Modern Quantum Mechanics Revised Edition

J. J. Sakurai

Late, University of California, Los Angeles

San Fu Tuan, Editor

University of Hawaii, Manoa

Addison-Wesley Publishing Company

Reading, Massachusetts • Menlo Park, California • New York
Don Mills, Ontario • Wokingham, England • Amsterdam • Bonn
Sydney • Singapore • Tokyo • Madrid • San Juan • Milan • Paris

Sponsoring Editor: Stuart W. Johnson
Assistant Editor: Jennifer Duggan
Senior Production Coordinator: Amy Willcutt
Manufacturing Manager: Roy Logan

Library of Congress Cataloging-in-Publication Data

Sakurai, J. J. (Jun John), 1933–1982.
Modern quantum mechanics / J. J. Sakurai ; San Fu Tuan, editor.—
Rev. ed.
p. cm.
Includes bibliographical references and index.
ISBN 0-201-53929-2
1. Quantum theory. I. Tuan, San Fu, 1932– . II. Title.
QC174.12.S25 1994
530.1′2—dc20 93-17803
CIP

5 6 7 8 9 10–MA–9695

Foreword

J. J. Sakurai was always a very welcome guest here at CERN, for he was one of those rare theorists to whom the experimental facts are even more interesting than the theoretical game itself. Nevertheless, he delighted in theoretical physics and in its teaching, a subject on which he held strong opinions. He thought that much theoretical physics teaching was both too narrow and too remote from application: "...we see a number of sophisticated, yet uneducated, theoreticians who are conversant in the LSZ formalism of the Heisenberg field operators, but do not know why an excited atom radiates, or are ignorant of the quantum theoretic derivation of Rayleigh's law that accounts for the blueness of the sky." And he insisted that the student must be able to use what has been taught: "The reader who has read the book but cannot do the exercises has learned nothing."

He put these principles to work in his fine book *Advanced Quantum Mechanics* (1967) and in *Invariance Principles and Elementary Particles* (1964), both of which have been very much used in the CERN library. This new book, *Modern Quantum Mechanics*, should be used even more, by a larger and less specialized group. The book combines breadth of interest with a thorough practicality. Its readers will find here what they need to know, with a sustained and successful effort to make it intelligible.

J. J. Sakurai's sudden death on November 1, 1982 left this book unfinished. Reinhold Bertlmann and I helped Mrs. Sakurai sort out her husband's papers at CERN. Among them we found a rough, handwritten version of most of the book and a large collection of exercises. Though only three chapters had been completely finished, it was clear that the bulk of the creative work had been done. It was also clear that much work remained to fill in gaps, polish the writing, and put the manuscript in order.

That the book is now finished is due to the determination of Noriko Sakurai and the dedication of San Fu Tuan. Upon her husband's death, Mrs. Sakurai resolved immediately that his last effort should not go to waste. With great courage and dignity she became the driving force behind the project, overcoming all obstacles and setting the high standards to be maintained. San Fu Tuan willingly gave his time and energy to the editing and completion of Sakurai's work. Perhaps only others close to the hectic field of high-energy theoretical physics can fully appreciate the sacrifice involved.

For me personally, J. J. had long been far more than just a particularly distinguished colleague. It saddens me that we will never again laugh together at physics and physicists and life in general, and that he will not see the success of his last work. But I am happy that it has been brought to fruition.

John S. Bell
CERN, Geneva

Preface to the Revised Edition

Since 1989 the Editor has enthusiastically pursued a revised edition of *Modern Quantum Mechanics* by his late great friend J. J. Sakurai, in order to extend this text's usefulness into the twenty-first century. Much consultation took place with the panel of Sakurai friends who helped with the original edition, but in particular with Professor Yasuo Hara of Tsukuba University and Professor Akio Sakurai of Kyoto Sangyo University in Japan.

The major motivation for this project is to revise the main text. There are three important additions and/or changes to the revised edition, which otherwise preserves the original version unchanged. These include a reworking of certain portions of Section 5.2 on time-independent perturbation theory for the degenerate case by Professor Kenneth Johnson of M.I.T., taking into account a subtle point that has not been properly treated by a number of texts on quantum mechanics in this country. Professor Roger Newton of Indiana University contributed refinements on lifetime broadening in Stark effect, additional explanations of phase shifts at resonances, the optical theorem, and on non-normalizable state. These appear as "remarks by the editor" or "editor's note" in the revised edition. Professor Thomas Fulton of the Johns Hopkins University reworked his Coulomb Scattering contribution (Section 7.13) so that it now appears as a shorter text portion emphasizing the physics, with the mathematical details relegated to Appendix C.

Though not a major part of the text, some additions were deemed necessary to take into account developments in quantum mechanics that have become prominent since November 1, 1982. To this end, two supplements are included at the end of the text. Supplement I is on adiabatic change and geometrical phase (popularized by M. V. Berry since 1983) and is actually an English translation of the supplement on this subject written by Professor Akio Sakurai for the Japanese version of *Modern Quantum Mechanics* (copyright © Yoshioka-Shoten Publishing of Kyoto). Supplement II is on non-exponential decays written by my colleague here, Professor Xerxes Tata, and read over by Professor E. C. G. Sudarshan of the University of Texas at Austin. Though non-exponential decays have a long history theoretically, experimental work on transition rates that tests indirectly such decays was done only in 1990. Introduction of additional material is of course a subjective matter on the part of the Editor; the readers will evaluate for themselves its appropriateness. Thanks to Professor Akio Sakurai, the revised edition has been "finely toothcombed" for misprint errors of the first ten printings of the original edition. My colleague, Professor Sandip Pakvasa, provided overall guidance and encouragement to me throughout this process of revision.

In addition to the acknowledgments above, my former students Li Ping, Shi Xiaohong, and Yasunaga Suzuki provided the sounding board for ideas on the revised edition when taking my graduate quantum mechanics course at the University of Hawaii during the spring of 1992. Suzuki provided the initial translation from Japanese of Supplement I as a course term paper. Dr. Andy Acker provided me with computer graphic assistance. The Department of Physics and Astronomy and particularly the High Energy Physics Group of the University of Hawaii at Manoa provided again both the facilities and a conducive atmosphere for me to carry out my editorial task. Finally I wish to express my gratitude to Physics (and sponsoring) Senior Editor, Stuart Johnson, and his Editorial Assistant, Jennifer Duggan, as well as Senior Production Coordinator Amy Willcutt, of Addison-Wesley for their encouragement and optimism that the revised edition will indeed materialize.

San Fu TUAN
Honolulu, Hawaii

J. J. Sakurai 1933–1982

In Memoriam

Jun John Sakurai was born in 1933 in Tokyo and came to the United States as a high school student in 1949. He studied at Harvard and at Cornell, where he received his Ph.D. in 1958. He was then appointed assistant professor of Physics at the University of Chicago, and became a full professor in 1964. He stayed at Chicago until 1970 when he moved to the University of California at Los Angeles, where he remained until his death. During his lifetime he wrote 119 articles in theoretical physics of elementary particles as well as several books and monographs on both quantum and particle theory.

The discipline of theoretical physics has as its principal aim the formulation of theoretical descriptions of the physical world that are at once concise and comprehensive. Because nature is subtle and complex, the pursuit of theoretical physics requires bold and enthusiastic ventures to the frontiers of newly discovered phenomena. This is an area in which Sakurai reigned supreme with his uncanny physical insight and intuition and also his ability to explain these phenomena in illuminating physical terms to the unsophisticated. One has but to read his very lucid textbooks on *Invariance Principles and Elementary Particles* and *Advanced Quantum Mechanics* as well as his reviews and summer school lectures to appreciate this. Without exaggeration I could say that much of what I did understand in particle physics came from these and from his articles and private tutoring.

When Sakurai was still a graduate student, he proposed what is now known as the V-A theory of weak interactions, independently of (and simultaneously with) Richard Feynman, Murray Gell-Mann, Robert Marshak, and George Sudarshan. In 1960 he published in *Annals of Physics* a prophetic paper, probably his single most important one. It was concerned with the first serious attempt to construct a theory of strong interactions based on Abelian and non-Abelian (Yang-Mills) gauge invariance. This seminal work induced theorists to attempt an understanding of the mechanisms of mass generation for gauge (vector) fields, now realized as the Higgs mechanism. Above all it stimulated the search for a realistic unification of forces under the gauge principle, now crowned with success in the celebrated Glashow-Weinberg-Salam unification of weak and electromagnetic forces. On the phenomenological side, Sakurai pursued and vigorously advocated the vector mesons dominance model of hadron dynamics. He was the first to discuss the mixing of ω and ϕ meson states. Indeed, he made numerous important contributions to particle physics phenomenology in a much more general sense, as his heart was always close to experimental activities.

I knew Jun John for more than 25 years, and I had the greatest admiration not only for his immense powers as a theoretical physicist but also for the warmth and generosity of his spirit. Though a graduate student himself at Cornell during 1957–1958, he took time from his own pioneering research in K-nucleon dispersion relations to help me (via extensive correspondence) with my Ph.D. thesis on the same subject at Berkeley. Both Sandip Pakvasa and I were privileged to be associated with one of his last papers on weak couplings of heavy quarks, which displayed once more his infectious and intuitive style of doing physics. It is of course gratifying to us in retrospect that Jun John counted this paper among the score of his published works that he particularly enjoyed.

The physics community suffered a great loss at Jun John Sakurai's death. The personal sense of loss is a severe one for me. Hence I am profoundly thankful for the opportunity to edit and complete his manuscript on *Modern Quantum Mechanics* for publication. In my faith no greater gift can be given me than an opportunity to show my respect and love for Jun John through meaningful service.

San Fu Tuan

Contents

Modern Quantum Mechanics

CHAPTER 1

Fundamental Concepts

The revolutionary change in our understanding of microscopic phenomena that took place during the first 27 years of the twentieth century is unprecedented in the history of natural sciences. Not only did we witness severe limitations in the validity of classical physics, but we found the alternative theory that replaced the classical physical theories to be far richer in scope and far richer in its range of applicability.

The most traditional way to begin a study of quantum mechanics is to follow the historical developments—Planck's radiation law, the Einstein-Debye theory of specific heats, the Bohr atom, de Broglie's matter waves, and so forth—together with careful analyses of some key experiments such as the Compton effect, the Franck-Hertz experiment, and the Davisson-Germer-Thompson experiment. In that way we may come to appreciate how the physicists in the first quarter of the twentieth century were forced to abandon, little by little, the cherished concepts of classical physics and how, despite earlier false starts and wrong turns, the great masters—Heisenberg, Schrödinger, and Dirac, among others—finally succeeded in formulating quantum mechanics as we know it today.

However, we do not follow the historical approach in this book. Instead, we start with an example that illustrates, perhaps more than any other example, the inadequacy of classical concepts in a fundamental way. We hope that by exposing the reader to a "shock treatment" at the onset, he or she may be attuned to what we might call the "quantum-mechanical way of thinking" at a very early stage.

1.1. THE STERN-GERLACH EXPERIMENT

The example we concentrate on in this section is the Stern-Gerlach experiment, originally conceived by O. Stern in 1921 and carried out in Frankfurt by him in collaboration with W. Gerlach in 1922. This experiment illustrates in a dramatic manner the necessity for a radical departure from the concepts of classical mechanics. In the subsequent sections the basic formalism of quantum mechanics is presented in a somewhat axiomatic manner but always with the example of the Stern-Gerlach experiment in the back of our minds. In a certain sense, a two-state system of the Stern-Gerlach type is the least classical, most quantum-mechanical system. A solid understanding of problems involving two-state systems will turn out to be rewarding to any serious student of quantum mechanics. It is for this reason that we refer repeatedly to two-state problems throughout this book.

Description of the Experiment

We now present a brief discussion of the Stern-Gerlach experiment, which is discussed in almost any book on modern physics.* First, silver (Ag) atoms are heated in an oven. The oven has a small hole through which some of the silver atoms escape. As shown in Figure 1.1, the beam goes through a collimator and is then subjected to an inhomogeneous magnetic field produced by a pair of pole pieces, one of which has a very sharp edge.

We must now work out the effect of the magnetic field on the silver atoms. For our purpose the following oversimplified model of the silver atom suffices. The silver atom is made up of a nucleus and 47 electrons, where 46 out of the 47 electrons can be visualized as forming a spherically symmetrical electron cloud with no net angular momentum. If we ignore the nuclear spin, which is irrelevant to our discussion, we see that the atom as a whole does have an angular momentum, which is due solely to the spin—intrinsic as opposed to orbital—angular momentum of the single 47th ($5s$) electron. The 47 electrons are attached to the nucleus, which is $\sim 2\times 10^5$ times heavier than the electron; as a result, the heavy atom as a whole possesses a magnetic moment equal to the spin magnetic moment of the 47th electron. In other words, the magnetic moment $\boldsymbol{\mu}$ of the atom is

*For an elementary but enlightening discussion of the Stern-Gerlach experiment, see French and Taylor (1978, 432–38).

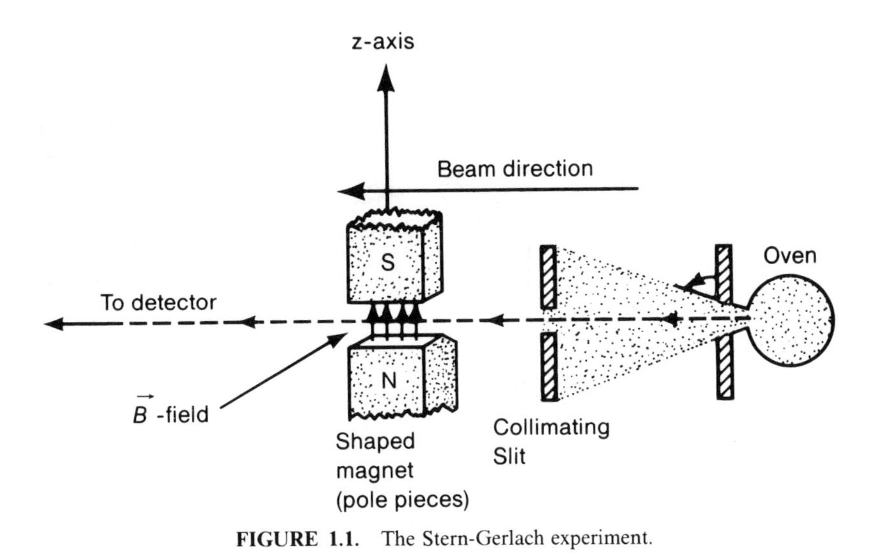

FIGURE 1.1. The Stern-Gerlach experiment.

proportional to the electron spin **S**,

$$\boldsymbol{\mu} \propto \mathbf{S}, \tag{1.1.1}$$

where the precise proportionality factor turns out to be $e/m_e c$ ($e < 0$ in this book) to an accuracy of about 0.2%.

Because the interaction energy of the magnetic moment with the magnetic field is just $-\boldsymbol{\mu}\cdot\mathbf{B}$, the z-component of the force experienced by the atom is given by

$$F_z = \frac{\partial}{\partial z}(\boldsymbol{\mu}\cdot\mathbf{B}) \simeq \mu_z \frac{\partial B_z}{\partial z}, \tag{1.1.2}$$

where we have ignored the components of **B** in directions other than the z-direction. Because the atom as a whole is very heavy, we expect that the classical concept of trajectory can be legitimately applied, a point which can be justified using the Heisenberg uncertainty principle to be derived later. With the arrangement of Figure 1.1, the $\mu_z > 0$ ($S_z < 0$) atom experiences a downward force, while the $\mu_z < 0$ ($S_z > 0$) atom experiences an upward force. The beam is then expected to get split according to the values of μ_z. In other words, the SG (Stern-Gerlach) apparatus "measures" the z-component of $\boldsymbol{\mu}$ or, equivalently, the z-component of **S** up to a proportionality factor.

The atoms in the oven are randomly oriented; there is no preferred direction for the orientation of $\boldsymbol{\mu}$. If the electron were like a classical spinning object, we would expect all values of μ_z to be realized between $|\boldsymbol{\mu}|$ and $-|\boldsymbol{\mu}|$. This would lead us to expect a continuous bundle of beams coming out of the SG apparatus, as shown in Figure 1.2a. Instead, what we

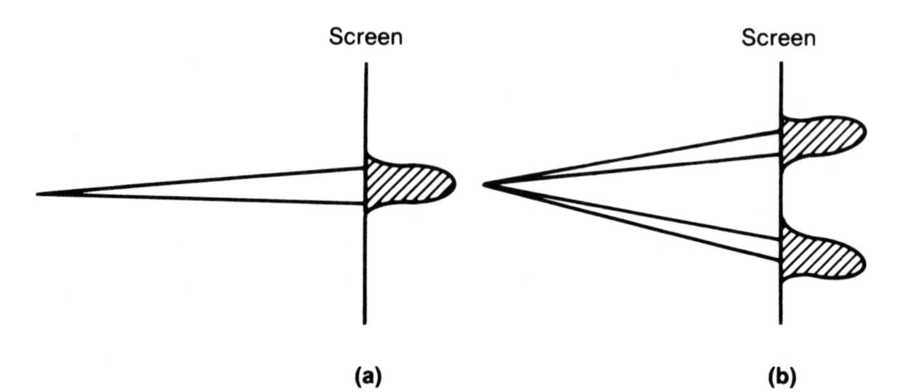

FIGURE 1.2. Beams from the SG apparatus; (a) is expected from classical physics, while (b) is actually observed.

experimentally observe is more like the situation in Figure 1.2b. In other words, the SG apparatus splits the original silver beam from the oven into *two distinct* components, a phenomenon referred to in the early days of quantum theory as "space quantization." To the extent that $\boldsymbol{\mu}$ can be identified within a proportionality factor with the electron spin $\mathbf{S}$, only two possible values of the z-component of $\mathbf{S}$ are observed to be possible, S_z up and S_z down, which we call S_z+ and S_z-. The two possible values of S_z are multiples of some fundamental unit of angular momentum; numerically it turns out that $S_z = \hbar/2$ and $-\hbar/2$, where

$$\begin{aligned}\hbar &= 1.0546\times 10^{-27}\,\text{erg-s}\\ &= 6.5822\times 10^{-16}\,\text{eV-s}\end{aligned} \tag{1.1.3}$$

This "quantization" of the electron spin angular momentum is the first important feature we deduce from the Stern-Gerlach experiment.

Of course, there is nothing sacred about the up-down direction or the z-axis. We could just as well have applied an inhomogeneous field in a horizontal direction, say in the x-direction, with the beam proceeding in the y-direction. In this manner we could have separated the beam from the oven into an S_x+ component and an S_x- component.

Sequential Stern-Gerlach Experiments

Let us now consider a sequential Stern-Gerlach experiment. By this we mean that the atomic beam goes through two or more SG apparatuses in sequence. The first arrangement we consider is relatively straightforward. We subject the beam coming out of the oven to the arrangement shown in Figure 1.3a, where SG$\hat{\mathbf{z}}$ stands for an apparatus with the inhomogeneous magnetic field in the z-direction, as usual. We then block the S_z- compo-

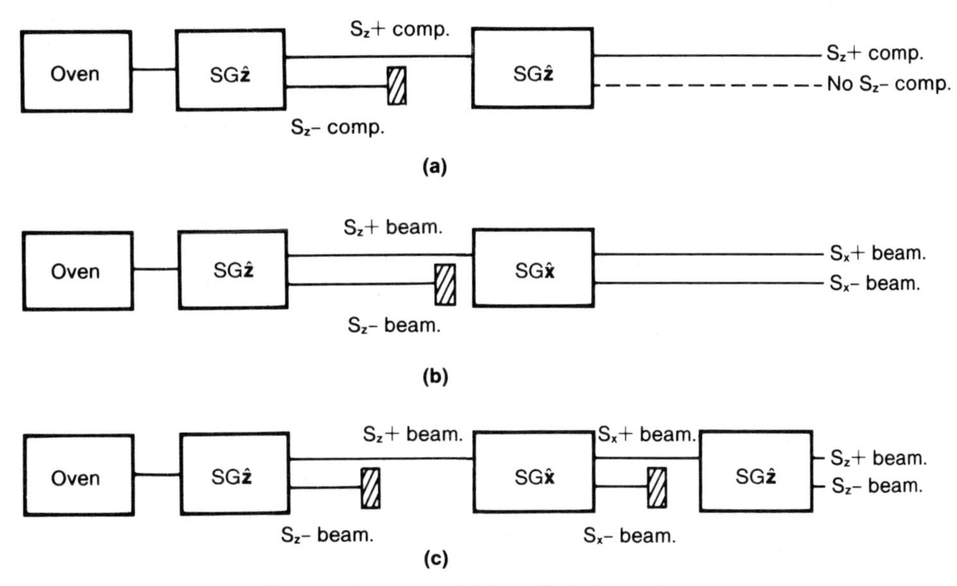

FIGURE 1.3. Sequential Stern-Gerlach experiments.

nent coming out of the first SG$\hat{\mathbf{z}}$ apparatus and let the remaining S_z+ component be subjected to another SG$\hat{\mathbf{z}}$ apparatus. This time there is only one beam component coming out of the second apparatus—just the S_z+ component. This is perhaps not so surprising; after all if the atom spins are up, they are expected to remain so, short of any external field that rotates the spins between the first and the second SG$\hat{\mathbf{z}}$ apparatuses.

A little more interesting is the arrangement shown in Figure 1.3b. Here the first SG apparatus is the same as before but the second one (SG$\hat{\mathbf{x}}$) has an inhomogeneous magnetic field in the x-direction. The S_z+ beam that enters the second apparatus (SG$\hat{\mathbf{x}}$) is now split into two components, an S_x+ component and an S_x- component, with equal intensities. How can we explain this? Does it mean that 50% of the atoms in the S_z+ beam coming out of the first apparatus (SG$\hat{\mathbf{z}}$) are made up of atoms characterized by both S_z+ and S_x+, while the remaining 50% have both S_z+ and S_x-? It turns out that such a picture runs into difficulty, as will be shown below.

We now consider a third step, the arrangement shown in Figure 1.3(c), which most dramatically illustrates the peculiarities of quantum-mechanical systems. This time we add to the arrangement of Figure 1.3b yet a third apparatus, of the SG$\hat{\mathbf{z}}$ type. It is observed experimentally that *two* components emerge from the third apparatus, not one; the emerging beams are seen to have *both* an S_z+ component and an S_z- component. This is a complete surprise because after the atoms emerged from the first apparatus, we made sure that the S_z- component was completely blocked. How is it possible that the S_z- component which, we thought, we eliminated earlier reappears? The model in which the atoms entering the third apparatus are visualized to have both S_z+ and S_x+ is clearly unsatisfactory.

This example is often used to illustrate that in quantum mechanics we cannot determine both S_z and S_x simultaneously. More precisely, we can say that the selection of the S_x+ beam by the second apparatus (SG$\hat{\mathbf{x}}$) completely destroys any *previous* information about S_z.

It is amusing to compare this situation with that of a spinning top in classical mechanics, where the angular momentum

$$\mathbf{L} = I\boldsymbol{\omega} \tag{1.1.4}$$

can be measured by determining the components of the angular-velocity vector $\boldsymbol{\omega}$. By observing how fast the object is spinning in which direction we can determine ω_x, ω_y, and ω_z simultaneously. The moment of inertia I is computable if we know the mass density and the geometric shape of the spinning top, so there is no difficulty in specifying both L_z and L_x in this classical situation.

It is to be clearly understood that the limitation we have encountered in determining S_z and S_x is not due to the incompetence of the experimentalist. By improving the experimental techniques we cannot make the S_z- component out of the third apparatus in Figure 1.3c disappear. The peculiarities of quantum mechanics are imposed upon us by the experiment itself. The limitation is, in fact, inherent in microscopic phenomena.

Analogy with Polarization of Light

Because this situation looks so novel, some analogy with a familiar classical situation may be helpful here. To this end we now digress to consider the polarization of light waves.

Consider a monochromatic light wave propagating in the z-direction. A linearly polarized (or plane polarized) light with a polarization vector in the x-direction, which we call for short an *x-polarized light*, has a space-time dependent electric field oscillating in the x-direction

$$\mathbf{E} = E_0\hat{\mathbf{x}}\cos(kz - \omega t). \tag{1.1.5}$$

Likewise, we may consider a y-polarized light, also propagating in the z-direction,

$$\mathbf{E} = E_0\hat{\mathbf{y}}\cos(kz - \omega t). \tag{1.1.6}$$

Polarized light beams of type (1.1.5) or (1.1.6) can be obtained by letting an unpolarized light beam go through a Polaroid filter. We call a filter that selects only beams polarized in the x-direction an *x-filter*. An x-filter, of course, becomes a y-filter when rotated by 90° about the propagation (z)

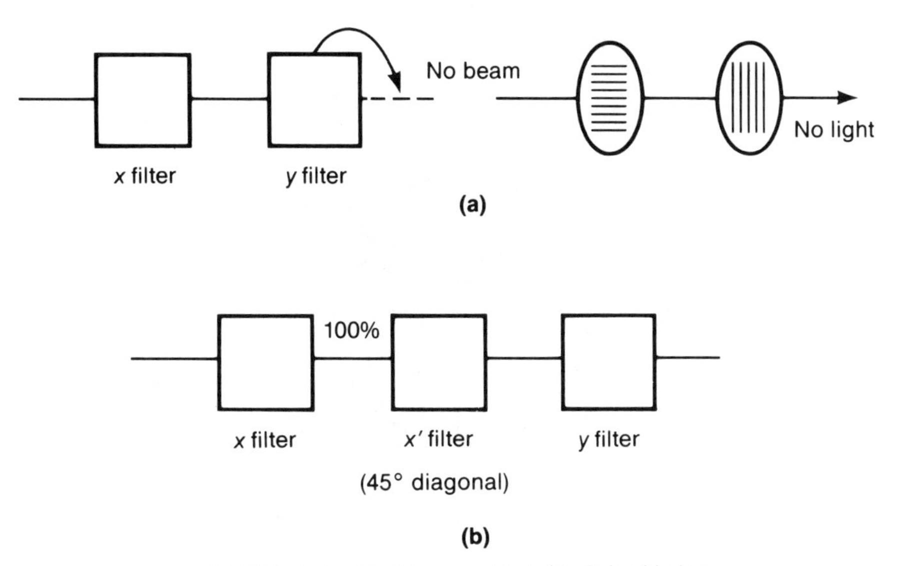

FIGURE 1.4. Light beams subjected to Polaroid filters.

direction. It is well known that when we let a light beam go through an x-filter and subsequently let it impinge on a y-filter, no light beam comes out provided, of course, we are dealing with 100% efficient Polaroids; see Figure 1.4a.

The situation is even more interesting if we insert between the x-filter and the y-filter yet another Polaroid that selects only a beam polarized in the direction—which we call the x'-direction—that makes an angle of 45° with the x-direction in the xy plane; see Figure 1.4b. This time, there is a light beam coming out of the y-filter despite the fact that right after the beam went through the x-filter it did not have any polarization component in the y-direction. In other words, once the x'-filter intervenes and selects the x'-polarized beam, it is immaterial whether the beam was previously x-polarized. The selection of the x'-polarized beam by the second Polaroid destroys any previous information on light polarization. Notice that this situation is quite analogous to the situation that we encountered earlier with the SG arrangement of Figure 1.3b, provided that the following correspondence is made:

$$\begin{aligned} S_z \pm \text{ atoms} &\leftrightarrow x\text{-},\ y\text{-polarized light} \\ S_x \pm \text{ atoms} &\leftrightarrow x'\text{-},\ y'\text{-polarized light,} \end{aligned} \tag{1.1.7}$$

where the x'- and the y'-axes are defined as in Figure 1.5.

Let us examine how we can quantitatively describe the behavior of 45°-polarized beams (x'- and y'-polarized beams) within the framework of

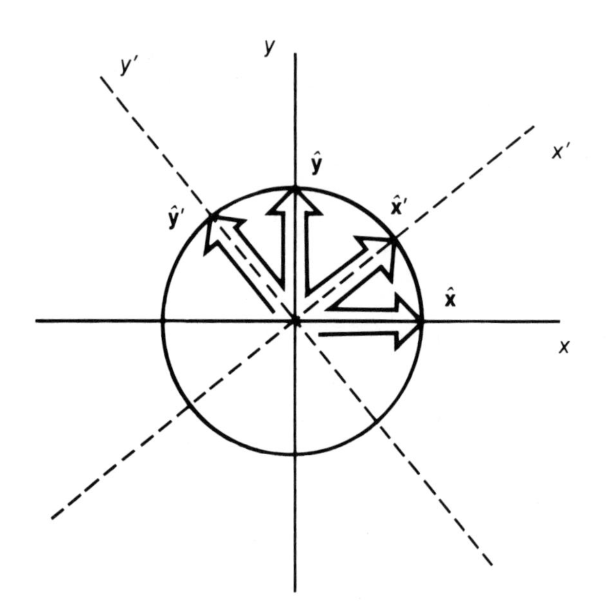

FIGURE 1.5. Orientations of the x'- and y'-axes.

classical electrodynamics. Using Figure 1.5 we obtain

$$E_0\hat{\mathbf{x}}'\cos(kz-\omega t) = E_0\left[\frac{1}{\sqrt{2}}\hat{\mathbf{x}}\cos(kz-\omega t)+\frac{1}{\sqrt{2}}\hat{\mathbf{y}}\cos(kz-\omega t)\right],$$
$$E_0\hat{\mathbf{y}}'\cos(kz-\omega t) = E_0\left[-\frac{1}{\sqrt{2}}\hat{\mathbf{x}}\cos(kz-\omega t)+\frac{1}{\sqrt{2}}\hat{\mathbf{y}}\cos(kz-\omega t)\right]. \tag{1.1.8}$$

In the triple-filter arrangement of Figure 1.4b the beam coming out of the first Polaroid is an $\hat{\mathbf{x}}$-polarized beam, which can be regarded as a linear combination of an x'-polarized beam and a y'-polarized beam. The second Polaroid selects the x'-polarized beam, which can in turn be regarded as a linear combination of an x-polarized and a y-polarized beam. And finally, the third Polaroid selects the y-polarized component.

Applying correspondence (1.1.7) from the sequential Stern-Gerlach experiment of Figure 1.3c, to the triple-filter experiment of Figure 1.4b suggests that we might be able to represent the spin state of a silver atom by some kind of vector in a new kind of two-dimensional vector space, an abstract vector space not to be confused with the usual two-dimensional (xy) space. Just as $\hat{\mathbf{x}}$ and $\hat{\mathbf{y}}$ in (1.1.8) are the base vectors used to decompose the polarization vector $\hat{\mathbf{x}}'$ of the $\hat{\mathbf{x}}'$-polarized light, it is reasonable to represent the S_x+ state by a vector, which we call a *ket* in the Dirac notation to be developed fully in the next section. We denote this vector by

$|S_x; +\rangle$ and write it as a linear combination of two base vectors, $|S_z; +\rangle$ and $|S_z; -\rangle$, which correspond to the $S_z +$ and the $S_z -$ states, respectively. So we may conjecture

$$|S_x; +\rangle \stackrel{?}{=} \frac{1}{\sqrt{2}}|S_z; +\rangle + \frac{1}{\sqrt{2}}|S_z; -\rangle \tag{1.1.9a}$$

$$|S_x; -\rangle \stackrel{?}{=} -\frac{1}{\sqrt{2}}|S_z; +\rangle + \frac{1}{\sqrt{2}}|S_z; -\rangle \tag{1.1.9b}$$

in analogy with (1.1.8). Later we will show how to obtain these expressions using the general formalism of quantum mechanics.

Thus the unblocked component coming out of the second (SG$\hat{x}$) apparatus of Figure 1.3c is to be regarded as a superposition of $S_z +$ and $S_z -$ in the sense of (1.1.9a). It is for this reason that two components emerge from the third (SG$\hat{z}$) apparatus.

The next question of immediate concern is, How are we going to represent the $S_y \pm$ states? Symmetry arguments suggest that if we observe an $S_z \pm$ beam going in the x-direction and subject it to an SG$\hat{\mathbf{y}}$ apparatus, the resulting situation will be very similar to the case where an $S_z \pm$ beam going in the y-direction is subjected to an SG$\hat{\mathbf{x}}$ apparatus. The kets for $S_y \pm$ should then be regarded as a linear combination of $|S_z; \pm\rangle$, but it appears from (1.1.9) that we have already used up the available possibilities in writing $|S_x; \pm\rangle$. How can our vector space formalism distinguish $S_y \pm$ states from $S_x \pm$ states?

An analogy with polarized light again rescues us here. This time we consider a circularly polarized beam of light, which can be obtained by letting a linearly polarized light pass through a quarter-wave plate. When we pass such a circularly polarized light through an x-filter or a y-filter, we again obtain either an x-polarized beam or a y-polarized beam of equal intensity. Yet everybody knows that the circularly polarized light is totally different from the 45°-linearly polarized (x'-polarized or y'-polarized) light.

Mathematically, how do we represent a circularly polarized light? A right circularly polarized light is nothing more than a linear combination of an x-polarized light and a y-polarized light, where the oscillation of the electric field for the y-polarized component is 90° out of phase with that of the x-polarized component:*

$$\mathbf{E} = E_0\left[\frac{1}{\sqrt{2}}\hat{\mathbf{x}}\cos(kz - \omega t) + \frac{1}{\sqrt{2}}\hat{\mathbf{y}}\cos\left(kz - \omega t + \frac{\pi}{2}\right)\right]. \tag{1.1.10}$$

It is more elegant to use complex notation by introducing $\boldsymbol{\epsilon}$ as follows:

$$\mathrm{Re}(\boldsymbol{\epsilon}) = \mathbf{E}/\mathbf{E}_0. \tag{1.1.11}$$

*Unfortunately, there is no unanimity in the definition of right versus left circularly polarized light in the literature.

For a right circularly polarized light, we can then write

$$\boldsymbol{\varepsilon} = \left[\frac{1}{\sqrt{2}} \hat{\mathbf{x}} e^{i(kz - \omega t)} + \frac{i}{\sqrt{2}} \hat{\mathbf{y}} e^{i(kz - \omega t)} \right], \tag{1.1.12}$$

where we have used $i = e^{i\pi/2}$.

We can make the following analogy with the spin states of silver atoms:

$$\begin{aligned} S_y + \text{ atom} &\leftrightarrow \text{ right circularly polarized beam}, \\ S_y - \text{ atom} &\leftrightarrow \text{ left circularly polarized beam}. \end{aligned} \tag{1.1.13}$$

Applying this analogy to (1.1.12), we see that if we are allowed to make the coefficients preceding base kets complex, there is no difficulty in accommodating the $S_y \pm$ atoms in our vector space formalism:

$$|S_y; \pm\rangle \stackrel{?}{=} \frac{1}{\sqrt{2}} |S_z; +\rangle \pm \frac{i}{\sqrt{2}} |S_z; -\rangle, \tag{1.1.14}$$

which are obviously different from (1.1.9). We thus see that the two-dimensional vector space needed to describe the spin states of silver atoms must be a *complex* vector space; an arbitrary vector in the vector space is written as a linear combination of the base vectors $|S_z; \pm\rangle$ with, in general, complex coefficients. The fact that the necessity of complex numbers is already apparent in such an elementary example is rather remarkable.

The reader must have noted by this time that we have deliberately avoided talking about photons. In other words, we have completely ignored the quantum aspect of light; nowhere did we mention the polarization states of individual photons. The analogy we worked out is between kets in an abstract vector space that describes the spin states of individual atoms with the polarization vectors of the *classical electromagnetic field*. Actually we could have made the analogy even more vivid by introducing the photon concept and talking about the probability of finding a circularly polarized photon in a linearly polarized state, and so forth; however, that is not needed here. Without doing so, we have already accomplished the main goal of this section: to introduce the idea that quantum-mechanical states are to be represented by vectors in an abstract complex vector space.*

1.2. KETS, BRAS, AND OPERATORS

In the preceding section we showed how analyses of the Stern-Gerlach experiment lead us to consider a complex vector space. In this and the

* The reader who is interested in grasping the basic concepts of quantum mechanics through a careful study of photon polarization may find Chapter 1 of Baym (1969) extremely illuminating.

following section we formulate the basic mathematics of vector spaces as used in quantum mechanics. Our notation throughout this book is the bra and ket notation developed by P. A. M. Dirac. The theory of linear vector spaces had, of course, been known to mathematicians prior to the birth of quantum mechanics, but Dirac's way of introducing vector spaces has many advantages, especially from the physicist's point of view.

Ket Space

We consider a complex vector space whose dimensionality is specified according to the nature of a physical system under consideration. In Stern-Gerlach–type experiments where the only quantum-mechanical degree of freedom is the spin of an atom, the dimensionality is determined by the number of alternative paths the atoms can follow when subjected to a SG apparatus; in the case of the silver atoms of the previous section, the dimensionality is just two, corresponding to the two possible values S_z can assume.* Later, in Section 1.6, we consider the case of continuous spectra—for example, the position (coordinate) or momentum of a particle—where the number of alternatives is nondenumerably infinite, in which case the vector space in question is known as a **Hilbert space** after D. Hilbert, who studied vector spaces in infinite dimensions.

In quantum mechanics a physical state, for example, a silver atom with a definite spin orientation, is represented by a **state vector** in a complex vector space. Following Dirac, we call such a vector a **ket** and denote it by $|\alpha\rangle$. This state ket is postulated to contain complete information about the physical state; everything we are allowed to ask about the state is contained in the ket. Two kets can be added:

$$|\alpha\rangle + |\beta\rangle = |\gamma\rangle. \tag{1.2.1}$$

The sum $|\gamma\rangle$ is just another ket. If we multiply $|\alpha\rangle$ by a complex number c, the resulting product $c|\alpha\rangle$ is another ket. The number c can stand on the left or on the right of a ket; it makes no difference:

$$c|\alpha\rangle = |\alpha\rangle c. \tag{1.2.2}$$

In the particular case where c is zero, the resulting ket is said to be a **null ket**.

One of the physics postulates is that $|\alpha\rangle$ and $c|\alpha\rangle$, with $c \neq 0$, represent the same physical state. In other words, only the "direction" in vector space is of significance. Mathematicians may prefer to say that we are here dealing with rays rather than vectors.

---

*For many physical systems the dimension of the state space is denumerably infinite. While we will usually indicate a finite number of dimensions, N, of the ket space, the results also hold for denumerably infinite dimensions.

An **observable**, such as momentum and spin components, can be represented by an **operator**, such as A, in the vector space in question. Quite generally, an operator acts on a ket *from the left*,

$$A\cdot(|\alpha\rangle) = A|\alpha\rangle, \tag{1.2.3}$$

which is yet another ket. There will be more on multiplication operations later.

In general, $A|\alpha\rangle$ is *not* a constant times $|\alpha\rangle$. However, there are particular kets of importance, known as **eigenkets** of operator A, denoted by

$$|a'\rangle, |a''\rangle, |a'''\rangle, \ldots \tag{1.2.4}$$

with the property

$$A|a'\rangle = a'|a'\rangle, A|a''\rangle = a''|a''\rangle, \ldots \tag{1.2.5}$$

where a', a'', ... are just numbers. Notice that applying A to an eigenket just reproduces the same ket apart from a multiplicative number. The set of numbers $\{a', a'', a''', \ldots\}$, more compactly denoted by $\{a'\}$, is called the set of **eigenvalues** of operator A. When it becomes necessary to order eigenvalues in a specific manner, $\{a^{(1)}, a^{(2)}, a^{(3)}, \ldots\}$ may be used in place of $\{a', a'', a''', \ldots\}$.

The physical state corresponding to an eigenket is called an **eigenstate**. In the simplest case of spin $\frac{1}{2}$ systems, the eigenvalue-eigenket relation (1.2.5) is expressed as

$$S_z|S_z;+\rangle = \frac{\hbar}{2}|S_z;+\rangle, \qquad S_z|S_z;-\rangle = -\frac{\hbar}{2}|S_z;-\rangle, \tag{1.2.6}$$

where $|S_z;\pm\rangle$ are eigenkets of operator S_z with eigenvalues $\pm\hbar/2$. Here we could have used just $|\hbar/2\rangle$ for $|S_z;+\rangle$ in conformity with the notation $|a'\rangle$, where an eigenket is labeled by its eigenvalue, but the notation $|S_z;\pm\rangle$, already used in the previous section, is more convenient here because we also consider eigenkets of S_x:

$$S_x|S_x;\pm\rangle = \pm\frac{\hbar}{2}|S_x;\pm\rangle. \tag{1.2.7}$$

We remarked earlier that the dimensionality of the vector space is determined by the number of alternatives in Stern-Gerlach–type experiments. More formally, we are concerned with an N-dimensional vector space spanned by the N eigenkets of observable A. Any arbitrary ket $|\alpha\rangle$ can be written as

$$|\alpha\rangle = \sum_{a'} c_{a'}|a'\rangle, \tag{1.2.8}$$

with $a', a'', \ldots$ up to $a^{(N)}$, where $c_{a'}$ is a complex coefficient. The question of the uniqueness of such an expansion will be postponed until we prove the orthogonality of eigenkets.

Bra Space and Inner Products

The vector space we have been dealing with is a ket space. We now introduce the notion of a **bra space**, a vector space "dual to" the ket space. We postulate that corresponding to every ket $|\alpha\rangle$ there exists a bra, denoted by $\langle\alpha|$, in this dual, or bra, space. The bra space is spanned by eigenbras $\{\langle a'|\}$ which correspond to the eigenkets $\{|a'\rangle\}$. There is a one-to-one correspondence between a ket space and a bra space:

$$|\alpha\rangle \overset{\text{DC}}{\leftrightarrow} \langle\alpha|$$

$$|a'\rangle, |a''\rangle, \ldots \overset{\text{DC}}{\leftrightarrow} \langle a'|, \langle a''|, \ldots \tag{1.2.9}$$

$$|\alpha\rangle + |\beta\rangle \overset{\text{DC}}{\leftrightarrow} \langle\alpha| + \langle\beta|$$

where DC stands for **dual correspondence**. Roughly speaking, we can regard the bra space as some kind of mirror image of the ket space.

The bra dual to $c|\alpha\rangle$ is postulated to be $c^*\langle\alpha|$, *not* $c\langle\alpha|$, which is a very important point. More generally, we have

$$c_\alpha|\alpha\rangle + c_\beta|\beta\rangle \overset{\text{DC}}{\leftrightarrow} c_\alpha^*\langle\alpha| + c_\beta^*\langle\beta|. \tag{1.2.10}$$

We now define the **inner product** of a bra and a ket.* The product is written as a bra standing on the left and a ket standing on the right, for example,

$$\langle\beta|\alpha\rangle = \underset{\text{bra }(c)\text{ ket}}{(\langle\beta|)\cdot(|\alpha\rangle)}. \tag{1.2.11}$$

This product is, in general, a complex number. Notice that in forming an inner product we always take one vector from the bra space and one vector from the ket space.

We postulate two fundamental properties of inner products. First,

$$\langle\beta|\alpha\rangle = \langle\alpha|\beta\rangle^*. \tag{1.2.12}$$

In other words, $\langle\beta|\alpha\rangle$ and $\langle\alpha|\beta\rangle$ are complex conjugates of each other. Notice that even though the inner product is, in some sense, analogous to the familiar scalar product $\mathbf{a}\cdot\mathbf{b}$, $\langle\beta|\alpha\rangle$ must be clearly distinguished from $\langle\alpha|\beta\rangle$; the analogous distinction is not needed in real vector space because $\mathbf{a}\cdot\mathbf{b}$ is equal to $\mathbf{b}\cdot\mathbf{a}$. Using (1.2.12) we can immediately deduce that $\langle\alpha|\alpha\rangle$ must be a real number. To prove this just let $\langle\beta| \to \langle\alpha|$.

---

*In the literature an inner product is often referred to as a *scalar product* because it is analogous to $\mathbf{a}\cdot\mathbf{b}$ in Euclidean space; in this book, however, we reserve the term *scalar* for a quantity invariant under rotations in the usual three-dimensional space.

The second postulate on inner products is

$$\langle\alpha|\alpha\rangle \geq 0, \tag{1.2.13}$$

where the equality sign holds only if $|\alpha\rangle$ is a *null ket*. This is sometimes known as the postulate of **positive definite metric**. From a physicist's point of view, this postulate is essential for the probabilistic interpretation of quantum mechanics, as will become apparent later.*

Two kets $|\alpha\rangle$ and $|\beta\rangle$ are said to be **orthogonal** if

$$\langle\alpha|\beta\rangle = 0, \tag{1.2.14}$$

even though in the definition of the inner product the bra $\langle\alpha|$ appears. The orthogonality relation (1.2.14) also implies, via (1.2.12),

$$\langle\beta|\alpha\rangle = 0. \tag{1.2.15}$$

Given a ket which is not a null ket, we can form a **normalized ket** $|\tilde{\alpha}\rangle$, where

$$|\tilde{\alpha}\rangle = \left(\frac{1}{\sqrt{\langle\alpha|\alpha\rangle}}\right)|\alpha\rangle, \tag{1.2.16}$$

with the property

$$\langle\tilde{\alpha}|\tilde{\alpha}\rangle = 1. \tag{1.2.17}$$

Quite generally, $\sqrt{\langle\alpha|\alpha\rangle}$ is known as the **norm** of $|\alpha\rangle$, analogous to the magnitude of vector $\sqrt{\mathbf{a}\cdot\mathbf{a}} = |\vec{\mathbf{a}}|$ in Euclidean vector space. Because $|\alpha\rangle$ and $c|\alpha\rangle$ represent the same physical state, we might as well require that the kets we use for physical states be normalized in the sense of (1.2.17).†

Operators

As we remarked earlier, observables like momentum and spin components are to be represented by operators that can act on kets. We can consider a more general class of operators that act on kets; they will be denoted by X, Y, and so forth, while A, B, and so on will be used for a restrictive class of operators that correspond to observables.

An operator acts on a ket from the left side,

$$X\cdot(|\alpha\rangle) = X|\alpha\rangle, \tag{1.2.18}$$

and the resulting product is another ket. Operators X and Y are said to be **equal**,

$$X = Y, \tag{1.2.19}$$

---

*Attempts to abandon this postulate led to physical theories with "indefinite metric." We shall not be concerned with such theories in this book.

†For eigenkets of observables with continuous spectra, different normalization conventions will be used; see Section 1.6.

if

$$X|\alpha\rangle = Y|\alpha\rangle \tag{1.2.20}$$

for an *arbitrary* ket in the ket space in question. Operator X is said to be the **null operator** if, for any *arbitrary* ket $|\alpha\rangle$, we have

$$X|\alpha\rangle = 0. \tag{1.2.21}$$

Operators can be added; addition operations are commutative and associative:

$$X + Y = Y + X, \tag{1.2.21a}$$

$$X + (Y + Z) = (X + Y) + Z. \tag{1.2.21b}$$

With the single exception of the time-reversal operator to be considered in Chapter 4, the operators that appear in this book are all linear, that is,

$$X(c_\alpha|\alpha\rangle + c_\beta|\beta\rangle) = c_\alpha X|\alpha\rangle + c_\beta X|\beta\rangle. \tag{1.2.22}$$

An operator X always acts on a bra from the *right* side

$$(\langle\alpha|)\cdot X = \langle\alpha|X, \tag{1.2.23}$$

and the resulting product is another bra. The ket $X|\alpha\rangle$ and the bra $\langle\alpha|X$ are, in general, *not* dual to each other. We define the symbol $X^\dagger$ as

$$X|\alpha\rangle \overset{\text{DC}}{\leftrightarrow} \langle\alpha|X^\dagger. \tag{1.2.24}$$

The operator $X^\dagger$ is called the **Hermitian adjoint**, or simply the adjoint, of X. An operator X is said to be Hermitian if

$$X = X^\dagger. \tag{1.2.25}$$

Multiplication

Operators X and Y can be multiplied. Multiplication operations are, in general, *noncommutative*, that is,

$$XY \neq YX. \tag{1.2.26}$$

Multiplication operations are, however, associative:

$$X(YZ) = (XY)Z = XYZ. \tag{1.2.27}$$

We also have

$$X(Y|\alpha\rangle) = (XY)|\alpha\rangle = XY|\alpha\rangle, \qquad (\langle\beta|X)Y = \langle\beta|(XY) = \langle\beta|XY. \tag{1.2.28}$$

Notice that

$$(XY)^\dagger = Y^\dagger X^\dagger \tag{1.2.29}$$

because

$$XY|\alpha\rangle = X(Y|\alpha\rangle) \overset{\text{DC}}{\leftrightarrow} (\langle\alpha|Y^\dagger)X^\dagger = \langle\alpha|Y^\dagger X^\dagger. \tag{1.2.30}$$

So far, we have considered the following products: $\langle\beta|\alpha\rangle$, $X|\alpha\rangle$, $\langle\alpha|X$, and XY. Are there other products we are allowed to form? Let us multiply $|\beta\rangle$ and $\langle\alpha|$, in that order. The resulting product

$$(|\beta\rangle)\cdot(\langle\alpha|) = |\beta\rangle\langle\alpha| \tag{1.2.31}$$

is known as the **outer product** of $|\beta\rangle$ and $\langle\alpha|$. We will emphasize in a moment that $|\beta\rangle\langle\alpha|$ is to be regarded as an operator; hence it is fundamentally different from the inner product $\langle\beta|\alpha\rangle$, which is just a number.

There are also "illegal products." We have already mentioned that an operator must stand on the left of a ket or on the right of a bra. In other words, $|\alpha\rangle X$ and $X\langle\alpha|$ are examples of illegal products. They are neither kets, nor bras, nor operators; they are simply nonsensical. Products like $|\alpha\rangle|\beta\rangle$ and $\langle\alpha|\langle\beta|$ are also illegal when $|\alpha\rangle$ and $|\beta\rangle$ ($\langle\alpha|$ and $\langle\beta|$) are ket (bra) vectors belonging to the same ket (bra) space.*

The Associative Axiom

As is clear from (1.2.27), multiplication operations among operators are associative. Actually the associative property is postulated to hold quite generally as long as we are dealing with "legal" multiplications among kets, bras, and operators. Dirac calls this important postulate the **associative axiom of multiplication**.

To illustrate the power of this axiom let us first consider an outer product acting on a ket:

$$(|\beta\rangle\langle\alpha|)\cdot|\gamma\rangle. \tag{1.2.32}$$

Because of the associative axiom, we can regard this equally well as

$$|\beta\rangle\cdot(\langle\alpha|\gamma\rangle), \tag{1.2.33}$$

where $\langle\alpha|\gamma\rangle$ is just a number. So the outer product acting on a ket is just another ket; in other words, $|\beta\rangle\langle\alpha|$ can be regarded as an operator. Because (1.2.32) and (1.2.33) are equal, we may as well omit the dots and let $|\beta\rangle\langle\alpha|\gamma\rangle$ stand for the operator $|\beta\rangle\langle\alpha|$ acting on $|\gamma\rangle$ *or*, equivalently, the number $\langle\alpha|\gamma\rangle$ multiplying $|\beta\rangle$. (On the other hand, if (1.2.33) is written as $(\langle\alpha|\gamma\rangle)\cdot|\beta\rangle$, we cannot afford to omit the dot and brackets because the

*Later in the book we will encounter products like $|\alpha\rangle|\beta\rangle$, which are more appropriately written as $|\alpha\rangle\otimes|\beta\rangle$, but in such cases $|\alpha\rangle$ and $|\beta\rangle$ always refer to kets from *different* vector spaces. For instance, the first ket belongs to the vector space for electron spin, the second ket to the vector space for electron orbital angular momentum; or the first ket lies in the vector space of particle 1, the second ket in the vector space of particle 2, and so forth.

resulting expression would look illegal.) Notice that the operator $|\beta\rangle\langle\alpha|$ rotates $|\gamma\rangle$ into the direction of $|\beta\rangle$. It is easy to see that if

$$X = |\beta\rangle\langle\alpha|, \tag{1.2.34}$$

then

$$X^\dagger = |\alpha\rangle\langle\beta|, \tag{1.2.35}$$

which is left as an exercise.

In a second important illustration of the associative axiom, we note that

$$\underset{\text{bra}}{(\langle\beta|)} \cdot \underset{\text{ket}}{(X|\alpha\rangle)} = \underset{\text{bra}}{(\langle\beta|X)} \cdot \underset{\text{ket}}{(|\alpha\rangle)}. \tag{1.2.36}$$

Because the two sides are equal, we might as well use the more compact notation

$$\langle\beta|X|\alpha\rangle \tag{1.2.37}$$

to stand for either side of (1.2.36). Recall now that $\langle\alpha|X^\dagger$ is the bra that is dual to $X|\alpha\rangle$, so

$$\begin{aligned}\langle\beta|X|\alpha\rangle &= \langle\beta|\cdot(X|\alpha\rangle)\\ &= \{(\langle\alpha|X^\dagger)\cdot|\beta\rangle\}^*\\ &= \langle\alpha|X^\dagger|\beta\rangle^*,\end{aligned} \tag{1.2.38}$$

where, in addition to the associative axiom, we used the fundamental property of the inner product (1.2.12). For a *Hermitian* X we have

$$\langle\beta|X|\alpha\rangle = \langle\alpha|X|\beta\rangle^*. \tag{1.2.39}$$

1.3. BASE KETS AND MATRIX REPRESENTATIONS

Eigenkets of an Observable

Let us consider the eigenkets and eigenvalues of a Hermitian operator A. We use the symbol A, reserved earlier for an observable, because in quantum mechanics Hermitian operators of interest quite often turn out to be the operators representing some physical observables.

We begin by stating an important theorem:

Theorem. *The eigenvalues of a Hermitian operator A are real; the eigenkets of A corresponding to different eigenvalues are orthogonal.*

Proof. First, recall that

$$A|a'\rangle = a'|a'\rangle. \tag{1.3.1}$$

Because A is Hermitian, we also have

$$\langle a''|A = a''^*\langle a''|, \tag{1.3.2}$$

where $a', a'' \ldots$ are eigenvalues of A. If we multiply both sides of (1.3.1) by $\langle a''|$ on the left, both sides of (1.3.2) by $|a'\rangle$ on the right, and subtract, we obtain

$$(a' - a''^*)\langle a''|a'\rangle = 0. \tag{1.3.3}$$

Now a' and a'' can be taken to be either the same or different. Let us first choose them to be the same; we then deduce the reality condition (the first half of the theorem)

$$a' = a'^*, \tag{1.3.4}$$

where we have used the fact that $|a'\rangle$ is not a null ket. Let us now assume a' and a'' to be different. Because of the just-proved reality condition, the difference $a' - a''^*$ that appears in (1.3.3) is equal to $a' - a''$, which cannot vanish, by assumption. The inner product $\langle a''|a'\rangle$ must then vanish:

$$\langle a''|a'\rangle = 0, \quad (a' \neq a''), \tag{1.3.5}$$

which proves the orthogonality property (the second half of the theorem). □

We expect on physical grounds that an observable has real eigenvalues, a point that will become clearer in the next section, where measurements in quantum mechanics will be discussed. The theorem just proved guarantees the reality of eigenvalues whenever the operator is Hermitian. That is why we talk about Hermitian observables in quantum mechanics.

It is conventional to normalize $|a'\rangle$ so the $\{|a'\rangle\}$ form a **orthonormal** set:

$$\langle a''|a'\rangle = \delta_{a''a'}. \tag{1.3.6}$$

We may logically ask, Is this set of eigenkets complete? Since we started our discussion by asserting that the whole ket space is spanned by the eigenkets of A, the eigenkets of A must therefore form a complete set by *construction* of our ket space.*

Eigenkets as Base Kets

We have seen that the normalized eigenkets of A form a complete orthonormal set. An arbitrary ket in the ket space can be expanded in terms

---

*The astute reader, already familiar with wave mechanics, may point out that the completeness of eigenfunctions we use can be proved by applying the Sturm-Liouville theory to the Schrödinger wave equation. But to "derive" the Schrödinger wave equation from our fundamental postulates, the completeness of the position eigenkets must be assumed.

of the eigenkets of A. In other words, the eigenkets of A are to be used as base kets in much the same way as a set of mutually orthogonal unit vectors is used as base vectors in Euclidean space.

Given an arbitrary ket $|\alpha\rangle$ in the ket space spanned by the eigenkets of A, let us attempt to expand it as follows:

$$|\alpha\rangle = \sum_{a'} c_{a'}|a'\rangle. \tag{1.3.7}$$

Multiplying $\langle a''|$ on the left and using the orthonormality property (1.3.6), we can immediately find the expansion coefficient,

$$c_{a'} = \langle a'|\alpha\rangle. \tag{1.3.8}$$

In other words, we have

$$|\alpha\rangle = \sum_{a'} |a'\rangle\langle a'|\alpha\rangle, \tag{1.3.9}$$

which is analogous to an expansion of a vector $\mathbf{V}$ in (real) Euclidean space:

$$\mathbf{V} = \sum_{i} \hat{\mathbf{e}}_i(\hat{\mathbf{e}}_i \cdot \mathbf{V}), \tag{1.3.10}$$

where $\{\hat{\mathbf{e}}_i\}$ form an orthogonal set of unit vectors. We now recall the associative axiom of multiplication: $|a'\rangle\langle a'|\alpha\rangle$ can be regarded either as the number $\langle a'|\alpha\rangle$ multiplying $|a'\rangle$ or, equivalently, as the operator $|a'\rangle\langle a'|$ acting on $|\alpha\rangle$. Because $|\alpha\rangle$ in (1.3.9) is an arbitrary ket, we must have

$$\sum_{a'} |a'\rangle\langle a'| = 1, \tag{1.3.11}$$

where the 1 on the right-hand side is to be understood as the identity *operator*. Equation (1.3.11) is known as the **completeness relation** or **closure**.

It is difficult to oveiestimate the usefulness of (1.3.11). Given a chain of kets, operators, or bras multiplied in legal orders, we can insert, in any place at our convenience, the identity operator written in form (1.3.11). Consider, for example $\langle\alpha|\alpha\rangle$; by inserting the identity operator between $\langle\alpha|$ and $|\alpha\rangle$, we obtain

$$\begin{aligned}\langle\alpha|\alpha\rangle &= \langle\alpha|\cdot\left(\sum_{a'}|a'\rangle\langle a'|\right)\cdot|\alpha\rangle \\ &= \sum_{a'}|\langle a'|\alpha\rangle|^2\end{aligned} \tag{1.3.12}$$

This, incidentally, shows that if $|\alpha\rangle$ is normalized, then the expansion coefficients in (1.3.7) must satisfy

$$\sum_{a'}|c_{a'}|^2 = \sum_{a'}|\langle a'|\alpha\rangle|^2 = 1. \tag{1.3.13}$$

Let us now look at $|a'\rangle\langle a'|$ that appears in (1.3.11). Since this is an outer product, it must be an operator. Let it operate on $|\alpha\rangle$:

$$(|a'\rangle\langle a'|)\cdot|\alpha\rangle = |a'\rangle\langle a'|\alpha\rangle = c_{a'}|a'\rangle. \tag{1.3.14}$$

We see that $|a'\rangle\langle a'|$ selects that portion of the ket $|\alpha\rangle$ parallel to $|a'\rangle$, so $|a'\rangle\langle a'|$ is known as the **projection operator** along the base ket $|a'\rangle$ and is denoted by $\Lambda_{a'}$:

$$\Lambda_{a'} \equiv |a'\rangle\langle a'|. \tag{1.3.15}$$

The completeness relation (1.3.11) can now be written as

$$\sum_{a'} \Lambda_{a'} = 1. \tag{1.3.16}$$

Matrix Representations

Having specified the base kets, we now show how to represent an operator, say X, by a square matrix. First, using (1.3.11) twice, we write the operator X as

$$X = \sum_{a''}\sum_{a'} |a''\rangle\langle a''|X|a'\rangle\langle a'|. \tag{1.3.17}$$

There are altogether N^2 numbers of form $\langle a''|X|a'\rangle$, where N is the dimensionality of the ket space. We may arrange them into an $N \times N$ square matrix such that the column and row indices appear as follows:

$$\underset{\text{row}}{\langle a''|}X\underset{\text{column}}{|a'\rangle}. \tag{1.3.18}$$

Explicitly we may write the matrix as

$$X \doteq \begin{pmatrix} \langle a^{(1)}|X|a^{(1)}\rangle & \langle a^{(1)}|X|a^{(2)}\rangle & \cdots \\ \langle a^{(2)}|X|a^{(1)}\rangle & \langle a^{(2)}|X|a^{(2)}\rangle & \cdots \\ \vdots & \vdots & \ddots \end{pmatrix}, \tag{1.3.19}$$

where the symbol $\doteq$ stands for "is represented by."*

Using (1.2.38), we can write

$$\langle a''|X|a'\rangle = \langle a'|X^\dagger|a''\rangle^*. \tag{1.3.20}$$

At last, the Hermitian adjoint operation, originally defined by (1.2.24), has been related to the (perhaps more familiar) concept of *complex conjugate transposed*. If an operator B is Hermitian, we have

$$\langle a''|B|a'\rangle = \langle a'|B|a''\rangle^*. \tag{1.3.21}$$

---

*We do not use the equality sign here because the particular form of a matrix representation depends on the particular choice of base kets used. The operator is different from a representation of the operator just as the actress is different from a poster of the actress.

The way we arranged $\langle a''|X|a'\rangle$ into a square matrix is in conformity with the usual rule of matrix multiplication. To see this just note that the matrix representation of the operator relation

$$Z = XY \tag{1.3.22}$$

reads

$$\begin{aligned}\langle a''|Z|a'\rangle &= \langle a''|XY|a'\rangle \\ &= \sum_{a'''} \langle a''|X|a'''\rangle\langle a'''|Y|a'\rangle.\end{aligned} \tag{1.3.23}$$

Again, all we have done is to insert the identity operator, written in form (1.3.11), between X and Y!

Let us now examine how the ket relation

$$|\gamma\rangle = X|\alpha\rangle \tag{1.3.24}$$

can be represented using our base kets. The expansion coefficients of $|\gamma\rangle$ can be obtained by multiplying $\langle a'|$ on the left:

$$\begin{aligned}\langle a'|\gamma\rangle &= \langle a'|X|\alpha\rangle \\ &= \sum_{a''} \langle a'|X|a''\rangle\langle a''|\alpha\rangle.\end{aligned} \tag{1.3.25}$$

But this can be seen as an application of the rule for multiplying a square matrix with a column matrix representing once the expansion coefficients of $|\alpha\rangle$ and $|\gamma\rangle$ arrange themselves to form column matrices as follows:

$$|\alpha\rangle \doteq \begin{pmatrix} \langle a^{(1)}|\alpha\rangle \\ \langle a^{(2)}|\alpha\rangle \\ \langle a^{(3)}|\alpha\rangle \\ \vdots \end{pmatrix}, \qquad |\gamma\rangle \doteq \begin{pmatrix} \langle a^{(1)}|\gamma\rangle \\ \langle a^{(2)}|\gamma\rangle \\ \langle a^{(3)}|\gamma\rangle \\ \vdots \end{pmatrix}. \tag{1.3.26}$$

Likewise, given

$$\langle\gamma| = \langle\alpha|X, \tag{1.3.27}$$

we can regard

$$\langle\gamma|a'\rangle = \sum_{a''} \langle\alpha|a''\rangle\langle a''|X|a'\rangle. \tag{1.3.28}$$

So a bra is represented by a row matrix as follows:

$$\begin{aligned}\langle\gamma| &\doteq (\langle\gamma|a^{(1)}\rangle, \langle\gamma|a^{(2)}\rangle, \langle\gamma|a^{(3)}\rangle, \ldots) \\ &= (\langle a^{(1)}|\gamma\rangle^*, \langle a^{(2)}|\gamma\rangle^*, \langle a^{(3)}|\gamma\rangle^*, \ldots).\end{aligned} \tag{1.3.29}$$

Note the appearance of complex conjugation when the elements of the column matrix are written as in (1.3.29). The inner product $\langle\beta|\alpha\rangle$ can be written as the product of the row matrix representing $\langle\beta|$ with the column matrix representing $|\alpha\rangle$:

$$\langle\beta|\alpha\rangle = \sum_{a'} \langle\beta|a'\rangle\langle a'|\alpha\rangle$$
$$= \left(\langle a^{(1)}|\beta\rangle^*, \langle a^{(2)}|\beta\rangle^*, \ldots\right) \begin{pmatrix} \langle a^{(1)}|\alpha\rangle \\ \langle a^{(2)}|\alpha\rangle \\ \vdots \end{pmatrix} \tag{1.3.30}$$

If we multiply the row matrix representing $\langle\alpha|$ with the column matrix representing $|\beta\rangle$, then we obtain just the complex conjugate of the preceding expression, which is consistent with the fundamental property of the inner product (1.2.12). Finally, the matrix representation of the outer product $|\beta\rangle\langle\alpha|$ is easily seen to be

$$|\beta\rangle\langle\alpha| \doteq \begin{pmatrix} \langle a^{(1)}|\beta\rangle\langle a^{(1)}|\alpha\rangle^* & \langle a^{(1)}|\beta\rangle\langle a^{(2)}|\alpha\rangle^* & \ldots \\ \langle a^{(2)}|\beta\rangle\langle a^{(1)}|\alpha\rangle^* & \langle a^{(2)}|\beta\rangle\langle a^{(2)}|\alpha\rangle^* & \ldots \\ \vdots & \vdots & \ddots \end{pmatrix}. \tag{1.3.31}$$

The matrix representation of an observable A becomes particularly simple if the eigenkets of A themselves are used as the base kets. First, we have

$$A = \sum_{a''}\sum_{a'} |a''\rangle\langle a''|A|a'\rangle\langle a'|. \tag{1.3.32}$$

But the square matrix $\langle a''|A|a'\rangle$ is obviously diagonal,

$$\langle a''|A|a'\rangle = \langle a'|A|a'\rangle\delta_{a'a''} = a'\delta_{a'a''}, \tag{1.3.33}$$

so

$$A = \sum_{a'} a'|a'\rangle\langle a'|$$
$$= \sum_{a'} a'\Lambda_{a'}. \tag{1.3.34}$$

Spin $\frac{1}{2}$ Systems

It is here instructive to consider the special case of spin $\frac{1}{2}$ systems. The base kets used are $|S_z;\pm\rangle$, which we denote, for brevity, as $|\pm\rangle$. The simplest operator in the ket space spanned by $|\pm\rangle$ is the identity operator, which, according to (1.3.11), can be written as

$$1 = |+\rangle\langle+| + |-\rangle\langle-|. \tag{1.3.35}$$

According to (1.3.34), we must be able to write S_z as

$$S_z = (\hbar/2)\left[(|+\rangle\langle+|) - (|-\rangle\langle-|)\right]. \tag{1.3.36}$$

The eigenket-eigenvalue relation

$$S_z|\pm\rangle = \pm(\hbar/2)|\pm\rangle \tag{1.3.37}$$

immediately follows from the orthonormality property of $|\pm\rangle$.

It is also instructive to look at two other operators,

$$S_+ \equiv \hbar|+\rangle\langle -|, \qquad S_- \equiv \hbar|-\rangle\langle +|, \tag{1.3.38}$$

which are both seen to be *non*-Hermitian. The operator S_+, acting on the spin-down ket $|-\rangle$, turns $|-\rangle$ into the spin-up ket $|+\rangle$ multiplied by $\hbar$. On the other hand, the spin-up ket $|+\rangle$, when acted upon by S_+, becomes a null ket. So the physical interpretation of S_+ is that it raises the spin component by one unit of $\hbar$; if the spin component cannot be raised any further, we automatically get a null state. Likewise, S_- can be interpreted as an operator that lowers the spin component by one unit of $\hbar$. Later we will show that $S_\pm$ can be written as $S_x \pm iS_y$.

In constructing the matrix representations of the angular momentum operators, it is customary to label the column (row) indices in *descending* order of angular momentum components, that is, the first entry corresponds to the maximum angular momentum component, the second, the next highest, and so forth. In our particular case of spin $\frac{1}{2}$ systems, we have

$$|+\rangle \doteq \begin{pmatrix} 1 \\ 0 \end{pmatrix}, \qquad |-\rangle \doteq \begin{pmatrix} 0 \\ 1 \end{pmatrix}, \tag{1.3.39a}$$

$$S_z \doteq \frac{\hbar}{2}\begin{pmatrix} 1 & 0 \\ 0 & -1 \end{pmatrix}, \qquad S_+ \doteq \hbar\begin{pmatrix} 0 & 1 \\ 0 & 0 \end{pmatrix}, \qquad S_- \doteq \hbar\begin{pmatrix} 0 & 0 \\ 1 & 0 \end{pmatrix}. \tag{1.3.39b}$$

We will come back to these explicit expressions when we discuss the Pauli two-component formalism in Chapter 3.

1.4. MEASUREMENTS, OBSERVABLES, AND THE UNCERTAINTY RELATIONS

Measurements

Having developed the mathematics of ket spaces, we are now in a position to discuss the quantum theory of measurement processes. This is not a particularly easy subject for beginners, so we first turn to the words of the great master, P. A. M. Dirac, for guidance (Dirac 1958, 36): "A measurement always causes the system to jump into an eigenstate of the dynamical variable that is being measured." What does all this mean? We interpret Dirac's words as follows: Before a measurement of observable A is made, the system is assumed to be represented by some linear combination

$$|\alpha\rangle = \sum_{a'} c_{a'}|a'\rangle = \sum_{a'}|a'\rangle\langle a'|\alpha\rangle. \tag{1.4.1}$$

When the measurement is performed, the system is "thrown into" one of the eigenstates, say $|a'\rangle$ of observable A. In other words,

$$|\alpha\rangle \xrightarrow{A \text{ measurement}} |a'\rangle. \tag{1.4.2}$$

For example, a silver atom with an arbitrary spin orientation will change into either $|S_z;+\rangle$ or $|S_z;-\rangle$ when subjected to a SG apparatus of type SG$\hat{z}$. Thus *a measurement usually changes the state*. The only exception is when the state is already in one of the eigenstates of the observable being measured, in which case

$$|a'\rangle \xrightarrow{A \text{ measurement}} |a'\rangle \tag{1.4.3}$$

with certainty, as will be discussed further. When the measurement causes $|\alpha\rangle$ to change into $|a'\rangle$, it is said that A is measured to be a'. It is in this sense that the result of a measurement yields one of the eigenvalues of the observable being measured.

Given (1.4.1), which is the state ket of a physical system before the measurement, we do not know in advance into which of the various $|a'\rangle$'s the system will be thrown as the result of the measurement. We do postulate, however, that the probability for jumping into some particular $|a'\rangle$ is given by

$$\text{Probability for } a' = |\langle a'|\alpha\rangle|^2, \tag{1.4.4}$$

provided that $|\alpha\rangle$ is normalized.

Although we have been talking about a single physical system, to determine probability (1.4.4) empirically, we must consider a great number of measurements performed on an ensemble—that is, a collection—of identically prepared physical systems, all characterized by the same ket $|\alpha\rangle$. Such an ensemble is known as a **pure ensemble**. (We will say more about ensembles in Chapter 3.) As an example, a beam of silver atoms which survive the first SG$\hat{z}$ apparatus of Figure 1.3 with the S_z- component blocked is an example of a pure ensemble because every member atom of the ensemble is characterized by $|S_z;+\rangle$.

The probabilistic interpretation (1.4.4) for the squared inner product $|\langle a'|\alpha\rangle|^2$ is one of the fundamental postulates of quantum mechanics, so it cannot be proven. Let us note, however, that it makes good sense in extreme cases. Suppose the state ket is $|a'\rangle$ itself even before a measurement is made; then according to (1.4.4), the probability for getting a'—or, more precisely, for being thrown into $|a'\rangle$—as the result of the measurement is predicted to be 1, which is just what we expect. By measuring A once again, we, of course, get $|a'\rangle$ only; quite generally, repeated measurements of the same observable in succession yield the same result.* If, on the other hand, we are interested in the probability for the system initially characterized by $|a'\rangle$ to be thrown into some other eigenket $|a''\rangle$ with $a'' \neq a'$, then (1.4.4) gives zero because of the orthogonality between $|a'\rangle$ and $|a''\rangle$. From the point of view of measurement theory, orthogonal kets correspond to mutually exclusive alternatives; for example, if a spin $\frac{1}{2}$ system is in $|S_z; +\rangle$, it is not in $|S_z; -\rangle$ with certainty.

Quite generally, the probability for anything must be nonnegative. Furthermore, the probabilities for the various alternative possibilities must add up to unity. Both of these expectations are met by our probability postulate (1.4.4).

We define the **expectation value** of A taken with respect to state $|\alpha\rangle$ as

$$\langle A\rangle \equiv \langle\alpha|A|\alpha\rangle. \tag{1.4.5}$$

To make sure that we are referring to state $|\alpha\rangle$, the notation $\langle A\rangle_\alpha$ is sometimes used. Equation (1.4.5) is a definition; however, it agrees with our intuitive notion of *average measured value* because it can be written as

$$\begin{aligned}\langle A\rangle &= \sum_{a'}\sum_{a''}\langle\alpha|a''\rangle\langle a''|A|a'\rangle\langle a'|\alpha\rangle \\ &= \sum_{a'} \underset{\text{measured value } a'}{\underset{\uparrow}{a'}} \qquad \underbrace{|\langle a'|\alpha\rangle|^2}_{\text{probability for obtaining } a'}\end{aligned} \tag{1.4.6}$$

It is very important not to confuse eigenvalues with expectation values. For example, the expectation value of S_z for spin $\frac{1}{2}$ systems can assume *any* real value between $-\hbar/2$ and $+\hbar/2$, say $0.273\hbar$; in contrast, the eigenvalue of S_z assumes only two values, $\hbar/2$ and $-\hbar/2$.

To clarify further the meaning of measurements in quantum mechanics we introduce the notion of a **selective measurement**, or *filtration*. In Section 1.1 we considered a Stern-Gerlach arrangement where we let only one of the spin components pass out of the apparatus while we completely blocked the other component. More generally, we imagine a measurement process with a device that selects only one of the eigenkets of A, say $|a'\rangle$, and rejects all others; see Figure 1.6. This is what we mean by a selective measurement; it is also called filtration because only one of the A eigenkets filters through the ordeal. Mathematically we can say that such a selective

*Here successive measurements must be carried out immediately afterward. This point will become clear when we discuss the time evolution of a state ket in Chapter 2.

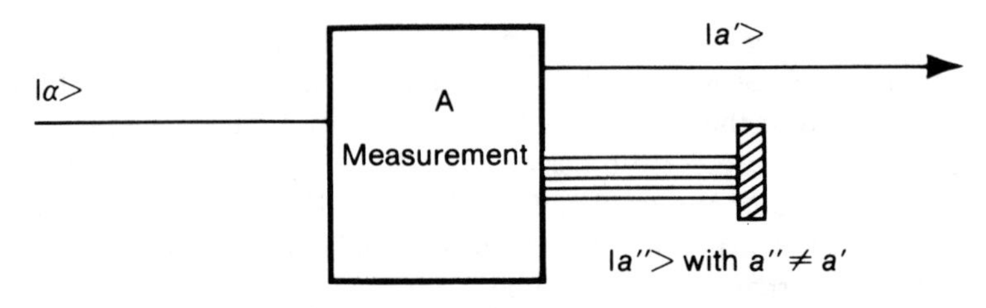

FIGURE 1.6. Selective measurement.

measurement amounts to applying the projection operator $\Lambda_{a'}$ to $|\alpha\rangle$:

$$\Lambda_{a'}|\alpha\rangle = |a'\rangle\langle a'|\alpha\rangle. \tag{1.4.7}$$

J. Schwinger has developed a formalism of quantum mechanics based on a thorough examination of selective measurements. He introduces a measurement symbol $M(a')$ in the beginning, which is identical to $\Lambda_{a'}$ or $|a'\rangle\langle a'|$ in our notation, and deduces a number of properties of $M(a')$ (and also of $M(b', a')$ which amount to $|b'\rangle\langle a'|$) by studying the outcome of various Stern-Gerlach–type experiments. In this way he motivates the entire mathematics of kets, bras, and operators. In this book we do not follow Schwinger's path; the interested reader may consult Gottfried's book. (Gottfried 1966, 192–9).

Spin $\frac{1}{2}$ Systems, Once Again

Before proceeding with a general discussion of observables, we once again consider spin $\frac{1}{2}$ systems. This time we show that the results of sequential Stern-Gerlach experiments, when combined with the postulates of quantum mechanics discussed so far, are sufficient to determine not only the $S_{x,y}$ eigenkets, $|S_x; \pm\rangle$ and $|S_y; \pm\rangle$, but also the operators S_x and S_y themselves.

First, we recall that when the S_x+ beam is subjected to an apparatus of type SG$\hat{z}$, the beam splits into two components with equal intensities. This means that the probability for the S_x+ state to be thrown into $|S_z; \pm\rangle$, simply denoted as $|\pm\rangle$, is $\frac{1}{2}$ each; hence,

$$|\langle +|S_x; +\rangle| = |\langle -|S_x; +\rangle| = \frac{1}{\sqrt{2}}. \tag{1.4.8}$$

We can therefore construct the S_x+ ket as follows:

$$|S_x; +\rangle = \frac{1}{\sqrt{2}}|+\rangle + \frac{1}{\sqrt{2}}e^{i\delta_1}|-\rangle, \tag{1.4.9}$$

with δ_1 real. In writing (1.4.9) we have used the fact that the *overall* phase (common to both $|+\rangle$ and $|-\rangle$) of a state ket is immaterial; the coefficient of $|+\rangle$ can be chosen to be real and positive by convention. The S_x- ket must be orthogonal to the S_x+ ket because the S_x+ alternative and S_x- alternative are mutually exclusive. This orthogonality requirement leads to

$$|S_x;-\rangle = \frac{1}{\sqrt{2}}|+\rangle - \frac{1}{\sqrt{2}}e^{i\delta_1}|-\rangle, \tag{1.4.10}$$

where we have, again, chosen the coefficient of $|+\rangle$ to be real and positive by convention. We can now construct the operator S_x using (1.3.34) as follows:

$$\begin{aligned} S_x &= \frac{\hbar}{2}[(|S_x;+\rangle\langle S_x;+|)-(|S_x;-\rangle\langle S_x;-|)] \\ &= \frac{\hbar}{2}\left[e^{-i\delta_1}(|+\rangle\langle -|)+e^{i\delta_1}(|-\rangle\langle +|)\right]. \end{aligned} \tag{1.4.11}$$

Notice that the S_x we have constructed is Hermitian, just as it must be. A similar argument with S_x replaced by S_y leads to

$$|S_y;\pm\rangle = \frac{1}{\sqrt{2}}|+\rangle \pm \frac{1}{\sqrt{2}}e^{i\delta_2}|-\rangle, \tag{1.4.12}$$

$$S_y = \frac{\hbar}{2}\left[e^{-i\delta_2}(|+\rangle\langle -|)+e^{i\delta_2}(|-\rangle\langle +|)\right]. \tag{1.4.13}$$

Is there any way of determining δ_1 and δ_2? Actually there is one piece of information we have not yet used. Suppose we have a beam of spin $\frac{1}{2}$ atoms moving in the z-direction. We can consider a sequential Stern-Gerlach experiment with SG$\hat{x}$ followed by SG$\hat{y}$. The results of such an experiment are completely analogous to the earlier case leading to (1.4.8):

$$|\langle S_y;\pm|S_x;+\rangle| = |\langle S_y;\pm|S_x;-\rangle| = \frac{1}{\sqrt{2}}, \tag{1.4.14}$$

which is not surprising in view of the invariance of physical systems under rotations. Inserting (1.4.10) and (1.4.12) into (1.4.14), we obtain

$$\frac{1}{2}|1 \pm e^{i(\delta_1-\delta_2)}| = \frac{1}{\sqrt{2}}, \tag{1.4.15}$$

which is satisfied only if

$$\delta_2 - \delta_1 = \pi/2 \quad \text{or} \quad -\pi/2. \tag{1.4.16}$$

We thus see that the matrix elements of S_x and S_y cannot all be real. If the S_x matrix elements are real, the S_y matrix elements must be purely imaginary (and vice versa). Just from this extremely simple example, the introduction of complex numbers is seen to be an essential feature in quantum mechanics. It is convenient to take the S_x matrix elements to be real* and

*This can always be done by adjusting arbitrary phase factors in the definition of $|+\rangle$ and $|-\rangle$. This point will become clearer in Chapter 3, where the behavior of $|\pm\rangle$ under rotations will be discussed.

set $\delta_1 = 0$; if we were to choose $\delta_1 = \pi$, the positive x-axis would be oriented in the opposite direction. The second phase angle δ_2 must then be $-\pi/2$ or $\pi/2$. The fact that there is still an ambiguity of this kind is not surprising. We have not yet specified whether the coordinate system we are using is right-handed or left-handed; given the x- and the z-axes there is still a twofold ambiguity in the choice of the positive y-axis. Later we will discuss angular momentum as a generator of rotations using the right-handed coordinate system; it can then be shown that $\delta_2 = \pi/2$ is the correct choice.

To summarize, we have

$$|S_x; \pm\rangle = \frac{1}{\sqrt{2}}|+\rangle \pm \frac{1}{\sqrt{2}}|-\rangle, \tag{1.4.17a}$$

$$|S_y; \pm\rangle = \frac{1}{\sqrt{2}}|+\rangle \pm \frac{i}{\sqrt{2}}|-\rangle, \tag{1.4.17b}$$

and

$$S_x = \frac{\hbar}{2}\left[(|+\rangle\langle -|) + (|-\rangle\langle +|)\right], \tag{1.4.18a}$$

$$S_y = \frac{\hbar}{2}\left[-i(|+\rangle\langle -|) + i(|-\rangle\langle +|)\right]. \tag{1.4.18b}$$

The $S_x \pm$ and $S_y \pm$ eigenkets given here are seen to be in agreement with our earlier guesses (1.1.9) and (1.1.14) based on an analogy with linearly and circularly polarized light. (Note, in this comparison, that only the relative phase between the $|+\rangle$ and $|-\rangle$ components is of physical significance.) Furthermore, the non-Hermitian $S_\pm$ operators defined by (1.3.38) can now be written as

$$S_\pm = S_x \pm iS_y. \tag{1.4.19}$$

The operators S_x and S_y, together with S_z given earlier, can be readily shown to satisfy the commutation relations

$$\left[S_i, S_j\right] = i\varepsilon_{ijk}\hbar S_k, \tag{1.4.20}$$

and the anticommutation relations

$$\{S_i, S_j\} = \frac{1}{2}\hbar^2\delta_{ij}, \tag{1.4.21}$$

where the commutator [,] and the anticommutator { , } are defined by

$$[A, B] \equiv AB - BA, \tag{1.4.22a}$$

$$\{A, B\} \equiv AB + BA. \tag{1.4.22b}$$

The commutation relations in (1.4.20) will be recognized as the simplest realization of the angular momentum commutation relations, whose significance will be discussed in detail in Chapter 3. In contrast, the anticommutation relations in (1.4.21) turn out to be a *special* property of spin $\frac{1}{2}$ systems.

We can also define the operator $\mathbf{S}\cdot\mathbf{S}$, or $\mathbf{S}^2$ for short, as follows:

$$\mathbf{S}^2 \equiv S_x^2 + S_y^2 + S_z^2. \tag{1.4.23}$$

Because of (1.4.21), this operator turns out to be just a constant multiple of the identity operator

$$\mathbf{S}^2 = \left(\frac{3}{4}\right)\hbar^2. \tag{1.4.24}$$

We obviously have

$$[\mathbf{S}^2, S_i] = 0. \tag{1.4.25}$$

As will be shown in Chapter 3, for spins higher than $\frac{1}{2}$, $\mathbf{S}^2$ is no longer a multiple of the identity operator; however, (1.4.25) still holds.

Compatible Observables

Returning now to the general formalism, we will discuss compatible versus incompatible observables. Observables A and B are defined to be **compatible** when the corresponding operators commute,

$$[A, B] = 0, \tag{1.4.26}$$

and **incompatible** when

$$[A, B] \neq 0, \tag{1.4.27}$$

For example, $\mathbf{S}^2$ and S_z are compatible observables, while S_x and S_z are incompatible observables.

Let us first consider the case of compatible observables A and B. As usual, we assume that the ket space is spanned by the eigenkets of A. We may also regard the same ket space as being spanned by the eigenkets of B. We now ask, How are the A eigenkets related to the B eigenkets when A and B are compatible observables?

Before answering this question we must touch upon a very important point we have bypassed earlier—the concept of *degeneracy*. Suppose there are two (or more) linearly independent eigenkets of A having the same eigenvalue; then the eigenvalues of the two eigenkets are said to be **degenerate**. In such a case the notation $|a'\rangle$ that labels the eigenket by its eigenvalue alone does not give a complete description; furthermore, we may recall that our earlier theorem on the orthogonality of different eigenkets was proved under the assumption of no degeneracy. Even worse, the whole concept that the ket space is spanned by $\{|a'\rangle\}$ appears to run into difficulty when the dimensionality of the ket space is larger than the number of distinct eigenvalues of A. Fortunately, in practical applications in quantum mechanics, it is usually the case that in such a situation the eigenvalues of *some other* commuting observable, say B, can be used to label the degenerate eigenkets.

Now we are ready to state an important theorem.

Theorem. *Suppose that A and B are compatible observables, and the eigenvalues of A are nondegenerate. Then the matrix elements $\langle a''|B|a'\rangle$ are all diagonal. (Recall here that the matrix elements of A are already diagonal if $\{|a'\rangle\}$ are used as the base kets.)*

Proof. The proof of this important theorem is extremely simple. Using the definition (1.4.26) of compatible observables, we observe that

$$\langle a''|[A, B]|a'\rangle = (a'' - a')\langle a''|B|a'\rangle = 0. \tag{1.4.28}$$

So $\langle a''|B|a'\rangle$ must vanish unless $a' = a''$, which proves our assertion. □

We can write the matrix elements of B as

$$\langle a''|B|a'\rangle = \delta_{a'a''}\langle a'|B|a'\rangle. \tag{1.4.29}$$

So both A and B can be represented by diagonal matrices with the *same* set of base kets. Using (1.3.17) and (1.4.29) we can write B as

$$B = \sum_{a''} |a''\rangle\langle a''|B|a''\rangle\langle a''|. \tag{1.4.30}$$

Suppose that this operator acts on an eigenket of A:

$$B|a'\rangle = \sum_{a''} |a''\rangle\langle a''|B|a''\rangle\langle a''|a'\rangle = (\langle a'|B|a'\rangle)|a'\rangle. \tag{1.4.31}$$

But this is nothing other than the eigenvalue equation for the operator B with eigenvalue

$$b' \equiv \langle a'|B|a'\rangle. \tag{1.4.32}$$

The ket $|a'\rangle$ is therefore a **simultaneous eigenket** of A and B. Just to be impartial to both operators, we may use $|a', b'\rangle$ to characterize this simultaneous eigenket.

We have seen that compatible observables have simultaneous eigenkets. Even though the proof given is for the case where the A eigenkets are nondegenerate, the statement holds even if there is an n-fold degeneracy, that is,

$$A|a'^{(i)}\rangle = a'|a'^{(i)}\rangle \quad \text{for } i = 1, 2, \ldots, n \tag{1.4.33}$$

where $|a'^{(i)}\rangle$ are n mutually orthonormal eigenkets of A, all with the same eigenvalue a'. To see this, all we need to do is construct appropriate linear combinations of $|a'^{(i)}\rangle$ that diagonalize the B operator by following the diagonalization procedure to be discussed in Section 1.5.

A simultaneous eigenket of A and B, denoted by $|a', b'\rangle$, has the property

$$A|a', b'\rangle = a'|a', b'\rangle, \tag{1.4.34a}$$

$$B|a', b'\rangle = b'|a', b'\rangle. \tag{1.4.34b}$$

When there is no degeneracy, this notation is somewhat superfluous because it is clear from (1.4.32) that if we specify a', we necessarily know the b' that appears in $|a', b'\rangle$. The notation $|a', b'\rangle$ is much more powerful when there are degeneracies. A simple example may be used to illustrate this point.

Even though a complete discussion of orbital angular momentum will not appear in this book until Chapter 3, the reader may be familiar from his or her earlier training in elementary wave mechanics that the eigenvalues of $\mathbf{L}^2$ (orbital angular momentum squared) and L_z (the z-component of orbital angular momentum) are $\hbar^2 l(l+1)$ and $m_l \hbar$, respectively, with l an integer and $m_l = -l, -l+1, \ldots, +l$. To characterize an orbital angular momentum state completely, it is necessary to specify *both* l and m_l. For example, if we just say $l=1$, the m_l value can still be 0, $+1$, or -1; if we just say $m_l = 1$, l can be 1, 2, 3, 4, and so on. Only by specifying *both* l and m_l do we succeed in uniquely characterizing the orbital angular momentum state in question. Quite often a **collective index** K' is used to stand for (a', b'), so that

$$|K'\rangle = |a', b'\rangle. \tag{1.4.35}$$

We can obviously generalize our considerations to a situation where there are several (more than two) mutually compatible observables, namely,

$$[A, B] = [B, C] = [A, C] = \cdots = 0. \tag{1.4.36}$$

Assume that we have found a **maximal** set of commuting observables; that is, we cannot add any more observables to our list without violating (1.4.36). The eigenvalues of individual operators $A, B, C, \ldots$ may have degeneracies, but if we specify a combination $(a', b', c', \ldots)$, then the corresponding simultaneous eigenket of $A, B, C, \ldots$ is uniquely specified. We can again use a collective index K' to stand for $(a', b', c', \ldots)$. The orthonormality relation for

$$|K'\rangle = |a', b', c', \ldots\rangle \tag{1.4.37}$$

reads

$$\langle K''|K'\rangle = \delta_{K'K''} = \delta_{aa'}\delta_{bb'}\delta_{cc'} \cdots, \tag{1.4.38}$$

while the completeness relation, or closure, can be written as

$$\sum_{K'} |K'\rangle\langle K'| = \sum_{a'}\sum_{b'}\sum_{c'} \cdots |a', b', c', \ldots\rangle\langle a', b', c', \ldots| = 1. \tag{1.4.39}$$

We now consider measurements of A and B when they are compatible observables. Suppose we measure A first and obtain result a'. Subsequently, we may measure B and get result b'. Finally we measure A again. It follows from our measurement formalism that the third measurement always gives a' with certainty, that is, the second (B) measurement does not destroy the previous information obtained in the first (A) measurement. This is rather obvious when the eigenvalues of A are nondegenerate:

$$|\alpha\rangle \xrightarrow{A \text{ measurement}} |a', b'\rangle \xrightarrow{B \text{ measurement}} |a', b'\rangle \xrightarrow{A \text{ measurement}} |a', b'\rangle. \tag{1.4.40}$$

When there is degeneracy, the argument goes as follows: After the first (A) measurement, which yields a', the system is thrown into some linear combination

$$\sum_i^n c_{a'}^{(i)} |a', b^{(i)}\rangle, \tag{1.4.41}$$

where n is the degree of degeneracy and the kets $|a', b^{(i)}\rangle$ all have the same eigenvalue a' as far as operator A is concerned. The second (B) measurement may select just one of the terms in the linear combination (1.4.41), say, $|a', b^{(j)}\rangle$, but the third (A) measurement applied to it still yields a'. Whether or not there is degeneracy, A measurements and B measurements do not interfere. The term *compatible* is indeed deemed appropriate.

Incompatible Observables

We now turn to incompatible observables, which are more nontrivial. The first point to be emphasized is that incompatible observables do not have a complete set of simultaneous eigenkets. To show this let us assume the converse to be true. There would then exist a set of simultaneous eigenkets with property (1.4.34a) and (1.4.34b). Clearly,

$$AB|a', b'\rangle = Ab'|a', b'\rangle = a'b'|a', b'\rangle. \tag{1.4.42}$$

Likewise,

$$BA|a', b'\rangle = Ba'|a', b'\rangle = a'b'|a', b'\rangle; \tag{1.4.43}$$

hence,

$$AB|a', b'\rangle = BA|a', b'\rangle, \tag{1.4.44}$$

and thus $[A, B] = 0$ in contradiction to the assumption. So in general, $|a', b'\rangle$ does not make sense for incompatible observables. There is, however, an interesting exception; it may happen that there exists a subspace of the ket space such that (1.4.44) holds for all elements of this subspace, even though A and B are incompatible. An example from the theory of orbital angular momentum may be helpful here. Suppose we consider an $l = 0$ state (s-state). Even though L_x and L_z do *not* commute, this state *is* a simultaneous eigenstate of L_x and L_z (with eigenvalue zero for both operators). The subspace in this case is one-dimensional.

We already encountered some of the peculiarities associated with incompatible observables when we discussed sequential Stern-Gerlach experiments in Section 1.1. We now give a more general discussion of experiments of that type. Consider the sequence of selective measurements shown in Figure 1.7(a). The first (A) filter selects some particular $|a'\rangle$ and rejects all others, the second (B) filter selects some particular $|b'\rangle$ and rejects all others, and the third (C) filter selects some particular $|c'\rangle$ and rejects all others. We are interested in the probability of obtaining $|c'\rangle$ when the beam coming out of the first filter is normalized to unity. Because the probabilities are multiplicative, we obviously have

$$|\langle c'|b'\rangle|^2 |\langle b'|a'\rangle|^2. \tag{1.4.45}$$

Now let us sum over b' to consider the total probability for going through all possible b' routes. Operationally this means that we first record the probability of obtaining c' with all but the first b' route blocked, then we repeat the procedure with all but the second b' blocked, and so on; then we sum the probabilities at the end and obtain

$$\sum_{b'} |\langle c'|b'\rangle|^2 |\langle b'|a'\rangle|^2 = \sum_{b'} \langle c'|b'\rangle\langle b'|a'\rangle\langle a'|b'\rangle\langle b'|c'\rangle. \tag{1.4.46}$$

We now compare this with a different arrangement, where the B filter is absent (or not operative); see Figure 1.7b. Clearly, the probability is

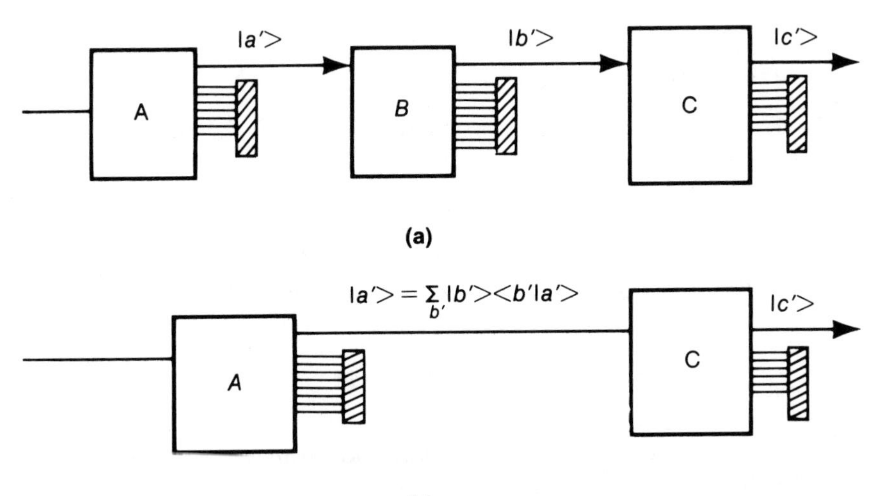

FIGURE 1.7. Sequential selective measurements.

just $|\langle c'|a'\rangle|^2$, which can also be written as follows:

$$|\langle c'|a'\rangle|^2 = |\sum_{b'} \langle c'|b'\rangle\langle b'|a'\rangle|^2 = \sum_{b'}\sum_{b''}\langle c'|b'\rangle\langle b'|a'\rangle\langle a'|b''\rangle\langle b''|c'\rangle. \tag{1.4.47}$$

Notice that expressions (1.4.46) and (1.4.47) are different! This is remarkable because in both cases the pure $|a'\rangle$ beam coming out of the first (A) filter can be regarded as being made up of the B eigenkets

$$|a'\rangle = \sum_{b'} |b'\rangle\langle b'|a'\rangle, \tag{1.4.48}$$

where the sum is over all possible values of b'. The crucial point to be noted is that the result coming out of the C filter depends on whether or not B measurements have actually been carried out. In the first case we experimentally ascertain which of the B eigenvalues are actually realized; in the second case, we merely imagine $|a'\rangle$ to be built up of the various $|b'\rangle$'s in the sense of (1.4.48). Put in another way, actually recording the probabilities of going through the various b' routes makes all the difference even though we sum over b' afterwards. Here lies the heart of quantum mechanics.

Under what conditions do the two expressions become equal? It is left as an exercise for the reader to show that for this to happen, in the absence of degeneracy, it is sufficient that

$$[A, B] = 0 \quad \text{or} \quad [B, C] = 0. \tag{1.4.49}$$

In other words, the peculiarity we have illustrated is characteristic of incompatible observables.

The Uncertainty Relation

The last topic to be discussed in this section is the uncertainty relation. Given an observable A, we define an **operator**

$$\Delta A \equiv A - \langle A\rangle, \tag{1.4.50}$$

where the expectation value is to be taken for a certain physical state under consideration. The expectation value of $(\Delta A)^2$ is known as the **dispersion** of A. Because we have

$$\langle(\Delta A)^2\rangle = \langle(A^2 - 2A\langle A\rangle + \langle A\rangle^2)\rangle = \langle A^2\rangle - \langle A\rangle^2, \tag{1.4.51}$$

the last line of (1.4.51) may be taken as an alternative definition of dispersion. Sometimes the terms **variance** and **mean square deviation** are used for the same quantity. Clearly, the dispersion vanishes when the state in question is an eigenstate of A. Roughly speaking, the dispersion of an observable characterizes "fuzziness." For example, for the S_z+ state of a spin $\frac{1}{2}$ system, the dispersion of S_x can be computed to be

$$\langle S_x^2 \rangle - \langle S_x \rangle^2 = \hbar^2/4. \tag{1.4.52}$$

In contrast the dispersion $\langle \Delta S_z)^2 \rangle$ obviously vanishes for the $S_z +$ state. So, for the $S_z +$ state, S_z is "sharp"—a vanishing dispersion for S_z—while S_x is fuzzy.

We now state the uncertainty relation, which is the generalization of the well-known x-p uncertainty relation to be discussed in Section 1.6. Let A and B be observables. Then for any state we must have the following inequality:

$$\langle (\Delta A)^2 \rangle \langle (\Delta B)^2 \rangle \geq \frac{1}{4} |\langle [A, B] \rangle|^2. \tag{1.4.53}$$

To prove this we first state three lemmas.

Lemma 1. *The Schwarz inequality*

$$\langle \alpha | \alpha \rangle \langle \beta | \beta \rangle \geq |\langle \alpha | \beta \rangle|^2, \tag{1.4.54}$$

which is analogous to

$$|\mathbf{a}|^2 |\mathbf{b}|^2 \geq |\mathbf{a} \cdot \mathbf{b}|^2 \tag{1.4.55}$$

in real Euclidian space.

Proof. First note

$$(\langle \alpha | + \lambda^* \langle \beta |) \cdot (|\alpha\rangle + \lambda |\beta\rangle) \geq 0, \tag{1.4.56}$$

where λ can be any complex number. This inequality must hold when λ is set equal to $-\langle \beta | \alpha \rangle / \langle \beta | \beta \rangle$:

$$\langle \alpha | \alpha \rangle \langle \beta | \beta \rangle - |\langle \alpha | \beta \rangle|^2 \geq 0, \tag{1.4.57}$$

which is the same as (1.4.54). □

Lemma 2. *The expectation value of a Hermitian operator is purely real.*

Proof. The proof is trivial—just use (1.3.21). □

Lemma 3. *The expectation value of an anti-Hermitian operator, defined by $C = -C^\dagger$, is purely imaginary.*

Proof. The proof is also trivial. □

Armed with these lemmas, we are in a position to prove the uncertainty relation (1.4.53). Using Lemma 1 with

$$\begin{aligned} |\alpha\rangle &= \Delta A | \ \rangle, \\ |\beta\rangle &= \Delta B | \ \rangle, \end{aligned} \tag{1.4.58}$$

where the blank ket $| \ \rangle$ emphasizes the fact that our consideration may be applied to *any* ket, we obtain

$$\langle(\Delta A)^2\rangle\langle(\Delta B)^2\rangle \geq |\langle\Delta A \Delta B\rangle|^2, \tag{1.4.59}$$

where the Hermiticity of ΔA and ΔB has been used. To evaluate the right-hand side of (1.4.59), we note

$$\Delta A \Delta B = \frac{1}{2}[\Delta A, \Delta B] + \frac{1}{2}\{\Delta A, \Delta B\}, \tag{1.4.60}$$

where the commutator $[\Delta A, \Delta B]$, which is equal to $[A, B]$, is clearly anti-Hermitian

$$([A, B])^\dagger = (AB - BA)^\dagger = BA - AB = -[A, B]. \tag{1.4.61}$$

In contrast, the anticommutator $\{\Delta A, \Delta B\}$ is obviously Hermitian, so

$$\langle\Delta A \Delta B\rangle = \frac{1}{2}\underbrace{\langle[A, B]\rangle}_{\text{purely imaginary}} + \frac{1}{2}\underbrace{\langle\{\Delta A, \Delta B\}\rangle}_{\text{purely real}}, \tag{1.4.62}$$

where Lemmas 2 and 3 have been used. The right-hand side of (1.4.59) now becomes

$$|\langle\Delta A \Delta B\rangle|^2 = \frac{1}{4}|\langle[A, B]\rangle|^2 + \frac{1}{4}|\langle\{\Delta A, \Delta B\}\rangle|^2. \tag{1.4.63}$$

The proof of (1.4.53) is now complete because the omission of the second (the anticommutator) term of (1.4.63) can only make the inequality relation stronger.*

Applications of the uncertainty relation to spin $\frac{1}{2}$ systems will be left as exercises. We come back to this topic when we discuss the fundamental x-p commutation relation in Section 1.6.

1.5. CHANGE OF BASIS

Transformation Operator

Suppose we have two incompatible observables A and B. The ket space in question can be viewed as being spanned either by the set $\{|a'\rangle\}$ or by the set $\{|b'\rangle\}$. For example, for spin $\frac{1}{2}$ systems $|S_z \pm\rangle$ may be used as our base kets; alternatively, $|S_x \pm\rangle$ may be used as our base kets. The two different sets of base kets, of course, span the same ket space. We are interested in finding out how the two descriptions are related. Changing the

---

*In the literature most authors use ΔA for our $\sqrt{\langle(\Delta A)^2\rangle}$ so the uncertainty relation is written as $\Delta A \Delta B \geq \frac{1}{2}|\langle[A, B]\rangle|$. In this book, however, ΔA and ΔB are to be understood as operators [see (1.4.50)], not numbers.

set of base kets is referred to as a **change of basis** or a **change of representation**. The basis in which the base eigenkets are given by $\{|a'\rangle\}$ is called the A representation or, sometimes, the A diagonal representation because the square matrix corresponding to A is diagonal in this basis.

Our basic task is to construct a transformation operator that connects the old orthonormal set $\{|a'\rangle\}$ and the new orthonormal set $\{|b'\rangle\}$. To this end, we first show the following.

Theorem. *Given two sets of base kets, both satisfying orthonormality and completeness, there exists a unitary operator U such that*

$$|b^{(1)}\rangle = U|a^{(1)}\rangle, |b^{(2)}\rangle = U|a^{(2)}\rangle, \ldots, |b^{(N)}\rangle = U|a^{(N)}\rangle. \qquad (1.5.1)$$

By a **unitary operator** *we mean an operator fulfilling the conditions*

$$U^\dagger U = 1 \qquad (1.5.2)$$

as well as

$$UU^\dagger = 1. \qquad (1.5.3)$$

Proof. We prove this theorem by explicit construction. We assert that the operator

$$U = \sum_k |b^{(k)}\rangle\langle a^{(k)}| \qquad (1.5.4)$$

will do the job and we apply this U to $|a^{(l)}\rangle$. Clearly,

$$U|a^{(l)}\rangle = |b^{(l)}\rangle \qquad (1.5.5)$$

is guaranteed by the orthonormality of $\{|a'\rangle\}$. Furthermore, U is unitary:

$$U^\dagger U = \sum_k \sum_l |a^{(l)}\rangle\langle b^{(l)}|b^{(k)}\rangle\langle a^{(k)}| = \sum_k |a^{(k)}\rangle\langle a^{(k)}| = 1, \qquad (1.5.6)$$

where we have used the orthonormality of $\{|b'\rangle\}$ and the completeness of $\{|a'\rangle\}$. We obtain relation (1.5.3) in an analogous manner. □

Transformation Matrix

It is instructive to study the matrix representation of the U operator in the old $\{|a'\rangle\}$ basis. We have

$$\langle a^{(k)}|U|a^{(l)}\rangle = \langle a^{(k)}|b^{(l)}\rangle, \qquad (1.5.7)$$

which is obvious from (1.5.5). In other words, the matrix elements of the U operator are built up of the inner products of old base bras and new base kets. We recall that the rotation matrix in three dimensions that changes one set of unit base vectors $(\hat{\mathbf{x}}, \hat{\mathbf{y}}, \hat{\mathbf{z}})$ into another set $(\hat{\mathbf{x}}', \hat{\mathbf{y}}', \hat{\mathbf{z}}')$ can be written as

(Goldstein 1980, 128–37, for example)

$$R = \begin{pmatrix} \hat{\mathbf{x}}\cdot\hat{\mathbf{x}}' & \hat{\mathbf{x}}\cdot\hat{\mathbf{y}}' & \hat{\mathbf{x}}\cdot\hat{\mathbf{z}}' \\ \hat{\mathbf{y}}\cdot\hat{\mathbf{x}}' & \hat{\mathbf{y}}\cdot\hat{\mathbf{y}}' & \hat{\mathbf{y}}\cdot\hat{\mathbf{z}}' \\ \hat{\mathbf{z}}\cdot\hat{\mathbf{x}}' & \hat{\mathbf{z}}\cdot\hat{\mathbf{y}}' & \hat{\mathbf{z}}\cdot\hat{\mathbf{z}}' \end{pmatrix}. \tag{1.5.8}$$

The square matrix made up of $\langle a^{(k)}|U|a^{(l)}\rangle$ is referred to as the **transformation matrix** from the $\{|a'\rangle\}$ basis to the $\{|b'\rangle\}$ basis.

Given an arbitrary ket $|\alpha\rangle$ whose expansion coefficients $\langle a'|\alpha\rangle$ are known in the old basis,

$$|\alpha\rangle = \sum_{a'} |a'\rangle\langle a'|\alpha\rangle, \tag{1.5.9}$$

how can we obtain $\langle b'|\alpha\rangle$, the expansion coefficients in the new basis? The answer is very simple: Just multiply (1.5.9) (with a' replaced by $a^{(l)}$ to avoid confusion) by $\langle b^{(k)}|$

$$\langle b^{(k)}|\alpha\rangle = \sum_l \langle b^{(k)}|a^{(l)}\rangle\langle a^{(l)}|\alpha\rangle = \sum_l \langle a^{(k)}|U^\dagger|a^{(l)}\rangle\langle a^{(l)}|\alpha\rangle. \tag{1.5.10}$$

In matrix notation, (1.5.10) states that the column matrix for $|\alpha\rangle$ in the new basis can be obtained just by applying the square matrix $U^\dagger$ to the column matrix in the old basis:

$$(\text{New}) = (U^\dagger)(\text{old}). \tag{1.5.11}$$

The relationships between the old matrix elements and the new matrix elements are also easy to obtain:

$$\begin{aligned} \langle b^{(k)}|X|b^{(l)}\rangle &= \sum_m \sum_n \langle b^{(k)}|a^{(m)}\rangle\langle a^{(m)}|X|a^{(n)}\rangle\langle a^{(n)}|b^{(l)}\rangle \\ &= \sum_m \sum_n \langle a^{(k)}|U^\dagger|a^{(m)}\rangle\langle a^{(m)}|X|a^{(n)}\rangle\langle a^{(n)}|U|a^{(l)}\rangle. \end{aligned} \tag{1.5.12}$$

This is simply the well-known formula for a **similarity transformation** in matrix algebra,

$$X' = U^\dagger X U. \tag{1.5.13}$$

The **trace** of an operator X is defined as the sum of diagonal elements:

$$\text{tr}(X) = \sum_{a'} \langle a'|X|a'\rangle. \tag{1.5.14}$$

Even though a particular set of base kets is used in the definition, $\text{tr}(X)$

turns out to be independent of representation, as shown:

$$\begin{aligned}\sum_{a'}\langle a'|X|a'\rangle &= \sum_{a'}\sum_{b'}\sum_{b''}\langle a'|b'\rangle\langle b'|X|b''\rangle\langle b''|a'\rangle \\ &= \sum_{b'}\sum_{b''}\langle b''|b'\rangle\langle b'|X|b''\rangle \\ &= \sum_{b'}\langle b'|X|b'\rangle. \end{aligned} \tag{1.5.15}$$

We can also prove

$$\operatorname{tr}(XY) = \operatorname{tr}(YX), \tag{1.5.16a}$$

$$\operatorname{tr}(U^\dagger X U) = \operatorname{tr}(X), \tag{1.5.16b}$$

$$\operatorname{tr}(|a'\rangle\langle a''|) = \delta_{a'a''}, \tag{1.5.16c}$$

$$\operatorname{tr}(|b'\rangle\langle a'|) = \langle a'|b'\rangle. \tag{1.5.16d}$$

Diagonalization

So far we have not discussed how to find the eigenvalues and eigenkets of an operator B whose matrix elements in the old $\{|a'\rangle\}$ basis are assumed to be known. This problem turns out to be equivalent to that of finding the unitary matrix that diagonalizes B. Even though the reader may already be familiar with the diagonalization procedure in matrix algebra, it is worth working out this problem using the Dirac bra-ket notation.

We are interested in obtaining the eigenvalue b' and the eigenket $|b'\rangle$ with the property

$$B|b'\rangle = b'|b'\rangle. \tag{1.5.17}$$

First, we rewrite this as

$$\sum_{a'}\langle a''|B|a'\rangle\langle a'|b'\rangle = b'\langle a''|b'\rangle. \tag{1.5.18}$$

When $|b'\rangle$ in (1.5.17) stands for the lth eigenket of operator B, we can write (1.5.18) in matrix notation as follows:

$$\begin{pmatrix} B_{11} & B_{12} & B_{13} & \cdots \\ B_{21} & B_{22} & B_{23} & \cdots \\ \vdots & \vdots & \vdots & \ddots \end{pmatrix}\begin{pmatrix} C_1^{(l)} \\ C_2^{(l)} \\ \vdots \end{pmatrix} = b^{(l)}\begin{pmatrix} C_1^{(l)} \\ C_2^{(l)} \\ \vdots \end{pmatrix}, \tag{1.5.19}$$

with

$$B_{ij} = \langle a^{(i)}|B|a^{(j)}\rangle, \tag{1.5.20a}$$

and

$$C_k^{(l)} = \langle a^{(k)}|b^{(l)}\rangle, \tag{1.5.20b}$$

where i, j, k run up to N, the dimensionality of the ket space. As we know from linear algebra, nontrivial solutions for $C_k^{(l)}$ are possible only if the characteristic equation

$$\det(B - \lambda 1) = 0 \tag{1.5.21}$$

is satisfied. This is an Nth order algebraic equation for λ, and the N roots obtained are to be identified with the various $b^{(l)}$'s we are trying to determine. Knowing $b^{(l)}$ we can solve for the corresponding $C_k^{(l)}$'s up to an overall constant to be determined from the normalization condition. Comparing (1.5.20b) with (1.5.7), we see that the $C_k^{(l)}$'s are just the elements of the unitary matrix involved in the change of basis $\{|a'\rangle\} \to \{|b'\rangle\}$.

For this procedure the Hermiticity of B is important. For example, consider S_+ defined by (1.3.38) or (1.4.19). This operator is obviously non-Hermitian. The corresponding matrix, which reads in the S_z basis as

$$S_+ \doteq \hbar \begin{pmatrix} 0 & 1 \\ 0 & 0 \end{pmatrix}, \tag{1.5.22}$$

cannot be diagonalized by any unitary matrix. In Chapter 2 we will encounter eigenkets of a non-Hermitian operator in connection with a coherent state of a simple harmonic oscillator. Such eigenkets, however, are known *not* to form a complete orthonormal set, and the formalism we have developed in this section cannot be immediately applied.

Unitary Equivalent Observables

We conclude this section by discussing a remarkable theorem on the unitary transform of an observable.

Theorem. *Consider again two sets of orthonormal basis $\{|a'\rangle\}$ and $\{|b'\rangle\}$ connected by the U operator* (1.5.4). *Knowing U, we may construct a* **unitary transform** *of A, UAU^{-1}; then A and UAU^{-1} are said to be* **unitary equivalent observables.** *The eigenvalue equation for A,*

$$A|a^{(l)}\rangle = a^{(l)}|a^{(l)}\rangle, \tag{1.5.23}$$

clearly implies that

$$UAU^{-1}U|a^{(l)}\rangle = a^{(l)}U|a^{(l)}\rangle. \tag{1.5.24}$$

But this can be rewritten as

$$(UAU^{-1})|b^{(l)}\rangle = a^{(l)}|b^{(l)}\rangle. \tag{1.5.25}$$

This deceptively simple result is quite profound. It tells us that the $|b'\rangle$'s are eigenkets of UAU^{-1} with *exactly the same eigenvalues* as the A

eigenvalues. In other words, *unitary equivalent observables have identical spectra.*

The eigenket $|b^{(l)}\rangle$, by definition, satisfies the relationship

$$B|b^{(l)}\rangle = b^{(l)}|b^{(l)}\rangle. \tag{1.5.26}$$

Comparing (1.5.25) and (1.5.26), we infer that B and UAU^{-1} are simultaneously diagonalizable. A natural question is, is UAU^{-1} the same as B itself? The answer quite often is yes in cases of physical interest. Take, for example, S_x and S_z. They are related by a unitary operator, which, as we will discuss in Chapter 3, is actually the rotation operator around the y-axis by angle $\pi/2$. In this case S_x itself is the unitary transform of S_z. Because we know that S_x and S_z exhibit the same set of eigenvalues—namely, $+\hbar/2$ and $-\hbar/2$—we see that our theorem holds in this particular example.

1.6. POSITION, MOMENTUM, AND TRANSLATION

Continuous Spectra

The observables considered so far have all been assumed to exhibit discrete eigenvalue spectra. In quantum mechanics, however, there are observables with continuous eigenvalues. Take, for instance, p_z, the z-component of momentum. In quantum mechanics this is again represented by a Hermitian operator. In contrast to S_z, however, the eigenvalues of p_z (in appropriate units) can assume any real value between $-\infty$ and ∞.

The rigorous mathematics of a vector space spanned by eigenkets that exhibit a continuous spectrum is rather treacherous. The dimensionality of such a space is obviously infinite. Fortunately, many of the results we worked out for a finite-dimensional vector space with discrete eigenvalues can immediately be generalized. In places where straightforward generalizations do not hold, we indicate danger signals.

We start with the analogue of eigenvalue equation (1.2.5), which, in the continuous-spectrum case, is written as

$$\xi|\xi'\rangle = \xi'|\xi'\rangle, \tag{1.6.1}$$

where ξ is an operator and ξ' is simply a number. The ket $|\xi'\rangle$ is, in other words, an eigenket of operator ξ with eigenvalue ξ', just as $|a'\rangle$ is an eigenket of operator A with eigenvalue a'.

In pursuing this analogy we replace the Kronecker symbol by Dirac's δ-function—a discrete sum over the eigenvalues $\{a'\}$ by an integral over the *continuous variable* ξ'—so

$$\langle a'|a''\rangle = \delta_{a'a''} \rightarrow \langle \xi'|\xi''\rangle = \delta(\xi' - \xi''), \tag{1.6.2a}$$

$$\sum_{a'} |a'\rangle\langle a'| = 1 \rightarrow \int d\xi' |\xi'\rangle\langle \xi'| = 1, \tag{1.6.2b}$$

$$|\alpha\rangle = \sum_{a'} |a'\rangle\langle a'|\alpha\rangle \rightarrow |\alpha\rangle = \int d\xi' |\xi'\rangle\langle \xi'|\alpha\rangle, \tag{1.6.2c}$$

$$\sum_{a'} |\langle a'|\alpha\rangle|^2 = 1 \rightarrow \int d\xi' |\langle \xi'|\alpha\rangle|^2 = 1, \tag{1.6.2d}$$

$$\langle \beta|\alpha\rangle = \sum_{a'} \langle \beta|a'\rangle\langle a'|\alpha\rangle \rightarrow \langle \beta|\alpha\rangle = \int d\xi' \langle \beta|\xi'\rangle\langle \xi'|\alpha\rangle, \tag{1.6.2e}$$

$$\langle a''|A|a'\rangle = a'\delta_{a'a''} \rightarrow \langle \xi''|\xi|\xi'\rangle = \xi'\delta(\xi'' - \xi'). \tag{1.6.2f}$$

Notice in particular how the completeness relation (1.6.2b) is used to obtain (1.6.2c) and (1.6.2e).

Position Eigenkets and Position Measurements

In Section 1.4 we emphasized that a measurement in quantum mechanics is essentially a filtering process. To extend this idea to measurements of observables exhibiting continuous spectra it is best to work with a specific example. To this end we consider the position (or coordinate) operator in one dimension.

The eigenkets $|x'\rangle$ of the position operator x satisfying

$$x|x'\rangle = x'|x'\rangle \tag{1.6.3}$$

are postulated to form a complete set. Here x' is just a number with the dimension of length 0.23 cm, for example, while x is an operator. The state ket for an arbitrary physical state can be expanded in terms of $\{|x'\rangle\}$:

$$|\alpha\rangle = \int_{-\infty}^{\infty} dx' |x'\rangle\langle x'|\alpha\rangle. \tag{1.6.4}$$

We now consider a highly idealized selective measurement of the position observable. Suppose we place a very tiny detector that clicks only when the particle is precisely at x' and nowhere else. Immediately after the detector clicks, we can say that the state in question is represented by $|x'\rangle$. In other words, when the detector clicks, $|\alpha\rangle$ abruptly "jumps into" $|x'\rangle$ in much the same way as an arbitrary spin state jumps into the S_z+ (or S_z-) state when subjected to an SG apparatus of the S_z type.

In practice the best the detector can do is to locate the particle within a narrow interval around x'. A realistic detector clicks when a particle is observed to be located within some narrow range $(x' - \Delta/2, x' + \Delta/2)$.

When a count is registered in such a detector, the state ket changes abruptly as follows:

$$|\alpha\rangle = \int_{-\infty}^{\infty} dx''|x''\rangle\langle x''|\alpha\rangle \xrightarrow{\text{measurement}} \int_{x'-\Delta/2}^{x'+\Delta/2} dx''|x''\rangle\langle x''|\alpha\rangle. \quad (1.6.5)$$

Assuming that $\langle x''|\alpha\rangle$ does not change appreciably within the narrow interval, the probability for the detector to click is given by

$$|\langle x'|\alpha\rangle|^2 dx', \quad (1.6.6)$$

where we have written dx' for Δ. This is analogous to $|\langle a'|\alpha\rangle|^2$ for the probability for $|\alpha\rangle$ to be thrown into $|a'\rangle$ when A is measured. The probability of recording the particle *somewhere* between $-\infty$ and ∞ is given by

$$\int_{-\infty}^{\infty} dx'|\langle x'|\alpha\rangle|^2, \quad (1.6.7)$$

which is normalized to unity if $|\alpha\rangle$ is normalized:

$$\langle\alpha|\alpha\rangle = 1 \Rightarrow \int_{-\infty}^{\infty} dx'\langle\alpha|x'\rangle\langle x'|\alpha\rangle = 1. \quad (1.6.8)$$

The reader familiar with wave mechanics may have recognized by this time that $\langle x'|\alpha\rangle$ is the wave function for the physical state represented by $|\alpha\rangle$. We will say more about this identification of the expansion coefficient with the x-representation of the wave function in Section 1.7.

The notion of a position eigenket can be extended to three dimensions. It is assumed in nonrelativistic quantum mechanics that the position eigenkets $|\mathbf{x}'\rangle$ are complete. The state ket for a particle with internal degrees of freedom, such as spin, ignored can therefore be expanded in terms of $\{|\mathbf{x}'\rangle\}$ as follows:

$$|\alpha\rangle = \int d^3x'|\mathbf{x}'\rangle\langle\mathbf{x}'|\alpha\rangle, \quad (1.6.9)$$

where $\mathbf{x}'$ stands for x', y', and z'; in other words, $|\mathbf{x}'\rangle$ is a *simultaneous* eigenket of the observables x, y, and z in the sense of Section 1.4:

$$|\mathbf{x}'\rangle \equiv |x', y', z'\rangle, \quad (1.6.10a)$$

$$x|\mathbf{x}'\rangle = x'|\mathbf{x}'\rangle, \qquad y|\mathbf{x}'\rangle = y'|\mathbf{x}'\rangle, \qquad z|\mathbf{x}'\rangle = z'|\mathbf{x}'\rangle. \quad (1.6.10b)$$

To be able to consider such a simultaneous eigenket at all, we are implicitly assuming that the three components of the position vector can be measured simultaneously to arbitrary degrees of accuracy; hence, we must have

$$[x_i, x_j] = 0, \quad (1.6.11)$$

where x_1, x_2, and x_3 stand for x, y, and z, respectively.

Translation

We now introduce the very important concept of translation, or spatial displacement. Suppose we start with a state that is well localized around $\mathbf{x}'$. Let us consider an operation that changes this state into another well-localized state, this time around $\mathbf{x}'+d\mathbf{x}'$ with everything else (for example, the spin direction) unchanged. Such an operation is defined to be an **infinitesimal translation** by $d\mathbf{x}'$, and the operator that does the job is denoted by $\mathscr{T}(d\mathbf{x}')$:

$$\mathscr{T}(d\mathbf{x}')|\mathbf{x}'\rangle = |\mathbf{x}'+d\mathbf{x}'\rangle, \tag{1.6.12}$$

where a possible arbitrary phase factor is set to unity by convention. Notice that the right-hand side of (1.6.12) is again a position eigenket, but this time with eigenvalue $\mathbf{x}'+\mathrm{d}\mathbf{x}'$. Obviously $|\mathbf{x}'\rangle$ is *not* an eigenket of the infinitesimal translation operator.

By expanding an arbitrary state ket $|\alpha\rangle$ in terms of the position eigenkets we can examine the effect of infinitesimal translation on $|\alpha\rangle$:

$$|\alpha\rangle \to \mathscr{T}(d\mathbf{x}')|\alpha\rangle = \mathscr{T}(d\mathbf{x}')\int d^3x'|\mathbf{x}'\rangle\langle\mathbf{x}'|\alpha\rangle = \int d^3x'|\mathbf{x}'+d\mathbf{x}'\rangle\langle\mathbf{x}'|\alpha\rangle. \tag{1.6.13}$$

We also write the right-hand side of (1.6.13) as

$$\int d^3x'|\mathbf{x}'+d\mathbf{x}'\rangle\langle\mathbf{x}'|\alpha\rangle = \int d^3x'|\mathbf{x}'\rangle\langle\mathbf{x}'-d\mathbf{x}'|\alpha\rangle \tag{1.6.14}$$

because the integration is over all space and $\mathbf{x}'$ is just an integration variable. This shows that the wave function of the translated state $\mathscr{T}(d\mathbf{x}')|\alpha\rangle$ is obtained by substituting $\mathbf{x}'-d\mathbf{x}'$ for $\mathbf{x}'$ in $\langle\mathbf{x}'|\alpha\rangle$.

There is an equivalent approach to translation that is often treated in the literature. Instead of considering an infinitesimal translation of the physical system itself, we consider a change in the coordinate system being used such that the origin is shifted in the *opposite* direction, $-d\mathbf{x}'$. Physically, in this alternative approach we are asking how the *same* state ket would look to another observer whose coordinate system is shifted by $-d\mathbf{x}'$. In this book we try not to use this approach. Obviously it is important that we do not mix the two approaches!

We now list the properties of the infinitesimal translation operator $\mathscr{T}(d\mathbf{x}')$. The first property we demand is the unitarity property imposed by probability conservation. It is reasonable to require that if the ket $|\alpha\rangle$ is normalized to unity, the translated ket $\mathscr{T}(d\mathbf{x}')|\alpha\rangle$ also be normalized to unity, so

$$\langle\alpha|\alpha\rangle = \langle\alpha|\mathscr{T}^\dagger(d\mathbf{x}')\mathscr{T}(d\mathbf{x}')|\alpha\rangle. \tag{1.6.15}$$

This condition is guaranteed by demanding that the infinitesimal translation be unitary:

$$\mathscr{T}^{\dagger}(d\mathbf{x}')\mathscr{T}(d\mathbf{x}')=1. \tag{1.6.16}$$

Quite generally, the norm of a ket is preserved under unitary transformations. For the second property, suppose we consider two successive infinitesimal translations—first by $d\mathbf{x}'$ and subsequently by $d\mathbf{x}''$, where $d\mathbf{x}'$ and $d\mathbf{x}''$ need not be in the same direction. We expect the net result to be just a *single* translation operation by the vector sum $d\mathbf{x}'+d\mathbf{x}''$, so we demand that

$$\mathscr{T}(d\mathbf{x}'')\mathscr{T}(d\mathbf{x}')=\mathscr{T}(d\mathbf{x}'+d\mathbf{x}''). \tag{1.6.17}$$

For the third property, suppose we consider a translation in the opposite direction; we expect the opposite-direction translation to be the same as the inverse of the original translation:

$$\mathscr{T}(-d\mathbf{x}')=\mathscr{T}^{-1}(d\mathbf{x}'). \tag{1.6.18}$$

For the fourth property, we demand that as $d\mathbf{x}'\to 0$, the translation operation reduce to the identity operation

$$\lim_{d\mathbf{x}'\to 0}\mathscr{T}(d\mathbf{x}')=1 \tag{1.6.19}$$

and that the difference between $\mathscr{T}(d\mathbf{x}')$ and the identity operator be of first order in $d\mathbf{x}'$.

We now demonstrate that if we take the infinitesimal translation operator to be

$$\mathscr{T}(d\mathbf{x}')=1-i\mathbf{K}\cdot d\mathbf{x}', \tag{1.6.20}$$

where the components of $\mathbf{K}$, K_x, K_y, and K_z, are **Hermitian operators**, then all the properties listed are satisfied. The first property, the unitarity of $\mathscr{T}(\mathrm{d}\mathbf{x}')$, is checked as follows:

$$\begin{aligned}\mathscr{T}^{\dagger}(d\mathbf{x}')\mathscr{T}(d\mathbf{x}') &= (1+i\mathbf{K}^{\dagger}\cdot d\mathbf{x}')(1-i\mathbf{K}\cdot d\mathbf{x}')\\ &=1-i(\mathbf{K}-\mathbf{K}^{\dagger})\cdot d\mathbf{x}'+0\left[(d\mathbf{x}')^2\right]\\ &\simeq 1,\end{aligned} \tag{1.6.21}$$

where terms of second order in $d\mathbf{x}'$ have been ignored for an infinitesimal translation. The second property [(1.6.17)] can also be proved as follows:

$$\begin{aligned}\mathscr{T}(d\mathbf{x}'')\mathscr{T}(d\mathbf{x}') &= (1-i\mathbf{K}\cdot d\mathbf{x}'')(1-i\mathbf{K}\cdot d\mathbf{x}')\\ &\simeq 1-i\mathbf{K}\cdot(d\mathbf{x}'+d\mathbf{x}'')\\ &=\mathscr{T}(d\mathbf{x}'+d\mathbf{x}'').\end{aligned} \tag{1.6.22}$$

The third and fourth properties are obviously satisfied by (1.6.20).

Accepting (1.6.20) to be the correct form for $\mathscr{T}(d\mathbf{x}')$, we are in a position to derive an extremely fundamental relation between the $\mathbf{K}$ operator and the $\mathbf{x}$ operator. First, note that

$$\mathbf{x}\mathscr{T}(d\mathbf{x}')|\mathbf{x}'\rangle = \mathbf{x}|\mathbf{x}'+d\mathbf{x}'\rangle = (\mathbf{x}'+d\mathbf{x}')|\mathbf{x}'+d\mathbf{x}'\rangle \tag{1.6.23a}$$

and

$$\mathscr{T}(d\mathbf{x}')\mathbf{x}|\mathbf{x}'\rangle = \mathbf{x}'\mathscr{T}(d\mathbf{x}')|\mathbf{x}'\rangle = \mathbf{x}'|\mathbf{x}'+d\mathbf{x}'\rangle; \tag{1.6.23b}$$

hence,

$$[\mathbf{x}, \mathscr{T}(d\mathbf{x}')]|\mathbf{x}'\rangle = d\mathbf{x}'|\mathbf{x}'+d\mathbf{x}'\rangle \simeq d\mathbf{x}'|\mathbf{x}'\rangle, \tag{1.6.24}$$

where the error made in writing the last line of (1.6.24) is of second order in $d\mathbf{x}'$. Now $|\mathbf{x}'\rangle$ can be *any* position eigenket, and the position eigenkets are known to form a complete set. We must therefore have an **operator identity**

$$[\mathbf{x}, \mathscr{T}(d\mathbf{x}')] = d\mathbf{x}', \tag{1.6.25}$$

or

$$-i\mathbf{x}\mathbf{K}\cdot d\mathbf{x}' + i\mathbf{K}\cdot d\mathbf{x}'\mathbf{x} = d\mathbf{x}', \tag{1.6.26}$$

where on the right-hand sides of (1.6.25) and (1.6.26) $d\mathbf{x}'$ is understood to be the number $d\mathbf{x}'$ multiplied by the identity operator in the ket space spanned by $|\mathbf{x}'\rangle$. By choosing $d\mathbf{x}'$ in the direction of $\hat{\mathbf{x}}_j$ and forming the scalar product with $\hat{\mathbf{x}}_i$, we obtain

$$[x_i, K_j] = i\delta_{ij}, \tag{1.6.27}$$

where again δ_{ij} is understood to be multiplied by the identity operator.

Momentum as a Generator of Translation

Equation (1.6.27) is the fundamental commutation relation between the position operators x, y, z and the K operators K_x, K_y, K_z. Remember that so far the K operator is *defined* in terms of the infinitesimal translation operator by (1.6.20). What is the physical significance we can attach to $\mathbf{K}$?

J. Schwinger, lecturing on quantum mechanics, once remarked, "... for fundamental properties we will borrow only names from classical physics." In the present case we would like to borrow from classical mechanics the notion that momentum is the generator of an infinitesimal translation. An infinitesimal translation in classical mechanics can be regarded as a canonical transformation,

$$\mathbf{x}_{\text{new}} \equiv \mathbf{X} = \mathbf{x} + d\mathbf{x}, \qquad \mathbf{p}_{\text{new}} \equiv \mathbf{P} = \mathbf{p}, \tag{1.6.28}$$

obtainable from the generating function (Goldstein 1980, 395 and 411)

$$F(\mathbf{x}, \mathbf{P}) = \mathbf{x}\cdot\mathbf{P} + \mathbf{p}\cdot d\mathbf{x}, \tag{1.6.29}$$

where $\mathbf{p}$ and $\mathbf{P}$ refer to the corresponding momenta.

This equation has a striking similarity to the infinitesimal translation operator (1.6.20) in quantum mechanics, particularly if we recall that $\mathbf{x}\cdot\mathbf{P}$ in

(1.6.29) is the generating function for the identity transformation ($\mathbf{X} = \mathbf{x}, \mathbf{P} = \mathbf{p}$). We are therefore led to speculate that the operator $\mathbf{K}$ is in some sense related to the momentum operator in quantum mechanics.

Can the K operator be identified with the momentum operator itself? Unfortunately the dimension is all wrong; the K operator has the dimension of 1/length because $\mathbf{K}\cdot d\mathbf{x}'$ must be dimensionless. But it appears legitimate to set

$$\mathbf{K} = \frac{\mathbf{p}}{\text{universal constant with the dimension of action}}. \tag{1.6.30}$$

From the fundamental postulates of quantum mechanics there is no way to determine the actual numerical value of the universal constant. Rather, this constant is needed here because, historically, classical physics was developed before quantum mechanics using units convenient for describing macroscopic quantities—the circumference of the earth, the mass of 1 cc of water, the duration of a mean solar day, and so forth. Had microscopic physics been formulated before macroscopic physics, the physicists would have almost certainly chosen the basic units in such a way that the universal constant appearing in (1.6.30) would be unity.

An analogy from electrostatics may be helpful here. The interaction energy between two particles of charge e separated at a distance r is proportional to e^2/r; in unrationalized Gaussian units, the proportionality factor is just 1, but in rationalized mks units, which may be more convenient for electrical engineers, the proportionality factor is $1/4\pi\varepsilon_o$.

The universal constant that appears in (1.6.30) turns out to be the same as the constant $\hbar$ that appears in L. de Broglie's relation, written in 1924,

$$\frac{2\pi}{\lambda} = \frac{p}{\hbar}, \tag{1.6.31}$$

where λ is the wavelength of a "particle wave." In other words, the K operator is the quantum mechanical operator that corresponds to the wave number—that is, 2π times the reciprocal wavelength, usually denoted by k. With this identification the infinitesimal translation operator $\mathscr{T}(d\mathbf{x}')$ reads

$$\mathscr{T}(d\mathbf{x}') = 1 - i\mathbf{p}\cdot d\mathbf{x}'/\hbar, \tag{1.6.32}$$

where $\mathbf{p}$ is the momentum operator. The commutation relation (1.6.27) now becomes

$$[x_i, p_j] = i\hbar\delta_{ij}. \tag{1.6.33}$$

The commutation relations (1.6.33) imply, for example, that x and p_x (but not x and p_y) are incompatible observables. It is therefore impossible to find simultaneous eigenkets of x and p_x. The general formalism of

Section 1.4 can be applied here to obtain the **position-momentum uncertainty relation** of W. Heisenberg:

$$\langle(\Delta x)^2\rangle\langle(\Delta p_x)^2\rangle \geq \hbar^2/4. \tag{1.6.34}$$

Some applications of (1.6.34) will appear in Section 1.7.

So far we have concerned ourselves with infinitesimal translations. A finite translation—that is, a spatial displacement by a finite amount—can be obtained by successively compounding infinitesimal translations. Let us consider a finite translation in the x-direction by an amount $\Delta x'$:

$$\mathscr{T}(\Delta x'\hat{\mathbf{x}})|\mathbf{x}'\rangle = |\mathbf{x}' + \Delta x'\hat{\mathbf{x}}\rangle. \tag{1.6.35}$$

By compounding N infinitesimal translations, each of which is characterized by a spatial displacement $\Delta x'/N$ in the x-direction, and letting $N \to \infty$, we obtain

$$\begin{aligned}\mathscr{T}(\Delta x'\hat{\mathbf{x}}) &= \lim_{N\to\infty}\left(1 - \frac{ip_x\Delta x'}{N\hbar}\right)^N \\ &= \exp\left(-\frac{ip_x\Delta x'}{\hbar}\right).\end{aligned} \tag{1.6.36}$$

Here $\exp(-ip_x\Delta x'/\hbar)$ is understood to be a function of the *operator* p_x; generally, for any operator X we have

$$\exp(X) \equiv 1 + X + \frac{X^2}{2!} + \cdots. \tag{1.6.37}$$

A fundamental property of translations is that successive translations in different directions, say in the x- and y-directions, commute. We see this clearly in Figure 1.8; in shifting from A and B it does not matter whether we go via C or via D. Mathematically,

$$\begin{aligned}\mathscr{T}(\Delta y'\hat{\mathbf{y}})\mathscr{T}(\Delta x'\hat{\mathbf{x}}) &= \mathscr{T}(\Delta x'\hat{\mathbf{x}} + \Delta y'\hat{\mathbf{y}}), \\ \mathscr{T}(\Delta x'\hat{\mathbf{x}})\mathscr{T}(\Delta y'\hat{\mathbf{y}}) &= \mathscr{T}(\Delta x'\hat{\mathbf{x}} + \Delta y'\hat{\mathbf{y}}).\end{aligned} \tag{1.6.38}$$

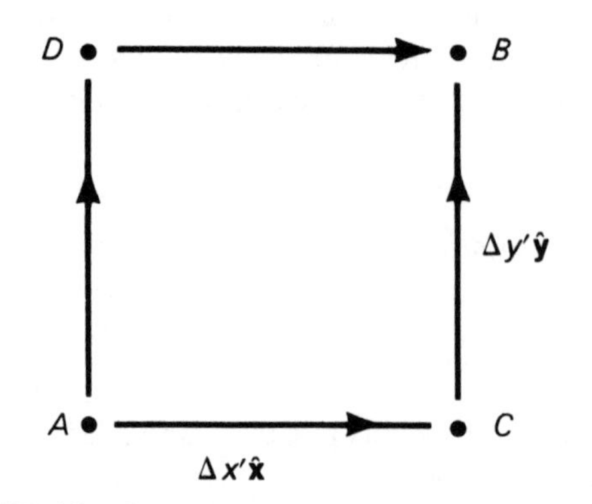

FIGURE 1.8. Successive translations in different directions.

This point is not so trivial as it may appear; we will show in Chapter 3 that rotations about different axes do *not* commute. Treating $\Delta x'$ and $\Delta y'$ up to second order, we obtain

$$[\mathscr{T}(\Delta y'\hat{\mathbf{y}}), \mathscr{T}(\Delta x'\hat{\mathbf{x}})] = \left[\left(1 - \frac{ip_y \Delta y'}{\hbar} - \frac{p_y^2(\Delta y')^2}{2\hbar^2} + \cdots\right), \left(1 - \frac{ip_x \Delta x'}{\hbar} - \frac{p_x^2(\Delta x')^2}{2\hbar^2} + \cdots\right)\right]$$
$$\simeq -\frac{(\Delta x')(\Delta y')[p_y, p_x]}{\hbar^2}. \quad (1.6.39)$$

Because $\Delta x'$ and $\Delta y'$ are arbitrary, requirement (1.6.38), or

$$[\mathscr{T}(\Delta y'\hat{\mathbf{y}}), \mathscr{T}(\Delta x'\hat{\mathbf{x}})] = 0, \quad (1.6.40)$$

immediately leads to

$$[p_x, p_y] = 0, \quad (1.6.41)$$

or, more generally,

$$[p_i, p_j] = 0. \quad (1.6.42)$$

This commutation relation is a direct consequence of the fact that translations in different directions commute. Whenever the generators of transformations commute, the corresponding group is said to be **Abelian**. The translation group in three dimensions is Abelian.

Equation (1.6.42) implies that p_x, p_y, and p_z are mutually compatible observables. We can therefore conceive of a simultaneous eigenket of p_x, p_y, p_z, namely,

$$|\mathbf{p}'\rangle \equiv |p_x', p_y', p_z'\rangle, \quad (1.6.43a)$$

$$p_x|\mathbf{p}'\rangle = p_x'|\mathbf{p}'\rangle, \; p_y|\mathbf{p}'\rangle = p_y'|\mathbf{p}'\rangle, \; p_z|\mathbf{p}'\rangle = p_z'|\mathbf{p}'\rangle. \quad (1.6.43b)$$

It is instructive to work out the effect of $\mathscr{T}(d\mathbf{x}')$ on such a momentum eigenket:

$$\mathscr{T}(d\mathbf{x}')|\mathbf{p}'\rangle = \left(1 - \frac{i\mathbf{p}\cdot d\mathbf{x}'}{\hbar}\right)|\mathbf{p}'\rangle = \left(1 - \frac{i\mathbf{p}'\cdot d\mathbf{x}'}{\hbar}\right)|\mathbf{p}'\rangle. \quad (1.6.44)$$

We see that the momentum eigenket remains the same even though it suffers a slight phase change, so unlike $|\mathbf{x}'\rangle$, $|\mathbf{p}'\rangle$ *is* an eigenket of $\mathscr{T}(d\mathbf{x}')$, which we anticipated because

$$[\mathbf{p}, \mathscr{T}(d\mathbf{x}')] = 0. \quad (1.6.45)$$

Notice, however, that the eigenvalue of $\mathscr{T}(d\mathbf{x}')$ is complex; we do not expect a real eigenvalue here because $\mathscr{T}(d\mathbf{x}')$, though unitary, is not Hermitian.

The Canonical Commutation Relations

We summarize the commutator relations we inferred by studying the properties of translation:

$$[x_i, x_j] = 0, \qquad [p_i, p_j] = 0, \qquad [x_i, p_j] = i\hbar\delta_{ij}. \tag{1.6.46}$$

These relations form the cornerstone of quantum mechanics; in his book, P. A. M. Dirac calls them the "fundamental quantum conditions." More often they are known as the **canonical commutation relations**, or the **fundamental commutation relations**.

Historically it was W. Heisenberg who, in 1925, showed that the combination rule for atomic transition lines known at that time could best be understood if one associated arrays of numbers obeying certain multiplication rules with these frequencies. Immediately afterward M. Born and P. Jordan pointed out that Heisenberg's multiplication rules are essentially those of matrix algebra, and a theory was developed based on the matrix analogues of (1.6.46), which is now known as **matrix mechanics.***

Also in 1925, P. A. M. Dirac observed that the various quantum-mechanical relations can be obtained from the corresponding classical relations just by replacing classical Poisson brackets by commutators, as follows:

$$[\ ,\ ]_{\text{classical}} \rightarrow \frac{[\ ,\ ]}{i\hbar}, \tag{1.6.47}$$

where we may recall that the classical Poisson brackets are defined for functions of q's and p's as

$$[A(q,p), B(q,p)]_{\text{classical}} \equiv \sum_s \left(\frac{\partial A}{\partial q_s}\frac{\partial B}{\partial p_s} - \frac{\partial A}{\partial p_s}\frac{\partial B}{\partial q_s} \right). \tag{1.6.48}$$

For example, in classical mechanics, we have

$$[x_i, p_j]_{\text{classical}} = \delta_{ij}, \tag{1.6.49}$$

which in quantum mechanics turns into (1.6.33).

Dirac's rule (1.6.47) is plausible because the classical Poisson brackets and quantum-mechanical commutators satisfy similar algebraic properties. In particular, the following relations can be proved regardless of whether [,] is understood as a classical Poisson bracket or as a quantum-

*Appropriately, $pq - qp = h/2\pi i$ is inscribed on the gravestone of M. Born in Göttingen.

mechanical commutator:

$$[A, A] = 0 \tag{1.6.50a}$$

$$[A, B] = -[B, A] \tag{1.6.50b}$$

$$[A, c] = 0 \qquad (c \text{ is just a number}) \tag{1.6.50c}$$

$$[A+B, C] = [A, C] + [B, C] \tag{1.6.50d}$$

$$[A, BC] = [A, B]C + B[A, C] \tag{1.6.50e}$$

$$[A,[B,C]] + [B,[C,A]] + [C,[A,B]] = 0, \tag{1.6.50f}$$

where the last relation is known as the **Jacobi identity**.* However, there are important differences. First, the dimension of the classical Poisson bracket differs from that of the quantum-mechanical commutator because of the differentiations with respect to q and p appearing in (1.6.48). Second, the Poisson bracket of real functions of q's and p's is purely real, while the commutator of two Hermitian operators is anti-Hermitian (see Lemma 3 of Section 1.4). To take care of these differences the factor $i\hbar$ is inserted in (1.6.47).

We have deliberately avoided exploiting Dirac's analogy in obtaining the canonical commutation relations. Our approach to the commutation relations is based solely on (1) the properties of translations and (2) the identification of the generator of translation with the momentum operator modulo a universal constant with the dimension of action. We believe that this approach is more powerful because it can be generalized to situations where observables have no classical analogues. For example, the spin-angular-momentum components we encountered in Section 1.4 having nothing to do with the p's and q's of classical mechanics; yet, as we will show in Chapter 3, the spin-angular-momentum commutation relations can be derived using the properties of rotations just as we derived the canonical commutation relations using the properties of translations.

1.7. WAVE FUNCTIONS IN POSITION AND MOMENTUM SPACE

Position-Space Wave Function

In this section we present a systematic study of the properties of wave functions in both position and momentum space. For simplicity let us return to the one-dimensional case. The base kets used are the position kets

*It is amusing that the Jacobi identity in quantum mechanics is much easier to prove than its classical analogue.

satisfying

$$x|x'\rangle = x'|x'\rangle, \tag{1.7.1}$$

normalized in such a way that the orthogonality condition reads

$$\langle x''|x'\rangle = \delta(x'' - x'). \tag{1.7.2}$$

We have already remarked that the ket representing a physical state can be expanded in terms of $|x'\rangle$,

$$|\alpha\rangle = \int dx' |x'\rangle\langle x'|\alpha\rangle, \tag{1.7.3}$$

and that the expansion coefficient $\langle x'|\alpha\rangle$ is interpreted in such a way that

$$|\langle x'|\alpha\rangle|^2\, dx' \tag{1.7.4}$$

is the probability for the particle to be found in a narrow interval dx' around x'. In our formalism the inner product $\langle x'|\alpha\rangle$ is what is usually referred to as the **wave function** $\psi_\alpha(x')$ for state $|\alpha\rangle$:

$$\langle x'|\alpha\rangle = \psi_\alpha(x'). \tag{1.7.5}$$

In elementary wave mechanics the probabilistic interpretations for the expansion coefficient $c_{a'}$ $(=\langle a'|\alpha\rangle)$ and for the wave function $\psi_\alpha(x')$ $(=\langle x'|\alpha\rangle)$ are often presented as separate postulates. One of the major advantages of our formalism, originally due to Dirac, is that the two kinds of probabilistic interpretations are unified; $\psi_\alpha(x')$ *is* an expansion coefficient [see (1.7.3)] in much the same way as $c_{a'}$ is. By following the footsteps of Dirac we come to appreciate the unity of quantum mechanics.

Consider the inner product $\langle\beta|\alpha\rangle$. Using the completeness of $|x'\rangle$, we have

$$\begin{aligned}\langle\beta|\alpha\rangle &= \int dx' \langle\beta|x'\rangle\langle x'|\alpha\rangle \\ &= \int dx' \psi_\beta^*(x')\psi_\alpha(x'),\end{aligned} \tag{1.7.6}$$

so $\langle\beta|\alpha\rangle$ characterizes the overlap between the two wave functions. Note that we are not defining $\langle\beta|\alpha\rangle$ as the overlap integral; the identification of $\langle\beta|\alpha\rangle$ with the overlap integral *follows* from our completeness postulate for $|x'\rangle$. The more general interpretation of $\langle\beta|\alpha\rangle$, *independent of representations*, is that it represents the probability amplitude for state $|\alpha\rangle$ to be found in state $|\beta\rangle$.

This time let us interpret the expansion

$$|\alpha\rangle = \sum_{a'} |a'\rangle\langle a'|\alpha\rangle \tag{1.7.7}$$

using the language of wave functions. We just multiply both sides of (1.7.7) by the position eigenbra $\langle x'|$ on the left. Thus

$$\langle x'|\alpha\rangle = \sum_{a'} \langle x'|a'\rangle\langle a'|\alpha\rangle. \tag{1.7.8}$$

In the usual notation of wave mechanics this is recognized as

$$\psi_\alpha(x') = \sum_{a'} c_{a'} u_{a'}(x'),$$

where we have introduced an **eigenfunction** of operator A with eigenvalue a':

$$u_{a'}(x') = \langle x'|a'\rangle. \tag{1.7.9}$$

Let us now examine how $\langle\beta|A|\alpha\rangle$ can be written using the wave functions for $|\alpha\rangle$ and $|\beta\rangle$. Clearly, we have

$$\begin{aligned}\langle\beta|A|\alpha\rangle &= \int dx' \int dx''\, \langle\beta|x'\rangle\langle x'|A|x''\rangle\langle x''|\alpha\rangle \\ &= \int dx' \int dx''\, \psi_\beta^*(x')\langle x'|A|x''\rangle\psi_\alpha(x''). \end{aligned} \tag{1.7.10}$$

So to be able to evaluate $\langle\beta|A|\alpha\rangle$, we must know the matrix element $\langle x'|A|x''\rangle$, which is, in general, a function of the two variables x' and x''.

An enormous simplification takes place if observable A is a function of the position operator x. In particular, consider

$$A = x^2, \tag{1.7.11}$$

which actually appears in the Hamiltonian for the simple harmonic oscillator problem to be discussed in Chapter 2. We have

$$\langle x'|x^2|x''\rangle = (\langle x'|)\cdot(x''^2|x''\rangle) = x'^2\delta(x'-x''), \tag{1.7.12}$$

where we have used (1.7.1) and (1.7.2). The double integral (1.7.10) is now reduced to a *single* integral:

$$\begin{aligned}\langle\beta|x^2|\alpha\rangle &= \int dx' \langle\beta|x'\rangle x'^2\langle x'|\alpha\rangle \\ &= \int dx'\, \psi_\beta^*(x')x'^2\psi_\alpha(x'). \end{aligned} \tag{1.7.13}$$

In general,

$$\langle\beta|f(x)|\alpha\rangle = \int dx'\, \psi_\beta^*(x')f(x')\psi_\alpha(x'). \tag{1.7.14}$$

Note that the $f(x)$ on the left-hand side of (1.7.14) is an operator, while the $f(x')$ on the right-hand side is not an operator.

Momentum Operator in the Position Basis

We now examine how the momentum operator may look in the x-basis—that is, in the representation where the position eigenkets are used as base kets. Our starting point is the definition of momentum as the generator of infinitesimal translations:

$$\begin{aligned}\left(1-\frac{ip\Delta x'}{\hbar}\right)|\alpha\rangle &= \int dx' \mathscr{T}(\Delta x')|x'\rangle\langle x'|\alpha\rangle \\ &= \int dx'|x'+\Delta x'\rangle\langle x'|\alpha\rangle \\ &= \int dx'|x'\rangle\langle x'-\Delta x'|\alpha\rangle \\ &= \int dx'|x'\rangle\left(\langle x'|\alpha\rangle - \Delta x'\frac{\partial}{\partial x'}\langle x'|\alpha\rangle\right). \end{aligned} \quad (1.7.15)$$

Comparison of both sides yields

$$p|\alpha\rangle = \int dx'|x'\rangle\left(-i\hbar\frac{\partial}{\partial x'}\langle x'|\alpha\rangle\right) \quad (1.7.16)$$

or

$$\langle x'|p|\alpha\rangle = -i\hbar\frac{\partial}{\partial x'}\langle x'|\alpha\rangle, \quad (1.7.17)$$

where we have used the orthogonality property (1.7.2). For the matrix element p in the x-representation, we obtain

$$\langle x'|p|x''\rangle = -i\hbar\frac{\partial}{\partial x'}\delta(x'-x''). \quad (1.7.18)$$

From (1.7.16) we get a very important identity:

$$\begin{aligned}\langle\beta|p|\alpha\rangle &= \int dx'\langle\beta|x'\rangle\left(-i\hbar\frac{\partial}{\partial x'}\langle x'|\alpha\rangle\right) \\ &= \int dx'\psi_\beta^*(x')\left(-i\hbar\frac{\partial}{\partial x'}\right)\psi_\alpha(x'). \end{aligned} \quad (1.7.19)$$

In our formalism (1.7.19) is not a postulate; rather, it has been *derived* using the basic properties of momentum. By repeatedly applying (1.7.17), we can also obtain

$$\langle x'|p^n|\alpha\rangle = (-i\hbar)^n\frac{\partial^n}{\partial x'^n}\langle x'|\alpha\rangle, \quad (1.7.20)$$

$$\langle\beta|p^n|\alpha\rangle = \int dx'\psi_\beta^*(x')(-i\hbar)^n\frac{\partial^n}{\partial x'^n}\psi_\alpha(x'). \quad (1.7.21)$$

Momentum-Space Wave Function

So far we have worked exclusively in the x-basis. There is actually a complete symmetry between x and p—apart from occasional minus signs—which we can infer from the canonical commutation relations. Let us now work in the p-basis, that is, in the momentum representation.

For simplicity we continue working in one-space. The base eigenkets in the p-basis specify

$$p|p'\rangle = p'|p'\rangle \tag{1.7.22}$$

and

$$\langle p'|p''\rangle = \delta(p'-p''). \tag{1.7.23}$$

The momentum eigenkets $\{|p'\rangle\}$ span the ket space in much the same way as the position eigenkets $\{|x'\rangle\}$. An arbitrary state ket $|\alpha\rangle$ can therefore be expanded as follows:

$$|\alpha\rangle = \int dp'|p'\rangle\langle p'|\alpha\rangle. \tag{1.7.24}$$

We can give a probabilistic interpretation for the expansion coefficient $\langle p'|\alpha\rangle$; the probability that a measurement of p gives eigenvalue p' within a narrow interval dp' is $|\langle p'|\alpha\rangle|^2 dp'$. It is customary to call $\langle p'|\alpha\rangle$ the **momentum-space wave function**; the notation $\phi_\alpha(p')$ is often used:

$$\langle p'|\alpha\rangle = \phi_\alpha(p'). \tag{1.7.25}$$

If $|\alpha\rangle$ is normalized, we obtain

$$\int dp'\langle\alpha|p'\rangle\langle p'|\alpha\rangle = \int dp'|\phi_\alpha(p')|^2 = 1. \tag{1.7.26}$$

Let us now establish the connection between the x-representation and the p-representation. We recall that in the case of the discrete spectra, the change of basis from the old set $\{|a'\rangle\}$ to the new set $\{|b'\rangle\}$ is characterized by the transformation matrix (1.5.7). Likewise, we expect that the desired information is contained in $\langle x'|p'\rangle$, which is a function of x' and p', usually called the **transformation function** from the x-representation to the p-representation. To derive the explicit form of $\langle x'|p'\rangle$, first recall (1.7.17); letting $|\alpha\rangle$ be the momentum eigenket $|p'\rangle$, we obtain

$$\langle x'|p|p'\rangle = -i\hbar\frac{\partial}{\partial x'}\langle x'|p'\rangle \tag{1.7.27}$$

or

$$p'\langle x'|p'\rangle = -i\hbar\frac{\partial}{\partial x'}\langle x'|p'\rangle. \tag{1.7.28}$$

The solution to this differential equation for $\langle x'|p'\rangle$ is

$$\langle x'|p'\rangle = N\exp\left(\frac{ip'x'}{\hbar}\right), \tag{1.7.29}$$

where N is the normalization constant to be determined in a moment. Even though the transformation function $\langle x'|p'\rangle$ is a function of two variables, x' and p', we can temporarily regard it as a function of x' with p' fixed. It can then be viewed as the probability amplitude for the momentum eigenstate specified by p' to be found at position x'; in other words, it is just the wave function for the momentum eigenstate $|p'\rangle$, often referred to as the momentum eigenfunction (still in the x-space). So (1.7.29) simply says that the wave function of a momentum eigenstate is a plane wave. It is amusing that we have obtained this plane-wave solution without solving the Schrödinger equation (which we have not yet written down).

To get the normalization constant N let us first consider

$$\langle x'|x''\rangle = \int dp' \langle x'|p'\rangle\langle p'|x''\rangle. \tag{1.7.30}$$

The left-hand side is just $\delta(x'-x'')$; the right-hand side can be evaluated using the explicit form of $\langle x'|p'\rangle$:

$$\begin{aligned}\delta(x'-x'') &= |N|^2 \int dp' \exp\left[\frac{ip'(x'-x'')}{\hbar}\right] \\ &= 2\pi\hbar|N|^2\delta(x'-x''). \end{aligned} \tag{1.7.31}$$

Choosing N to be purely real and positive by convention, we finally have

$$\langle x'|p'\rangle = \frac{1}{\sqrt{2\pi\hbar}}\exp\left(\frac{ip'x'}{\hbar}\right). \tag{1.7.32}$$

We can now demonstrate how the position-space wave function is related to the momentum-space wave function. All we have to do is rewrite

$$\langle x'|\alpha\rangle = \int dp' \langle x'|p'\rangle\langle p'|\alpha\rangle \tag{1.7.33a}$$

and

$$\langle p'|\alpha\rangle = \int dx' \langle p'|x'\rangle\langle x'|\alpha\rangle \tag{1.7.33b}$$

as

$$\psi_\alpha(x') = \left[\frac{1}{\sqrt{2\pi\hbar}}\right]\int dp' \exp\left(\frac{ip'x'}{\hbar}\right)\phi_\alpha(p') \tag{1.7.34a}$$

and

$$\phi_\alpha(p') = \left[\frac{1}{\sqrt{2\pi\hbar}}\right]\int dx' \exp\left(\frac{-ip'x'}{\hbar}\right)\psi_\alpha(x'). \tag{1.7.34b}$$

This pair of equations is just what one expects from Fourier's inversion theorem. Apparently the mathematics we have developed somehow "knows" Fourier's work on integral transforms.

Gaussian Wave Packets

It is instructive to look at a physical example to illustrate our basic formalism. We consider what is known as a **Gaussian wave packet**, whose x-space wave function is given by

$$\langle x'|\alpha\rangle = \left[\frac{1}{\pi^{1/4}\sqrt{d}}\right]\exp\left[ikx' - \frac{x'^2}{2d^2}\right]. \tag{1.7.35}$$

This is a plane wave with wave number k modulated by a Gaussian profile centered on the origin. The probability of observing the particle vanishes very rapidly for $|x'| > d$; more quantitatively, the probability density $|\langle x'|\alpha\rangle|^2$ has a Gaussian shape with width d.

We now compute the expectation values of x, x^2, p, and p^2. The expectation value of x is clearly zero by symmetry:

$$\langle x\rangle = \int_{-\infty}^{\infty} dx'\langle\alpha|x'\rangle x'\langle x'|\alpha\rangle = \int_{-\infty}^{\infty} dx'|\langle x'|\alpha\rangle|^2 x' = 0. \tag{1.7.36}$$

For x^2 we obtain

$$\begin{aligned}\langle x^2\rangle &= \int_{-\infty}^{\infty} dx' x'^2|\langle x'|\alpha\rangle|^2\\ &= \left(\frac{1}{\sqrt{\pi}\,\mathrm{d}}\right)\int_{-\infty}^{\infty} dx' x'^2 \exp\left[\frac{-x'^2}{d^2}\right]\\ &= \frac{d^2}{2},\end{aligned} \tag{1.7.37}$$

which leads to

$$\left\langle(\Delta x)^2\right\rangle = \langle x^2\rangle - \langle x\rangle^2 = \frac{d^2}{2} \tag{1.7.38}$$

for the dispersion of the position operator. The expectation values of p and p^2 can also be computed as follows:

$$\langle p\rangle = \hbar k \tag{1.7.39a}$$

$$\langle p^2\rangle = \frac{\hbar^2}{2d^2} + \hbar^2 k^2, \tag{1.7.39b}$$

which is left as an exercise. The momentum dispersion is therefore given by

$$\left\langle(\Delta p)^2\right\rangle = \langle p^2\rangle - \langle p\rangle^2 = \frac{\hbar^2}{2d^2}. \tag{1.7.40}$$

Armed with (1.7.38) and (1.7.40), we can check the Heisenberg uncertainty relation (1.6.34); in this case the uncertainty product is given by

$$\left\langle (\Delta x)^2 \right\rangle \left\langle (\Delta p)^2 \right\rangle = \frac{\hbar^2}{4}, \tag{1.7.41}$$

independent of d, so for a Gaussian wave packet we actually have an *equality* relation rather than the more general inequality relation (1.6.34). For this reason a Gaussian wave packet is often called a *minimum uncertainty wave packet*.

We now go to momentum space. By a straightforward integration—just completing the square in the exponent—we obtain

$$\begin{aligned}\langle p'|\alpha\rangle &= \left(\frac{1}{\sqrt{2\pi\hbar}}\right)\left(\frac{1}{\pi^{1/4}\sqrt{d}}\right)\int_{-\infty}^{\infty} dx' \exp\left(\frac{-ip'x'}{\hbar} + ikx' - \frac{x'^2}{2d^2}\right)\\ &= \sqrt{\frac{d}{\hbar\sqrt{\pi}}}\exp\left[\frac{-(p'-\hbar k)^2 d^2}{2\hbar^2}\right].\end{aligned} \tag{1.7.42}$$

This momentum-space wave function provides an alternative method for obtaining $\langle p \rangle$ and $\langle p^2 \rangle$, which is also left as an exercise.

The probability of finding the particle with momentum p' is Gaussian (in momentum space) centered on $\hbar k$, just as the probability of finding the particle at x' is Gaussian (in position space) centered on zero. Furthermore, the widths of the two Gaussians are inversely proportional to each other, which is just another way of expressing the constancy of the uncertainty product $\langle (\Delta x)^2 \rangle \langle \Delta p)^2 \rangle$ explicitly computed in (1.7.41). The wider the spread in the p-space, the narrower the spread in the x-space, and vice versa.

As an extreme example, suppose we let $d \to \infty$. The position-space wave function (1.7.35) then becomes a plane wave extending over all space; the probability of finding the particle is just constant, independent of x'. In contrast, the momentum-space wave function is δ-function-like and is sharply peaked at $\hbar k$. In the opposite extreme, by letting $d \to 0$, we obtain a position-space wave function localized like the δ-function, but the momentum-space wave function (1.7.42) is just constant, independent of p'.

We have seen that an extremely well localized (in the x-space) state is to be regarded as a superposition of momentum eigenstates with all possible values of momenta. Even those momentum eigenstates whose momenta are comparable to or exceed mc must be included in the superposition. However, at such high values of momentum, a description based on nonrelativistic quantum mechanics is bound to break down.* Despite this limitation

*It turns out that the concept of a localized state in relativistic quantum mechanics is far more intricate because of the possibility of "negative energy states," or pair creation (Sakurai 1967, 118–19).

our formalism, based on the existence of the position eigenket $|x'\rangle$, has a wide domain of applicability.

Generalization to Three Dimensions

So far in this section we have worked exclusively in one-space for simplicity, but everything we have done can be generalized to three-space, if the necessary changes are made. The base kets to be used can be taken as either the position eigenkets satisfying

$$\mathbf{x}|\mathbf{x}'\rangle = \mathbf{x}'|\mathbf{x}'\rangle \tag{1.7.43}$$

or the momentum eigenkets satisfying

$$\mathbf{p}|\mathbf{p}'\rangle = \mathbf{p}'|\mathbf{p}'\rangle. \tag{1.7.44}$$

They obey the normalization conditions

$$\langle\mathbf{x}'|\mathbf{x}''\rangle = \delta^3(\mathbf{x}'-\mathbf{x}'') \tag{1.7.45a}$$

and

$$\langle\mathbf{p}'|\mathbf{p}''\rangle = \delta^3(\mathbf{p}'-\mathbf{p}''), \tag{1.7.45b}$$

where δ^3 stands for the three-dimensional δ-function

$$\delta^3(\mathbf{x}'-\mathbf{x}'') = \delta(x'-x'')\delta(y'-y'')\delta(z'-z''). \tag{1.7.46}$$

The completeness relations read

$$\int d^3x'|\mathbf{x}'\rangle\langle\mathbf{x}'| = 1 \tag{1.7.47a}$$

and

$$\int d^3p'|\mathbf{p}'\rangle\langle\mathbf{p}'| = 1, \tag{1.7.47b}$$

which can be used to expand an arbitrary state ket:

$$|\alpha\rangle = \int d^3x'|\mathbf{x}'\rangle\langle\mathbf{x}'|\alpha\rangle, \tag{1.7.48a}$$

$$|\alpha\rangle = \int d^3p'|\mathbf{p}'\rangle\langle\mathbf{p}'|\alpha\rangle. \tag{1.7.48b}$$

The expansion coefficients $\langle\mathbf{x}'|\alpha\rangle$ and $\langle\mathbf{p}'|\alpha\rangle$ are identified with the wave functions $\psi_\alpha(\mathbf{x}')$ and $\phi_\alpha(\mathbf{p}')$ in position and momentum space, respectively.

The momentum operator, when taken between $|\beta\rangle$ and $|\alpha\rangle$, becomes

$$\langle\beta|\mathbf{p}|\alpha\rangle = \int d^3x'\psi_\beta^*(\mathbf{x}')(-i\hbar\nabla)\psi_\alpha(\mathbf{x}'). \tag{1.7.49}$$

The transformation function analogous to (1.7.32) is

$$\langle \mathbf{x}'|\mathbf{p}'\rangle = \left[\frac{1}{(2\pi\hbar)^{3/2}}\right]\exp\left(\frac{i\mathbf{p}'\cdot\mathbf{x}'}{\hbar}\right), \tag{1.7.50}$$

so that

$$\psi_\alpha(\mathbf{x}') = \left[\frac{1}{(2\pi\hbar)^{3/2}}\right]\int d^3p' \exp\left(\frac{i\mathbf{p}'\cdot\mathbf{x}'}{\hbar}\right)\phi_\alpha(\mathbf{p}') \tag{1.7.51a}$$

and

$$\phi_\alpha(\mathbf{p}') = \left[\frac{1}{(2\pi\hbar)^{3/2}}\right]\int d^3x' \exp\left(\frac{-i\mathbf{p}'\cdot\mathbf{x}'}{\hbar}\right)\psi_\alpha(\mathbf{x}'). \tag{1.7.51b}$$

It is interesting to check the dimension of the wave functions. In one-dimensional problems the normalization requirement (1.6.8) implies that $|\langle x'|\alpha\rangle|^2$ has the dimension of inverse length, so the wave function itself must have the dimension of $(\text{length})^{-1/2}$. In contrast, the wave function in three-dimensional problems must have the dimension of $(\text{length})^{-3/2}$ because $|\langle \mathbf{x}'|\alpha\rangle|^2$ integrated over all spatial volume must be unity (dimensionless).

Problems

1. Prove

$$[AB, CD] = -AC\{D, B\} + A\{C, B\}D - C\{D, A\}B + \{C, A\}DB.$$

2. Suppose a 2×2 matrix X (not necessarily Hermitian, nor unitary) is written as

$$X = a_0 + \boldsymbol{\sigma}\cdot\mathbf{a},$$

where a_0 and $a_{1,2,3}$ are numbers.
   a. How are a_0 and a_k $(k = 1,2,3)$ related to $\mathrm{tr}(X)$ and $\mathrm{tr}(\sigma_k X)$?
   b. Obtain a_0 and a_k in terms of the matrix elements X_{ij}.
3. Show that the determinant of a 2×2 matrix $\boldsymbol{\sigma}\cdot\mathbf{a}$ is invariant under

$$\boldsymbol{\sigma}\cdot\mathbf{a} \to \boldsymbol{\sigma}\cdot\mathbf{a}' \equiv \exp\left(\frac{i\boldsymbol{\sigma}\cdot\hat{\mathbf{n}}\phi}{2}\right)\boldsymbol{\sigma}\cdot\mathbf{a}\exp\left(\frac{-i\boldsymbol{\sigma}\cdot\hat{\mathbf{n}}\phi}{2}\right).$$

   Find a'_k in terms of a_k when $\hat{\mathbf{n}}$ is in the positive z-direction and interpret your result.
4. Using the rules of bra-ket algebra, prove or evaluate the following:
   a. $\mathrm{tr}(XY) = \mathrm{tr}(YX)$, where X and Y are operators;
   b. $(XY)^\dagger = Y^\dagger X^\dagger$, where X and Y are operators;
   c. $\exp[if(A)] = ?$ in ket-bra form, where A is a Hermitian operator whose eigenvalues are known;
   d. $\sum_{a'}\psi^*_{a'}(\mathbf{x}')\psi_{a'}(\mathbf{x}'')$, where $\psi_{a'}(\mathbf{x}') = \langle \mathbf{x}'|a'\rangle$.

5. a. Consider two kets $|\alpha\rangle$ and $|\beta\rangle$. Suppose $\langle a'|\alpha\rangle$, $\langle a''|\alpha\rangle,\ldots$ and $\langle a'|\beta\rangle$, $\langle a''|\beta\rangle,\ldots$ are all known, where $|a'\rangle$, $|a''\rangle,\ldots$ form a complete set of base kets. Find the matrix representation of the operator $|\alpha\rangle\langle\beta|$ in that basis.
   b. We now consider a spin $\frac{1}{2}$ system and let $|\alpha\rangle$ and $|\beta\rangle$ be $|s_z = \hbar/2\rangle$ and $|s_x = \hbar/2\rangle$, respectively. Write down explicitly the square matrix that corresponds to $|\alpha\rangle\langle\beta|$ in the usual (s_z diagonal) basis.
6. Suppose $|i\rangle$ and $|j\rangle$ are eigenkets of some Hermitian operator A. Under what condition can we conclude that $|i\rangle + |j\rangle$ is also an eigenket of A? Justify your answer.
7. Consider a ket space spanned by the eigenkets $\{|a'\rangle\}$ of a Hermitian operator A. There is no degeneracy.
   a. Prove that

$$\prod_{a'}(A - a')$$

   is the null operator.
   b. What is the significance of

$$\prod_{a''\neq a'}\frac{(A - a'')}{(a' - a'')}?$$

   c. Illustrate (a) and (b) using A set equal to S_z of a spin $\frac{1}{2}$ system.
8. Using the orthonormality of $|+\rangle$ and $|-\rangle$, prove

$$[S_i, S_j] = i\varepsilon_{ijk}\hbar S_k, \qquad \{S_i, S_j\} = \left(\frac{\hbar^2}{2}\right)\delta_{ij},$$

   where

$$S_x = \frac{\hbar}{2}(|+\rangle\langle-| + |-\rangle\langle+|), \qquad S_y = \frac{i\hbar}{2}(-|+\rangle\langle-| + |-\rangle\langle+|),$$

$$S_z = \frac{\hbar}{2}(|+\rangle\langle+| - |-\rangle\langle-|).$$

9. Construct $|\mathbf{S}\cdot\hat{\mathbf{n}}; +\rangle$ such that

$$\mathbf{S}\cdot\hat{\mathbf{n}}|\mathbf{S}\cdot\hat{\mathbf{n}}; +\rangle = \left(\frac{\hbar}{2}\right)|\mathbf{S}\cdot\hat{\mathbf{n}}; +\rangle$$

   where $\hat{\mathbf{n}}$ is characterized by the angles shown in the figure. Express your answer as a linear combination of $|+\rangle$ and $|-\rangle$. [*Note*: The answer is

$$\cos\left(\frac{\beta}{2}\right)|+\rangle + \sin\left(\frac{\beta}{2}\right)e^{i\alpha}|-\rangle.$$

   But do not just verify that this answer satisfies the above eigenvalue equation. Rather, treat the problem as a straightforward eigenvalue

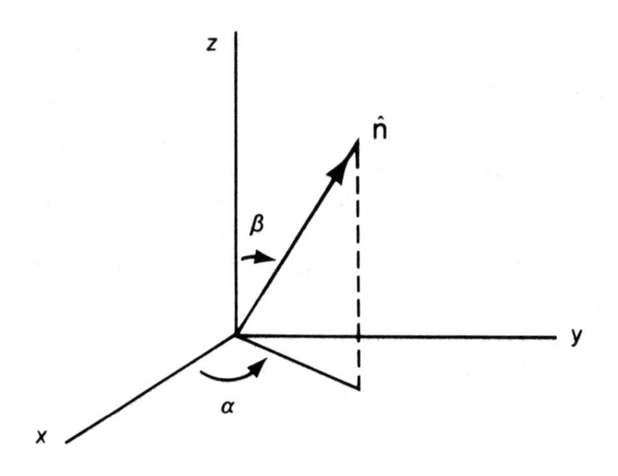

problem. Also do not use rotation operators, which we will introduce later in this book.]

10. The Hamiltonian operator for a two-state system is given by
$$H = a(|1\rangle\langle 1| - |2\rangle\langle 2| + |1\rangle\langle 2| + |2\rangle\langle 1|),$$
where a is a number with the dimension of energy. Find the energy eigenvalues and the corresponding energy eigenkets (as linear combinations of $|1\rangle$ and $|2\rangle$).

11. A two-state system is characterized by the Hamiltonian
$$H = H_{11}|1\rangle\langle 1| + H_{22}|2\rangle\langle 2| + H_{12}[|1\rangle\langle 2| + |2\rangle\langle 1|]$$
where H_{11}, H_{22}, and H_{12} are real numbers with the dimension of energy, and $|1\rangle$ and $|2\rangle$ are eigenkets of some observable ($\neq H$). Find the energy eigenkets and corresponding energy eigenvalues. Make sure that your answer makes good sense for $H_{12} = 0$. (You need not solve this problem from scratch. The following fact may be used without proof:
$$(\mathbf{S}\cdot\hat{\mathbf{n}})|\hat{\mathbf{n}}; +\rangle = \frac{\hbar}{2}|\hat{\mathbf{n}}; +\rangle,$$
with $|\hat{\mathbf{n}}; +\rangle$ given by
$$|\hat{\mathbf{n}}; +\rangle = \cos\frac{\beta}{2}|+\rangle + e^{i\alpha}\sin\frac{\beta}{2}|-\rangle,$$
where β and α are the polar and azimuthal angles, respectively, that characterize $\hat{\mathbf{n}}$.)

12. A spin $\frac{1}{2}$ system is known to be in an eigenstate of $\mathbf{S}\cdot\hat{\mathbf{n}}$ with eigenvalue $\hbar/2$, where $\hat{\mathbf{n}}$ is a unit vector lying in the xz-plane that makes an angle γ with the positive z-axis.

see Notes 10/13/97

a. Suppose S_x is measured. What is the probability of getting $+\hbar/2$?
b. Evaluate the dispersion in S_x, that is,

$$\langle (S_x - \langle S_x \rangle)^2 \rangle.$$

(For your own peace of mind check your answers for the special cases $\gamma = 0$, $\pi/2$, and π.)

13. A beam of spin $\frac{1}{2}$ atoms goes through a series of Stern-Gerlach-type measurements as follows:
a. The first measurement accepts $s_z = \hbar/2$ atoms and rejects $s_z = -\hbar/2$ atoms.
b. The second measurement accepts $s_n = \hbar/2$ atoms and rejects $s_n = -\hbar/2$ atoms, where s_n is the eigenvalue of the operator $\mathbf{S}\cdot\hat{\mathbf{n}}$, with $\hat{\mathbf{n}}$ making an angle β in the xz-plane with respect to the z-axis.
c. The third measurement accepts $s_z = -\hbar/2$ atoms and rejects $s_z = \hbar/2$ atoms.

What is the intensity of the final $s_z = -\hbar/2$ beam when the $s_z = \hbar/2$ beam surviving the first measurement is normalized to unity? How must we orient the second measuring apparatus if we are to maximize the intensity of the final $s_z = -\hbar/2$ beam?

14. A certain observable in quantum mechanics has a 3×3 matrix representation as follows:

$$\frac{1}{\sqrt{2}}\begin{pmatrix} 0 & 1 & 0 \\ 1 & 0 & 1 \\ 0 & 1 & 0 \end{pmatrix}.$$

a. Find the normalized eigenvectors of this observable and the corresponding eigenvalues. Is there any degeneracy?
b. Give a physical example where all this is relevant.

15. Let A and B be observables. Suppose the simultaneous eigenkets of A and B $\{|a', b'\rangle\}$ form a *complete* orthonormal set of base kets. Can we always conclude that

$$[A, B] = 0?$$

If your answer is yes, prove the assertion. If your answer is no, give a counterexample.

16. Two Hermitian operators anticommute:

$$\{A, B\} = AB + BA = 0.$$

Is it possible to have a simultaneous (that is, common) eigenket of A and B? Prove or illustrate your assertion.

17. Two observables A_1 and A_2, which do not involve time explicitly, are known not to commute,

$$[A_1, A_2] \neq 0,$$

yet we also know that A_1 and A_2 both commute with the Hamiltonian:

$$[A_1, H]=0, \qquad [A_2, H]=0.$$

Prove that the energy eigenstates are, in general, degenerate. Are there exceptions? As an example, you may think of the central-force problem $H=\mathbf{p}^2/2m+V(r)$, with $A_1 \to L_z$, $A_2 \to L_x$.

18. a. The simplest way to derive the Schwarz inequality goes as follows. First, observe

$$(\langle\alpha|+\lambda^*\langle\beta|)\cdot(|\alpha\rangle+\lambda|\beta\rangle)\geq 0$$

for any complex number λ; then choose λ in such a way that the preceding inequality reduces to the Schwarz inequality.

b. Show that the equality sign in the generalized uncertainty relation holds if the state in question satisfies

$$\Delta A|\alpha\rangle=\lambda\,\Delta B|\alpha\rangle$$

with λ purely *imaginary*.

c. Explicit calculations using the usual rules of wave mechanics show that the wave function for a Gaussian wave packet given by

$$\langle x'|\alpha\rangle=(2\pi d^2)^{-1/4}\exp\left[\frac{i\langle p\rangle x'}{\hbar}-\frac{(x'-\langle x\rangle)^2}{4d^2}\right]$$

satisfies the minimum uncertainty relation

$$\sqrt{\langle(\Delta x)^2\rangle}\sqrt{\langle(\Delta p)^2\rangle}=\frac{\hbar}{2}.$$

Prove that the requirement

$$\langle x'|\Delta x|\alpha\rangle=(\text{imaginary number})\langle x'|\Delta p|\alpha\rangle$$

is indeed satisfied for such a Gaussian wave packet, in agreement with (b).

19. a. Compute

$$\langle(\Delta S_x)^2\rangle\equiv\langle S_x^2\rangle-\langle S_x\rangle^2,$$

where the expectation value is taken for the S_z+ state. Using your result, check the generalized uncertainty relation

$$\langle(\Delta A)^2\rangle\langle(\Delta B)^2\rangle\geq\frac{1}{4}|\langle[A,B]\rangle|^2,$$

with $A\to S_x$, $B\to S_y$.

b. Check the uncertainty relation with $A\to S_x$, $B\to S_y$ for the S_x+ state.

20. Find the linear combination of $|+\rangle$ and $|-\rangle$ kets that maximizes the

uncertainty product

$$\langle(\Delta S_x)^2\rangle\langle(\Delta S_y)^2\rangle.$$

Verify explicitly that for the linear combination you found, the uncertainty relation for S_x and S_y is not violated.

21. Evaluate the x-p uncertainty product $\langle(\Delta x)^2\rangle\langle(\Delta p)^2\rangle$ for a one-dimensional particle confined between two rigid walls

$$V = \begin{cases} 0 & \text{for } 0 < x < a, \\ \infty & \text{otherwise.} \end{cases}$$

Do this for both the ground and excited states.

22. Estimate the rough order of magnitude of the length of time that an ice pick can be balanced on its point if the only limitation is that set by the Heisenberg uncertainty principle. Assume that the point is sharp and that the point and the surface on which it rests are hard. You may make approximations which do not alter the general order of magnitude of the result. Assume reasonable values for the dimensions and weight of the ice pick. Obtain an approximate numerical result and express it *in seconds*.

23. Consider a three-dimensional ket space. If a certain set of orthonormal kets—say, $|1\rangle$, $|2\rangle$, and $|3\rangle$—are used as the base kets, the operators A and B are represented by

$$A \doteq \begin{pmatrix} a & 0 & 0 \\ 0 & -a & 0 \\ 0 & 0 & -a \end{pmatrix}, \qquad B \doteq \begin{pmatrix} b & 0 & 0 \\ 0 & 0 & -ib \\ 0 & ib & 0 \end{pmatrix}$$

with a and b both real.

a. Obviously A exhibits a degenerate spectrum. Does B also exhibit a degenerate spectrum?

b. Show that A and B commute.

c. Find a new set of orthonormal kets which are simultaneous eigenkets of both A and B. Specify the eigenvalues of A and B for each of the three eigenkets. Does your specification of eigenvalues completely characterize each eigenket?

24. a. Prove that $(1/\sqrt{2})(1 + i\sigma_x)$ acting on a two-component spinor can be regarded as the matrix representation of the rotation operator about the x-axis by angle $-\pi/2$. (The minus sign signifies that the rotation is clockwise.)

b. Construct the matrix representation of S_z when the eigenkets of S_y are used as base vectors.

25. Some authors define an *operator* to be real when every member of its matrix elements $\langle b'|A|b''\rangle$ is real in some representation ($\{|b'\rangle\}$ basis in this case). Is this concept representation independent, that is, do the

matrix elements remain real even if some basis other than $\{|b'\rangle\}$ is used? Check your assertion using familiar operators such as S_y and S_z (see Problem 24) or x and p_x.

26. Construct the transformation matrix that connects the S_z diagonal basis to the S_x diagonal basis. Show that your result is consistent with the general relation

$$U = \sum_r |b^{(r)}\rangle\langle a^{(r)}|.$$

27. a. Suppose that $f(A)$ is a function of a Hermitian operator A with the property $A|a'\rangle = a'|a'\rangle$. Evaluate $\langle b''|f(A)|b'\rangle$ when the transformation matrix from the a' basis to the b' basis is known.

    b. Using the continuum analogue of the result obtained in (a), evaluate

$$\langle \mathbf{p}''|F(r)|\mathbf{p}'\rangle.$$

    Simplify your expression as far as you can. Note that r is $\sqrt{x^2+y^2+z^2}$, where x, y, and z are *operators*.

28. a. Let x and p_x be the coordinate and linear momentum in one dimension. Evaluate the classical Poisson bracket

$$[x, F(p_x)]_{\text{classical}}.$$

    b. Let x and p_x be the corresponding quantum-mechanical operators this time. Evaluate the commutator

$$\left[x, \exp\left(\frac{ip_x a}{\hbar}\right)\right].$$

    c. Using the result obtained in (b), prove that

$$\exp\left(\frac{ip_x a}{\hbar}\right)|x'\rangle, \qquad (x|x'\rangle = x'|x'\rangle)$$

    is an eigenstate of the coordinate operator x. What is the corresponding eigenvalue?

29. a. On page 247, Gottfried (1966) states that

$$[x_i, G(\mathbf{p})] = i\hbar\frac{\partial G}{\partial p_i}, \qquad [p_i, F(\mathbf{x})] = -i\hbar\frac{\partial F}{\partial x_i}$$

    can be "easily derived" from the fundamental commutation relations for all functions of F and G that can be expressed as power series in their arguments. Verify this statement.

    b. Evaluate $[x^2, p^2]$. Compare your result with the classical Poisson bracket $[x^2, p^2]_{\text{classical}}$.

30. The translation operator for a finite (spatial) displacement is given by

$$\mathscr{T}(\mathbf{l}) = \exp\left(\frac{-i\mathbf{p}\cdot\mathbf{l}}{\hbar}\right),$$

where $\mathbf{p}$ is the momentum *operator*.

a. Evaluate

$$[x_i, \mathscr{T}(\mathbf{l})].$$

b. Using (a) (or otherwise), demonstrate how the expectation value $\langle\mathbf{x}\rangle$ changes under translation.

31. In the main text we discussed the effect of $\mathscr{T}(d\mathbf{x}')$ on the position and momentum eigenkets and on a more general state ket $|\alpha\rangle$. We can also study the behavior of expectation values $\langle\mathbf{x}\rangle$ and $\langle\mathbf{p}\rangle$ under infinitesimal translation. Using (1.6.25), (1.6.45), and $|\alpha\rangle \to \mathscr{T}(d\mathbf{x}')|\alpha\rangle$ only, prove $\langle\mathbf{x}\rangle \to \langle\mathbf{x}\rangle + d\mathbf{x}', \langle\mathbf{p}\rangle \to \langle\mathbf{p}\rangle$ under infinitesimal translation.

32. a. Verify (1.7.39a) and (1.7.39b) for the expectation value of p and p^2 from the Gaussian wave packet (1.7.35).

b. Evaluate the expectation value of p and p^2 using the momentum-space wave function (1.7.42).

33. a. Prove the following:

$$\text{(i)} \qquad \langle p'|x|\alpha\rangle = i\hbar\frac{\partial}{\partial p'}\langle p'|\alpha\rangle,$$

$$\text{(ii)} \qquad \langle\beta|x|\alpha\rangle = \int dp'\,\phi_\beta^*(p')i\hbar\frac{\partial}{\partial p'}\phi_\alpha(p'),$$

where $\phi_\alpha(p') = \langle p'|\alpha\rangle$ and $\phi_\beta(p') = \langle p'|\beta\rangle$ are momentum-space wave functions.

b. What is the physical significance of

$$\exp\left(\frac{ix\Xi}{\hbar}\right),$$

where x is the position operator and Ξ is some number with the dimension of momentum? Justify your answer.

CHAPTER 2

Quantum Dynamics

So far we have not discussed how physical systems change with time. This chapter is devoted exclusively to the dynamic development of state kets and/or observables. In other words, we are concerned here with the quantum mechanical analogue of Newton's (or Lagrange's or Hamilton's) equations of motion.

2.1. TIME EVOLUTION AND THE SCHRÖDINGER EQUATION

The first important point we should keep in mind is that time is just a parameter in quantum mechanics, *not* an operator. In particular, time is not an observable in the language of the previous chapter. It is nonsensical to talk about the time operator in the same sense as we talk about the position operator. Ironically, in the historical development of wave mechanics both L. de Broglie and E. Schrödinger were guided by a kind of covariant analogy between energy and time on the one hand and momentum and position (spatial coordinate) on the other. Yet when we now look at quantum mechanics in its finished form, there is no trace of a symmetrical treatment between time and space. The relativistic quantum theory of fields does treat the time and space coordinates on the same footing, but it does so only at the expense of demoting position from the status of being an observable to that of being just a parameter.

Time Evolution Operator

Our basic concern in this section is, How does a state ket change with time? Suppose we have a physical system whose state ket at t_0 is represented by $|\alpha\rangle$. At later times, we do not, in general, expect the system to remain in the same state $|\alpha\rangle$. Let us denote the ket corresponding to the state at some later time by

$$|\alpha, t_0; t\rangle, \quad (t > t_0), \tag{2.1.1}$$

where we have written α, t_0 to remind ourselves that the system *used to be* in state $|\alpha\rangle$ at some earlier reference time t_0. Because time is assumed to be a continuous parameter, we expect

$$\lim_{t \to t_0} |\alpha, t_0; t\rangle = |\alpha\rangle \tag{2.1.2}$$

and we may as well use a shorthand notation,

$$|\alpha, t_0; t_0\rangle = |\alpha, t_0\rangle, \tag{2.1.3}$$

for this. Our basic task is to study the time evolution of a state ket:

$$|\alpha, t_0\rangle = |\alpha\rangle \xrightarrow{\text{time evolution}} |\alpha, t_0; t\rangle. \tag{2.1.4}$$

Put in another way, we are interested in asking how the state ket changes under a time displacement $t_0 \to t$.

As in the case of translation, the two kets are related by an operator which we call the **time-evolution operator** $\mathscr{U}(t, t_0)$:

$$|\alpha, t_0; t\rangle = \mathscr{U}(t, t_0)|\alpha, t_0\rangle. \tag{2.1.5}$$

What are some of the properties we would like to ascribe to the time-evolution operator? The first important property is the unitary requirement for $\mathscr{U}(t, t_0)$ that follows from probability conservation. Suppose that at t_0 the state ket is expanded in terms of the eigenkets of some observable A:

$$|\alpha, t_0\rangle = \sum_{a'} c_{a'}(t_0)|a'\rangle. \tag{2.1.6}$$

Likewise, at some later time, we have

$$|\alpha, t_0; t\rangle = \sum_{a'} c_{a'}(t)|a'\rangle. \tag{2.1.7}$$

In general, we do not expect the modulus of the individual expansion coefficient to remain the same:*

$$|c_{a'}(t)| \neq |c_{a'}(t_0)|. \tag{2.1.8}$$

For instance, consider a spin $\frac{1}{2}$ system with its spin magnetic moment

---

*We later show, however, that if the Hamiltonian commutes with A, then $|c_{a'}(t)|$ is indeed equal to $|c_{a'}(t_0)|$.

subjected to a uniform magnetic field in the z-direction. To be specific, suppose that at t_0 the spin is in the positive x-direction; that is, the system is found in an eigenstate of S_x with eigenvalue $\hbar/2$. As time goes on, the spin precesses in the xy-plane, as will be quantitatively demonstrated later in this section. This means that the probability for observing S_x+ is no longer unity at $t>t_0$; there is a finite probability for observing S_x- as well. Yet the *sum* of the probabilities for S_x+ and S_x- remains unity at all times. Generally, in the notation of (2.1.6) and (2.1.7), we must have

$$\sum_{a'} |c_{a'}(t_0)|^2 = \sum_{a'} |c_{a'}(t)|^2 \tag{2.1.9}$$

despite (2.1.8) for the individual expansion coefficients. Stated another way, if the state ket is initially normalized to unity, it must remain normalized to unity at all later times:

$$\langle \alpha, t_0 | \alpha, t_0 \rangle = 1 \Rightarrow \langle \alpha, t_0; t | \alpha, t_0; t \rangle = 1. \tag{2.1.10}$$

As in the translation case, this property is guaranteed if the time-evolution operator is taken to be unitary. For this reason we take unitarity,

$$\mathscr{U}^\dagger(t, t_0)\mathscr{U}(t, t_0) = 1, \tag{2.1.11}$$

to be one of the fundamental properties of the $\mathscr{U}$ operator. It is no coincidence that many authors regard unitarity as being synonymous with probability conservation.

Another feature we require of the $\mathscr{U}$ operator is the composition property:

$$\mathscr{U}(t_2, t_0) = \mathscr{U}(t_2, t_1)\mathscr{U}(t_1, t_0), \quad (t_2 > t_1 > t_0). \tag{2.1.12}$$

This equation says that if we are interested in obtaining time evolution from t_0 to t_2, then we can obtain the same result by first considering time evolution from t_0 to t_1, then from t_1 to t_2—a reasonable requirement. Note that we read (2.1.12) from right to left!

It also turns out to be advantageous to consider an infinitesimal time-evolution operator $\mathscr{U}(t_0 + dt, t_0)$:

$$|\alpha, t_0; t_0 + dt\rangle = \mathscr{U}(t_0 + dt, t_0)|\alpha, t_0\rangle. \tag{2.1.13}$$

Because of continuity [see (2.1.2)], the infinitesimal time-evolution operator must reduce to the identity operator as dt goes to zero,

$$\lim_{dt \to 0} \mathscr{U}(t_0 + dt, t_0) = 1, \tag{2.1.14}$$

and as in the translation case, we expect the difference between $\mathscr{U}(t_0 + dt, t_0)$ and 1 to be of first order in dt.

We assert that all these requirements are satisfied by

$$\mathscr{U}(t_0 + dt, t_0) = 1 - i\Omega\, dt, \tag{2.1.15}$$

where Ω is a Hermitian operator,*

$$\Omega^\dagger = \Omega. \tag{2.1.16}$$

With (2.1.15) the infinitesimal time-displacement operator satisfies the composition property

$$\mathscr{U}(t_0 + dt_1 + dt_2, t_0) = \mathscr{U}(t_0 + dt_1 + dt_2, t_0 + dt_1)\mathscr{U}(t_0 + dt_1, t_0); \tag{2.1.17}$$

it differs from the identity operator by a term of order dt. The unitarity property can also be checked as follows:

$$\mathscr{U}^\dagger(t_0 + dt, t_0)\mathscr{U}(t_0 + dt, t_0) = (1 + i\Omega^\dagger dt)(1 - i\Omega\, dt) \simeq 1, \tag{2.1.18}$$

to the extent that terms of order $(dt)^2$ or higher can be ignored.

The operator Ω has the dimension of frequency or inverse time. Is there any familiar observable with the dimension of frequency? We recall that in the old quantum theory, angular frequency ω is postulated to be related to energy by the Planck-Einstein relation

$$E = \hbar\omega. \tag{2.1.19}$$

Let us now borrow from classical mechanics the idea that the Hamiltonian is the generator of time evolution (Goldstein 1980, 407–8). It is then natural to relate Ω to the Hamiltonian operator H:

$$\Omega = \frac{H}{\hbar}. \tag{2.1.20}$$

To sum up, the infinitesimal time-evolution operator is written as

$$\mathscr{U}(t_0 + dt, t_0) = 1 - \frac{iH\,dt}{\hbar}, \tag{2.1.21}$$

where H, the Hamiltonian operator, is assumed to be Hermitian. The reader may ask whether the $\hbar$ introduced here is the same as the $\hbar$ that appears in the expression for the translation operator (1.6.32). This question can be answered by comparing the quantum-mechanical equation of motion we derive later with the classical equation of motion. It turns out that unless the two $\hbar$'s are taken to be the same, we are unable to obtain a relation like

$$\frac{d\mathbf{x}}{dt} = \frac{\mathbf{p}}{m} \tag{2.1.22}$$

as the classical limit of the corresponding quantum-mechanical relation.

The Schrödinger Equation

We are now in a position to derive the fundamental differential equation for the time-evolution operator $\mathscr{U}(t, t_0)$. We exploit the composi-

*If the Ω operator depends on time explicitly, it must be evaluated at t_0.

tion property of the time-evolution operator by letting $t_1 \to t,\ t_2 \to t + dt$ in (2.1.12):

$$\mathscr{U}(t+dt, t_0) = \mathscr{U}(t+dt, t)\mathscr{U}(t, t_0) = \left(1 - \frac{iH\,dt}{\hbar}\right)\mathscr{U}(t, t_0), \tag{2.1.23}$$

where the time difference $t - t_0$ need not be infinitesimal. We have

$$\mathscr{U}(t+dt, t_0) - \mathscr{U}(t, t_0) = -i\left(\frac{H}{\hbar}\right) dt\,\mathscr{U}(t, t_0), \tag{2.1.24}$$

which can be written in differential equation form:

$$i\hbar\frac{\partial}{\partial t}\mathscr{U}(t, t_0) = H\mathscr{U}(t, t_0). \tag{2.1.25}$$

This is the **Schrödinger equation for the time-evolution operator**. Everything that has to do with time development follows from this fundamental equation.

Equation (2.1.25) immediately leads to the Schrödinger equation for a state ket. Multiplying both sides of (2.1.25) by $|\alpha, t_0\rangle$ on the right, we obtain

$$i\hbar\frac{\partial}{\partial t}\mathscr{U}(t, t_0)|\alpha, t_0\rangle = H\mathscr{U}(t, t_0)|\alpha, t_0\rangle. \tag{2.1.26}$$

But $|\alpha, t_0\rangle$ does not depend on t, so this is the same as

$$i\hbar\frac{\partial}{\partial t}|\alpha, t_0; t\rangle = H|\alpha, t_0; t\rangle, \tag{2.1.27}$$

where (2.1.5) has been used.

If we are given $\mathscr{U}(t, t_0)$ and, in addition, know how $\mathscr{U}(t, t_0)$ acts on the initial state ket $|\alpha, t_0\rangle$, it is not necessary to bother with the Schrödinger equation for the state ket (2.1.27). All we have to do is apply $\mathscr{U}(t, t_0)$ to $|\alpha, t_0\rangle$; in this manner we can obtain a state ket at any t. Our first task is therefore to derive formal solutions to the Schrödinger equation for the time evolution operator (2.1.25). There are three cases to be treated separately:

Case 1. The Hamiltonian operator is independent of time. By this we mean that even when the parameter t is changed, the H operator remains unchanged. The Hamiltonian for a spin-magnetic moment interacting with a time-independent magnetic field is an example of this. The solution to (2.1.25) in such a case is given by

$$\mathscr{U}(t, t_0) = \exp\left[\frac{-iH(t-t_0)}{\hbar}\right]. \tag{2.1.28}$$

To prove this let us expand the exponential as follows:

$$\exp\left[\frac{-iH(t-t_0)}{\hbar}\right] = 1 - \frac{iH(t-t_0)}{\hbar} + \left[\frac{(-i)^2}{2}\right]\left[\frac{H(t-t_0)}{\hbar}\right]^2 + \cdots. \tag{2.1.29}$$

Because the time derivative of this expansion is given by

$$\frac{\partial}{\partial t}\exp\left[\frac{-iH(t-t_0)}{\hbar}\right]=\frac{-iH}{\hbar}+\left[\frac{(-i)^2}{2}\right]2\left(\frac{H}{\hbar}\right)^2(t-t_0)+\cdots, \tag{2.1.30}$$

expression (2.1.28) obviously satisfies differential equation (2.1.25). The boundary condition is also satisfied because as $t \to t_0$, (2.1.28) reduces to the identity operator. An alternative way to obtain (2.1.28) is to compound successively infinitesimal time-evolution operators just as we did to obtain (1.6.36) for finite translation:

$$\lim_{N\to\infty}\left[1-\frac{(iH/\hbar)(t-t_0)}{N}\right]^N=\exp\left[\frac{-iH(t-t_0)}{\hbar}\right]. \tag{2.1.31}$$

Case 2. The Hamiltonian operator H is time-dependent but the H's at different times commute. As an example, let us consider the spin-magnetic moment subjected to a magnetic field whose strength varies with time but whose direction is always unchanged. The formal solution to (2.1.25) in this case is

$$\mathscr{U}(t,t_0)=\exp\left[-\left(\frac{i}{\hbar}\right)\int_{t_0}^{t}dt'H(t')\right]. \tag{2.1.32}$$

This can be proved in a similar way. We simply replace $H(t-t_0)$ in (2.1.29) and (2.1.30) by $\int_{t_0}^{t}dt'H(t')$.

Case 3. The H's at different times do *not* commute. Continuing with the example involving spin-magnetic moment, we suppose, this time, that the magnetic field direction also changes with time: at $t=t_1$ in the x-direction, at $t=t_2$ in the y-direction, and so forth. Because S_x and S_y do not commute, $H(t_1)$ and $H(t_2)$, which go like $\mathbf{S}\cdot\mathbf{B}$, do not commute either. The formal solution in such a situation is given by

$$\mathscr{U}(t,t_0)=1+\sum_{n=1}^{\infty}\left(\frac{-i}{\hbar}\right)^n\int_{t_0}^{t}dt_1\int_{t_0}^{t_1}dt_2\cdots\int_{t_0}^{t_{n-1}}dt_n\,H(t_1)H(t_2)\cdots H(t_n), \tag{2.1.33}$$

which is sometimes known as the **Dyson series**, after F. J. Dyson, who developed a perturbation expansion of this form in quantum field theory. We do not prove (2.1.33) now because the proof is very similar to the one presented in Chapter 5 for the time-evolution operator in the interaction picture.

In elementary applications, only case 1 is of practical interest. In the remaining part of this chapter we assume that the H operator is time-independent. We will encounter time-dependent Hamiltonians in Chapter 5.

Energy Eigenkets

To be able to evaluate the effect of the time-evolution operator (2.1.28) on a general initial ket $|\alpha\rangle$, we must first know how it acts on the base kets used in expanding $|\alpha\rangle$. This is particularly straightforward if the base kets used are eigenkets of A such that

$$[A, H] = 0; \tag{2.1.34}$$

then the eigenkets of A are also eigenkets of H, called **energy eigenkets**, whose eigenvalues are denoted by $E_{a'}$:

$$H|a'\rangle = E_{a'}|a'\rangle. \tag{2.1.35}$$

We can now expand the time-evolution operator in terms of $|a'\rangle\langle a'|$. Taking $t_0 = 0$ for simplicity, we obtain

$$\begin{aligned}\exp\left(\frac{-iHt}{\hbar}\right) &= \sum_{a'}\sum_{a''}|a''\rangle\langle a''|\exp\left(\frac{-iHt}{\hbar}\right)|a'\rangle\langle a'| \\ &= \sum_{a'}|a'\rangle\exp\left(\frac{-iE_{a'}t}{\hbar}\right)\langle a'|.\end{aligned} \tag{2.1.36}$$

The time-evolution operator written in this form enables us to solve any initial-value problem once the expansion of the initial ket in terms of $\{|a'\rangle\}$ is known. As an example, suppose that the initial ket expansion reads

$$|\alpha, t_0 = 0\rangle = \sum_{a'}|a'\rangle\langle a'|\alpha\rangle = \sum_{a'}c_{a'}|a'\rangle. \tag{2.1.37}$$

We then have

$$|\alpha, t_0 = 0; t\rangle = \exp\left(\frac{-iHt}{\hbar}\right)|\alpha, t_0 = 0\rangle = \sum_{a'}|a'\rangle\langle a'|\alpha\rangle\exp\left(\frac{-iE_{a'}t}{\hbar}\right). \tag{2.1.38}$$

In other words, the expansion coefficient changes with time as

$$c_{a'}(t=0) \to c_{a'}(t) = c_{a'}(t=0)\exp\left(\frac{-iE_{a'}t}{\hbar}\right) \tag{2.1.39}$$

with its modulus unchanged. Notice that the relative phases among various components do vary with time because the oscillation frequencies are different.

A special case of interest is where the initial state happens to be one of $\{|a'\rangle\}$ itself. We have

$$|\alpha, t_0 = 0\rangle = |a'\rangle \tag{2.1.40}$$

initially, and at a later time

$$|\alpha, t_0 = 0; t\rangle = |a'\rangle \exp\left(\frac{-iE_{a'}t}{\hbar}\right), \tag{2.1.41}$$

so if the system is initially a simultaneous eigenstate of A and H, it remains so at all times. The most that can happen is the phase modulation, $\exp(-iE_{a'}t/\hbar)$. It is in this sense that an observable compatible with H [see (2.1.34)] is a *constant of the motion*. We will encounter this connection once again in a different form when we discuss the Heisenberg equation of motion.

In the foregoing discussion the basic task in quantum dynamics is reduced to finding an observable that commutes with H and evaluating its eigenvalues. Once that is done, we expand the initial ket in terms of the eigenkets of that observable and just apply the time-evolution operator. This last step merely amounts to changing the phase of each expansion coefficient, as indicated by (2.1.39).

Even though we worked out the case where there is just one observable A that commutes with H, our considerations can easily be generalized when there are several mutually compatible observables all also commuting with H:

$$\begin{aligned}[A, B] = [B, C] = [A, C] = \cdots = 0,\\ [A, H] = [B, H] = [C, H] = \cdots = 0.\end{aligned} \tag{2.1.42}$$

Using the collective index notation of Section 1.4 [see (1.4.37)], we have

$$\exp\left(\frac{-iHt}{\hbar}\right) = \sum_{K'} |K'\rangle \exp\left(\frac{-iE_{K'}t}{\hbar}\right)\langle K'|, \tag{2.1.43}$$

where $E_{K'}$ is uniquely specified once $a', b', c', \ldots$ are specified. It is therefore of fundamental importance to find *a complete set of mutually compatible observables that also commute with H*. Once such a set is found, we express the initial ket as a superposition of the simultaneous eigenkets of $A, B, C, \ldots$ and H. The final step is just to apply the time-evolution operator, written as (2.1.43). In this manner we can solve the most general initial-value problem with a time-independent H.

Time Dependence of Expectation Values

It is instructive to study how the expectation value of an observable changes as a function of time. Suppose that at $t = 0$ the initial state is one of the eigenstates of an observable A that commutes with H, as in (2.1.40). We now look at the expectation value of some other observable B, which need not commute with A nor with H. Because at a later time we have

$$|a', t_0 = 0; t\rangle = \mathscr{U}(t, 0)|a'\rangle \tag{2.1.44}$$

for the state ket, $\langle B \rangle$ is given by

$$\begin{aligned}\langle B \rangle &= (\langle a'|\mathscr{U}^\dagger(t,0))\cdot B \cdot (\mathscr{U}(t,0)|a'\rangle)\\ &= \langle a'|\exp\left(\frac{iE_{a'}t}{\hbar}\right) B \exp\left(\frac{-iE_{a'}t}{\hbar}\right)|a'\rangle\\ &= \langle a'|B|a'\rangle, \end{aligned} \tag{2.1.45}$$

which is *independent of t*. So the expectation value of an observable taken with respect to an energy eigenstate does not change with time. For this reason an energy eigenstate is often referred to as a **stationary state**.

The situation is more interesting when the expectation value is taken with respect to a *superposition* of energy eigenstates, or a **nonstationary state**. Suppose that initially we have

$$|\alpha, t_0 = 0\rangle = \sum_{a'} c_{a'}|a'\rangle. \tag{2.1.46}$$

We easily compute the expectation value of B to be

$$\begin{aligned}\langle B \rangle &= \left[\sum_{a'} c_{a'}^* \langle a'|\exp\left(\frac{iE_{a'}t}{\hbar}\right)\right]\cdot B \cdot \left[\sum_{a''} c_{a''}\exp\left(\frac{-iE_{a''}t}{\hbar}\right)|a''\rangle\right]\\ &= \sum_{a'}\sum_{a''} c_{a'}^* c_{a''}\langle a'|B|a''\rangle \exp\left[\frac{-i(E_{a''}-E_{a'})t}{\hbar}\right]. \end{aligned} \tag{2.1.47}$$

So this time the expectation value consists of oscillating terms whose angular frequencies are determined by N. Bohr's frequency condition

$$\omega_{a''a'} = \frac{(E_{a''} - E_{a'})}{\hbar}. \tag{2.1.48}$$

Spin Precession

It is appropriate to treat an example here. We consider an extremely simple system which, however, illustrates the basic formalism we have developed.

We start with a Hamiltonian of a spin $\frac{1}{2}$ system with magnetic moment $e\hbar/2m_e c$ subjected to an external magnetic field $\mathbf{B}$:

$$H = -\left(\frac{e}{m_e c}\right)\mathbf{S}\cdot\mathbf{B} \tag{2.1.49}$$

($e < 0$ for the electron). Furthermore, we take $\mathbf{B}$ to be a static, uniform magnetic field in the z-direction. We can then write H as

$$H = -\left(\frac{eB}{m_e c}\right)S_z. \tag{2.1.50}$$

Because S_z and H differ just by a multiplicative constant, they obviously commute. The S_z eigenstates are also energy eigenstates, and the corresponding energy eigenvalues are

$$E_{\pm} = \mp \frac{e\hbar B}{2m_e c}, \quad \text{for } S_z \pm . \tag{2.1.51}$$

It is convenient to define ω in such a way that the difference in the two energy eigenvalues is $\hbar\omega$:

$$\omega \equiv \frac{|e|B}{m_e c}. \tag{2.1.52}$$

We can then rewrite the H operator simply as

$$H = \omega S_z. \tag{2.1.53}$$

All the information on time development is contained in the time-evolution operator

$$\mathscr{U}(t,0) = \exp\left(\frac{-i\omega S_z t}{\hbar}\right). \tag{2.1.54}$$

We apply this to the initial state. The base kets we must use in expanding the initial ket are obviously the S_z eigenkets, $|+\rangle$ and $|-\rangle$, which are also energy eigenkets. Suppose that at $t = 0$ the system is characterized by

$$|\alpha\rangle = c_+|+\rangle + c_-|-\rangle. \tag{2.1.55}$$

Upon applying (2.1.54), we see that the state ket at some later time is

$$|\alpha, t_0 = 0; t\rangle = c_+ \exp\left(\frac{-i\omega t}{2}\right)|+\rangle + c_- \exp\left(\frac{+i\omega t}{2}\right)|-\rangle, \tag{2.1.56}$$

where we have used

$$H|\pm\rangle = \left(\frac{\pm\hbar\omega}{2}\right)|\pm\rangle. \tag{2.1.57}$$

Specifically, let us suppose that the initial ket $|\alpha\rangle$ represents the spin-up (or, more precisely, S_z+) state $|+\rangle$, which means that

$$c_+ = 1, \qquad c_- = 0. \tag{2.1.58}$$

At a later time, (2.1.56) tells us that it is still in the spin-up state, which is no surprise because this is a stationary state.

Next, let us suppose that initially the system is in the S_x+ state. Comparing (1.4.17a) with (2.1.55), we see that

$$c_+ = c_- = \frac{1}{\sqrt{2}}. \tag{2.1.59}$$

It is straightforward to work out the probabilities for the system to be found in the $S_x \pm$ state at some later time t:

$$
\begin{aligned}
|\langle S_x \pm | \alpha, t_0 = 0; t\rangle|^2 &= \left| \left[\left(\frac{1}{\sqrt{2}}\right)\langle + | \pm \left(\frac{1}{\sqrt{2}}\right)\langle - | \right] \cdot \left[\left(\frac{1}{\sqrt{2}}\right)\exp\left(\frac{-i\omega t}{2}\right)|+\rangle \right.\right. \\
&\quad \left.\left. + \left(\frac{1}{\sqrt{2}}\right)\exp\left(\frac{+i\omega t}{2}\right)|-\rangle \right] \right|^2 \\
&= \left| \frac{1}{2}\exp\left(\frac{-i\omega t}{2}\right) \pm \frac{1}{2}\exp\left(\frac{+i\omega t}{2}\right) \right|^2 \\
&= \begin{cases} \cos^2 \dfrac{\omega t}{2}, & \text{for } S_x +, \qquad (2.1.60a) \\ \sin^2 \dfrac{\omega t}{2}, & \text{for } S_x -. \qquad (2.1.60b) \end{cases}
\end{aligned}
$$

Even though the spin is initially in the positive x-direction, the magnetic field in the z-direction causes it to rotate; as a result, we obtain a finite probability for finding $S_x -$ at some later time. The sum of the two probabilities is seen to be unity at all times, in agreement with the unitarity property of the time-evolution operator.

Using (1.4.6), we can write the expectation value of S_x as

$$
\begin{aligned}
\langle S_x \rangle &= \left(\frac{\hbar}{2}\right)\cos^2\left(\frac{\omega t}{2}\right) + \left(\frac{-\hbar}{2}\right)\sin^2\left(\frac{\omega t}{2}\right) \\
&= \left(\frac{\hbar}{2}\right)\cos \omega t, \qquad (2.1.61)
\end{aligned}
$$

so this quantity oscillates with an angular frequency corresponding to the difference of the two energy eigenvalues divided by $\hbar$, in agreement with our general formula (2.1.47). Similar exercises with S_y and S_z show that

$$\langle S_y \rangle = \left(\frac{\hbar}{2}\right)\sin \omega t \qquad (2.1.62a)$$

and

$$\langle S_z \rangle = 0. \qquad (2.1.62b)$$

Physically this means that the spin precesses in the xy-plane. We will comment further on spin precession when we discuss rotation operators in Chapter 3.

Correlation Amplitude and the Energy-Time Uncertainty Relation

We conclude this section by asking how state kets at different times are correlated with each other. Suppose the initial state ket at $t = 0$ of a physical system is given by $|\alpha\rangle$. With time it changes into $|\alpha, t_0 = 0; t\rangle$, which we obtain by applying the time-evolution operator. We are concerned with the extent to which the state ket at a later time t is similar to the state ket at $t=0$; we therefore construct the inner product between the two state kets at different times:

$$\begin{aligned} C(t) &\equiv \langle \alpha | \alpha, t_0 = 0; t \rangle \\ &= \langle \alpha | \mathscr{U}(t,0) | \alpha \rangle, \end{aligned} \tag{2.1.63}$$

which is known as the **correlation amplitude**. The modulus of $C(t)$ provides a quantitative measure of the "resemblance" between the state kets at different times.

As an extreme example, consider the very special case where the initial ket $|\alpha\rangle$ is an eigenket of H; we then have

$$C(t) = \langle a' | a', t_0 = 0; t \rangle = \exp\left(\frac{-iE_{a'}t}{\hbar} \right), \tag{2.1.64}$$

so the modulus of the correlation amplitude is unity at all times—which is not surprising for a stationary state. In the more general situation where the initial ket is represented by a superposition of $\{|a'\rangle\}$, as in (2.1.37), we have

$$\begin{aligned} C(t) &= \left(\sum_{a'} c_{a'}^* \langle a' | \right) \left[\sum_{a''} c_{a''} \exp\left(\frac{-iE_{a''}t}{\hbar} \right) |a''\rangle \right] \\ &= \sum_{a'} |c_{a'}|^2 \exp\left(\frac{-iE_{a'}t}{\hbar} \right). \end{aligned} \tag{2.1.65}$$

As we sum over many terms with oscillating time dependence of different frequencies, a strong cancellation is possible for moderately large values of t. We expect the correlation amplitude that starts with unity at $t=0$ to decrease in magnitude with time.

To estimate (2.1.65) in a more concrete manner, let us suppose that the state ket can be regarded as a superposition of so many energy eigenkets with similar energies that we can regard them as exhibiting essentially a quasi-continuous spectrum. It is then legitimate to replace the sum by the integral

$$\sum_{a'} \to \int dE \rho(E), \qquad c_{a'} \to g(E)\Big|_{E \simeq E_{a'}}, \tag{2.1.66}$$

where $\rho(E)$ characterizes the density of energy eigenstates. Expression (2.1.65) now becomes

$$C(t) = \int dE\, |g(E)|^2 \rho(E) \exp\left(\frac{-iEt}{\hbar} \right), \tag{2.1.67}$$

subject to the normalization condition

$$\int dE\, |g(E)|^2 \rho(E) = 1. \tag{2.1.68}$$

In a realistic physical situation $|g(E)|^2\rho(E)$ may be peaked around $E = E_0$ with width ΔE. Writing (2.1.67) as

$$C(t) = \exp\left(\frac{-iE_0 t}{\hbar}\right)\int dE|g(E)|^2\rho(E)\exp\left[\frac{-i(E-E_0)t}{\hbar}\right], \tag{2.1.69}$$

we see that as t becomes large, the integrand oscillates very rapidly unless the energy interval $|E - E_0|$ is small compared with $\hbar/t$. If the interval for which $|E - E_0| \simeq \hbar/t$ holds is much narrower than ΔE—the width of $|g(E)|^2\rho(E)$—we get essentially no contribution to $C(t)$ because of strong cancellations. The characteristic time at which the modulus of the correlation amplitude starts becoming appreciably different from 1 is given by

$$t \simeq \frac{\hbar}{\Delta E}. \tag{2.1.70}$$

Even though this equation is obtained for a superposition state with a quasi-continuous energy spectrum, it also makes sense for a two-level system; in the spin-precession problem considered earlier, the state ket, which is initially $|S_x+\rangle$, starts losing its identity after $\sim 1/\omega = \hbar/(E_+ - E_-)$, as is evident from (2.1.60).

To summarize, as a result of time evolution the state ket of a physical system ceases to retain its original form after a time interval of order $\hbar/\Delta E$. In the literature this point is often said to illustrate the *time-energy uncertainty relation*

$$\Delta t \Delta E \simeq \hbar. \tag{2.1.71}$$

However, it is to be clearly understood that this time-energy uncertainty relation is of a very different nature from the uncertainty relation between two incompatible observables discussed in Section 1.4. In Chapter 5 we will come back to (2.1.71) in connection with time-dependent perturbation theory.

2.2. THE SCHRÖDINGER VERSUS THE HEISENBERG PICTURE

Unitary Operators

In the previous section we introduced the concept of time development by considering the time-evolution operator that affects state kets; that approach to quantum dynamics is known as the **Schrödinger picture**. There is another formulation of quantum dynamics where observables, rather than state kets, vary with time; this second approach is known as the **Heisenberg picture**. Before discussing the differences between the two approaches in detail, we digress to make some general comments on unitary operators.

Unitary operators are used for many different purposes in quantum mechanics. In this book we introduced (Section 1.5) an operator satisfying the unitarity property. In that section we were concerned with the question of how the base kets in one representation are related to those in some other representations. The state kets themselves are assumed not to change as we switch to a different set of base kets even though the numerical values of the expansion coefficients for $|\alpha\rangle$ are, of course, different in different representations. Subsequently we introduced two unitary operators that actually change the state kets, the translation operator of Section 1.6 and the time-evolution operator of Section 2.1. We have

$$|\alpha\rangle \rightarrow U|\alpha\rangle, \tag{2.2.1}$$

where U may stand for $\mathscr{T}(d\mathbf{x})$ or $\mathscr{U}(t, t_0)$. Here $U|\alpha\rangle$ is the state ket corresponding to a physical system that actually has undergone translation or time evolution.

It is important to keep in mind that under a unitary transformation that changes the state kets, the inner product of a state bra and a state ket remains unchanged:

$$\langle\beta|\alpha\rangle \rightarrow \langle\beta|U^\dagger U|\alpha\rangle = \langle\beta|\alpha\rangle. \tag{2.2.2}$$

Using the fact that these transformations affect the state kets but not operators, we can infer how $\langle\beta|X|\alpha\rangle$ must change:

$$\langle\beta|X|\alpha\rangle \rightarrow (\langle\beta|U^\dagger)\cdot X\cdot(U|\alpha\rangle) = \langle\beta|U^\dagger XU|\alpha\rangle. \tag{2.2.3}$$

We now make a very simple mathematical observation that follows from the associative axiom of multiplication:

$$(\langle\beta|U^\dagger)\cdot X\cdot(U|\alpha\rangle) = \langle\beta|\cdot(U^\dagger XU)\cdot|\alpha\rangle. \tag{2.2.4}$$

Is there any physics in this observation? This mathematical identity suggests two approaches to unitary transformations:

Approach 1:

$$|\alpha\rangle \rightarrow U|\alpha\rangle, \quad \text{with operators unchanged,} \tag{2.2.5a}$$

Approach 2:

$$X \rightarrow U^\dagger XU, \quad \text{with state kets unchanged.} \tag{2.2.5b}$$

In classical physics we do not introduce state kets, yet we talk about translation, time evolution, and the like. This is possible because these operations actually change quantities such as $\mathbf{x}$ and $\mathbf{L}$, which are observables of classical mechanics. We therefore conjecture that a closer connection with classical physics may be established if we follow approach 2.

A simple example may be helpful here. We go back to the infinitesimal translation operator $\mathscr{T}(d\mathbf{x}')$. The formalism presented in Section 1.6 is based on approach 1; $\mathscr{T}(d\mathbf{x}')$ affects the state kets, not the position operator:

$$|\alpha\rangle \to \left(1 - \frac{i\mathbf{p}\cdot d\mathbf{x}'}{\hbar}\right)|\alpha\rangle,$$
$$\mathbf{x} \to \mathbf{x}. \tag{2.2.6}$$

In contrast, if we follow approach 2, we obtain

$$\begin{aligned}|\alpha\rangle &\to |\alpha\rangle,\\ \mathbf{x} &\to \left(1 + \frac{i\mathbf{p}\cdot d\mathbf{x}'}{\hbar}\right)\mathbf{x}\left(1 - \frac{i\mathbf{p}\cdot d\mathbf{x}'}{\hbar}\right)\\ &= \mathbf{x} + \left(\frac{i}{\hbar}\right)[\mathbf{p}\cdot d\mathbf{x}', \mathbf{x}]\\ &= \mathbf{x} + d\mathbf{x}'.\end{aligned} \tag{2.2.7}$$

We leave it as an exercise for the reader to show that both approaches lead to the same result for the expectation value of $\mathbf{x}$:

$$\langle \mathbf{x}\rangle \to \langle \mathbf{x}\rangle + \langle d\mathbf{x}'\rangle. \tag{2.2.8}$$

State Kets and Observables in the Schrödinger and the Heisenberg Pictures

We now return to the time-evolution operator $\mathscr{U}(t, t_0)$. In the previous section we examined how state kets evolve with time. This means that we were following approach 1, known as the **Schrödinger picture** when applied to time evolution. Alternatively we may follow approach 2, known as the **Heisenberg picture** when applied to time evolution.

In the Schrödinger picture the operators corresponding to observables like x, p_y, and S_z are fixed in time, while state kets vary with time, as indicated in the previous section. In contrast, in the Heisenberg picture the operators corresponding to observables vary with time; the state kets are fixed, frozen so to speak, at what they were at t_0. It is convenient to set t_0 in $\mathscr{U}(t, t_0)$ to zero for simplicity and work with $\mathscr{U}(t)$, which is defined by

$$\mathscr{U}(t, t_0 = 0) \equiv \mathscr{U}(t) = \exp\left(\frac{-iHt}{\hbar}\right). \tag{2.2.9}$$

Motivated by (2.2.5b) of approach 2, we define the Heisenberg picture observable by

$$A^{(H)}(t) \equiv \mathscr{U}^\dagger(t) A^{(S)} \mathscr{U}(t), \tag{2.2.10}$$

where the superscripts H and S stand for Heisenberg and Schrödinger, respectively. At $t = 0$, the Heisenberg picture observable and the corresponding Schrödinger picture observable coincide:

$$A^{(H)}(0) = A^{(S)}. \tag{2.2.11}$$

The state kets also coincide between the two pictures at $t = 0$; at later t the Heisenberg picture state ket is frozen to what it was at $t = 0$:

$$|\alpha, t_0 = 0; t\rangle_H = |\alpha, t_0 = 0\rangle, \tag{2.2.12}$$

independent of t. This is in dramatic contrast with the Schrödinger-picture state ket,

$$|\alpha, t_0 = 0; t\rangle_S = \mathscr{U}(t)|\alpha, t_0 = 0\rangle. \tag{2.2.13}$$

The expectation value $\langle A \rangle$ is obviously the same in both pictures:

$$\begin{aligned} {}_S\langle \alpha, t_0 = 0; t|A^{(S)}|\alpha, t_0 = 0; t\rangle_S &= \langle \alpha, t_0 = 0|\mathscr{U}^\dagger A^{(S)}\mathscr{U}|\alpha, t_0 = 0\rangle \\ &= {}_H\langle \alpha, t_0 = 0; t|A^{(H)}(t)|\alpha, t_0 = 0; t\rangle_H. \end{aligned} \tag{2.2.14}$$

The Heisenberg Equation of Motion

We now derive the fundamental equation of motion in the Heisenberg picture. Assuming that $A^{(S)}$ does not depend explicitly on time, which is the case in most physical situations of interest, we obtain [by differentiating (2.2.10)]

$$\begin{aligned} \frac{dA^{(H)}}{dt} &= \frac{\partial \mathscr{U}^\dagger}{\partial t} A^{(S)}\mathscr{U} + \mathscr{U}^\dagger A^{(S)} \frac{\partial \mathscr{U}}{\partial t} \\ &= -\frac{1}{i\hbar}\mathscr{U}^\dagger H \mathscr{U}\mathscr{U}^\dagger A^{(S)}\mathscr{U} + \frac{1}{i\hbar}\mathscr{U}^\dagger A^{(S)}\mathscr{U}\mathscr{U}^\dagger H\mathscr{U} \\ &= \frac{1}{i\hbar}[A^{(H)}, \mathscr{U}^\dagger H \mathscr{U}], \end{aligned} \tag{2.2.15}$$

where we have used [see (2.1.25)]

$$\frac{\partial \mathscr{U}}{\partial t} = \frac{1}{i\hbar} H\mathscr{U}, \tag{2.2.16a}$$

$$\frac{\partial \mathscr{U}^\dagger}{\partial t} = -\frac{1}{i\hbar}\mathscr{U}^\dagger H. \tag{2.2.16b}$$

Because H was originally introduced in the Schrödinger picture, we may be tempted to define

$$H^{(H)} = \mathscr{U}^\dagger H \mathscr{U} \tag{2.2.17}$$

in accordance with (2.2.10). But in elementary applications where $\mathscr{U}$ is given by (2.2.9), $\mathscr{U}$ and H obviously commute; as a result,

$$\mathscr{U}^\dagger H \mathscr{U} = H, \tag{2.2.18}$$

so it is all right to write (2.2.15) as

$$\frac{dA^{(H)}}{dt} = \frac{1}{i\hbar}[A^{(H)}, H]. \tag{2.2.19}$$

This equation is known as the **Heisenberg equation of motion**. Notice that we have derived it using the properties of the time-evolution operator and the defining equation for $A^{(H)}$.

It is instructive to compare (2.2.19) with the classical equation of motion in Poisson bracket form. In classical physics, for a function A of q's and p's that does not involve time explicitly, we have (Goldstein 1980, 405–6)

$$\frac{dA}{dt} = [A, H]_{\text{classical}}. \tag{2.2.20}$$

Again, we see that Dirac's quantization rule (1.6.47) leads to the correct equation in quantum mechanics. Indeed, historically (2.2.19) was first written by P. A. M. Dirac, who—with his characteristic modesty—called it the Heisenberg equation of motion. It is worth noting, however, that (2.2.19) makes sense whether or not $A^{(H)}$ has a classical analogue. For example, the spin operator in the Heisenberg picture satisfies

$$\frac{dS_i^{(H)}}{dt} = \frac{1}{i\hbar}\left[S_i^{(H)}, H\right], \tag{2.2.21}$$

which can be used to discuss spin precession, but this equation has no classical counterpart because S_z *cannot* be written as a function of q's and p's. Rather than insisting on Dirac's rule, (1.6.47), we may argue that for quantities possessing classical counterparts, the correct classical equation can be obtained from the corresponding quantum-mechanical equation via the ansatz,

$$\frac{[\ ,\ ]}{i\hbar} \to [\ ,\ ]_{\text{classical}}. \tag{2.2.22}$$

Classical mechanics can be derived from quantum mechanics, but the opposite is not true.*

Free Particles; Ehrenfest's Theorem

Whether we work in the Schrödinger picture or in the Heisenberg picture, to be able to use the equations of motion we must first learn how to construct the appropriate Hamiltonian operator. For a physical system with classical analogues, we assume the Hamiltonian to be of the same form as in classical physics; we merely replace the classical x_i's and p_i's by the corresponding operators in quantum mechanics. With this assumption we can reproduce the correct classical equations in the classical limit. Whenever

*In this book we follow the order: the Schrödinger picture → the Heisenberg picture → classical. For an enlightening treatment of the same subject in opposite order, classical → the Heisenberg picture → the Schrödinger picture, see Finkelstein (1973), 68–70 and 109.

an ambiguity arises because of noncommuting observables, we attempt to resolve it by requiring H to be Hermitian; for instance, we write the quantum-mechanical analogue of the classical product xp as $\frac{1}{2}(xp + px)$. When the physical system in question has no classical analogues, we can only guess the structure of the Hamiltonian operator. We try various forms until we get the Hamiltonian that leads to results agreeing with empirical observation.

In practical applications it is often necessary to evaluate the commutator of x_i (or p_i) with functions of x_j and p_j. To this end the following formulas are found to be useful:

$$[x_i, F(\mathbf{p})] = i\hbar \frac{\partial F}{\partial p_i} \tag{2.2.23a}$$

and

$$[p_i, G(\mathbf{x})] = -i\hbar \frac{\partial G}{\partial x_i}, \tag{2.2.23b}$$

where F and G are functions that can be expanded in powers of p_j's and x_j's, respectively. We can easily prove both formulas by repeatedly applying (1.6.50e).

We are now in a position to apply the Heisenberg equation of motion to a free particle of mass m. The Hamiltonian is taken to be of the same form as in classical mechanics:

$$H = \frac{\mathbf{p}^2}{2m} = \frac{\left(p_x^2 + p_y^2 + p_z^2\right)}{2m}. \tag{2.2.24}$$

We look at the observables p_i and x_i, which are understood to be the momentum and the position operator in the Heisenberg picture even though we omit the superscript (H). Because p_i commutes with any function of p_j's, we have

$$\frac{dp_i}{dt} = \frac{1}{i\hbar}[p_i, H] = 0. \tag{2.2.25}$$

Thus for a free particle, the momentum operator is a constant of the motion, which means that $p_i(t)$ is the same as $p_i(0)$ at all times. Quite generally, it is evident from the Heisenberg equation of motion (2.2.19) that whenever $A^{(H)}$ commutes with the Hamiltonian, $A^{(H)}$ is a constant of the motion. Next,

$$\begin{aligned} \frac{dx_i}{dt} &= \frac{1}{i\hbar}[x_i, H] = \frac{1}{i\hbar}\frac{1}{2m}i\hbar\frac{\partial}{\partial p_i}\left(\sum_{j=1}^{3} p_j^2\right) \\ &= \frac{p_i}{m} = \frac{p_i(0)}{m}, \end{aligned} \tag{2.2.26}$$

where we have taken advantage of (2.2.23a), so we have the solution

$$x_i(t) = x_i(0) + \left(\frac{p_i(0)}{m}\right)t, \tag{2.2.27}$$

which is reminiscent of the classical trajectory equation for a uniform rectilinear motion. It is important to note that even though we have

$$[x_i(0), x_j(0)] = 0 \tag{2.2.28}$$

at equal times, the commutator of the x_i's at *different* times does *not* vanish; specifically,

$$[x_i(t), x_i(0)] = \left[\frac{p_i(0)t}{m}, x_i(0)\right] = \frac{-i\hbar t}{m}. \tag{2.2.29}$$

Applying the uncertainty relation (1.4.53) to this commutator, we obtain

$$\langle(\Delta x_i)^2\rangle_t \langle(\Delta x_i)^2\rangle_{t=0} \geq \frac{\hbar^2 t^2}{4m^2}. \tag{2.2.30}$$

Among other things, this relation implies that even if the particle is well localized at $t = 0$, its position becomes more and more uncertain with time, a conclusion which can also be obtained by studying the time-evolution behavior of free-particle wave packets in wave mechanics.

We now add a potential $V(\mathbf{x})$ to our earlier free-particle Hamiltonian:

$$H = \frac{\mathbf{p}^2}{2m} + V(\mathbf{x}). \tag{2.2.31}$$

Here $V(\mathbf{x})$ is to be understood as a function of the x-, y-, and z-*operators*. Using (2.2.23b) this time, we obtain

$$\frac{dp_i}{dt} = \frac{1}{i\hbar}[p_i, V(\mathbf{x})] = -\frac{\partial}{\partial x_i}V(\mathbf{x}). \tag{2.2.32}$$

On the other hand, we see that

$$\frac{dx_i}{dt} = \frac{p_i}{m} \tag{2.2.33}$$

still holds because x_i commutes with the newly added term $V(\mathbf{x})$. We can use the Heisenberg equation of motion once again to deduce

$$\frac{d^2x_i}{dt^2} = \frac{1}{i\hbar}\left[\frac{dx_i}{dt}, H\right] = \frac{1}{i\hbar}\left[\frac{p_i}{m}, H\right]$$
$$= \frac{1}{m}\frac{dp_i}{dt}. \tag{2.2.34}$$

Combining this with (2.2.32), we finally obtain in vectorial form

$$m\frac{d^2\mathbf{x}}{dt^2} = -\nabla V(\mathbf{x}). \tag{2.2.35}$$

This is the quantum-mechanical analogue of Newton's second law. By taking the expectation values of both sides with respect to a Heisenberg state ket that does *not* move with time, we obtain

$$m\frac{d^2}{dt^2}\langle \mathbf{x}\rangle = \frac{d\langle \mathbf{p}\rangle}{dt} = -\langle \nabla V(\mathbf{x})\rangle. \tag{2.2.36}$$

This is known as the **Ehrenfest theorem** after P. Ehrenfest, who derived it in 1927 using the formalism of wave mechanics. When written in this expectation form, its validity is independent of whether we are using the Heisenberg or the Schrödinger picture; after all, the expectation values are the same in the two pictures. In contrast, the operator form (2.2.35) is meaningful only if we understand $\mathbf{x}$ and $\mathbf{p}$ to be Heisenberg-picture operators.

We note that in (2.2.36) the $\hbar$'s have completely disappeared. It is therefore not surprising that the center of a wave packet moves like a *classical* particle subjected to $V(\mathbf{x})$.

Base Kets and Transition Amplitudes

So far we have avoided asking how the base kets evolve in time. A common misconception is that as time goes on, all kets move in the Schrödinger picture and are stationary in the Heisenberg picture. This is *not* the case, as we will make clear shortly. The important point is to distinguish the behavior of state kets from that of base kets.

We started our discussion of ket spaces in Section 1.2 by remarking that the eigenkets of observables are to be used as base kets. What happens to the defining eigenvalue equation

$$A|a'\rangle = a'|a'\rangle \tag{2.2.37}$$

with time? In the Schrödinger picture, A does not change, so the base kets, obtained as the solutions to this eigenvalue equation at $t=0$, for instance, must remain unchanged. Unlike state kets, the base kets do *not* change in the Schrödinger picture.

The whole situation is very different in the Heisenberg picture, where the eigenvalue equation we must study is for the time-dependent operator

$$A^{(H)}(t) = \mathcal{U}^\dagger A(0)\mathcal{U}. \tag{2.2.38}$$

From (2.2.37) evaluated at $t=0$, when the two pictures coincide, we deduce

$$\mathcal{U}^\dagger A(0)\mathcal{U}\mathcal{U}^\dagger|a'\rangle = a'\mathcal{U}^\dagger|a'\rangle, \tag{2.2.39}$$

which implies an eigenvalue equation for $A^{(H)}$:

$$A^{(H)}\left(\mathcal{U}^\dagger|a'\rangle\right) = a'\left(\mathcal{U}^\dagger|a'\rangle\right). \tag{2.2.40}$$

If we continue to maintain the view that the eigenkets of observables form the base kets, then $\{\mathcal{U}^\dagger|a'\rangle\}$ must be used as the base kets in the Heisenberg picture. As time goes on, the Heisenberg-picture base kets, denoted by $|a', t\rangle_H$, move as follows:

$$|a', t\rangle_H = \mathscr{U}^\dagger |a'\rangle. \tag{2.2.41}$$

Because of the appearance of $\mathscr{U}^\dagger$ rather than $\mathscr{U}$ in (2.2.41), the Heisenberg-picture base kets are seen to rotate oppositely when compared with the Schrödinger-picture state kets; specifically, $|a', t\rangle_H$ satisfies the "wrong-sign Schrödinger equation"

$$i\hbar \frac{\partial}{\partial t} |a', t\rangle_H = -H |a', t\rangle_H. \tag{2.2.42}$$

As for the eigenvalues themselves, we see from (2.2.40) that they are unchanged with time. This is consistent with the theorem on unitary equivalent observables discussed in Section 1.5. Notice also the following expansion for $A^{(H)}(t)$ in terms of the base kets and bras of the Heisenberg picture:

$$\begin{aligned} A^{(H)}(t) &= \sum_{a'} |a', t\rangle_H a' {}_H\langle a', t| \\ &= \sum_{a'} \mathscr{U}^\dagger |a'\rangle a' \langle a'| \mathscr{U} \\ &= \mathscr{U}^\dagger A^{(S)} \mathscr{U}, \end{aligned} \tag{2.2.43}$$

which shows that everything is quite consistent provided that the Heisenberg base kets change as in (2.2.41).

We see that the expansion coefficients of a state ket in terms of base kets are the same in both pictures:

$$c_{a'}(t) = \underbrace{\langle a'|}_{\text{base bra}} \cdot \underbrace{(\mathscr{U}|\alpha, t_0 = 0\rangle)}_{\text{state ket}} \qquad \text{(the Schrödinger picture)} \tag{2.2.44a}$$

$$c_{a'}(t) = \underbrace{(\langle a'|\mathscr{U})}_{\text{base bra}} \cdot \underbrace{|\alpha, t_0 = 0\rangle}_{\text{state ket}} \qquad \text{(the Heisenberg picture).} \tag{2.2.44b}$$

Pictorially, we may say that the cosine of the angle between the state ket and the base ket is the same whether we rotate the state ket counterclockwise or the base ket clockwise. These considerations apply equally well to base kets that exhibit a continuous spectrum; in particular, the wave function $\langle \mathbf{x}'|\alpha\rangle$ can be regarded either as (1) the inner product of the stationary position eigenbra with the moving state ket (the Schrödinger picture) or as (2) the inner product of the moving position eigenbra with the stationary state ket (the Heisenberg picture). We will discuss the time dependence of the wave function in Section 2.4, where we will derive the celebrated wave equation of Schrödinger.

To illustrate further the equivalence between the two pictures, we study transition amplitudes, which will play a fundamental role in Section

TABLE 2.1. The Schrödinger Picture Versus the Heisenberg Picture

	Schrödinger picture	Heisenberg picture
State ket	Moving: (2.1.5), (2.1.27)	Stationary
Observable	Stationary	Moving: (2.2.10), (2.2.19)
Base ket	Stationary	Moving oppositely: (2.2.41), (2.2.42)

2.5. Suppose there is a physical system prepared at $t = 0$ to be in an eigenstate of observable A with eigenvalue a'. At some later time t we may ask, What is the probability amplitude, known as the **transition amplitude**, for the system to be found in an eigenstate of observable B with eigenvalue b'? Here A and B can be the same or different. In the Schrödinger picture the state ket at t is given by $\mathscr{U}|a'\rangle$, while the base kets $|a'\rangle$ and $|b'\rangle$ do not vary with time; so we have

$$\underbrace{\langle b'|}_{\text{base bra}} \cdot \underbrace{(\mathscr{U}|a'\rangle)}_{\text{state ket}} \tag{2.2.45}$$

for this transition amplitude. In contrast, in the Heisenberg picture the state ket is stationary, that is, it remains as $|a'\rangle$ at all times, but the base kets evolve oppositely. So the transition amplitude is

$$\underbrace{(\langle b'|\mathscr{U})}_{\text{base bra}} \cdot \underbrace{|a'\rangle}_{\text{state ket}} . \tag{2.2.46}$$

Obviously (2.2.45) and (2.2.46) are the same. They can both be written as

$$\langle b'|\mathscr{U}(t,0)|a'\rangle. \tag{2.2.47}$$

In some loose sense this is the transition amplitude for "going" from state $|a'\rangle$ to state $|b'\rangle$.

To conclude this section let us summarize the differences between the Schrödinger picture and the Heisenberg picture; see Table 2.1.

2.3. SIMPLE HARMONIC OSCILLATOR

The simple harmonic oscillator is one of the most important problems in quantum mechanics. From a pedagogical point of view it can be used to illustrate the basic concepts and methods in quantum mechanics. From a practical point of view it has applications in a variety of branches of modern physics—molecular spectroscopy, solid state physics, nuclear structure, quantum field theory, quantum optics, quantum statistical mechanics, and so forth. From a historical point of view it was M. Planck's proposal to associate discrete units of energy with radiation oscillators that led to the birth of quantum concepts. A thorough understanding of the properties of quantum-mechanical oscillators is indispensable for any serious student of modern physics.

Energy Eigenkets and Energy Eigenvalues

We begin our discussion with Dirac's elegant operator method, which is based on the earlier work of M. Born and N. Wiener, to obtain the energy eigenkets and energy eigenvalues of the simple harmonic oscillator. The basic Hamiltonian is

$$H = \frac{p^2}{2m} + \frac{m\omega^2 x^2}{2}, \tag{2.3.1}$$

where ω is the angular frequency of the classical oscillator related to the spring constant k in Hooke's law via $\omega = \sqrt{k/m}$. The operators x and p are, of course, Hermitian. It is convenient to define two non-Hermitian operators,

$$a = \sqrt{\frac{m\omega}{2\hbar}}\left(x + \frac{ip}{m\omega}\right), \qquad a^\dagger = \sqrt{\frac{m\omega}{2\hbar}}\left(x - \frac{ip}{m\omega}\right), \tag{2.3.2}$$

known as the **annihilation operator** and the **creation operator**, respectively, for reasons that will become evident shortly. Using the canonical commutation relations, we readily obtain

$$[a, a^\dagger] = \left(\frac{1}{2\hbar}\right)(-i[x, p] + i[p, x]) = 1. \tag{2.3.3}$$

We also define the number operator

$$N = a^\dagger a, \tag{2.3.4}$$

which is obviously Hermitian. It is straightforward to show that

$$\begin{aligned} a^\dagger a &= \left(\frac{m\omega}{2\hbar}\right)\left(x^2 + \frac{p^2}{m^2\omega^2}\right) + \left(\frac{i}{2\hbar}\right)[x, p] \\ &= \frac{H}{\hbar\omega} - \frac{1}{2}, \end{aligned} \tag{2.3.5}$$

so we have an important relation between the number operator and the Hamiltonian operator:

$$H = \hbar\omega\left(N + \tfrac{1}{2}\right). \tag{2.3.6}$$

Because H is just a linear function of N, N can be diagonalized simultaneously with H. We denote an energy eigenket of N by its eigenvalue n, so

$$N|n\rangle = n|n\rangle. \tag{2.3.7}$$

We will later show that n must be a nonnegative integer. Because of (2.3.6) we also have

$$H|n\rangle = \left(n + \tfrac{1}{2}\right)\hbar\omega|n\rangle, \tag{2.3.8}$$

which means that the energy eigenvalues are given by

$$E_n = \left(n + \tfrac{1}{2}\right)\hbar\omega. \tag{2.3.9}$$

To appreciate the physical significance of a, $a^\dagger$, and N, let us first note that

$$[N, a] = [a^\dagger a, a] = a^\dagger[a, a] + [a^\dagger, a]a = -a, \tag{2.3.10}$$

where we have used (2.3.3). Likewise, we can derive

$$[N, a^\dagger] = a^\dagger. \tag{2.3.11}$$

As a result, we have

$$\begin{aligned} Na^\dagger|n\rangle &= \left([N, a^\dagger] + a^\dagger N\right)|n\rangle \\ &= (n+1)a^\dagger|n\rangle \end{aligned} \tag{2.3.12a}$$

and

$$\begin{aligned} Na|n\rangle &= \left([N, a] + aN\right)|n\rangle \\ &= (n-1)a|n\rangle. \end{aligned} \tag{2.3.12b}$$

These relations imply that $a^\dagger|n\rangle(a|n\rangle)$ is also an eigenket of N with eigenvalue increased (decreased) by one. Because the increase (decrease) of n by one amounts to the creation (annihilation) of one quantum unit of energy $\hbar\omega$, the term *creation operator* (*annihilation operator*) for $a^\dagger(a)$ is deemed appropriate.

Equation (2.3.12b) implies that $a|n\rangle$ and $|n-1\rangle$ are the same up to a multiplicative constant. We write

$$a|n\rangle = c|n-1\rangle, \tag{2.3.13}$$

where c is a numerical constant to be determined from the requirement that both $|n\rangle$ and $|n-1\rangle$ be normalized. First, note that

$$\langle n|a^\dagger a|n\rangle = |c|^2. \tag{2.3.14}$$

We can evaluate the left-hand side of (2.3.14) by noting that $a^\dagger a$ is just the number operator, so

$$n = |c|^2. \tag{2.3.15}$$

Taking c to be real and positive by convention, we finally obtain

$$a|n\rangle = \sqrt{n}\,|n-1\rangle. \tag{2.3.16}$$

Similarly, it is easy to show that

$$a^\dagger|n\rangle = \sqrt{n+1}\,|n+1\rangle. \tag{2.3.17}$$

Suppose that we keep on applying the annihilation operator a to both sides of (2.3.16):

$$\begin{aligned} a^2|n\rangle &= \sqrt{n(n-1)}\,|n-2\rangle, \\ a^3|n\rangle &= \sqrt{n(n-1)(n-2)}\,|n-3\rangle, \\ &\vdots \end{aligned} \tag{2.3.18}$$

We can obtain numerical operator eigenkets with smaller and smaller n until the sequence terminates, which is bound to happen whenever we start with a positive integer n. One may argue that if we start with a noninteger n, the sequence will not terminate, leading to eigenkets with a negative value of n. But we also have the positivity requirement for the norm of $a|n\rangle$:

$$n = \langle n|N|n\rangle = (\langle n|a^\dagger)\cdot(a|n\rangle) \geq 0, \tag{2.3.19}$$

which implies that n can never be negative! So we conclude that the sequence must terminate with $n = 0$ and that the allowed values of n are nonnegative integers.

Because the smallest possible value of n is zero, the ground state of the harmonic oscillator has

$$E_0 = \tfrac{1}{2}\hbar\omega. \tag{2.3.20}$$

We can now successively apply the creation operator $a^\dagger$ to the ground state $|0\rangle$. Using (2.3.17), we obtain

$$\begin{aligned}
|1\rangle &= a^\dagger|0\rangle,\\
|2\rangle &= \left(\frac{a^\dagger}{\sqrt{2}}\right)|1\rangle = \left[\frac{(a^\dagger)^2}{\sqrt{2}}\right]|0\rangle,\\
|3\rangle &= \left(\frac{a^\dagger}{\sqrt{3}}\right)|2\rangle = \left[\frac{(a^\dagger)^3}{\sqrt{3!}}\right]|0\rangle,\\
&\;\vdots\\
|n\rangle &= \left[\frac{(a^\dagger)^n}{\sqrt{n!}}\right]|0\rangle.
\end{aligned} \tag{2.3.21}$$

In this way we have succeeded in constructing simultaneous eigenkets of N and H with energy eigenvalues

$$E_n = \left(n + \tfrac{1}{2}\right)\hbar\omega, \quad (n = 0, 1, 2, 3, \ldots). \tag{2.3.22}$$

From (2.3.16), (2.3.17), and the orthonormality requirement for $\{|n\rangle\}$, we obtain the matrix elements

$$\langle n'|a|n\rangle = \sqrt{n}\,\delta_{n',n-1}, \qquad \langle n'|a^\dagger|n\rangle = \sqrt{n+1}\,\delta_{n',n+1}. \tag{2.3.23}$$

Using these together with

$$x = \sqrt{\frac{\hbar}{2m\omega}}\,(a + a^\dagger), \qquad p = i\sqrt{\frac{m\hbar\omega}{2}}\,(-a + a^\dagger), \tag{2.3.24}$$

we derive the matrix elements of the x and p operators:

$$\langle n'|x|n\rangle = \sqrt{\frac{\hbar}{2m\omega}}\left(\sqrt{n}\,\delta_{n',n-1} + \sqrt{n+1}\,\delta_{n',n+1}\right), \tag{2.3.25a}$$

$$\langle n'|p|n\rangle = i\sqrt{\frac{m\hbar\omega}{2}}\left(-\sqrt{n}\,\delta_{n',n-1} + \sqrt{n+1}\,\delta_{n',n+1}\right). \tag{2.3.25b}$$

Notice that neither x nor p is diagonal in the N-representation we are using. This is not surprising because x and p, like a and $a^\dagger$, do not commute with N.

The operator method can also be used to obtain the energy eigenfunctions in position space. Let us start with the ground state defined by

$$a|0\rangle = 0, \tag{2.3.26}$$

which, in the x-representation, reads

$$\langle x'|a|0\rangle = \sqrt{\frac{m\omega}{2\hbar}}\langle x'|\left(x + \frac{ip}{m\omega}\right)|0\rangle = 0. \tag{2.3.27}$$

Recalling (1.7.17), we can regard this as a differential equation for the ground-state wave function $\langle x'|0\rangle$:

$$\left(x' + x_0^2 \frac{d}{dx'}\right)\langle x'|0\rangle = 0, \tag{2.3.28}$$

where we have introduced

$$x_0 \equiv \sqrt{\frac{\hbar}{m\omega}}, \tag{2.3.29}$$

which sets the length scale of the oscillator. We see that the normalized solution to (2.3.28) is

$$\langle x'|0\rangle = \left(\frac{1}{\pi^{1/4}\sqrt{x_0}}\right)\exp\left[-\frac{1}{2}\left(\frac{x'}{x_0}\right)^2\right]. \tag{2.3.30}$$

We can also obtain the energy eigenfunctions for excited states by evaluating

$$\langle x'|1\rangle = \langle x'|a^\dagger|0\rangle = \left(\frac{1}{\sqrt{2}x_0}\right)\left(x' - x_0^2\frac{d}{dx'}\right)\langle x'|0\rangle,$$

$$\langle x'|2\rangle = \left(\frac{1}{\sqrt{2}}\right)\langle x'|(a^\dagger)^2|0\rangle = \left(\frac{1}{\sqrt{2!}}\right)\left(\frac{1}{\sqrt{2}x_0}\right)^2\left(x' - x_0^2\frac{d}{dx'}\right)^2\langle x'|0\rangle, \ldots \tag{2.3.31}$$

In general, we obtain

$$\langle x'|n\rangle = \left(\frac{1}{\pi^{1/4}\sqrt{2^n n!}}\right)\left(\frac{1}{x_0^{n+1/2}}\right)\left(x' - x_0^2\frac{d}{dx'}\right)^n \exp\left[-\frac{1}{2}\left(\frac{x'}{x_0}\right)^2\right]. \tag{2.3.32}$$

It is instructive to look at the expectation values of x^2 and p^2 for the ground state. First, note that

$$x^2 = \left(\frac{\hbar}{2m\omega}\right)(a^2 + a^{\dagger 2} + a^\dagger a + aa^\dagger). \tag{2.3.33}$$

When we take the expectation value of x^2, only the last term in (2.3.33) yields a nonvanishing contribution:

$$\langle x^2 \rangle = \frac{\hbar}{2m\omega} = \frac{x_0^2}{2}. \tag{2.3.34}$$

Likewise,

$$\langle p^2 \rangle = \frac{\hbar m\omega}{2}. \tag{2.3.35}$$

It follows that the expectation values of the kinetic and the potential energies are, respectively,

$$\left\langle \frac{p^2}{2m} \right\rangle = \frac{\hbar\omega}{4} = \frac{\langle H \rangle}{2} \quad \text{and} \quad \left\langle \frac{m\omega^2 x^2}{2} \right\rangle = \frac{\hbar\omega}{4} = \frac{\langle H \rangle}{2}, \tag{2.3.36}$$

as expected from the virial theorem. From (2.3.25a) and (2.3.25b), it follows that

$$\langle x \rangle = \langle p \rangle = 0, \tag{2.3.37}$$

which also holds for the excited states. We therefore have

$$\langle (\Delta x)^2 \rangle = \langle x^2 \rangle = \frac{\hbar}{2m\omega} \quad \text{and} \quad \langle (\Delta p)^2 \rangle = \langle p^2 \rangle = \frac{\hbar m\omega}{2}, \tag{2.3.38}$$

and we see that the uncertainty relation is satisfied in the minimum uncertainty product form:

$$\langle (\Delta x)^2 \rangle \langle (\Delta p)^2 \rangle = \frac{\hbar^2}{4}. \tag{2.3.39}$$

This is not surprising because the ground-state wave function has a Gaussian shape. In contrast, the uncertainty products for the excited states are larger:

$$\langle (\Delta x)^2 \rangle \langle (\Delta p)^2 \rangle = \left(n + \tfrac{1}{2}\right)^2 \hbar^2, \tag{2.3.40}$$

as the reader may easily verify.

Time Development of the Oscillator

So far we have not discussed the time evolution of oscillator state kets nor of observables like x and p. Everything we have done is supposed to hold at some instant of time, say at $t = 0$; the operators x, p, a, and $a^\dagger$ are to be regarded either as Schrödinger-picture operators (at all t) or as Heisenberg-picture operators at $t = 0$. In the remaining part of this section, we work exclusively in the Heisenberg picture, which means that x, p, a, and $a^\dagger$ are all time-dependent even though we do not explicitly write $x^{(H)}(t)$, and so forth.

The Heisenberg equations of motion for p and x are, from (2.2.32) and (2.2.33),

$$\frac{dp}{dt} = -m\omega^2 x \tag{2.3.41a}$$

and

$$\frac{dx}{dt} = \frac{p}{m}. \tag{2.3.41b}$$

This pair of coupled differential equations is equivalent to two uncoupled differential equations for a and $a^\dagger$, namely,

$$\frac{da}{dt} = \sqrt{\frac{m\omega}{2\hbar}}\left(\frac{p}{m} - i\omega x\right) = -i\omega a \tag{2.3.42a}$$

and

$$\frac{da^\dagger}{dt} = i\omega a^\dagger, \tag{2.3.42b}$$

whose solutions are

$$a(t) = a(0)\exp(-i\omega t) \quad \text{and} \quad a^\dagger(t) = a^\dagger(0)\exp(i\omega t). \tag{2.3.43}$$

Incidentally, these relations explicitly show that N and H are *time-independent* operators even in the Heisenberg picture, as they must be. In terms of x and p, we can rewrite (2.3.43) as

$$\begin{aligned} x(t) + \frac{ip(t)}{m\omega} &= x(0)\exp(-i\omega t) + i\left[\frac{p(0)}{m\omega}\right]\exp(-i\omega t), \\ x(t) - \frac{ip(t)}{m\omega} &= x(0)\exp(i\omega t) - i\left[\frac{p(0)}{m\omega}\right]\exp(i\omega t). \end{aligned} \tag{2.3.44}$$

Equating the Hermitian and anti-Hermitian parts of both sides separately, we deduce

$$x(t) = x(0)\cos\omega t + \left[\frac{p(0)}{m\omega}\right]\sin\omega t \tag{2.3.45a}$$

and

$$p(t) = -m\omega x(0)\sin\omega t + p(0)\cos\omega t. \tag{2.3.45b}$$

These look the same as the classical equations of motion. We see that the x and p operators "oscillate" just like their classical analogues.

For pedagogical reasons we now present an alternative derivation of (2.3.45a). Instead of solving the Heisenberg equation of motion, we attempt to evaluate

$$x(t) = \exp\left(\frac{iHt}{\hbar}\right)x(0)\exp\left(\frac{-iHt}{\hbar}\right). \tag{2.3.46}$$

To this end we record a very useful formula:

$$\exp(iG\lambda)A\exp(-iG\lambda) = A + i\lambda[G,A] + \left(\frac{i^2\lambda^2}{2!}\right)[G,[G,A]] + \cdots + \left(\frac{i^n\lambda^n}{n!}\right)[G,[G,[G,\ldots[G,A]]]\ldots] + \cdots, \tag{2.3.47}$$

where G is a Hermitian operator and λ is a real parameter. We leave the proof of this formula, known as the **Baker-Hausdorff lemma** as an exercise. Applying this formula to (2.3.46), we obtain

$$\exp\left(\frac{iHt}{\hbar}\right)x(0)\exp\left(\frac{-iHt}{\hbar}\right) = x(0)+\left(\frac{it}{\hbar}\right)[H,x(0)]+\left(\frac{i^2t^2}{2!\hbar^2}\right)[H,[H,x(0)]]+\cdots. \tag{2.3.48}$$

Each term on the right-hand side can be reduced to either x or p by repeatedly using

$$[H,x(0)] = \frac{-i\hbar p(0)}{m} \tag{2.3.49a}$$

and

$$[H,p(0)] = i\hbar m\omega^2 x(0). \tag{2.3.49b}$$

Thus

$$\begin{aligned}\exp\left(\frac{iHt}{\hbar}\right)x(0)\exp\left(\frac{-iHt}{\hbar}\right) &= x(0)+\left[\frac{p(0)}{m}\right]t-\left(\frac{1}{2!}\right)t^2\omega^2x(0) \\ &\quad -\left(\frac{1}{3!}\right)\frac{t^3\omega^2p(0)}{m}+\cdots \\ &= x(0)\cos\omega t+\left[\frac{p(0)}{m\omega}\right]\sin\omega t,\end{aligned} \tag{2.3.50}$$

in agreement with (2.3.45a).

From (2.3.45a) and (2.3.45b), one may be tempted to conclude that $\langle x\rangle$ and $\langle p\rangle$ always oscillate with angular frequency ω. However, this inference is not correct. Take any energy eigenstate characterized by a definite value of n; the expectation value $\langle n|x(t)|n\rangle$ vanishes because the operators $x(0)$ and $p(0)$ change n by ± 1 and $|n\rangle$ and $|n\pm 1\rangle$ are orthogonal. This point is also obvious from our earlier conclusion (see Section 2.1) that the expectation value of an observable taken with respect to a stationary state does not vary with time. To observe oscillations reminiscent of the classical oscillator, we must look at a *superposition* of energy eigenstates such as

$$|\alpha\rangle = c_0|0\rangle + c_1|1\rangle. \tag{2.3.51}$$

The expectation value of $x(t)$ taken with respect to (2.3.51) does oscillate, as the reader may readily verify.

We have seen that an energy eigenstate does not behave like the classical oscillator—in the sense of oscillating expectation values for x and p—no matter how large n may be. We may logically ask, How can we construct a superposition of energy eigenstates that most closely imitates the classical oscillator? In wave-function language, we want a wave packet that bounces back and forth without spreading in shape. It turns out that a *coherent state* defined by the eigenvalue equation for the non-Hermitian annihilation operator a,

$$a|\lambda\rangle = \lambda|\lambda\rangle, \tag{2.3.52}$$

with, in general, a complex eigenvalue λ does the desired job. The coherent state has many other remarkable properties:

1. When expressed as a superposition of energy (or N) eigenstates,

$$|\lambda\rangle = \sum_{n=0}^{\infty} f(n)|n\rangle, \tag{2.3.53}$$

the distribution of $|f(n)|^2$ with respect to n is of the Poisson type about some mean value $\bar{n}$:

$$|f(n)|^2 = \left(\frac{\bar{n}^n}{n!}\right)\exp(-\bar{n}). \tag{2.3.54}$$

2. It can be obtained by translating the oscillator ground state by some finite distance.
3. It satisfies the minimum uncertainty product relation at all times.

A systematic study of coherent states, pioneered by R. Glauber, is very rewarding; the reader is urged to work out an exercise on this subject at the end of this chapter.*

2.4. SCHRÖDINGER'S WAVE EQUATION

Time-Dependent Wave Equation

We now turn to the Schrödinger picture and examine the time evolution of $|\alpha, t_0; t\rangle$ in the x-representation. In other words, our task is to study the behavior of the wave function

$$\psi(\mathbf{x}', t) = \langle \mathbf{x}'|\alpha, t_0; t\rangle \tag{2.4.1}$$

as a function of time, where $|\alpha, t_0; t\rangle$ is a state ket in the Schrödinger

*For applications to laser physics, see Sargent, Scully, and Lamb (1974).

picture at time t, and $\langle \mathbf{x}'|$ is a time-independent position eigenbra with eigenvalue $\mathbf{x}'$. The Hamiltonian operator is taken to be

$$H = \frac{\mathbf{p}^2}{2m} + V(\mathbf{x}). \tag{2.4.2}$$

The potential $V(\mathbf{x})$ is a Hermitian operator; it is also local in the sense that in the $\mathbf{x}$-representation we have

$$\langle \mathbf{x}''|V(\mathbf{x})|\mathbf{x}'\rangle = V(\mathbf{x}')\delta^3(\mathbf{x}' - \mathbf{x}''), \tag{2.4.3}$$

where $V(\mathbf{x}')$ is a real function of $\mathbf{x}'$. Later in this book we will consider more-complicated Hamiltonians—a time-dependent potential $V(\mathbf{x}, t)$; a nonlocal but separable potential where the right-hand side of (2.4.3) is replaced by $v_1(\mathbf{x}'')v_2(\mathbf{x}')$; a momentum-dependent interaction of the form $\mathbf{p}\cdot\mathbf{A} + \mathbf{A}\cdot\mathbf{p}$, where $\mathbf{A}$ is the vector potential in electrodynamics, and so on.

We now derive Schrödinger's time-dependent wave equation. We first write the Schrödinger equation for a state ket (2.1.27) in the $\mathbf{x}$-representation:

$$i\hbar \frac{\partial}{\partial t}\langle \mathbf{x}'|\alpha, t_0; t\rangle = \langle \mathbf{x}'|H|\alpha, t_0; t\rangle, \tag{2.4.4}$$

where we have used the fact that the position eigenbras in the Schrödinger picture do not change with time. Using (1.7.20), we can write the kinetic-energy contribution to the right-hand side of (2.4.4) as

$$\left\langle \mathbf{x}' \left| \frac{\mathbf{p}^2}{2m} \right| \alpha, t_0; t \right\rangle = -\left(\frac{\hbar^2}{2m}\right)\nabla'^2\langle \mathbf{x}'|\alpha, t_0; t\rangle. \tag{2.4.5}$$

As for $V(\mathbf{x})$, we simply use

$$\langle \mathbf{x}'|V(\mathbf{x}) = \langle \mathbf{x}'|V(\mathbf{x}'), \tag{2.4.6}$$

where $V(\mathbf{x}')$ is no longer an operator. Combining everything, we deduce

$$i\hbar \frac{\partial}{\partial t}\langle \mathbf{x}'|\alpha, t_0; t\rangle = -\left(\frac{\hbar^2}{2m}\right)\nabla'^2\langle \mathbf{x}'|\alpha, t_0; t\rangle + V(\mathbf{x}')\langle \mathbf{x}'|\alpha, t_0; t\rangle, \tag{2.4.7}$$

which we recognize to be the celebrated time-dependent wave equation of E. Schrödinger, usually written as

$$i\hbar \frac{\partial}{\partial t}\psi(\mathbf{x}', t) = -\left(\frac{\hbar^2}{2m}\right)\nabla'^2\psi(\mathbf{x}', t) + V(\mathbf{x}')\psi(\mathbf{x}', t). \tag{2.4.8}$$

The quantum mechanics based on wave equation (2.4.8) is known as **wave mechanics**. This equation is, in fact, the starting point of many textbooks on quantum mechanics. In our formalism, however, this is just the Schrödinger equation for a state ket written explicitly in the x-basis when the Hamiltonian operator is taken to be (2.4.2).

The Time-Independent Wave Equation

We now derive the partial differential equation satisfied by energy eigenfunctions. We showed in Section 2.1 that the time dependence of a stationary state is given by $\exp(-iE_{a'}t/\hbar)$. This enables us to write its wave function as

$$\langle \mathbf{x}'|a', t_0; t\rangle = \langle \mathbf{x}'|a'\rangle \exp\left(\frac{-iE_{a'}t}{\hbar}\right), \tag{2.4.9}$$

where it is understood that initially the system is prepared in a simultaneous eigenstate of A and H with eigenvalues a' and $E_{a'}$, respectively. Let us now substitute (2.4.9) into the time-dependent Schrödinger equation (2.4.7). We are then led to

$$-\left(\frac{\hbar^2}{2m}\right)\nabla'^2\langle \mathbf{x}'|a'\rangle + V(\mathbf{x}')\langle \mathbf{x}'|a'\rangle = E_{a'}\langle \mathbf{x}'|a'\rangle. \tag{2.4.10}$$

This partial differential equation is satisfied by the energy eigenfunction $\langle \mathbf{x}'|a'\rangle$ with energy eigenvalue $E_{a'}$. Actually, in wave mechanics where the Hamiltonian operator is given as a function of $\mathbf{x}$ and $\mathbf{p}$, as in (2.4.2), it is not necessary to refer explicitly to observable A that commutes with H because we can always choose A to be that function of the observables $\mathbf{x}$ and $\mathbf{p}$ which coincides with H itself. We may therefore omit reference to a' and simply write (2.4.10) as the partial differential equation to be satisfied by the energy eigenfunction $u_E(\mathbf{x}')$:

$$-\left(\frac{\hbar^2}{2m}\right)\nabla'^2 u_E(\mathbf{x}') + V(\mathbf{x}')u_E(\mathbf{x}') = Eu_E(\mathbf{x}'). \tag{2.4.11}$$

This is the **time-independent wave equation** of E. Schrödinger—announced in the first of four monumental papers, all written in the first half of 1926—that laid the foundations of wave mechanics. In the same paper he immediately applied (2.4.11) to derive the energy spectrum of the hydrogen atom.

To solve (2.4.11) some boundary condition has to be imposed. Suppose we seek a solution to (2.4.11) with

$$E < \lim_{|\mathbf{x}'|\to\infty} V(\mathbf{x}'), \tag{2.4.12}$$

where the inequality relation is to hold for $|\mathbf{x}'| \to \infty$ in any direction. The appropriate boundary condition to be used in this case is

$$u_E(\mathbf{x}') \to 0 \quad \text{as } |\mathbf{x}'| \to \infty. \tag{2.4.13}$$

Physically this means that the particle is bound or confined within a finite region of space. We know from the theory of partial differential equations that (2.4.11) subject to boundary condition (2.4.13) allows nontrivial solutions only for a discrete set of values of E. It is in this sense that the time-independent Schrödinger equation (2.4.11) yields the *quantization of energy levels*.* Once the partial differential equation (2.4.11) is written, the problem of finding the energy levels of microscopic physical systems is as straightforward as that of finding the characteristic frequencies of vibrating strings or membranes. In both cases we solve boundary-value problems in mathematical physics.

A short digression on the history of quantum mechanics is in order here. The fact that exactly soluble eigenvalue problems in the theory of partial differential equations can also be treated using matrix methods was already known to mathematicians in the first quarter of the twentieth century. Furthermore, theoretical physicists like M. Born frequently consulted great mathematicians of the day—D. Hilbert and H. Weyl, in particular. Yet when matrix mechanics was born in the summer of 1925, it did not immediately occur to the theoretical physicists or to the mathematicians to reformulate it using the language of partial differential equations. Six months after Heisenberg's pioneering paper, wave mechanics was proposed by Schrödinger. However, a close inspection of his papers shows that he was not at all influenced by the earlier works of Heisenberg, Born, and Jordan. Instead, the train of reasoning that led Schrödinger to formulate wave mechanics has its roots in W. R. Hamilton's analogy between optics and mechanics, on which we will comment later, and the particle-wave hypothesis of L. de Broglie. Once wave mechanics was formulated, many people, including Schrödinger himself, showed the equivalence between wave mechanics and matrix mechanics.

It is assumed that the reader of this book has some experience in solving the time-dependent and time-independent wave equations. He or she should be familiar with the time evolution of a Gaussian wave packet in a force-free region; should be able to solve one-dimensional transmission-reflection problems involving a rectangular potential barrier, and the like; should have seen derived some simple solutions of the time-independent wave equation—a particle in a box, a particle in a square well, the simple harmonic oscillator, the hydrogen atom, and so on—and should also be familiar with some general properties of the energy eigenfunctions and energy eigenvalues, such as (1) the fact that the energy levels exhibit a discrete or continuous spectrum depending on whether or not (2.4.12) is satisfied and (2) the property that the energy eigenfunction in one dimension is *sinusoidal* or *damped* depending on whether $E - V(\mathbf{x}')$ is positive or negative. In this book we will not cover these topics. A brief summary of elementary solutions to Schrödinger's equations is presented in Appendix A.

---

*Schrödinger's paper that announced (2.4.11) is appropriately entitled *Quantisierung als Eigenwertproblem* (*Quantization as an Eigenvalue Problem*).

Interpretations of the Wave Function

We now turn to discussions of the physical interpretations of the wave function. In Section 1.7 we commented on the probabilistic interpretation of $|\psi|^2$ that follows from the fact that $\langle \mathbf{x}'|\alpha, t_0; t\rangle$ is to be regarded as an expansion coefficient of $|\alpha, t_0; t\rangle$ in terms of the position eigenkets $\{|\mathbf{x}'\rangle\}$. The quantity $\rho(\mathbf{x}', t)$ defined by

$$\rho(\mathbf{x}', t) = |\psi(\mathbf{x}', t)|^2 = |\langle \mathbf{x}'|\alpha, t_0; t\rangle|^2 \tag{2.4.14}$$

is therefore regarded as the **probability density** in wave mechanics. Specifically, when we use a detector that ascertains the presence of the particle within a small volume element d^3x' around $\mathbf{x}'$, the probability of recording a positive result at time t is given by $\rho(\mathbf{x}', t)\, d^3x'$.

In the remainder of this section we use $\mathbf{x}$ for $\mathbf{x}'$ because the position operator will not appear. Using Schrödinger's time-dependent wave equation, it is straightforward to derive the continuity equation

$$\frac{\partial \rho}{\partial t} + \nabla\cdot\mathbf{j} = 0, \tag{2.4.15}$$

where $\rho(\mathbf{x}, t)$ stands for $|\psi|^2$ as before, and $\mathbf{j}(\mathbf{x}, t)$, known as the **probability flux**, is given by

$$\begin{aligned}\mathbf{j}(\mathbf{x}, t) &= -\left(\frac{i\hbar}{2m}\right)\left[\psi^*\nabla\psi - (\nabla\psi^*)\psi\right]\\ &= \left(\frac{\hbar}{m}\right)\mathrm{Im}(\psi^*\nabla\psi).\end{aligned} \tag{2.4.16}$$

The reality of the potential V (or the Hermiticity of the V operator) has played a crucial role in our obtaining this result. Conversely, a complex potential can phenomenologically account for the disappearance of a particle; such a potential is often used for nuclear reactions where incident particles get absorbed by nuclei.

We may intuitively expect that the probability flux $\mathbf{j}$ is related to momentum. This is indeed the case for $\mathbf{j}$ *integrated over all space*. From (2.4.16) we obtain

$$\int d^3x\, \mathbf{j}(\mathbf{x}, t) = \frac{\langle \mathbf{p}\rangle_t}{m}, \tag{2.4.17}$$

where $\langle \mathbf{p}\rangle_t$ is the expectation value of the momentum operator at time t.

Equation (2.4.15) is reminiscent of the continuity equation in fluid dynamics that characterizes a hydrodynamic flow of a fluid in a source-free, sink-free region. Indeed, historically Schrödinger was first led to interpret $|\psi|^2$ as the actual matter density, or $e|\psi|^2$ as the actual electric charge density. If we adopt such a view, we are led to face some bizarre consequences.

A typical argument for a position measurement might go as follows. An atomic electron is to be regarded as a continuous distribution of matter filling up a finite region of space around the nucleus; yet, when a measurement is made to make sure that the electron is at some particular point, this continuous distribution of matter suddenly shrinks to a pointlike particle with no spatial extension. The more satisfactory *statistical interpretation* of $|\psi|^2$ as the probability density was first given by M. Born.

To understand the physical significance of the wave function, let us write it as

$$\psi(\mathbf{x},t)=\sqrt{\rho(\mathbf{x},t)}\exp\left[\frac{iS(\mathbf{x},t)}{\hbar}\right], \tag{2.4.18}$$

with S real and $\rho>0$, which can always be done for any complex function of $\mathbf{x}$ and t. The meaning of ρ has already been given. What is the physical interpretation of S? Noting

$$\psi^*\nabla\psi=\sqrt{\rho}\,\nabla(\sqrt{\rho})+\left(\frac{i}{\hbar}\right)\rho\nabla S, \tag{2.4.19}$$

we can write the probability flux as [see (2.4.16)]

$$\mathbf{j}=\frac{\rho\nabla S}{m}. \tag{2.4.20}$$

We now see that there is more to the wave function than the fact that $|\psi|^2$ is the probability density; the gradient of the phase S contains a vital piece of information. From (2.4.20) we see that the *spatial variation of the phase* of the wave function characterizes the probability flux; the stronger the phase variation, the more intense the flux. The direction of $\mathbf{j}$ at some point $\mathbf{x}$ is seen to be normal to the surface of a constant phase that goes through that point. In the particularly simple example of a plane wave (a momentum eigenfunction)

$$\psi(\mathbf{x},t)\propto\exp\left(\frac{i\mathbf{p}\cdot\mathbf{x}}{\hbar}-\frac{iEt}{\hbar}\right), \tag{2.4.21}$$

where $\mathbf{p}$ stands for the eigenvalue of the momentum operator. All this is evident because

$$\nabla S=\mathbf{p}. \tag{2.4.22}$$

More generally, it is tempting to regard $\nabla S/m$ as some kind of "velocity,"

$$\text{“}\mathbf{v}\text{”}=\frac{\nabla S}{m}, \tag{2.4.23}$$

and to write the continuity equation (2.4.15) as

$$\frac{\partial\rho}{\partial t}+\nabla\cdot(\rho\text{“}\mathbf{v}\text{”})=0, \tag{2.4.24}$$

just as in fluid dynamics. However, we would like to caution the reader against a too literal interpretation of $\mathbf{j}$ as ρ times the velocity defined at every point in space, because a simultaneous precision measurement of position and velocity would necessarily violate the uncertainty principle.

The Classical Limit

We now discuss the classical limit of wave mechanics. First, we substitute ψ written in form (2.4.18) into both sides of the time-dependent wave equation. Straightforward differentiations lead to

$$-\left(\frac{\hbar^2}{2m}\right) \times\left[\nabla^2\sqrt{\rho}+\left(\frac{2i}{\hbar}\right)(\nabla\sqrt{\rho})\cdot(\nabla S)-\left(\frac{1}{\hbar^2}\right)\sqrt{\rho}|\nabla S|^2+\left(\frac{i}{\hbar}\right)\sqrt{\rho}\,\nabla^2 S\right]+\sqrt{\rho}\,V = i\hbar\left[\frac{\partial\sqrt{\rho}}{\partial t}+\left(\frac{i}{\hbar}\right)\sqrt{\rho}\,\frac{\partial S}{\partial t}\right]. \tag{2.4.25}$$

So far everything has been exact. Let us suppose now that $\hbar$ can, in some sense, be regarded as a small quantity. The precise physical meaning of this approximation, to which we will come back later, is not evident now, but let us assume

$$\hbar|\nabla^2 S| \ll |\nabla S|^2, \tag{2.4.26}$$

and so forth. We can then collect terms in (2.4.25) that do not explicitly contain $\hbar$ to obtain a nonlinear partial differential equation for S:

$$\frac{1}{2m}|\nabla S(\mathbf{x},t)|^2+V(\mathbf{x})+\frac{\partial S(\mathbf{x},t)}{\partial t}=0. \tag{2.4.27}$$

We recognize this to be the **Hamilton-Jacobi equation** in classical mechanics, first written in 1836, where $S(\mathbf{x},t)$ stands for Hamilton's principal function. So, not surprisingly, in the $\hbar \to 0$ limit, classical mechanics is contained in Schrödinger's wave mechanics. We have a semiclassical interpretation of the phase of the wave function: $\hbar$ times the phase is equal to Hamilton's principal function provided that $\hbar$ can be regarded as a small quantity.

Let us now look at a stationary state with time dependence $\exp(-iEt/\hbar)$. This time dependence is anticipated from the fact that for a classical system with a constant Hamiltonian, Hamilton's principal function S is separable:

$$S(x,t)=W(x)-Et, \tag{2.4.28}$$

where $W(x)$ is called **Hamilton's characteristic function** (Goldstein 1980, 445–46). As time goes on, a surface of a constant S advances in much the same way as a surface of a constant phase in wave optics—a "wave front" —advances. The momentum in the classical Hamilton-Jacobi theory is given by

$$\mathbf{p}_{\text{class}} = \nabla S = \nabla W, \tag{2.4.29}$$

which is consistent with our earlier identification of $\nabla S/m$ with some kind of velocity. In classical mechanics the velocity vector is tangential to the particle trajectory, and as a result we can trace the trajectory by following continuously the direction of the velocity vector. The particle trajectory is like a ray in geometric optics because the ∇S that traces the trajectory is normal to the wave front defined by a constant S. In this sense geometrical optics is to wave optics what classical mechanics is to wave mechanics.

One might wonder, in hindsight, why this optical-mechanical analogy was not fully exploited in the nineteenth century. The reason is that there was no motivation for regarding Hamilton's principal function as the phase of some traveling wave; the wave nature of a material particle did not become apparent until the 1920s. Besides, the basic unit of action $\hbar$, which must enter into (2.4.18) for dimensional reasons, was missing in the physics of the nineteenth century.

Semiclassical (WKB) Approximation

Let us now restrict ourselves to one dimension and obtain an approximate stationary-state solution to Schrödinger's wave equation. This can easily be accomplished by noting that the corresponding solution of the classical Hamilton-Jacobi equation is

$$\begin{aligned} S(x,t) &= W(x) - Et \\ &= \pm \int^{x} dx' \sqrt{2m[E - V(x')]} - Et. \end{aligned} \tag{2.4.30}$$

For a stationary state we must have

$$\frac{\partial \rho}{\partial t} = 0, \quad (\text{all } x), \tag{2.4.31}$$

which, because of the continuity equation [see (2.4.24)]

$$\frac{\partial \rho}{\partial t} + \frac{1}{m}\frac{\partial}{\partial x}\left(\rho \frac{\partial S}{\partial x}\right) = 0 \tag{2.4.32}$$

implies

$$\rho \frac{dW}{dx} = \pm \rho \sqrt{2m[E - V(x)]} = \text{constant}; \tag{2.4.33}$$

hence,

$$\sqrt{\rho} = \frac{\text{constant}}{[E - V(x)]^{1/4}} \propto \frac{1}{\sqrt{v_{\text{classical}}}}. \tag{2.4.34}$$

This makes good sense because classically the probability for finding the particle at a given point should be inversely proportional to the velocity. Combining everything, we obtain an approximate solution:

$$\psi(x,t) \simeq \left\{\frac{\text{constant}}{[E-V(x)]^{1/4}}\right\} \times \exp\left[\pm\left(\frac{i}{\hbar}\right)\int^x dx'\sqrt{2m[E-V(x')]} - \frac{iEt}{\hbar}\right]. \quad (2.4.35)$$

This is known as the WKB solution, after G. Wentzel, A. Kramers, and L. Brillouin.*

Having obtained (2.4.35), let us return to the question of what we mean when we say that $\hbar$ is small. Our derivation of the Hamilton-Jacobi equation from the Schrödinger equation rested on (2.4.26), which, in one-dimensional problems, is equivalent to

$$\hbar\left|\frac{d^2W}{dx^2}\right| \ll \left|\frac{dW}{dx}\right|^2. \quad (2.4.36)$$

In terms of the de Broglie wavelength divided by 2π, this condition amounts to

$$\lambda\!\!\!^{-} = \frac{\hbar}{\sqrt{2m[E-V(x)]}} \ll \frac{2[E-V(x)]}{|dV/dx|}. \quad (2.4.37)$$

In other words, $\lambda\!\!\!^{-}$ must be small compared with the characteristic distance over which the potential varies appreciably. Roughly speaking, the potential must be essentially constant over many wavelengths. Thus we see that the semiclassical picture is reliable *in the short-wavelength limit.*

Solution (2.4.35) has been derived for the classically allowed region where $E-V(x)$ is positive. We now consider the classically forbidden region where $E-V(x)$ is negative. The classical Hamilton-Jacobi theory does not make sense in this case, so our approximate solution (2.4.35), which is valid for $E>V$, must therefore be modified. Fortunately an analogous solution exists in the $E<V$ region; by direct substitution we can check that

$$\psi(\mathbf{x},t) \simeq \left\{\frac{\text{constant}}{[V(x)-E]^{1/4}}\right\}\exp\left[\pm\left(\frac{1}{\hbar}\right)\int^x dx'\sqrt{2m[V(x')-E]} - \frac{iEt}{\hbar}\right] \quad (2.4.38)$$

*A similar technique was used earlier by H. Jeffreys; this solution is referred to as the JWKB solution in some English books.

satisfies the wave equation provided that $\hbar/\sqrt{2m(V-E)}$ is small compared with the characteristic distance over which the potential varies.

Neither (2.4.35) nor (2.4.38) makes sense near the classical turning point defined by the value of x for which

$$V(x) = E \tag{2.4.39}$$

because $\lambda\!\!\!^{-}$ (or its purely imaginary analogue) becomes infinite at that point, leading to a violent violation of (2.4.37). In fact, it is a nontrivial task to match the two solutions across the classical turning point. The standard procedure is based on the following steps:

1. Make a linear approximation to the potential $V(x)$ near the turning point x_0, defined by the root of (2.4.39).
2. Solve the differential equation

$$\frac{d^2 u_E}{dx^2} - \left(\frac{2m}{\hbar^2}\right)\left(\frac{dV}{dx}\right)_{x=x_0}(x - x_0)u_E = 0 \tag{2.4.40}$$

   exactly to obtain a third solution involving the Bessel function of order $\pm\frac{1}{3}$, valid near x_0.
3. Match this solution to the other two solutions by choosing appropriately various constants of integration.

We do not discuss these steps in detail, as they are discussed in many places (Schiff 1968, 268–76, for example). Instead, we content ourselves to present the results of such an analysis for a potential well, schematically shown in Figure 2.1, with two turning points, x_1 and x_2. The wave function must behave like (2.4.35) in region II and like (2.4.38) in regions I and III. The correct matching from region I into region II can be shown to be

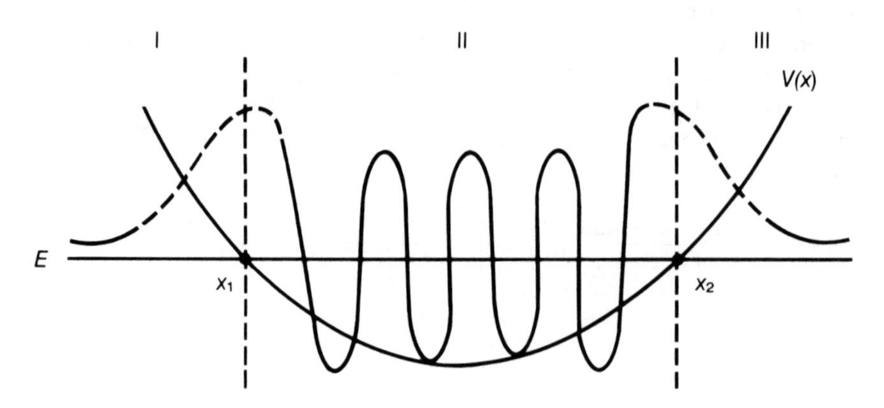

FIGURE 2.1. Schematic diagram for behavior of wave function $u_E(x)$ in potential well $V(x)$ with turning points x_1 and x_2.

accomplished by choosing the integration constants in such a way that*

$$\left\{\frac{1}{[V(x)-E]^{1/4}}\right\}\exp\left[-\left(\frac{1}{\hbar}\right)\int_x^{x_1} dx'\sqrt{2m[V(x')-E]}\right]$$
$$\rightarrow \left\{\frac{2}{[E-V(x)]^{1/4}}\right\}\cos\left[\left(\frac{1}{\hbar}\right)\int_{x_1}^{x} dx'\sqrt{2m[E-V(x')]}-\frac{\pi}{4}\right].$$
$$(2.4.41)$$

Likewise, from region III into region II we have

$$\left\{\frac{1}{[V(x)-E]^{1/4}}\right\}\exp\left[-\left(\frac{1}{\hbar}\right)\int_{x_2}^{x} dx'\sqrt{2m[V(x')-E]}\right]$$
$$\rightarrow \left\{\frac{2}{[E-V(x)]^{1/4}}\right\}\cos\left[-\left(\frac{1}{\hbar}\right)\int_{x}^{x_2} dx'\sqrt{2m[E-V(x')]}+\frac{\pi}{4}\right].$$
$$(2.4.42)$$

The uniqueness of the wave function in region II implies that the arguments of the cosine in (2.4.41) and (2.4.42) must differ at most by an integer multiple of π [not of 2π, because the signs of both sides of (2.4.42) can be reversed]. In this way we obtain a very interesting consistency condition,

$$\int_{x_1}^{x_2} dx\sqrt{2m[E-V(x)]} = \left(n+\tfrac{1}{2}\right)\pi\hbar \quad (n=0,1,2,3,\ldots). \quad (2.4.43)$$

Apart from the difference between $n+\frac{1}{2}$ and n, this equation is simply the quantization condition of the old quantum theory due to A. Sommerfeld and W. Wilson, originally written in 1915 as

$$\oint p\,dq = nh, \quad (2.4.44)$$

where h is Planck's h, not Dirac's $\hbar$, and the integral is evaluated over one whole period of classical motion, from x_1 to x_2 *and* back.

Equation (2.4.43) can be used to obtain approximate expressions for the energy levels of a particle confined in a potential well. As an example, we consider the energy spectrum of a ball bouncing up and down over a

*A quick way to understand the appearance of $\pi/4$ in (2.4.41) is as follows. The correctly matched pair of solutions in region I and region II, of the form (2.4.38) and (2.4.35), respectively, must be related to each other by analytic continuation in a *complex* x-plane. An excursion into the complex x-plane enables us to change the sign of $V(x)-E \simeq (dV/dx)_{x=x_1}(x-x_1)$ without going through the dangerous region with $x \simeq x_1$. We now observe that under such an analytic continuation, the $1/[V(x)-E]^{1/4}$ factor in (2.4.38) picks up a phase: $[(dV/dx)_{x=x_1}(x-x_1)]^{1/4} \rightarrow [(dV/dx)_{x=x_1}(x_1-x)]^{1/4}e^{i\pi/4}$.

hard surface:

$$V = \begin{cases} mgx, & \text{for } x > 0 \\ \infty, & \text{for } x < 0, \end{cases} \tag{2.4.45}$$

where x stands for the height of the ball measured from the hard surface. One might be tempted to use (2.4.43) directly with

$$x_1 = 0, \qquad x_2 = \frac{E}{mg}, \tag{2.4.46}$$

which are the classical turning points of this problem. We note, however, that (2.4.43) was derived under the assumption that the WKB wave function "leaks into" the $x < x_1$ region, while in our problem the wave function must strictly vanish for $x \le x_1 = 0$. A much more satisfactory approach to this problem is to consider the *odd-parity solutions*—guaranteed to vanish at $x = 0$—of a modified problem defined by

$$V(x) = mg|x| \quad (-\infty < x < \infty) \tag{2.4.47}$$

whose turning points are

$$x_1 = -\frac{E}{mg}, \qquad x_2 = \frac{E}{mg}. \tag{2.4.48}$$

The energy spectrum of the odd-parity states for this modified problem must clearly be the same as that of the original problem. The quantization condition then becomes

$$\int_{-E/mg}^{E/mg} dx\sqrt{2m(E - mg|x|)} = \left(n_{\text{odd}} + \tfrac{1}{2}\right)\pi\hbar, \quad (n_{\text{odd}} = 1,3,5,\ldots) \tag{2.4.49}$$

or, equivalently,

$$\int_0^{E/mg} dx\sqrt{2m(E - mgx)} = \left(n - \tfrac{1}{4}\right)\pi\hbar, \quad (n = 1,2,3,4,\ldots). \tag{2.4.50}$$

This integral is elementary, and we obtain

$$E_n = \left\{\frac{\left[3\left(n - \frac{1}{4}\right)\pi\right]^{2/3}}{2}\right\}\left(mg^2\hbar^2\right)^{1/3} \tag{2.4.51}$$

for the quantized energy levels of the bouncing ball.

It is known that this problem is soluble analytically without any approximation. The energy eigenvalues turn out to be expressible in terms of the zeros of the Airy function

$$Ai(-\lambda_n) = 0 \tag{2.4.52}$$

TABLE 2.2. The Quantized Energies of a Bouncing Ball in Units of $(mg^2\hbar^2/2)^{1/3}$

n	WKB	Exact
1	2.320	2.338
2	4.082	4.088
3	5.517	5.521
4	6.784	6.787
5	7.942	7.944
6	9.021	9.023
7	10.039	10.040
8	11.008	11.009
9	11.935	11.936
10	12.828	12.829

as

$$E_n = \left(\frac{\lambda_n}{2^{1/3}}\right)\left(mg^2\hbar^2\right)^{1/3}. \tag{2.4.53}$$

The two approaches are compared numerically in Table 2.2 for the first 10 energy levels. We see that agreement is excellent even for small values of n and essentially exact for $n \simeq 10$.

The quantum-theoretical treatment of a bouncing ball may appear to have little to do with the real world. It turns out, however, that a potential of type (2.4.45) is actually of practical interest in studying the energy spectrum of a quark-antiquark bound system, called **quarkonium**. To go from a bouncing ball to a quarkonium, the x in (2.4.45) is replaced by the quark-antiquark separation distance r. The analogue of the downward gravitational force mg is a constant (that is, r-independent) force believed to be operative between a quark and an antiquark. This force is empirically estimated to be in the neighborhood of

$$1\ \text{GeV/fm} \simeq 1.6\times 10^5\ \text{N}, \tag{2.4.54}$$

which corresponds to about 16 tons. This contrasts with the gravitational force of 0.98 N on a ball of 0.1 kg.

2.5. PROPAGATORS AND FEYNMAN PATH INTEGRALS

Propagators in Wave Mechanics

In Section 2.1 we showed how the most general time-evolution problem with a time-independent Hamiltonian can be solved once we expand the initial ket in terms of the eigenkets of an observable that commutes with H. Let us translate this statement into the language of wave mechanics. We start with

$$|\alpha, t_0; t\rangle = \exp\left[\frac{-iH(t-t_0)}{\hbar}\right]|\alpha, t_0\rangle$$

$$= \sum_{a'} |a'\rangle\langle a'|\alpha, t_0\rangle \exp\left[\frac{-iE_{a'}(t-t_0)}{\hbar}\right]. \tag{2.5.1}$$

Multiplying both sides by $\langle \mathbf{x}'|$ on the left, we have

$$\langle \mathbf{x}'|\alpha, t_0; t\rangle = \sum_{a'} \langle \mathbf{x}'|a'\rangle\langle a'|\alpha, t_0\rangle \exp\left[\frac{-iE_{a'}(t-t_0)}{\hbar}\right], \tag{2.5.2}$$

which is of the form

$$\psi(\mathbf{x}', t) = \sum_{a'} c_{a'}(t_0) u_{a'}(\mathbf{x}') \exp\left[\frac{-iE_{a'}(t-t_0)}{\hbar}\right], \tag{2.5.3}$$

with

$$u_{a'}(\mathbf{x}') = \langle \mathbf{x}'|a'\rangle \tag{2.5.4}$$

standing for the eigenfunction of operator A with eigenvalue a'. Note also that

$$\langle a'|\alpha, t_0\rangle = \int d^3x' \langle a'|\mathbf{x}'\rangle\langle \mathbf{x}'|\alpha, t_0\rangle, \tag{2.5.5}$$

which we recognize as the usual rule in wave mechanics for getting the expansion coefficients of the initial state:

$$c_{a'}(t_0) = \int d^3x' u_{a'}^*(\mathbf{x}') \psi(\mathbf{x}', t_0). \tag{2.5.6}$$

All this should be straightforward and familiar. Now (2.5.2) together with (2.5.5) can also be visualized as some kind of integral operator acting on the initial wave function to yield the final wave function:

$$\psi(\mathbf{x}'', t) = \int d^3x' K(\mathbf{x}'', t; \mathbf{x}', t_0) \psi(\mathbf{x}', t_0). \tag{2.5.7}$$

Here the kernel of the integral operator, known as the **propagator** in wave mechanics, is given by

$$K(\mathbf{x}'', t; \mathbf{x}', t_0) = \sum_{a'} \langle \mathbf{x}''|a'\rangle\langle a'|\mathbf{x}'\rangle \exp\left[\frac{-iE_{a'}(t-t_0)}{\hbar}\right]. \tag{2.5.8}$$

In any given problem the propagator depends only on the potential and is independent of the initial wave function. It can be constructed once the energy eigenfunctions and their eigenvalues are given.

Clearly, the time evolution of the wave function is completely predicted if $K(\mathbf{x}'', t; \mathbf{x}', t_0)$ is known and $\psi(\mathbf{x}', t_0)$ is given initially. In this sense Schrödinger's wave mechanics is a *perfectly causal theory*. The time development of a wave function subjected to some potential is as "deterministic" as anything else in classical mechanics *provided that the system is left undisturbed*. The only peculiar feature, if any, is that when a measurement intervenes, the wave function changes abruptly, in an uncontrollable way, into one of the eigenfunctions of the observable being measured.

There are two properties of the propagator worth recording here. First, for $t > t_0$, $K(\mathbf{x}'', t; \mathbf{x}', t_0)$ satisfies Schrödinger's time-dependent wave equation in the variables $\mathbf{x}''$ and t, with $\mathbf{x}'$ and t_0 fixed. This is evident from (2.5.8) because $\langle \mathbf{x}''|a'\rangle \exp[-iE_{a'}(t-t_0)/\hbar]$, being the wave function corresponding to $\mathscr{U}(t, t_0)|a'\rangle$, satisfies the wave equation. Second,

$$\lim_{t \to t_0} K(\mathbf{x}'', t; \mathbf{x}', t_0) = \delta^3(\mathbf{x}'' - \mathbf{x}'), \tag{2.5.9}$$

which is also obvious; as $t \to t_0$, because of the completeness of $\{|a'\rangle\}$, sum (2.5.8) just reduces to $\langle \mathbf{x}''|\mathbf{x}'\rangle$.

Because of these two properties, the propagator (2.5.8), regarded as a function of $\mathbf{x}''$, is simply the wave function at t of a particle which was localized *precisely* at $\mathbf{x}'$ at some earlier time t_0. Indeed, this interpretation follows, perhaps more elegantly, from noting that (2.5.8) can also be written as

$$K(\mathbf{x}'', t; \mathbf{x}', t_0) = \langle \mathbf{x}''| \exp\left[\frac{-iH(t-t_0)}{\hbar}\right] |\mathbf{x}'\rangle, \tag{2.5.10}$$

where the time-evolution operator acting on $|\mathbf{x}'\rangle$ is just the state ket at t of a system that was localized precisely at $\mathbf{x}'$ at time t_0 ($< t$). If we wish to solve a more general problem where the initial wave function extends over a finite region of space, all we have to do is multiply $\psi(\mathbf{x}', t_0)$ by the propagator $K(\mathbf{x}'', t; \mathbf{x}', t_0)$ and integrate over all space (that is, over $\mathbf{x}'$). In this manner we can add the various contributions from different positions ($\mathbf{x}'$). This situation is analogous to one in electrostatics; if we wish to find the electrostatic potential due to a general charge distribution $\rho(\mathbf{x}')$, we first solve the point-charge problem, multiply the point-charge solution with the charge distribution, and integrate:

$$\phi(\mathbf{x}) = \int d^3x' \frac{\rho(\mathbf{x}')}{|\mathbf{x} - \mathbf{x}'|}. \tag{2.5.11}$$

The reader familiar with the theory of the Green's functions must have recognized by this time that the propagator is simply the Green's function for the time-dependent wave equation satisfying

$$\left[-\left(\frac{\hbar^2}{2m}\right)\nabla''^2 + V(\mathbf{x}'') - i\hbar\frac{\partial}{\partial t}\right] K(\mathbf{x}'', t; \mathbf{x}', t_0) = -i\hbar\delta^3(\mathbf{x}'' - \mathbf{x}')\delta(t - t_0) \tag{2.5.12}$$

with the boundary condition

$$K(\mathbf{x}'', t; \mathbf{x}', t_0) = 0, \quad \text{for } t < t_0. \tag{2.5.13}$$

The delta function $\delta(t - t_0)$ is needed on the right-hand side of (2.5.12) because K varies discontinuously at $t = t_0$.

The particular form of the propagator is, of course, dependent on the particular potential to which the particle is subjected. Consider, as an example, a free particle in one dimension. The obvious observable that commutes with H is momentum; $|p'\rangle$ is a simultaneous eigenket of the operators p and H:

$$p|p'\rangle = p'|p'\rangle \qquad H|p'\rangle = \left(\frac{p'^2}{2m}\right)|p'\rangle. \tag{2.5.14}$$

The momentum eigenfunction is just the transformation function of Section 1.7 [see (1.7.32)] which is of the plane-wave form. Combining everything, we have

$$K(x'', t; x', t_0) = \left(\frac{1}{2\pi\hbar}\right)\int_{-\infty}^{\infty} dp' \exp\left[\frac{ip'(x''-x')}{\hbar} - \frac{ip'^2(t-t_0)}{2m\hbar}\right]. \tag{2.5.15}$$

The integral can be evaluated by completing the square in the exponent. Here we simply record the result:

$$K(x'', t; x', t_0) = \sqrt{\frac{m}{2\pi i\hbar(t-t_0)}} \exp\left[\frac{im(x''-x')^2}{2\hbar(t-t_0)}\right]. \tag{2.5.16}$$

This expression may be used, for example, to study how a Gaussian wave packet spreads out as a function of time.

For the simple harmonic oscillator, where the wave function of an energy eigenstate is given by

$$u_n(x)\exp\left(\frac{-iE_n t}{\hbar}\right) = \left(\frac{1}{2^{n/2}\sqrt{n!}}\right)\left(\frac{m\omega}{\pi\hbar}\right)^{1/4} \exp\left(\frac{-m\omega x^2}{2\hbar}\right) \times H_n\left(\sqrt{\frac{m\omega}{\hbar}}\,x\right)\exp\left[-i\omega\left(n+\frac{1}{2}\right)t\right], \tag{2.5.17}$$

the propagator is given by

$$K(x'', t; x', t_0) = \sqrt{\frac{m\omega}{2\pi i\hbar \sin[\omega(t-t_0)]}} \exp\left[\left\{\frac{im\omega}{2\hbar\sin[\omega(t-t_0)]}\right\} \times \left\{(x''^2 + x'^2)\cos[\omega(t-t_0)] - 2x''x'\right\}\right]. \tag{2.5.18}$$

One way to prove this is to use

$$\left(\frac{1}{\sqrt{1-\zeta^2}}\right)\exp\left[\frac{-(\xi^2+\eta^2-2\xi\eta\zeta)}{(1-\zeta^2)}\right]$$
$$= \exp[-(\xi^2+\eta^2)] \sum_{n=0} \left(\frac{\zeta^n}{2^n n!}\right) H_n(\xi) H_n(\eta), \qquad (2.5.19)$$

which is found in books on special functions (Morse and Feshbach 1953, 786). It can also be obtained using the $a, a^\dagger$ operator method (Saxon 1968, 144–45) or, alternatively, the path-integral method to be described later. Notice that (2.5.18) is a periodic function of t with angular frequency ω, the classical oscillator frequency. This means, among other things, that a particle initially localized precisely at x' will return to its original position with certainty at $2\pi/\omega$ ($4\pi/\omega$, and so forth) later.

Certain space and time integrals derivable from $K(\mathbf{x}'', t; \mathbf{x}', t_0)$ are of considerable interest. Without loss of generality we set $t_0 = 0$ in the following. The first integral we consider is obtained by setting $\mathbf{x}'' = \mathbf{x}'$ and integrating over all space. We have

$$\begin{aligned} G(t) &\equiv \int d^3x' K(\mathbf{x}', t; \mathbf{x}', 0) \\ &= \int d^3x' \sum_{a'} |\langle \mathbf{x}'|a'\rangle|^2 \exp\left(\frac{-iE_{a'}t}{\hbar}\right) \\ &= \sum_{a'} \exp\left(\frac{-iE_{a'}t}{\hbar}\right). \end{aligned} \qquad (2.5.20)$$

This result is anticipated; recalling (2.5.10),we observe that setting $\mathbf{x}' = \mathbf{x}''$ and integrating are equivalent to taking the trace of the time-evolution operator in the $\mathbf{x}$-representation. But the trace is independent of representations; it can be evaluated more readily using the $\{|a'\rangle\}$ basis where the time-evolution operator is diagonal, which immediately leads to the last line of (2.5.20). Now we see that (2.5.20) is just the "sum over states," reminiscent of the partition function in statistical mechanics. In fact, if we analytically continue in the t variable and make t purely imaginary, with β defined by

$$\beta = \frac{it}{\hbar} \qquad (2.5.21)$$

real and positive, we can identify (2.5.20) with the partition function itself:

$$Z = \sum_{a'} \exp(-\beta E_{a'}). \qquad (2.5.22)$$

For this reason some of the techniques encountered in studying propagators in quantum mechanics are also useful in statistical mechanics.

Next, let us consider the Laplace-Fourier transform of $G(t)$:

$$\begin{aligned}\tilde{G}(E) &\equiv -i\int_0^\infty dt G(t)\exp(iEt/\hbar)/\hbar \\ &= -i\int_0^\infty dt \sum_{a'} \exp(-iE_{a'}t/\hbar)\exp(iEt/\hbar)/\hbar. \end{aligned} \tag{2.5.23}$$

The integrand here oscillates indefinitely. But we can make the integral meaningful by letting E acquire a small positive imaginary part:

$$E \to E + i\varepsilon. \tag{2.5.24}$$

We then obtain, in the limit $\varepsilon \to 0$,

$$\tilde{G}(E) = \sum_{a'} \frac{1}{E - E_{a'}}. \tag{2.5.25}$$

Observe now that the complete energy spectrum is exhibited as simple poles of $\tilde{G}(E)$ in the complex E-plane. If we wish to know the energy spectrum of a physical system, it is sufficient to study the analytic properties of $\tilde{G}(E)$.

Propagator as a Transition Amplitude

To gain further insight into the physical meaning of the propagator, we wish to relate it to the concept of transition amplitudes introduced in Section 2.2. But first, recall that the wave function which is the inner product of the fixed position bra $\langle \mathbf{x}'|$ with the moving state ket $|\alpha, t_0; t\rangle$ can also be regarded as the inner product of the Heisenberg-picture position bra $\langle \mathbf{x}', t|$, which moves "oppositely" with time, with the Heisenberg-picture state ket $|\alpha, t_0\rangle$, which is fixed in time. Likewise, the propagator can also be written as

$$\begin{aligned} K(\mathbf{x}'', t; \mathbf{x}', t_0) &= \sum_{a'} \langle \mathbf{x}''|a'\rangle\langle a'|\mathbf{x}'\rangle \exp\left[\frac{-iE_{a'}(t-t_0)}{\hbar}\right] \\ &= \sum_{a'} \langle \mathbf{x}''|\exp\left(\frac{-iHt}{\hbar}\right)|a'\rangle\langle a'|\exp\left(\frac{iHt_0}{\hbar}\right)|\mathbf{x}'\rangle \\ &= \langle \mathbf{x}'', t|\mathbf{x}', t_0\rangle, \end{aligned} \tag{2.5.26}$$

where $|\mathbf{x}', t_0\rangle$ and $\langle \mathbf{x}'', t|$ are to be understood as an eigenket and an eigenbra of the position operator in the Heisenberg picture. In Section 2.1 we showed that $\langle b', t|a'\rangle$, in the Heisenberg-picture notation, is the probability amplitude for a system originally prepared to be an eigenstate of A with eigenvalue a' at some initial time $t_0 = 0$ to be found at a later time t in an eigenstate of B with eigenvalue b', and we called it the transition amplitude for going from state $|a'\rangle$ to state $|b'\rangle$. Because there is nothing special about the choice of t_0—only the time difference $t - t_0$ is relevant—we can identify $\langle \mathbf{x}'', t|\mathbf{x}', t_0\rangle$ as the probability amplitude for the particle prepared at t_0 with position eigenvalue $\mathbf{x}'$ to be found at a later time t at $\mathbf{x}''$. Roughly speaking, $\langle \mathbf{x}'', t|\mathbf{x}', t_0\rangle$ is the amplitude for the particle to go from a space-time point $(\mathbf{x}', t_0)$ to another space-time point $(\mathbf{x}'', t)$, so the term *transition amplitude* for this expression is quite appropriate. This interpretation is, of course, in complete accord with the interpretation we gave earlier for $K(\mathbf{x}'', t; \mathbf{x}', t_0)$.

Yet another way to interpret $\langle \mathbf{x}'', t|\mathbf{x}', t_0\rangle$ is as follows. As we emphasized earlier, $|\mathbf{x}', t_0\rangle$ is the position eigenket at t_0 with the eigenvalue $\mathbf{x}'$ in the Heisenberg picture. Because at any given time the Heisenberg-picture eigenkets of an observable can be chosen as base kets, we can regard $\langle \mathbf{x}'', t|\mathbf{x}', t_0\rangle$ as the transformation function that connects the two sets of base kets at *different* times. So in the Heisenberg picture, time evolution can be viewed as a *unitary transformation*, in the sense of changing bases, that connects one set of base kets formed by $\{|\mathbf{x}', t_0\rangle\}$ to another formed by $\{|\mathbf{x}'', t\rangle\}$. This is reminiscent of classical physics, in which the time development of a classical dynamic variable such as $\mathbf{x}(t)$ is viewed as a canonical (or contact) transformation generated by the classical Hamiltonian (Goldstein 1980, 407–8).

It turns out to be convenient to use a notation that treats the space and time coordinates more symmetrically. To this end we write $\langle \mathbf{x}'', t''|\mathbf{x}', t'\rangle$ in place of $\langle \mathbf{x}'', t|\mathbf{x}', t_0\rangle$. Because at any given time the position kets in the Heisenberg picture form a complete set, it is legitimate to insert the identity operator written as

$$\int d^3x''|\mathbf{x}'', t''\rangle\langle \mathbf{x}'', t''| = 1 \tag{2.5.27}$$

at any place we desire. For example, consider the time evolution from t' to t'''; by dividing the time interval (t', t''') into two parts, (t', t'') and (t'', t'''), we have

$$\langle \mathbf{x}''', t'''|\mathbf{x}', t'\rangle = \int d^3x''\langle \mathbf{x}''', t'''|\mathbf{x}'', t''\rangle\langle \mathbf{x}'', t''|\mathbf{x}', t'\rangle,$$
$$(t''' > t'' > t'). \tag{2.5.28}$$

We call this the **composition property** of the transition amplitude.* Clearly, we can divide the time interval into as many smaller subintervals as we wish. We have

$$\langle \mathbf{x}'''', t''''|\mathbf{x}', t'\rangle = \int d^3x''' \int d^3x''\langle \mathbf{x}'''', t''''|\mathbf{x}''', t'''\rangle\langle \mathbf{x}''', t'''|\mathbf{x}'', t''\rangle$$
$$\times \langle \mathbf{x}'', t''|\mathbf{x}', t'\rangle, \quad (t'''' > t''' > t'' > t'), \tag{2.5.29}$$

*The analogue of (2.5.28) in probability theory is known as the Chapman-Kolmogoroff equation, and in diffusion theory, the Smoluchowsky equation.

and so on. If we somehow guess the form of $\langle \mathbf{x}'', t'' | \mathbf{x}', t' \rangle$ for an *infinitesimal* time interval (between t' and $t'' = t' + dt$), we should be able to obtain the amplitude $\langle \mathbf{x}'', t'' | \mathbf{x}', t' \rangle$ for a finite time interval by compounding the appropriate transition amplitudes for infinitesimal time intervals in a manner analogous to (2.5.29). This kind of reasoning leads to an *independent formulation* of quantum mechanics due to R. P. Feynman, published in 1948, to which we now turn our attention.

Path Integrals as the Sum Over Paths

Without loss of generality we restrict ourselves to one-dimensional problems. Also, we avoid awkward expressions like

$$\underbrace{x'''' \cdots x'''}_{N \text{ times}}$$

by using notation such as x_N. With this notation we consider the transition amplitude for a particle going from the initial space-time point (x_1, t_1) to the final space-time point (x_N, t_N). The entire time interval between t_1 and t_N is divided into $N-1$ equal parts:

$$t_j - t_{j-1} = \Delta t = \frac{(t_N - t_1)}{(N-1)}. \tag{2.5.30}$$

Exploiting the composition property, we obtain

$$\begin{aligned} \langle x_N, t_N | x_1, t_1 \rangle = \int dx_{N-1} \int dx_{N-2} \cdots \int dx_2 \langle x_N, t_N | x_{N-1}, t_{N-1} \rangle \\ \times \langle x_{N-1}, t_{N-1} | x_{N-2}, t_{N-2} \rangle \cdots \langle x_2, t_2 | x_1, t_1 \rangle. \end{aligned} \tag{2.5.31}$$

To visualize this pictorially, we consider a space-time plane, as shown in Figure 2.2. The initial and final space-time points are fixed to be (x_1, t_1) and (x_N, t_N), respectively. For each time segment, say between t_{n-1} and t_n, we are instructed to consider the transition amplitude to go from (x_{n-1}, t_{n-1}) to (x_n, t_n); we then integrate over $x_2, x_3, \ldots, x_{N-1}$. This means that we must *sum over all possible paths* in the space-time plane with the end points fixed.

Before proceeding further, it is profitable to review here how paths appear in classical mechanics. Suppose we have a particle subjected to a force field derivable from a potential $V(x)$. The *classical* Lagrangian is written as

$$L_{\text{classical}}(x, \dot{x}) = \frac{m\dot{x}^2}{2} - V(x). \tag{2.5.32}$$

Given this Lagrangian with the end points (x_1, t_1) and (x_N, t_N) specified, we do *not* consider just any path joining (x_1, t_1) and (x_N, t_N) in classical mechanics. On the contrary, there exists a *unique path* that corresponds to

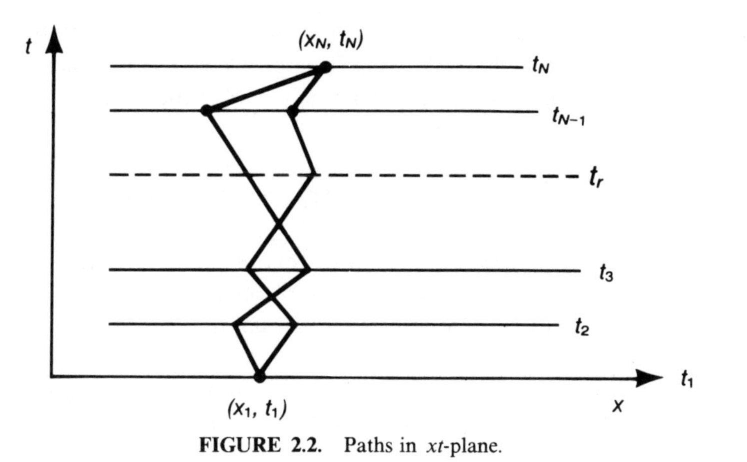

FIGURE 2.2. Paths in *xt*-plane.

the actual motion of the classical particle. For example, given

$$V(x) = mgx, \qquad (x_1, t_1) = (h, 0), \qquad (x_N, t_N) = \left(0, \sqrt{\frac{2h}{g}}\right), \tag{2.5.33}$$

where h may stand for the height of the Leaning Tower of Pisa, the classical path in the *xt*-plane can *only* be

$$x = h - \frac{gt^2}{2}. \tag{2.5.34}$$

More generally, according to Hamilton's principle, the unique path is that which minimizes the action, defined as the time integral of the classical Lagrangian:

$$\delta \int_{t_1}^{t_2} dt \, L_{\text{classical}}(x, \dot{x}) = 0, \tag{2.5.35}$$

from which Lagrange's equation of motion can be obtained.

Feynman's Formulation

The basic difference between classical mechanics and quantum mechanics should now be apparent. In classical mechanics a definite path in the *xt*-plane is associated with the particle's motion; in contrast, in quantum mechanics all possible paths must play roles including those which do not bear any resemblance to the classical path. Yet we must somehow be able to reproduce classical mechanics in a smooth manner in the limit $\hbar \to 0$. How are we to accomplish this?

As a young graduate student at Princeton University, R. P. Feynman tried to attack this problem. In looking for a possible clue, he was said to be intrigued by a mysterious remark in Dirac's book which, in our notation, amounts to the following statement:

$$\exp\left[i\int_{t_1}^{t_2}\frac{dt\,L_{\text{classical}}(x,\dot{x})}{\hbar}\right] \quad \text{corresponds to} \quad \langle x_2, t_2|x_1, t_1\rangle.$$

Feynman attempted to make sense out of this remark. Is "corresponds to" the same thing as "is equal to" or "is proportional to"? In so doing he was led to formulate a space-time approach to quantum mechanics based on *path integrals.*

In Feynman's formulation the classical action plays a very important role. For compactness, we introduce a new notation:

$$S(n, n-1) \equiv \int_{t_{n-1}}^{t_n} dt\,L_{\text{classical}}(x,\dot{x}). \tag{2.5.36}$$

Because $L_{\text{classical}}$ is a function of x and $\dot{x}$, $S(n, n-1)$ is defined only after a definite path is specified along which the integration is to be carried out. So even though the path dependence is not explicit in this notation, it is understood that we are considering a particular path in evaluating the integral. Imagine now that we are following some prescribed path. We concentrate our attention on a small segment along that path, say between (x_{n-1}, t_{n-1}) and (x_n, t_n). According to Dirac, we are instructed to associate $\exp[iS(n, n-1)/\hbar]$ with that segment. Going along the definite path we are set to follow, we successively multiply expressions of this type to obtain

$$\prod_{n=2}^{N}\exp\left[\frac{iS(n,n-1)}{\hbar}\right] = \exp\left[\left(\frac{i}{\hbar}\right)\sum_{n=2}^{N}S(n,n-1)\right] = \exp\left[\frac{iS(N,1)}{\hbar}\right]. \tag{2.5.37}$$

This does not yet give $\langle x_N, t_N|x_1, t_1\rangle$; rather, this equation is the contribution to $\langle x_N, t_N|x_1, t_1\rangle$ arising from the particular path we have considered. We must still integrate over $x_2, x_3, \ldots, x_{N-1}$. At the same time, exploiting the composition property, we let the time interval between t_{n-1} and t_n be infinitesimally small. Thus our candidate expression for $\langle x_N, t_N|x_1, t_1\rangle$ may be written, in some loose sense, as

$$\langle x_N, t_N|x_1, t_1\rangle \sim \sum_{\text{all paths}}\exp\left[\frac{iS(N,1)}{\hbar}\right], \tag{2.5.38}$$

where the sum is to be taken over an innumerably infinite set of paths!

Before presenting a more precise formulation, let us see whether considerations along this line make sense in the classical limit. As $\hbar \to 0$, the exponential in (2.5.38) oscillates very violently, so there is a tendency for cancellation among various contributions from neighboring paths. This is because $\exp[iS/\hbar]$ for some definite path and $\exp[iS/\hbar]$ for a slightly different path have very different phases because of the smallness of $\hbar$. So most paths do *not* contribute when $\hbar$ is regarded as a small quantity. However, there is an important exception.

Suppose that we consider a path that satisfies

$$\delta S(N,1)=0, \tag{2.5.39}$$

where the change in S is due to a slight deformation of the path with the end points fixed. This is precisely the classical path by virtue of Hamilton's principle. We denote the S that satisfies (2.5.39) by S_{min}. We now attempt to deform the path a little bit from the classical path. The resulting S is still equal to S_{min} to first order in deformation. This means that the phase of $\exp[iS/\hbar]$ does not vary very much as we deviate slightly from the classical path even if $\hbar$ is small. As a result, as long as we stay near the classical path, constructive interference between neighboring paths is possible. In the $\hbar \to 0$ limit, the major contributions must then arise from a very narrow strip (or a tube in higher dimensions) containing the classical path, as shown in Figure 2.3. Our (or Feynman's) guess based on Dirac's mysterious remark makes good sense because the classical path gets singled out in the $\hbar \to 0$ limit.

To formulate Feynman's conjecture more precisely, let us go back to $\langle x_n, t_n | x_{n-1}, t_{n-1} \rangle$, where the time difference $t_n - t_{n-1}$ is assumed to be infinitesimally small. We write

$$\langle x_n, t_n | x_{n-1}, t_{n-1} \rangle = \left[\frac{1}{w(\Delta t)}\right] \exp\left[\frac{iS(n,n-1)}{\hbar}\right], \tag{2.5.40}$$

where we evaluate $S(n,n-1)$ in a moment in the $\Delta t \to 0$ limit. Notice that we have inserted a weight factor, $1/w(\Delta t)$, which is assumed to depend only on the time interval $t_n - t_{n-1}$ and not on $V(x)$. That such a factor is needed is clear from dimensional considerations; according to the way we

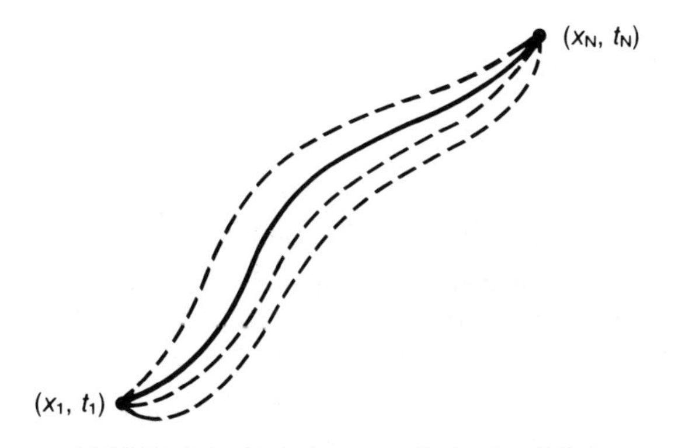

FIGURE 2.3. Paths important in the $\hbar \to 0$ limit.

normalized our position eigenkets, $\langle x_n, t_n | x_{n-1}, t_{n-1}\rangle$ must have the dimension of 1/length.

We now look at the exponential in (2.5.40). Our task is to evaluate the $\Delta t \to 0$ limit of $S(n, n-1)$. Because the time interval is so small, it is legitimate to make a straight-line approximation to the path joining (x_{n-1}, t_{n-1}) and (x_n, t_n) as follows:

$$\begin{aligned} S(n, n-1) &= \int_{t_{n-1}}^{t_n} dt \left[\frac{m\dot{x}^2}{2} - V(x) \right] \\ &= \Delta t \left\{ \left(\frac{m}{2}\right) \left[\frac{(x_n - x_{n-1})}{\Delta t} \right]^2 - V\left(\frac{(x_n + x_{n-1})}{2} \right) \right\}. \end{aligned} \tag{2.5.41}$$

As an example, we consider specifically the free-particle case, $V = 0$. Equation (2.5.40) now becomes

$$\langle x_n, t_n | x_{n-1}, t_{n-1}\rangle = \left[\frac{1}{w(\Delta t)} \right] \exp\left[\frac{im(x_n - x_{n-1})^2}{2\hbar \Delta t} \right]. \tag{2.5.42}$$

We see that the exponent appearing here is completely identical to the one in the expression for the free-particle propagator (2.5.16). The reader may work out a similar comparison for the simple harmonic oscillator.

We remarked earlier that the weight factor $1/w(\Delta t)$ appearing in (2.5.40) is assumed to be independent of $V(x)$, so we may as well evaluate it for the free particle. Noting the orthonormality, in the sense of δ-function, of Heisenberg-picture position eigenkets at equal times,

$$\langle x_n, t_n | x_{n-1}, t_{n-1}\rangle \big|_{t_n = t_{n-1}} = \delta(x_n - x_{n-1}), \tag{2.5.43}$$

we obtain

$$\frac{1}{w(\Delta t)} = \sqrt{\frac{m}{2\pi i \hbar \Delta t}}, \tag{2.5.44}$$

where we have used

$$\int_{-\infty}^{\infty} d\xi \exp\left(\frac{im\xi^2}{2\hbar \Delta t} \right) = \sqrt{\frac{2\pi i \hbar \Delta t}{m}} \tag{2.5.45a}$$

and

$$\lim_{\Delta t \to 0} \sqrt{\frac{m}{2\pi i \hbar \Delta t}} \exp\left(\frac{im\xi^2}{2\hbar \Delta t} \right) = \delta(\xi). \tag{2.5.45b}$$

This weight factor is, of course, anticipated from the expression for the free-particle propagator (2.5.16).

To summarize, as $\Delta t \to 0$, we are led to

$$\langle x_n, t_n | x_{n-1}, t_{n-1}\rangle = \sqrt{\frac{m}{2\pi i \hbar \Delta t}} \exp\left[\frac{iS(n, n-1)}{\hbar} \right]. \tag{2.5.46}$$

The final expression for the transition amplitude with $t_N - t_1$ finite is

$$\langle x_N, t_N | x_1, t_1 \rangle = \lim_{N \to \infty} \left(\frac{m}{2\pi i \hbar \Delta t} \right)^{(N-1)/2} \times \int dx_{N-1} \int dx_{N-2} \cdots \int dx_2 \prod_{n=2}^{N} \exp\left[\frac{iS(n, n-1)}{\hbar} \right], \tag{2.5.47}$$

where the $N \to \infty$ limit is taken with x_N and t_N fixed. It is customary here to define a new kind of multidimensional (in fact, infinite-dimensional) integral operator

$$\int_{x_1}^{x_N} \mathscr{D}[x(t)] \equiv \lim_{N \to \infty} \left(\frac{m}{2\pi i \hbar \Delta t} \right)^{(N-1)/2} \int dx_{N-1} \int dx_{N-2} \cdots \int dx_2 \tag{2.5.48}$$

and write (2.5.47) as

$$\langle x_N, t_N | x_1, t_1 \rangle = \int_{x_1}^{x_N} \mathscr{D}[x(t)] \exp\left[i \int_{t_1}^{t_N} dt \frac{L_{\text{classical}}(x, \dot{x})}{\hbar} \right]. \tag{2.5.49}$$

This expression is known as **Feynman's path integral**. Its meaning as the sum over all possible paths should be apparent from (2.5.47).

Our steps leading to (2.5.49) are not meant to be a derivation. Rather, we (or Feynman) have attempted a new formulation of quantum mechanics based on the concept of paths, motivated by Dirac's mysterious remark. The only ideas we borrowed from the conventional form of quantum mechanics are (1) the superposition principle (used in summing the contributions from various alternate paths), (2) the composition property of the transition amplitude, and (3) classical correspondence in the $\hbar \to 0$ limit.

Even though we obtained the same result as the conventional theory for the free-particle case, it is now obvious, from what we have done so far, that Feynman's formulation is completely equivalent to Schrödinger's wave mechanics. We conclude this section by proving that Feynman's expression for $\langle x_N, t_N | x_1, t_1 \rangle$ indeed satisfies Schrödinger's time-dependent wave equation in the variables x_N, t_N, just as the propagator defined by (2.5.8).

We start with

$$\begin{aligned} \langle x_N, t_N | x_1, t_1 \rangle &= \int dx_{N-1} \langle x_N, t_N | x_{N-1}, t_{N-1} \rangle \langle x_{N-1}, t_{N-1} | x_1, t_1 \rangle \\ &= \int_{-\infty}^{\infty} dx_{N-1} \sqrt{\frac{m}{2\pi i \hbar \Delta t}} \exp\left[\left(\frac{im}{2\hbar} \right) \frac{(x_N - x_{N-1})^2}{\Delta t} - \frac{iV \Delta t}{\hbar} \right] \\ &\quad \times \langle x_{N-1}, t_{N-1} | x_1, t_1 \rangle, \end{aligned} \tag{2.5.50}$$

where we have assumed $t_N - t_{N-1}$ to be infinitesimal. Introducing

$$\xi = x_N - x_{N-1} \tag{2.5.51}$$

and letting $x_N \to x$ and $t_N \to t + \Delta t$, we obtain

$$\langle x, t+\Delta t | x_1, t_1 \rangle = \sqrt{\frac{m}{2\pi i \hbar \Delta t}} \int_{-\infty}^{\infty} d\xi \exp\left(\frac{im\xi^2}{2\hbar \Delta t} - \frac{iV\Delta t}{\hbar} \right) \langle x - \xi, t | x_1, t_1 \rangle . \tag{2.5.52}$$

As is evident from (2.5.45b), in the limit $\Delta t \to 0$, the major contribution to this integral comes from the $\xi \simeq 0$ region. It is therefore legitimate to expand $\langle x - \xi, t | x_1, t_1 \rangle$ in powers of ξ. We also expand $\langle x, t+\Delta t | x_1, t_1 \rangle$ and $\exp(-iV\Delta t/\hbar)$ in powers of Δt, so

$$\begin{aligned} \langle x, t | x_1, t_1 \rangle + \Delta t \frac{\partial}{\partial t} \langle x, t | x_1, t_1 \rangle & \\ = \sqrt{\frac{m}{2\pi i \hbar \Delta t}} \int_{-\infty}^{\infty} & d\xi \exp\left(\frac{im\xi^2}{2\hbar \Delta t} \right) \left(1 - \frac{iV\Delta t}{\hbar} + \cdots \right) \\ & \times \left[\langle x, t | x_1, t_1 \rangle + \left(\frac{\xi^2}{2} \right) \frac{\partial^2}{\partial x^2} \langle x, t | x_1, t_1 \rangle + \cdots \right], \end{aligned} \tag{2.5.53}$$

where we have dropped a term linear in ξ because it vanishes when integrated with respect to ξ. The $\langle x, t | x_1, t_1 \rangle$ term on the left-hand side just matches the leading term on the right-hand side because of (2.5.45a). Collecting terms first order in Δt, we obtain

$$\begin{aligned} \Delta t \frac{\partial}{\partial t} \langle x, t | x_1, t_1 \rangle = & \left(\sqrt{\frac{m}{2\pi i \hbar \Delta t}} \right) (\sqrt{2\pi}) \left(\frac{i\hbar \Delta t}{m} \right)^{3/2} \frac{1}{2} \frac{\partial^2}{\partial x^2} \langle x, t | x_1, t_1 \rangle \\ & - \left(\frac{i}{\hbar} \right) \Delta t \, V \langle x, t | x_1, t_1 \rangle , \end{aligned} \tag{2.5.54}$$

where we have used

$$\int_{-\infty}^{\infty} d\xi \, \xi^2 \exp\left(\frac{im\xi^2}{2\hbar \Delta t} \right) = \sqrt{2\pi} \left(\frac{i\hbar \Delta t}{m} \right)^{3/2}, \tag{2.5.55}$$

obtained by differentiating (2.5.45a) with respect to Δt. In this manner we see that $\langle x, t | x_1, t_1 \rangle$ satisfies Schrödinger's time-dependent wave equation:

$$i\hbar \frac{\partial}{\partial t} \langle x, t | x_1, t_1 \rangle = - \left(\frac{\hbar^2}{2m} \right) \frac{\partial^2}{\partial x^2} \langle x, t | x_1, t_1 \rangle + V \langle x, t | x_1, t_1 \rangle . \tag{2.5.56}$$

Thus we can conclude that $\langle x, t | x_1, t_1 \rangle$ constructed according to Feynman's prescription is the same as the propagator in Schrödinger's wave mechanics.

Feynman's space-time approach based on path integrals is not too convenient for attacking practical problems in nonrelativistic quantum mechanics. Even for the simple harmonic oscillator it is rather cumbersome to evaluate explicitly the relevant path integral.* However, his approach is extremely gratifying from a conceptual point of view. By imposing a certain set of sensible requirements on a physical theory, we are inevitably led to a formalism equivalent to the usual formulation of quantum mechanics. It makes us wonder whether it is at all possible to construct a sensible alternative theory that is equally successful in accounting for microscopic phenomena.

Methods based on path integrals have been found to be very powerful in other branches of modern physics, such as quantum field theory and statistical mechanics. In this book the path-integral method will appear again when we discuss the Aharonov-Bohm effect.†

2.6. POTENTIALS AND GAUGE TRANSFORMATIONS

Constant Potentials

In classical mechanics it is well known that the zero point of the potential energy is of no physical significance. The time development of dynamic variables such as $\mathbf{x}(t)$ and $\mathbf{L}(t)$ is independent of whether we use $V(\mathbf{x})$ or $V(\mathbf{x})+V_0$ with V_0 constant both in space and time. The force that appears in Newton's second law depends only on the gradient of the potential; an additive constant is clearly irrelevant. What is the analogous situation in quantum mechanics?

We look at the time evolution of a Schrödinger-picture state ket subject to some potential. Let $|\alpha, t_0; t\rangle$ be a state ket in the presence of $V(\mathbf{x})$, and $\widetilde{|\alpha, t_0; t\rangle}$, the corresponding state ket appropriate for

$$\tilde{V}(\mathbf{x}) = V(\mathbf{x}) + V_0. \tag{2.6.1}$$

To be precise let us agree that the initial conditions are such that both kets coincide with $|\alpha\rangle$ at $t = t_0$. If they represent the same physical situation, this can always be done by a suitable choice of the phase. Recalling that the state ket at t can be obtained by applying the time-evolution operator

*The reader is challenged to solve the simple harmonic oscillator problem using the Feynman path integral method in Problem 2-31.

†The reader who is interested in the fundamentals and applications of path integrals may consult Feynman and Hibbs 1965.

$\mathscr{U}(t, t_0)$ to the state ket at t_0, we obtain

$$\widetilde{|\alpha, t_0; t\rangle} = \exp\left[-i\left(\frac{\mathbf{p}^2}{2m} + V(x) + V_0\right)\frac{(t-t_0)}{\hbar}\right]|\alpha\rangle$$
$$= \exp\left[\frac{-iV_0(t-t_0)}{\hbar}\right]|\alpha, t_0; t\rangle. \tag{2.6.2}$$

In other words, the ket computed under the influence of $\tilde{V}$ has a time dependence different only by a phase factor $\exp[-iV_0(t-t_0)/\hbar]$. For stationary states, this means that if the time dependence computed with $V(\mathbf{x})$ is $\exp[-iE(t-t_0)/\hbar]$, then the corresponding time dependence computed with $V(\mathbf{x})+V_0$ is $\exp[-i(E+V_0)(t-t_0)/\hbar]$. In other words, the use of $\tilde{V}$ in place of V just amounts to the following change:

$$E \to E + V_0, \tag{2.6.3}$$

which the reader probably guessed immediately. Observable effects such as the time evolution of expectation values of $\langle \mathbf{x} \rangle$ and $\langle \mathbf{S} \rangle$ always depend on energy *differences* [see (2.1.47)]; the Bohr frequencies that characterize the sinusoidal time dependence of expectation values are the same whether we use $V(\mathbf{x})$ or $V(\mathbf{x})+V_0$. In general, there can be no difference in the expectation values of observables if every state ket in the world is multiplied by a common factor $\exp[-iV_0(t-t_0)/\hbar]$.

Trivial as it may seem, we see here the first example of a class of transformations known as **gauge transformations**. The change in our convention for the zero-point energy of the potential

$$V(\mathbf{x}) \to V(\mathbf{x}) + V_0 \tag{2.6.4}$$

must be accompanied by a change in the state ket

$$|\alpha, t_0; t\rangle \to \exp\left[\frac{-iV_0(t-t_0)}{\hbar}\right]|\alpha, t_0; t\rangle. \tag{2.6.5}$$

Of course, this change implies the following change in the wave function:

$$\psi(\mathbf{x}', t) \to \exp\left[\frac{-iV_0(t-t_0)}{\hbar}\right]\psi(\mathbf{x}', t). \tag{2.6.6}$$

Next we consider V_0 that is spatially uniform but dependent on time. We then easily see that the analogue of (2.6.5) is

$$|\alpha, t_0; t\rangle \to \exp\left[-i\int_{t_0}^{t} dt' \frac{V_0(t')}{\hbar}\right]|\alpha, t_0; t\rangle. \tag{2.6.7}$$

Physically, the use of $V(\mathbf{x})+V_0(t)$ in place of $V(\mathbf{x})$ simply means that we are choosing a new zero point of the energy scale at each instant of time.

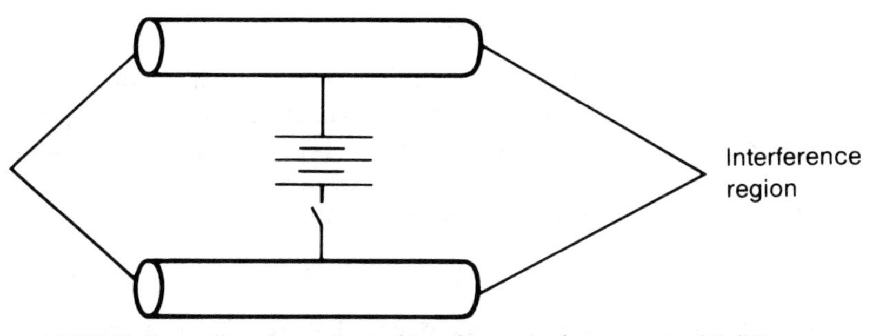

FIGURE 2.4. Quantum-mechanical interference to detect a potential difference.

Even though the choice of the absolute scale of the potential is arbitrary, *potential differences* are of nontrivial physical significance and, in fact, can be detected in a very striking way. To illustrate this point, let us consider the arrangement shown in Figure 2.4. A beam of charged particles is split into two parts, each of which enters a metallic cage. If we so desire, we can maintain a finite potential difference between the two cages by turning on a switch, as shown. A particle in the beam can be visualized as a wave packet whose dimension is much smaller than the dimension of the cage. Suppose we switch on the potential difference only after the wave packets enter the cages and switch it off before the wave packets leave the cages. The particle in the cage experiences *no force* because inside the cage the potential is spatially uniform; hence no electric field is present. Now let us recombine the two beam components in such a way that they meet in the interference region of Figure 2.4. Because of the existence of the potential, each beam component suffers a phase change, as indicated by (2.6.7). As a result, there is an observable interference term in the beam intensity in the interference region, namely,

$$\cos(\phi_1 - \phi_2), \qquad \sin(\phi_1 - \phi_2), \tag{2.6.8}$$

where

$$\phi_1 - \phi_2 = \left(\frac{1}{\hbar}\right)\int_{t_i}^{t_f} dt\,[V_2(t) - V_1(t)]. \tag{2.6.9}$$

So despite the fact that the particle experiences no force, there is an observable effect that depends on whether $V_2(t) - V_1(t)$ has been applied. Notice that this effect is *purely quantum mechanical*; in the limit $\hbar \to 0$, the interesting interference effect gets washed out because the oscillation of the cosine becomes infinitely rapid.*

*This gedanken experiment is the Minkowski-rotated form of the Aharonov-Bohm experiment to be discussed later in this section.

Gravity in Quantum Mechanics

There is an experiment that exhibits in a striking manner how a gravitational effect appears in quantum mechanics. Before describing it, we first comment on the role of gravity in both classical and quantum mechanics.

Consider the classical equation of motion for a purely falling body:

$$m\ddot{\mathbf{x}} = -m\nabla\Phi_{\text{grav}} = -mg\hat{\mathbf{z}}. \tag{2.6.10}$$

The mass term drops out; so in the absence of air resistance, a feather and a stone would behave in the same way—à la Galileo—under the influence of gravity. This is, of course, a direct consequence of the equality of the gravitational and the inertial masses. Because the mass does not appear in the equation of a particle trajectory, gravity in classical mechanics is often said to be a purely geometric theory.

The situation is rather different in quantum mechanics. In the wave-mechanical formulation, the analogue of (2.6.10) is

$$\left[-\left(\frac{\hbar^2}{2m}\right)\nabla^2 + m\Phi_{\text{grav}}\right]\psi = i\hbar\frac{\partial\psi}{\partial t}. \tag{2.6.11}$$

The mass no longer cancels; instead it appears in the combination $\hbar/m$, so in a problem where $\hbar$ appears, m is also expected to appear. We can see this point also using the Feynman path-integral formulation of a falling body based on

$$\langle\mathbf{x}_n, t_n|\mathbf{x}_{n-1}, t_{n-1}\rangle = \sqrt{\frac{m}{2\pi i\hbar\Delta t}}\exp\left[i\int_{t_{n-1}}^{t_n} dt\,\frac{\left(\frac{1}{2}m\dot{\mathbf{x}}^2 - mgz\right)}{\hbar}\right],$$

$$(t_n - t_{n-1} = \Delta t \to 0). \tag{2.6.12}$$

Here again we see that m appears in the combination $m/\hbar$. This is in sharp contrast with Hamilton's classical approach, based on

$$\delta\int_{t_1}^{t_2} dt\left(\frac{m\dot{\mathbf{x}}^2}{2} - mgz\right) = 0, \tag{2.6.13}$$

where m can be eliminated in the very beginning.

Starting with the Schrödinger equation (2.6.11), we may derive the Ehrenfest theorem

$$\frac{d^2}{dt^2}\langle\mathbf{x}\rangle = -g\hat{\mathbf{z}}. \tag{2.6.14}$$

However, $\hbar$ does not appear here, nor does m. To see a *nontrivial* quantum-mechanical effect of gravity, we must study effects in which $\hbar$ appears explicitly—and consequently where we expect the mass to appear—in contrast with purely gravitational phenomena in classical mechanics.

Until 1975, there had been no direct experiment that established the presence of the $m\Phi_{\text{grav}}$ term in (2.6.11). To be sure, a free fall of an elementary particle had been observed, but the classical equation of motion —or the Ehrenfest theorem (2.6.14), where $\hbar$ does not appear—sufficed to account for this. The famous "weight of photon" experiment of V. Pound and collaborators did not test gravity in the quantum domain either because they measured a frequency shift where $\hbar$ does not explicitly appear.

On the microscopic scale, gravitational forces are too weak to be readily observable. To appreciate the difficulty involved in seeing gravity in bound-state problems, let us consider the ground state of an electron and a neutron bound by gravitational forces. This is the gravitational analogue of the hydrogen atom, where an electron and a proton are bound by Coulomb forces. At the same distance, the gravitational force between the electron and the neutron is weaker than the Coulomb force between the electron and the proton by a factor of $\sim 2\times 10^{39}$. The Bohr radius involved here can be obtained simply:

$$a_0 = \frac{\hbar^2}{e^2 m_e} \to \frac{\hbar^2}{G_N m_e^2 m_n}, \tag{2.6.15}$$

where G_N is Newton's gravitational constant. If we substitute numbers in the equation, the Bohr radius of this gravitationally bound system turns out to be $\sim 10^{31}$ cm, or $\sim 10^{13}$ light years, which is larger than the estimated radius of the universe by a few orders of magnitude!

We now discuss a remarkable phenomenon known as **gravity-induced quantum interference**. A nearly monoenergetic beam of particles—in practice, thermal neutrons—is split into two parts and then brought together as shown in Figure 2.5. In actual experiments the neutron beam is split and bent by silicon crystals, but the details of this beautiful art of neutron interferometry do not concern us here. Because the size of the wave packet can be assumed to be much smaller than the macroscopic dimension of the loop formed by the two alternate paths, we can apply the concept of a classical trajectory. Let us first suppose that path $A \to B \to D$ and path $A \to C \to D$ lie in a horizontal plane. Because the absolute zero of the potential due to gravity is of no significance, we can set $V = 0$ for any phenomenon that takes place in this plane; in other words, it is legitimate to ignore gravity altogether. The situation is very different if the plane formed by the two alternate paths is rotated around segment AC by angle δ. This time the potential at level BD is higher than that at level AC by $mgl_2 \sin\delta$, which means that the state ket associated with path BD "rotates faster." This leads to a gravity-induced phase difference between the amplitudes for the two wave packets arriving at D. Actually there is also a gravity-induced phase change associated with AB and also with CD, but the effects cancel as we compare the two alternate paths. The net result is that the wave packet

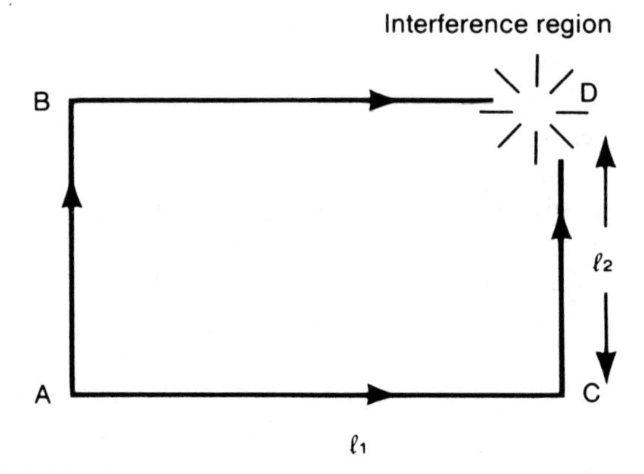

FIGURE 2.5. Experiment to detect gravity-induced quantum interference.

arriving at D via path ABD suffers a phase change

$$\exp\left[\frac{-im_n g l_2 \sin\delta\, T}{\hbar}\right] \tag{2.6.16}$$

relative to that of the wave packet arriving at D via path ACD, where T is the time spent for the wave packet to go from B to D (or from A to C) and m_n, the neutron mass. We can control this phase difference by rotating the plane of Figure 2.5; δ can change from 0 to $\pi/2$, or from 0 to $-\pi/2$. Expressing the time spent T, or $l_1/v_{\text{wave packet}}$, in terms of $\lambda\!\!\!^-$, the de Broglie wavelength of the neutron, we obtain the following expression for the phase difference:

$$\phi_{ABD} - \phi_{ACD} = -\frac{\left(m_n^2 g l_1 l_2 \lambda\!\!\!^- \sin\delta\right)}{\hbar^2}. \tag{2.6.17}$$

In this manner we predict an observable interference effect that depends on angle δ, which is reminiscent of fringes in Michelson-type interferometers in optics.

An alternative, more wave-mechanical way to understand (2.6.17) follows. Because we are concerned with a time-independent potential, the sum of the kinetic energy and the potential energy is constant:

$$\frac{\mathbf{p}^2}{2m} + mgz = E. \tag{2.6.18}$$

The difference in height between level BD and level AC implies a slight difference in $\mathbf{p}$, or $\lambda\!\!\!^-$. As a result, there is an accumulation of phase differences due to the $\lambda\!\!\!^-$ difference. It is left as an exercise to show that this wave-mechanical approach also leads to result (2.6.17).

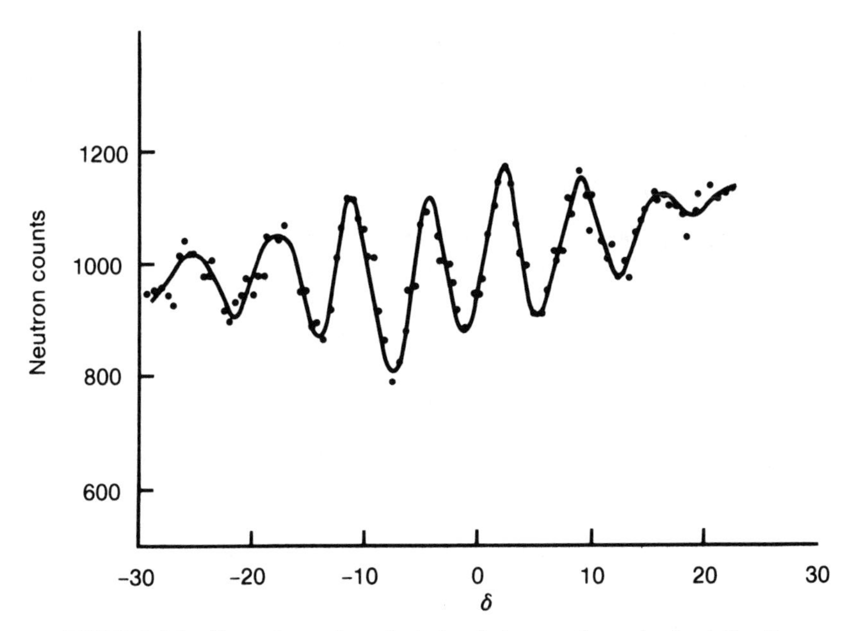

FIGURE 2.6. Dependence of gravity-induced phase on the angle of rotation δ.

What is interesting about expression (2.6.17) is that its magnitude is neither too small nor too large; it is just right for this interesting effect to be detected with thermal neutrons traveling through paths of "table-top" dimensions. For $\lambda = 1.42$ Å (comparable to interatomic spacing in silicon) and $l_1 l_2 = 10$ cm^2, we obtain 55.6 for $m_n^2 g l_1 l_2 \lambda\!\!\!^{-} / \hbar^2$. As we rotate the loop plane gradually by 90°, we predict the intensity in the interference region to exhibit a series of maxima and minima; quantitatively we should see $55.6/2\pi \simeq 9$ oscillations. It is extraordinary that such an effect has indeed been observed experimentally; see Figure 2.6 taken from a 1975 experiment of R. Colella, A. Overhauser, and S. A. Werner. The phase shift due to gravity is seen to be verified to well within 1%.

We emphasize that this effect is purely quantum mechanical because as $\hbar \to 0$, the interference pattern gets washed out. The gravitational potential has been shown to enter into the Schrödinger equation just as expected. This experiment also shows that gravity is not purely geometric at the quantum level because the effect depends on $(m/\hbar)^2$.*

*However, this does not imply that the equivalence principle is unimportant in understanding an effect of this sort. If the gravitational mass (m_{grav}) and inertial mass (m_{inert}) were unequal, $(m/\hbar)^2$ would have to be replaced by $m_{\text{grav}} m_{\text{inert}}/\hbar^2$. The fact that we could correctly predict the interference pattern without making a distinction between m_{grav} and m_{inert} shows some support for the equivalence principle at the quantum level.

Gauge Transformations in Electromagnetism

Let us now turn to potentials that appear in electromagnetism. We consider an electric and a magnetic field derivable from the time-independent scalar and vector potential, $\phi(\mathbf{x})$ and $\mathbf{A}(\mathbf{x})$:

$$\mathbf{E} = -\nabla\phi, \qquad \mathbf{B} = \nabla \times \mathbf{A}. \tag{2.6.19}$$

The Hamiltonian for a particle of electric charge e ($e < 0$ for the electron) subjected to the electromagnetic field is taken from classical physics to be

$$H = \frac{1}{2m}\left(\mathbf{p} - \frac{e\mathbf{A}}{c}\right)^2 + e\phi. \tag{2.6.20}$$

In quantum mechanics ϕ and $\mathbf{A}$ are understood to be functions of the position *operator* $\mathbf{x}$ of the charged particle. Because $\mathbf{p}$ and $\mathbf{A}$ do not commute, some care is needed in interpreting (2.6.20). The safest procedure is to write

$$\left(\mathbf{p} - \frac{e\mathbf{A}}{c}\right)^2 \rightarrow p^2 - \left(\frac{e}{c}\right)(\mathbf{p}\cdot\mathbf{A} + \mathbf{A}\cdot\mathbf{p}) + \left(\frac{e}{c}\right)^2 \mathbf{A}^2. \tag{2.6.21}$$

In this form the Hamiltonian is obviously Hermitian.

To study the dynamics of a charged particle subjected to ϕ and $\mathbf{A}$, let us first proceed in the Heisenberg picture. We can evaluate the time derivative of $\mathbf{x}$ in a straightforward manner as

$$\frac{dx_i}{dt} = \frac{[x_i, H]}{i\hbar} = \frac{(p_i - eA_i/c)}{m}, \tag{2.6.22}$$

which shows that the operator $\mathbf{p}$, defined in this book to be the generator of translation, is not the same as $m\,d\mathbf{x}/dt$. Quite often $\mathbf{p}$ is called **canonical momentum**, as distinguished from **kinematical** (or mechanical) **momentum**, denoted by Π:

$$\Pi \equiv m\frac{d\mathbf{x}}{dt} = \mathbf{p} - \frac{e\mathbf{A}}{c}. \tag{2.6.23}$$

Even though we have

$$[p_i, p_j] = 0 \tag{2.6.24}$$

for canonical momentum, the analogous commutator does not vanish for mechanical momentum. Instead we have

$$[\Pi_i, \Pi_j] = \left(\frac{i\hbar e}{c}\right)\varepsilon_{ijk}B_k, \tag{2.6.25}$$

as the reader may easily verify. Rewriting the Hamiltonian as

$$H = \frac{\Pi^2}{2m} + e\phi \tag{2.6.26}$$

and using the fundamental commutation relation, we can derive the quantum-mechanical version of the **Lorentz force**, namely,

$$m\frac{d^2\mathbf{x}}{dt^2} = \frac{d\mathbf{\Pi}}{dt} = e\left[\mathbf{E} + \frac{1}{2c}\left(\frac{d\mathbf{x}}{dt}\times\mathbf{B} - \mathbf{B}\times\frac{d\mathbf{x}}{dt}\right)\right]. \tag{2.6.27}$$

This then is Ehrenfest's theorem, written in the Heisenberg picture, for the charged particle in the presence of **E** and **B**.

We now study Schrödinger's wave equation with ϕ and **A**. Our first task is to sandwich H between $\langle\mathbf{x}'|$ and $|\alpha, t_0; t\rangle$. The only term with which we have to be careful is

$$\begin{aligned}
&\langle\mathbf{x}'|\left[\mathbf{p} - \frac{e\mathbf{A}(\mathbf{x})}{c}\right]^2|\alpha, t_0; t\rangle \\
&\qquad = \left[-i\hbar\nabla' - \frac{e\mathbf{A}(\mathbf{x}')}{c}\right]\langle\mathbf{x}'|\left[\mathbf{p} - \frac{e\mathbf{A}(\mathbf{x})}{c}\right]|\alpha, t_0; t\rangle \\
&\qquad = \left[-i\hbar\nabla' - \frac{e\mathbf{A}(\mathbf{x}')}{c}\right]\cdot\left[-i\hbar\nabla' - \frac{e\mathbf{A}(\mathbf{x}')}{c}\right]\langle\mathbf{x}'|\alpha, t_0; t\rangle.
\end{aligned} \tag{2.6.28}$$

It is important to emphasize that the first ∇' in the last line can differentiate *both* $\langle\mathbf{x}'|\alpha, t_0; t\rangle$ *and* $\mathbf{A}(\mathbf{x}')$. Combining everything, we have

$$\begin{aligned}
&\frac{1}{2m}\left[-i\hbar\nabla' - \frac{e\mathbf{A}(\mathbf{x}')}{c}\right]\cdot\left[-i\hbar\nabla' - \frac{e\mathbf{A}(\mathbf{x}')}{c}\right]\langle\mathbf{x}'|\alpha, t_0; t\rangle \\
&\qquad + e\phi(\mathbf{x}')\langle\mathbf{x}'|\alpha, t_0; t\rangle = i\hbar\frac{\partial}{\partial t}\langle\mathbf{x}'|\alpha, t_0; t\rangle.
\end{aligned} \tag{2.6.29}$$

From this expression we readily obtain the continuity equation

$$\frac{\partial\rho}{\partial t} + \nabla'\cdot\mathbf{j} = 0, \tag{2.6.30}$$

where ρ is $|\psi|^2$ as before, with $\langle\mathbf{x}'|\alpha, t_0; t\rangle$ written as ψ, but for the probability flux **j** we have

$$\mathbf{j} = \left(\frac{\hbar}{m}\right)\mathrm{Im}(\psi^*\nabla'\psi) - \left(\frac{e}{mc}\right)\mathbf{A}|\psi|^2, \tag{2.6.31}$$

which is just what we expect from the substitution

$$\nabla' \to \nabla' - \left(\frac{ie}{\hbar c}\right)\mathbf{A}. \tag{2.6.32}$$

Writing the wave function of $\sqrt{\rho}\,\exp(iS/\hbar)$ [see (2.4.18)], we obtain an alternative form for **j**, namely,

$$\mathbf{j} = \left(\frac{\rho}{m}\right)\left(\nabla S - \frac{e\mathbf{A}}{c}\right), \tag{2.6.33}$$

which is to be compared with (2.4.20). We will find this form to be convenient in discussing superconductivity, flux quantization, and so on. We also note that the space integral of $\mathbf{j}$ is the expectation value of kinematical momentum (not canonical momentum) apart from $1/m$:

$$\int d^3x'\mathbf{j} = \frac{\langle \mathbf{p} - e\mathbf{A}/c \rangle}{m} = \langle \mathbf{\Pi} \rangle / m. \tag{2.6.34}$$

We are now in a position to discuss the subject of **gauge transformations** in electromagnetism. First, consider

$$\phi \to \phi + \lambda, \qquad \mathbf{A} \to \mathbf{A}, \tag{2.6.35}$$

with λ constant, that is, independent of $\mathbf{x}$ and t. Both $\mathbf{E}$ and $\mathbf{B}$ obviously remain unchanged. This transformation just amounts to a change in the zero point of the energy scale, a possibility treated in the beginning of this section; we just replace V by $e\phi$. We have already discussed the accompanying change needed for the state ket [see (2.6.5)], so we do not dwell on this transformation any further.

Much more interesting is the transformation

$$\phi \to \phi, \qquad \mathbf{A} \to \mathbf{A} + \nabla\Lambda, \tag{2.6.36}$$

where Λ is a function of $\mathbf{x}$. The static electromagnetic fields $\mathbf{E}$ and $\mathbf{B}$ are unchanged under (2.6.36). Both (2.6.35) and (2.6.36) are special cases of

$$\phi \to \phi - \frac{1}{c}\frac{\partial\Lambda}{\partial t}, \qquad \mathbf{A} \to \mathbf{A} + \nabla\Lambda, \tag{2.6.37}$$

which leave $\mathbf{E}$ and $\mathbf{B}$, given by

$$\mathbf{E} = -\nabla\phi - \frac{1}{c}\frac{\partial \mathbf{A}}{\partial t}, \qquad \mathbf{B} = \nabla \times \mathbf{A}, \tag{2.6.38}$$

unchanged, but in the following we do not consider time-dependent fields and potentials. In the remaining part of this section the term *gauge transformation* refers to (2.6.36).

In classical physics observable effects such as the trajectory of a charged particle are independent of the gauge used, that is, of the particular choice of Λ we happen to adopt. Consider a charged particle in a uniform magnetic field in the z-direction

$$\mathbf{B} = B\hat{\mathbf{z}}. \tag{2.6.39}$$

This magnetic field may be derived from

$$A_x = \frac{-By}{2}, \qquad A_y = \frac{Bx}{2}, \qquad A_z = 0 \tag{2.6.40}$$

or also from

$$A_x = -By, \qquad A_y = 0, \qquad A_z = 0. \tag{2.6.41}$$

The second form is obtained from the first by

$$\mathbf{A} \to \mathbf{A} - \nabla\left(\frac{Bxy}{2}\right), \tag{2.6.42}$$

which is indeed of the form of (2.6.36). Regardless of which **A** we may use, the trajectory of the charged particle with a given set of initial conditions is the same; it is just a helix—a uniform circular motion when projected in the xy-plane, superposed with a uniform rectilinear motion in the z-direction. Yet if we look at p_x and p_y, the results are very different. For one thing, p_x is a constant of the motion when (2.6.41) is used but not when (2.6.40) is used.

Recall Hamilton's equations of motion:

$$\frac{dp_x}{dt} = -\frac{\partial H}{\partial x}, \qquad \frac{dp_y}{dt} = -\frac{\partial H}{\partial y}, \ldots. \tag{2.6.43}$$

In general, the canonical momentum **p** is *not* a gauge-invariant quantity; its numerical value depends on the particular gauge used even when we are referring to the same physical situation. In contrast, the *kinematic* momentum $\mathbf{\Pi}$, or $m d\mathbf{x}/dt$, that traces the trajectory of the particle *is* a gauge-invariant quantity, as one may explicitly verify. Because **p** and $m d\mathbf{x}/dt$ are related via (2.6.23), **p** must change to compensate for the change in **A** given by (2.6.42).

We now return to quantum mechanics. We believe that it is reasonable to demand that the expectation values in quantum mechanics behave in a manner similar to the corresponding classical quantities under gauge transformations, so $\langle \mathbf{x} \rangle$ and $\langle \mathbf{\Pi} \rangle$ are *not* to change under gauge transformations, while $\langle \mathbf{p} \rangle$ is expected to change.

Let us denote by $|\alpha\rangle$ the state ket in the presence of **A**; the state ket for the same physical situation when

$$\tilde{\mathbf{A}} = \mathbf{A} + \nabla\Lambda \tag{2.6.44}$$

is used in place of **A** is denoted by $|\tilde{\alpha}\rangle$. Here Λ, as well as **A**, is a function of the position operator **x**. Our basic requirements are

$$\langle \alpha|\mathbf{x}|\alpha\rangle = \langle \tilde{\alpha}|\mathbf{x}|\tilde{\alpha}\rangle \tag{2.6.45a}$$

and

$$\langle \alpha|\left(\mathbf{p} - \frac{e\mathbf{A}}{c}\right)|\alpha\rangle = \langle \tilde{\alpha}|\left(\mathbf{p} - \frac{e\tilde{\mathbf{A}}}{c}\right)|\tilde{\alpha}\rangle. \tag{2.6.45b}$$

In addition, we require, as usual, the norm of the state ket to be preserved:

$$\langle \alpha|\alpha\rangle = \langle \tilde{\alpha}|\tilde{\alpha}\rangle. \tag{2.6.46}$$

We must construct an operator $\mathscr{G}$ that relates $|\tilde{\alpha}\rangle$ to $|\alpha\rangle$:

$$|\tilde{\alpha}\rangle = \mathscr{G}|\alpha\rangle. \tag{2.6.47}$$

Invariance properties (2.6.45a) and (2.6.45b) are guaranteed if

$$\mathscr{G}^{\dagger}\mathbf{x}\mathscr{G} = \mathbf{x} \tag{2.6.48a}$$

and

$$\mathscr{G}^{\dagger}\left(\mathbf{p} - \frac{e\mathbf{A}}{c} - \frac{e\nabla\Lambda}{c}\right)\mathscr{G} = \mathbf{p} - \frac{e\mathbf{A}}{c}. \tag{2.6.48b}$$

We assert that

$$\mathscr{G} = \exp\left[\frac{ie\Lambda(\mathbf{x})}{\hbar c}\right] \tag{2.6.49}$$

will do the job. First, $\mathscr{G}$ is unitary, so (2.6.46) is all right. Second, (2.6.48a) is obviously satisfied because $\mathbf{x}$ commutes with any function of $\mathbf{x}$. As for (2.6.48b), just note that

$$\begin{aligned}\exp\left(\frac{-ie\Lambda}{\hbar c}\right)\mathbf{p}\exp\left(\frac{ie\Lambda}{\hbar c}\right) &= \exp\left(\frac{-ie\Lambda}{\hbar c}\right)\left[\mathbf{p}, \exp\left(\frac{ie\Lambda}{\hbar c}\right)\right] + \mathbf{p}\\ &= -\exp\left(\frac{-ie\Lambda}{\hbar c}\right)i\hbar\nabla\left[\exp\left(\frac{ie\Lambda}{\hbar c}\right)\right] + \mathbf{p}\\ &= \mathbf{p} + \frac{e\nabla\Lambda}{c},\end{aligned} \tag{2.6.50}$$

where we have used (2.2.23b).

The invariance of quantum mechanics under gauge transformations can also be demonstrated by looking directly at the Schrödinger equation. Let $|\alpha, t_0; t\rangle$ be a solution to the Schrödinger equation in the presence of $\mathbf{A}$:

$$\left[\frac{(\mathbf{p} - e\mathbf{A}/c)^2}{2m} + e\phi\right]|\alpha, t_0; t\rangle = i\hbar\frac{\partial}{\partial t}|\alpha, t_0; t\rangle. \tag{2.6.51}$$

The corresponding solution in the presence of $\tilde{\mathbf{A}}$ must satisfy

$$\left[\frac{(\mathbf{p} - e\mathbf{A}/c - e\nabla\Lambda/c)^2}{2m} + e\phi\right]\widetilde{|\alpha, t_0; t\rangle} = i\hbar\frac{\partial}{\partial t}\widetilde{|\alpha, t_0; t\rangle}. \tag{2.6.52}$$

We see that if the new ket is taken to be

$$\widetilde{|\alpha, t_0; t\rangle} = \exp\left(\frac{ie\Lambda}{\hbar c}\right)|\alpha, t_0; t\rangle \tag{2.6.53}$$

in accordance with (2.6.49), then the new Schrödinger equation (2.6.52) will be satisfied; all we have to note is that

$$\exp\left(\frac{-ie\Lambda}{\hbar c}\right)\left(\mathbf{p} - \frac{e\mathbf{A}}{c} - \frac{e\nabla\Lambda}{c}\right)^2\exp\left(\frac{ie\Lambda}{\hbar c}\right) = \left(\mathbf{p} - \frac{e\mathbf{A}}{c}\right)^2, \tag{2.6.54}$$

which follows from applying (2.6.50) twice.

Equation (2.6.53) also implies that the corresponding wave equations are related via

$$\tilde{\psi}(\mathbf{x}', t) = \exp\left[\frac{ie\Lambda(\mathbf{x}')}{\hbar c}\right]\psi(\mathbf{x}', t), \tag{2.6.55}$$

where $\Lambda(\mathbf{x}')$ is now a real function of the position vector eigenvalue $\mathbf{x}'$. This can, of course, be verified also by directly substituting (2.6.55) into Schrödinger's wave equation with $\mathbf{A}$ replaced by $\mathbf{A} + \nabla\Lambda$. In terms of ρ and S, we see that ρ is unchanged but S is modified as follows:

$$S \to S + \frac{e\Lambda}{c}. \tag{2.6.56}$$

This is highly satisfactory because we see that the probability flux given by (2.6.33) is then gauge invariant.

To summarize, when vector potentials in different gauges are used for the same physical situation, the corresponding state kets (or wave functions) must necessarily be different. However, only a simple change is needed; we can go from a gauge specified by $\mathbf{A}$ to another specified by $\mathbf{A} + \nabla\Lambda$ by merely multiplying the old ket (the old wave function) by $\exp[ie\Lambda(\mathbf{x})/\hbar c]$ ($\exp[ie\Lambda(\mathbf{x}')/\hbar c]$). The **canonical momentum**, defined as the generator of translation, is manifestly *gauge dependent* in the sense that its expectation value depends on the particular gauge chosen, while the **kinematic momentum** and the probability flux are *gauge invariant*.

The reader may wonder why invariance under (2.6.49) is called *gauge invariance*. This word is the translation of the German *Eichinvarianz*, where *Eich* means *gauge*. There is a historical anecdote that goes with the origin of this term.

Consider some function of position at $\mathbf{x}$: $F(\mathbf{x})$. At a neighboring point we obviously have

$$F(\mathbf{x} + d\mathbf{x}) \simeq F(\mathbf{x}) + (\nabla F)\cdot d\mathbf{x}. \tag{2.6.57}$$

But suppose we apply a scale change as we go from $\mathbf{x}$ to $\mathbf{x} + d\mathbf{x}$ as follows:

$$1\big|_{\text{at }\mathbf{x}} \to [1 + \boldsymbol{\Sigma}(\mathbf{x})\cdot d\mathbf{x}]\big|_{\text{at }\mathbf{x}+d\mathbf{x}}. \tag{2.6.58}$$

We must then rescale $F(\mathbf{x})$ as follows:

$$F(\mathbf{x} + d\mathbf{x})\big|_{\text{rescaled}} \simeq F(\mathbf{x}) + [(\nabla + \boldsymbol{\Sigma})F]\cdot d\mathbf{x} \tag{2.6.59}$$

instead of (2.6.57). The combination $\nabla + \boldsymbol{\Sigma}$ is similar to the gauge-invariant combination

$$\nabla - \left(\frac{ie}{\hbar c}\right)\mathbf{A} \tag{2.6.60}$$

encountered in (2.6.32) except for the absence of i. Historically, H. Weyl unsuccessfully attempted to construct a geometric theory of electromagnetism based on *Eichinvarianz* by identifying the scale function $\boldsymbol{\Sigma}(\mathbf{x})$ in

(2.6.58) and (2.6.59) with the vector potential **A** itself. With the birth of quantum mechanics, V. Fock and F. London realized the importance of the gauge-invariant combination (2.6.60), and they recalled Weyl's earlier work by comparing Σ with i times Λ. We are stuck with the term *gauge invariance* even though the quantum-mechanical analogue of (2.6.58)

$$1\Big|_{\text{at } \mathbf{x}} \to \left[1 - \left(\frac{ie}{\hbar c}\right)\mathbf{A}\cdot d\mathbf{x}\right]\Big|_{\text{at } \mathbf{x}+d\mathbf{x}} \tag{2.6.61}$$

would actually correspond to "phase change" rather than to "scale change."

The Aharonov-Bohm Effect

The use of vector potential in quantum mechanics has many far-reaching consequences, some of which we are now ready to discuss. We start with a relatively innocuous-looking problem.

Consider a hollow cylindrical shell, as shown in Figure 2.7a. We assume that a particle of charge e can be completely confined to the interior of the shell with rigid walls. The wave function is required to vanish on the inner ($\rho = \rho_a$) and outer ($\rho = \rho_b$) walls as well as at the top and the bottom. It is a straightforward boundary-value problem in mathematical physics to obtain the energy eigenvalues.

Let us now consider a modified arrangement where the cylindrical shell encloses a uniform magnetic field, as shown in Figure 2.7b. Specifi-

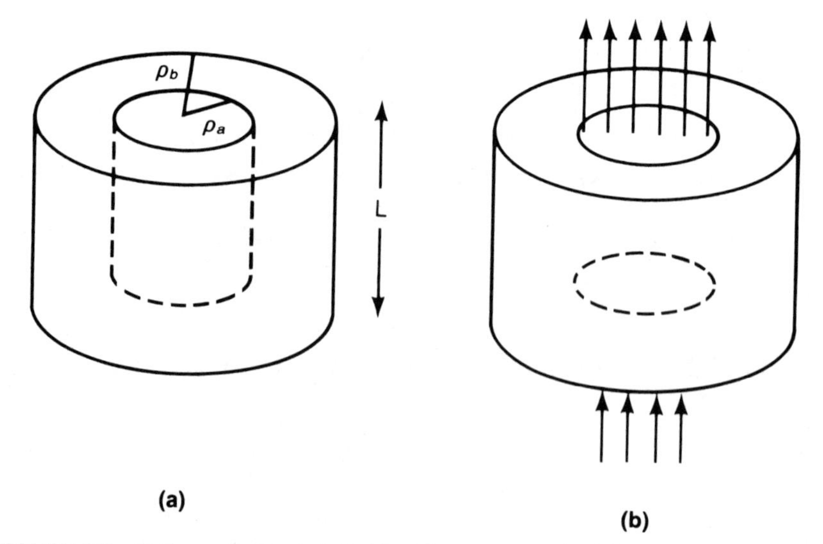

FIGURE 2.7. Hollow cylindrical shell (a) without a magnetic field, (b) with a uniform magnetic field.

cally, you may imagine fitting a very long solenoid into the hole in the middle in such a way that no magnetic field leaks into the region $\rho \geq \rho_a$. The boundary conditions for the wave function are taken to be the same as before; the walls are assumed to be just as rigid. Intuitively we may conjecture that the energy spectrum is unchanged because the region with $\mathbf{B} \neq 0$ is completely inaccessible to the charged particle trapped inside the shell. However, quantum mechanics tells us that this conjecture is *not* correct.

Even though the magnetic field vanishes in the interior, the vector potential $\mathbf{A}$ is nonvanishing there; using Stokes's theorem we can infer that the vector potential needed to produce the magnetic field $\mathbf{B}$ ($= B\hat{\mathbf{z}}$) is

$$\mathbf{A} = \left(\frac{B\rho_a^2}{2\rho} \right) \hat{\boldsymbol{\phi}}, \tag{2.6.62}$$

where $\hat{\boldsymbol{\phi}}$ is the unit vector in the direction of increasing azimuthal angle. In attempting to solve the Schrödinger equation to find the energy eigenvalues for this new problem, we need only to replace the gradient ∇ by $\nabla - (ie/\hbar c)\mathbf{A}$; we can accomplish this in cylindrical coordinates by replacing the partial derivative with respect to ϕ as follows:

$$\frac{\partial}{\partial \phi} \rightarrow \frac{\partial}{\partial \phi} - \left(\frac{ie}{\hbar c} \right) \frac{B\rho_a^2}{2}; \tag{2.6.63}$$

recall the expression for gradient in cylindrical coordinates:

$$\nabla = \hat{\boldsymbol{\rho}} \frac{\partial}{\partial \rho} + \hat{\mathbf{z}} \frac{\partial}{\partial z} + \hat{\boldsymbol{\phi}} \frac{1}{\rho} \frac{\partial}{\partial \phi}. \tag{2.6.64}$$

The replacement (2.6.63) results in an *observable* change in the energy spectrum, as the reader may verify explicitly. This is quite remarkable because the particle never "touches" the magnetic field; the Lorentz force the particle experiences is identically zero in this problem, yet the energy levels depend on whether or not the magnetic field is finite in the hole region inaccessible to the particle.

The problem we have just treated is the bound-state version of what is commonly referred to as the *Aharonov-Bohm effect*.* We are now in a position to discuss the original form of the Aharonov-Bohm effect itself. Consider a particle of charge e going above or below a very long impenetrable cylinder, as shown in Figure 2.8. Inside the cylinder is a magnetic field parallel to the cylinder axis, taken to be normal to the plane of Figure 2.8. So the particle paths above and below enclose a magnetic flux. Our object is to study how the probability of finding the particle in the interference region B depends on the magnetic flux.

*After a 1959 paper by Y. Aharonov and D. Bohm. Essentially the same effect was discussed 10 years earlier by W. Ehrenberg and R. E. Siday.

Even though this problem can be attacked by comparing the solutions to the Schrödinger equation in the presence and absence of **B**, for pedagogical reasons we prefer to use the Feynman path-integral method. Let $\mathbf{x}_1$ and $\mathbf{x}_N$ be typical points in source region A and interference region B, respectively. We recall from classical mechanics that the Lagrangian in the presence of the magnetic field can be obtained from that in the absence of the magnetic field, denoted by $L^{(0)}_{\text{classical}}$, as follows:

$$L^{(0)}_{\text{classical}} = \frac{m}{2}\left(\frac{d\mathbf{x}}{dt}\right)^2 \rightarrow L^{(0)}_{\text{classical}} + \frac{e}{c}\frac{d\mathbf{x}}{dt}\cdot\mathbf{A}. \tag{2.6.65}$$

The corresponding change in the action for some definite path segment going from $(\mathbf{x}_{n-1}, t_{n-1})$ to $(\mathbf{x}_n, t_n)$ is then given by

$$S^{(0)}(n, n-1) \rightarrow S^{(0)}(n, n-1) + \frac{e}{c}\int_{t_{n-1}}^{t_n} dt\left(\frac{d\mathbf{x}}{dt}\right)\cdot\mathbf{A}. \tag{2.6.66}$$

But this last integral can be written as

$$\frac{e}{c}\int_{t_{n-1}}^{t_n} dt\left(\frac{d\mathbf{x}}{dt}\right)\cdot\mathbf{A} = \frac{e}{c}\int_{\mathbf{x}_{n-1}}^{\mathbf{x}_n} \mathbf{A}\cdot d\mathbf{s}, \tag{2.6.67}$$

where $d\mathbf{s}$ is the differential line element along the path segment, so when we consider the entire contribution from $\mathbf{x}_1$ to $\mathbf{x}_N$, we have the following change:

$$\prod \exp\left[\frac{iS^{(0)}(n, n-1)}{\hbar}\right] \rightarrow \left\{\prod \exp\left[\frac{iS^{(0)}(n, n-1)}{\hbar}\right]\right\} \exp\left(\frac{ie}{\hbar c}\int_{\mathbf{x}_1}^{\mathbf{x}_N} \mathbf{A}\cdot d\mathbf{s}\right). \tag{2.6.68}$$

All this is for a particular path, such as going above the cylinder. We must still sum over all possible paths, which may appear to be a formidable task. Fortunately, we know from the theory of electromagnetism that the line integral $\int \mathbf{A}\cdot d\mathbf{s}$ is independent of paths, that is, it is dependent only on the end points, as long as the loop formed by a pair of different paths does not enclose a magnetic flux. As a result, the contributions due to $\mathbf{A} \neq 0$ to all paths going above the cylinder are given by a *common* phase factor;

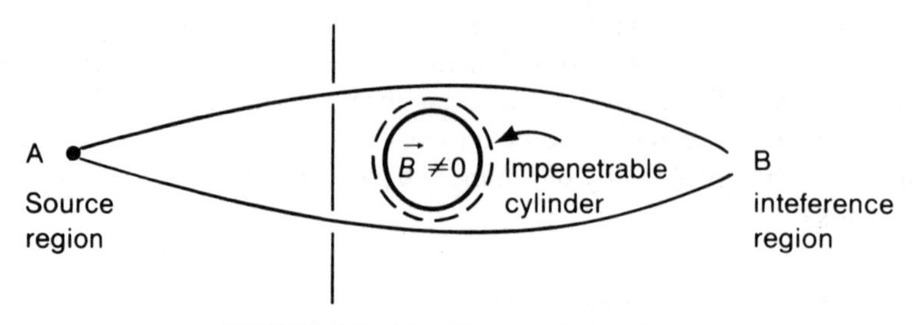

FIGURE 2.8. The Aharonov-Bohm effect.

similarly, the contributions from all paths going below the cylinder are multiplied by another common phase factor. In the path-integral notation we have, for the entire transition amplitude,

$$\int_{\text{above}} \mathscr{D}[\mathbf{x}(t)] \exp\left[\frac{iS^{(0)}(N,1)}{\hbar}\right] + \int_{\text{below}} \mathscr{D}[\mathbf{x}(t)] \exp\left[\frac{iS^{(0)}(N,1)}{\hbar}\right]$$

$$\to \int_{\text{above}} \mathscr{D}[\mathbf{x}(t)] \exp\left[\frac{iS^{(0)}(N,1)}{\hbar}\right] \left\{ \exp\left[\left(\frac{ie}{\hbar c}\right) \int_{\mathbf{x}_1}^{\mathbf{x}_N} \mathbf{A} \cdot d\mathbf{s}\right]_{\text{above}} \right\}$$

$$+ \int_{\text{below}} \mathscr{D}[\mathbf{x}(t)] \exp\left[\frac{iS^{(0)}(N,1)}{\hbar}\right] \left\{ \exp\left[\left(\frac{ie}{\hbar c}\right) \int_{\mathbf{x}_1}^{\mathbf{x}_N} \mathbf{A} \cdot d\mathbf{s}\right]_{\text{below}} \right\}. \tag{2.6.69}$$

The probability for finding the particle in the interference region B depends on the modulus squared of the entire transition amplitude and hence on the phase difference between the contribution from the paths going above and below. The phase difference due to the presence of **B** is just

$$\left[\left(\frac{e}{\hbar c}\right) \int_{\mathbf{x}_1}^{\mathbf{x}_N} \mathbf{A} \cdot d\mathbf{s}\right]_{\text{above}} - \left[\left(\frac{e}{\hbar c}\right) \int_{\mathbf{x}_1}^{\mathbf{x}_N} \mathbf{A} \cdot d\mathbf{s}\right]_{\text{below}} = \left(\frac{e}{\hbar c}\right) \oint \mathbf{A} \cdot d\mathbf{s}$$

$$= \left(\frac{e}{\hbar c}\right) \Phi_B, \tag{2.6.70}$$

where Φ_B stands for the magnetic flux inside the impenetrable cylinder. This means that as we change the magnetic field strength, there is a sinusoidal component in the probability for observing the particle in region B with a period given by a *fundamental unit of magnetic flux*, namely,

$$\frac{2\pi \hbar c}{|e|} = 4.135 \times 10^{-7} \quad \text{Gauss-cm}^2. \tag{2.6.71}$$

We emphasize that the interference effect discussed here is purely quantum mechanical. Classically, the motion of a charged particle is determined solely by Newton's second law supplemented by the force law of Lorentz. Here, as in the previous bound-state problem, the particle can never enter the region in which **B** is finite; the Lorentz force is identically zero in all regions where the particle wave function is finite. Yet there is a striking interference pattern that depends on the presence or absence of a magnetic field inside the impenetrable cylinder. This point has led some people to conclude that in quantum mechanics it is **A** rather than **B** that is fundamental. It is to be noted, however, that the observable effects in both examples depend only on Φ_B, which is directly expressible in terms of **B**. Experiments to verify the Aharonov-Bohm effect have been performed using a thin magnetized iron filament called a *whisker*.*

*One such recent experiment is that of A. Tonomura et al., *Phys. Rev. Lett.* **48** (1982): 1443.

Magnetic Monopole

We conclude this section with one of the most remarkable predictions of quantum physics, which has yet to be verified experimentally. An astute student of classical electrodynamics may be struck by the fact that there is a strong symmetry between **E** and **B**, yet a magnetic charge—commonly referred to as a **magnetic monopole**—analogous to electric charge is peculiarly absent in Maxwell's equations. The source of a magnetic field observed in nature is either a moving electric charge or a static magnetic dipole, never a static magnetic charge. Instead of

$$\nabla\cdot\mathbf{B} = 4\pi\rho_M \tag{2.6.72}$$

analogous to

$$\nabla\cdot\mathbf{E} = 4\pi\rho, \tag{2.6.73}$$

$\nabla\cdot\mathbf{B}$ actually vanishes in the usual way of writing Maxwell's equations. Quantum mechanics does not predict that a magnetic monopole must exist. However, it unambiguously requires that if a magnetic monopole is ever found in nature, the magnitude of magnetic charge must be quantized in terms of e, $\hbar$, and c, as we now demonstrate.

Suppose there is a point magnetic monopole, situated at the origin, of strength e_M analogous to a point electric charge. The static magnetic field is then given by

$$\mathbf{B} = \left(\frac{e_M}{r^2}\right)\hat{\mathbf{r}}. \tag{2.6.74}$$

At first sight it may appear that the magnetic field (2.6.74) can be derived from

$$\mathbf{A} = \left[\frac{e_M(1-\cos\theta)}{r\sin\theta}\right]\hat{\boldsymbol{\phi}}. \tag{2.6.75}$$

Recall the expression for curl in spherical coordinates:

$$\nabla\times\mathbf{A} = \hat{\mathbf{r}}\left[\frac{1}{r\sin\theta}\frac{\partial}{\partial\theta}(A_\phi\sin\theta) - \frac{\partial A_\theta}{\partial\phi}\right]$$

$$+\hat{\boldsymbol{\theta}}\frac{1}{r}\left[\frac{1}{\sin\theta}\frac{\partial A_r}{\partial\phi} - \frac{\partial}{\partial r}(rA_\phi)\right] + \hat{\boldsymbol{\phi}}\frac{1}{r}\left[\frac{\partial}{\partial r}(rA_\theta) - \frac{\partial A_r}{\partial\theta}\right]. \tag{2.6.76}$$

But vector potential (2.6.75) has one difficulty—it is singular on the negative z-axis ($\theta = \pi$). In fact, it turns out to be impossible to construct a singularity-free potential valid everywhere for this problem. To see this we first note "Gauss's law"

$$\int_{\text{closed surface}} \mathbf{B}\cdot d\boldsymbol{\sigma} = 4\pi e_M \tag{2.6.77}$$

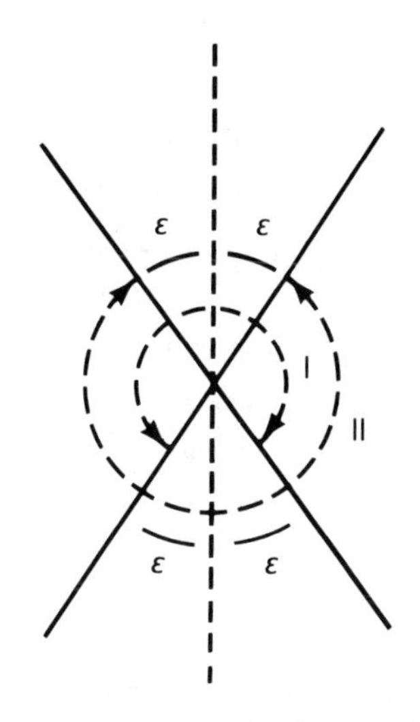

FIGURE 2.9. Regions of validity for the potentials $\mathbf{A}^{(\mathrm{I})}$ and $\mathbf{A}^{(\mathrm{II})}$.

for any surface boundary enclosing the origin at which the magnetic monopole is located. On the other hand, if **A** were nonsingular, we would have

$$\nabla \cdot (\nabla \times \mathbf{A}) = 0 \tag{2.6.78}$$

everywhere; hence,

$$\int_{\text{closed surface}} \mathbf{B} \cdot d\boldsymbol{\sigma} = \int_{\text{volume inside}} \nabla \cdot (\nabla \times \mathbf{A})\, d^3x = 0, \tag{2.6.79}$$

in contradiction with (2.6.77).

However, one might argue that because the vector potential is just a device for obtaining **B**, we need not insist on having a single expression for **A** valid everywhere. Suppose we construct a pair of potentials,

$$\mathbf{A}^{(\mathrm{I})} = \left[\frac{e_M(1-\cos\theta)}{r\sin\theta}\right]\hat{\boldsymbol{\phi}}, \qquad (\theta < \pi - \varepsilon) \tag{2.6.80a}$$

$$\mathbf{A}^{(\mathrm{II})} = -\left[\frac{e_M(1+\cos\theta)}{r\sin\theta}\right]\hat{\boldsymbol{\phi}}, \quad (\theta > \varepsilon), \tag{2.6.80b}$$

such that the potential $\mathbf{A}^{(\mathrm{I})}$ can be used everywhere except inside the cone defined by $\theta = \pi - \varepsilon$ around the negative z-axis; likewise, the potential $\mathbf{A}^{(\mathrm{II})}$ can be used everywhere except inside the cone $\theta = \varepsilon$ around the positive z-axis; see Figure 2.9. Together they lead to the correct expression for **B** everywhere.*

*An alternative approach to this problem uses $\mathbf{A}^{(\mathrm{I})}$ everywhere, but taking special care of the string of singularities, known as a **Dirac string**, along the negative z-axis.

Consider now what happens in the overlap region $\varepsilon < \theta < \pi - \varepsilon$, where we may use either $\mathbf{A}^{(\mathrm{I})}$ or $\mathbf{A}^{(\mathrm{II})}$. Because the two potentials lead to the same magnetic field, they must be related to each other by a gauge transformation. To find Λ appropriate for this problem we first note that

$$\mathbf{A}^{(\mathrm{II})} - \mathbf{A}^{(\mathrm{I})} = -\left(\frac{2e_M}{r\sin\theta}\right)\hat{\boldsymbol{\phi}}. \tag{2.6.81}$$

Recalling the expression for gradient in spherical coordinates,

$$\nabla\Lambda = \hat{\mathbf{r}}\frac{\partial\Lambda}{\partial r} + \hat{\boldsymbol{\theta}}\frac{1}{r}\frac{\partial\Lambda}{\partial\theta} + \hat{\boldsymbol{\phi}}\frac{1}{r\sin\theta}\frac{\partial\Lambda}{\partial\phi}, \tag{2.6.82}$$

we deduce that

$$\Lambda = -2e_M\phi \tag{2.6.83}$$

will do the job.

Next, we consider the wave function of an electrically charged particle of charge e subjected to magnetic field (2.6.74). As we emphasized earlier, the particular form of the wave function depends on the particular gauge used. In the overlap region where we may use either $\mathbf{A}^{(\mathrm{I})}$ or $\mathbf{A}^{(\mathrm{II})}$, the corresponding wave functions are, according to (2.6.55), related to each other by

$$\psi^{(\mathrm{II})} = \exp\left(\frac{-2iee_M\phi}{\hbar c}\right)\psi^{(\mathrm{I})}. \tag{2.6.84}$$

Wave functions $\psi^{(\mathrm{I})}$ and $\psi^{(\mathrm{II})}$ must each be *single-valued* because once we choose particular gauge, the expansion of the state ket in terms of the position eigenkets must be *unique*. After all, as we have repeatedly emphasized, the wave function is simply an expansion coefficient for the state ket in terms of the position eigenkets.

Let us now examine the behavior of wave function $\psi^{(\mathrm{II})}$ on the equator $\theta = \pi/2$ with some definite radius r, which is a constant. When we increase the azimuthal angle ϕ along the equator and go around once, say from $\phi = 0$ to $\phi = 2\pi$, $\psi^{(\mathrm{II})}$, as well as $\psi^{(\mathrm{I})}$, must return to its original value because each is single-valued. According to (2.6.84), this is possible only if

$$\frac{2ee_M}{\hbar c} = \pm N, \quad N = 0, \pm 1, \pm 2, \ldots. \tag{2.6.85}$$

So we reach a very far-reaching conclusion: The magnetic charges must be *quantized* in units of

$$\frac{\hbar c}{2|e|} \simeq \left(\frac{137}{2}\right)|e|. \tag{2.6.86}$$

The smallest magnetic charge possible is $\hbar c/2|e|$, where e is the electronic charge. It is amusing that once a magnetic monopole is assumed to exist, we can use (2.6.85) backward, so to speak, to explain why the electric charges are quantized, for example, why the proton charge cannot be 0.999972 times $|e|$.*

We repeat once again that quantum mechanics does not require magnetic monopoles to exist. However, it unambiguously predicts that a magnetic charge, if it is ever found in nature, must be quantized in units of $\hbar c/2|e|$. The quantization of magnetic charges in quantum mechanics was first shown in 1931 by P. A. M. Dirac. The derivation given here is due to T. T. Wu and C. N. Yang.

PROBLEMS

1. Consider the spin-precession problem discussed in the text. It can also be solved in the Heisenberg picture. Using the Hamiltonian

$$H = -\left(\frac{eB}{mc}\right)S_z = \omega S_z,$$

write the Heisenberg equations of motion for the time-dependent operators $S_x(t)$, $S_y(t)$, and $S_z(t)$. Solve them to obtain $S_{x,y,z}$ as functions of time.

2. Look again at the Hamiltonian of Chapter 1, Problem 11. Suppose the typist made an error and wrote H as

$$H = H_{11}|1\rangle\langle 1| + H_{22}|2\rangle\langle 2| + H_{12}|1\rangle\langle 2|.$$

What principle is now violated? Illustrate your point explicitly by attempting to solve the most general time-dependent problem using an illegal Hamiltonian of this kind. (You may assume $H_{11} = H_{22} = 0$ for simplicity.)

3. An electron is subject to a uniform, time-independent magnetic field of strength B in the positive z-direction. At $t = 0$ the electron is known to be in an eigenstate of $\mathbf{S}\cdot\hat{\mathbf{n}}$ with eigenvalue $\hbar/2$, where $\hat{\mathbf{n}}$ is a unit vector, lying in the xz-plane, that makes an angle β with the z-axis.
   a. Obtain the probability for finding the electron in the $s_x = \hbar/2$ state as a function of time.
   b. Find the expectation value of S_x as a function of time.
   c. For your own peace of mind show that your answers make good sense in the extreme cases (i) $\beta \to 0$ and (ii) $\beta \to \pi/2$.

4. Let $x(t)$ be the coordinate operator for a free particle in one dimension in the Heisenberg picture. Evaluate

$$[x(t), x(0)].$$

*Empirically the equality in magnitude between the electron charge and the proton charge is established to an accuracy of four parts in 10^{19}.

5. Consider a particle in one dimension whose Hamiltonian is given by

$$H = \frac{p^2}{2m} + V(x).$$

By calculating $[[H, x], x]$ prove

$$\sum_{a'} |\langle a''|x|a'\rangle|^2 (E_{a'} - E_{a''}) = \frac{\hbar^2}{2m},$$

where $|a'\rangle$ is an energy eigenket with eigenvalue $E_{a'}$.

6. Consider a particle in three dimensions whose Hamiltonian is given by

$$H = \frac{\mathbf{p}^2}{2m} + V(\mathbf{x}).$$

By calculating $[\mathbf{x}\cdot\mathbf{p}, H]$ obtain

$$\frac{d}{dt}\langle \mathbf{x}\cdot\mathbf{p}\rangle = \left\langle \frac{\mathbf{p}^2}{m} \right\rangle - \langle \mathbf{x}\cdot\nabla V\rangle.$$

To identify the preceding relation with the quantum-mechanical analogue of the virial theorem it is essential that the left-hand side vanish. Under what condition would this happen?

7. Consider a free-particle wave packet in one dimension. At $t = 0$ it satisfies the minimum uncertainty relation

$$\langle(\Delta x)^2\rangle\langle(\Delta p)^2\rangle = \frac{\hbar^2}{4} \quad (t = 0).$$

In addition, we know

$$\langle x\rangle = \langle p\rangle = 0 \quad (t = 0).$$

Using the Heisenberg picture, obtain $\langle(\Delta x)^2\rangle_t$ as a function of $t (t \geq 0)$ when $\langle(\Delta x)^2\rangle_{t=0}$ is given. (*Hint:* Take advantage of the property of the minimum-uncertainty wave packet you worked out in Chapter 1, Problem 18).

8. Let $|a'\rangle$ and $|a''\rangle$ be eigenstates of a Hermitian operator A with eigenvalues a' and a'', respectively ($a' \neq a''$). The Hamiltonian operator is given by

$$H = |a'\rangle\delta\langle a''| + |a''\rangle\delta\langle a'|,$$

where δ is just a real number.

   a. Clearly, $|a'\rangle$ and $|a''\rangle$ are not eigenstates of the Hamiltonian. Write down the eigenstates of the Hamiltonian. What are their energy eigenvalues?
   b. Suppose the system is known to be in state $|a'\rangle$ at $t = 0$. Write down the state vector in the Schrödinger picture for $t > 0$.

c. What is the probability for finding the system in $|a''\rangle$ for $t > 0$ if the system is known to be in state $|a'\rangle$ at $t = 0$?

d. Can you think of a physical situation corresponding to this problem?

9. A box containing a particle is divided into a right and left compartment by a thin partition. If the particle is known to be on the right (left) side with certainty, the state is represented by the position eigenket $|R\rangle(|L\rangle)$, where we have neglected spatial variations within each half of the box. The most general state vector can then be written as

$$|\alpha\rangle = |R\rangle\langle R|\alpha\rangle + |L\rangle\langle L|\alpha\rangle,$$

where $\langle R|\alpha\rangle$ and $\langle L|\alpha\rangle$ can be regarded as "wave functions." The particle can tunnel through the partition; this tunneling effect is characterized by the Hamiltonian

$$H = \Delta(|L\rangle\langle R| + |R\rangle\langle L|),$$

where Δ is a real number with the dimension of energy.

a. Find the normalized energy eigenkets. What are the corresponding energy eigenvalues?

b. In the Schrödinger picture the base kets $|R\rangle$ and $|L\rangle$ are fixed, and the state vector moves with time. Suppose the system is represented by $|\alpha\rangle$ as given above at $t = 0$. Find the state vector $|\alpha, t_0 = 0; t\rangle$ for $t > 0$ by applying the appropriate time-evolution operator to $|\alpha\rangle$.

c. Suppose at $t = 0$ the particle is on the right side with certainty. What is the probability for observing the particle on the left side as a function of time?

d. Write down the coupled Schrödinger equations for the wave functions $\langle R|\alpha, t_0 = 0; t\rangle$ and $\langle L|\alpha, t_0 = 0; t\rangle$. Show that the solutions to the coupled Schrödinger equations are just what you expect from (b).

e. Suppose the printer made an error and wrote H as

$$H = \Delta|L\rangle\langle R|.$$

By explicitly solving the most general time-evolution problem with this Hamiltonian, show that probability conservation is violated.

10. Using the one-dimensional simple harmonic oscillator as an example, illustrate the difference between the Heisenberg picture and the Schrödinger picture. Discuss in particular how (a) the dynamic variables x and p and (b) the most general state vector evolve with time in each of the two pictures.

11. Consider a particle subject to a one-dimensional simple harmonic oscillator potential. Suppose at $t = 0$ the state vector is given by

$$\exp\left(\frac{-ipa}{\hbar}\right)|0\rangle,$$

where p is the momentum operator and a is some number with dimension of length. Using the Heisenberg picture, evaluate the expectation value $\langle x\rangle$ for $t \geq 0$.

12. a. Write down the wave function (in coordinate space) for the state specified in Problem 11 at $t = 0$. You may use

$$\langle x'|0\rangle = \pi^{-1/4} x_0^{-1/2} \exp\left[-\frac{1}{2}\left(\frac{x'}{x_0}\right)^2\right], \quad \left(x_0 \equiv \left(\frac{\hbar}{m\omega}\right)^{1/2}\right).$$

b. Obtain a simple expression for the probability that the state is found in the ground state at $t = 0$. Does this probability change for $t > 0$?

13. Consider a one-dimensional simple harmonic oscillator.

a. Using

$$\left.\begin{matrix} a \\ a^\dagger \end{matrix}\right\} = \sqrt{\frac{m\omega}{2\hbar}}\left(x \pm \frac{ip}{m\omega}\right), \quad \left.\begin{matrix} a|n\rangle \\ a^\dagger|n\rangle \end{matrix}\right\} = \left\{\begin{matrix} \sqrt{n}\,|n-1\rangle \\ \sqrt{n+1}\,|n+1\rangle, \end{matrix}\right.$$

evaluate $\langle m|x|n\rangle$, $\langle m|p|n\rangle$, $\langle m|\{x,p\}|n\rangle$, $\langle m|x^2|n\rangle$, and $\langle m|p^2|n\rangle$.

b. Check that the virial theorem holds for the expectation values of the kinetic and the potential energy taken with respect to an energy eigenstate.

14. a. Using

$$\langle x'|p'\rangle = (2\pi\hbar)^{-1/2} e^{ip'x'/\hbar} \qquad \text{(one dimension)}$$

prove

$$\langle p'|x|\alpha\rangle = i\hbar\frac{\partial}{\partial p'}\langle p'|\alpha\rangle.$$

b. Consider a one-dimensional simple harmonic oscillator. Starting with the Schrödinger equation for the state vector, derive the Schrödinger equation for the *momentum-space* wave function. (Make sure to distinguish the operator p from the eigenvalue p'.) Can you guess the energy eigenfunctions in momentum space?

15. Consider a function, known as the **correlation function**, defined by

$$C(t) = \langle x(t)x(0)\rangle,$$

where $x(t)$ is the position operator in the Heisenberg picture. Evaluate the correlation function explicitly for the ground state of a one-dimensional simple harmonic oscillator.

16. Consider again a one-dimensional simple harmonic oscillator. Do the following algebraically, that is, without using wave functions.

a. Construct a linear combination of $|0\rangle$ and $|1\rangle$ such that $\langle x\rangle$ is as large as possible.

b. Suppose the oscillator is in the state constructed in (a) at $t = 0$. What is the state vector for $t > 0$ in the Schrödinger picture? Evaluate the expectation value $\langle x\rangle$ as a function of time for $t > 0$ using (i) the Schrödinger picture and (ii) the Heisenberg picture.

c. Evaluate $\langle(\Delta x)^2\rangle$ as a function of time using either picture.

17. Show for the one-dimensional simple harmonic oscillator

$$\langle 0|e^{ikx}|0\rangle = \exp\left[-k^2\langle 0|x^2|0\rangle/2\right],$$

where x is the position *operator*.

18. A coherent state of a one-dimensional simple harmonic oscillator is defined to be an eigenstate of the (non-Hermitian) annihilation operator a:

$$a|\lambda\rangle = \lambda|\lambda\rangle,$$

where λ is, in general, a complex number.

a. Prove that

$$|\lambda\rangle = e^{-|\lambda|^2/2} e^{\lambda a^\dagger}|0\rangle$$

is a normalized coherent state.

b. Prove the minimum uncertainty relation for such a state.

c. Write $|\lambda\rangle$ as

$$|\lambda\rangle = \sum_{n=0}^{\infty} f(n)|n\rangle.$$

Show that the distribution of $|f(n)|^2$ with respect to n is of the Poisson form. Find the most probable value of n, hence of E.

d. Show that a coherent state can also be obtained by applying the translation (finite-displacement) operator $e^{-ipl/\hbar}$ (where p is the momentum operator, and l is the displacement distance) to the ground state. (See also Gottfried 1966, 262–64.)

19. Let

$$J_\pm = \hbar a_\pm^\dagger a_\mp, \qquad J_z = \frac{\hbar}{2}\left(a_+^\dagger a_+ - a_-^\dagger a_-\right), \qquad N = a_+^\dagger a_+ + a_-^\dagger a_-$$

where $a_\pm$ and $a_\pm^\dagger$ are the annihilation and creation operators of two *independent* simple harmonic oscillator satisfying the usual simple harmonic oscillator commutation relations. Prove

$$[J_z, J_\pm] = \pm\hbar J_\pm, \qquad [\mathbf{J}^2, J_z] = 0, \qquad \mathbf{J}^2 = \left(\frac{\hbar^2}{2}\right) N\left[\left(\frac{N}{2}\right)+1\right].$$

20. Consider a particle of mass m subject to a one-dimensional potential of the following form:

$$V = \begin{cases} \dfrac{1}{2}kx^2 & \text{for } x > 0 \\ \infty & \text{for } x < 0. \end{cases}$$

a. What is the ground-state energy?

b. What is the expectation value $\langle x^2\rangle$ for the ground state?

21. A particle in one dimension is trapped between two rigid walls:

$$V(x) = \begin{cases} 0, & \text{for } 0 < x < L \\ \infty, & \text{for } x < 0, \quad x > L. \end{cases}$$

At $t = 0$ it is known to be exactly at $x = L/2$ with certainty. What are the *relative* probabilities for the particle to be found in various energy eigenstates? Write down the wave function for $t \geq 0$. (You need not worry about absolute normalization, convergence, and other mathematical subtleties.)

22. Consider a particle in one dimension bound to a fixed center by a δ-function potential of the form

$$V(x) = -v_0 \delta(x), \quad (v_0 \text{ real and positive}).$$

Find the wave function and the binding energy of the ground state. Are there excited bound states?

23. A particle of mass m in one dimension is bound to a fixed center by an attractive δ-function potential:

$$V(x) = -\lambda\delta(x), \quad (\lambda > 0).$$

At $t = 0$, the potential is suddenly switched off (that is, $V = 0$ for $t > 0$). Find the wave function for $t > 0$. (Be quantitative! But you need not attempt to evaluate an integral that may appear.)

24. A particle in one dimension $(-\infty < x < \infty)$ is subjected to a constant force derivable from

$$V = \lambda x, \quad (\lambda > 0).$$

   a. Is the energy spectrum continuous or discrete? Write down an approximate expression for the energy eigenfunction specified by E. Also sketch it crudely.
   b. Discuss briefly what changes are needed if V is replaced by

$$V = \lambda |x|.$$

25. Consider an electron confined to the *interior* of a hollow cylindrical shell whose axis coincides with the z-axis. The wave function is required to vanish on the inner and outer walls, $\rho = \rho_a$ and ρ_b, and also at the top and bottom, $z = 0$ and L.
   a. Find the energy eigenfunctions. (Do not bother with normalization.) Show that the energy eigenvalues are given by

$$E_{lmn} = \left(\frac{\hbar^2}{2m_e}\right)\left[k_{mn}^2 + \left(\frac{l\pi}{L}\right)^2\right] \quad (l = 1,2,3,\ldots, \; m = 0,1,2,\ldots),$$

   where k_{mn} is the nth root of the transcendental equation

$$J_m(k_{mn}\rho_b)N_m(k_{mn}\rho_a) - N_m(k_{mn}\rho_b)J_m(k_{mn}\rho_a) = 0.$$

   b. Repeat the same problem when there is a uniform magnetic field $\mathbf{B} = B\hat{\mathbf{z}}$ for $0 < \rho < \rho_a$. Note that the energy eigenvalues are influenced by the magnetic field even though the electron never "touches" the magnetic field.

c. Compare, in particular, the ground state of the $B = 0$ problem with that of the $B \neq 0$ problem. Show that if we require the ground-state energy to be unchanged in the presence of B, we obtain "flux quantization"

$$\pi\rho_a^2 B = \frac{2\pi N\hbar c}{e}, \quad (N = 0, \pm 1, \pm 2, \ldots).$$

26. Consider a particle moving in one dimension under the influence of a potential $V(x)$. Suppose its wave function can be written as $\exp[iS(x,t)/\hbar]$. Prove that $S(x,t)$ satisfies the classical Hamilton-Jacobi equation to the extent that $\hbar$ can be regarded as small in some sense. Show how one may obtain the correct wave function for a plane wave by starting with the solution of the classical Hamilton-Jacobi equation with $V(x)$ set equal to zero. Why do we get the exact wave function in this particular case?
27. Using spherical coordinates, obtain an expression for $\mathbf{j}$ for the ground and excited states of the hydrogen atom. Show, in particular, that for $m_l \neq 0$ states, there is a circulating flux in the sense that $\mathbf{j}$ is in the direction of increasing or decreasing ϕ, depending on whether m_l is positive or negative.
28. Derive (2.5.16) and obtain the three-dimensional generalization of (2.5.16).
29. Define the partition function as

$$Z = \int d^3x' K(\mathbf{x}', t; \mathbf{x}', 0)|_{\beta = it/\hbar},$$

as in (2.5.20)–(2.5.22). Show that the ground-state energy is obtained by taking

$$-\frac{1}{Z}\frac{\partial Z}{\partial \beta}, \quad (\beta \to \infty).$$

Illustrate this for a particle in a one-dimensional box.
30. The propagator in momentum space analogous to (2.5.26) is given by $\langle \mathbf{p}'', t|\mathbf{p}', t_0\rangle$. Derive an explicit expression for $\langle \mathbf{p}'', t|\mathbf{p}', t_0\rangle$ for the free-particle case.
31. a. Write down an expression for the classical action for a simple harmonic oscillator for a finite time interval.
    b. Construct $\langle x_n, t_n|x_{n-1}, t_{n-1}\rangle$ for a simple harmonic oscillator using Feynman's prescription for $t_n - t_{n-1} = \Delta t$ small. Keeping only terms up to order $(\Delta t)^2$, show that it is in complete agreement with the $t - t_0 \to 0$ limit of the propagator given by (2.5.26).
32. State the Schwinger action principle (see Finkelstein 1973, 155). Obtain the solution for $\langle x_2 t_2|x_1 t_1\rangle$ by integrating the Schwinger principle and compare it with the corresponding Feynman expression for $\langle x_2 t_2|x_1 t_1\rangle$. Describe the classical limits of these two expressions.

33. Show that the wave-mechanics approach to the gravity-induced problem discussed in Section 2.6 also leads to phase-difference expression (2.6.17).
34. a. Verify (2.6.25) and (2.6.27).
    b. Verify continuity equation (2.6.30) with $\mathbf{j}$ given by (2.6.31).
35. Consider the Hamiltonian of a spinless particle of charge e. In the presence of a static magnetic field, the interaction terms can be generated by

$$\mathbf{p}_{\text{operator}} \to \mathbf{p}_{\text{operator}} - \frac{e\mathbf{A}}{c},$$

where $\mathbf{A}$ is the appropriate vector potential. Suppose, for simplicity, that the magnetic field $\mathbf{B}$ is uniform in the positive z-direction. Prove that the above prescription indeed leads to the correct expression for the interaction of the orbital magnetic moment $(e/2mc)\mathbf{L}$ with the magnetic field $\mathbf{B}$. Show that there is also an extra term proportional to $B^2(x^2+y^2)$, and comment briefly on its physical significance.
36. An electron moves in the presence of a uniform magnetic field in the z-direction ($\mathbf{B} = B\hat{\mathbf{z}}$).
    a. Evaluate

$$[\Pi_x, \Pi_y],$$

where

$$\Pi_x \equiv p_x - \frac{eA_x}{c}, \qquad \Pi_y \equiv p_y - \frac{eA_y}{c}.$$

    b. By comparing the Hamiltonian and the commutation relation obtained in (a) with those of the one-dimensional oscillator problem, show how we can immediately write the energy eigenvalues as

$$E_{k,n} = \frac{\hbar^2 k^2}{2m} + \left(\frac{|eB|\hbar}{mc}\right)\left(n + \frac{1}{2}\right),$$

where $\hbar k$ is the continuous eigenvalue of the p_z operator and n is a nonnegative integer including zero.
37. Consider the neutron interferometer.

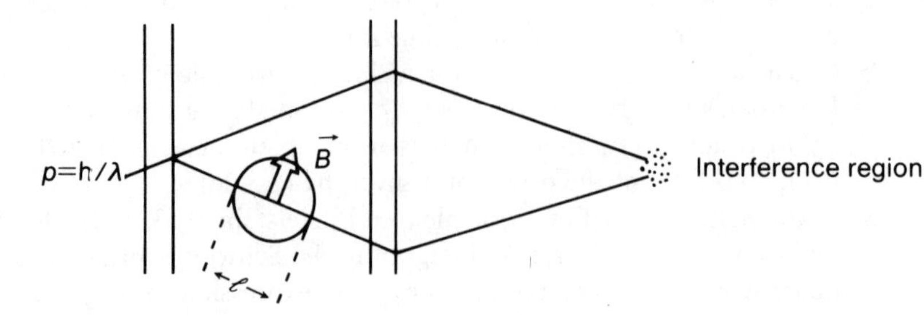

Prove that the difference in the magnetic fields that produce two successive maxima in the counting rates is given by

$$\Delta B = \frac{4\pi\hbar c}{|e| g_n \lambda\!\!\!^{-} l},$$

where g_n $(= -1.91)$ is the neutron magnetic moment in units of $-e\hbar/2m_n c$. [If you had solved this problem in 1967, you could have published it in *Physical Review Letters*!]

CHAPTER 3

Theory of Angular Momentum

This chapter is concerned with a systematic treatment of angular momentum and related topics. The importance of angular momentum in modern physics can hardly be overemphasized. A thorough understanding of angular momentum is essential in molecular, atomic, and nuclear spectroscopy; angular-momentum considerations play an important role in scattering and collision problems as well as in bound-state problems. Furthermore, angular-momentum concepts have important generalizations—isospin in nuclear physics, SU(3), SU(2)⊗U(1) in particle physics, and so forth.

3.1. ROTATIONS AND ANGULAR MOMENTUM COMMUTATION RELATIONS

Finite Versus Infinitesimal Rotations

We recall from elementary physics that rotations about the same axis commute, whereas rotations about different axes do not. For instance, a 30° rotation about the z-axis followed by a 60° rotation about the same z-axis is obviously equivalent to a 60° rotation followed by a 30° rotation, both about the same axis. However, let us consider a 90° rotation about the z-axis, denoted by $R_z(\pi/2)$, followed by a 90° rotation about the x-axis, denoted by $R_x(\pi/2)$; compare this with a 90° rotation about the x-axis followed by a 90° rotation about the z-axis. The net results are different, as we can see from Figure 3.1.

Our first basic task is to work out quantitatively the manner in which rotations about different axes *fail* to commute. To this end, we first recall how to represent rotations in three dimensions by 3×3 real, orthogonal matrices. Consider a vector $\mathbf{V}$ with components V_x, V_y, and V_z. When we rotate, the three components become some other set of numbers, V_x', V_y', and V_z'. The old and new components are related via a 3×3 orthogonal matrix R:

$$\begin{pmatrix} V_x' \\ V_y' \\ V_z' \end{pmatrix} = \begin{pmatrix} & & \\ & R & \\ & & \end{pmatrix} \begin{pmatrix} V_x \\ V_y \\ V_z \end{pmatrix}, \tag{3.1.1a}$$

$$RR^T = R^T R = 1, \tag{3.1.1b}$$

where the superscript T stands for a transpose of a matrix. It is a property of orthogonal matrices that

$$\sqrt{V_x^2 + V_y^2 + V_z^2} = \sqrt{V_x'^2 + V_y'^2 + V_z'^2} \tag{3.1.2}$$

is automatically satisfied.

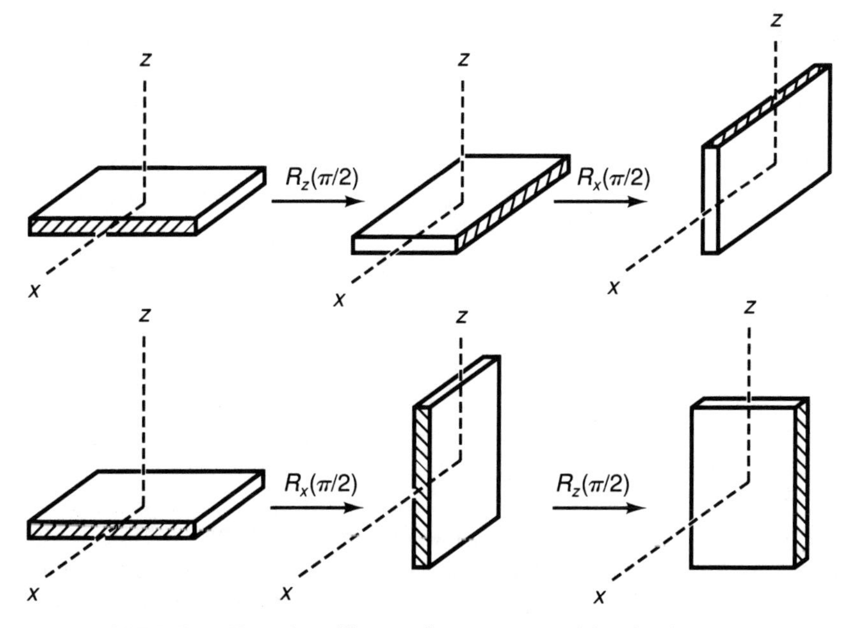

FIGURE 3.1. Example to illustrate the noncommutativity of finite rotations.

To be definite, we consider a rotation about the z-axis by angle ϕ. The convention we follow throughout this book is that a rotation operation affects a physical system itself, as in Figure 3.1, while the coordinate axes remain *unchanged*. The angle ϕ is taken to be positive when the rotation in question is counterclockwise in the xy-plane, as viewed from the positive z-side. If we associate a right-handed screw with such a rotation, a positive ϕ rotation around the z-axis means that the screw is advancing in the positive z-direction. With this convention, we easily verify that

$$R_z(\phi) = \begin{pmatrix} \cos\phi & -\sin\phi & 0 \\ \sin\phi & \cos\phi & 0 \\ 0 & 0 & 1 \end{pmatrix}. \tag{3.1.3}$$

Had we adopted a different convention, in which a physical system remained fixed but the coordinate axes rotated, this same matrix with a positive ϕ would have represented a *clockwise* rotation of the x- and y-axes, when viewed from the positive z-side. It is obviously important not to mix the two conventions! Some authors distinguish the two approaches by using "active rotations" for physical systems rotated and "passive rotations" for the coordinate axes rotated.

We are particularly interested in an infinitesimal form of R_z:

$$R_z(\varepsilon) = \begin{pmatrix} 1-\frac{\varepsilon^2}{2} & -\varepsilon & 0 \\ \varepsilon & 1-\frac{\varepsilon^2}{2} & 0 \\ 0 & 0 & 1 \end{pmatrix}, \tag{3.1.4}$$

where terms of order ε^3 and higher are ignored. Likewise, we have

$$R_x(\varepsilon) = \begin{pmatrix} 1 & 0 & 0 \\ 0 & 1-\frac{\varepsilon^2}{2} & -\varepsilon \\ 0 & \varepsilon & 1-\frac{\varepsilon^2}{2} \end{pmatrix} \tag{3.1.5a}$$

and

$$R_y(\varepsilon) = \begin{pmatrix} 1-\frac{\varepsilon^2}{2} & 0 & \varepsilon \\ 0 & 1 & 0 \\ -\varepsilon & 0 & 1-\frac{\varepsilon^2}{2} \end{pmatrix}, \tag{3.1.5b}$$

which may be read from (3.1.4) by cyclic permutations of x, y, z—that is, $x \to y$, $y \to z$, $z \to x$. Compare now the effect of a y-axis rotation followed by an x-axis rotation with that of an x-axis rotation followed by a y-axis rotation. Elementary matrix manipulations lead to

$$R_x(\varepsilon)R_y(\varepsilon) = \begin{pmatrix} 1-\dfrac{\varepsilon^2}{2} & 0 & \varepsilon \\ \varepsilon^2 & 1-\dfrac{\varepsilon^2}{2} & -\varepsilon \\ -\varepsilon & \varepsilon & 1-\varepsilon^2 \end{pmatrix} \tag{3.1.6a}$$

and

$$R_y(\varepsilon)R_x(\varepsilon) = \begin{pmatrix} 1-\dfrac{\varepsilon^2}{2} & \varepsilon^2 & \varepsilon \\ 0 & 1-\dfrac{\varepsilon^2}{2} & -\varepsilon \\ -\varepsilon & \varepsilon & 1-\varepsilon^2 \end{pmatrix}. \tag{3.1.6b}$$

From (3.1.6a) and (3.1.6b) we have the first important result: Infinitesimal rotations about different axes do commute if terms of order ε^2 and higher are ignored.* The second and even more important result concerns the manner in which rotations about different axes *fail* to commute when terms of order ε^2 are kept:

$$\begin{aligned} R_x(\varepsilon)R_y(\varepsilon) - R_y(\varepsilon)R_x(\varepsilon) &= \begin{pmatrix} 0 & -\varepsilon^2 & 0 \\ \varepsilon^2 & 0 & 0 \\ 0 & 0 & 0 \end{pmatrix} \\ &= R_z(\varepsilon^2) - 1, \end{aligned} \tag{3.1.7}$$

where all terms of order higher than ε^2 have been ignored throughout this derivation. We also have

$$1 = R_{\text{any}}(0) \tag{3.1.8}$$

where *any* stands for any rotation axis. Thus the final result can be written as

$$R_x(\varepsilon)R_y(\varepsilon) - R_y(\varepsilon)R_x(\varepsilon) = R_z(\varepsilon^2) - R_{\text{any}}(0). \tag{3.1.9}$$

This is an example of the commutation relations between rotation operations about different axes, which we will use later in deducing the angular-momentum commutation relations in quantum mechanics.

---

*Actually there is a familiar example of this in elementary mechanics. The angular velocity vector $\boldsymbol{\omega}$ that characterizes an infinitesimal change in rotation angle during an infinitesimal time interval follows the usual rule of vector addition, including commutativity of vector addition. However, we cannot ascribe a vectorial property to a *finite* angular change.

Infinitesimal Rotations in Quantum Mechanics

So far we have not used quantum-mechanical concepts. The matrix R is just a 3×3 orthogonal matrix acting on a vector $\mathbf{V}$ written in column matrix form. We must now understand how to characterize rotations in quantum mechanics.

Because rotations affect physical systems, the state ket corresponding to a rotated system is expected to look different from the state ket corresponding to the original unrotated system. Given a rotation operation R, characterized by a 3×3 orthogonal matrix R, we associate an operator $\mathscr{D}(R)$ in the appropriate ket space such that

$$|\alpha\rangle_R = \mathscr{D}(R)|\alpha\rangle, \tag{3.1.10}$$

where $|\alpha\rangle_R$ and $|\alpha\rangle$ stand for the kets of the rotated and original system, respectively.* Note that the 3×3 orthogonal matrix R acts on a column matrix made up of the three components of a classical vector, while the operator $\mathscr{D}(R)$ acts on state vectors in ket space. The matrix representation of $\mathscr{D}(R)$, which we will study in great detail in the subsequent sections, depends on the dimensionality N of the particular ket space in question. For $N=2$, appropriate for describing a spin $\frac{1}{2}$ system with no other degrees of freedom, $\mathscr{D}(R)$ is represented by a 2×2 matrix; for a spin 1 system, the appropriate representation is a 3×3 unitary matrix, and so on.

To construct the rotation operator $\mathscr{D}(R)$, it is again fruitful to examine first its properties under an infinitesimal rotation. We can almost guess how we must proceed by analogy. In both translations and time evolution, which we studied in Sections 1.6 and 2.1, respectively, the appropriate infinitesimal operators could be written as

$$U_\varepsilon = 1 - iG\varepsilon \tag{3.1.11}$$

with a Hermitian operator G. Specifically,

$$G \to \frac{p_x}{\hbar}, \qquad \varepsilon \to dx' \tag{3.1.12}$$

for an infinitesimal translation by a displacement dx' in the x-direction, and

$$G \to \frac{H}{\hbar}, \qquad \varepsilon \to dt \tag{3.1.13}$$

for an infinitesimal time evolution with time displacement dt. We know from classical mechanics that angular momentum is the generator of rotation in much the same way as momentum and Hamiltonian are the generators of translation and time evolution, respectively. We therefore *define* the angular momentum operator J_k in such a way that the operator for an infinitesimal rotation around the kth axis by angle $d\phi$ can be

*The symbol $\mathscr{D}$ stems from the German word *Drehung*, meaning *rotation*.

obtained by letting

$$G \to \frac{J_k}{\hbar}, \qquad \varepsilon \to d\phi \tag{3.1.14}$$

in (3.1.11). With J_k taken to be Hermitian, the infinitesimal rotation operator is guaranteed to be unitary and reduces to the identity operator in the limit $d\phi \to 0$. More generally, we have

$$\mathscr{D}(\hat{\mathbf{n}}, d\phi) = 1 - i\left(\frac{\mathbf{J}\cdot\hat{\mathbf{n}}}{\hbar}\right) d\phi \tag{3.1.15}$$

for a rotation about the direction characterized by a unit vector $\hat{\mathbf{n}}$ by an infinitesimal angle $d\phi$.

We stress that in this book we do not define the angular-momentum operator to be $\mathbf{x}\times\mathbf{p}$. This is important because spin angular momentum, to which our general formalism also applies, has nothing to do with x_i and p_j. Put in another way, in classical mechanics one can prove that the angular momentum defined to be $\mathbf{x}\times\mathbf{p}$ is the generator of a rotation; in contrast, in quantum mechanics we *define* $\mathbf{J}$ so that the operator for an infinitesimal rotation takes form (3.1.15).

A finite rotation can be obtained by compounding successively infinitesimal rotations about the same axis. For instance, if we are interested in a finite rotation about the z-axis by angle ϕ, we consider

$$\begin{aligned}\mathscr{D}_z(\phi) &= \lim_{N\to\infty}\left[1 - i\left(\frac{J_z}{\hbar}\right)\left(\frac{\phi}{N}\right)\right]^N \\ &= \exp\left(\frac{-iJ_z\phi}{\hbar}\right) \\ &= 1 - \frac{iJ_z\phi}{\hbar} - \frac{J_z^2\phi^2}{2\hbar^2} + \cdots .\end{aligned} \tag{3.1.16}$$

In order to obtain the angular-momentum commutation relations, we need one more concept. As we remarked earlier, for every rotation R represented by a 3×3 orthogonal matrix R there exists a rotation operator $\mathscr{D}(R)$ in the appropriate ket space. We further postulate that $\mathscr{D}(R)$ has the same group properties as R:

Identity: $$R\cdot 1 = R \Rightarrow \mathscr{D}(R)\cdot 1 = \mathscr{D}(R) \tag{3.1.17a}$$

Closure: $$R_1R_2 = R_3 \Rightarrow \mathscr{D}(R_1)\mathscr{D}(R_2) = \mathscr{D}(R_3) \tag{3.1.17b}$$

Inverses: $$RR^{-1} = 1 \Rightarrow \mathscr{D}(R)\mathscr{D}^{-1}(R) = 1 \tag{3.1.17c}$$

$$R^{-1}R = 1 \Rightarrow \mathscr{D}^{-1}(R)\mathscr{D}(R) = 1$$

Associativity: $$\begin{aligned}R_1(R_2R_3) &= (R_1R_2)R_3 = R_1R_2R_3 \\ &\Rightarrow \mathscr{D}(R_1)[\mathscr{D}(R_2)\mathscr{D}(R_3)] \\ &= [\mathscr{D}(R_1)\mathscr{D}(R_2)]\mathscr{D}(R_3) \\ &= \mathscr{D}(R_1)\mathscr{D}(R_2)\mathscr{D}(R_3).\end{aligned} \tag{3.1.17d}$$

Let us now return to the fundamental commutation relations for rotation operations (3.1.9) written in terms of the R matrices. Its rotation operator analogue would read

$$\left(1-\frac{iJ_x\varepsilon}{\hbar}-\frac{J_x^2\varepsilon^2}{2\hbar^2}\right)\left(1-\frac{iJ_y\varepsilon}{\hbar}-\frac{J_y^2\varepsilon^2}{2\hbar^2}\right)$$
$$-\left(1-\frac{iJ_y\varepsilon}{\hbar}-\frac{J_y^2\varepsilon^2}{2\hbar^2}\right)\left(1-\frac{iJ_x\varepsilon}{\hbar}-\frac{J_x^2\varepsilon^2}{2\hbar^2}\right)=1-\frac{iJ_z\varepsilon^2}{\hbar}-1. \quad (3.1.18)$$

Terms of order ε automatically drop out. Equating terms of order ε^2 on both sides of (3.1.18), we obtain

$$\left[J_x, J_y\right] = i\hbar J_z. \quad (3.1.19)$$

Repeating this kind of argument with rotations about other axes, we obtain

$$\left[J_i, J_j\right] = i\hbar\varepsilon_{ijk}J_k, \quad (3.1.20)$$

known as the **fundamental commutation relations of angular momentum**.

In general, when the generators of infinitesimal transformations do not commute, the corresponding group of operations is said to be **non-Abelian**. Because of (3.1.20), the rotation group in three dimensions is non-Abelian. In contrast, the translation group in three dimensions is Abelian because p_i and p_j commute even with $i \neq j$.

We emphasize that in obtaining the commutation relations (3.1.20) we have used the following two concepts:

1. J_k is the generator of rotation about the kth axis.
2. Rotations about different axes fail to commute.

It is no exaggeration to say that commutation relations (3.1.20) summarize in a compact manner *all* the basic properties of rotations in three dimensions.

3.2. SPIN $\frac{1}{2}$ SYSTEMS AND FINITE ROTATIONS

Rotation Operator for Spin $\frac{1}{2}$

The lowest number, N, of dimensions in which the angular-momentum commutation relations (3.1.20) are realized, is $N=2$. The reader has already checked in Problem 8 of Chapter 1 that the operators defined by

$$S_x = \left(\frac{\hbar}{2}\right)\{(|+\rangle\langle-|)+(|-\rangle\langle+|)\},$$
$$S_y = \left(\frac{i\hbar}{2}\right)\{-(|+\rangle\langle-|)+(|-\rangle\langle+|)\}, \quad (3.2.1)$$
$$S_z = \left(\frac{\hbar}{2}\right)\{(|+\rangle\langle+|)-(|-\rangle\langle-|)\}$$

satisfy commutation relations (3.1.20) with J_k replaced by S_k. It is not a priori obvious that nature takes advantage of the lowest dimensional realization of (3.1.20), but numerous experiments—from atomic spectroscopy to nuclear magnetic resonance—suffice to convince us that this is in fact the case.

Consider a rotation by a finite angle ϕ about the z-axis. If the ket of a spin $\frac{1}{2}$ system before rotation is given by $|\alpha\rangle$, the ket after rotation is given by

$$|\alpha\rangle_R = \mathscr{D}_z(\phi)|\alpha\rangle \tag{3.2.2}$$

with

$$\mathscr{D}_z(\phi) = \exp\left(\frac{-iS_z\phi}{\hbar}\right). \tag{3.2.3}$$

To see that this operator really rotates the physical system, let us look at its effect on $\langle S_x \rangle$. Under rotation this expectation value changes as follows:

$$\langle S_x \rangle \to {}_R\langle\alpha|S_x|\alpha\rangle_R = \langle\alpha|\mathscr{D}_z^\dagger(\phi)S_x\mathscr{D}_z(\phi)|\alpha\rangle. \tag{3.2.4}$$

We must therefore compute

$$\exp\left(\frac{iS_z\phi}{\hbar}\right)S_x\exp\left(\frac{-iS_z\phi}{\hbar}\right). \tag{3.2.5}$$

For pedagogical reasons we evaluate this in two different ways.

Derivation 1: Here we use the specific form of S_x given by (3.2.1). We then obtain for (3.2.5)

$$
\begin{aligned}
&\left(\frac{\hbar}{2}\right)\exp\left(\frac{iS_z\phi}{\hbar}\right)\{(|+\rangle\langle-|)+(|-\rangle\langle+|)\}\exp\left(\frac{-iS_z\phi}{\hbar}\right)\\
&\quad=\left(\frac{\hbar}{2}\right)\left(e^{i\phi/2}|+\rangle\langle-|e^{i\phi/2}+e^{-i\phi/2}|-\rangle\langle+|e^{-i\phi/2}\right)\\
&\quad=\frac{\hbar}{2}\left[\{(|+\rangle\langle-|)+(|-\rangle\langle+|)\}\cos\phi+i\{(|+\rangle\langle-|)-(|-\rangle\langle+|)\}\sin\phi\right]\\
&\quad=S_x\cos\phi-S_y\sin\phi.
\end{aligned} \tag{3.2.6}
$$

Derivation 2: Alternatively we may use formula (2.3.47) to evaluate (3.2.5):

$$
\begin{aligned}
\exp\left(\frac{iS_z\phi}{\hbar}\right)S_x\exp\left(\frac{-iS_z\phi}{\hbar}\right) &= S_x + \left(\frac{i\phi}{\hbar}\right)\underbrace{[S_z, S_x]}_{i\hbar S_y}\\
&\quad+\left(\frac{1}{2!}\right)\left(\frac{i\phi}{\hbar}\right)^2\underbrace{[S_z,\underbrace{[S_z,S_x]}_{i\hbar S_y}]}_{\hbar^2 S_x}+\left(\frac{1}{3!}\right)\left(\frac{i\phi}{\hbar}\right)^3\underbrace{[S_z,\underbrace{[S_z,[S_z,S_x]]}_{\hbar^2 S_x}]}_{i\hbar^3 S_y}+\cdots\\
&= S_x\left[1-\frac{\phi^2}{2!}+\cdots\right]-S_y\left[\phi-\frac{\phi^3}{3!}+\cdots\right]\\
&= S_x\cos\phi - S_y\sin\phi.
\end{aligned} \tag{3.2.7}
$$

Notice that in derivation 2 we used only the commutation relations for S_i, so this method can be generalized to rotations of systems with angular momentum higher than $\frac{1}{2}$.

For spin $\frac{1}{2}$, both methods give

$$\langle S_x \rangle \to {}_R\langle \alpha | S_x | \alpha \rangle_R = \langle S_x \rangle \cos\phi - \langle S_y \rangle \sin\phi, \tag{3.2.8}$$

where the expectation value without subscripts is understood to be taken with respect to the (old) unrotated system. Similarly,

$$\langle S_y \rangle \to \langle S_y \rangle \cos\phi + \langle S_x \rangle \sin\phi. \tag{3.2.9}$$

As for the expectation value of S_z, there is no change because S_z commutes with $\mathscr{D}_z(\phi)$:

$$\langle S_z \rangle \to \langle S_z \rangle. \tag{3.2.10}$$

Relations (3.2.8), (3.2.9), and (3.2.10) are quite reasonable. They show that rotation operator (3.2.3), when applied to the state ket, does rotate the expectation value of $\mathbf{S}$ around the z-axis by angle ϕ. In other words, the expectation value of the spin operator behaves as though it were a classical vector under rotation:

$$\langle S_k \rangle \to \sum_l R_{kl} \langle S_l \rangle, \tag{3.2.11}$$

where R_{kl} are the elements of the 3×3 orthogonal matrix R that specifies the rotation in question. It should be clear from our derivation 2 that this property is not restricted to the spin operator of spin $\frac{1}{2}$ systems. In general, we have

$$\langle J_k \rangle \to \sum_l R_{kl} \langle J_l \rangle \tag{3.2.12}$$

under rotation, where J_k are the generators of rotations satisfying the angular-momentum commutation relations (3.1.20). Later we will show that relations of this kind can be further generalized to any vector operator.

So far everything has been as expected. But now, be prepared for a surprise! We examine the effect of rotation operator (3.2.3) on a general ket,

$$|\alpha\rangle = |+\rangle\langle +|\alpha\rangle + |-\rangle\langle -|\alpha\rangle, \tag{3.2.13}$$

a little more closely. We see that

$$\exp\left(\frac{-iS_z\phi}{\hbar}\right)|\alpha\rangle = e^{-i\phi/2}|+\rangle\langle +|\alpha\rangle + e^{i\phi/2}|-\rangle\langle -|\alpha\rangle. \tag{3.2.14}$$

The appearance of the half-angle $\phi/2$ here has an extremely interesting consequence.

Let us consider a rotation by 2π. We then have

$$|\alpha\rangle_{R_z(2\pi)} \to -|\alpha\rangle. \tag{3.2.15}$$

So the ket for the 360° rotated state differs from the original ket by a minus sign. We would need a 720° ($\phi = 4\pi$) rotation to get back to the same ket with a *plus* sign. Notice that this minus sign disappears for the expectation value of **S** because **S** is sandwiched by $|\alpha\rangle$ and $\langle\alpha|$, both of which change sign. Will this minus sign ever be observable? We will give the answer to this interesting question after we discuss spin precession once again.

Spin Precession Revisited

We now treat the problem of spin precession, already discussed in Section 2.1, from a new point of view. We recall that the basic Hamiltonian of the problem is given by

$$H = -\left(\frac{e}{m_e c}\right)\mathbf{S}\cdot\mathbf{B} = \omega S_z, \tag{3.2.16}$$

where

$$\omega \equiv \frac{|e|B}{m_e c}. \tag{3.2.17}$$

The time-evolution operator based on this Hamiltonian is given by

$$\mathscr{U}(t,0) = \exp\left(\frac{-iHt}{\hbar}\right) = \exp\left(\frac{-iS_z\omega t}{\hbar}\right). \tag{3.2.18}$$

Comparing this equation with (3.2.3), we see that the time-evolution operator here is precisely the same as the rotation operator in (3.2.3) with ϕ set equal to ωt. In this manner we see immediately why this Hamiltonian causes spin precession. Paraphrasing (3.2.8), (3.2.9), and (3.2.10), we obtain

$$\langle S_x\rangle_t = \langle S_x\rangle_{t=0}\cos\omega t - \langle S_y\rangle_{t=0}\sin\omega t, \tag{3.2.19a}$$

$$\langle S_y\rangle_t = \langle S_y\rangle_{t=0}\cos\omega t + \langle S_x\rangle_{t=0}\sin\omega t, \tag{3.2.19b}$$

$$\langle S_z\rangle_t = \langle S_z\rangle_{t=0}. \tag{3.2.19c}$$

After $t = 2\pi/\omega$, the spin returns to its original direction.

This set of equations can be used to discuss the spin precession of a **muon**, an electronlike particle which, however, is 210 times as heavy. The muon magnetic moment can be determined from other experiments—for example, the hyperfine splitting in muonium, a bound state of a positive muon and an electron—to be $e\hbar/2m_\mu c$, just as expected from Dirac's relativistic theory of spin $\frac{1}{2}$ particles. (We will here neglect very small corrections that arise from quantum field theory effects). Knowing the magnetic moment we can predict the angular frequency of precession. So (3.2.19) can be and, in fact, has been checked experimentally. In practice, as the external magnetic field causes spin precession, the spin direction is analyzed by taking advantage of the fact that electrons from muon decay tend to be emitted preferentially in the direction opposite to the muon spin.

Let us now look at the time evolution of the state ket itself. Assuming that the initial ($t=0$) ket is given by (3.2.13), we obtain after time t

$$|\alpha, t_0=0; t\rangle = e^{-i\omega t/2}|+\rangle\langle+|\alpha\rangle + e^{+i\omega t/2}|-\rangle\langle-|\alpha\rangle. \quad (3.2.20)$$

Expression (3.2.20) acquires a minus sign at $t=2\pi/\omega$, and we must wait until $t=4\pi/\omega$ to get back to the original state ket with the same sign. To sum up, the period for the state ket is *twice* as long as the period for spin precession

$$\tau_{\text{precession}} = \frac{2\pi}{\omega}, \quad (3.2.21a)$$

$$\tau_{\text{state ket}} = \frac{4\pi}{\omega}. \quad (3.2.21b)$$

Neutron Interferometry Experiment to Study 2π Rotations

We now describe an experiment performed to detect the minus sign in (3.2.15). Quite clearly, if every state ket in the universe is multiplied by a minus sign, there will be no observable effect. The only way to detect the predicted minus sign is to make a comparison between an unrotated state and a rotated state. As in gravity-induced quantum interference, discussed in Section 2.6, we rely on the art of neutron interferometry to verify this extraordinary prediction of quantum mechanics.

A nearly monoenergetic beam of thermal neutrons is split into two parts—path A and path B; see Figure 3.2. Path A always goes through a magnetic-field-free region; in contrast, path B enters a small region where a static magnetic field is present. As a result, the neutron state ket going via path B suffers a phase change $e^{\mp i\omega T/2}$, where T is the time spent in the $\mathbf{B}\neq 0$ region and ω is the spin-precession frequency

$$\omega = \frac{g_n eB}{m_p c}, \quad (g_n \simeq -1.91) \quad (3.2.22)$$

for the neutron with a magnetic moment of $g_n e\hbar/2m_p c$, as we can see if we

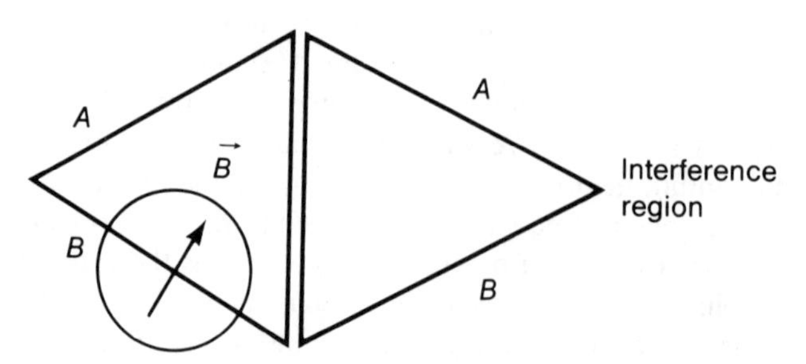

FIGURE 3.2. Experiment to study the predicted minus sign under a 2π rotation.

compare this with (3.2.17), which is appropriate for the electron with magnetic moment $e\hbar/2m_e c$. When path A and path B meet again in the interference region of Figure 3.2, the amplitude of the neutron arriving via path B is

$$c_2 = c_2(B=0)e^{\mp i\omega T/2}, \tag{3.2.23}$$

while the amplitude of the neutron arriving via path A is c_1, independent of **B**. So the intensity observable in the interference region must exhibit a sinusoidal variation

$$\cos\left(\frac{\mp\omega T}{2} + \delta\right), \tag{3.2.24}$$

where δ is the phase difference between c_1 and c_2 $(B=0)$. In practice, T, the time spent in the $B \neq 0$ region, is fixed but the precession frequency ω is varied by changing the strength of the magnetic field. The intensity in the interference region as a function of B is predicted to have a sinusoidal variation. If we call ΔB the difference in B needed to produce successive maxima, we can easily show that

$$\Delta B = \frac{4\pi\hbar c}{e g_n \lambda\!\!\!^{-} l}, \tag{3.2.25}$$

where l is the path length.

In deriving this formula we used the fact that a 4π rotation is needed for the state ket to return to the original ket with the same sign, as required by our formalism. If, on the other hand, our description of spin $\frac{1}{2}$ systems were incorrect and the ket were to return to its original ket with the same sign under a 2π rotation, the predicted value for ΔB would be just one-half of (3.2.25).

Two different groups have conclusively demonstrated experimentally that prediction (3.2.25) is correct to an accuracy of a fraction of a percent.* This is another triumph of quantum mechanics. The nontrivial prediction (3.2.15) has been experimentally established in a direct manner.

Pauli Two-Component Formalism

Manipulations with the state kets of spin $\frac{1}{2}$ systems can be conveniently carried out using the two-component spinor formalism introduced by W. Pauli in 1926. In Section 1.3 we learned how a ket (bra) can be represented by a column (row) matrix; all we have to do is arrange the expansion coefficients in terms of a certain specified set of base kets into a

*H. Rauch et al., *Phys. Lett.* **54A**, 425 (1975); S. A. Werner et al., *Phys. Rev. Lett.* **35** (1975), 1053.

column (row) matrix. In the spin $\frac{1}{2}$ case we have

$$|+\rangle \doteq \begin{pmatrix} 1 \\ 0 \end{pmatrix} \equiv \chi_+ \qquad |-\rangle \doteq \begin{pmatrix} 0 \\ 1 \end{pmatrix} \equiv \chi_-$$
$$\langle +| \doteq (1,0) = \chi_+^\dagger \qquad \langle -| \doteq (0,1) = \chi_-^\dagger \tag{3.2.26}$$

for the base kets and bras and

$$|\alpha\rangle = |+\rangle\langle +|\alpha\rangle + |-\rangle\langle -|\alpha\rangle \doteq \begin{pmatrix} \langle +|\alpha\rangle \\ \langle -|\alpha\rangle \end{pmatrix} \tag{3.2.27a}$$

and

$$\langle\alpha| = \langle\alpha|+\rangle\langle +| + \langle\alpha|-\rangle\langle -| \doteq (\langle\alpha|+\rangle, \langle\alpha|-\rangle) \tag{3.2.27b}$$

for an arbitrary state ket and the corresponding state bra. Column matrix (3.2.27a) is referred to as a **two-component spinor** and is written as

$$\chi = \begin{pmatrix} \langle +|\alpha\rangle \\ \langle -|\alpha\rangle \end{pmatrix} \equiv \begin{pmatrix} c_+ \\ c_- \end{pmatrix}$$
$$= c_+\chi_+ + c_-\chi_-, \tag{3.2.28}$$

where c_+ and c_- are, in general, complex numbers. For $\chi^\dagger$ we have

$$\chi^\dagger = (\langle\alpha|+\rangle, \langle\alpha|-\rangle) = (c_+^*, c_-^*). \tag{3.2.29}$$

The matrix elements $\langle \pm|S_k|+\rangle$ and $\langle \pm|S_k|-\rangle$, apart from $\hbar/2$, are to be set equal to those of 2×2 matrices σ_k, known as the **Pauli matrices.** We identify

$$\langle \pm|S_k|+\rangle \equiv \left(\frac{\hbar}{2}\right)(\sigma_k)_{\pm,+}, \qquad \langle \pm|S_k|-\rangle \equiv \left(\frac{\hbar}{2}\right)(\sigma_k)_{\pm,-}. \tag{3.2.30}$$

We can now write the expectation value $\langle S_k\rangle$ in terms of χ and σ_k:

$$\langle S_k\rangle = \langle\alpha|S_k|\alpha\rangle = \sum_{a'=+,-}\ \sum_{a''=+,-} \langle\alpha|a'\rangle\langle a'|S_k|a''\rangle\langle a''|\alpha\rangle$$
$$= \left(\frac{\hbar}{2}\right)\chi^\dagger\sigma_k\chi, \tag{3.2.31}$$

where the usual rule of matrix multiplication is used in the last line. Explicitly, we see from (3.2.1) together with (3.2.30) that

$$\sigma_1 = \begin{pmatrix} 0 & 1 \\ 1 & 0 \end{pmatrix}, \qquad \sigma_2 = \begin{pmatrix} 0 & -i \\ i & 0 \end{pmatrix}, \qquad \sigma_3 = \begin{pmatrix} 1 & 0 \\ 0 & -1 \end{pmatrix} \tag{3.2.32}$$

where the subscripts 1, 2, and 3 refer to x, y, and z, respectively.

We record some properties of the Pauli matrices. First,

$$\sigma_i^2 = 1 \tag{3.2.33a}$$
$$\sigma_i\sigma_j + \sigma_j\sigma_i = 0, \quad \text{for } i \neq j, \tag{3.2.33b}$$

where the right-hand side of (3.2.33a) is to be understood as the 2×2

identity matrix. These two relations are, of course, equivalent to the anticommutation relations

$$\{\sigma_i, \sigma_j\} = 2\delta_{ij}. \tag{3.2.34}$$

We also have the commutation relations

$$[\sigma_i, \sigma_j] = 2i\varepsilon_{ijk}\sigma_k, \tag{3.2.35}$$

which we see to be the explicit 2×2 matrix realizations of the angular-momentum commutation relations (3.1.20). Combining (3.2.34) and (3.2.35), we can obtain

$$\sigma_1\sigma_2 = -\sigma_2\sigma_1 = i\sigma_3 \dots . \tag{3.2.36}$$

Notice also that

$$\sigma_i^\dagger = \sigma_i, \tag{3.2.37a}$$

$$\det(\sigma_i) = -1, \tag{3.2.37b}$$

$$\mathrm{Tr}(\sigma_i) = 0. \tag{3.2.37c}$$

We now consider $\boldsymbol{\sigma}\cdot\mathbf{a}$, where $\mathbf{a}$ is a vector in three dimensions. This is actually to be understood as a 2×2 matrix. Thus

$$\begin{aligned}\boldsymbol{\sigma}\cdot\mathbf{a} &\equiv \sum_k a_k\sigma_k \\ &= \begin{pmatrix} +a_3 & a_1 - ia_2 \\ a_1 + ia_2 & -a_3 \end{pmatrix}.\end{aligned} \tag{3.2.38}$$

There is also a very important identity,

$$(\boldsymbol{\sigma}\cdot\mathbf{a})(\boldsymbol{\sigma}\cdot\mathbf{b}) = \mathbf{a}\cdot\mathbf{b} + i\boldsymbol{\sigma}\cdot(\mathbf{a}\times\mathbf{b}). \tag{3.2.39}$$

To prove this all we need are the anticommutation and commutation relations, (3.2.34) and (3.2.35), respectively:

$$\begin{aligned}\sum_j \sigma_j a_j \sum_k \sigma_k b_k &= \sum_j\sum_k \left(\frac{1}{2}\{\sigma_j, \sigma_k\} + \frac{1}{2}[\sigma_j, \sigma_k]\right) a_j b_k \\ &= \sum_j\sum_k (\delta_{jk} + i\varepsilon_{jkl}\sigma_l) a_j b_k \\ &= \mathbf{a}\cdot\mathbf{b} + i\boldsymbol{\sigma}\cdot(\mathbf{a}\times\mathbf{b}).\end{aligned} \tag{3.2.40}$$

If the components of $\mathbf{a}$ are real, we have

$$(\boldsymbol{\sigma}\cdot\mathbf{a})^2 = |\mathbf{a}|^2, \tag{3.2.41}$$

where $|\mathbf{a}|$ is the magnitude of the vector $\mathbf{a}$.

Rotations in the Two-Component Formalism

Let us now study the 2×2 matrix representation of the rotation operator $\mathscr{D}(\hat{\mathbf{n}}, \phi)$. We have

$$\exp\left(\frac{-i\mathbf{S}\cdot\hat{\mathbf{n}}\phi}{\hbar}\right) \doteq \exp\left(\frac{-i\boldsymbol{\sigma}\cdot\hat{\mathbf{n}}\phi}{2}\right). \tag{3.2.42}$$

Using

$$(\boldsymbol{\sigma}\cdot\hat{\mathbf{n}})^n = \begin{cases} 1 & \text{for } n \text{ even,} \\ \boldsymbol{\sigma}\cdot\hat{\mathbf{n}} & \text{for } n \text{ odd,} \end{cases} \tag{3.2.43}$$

which follows from (3.2.41), we can write

$$\begin{aligned} \exp\left(\frac{-i\boldsymbol{\sigma}\cdot\hat{\mathbf{n}}\phi}{2}\right) &= \left[1 - \frac{(\boldsymbol{\sigma}\cdot\hat{\mathbf{n}})^2}{2!}\left(\frac{\phi}{2}\right)^2 + \frac{(\boldsymbol{\sigma}\cdot\hat{\mathbf{n}})^4}{4!}\left(\frac{\phi}{2}\right)^4 - \cdots\right] \\ &\quad - i\left[(\boldsymbol{\sigma}\cdot\hat{\mathbf{n}})\frac{\phi}{2} - \frac{(\boldsymbol{\sigma}\cdot\hat{\mathbf{n}})^3}{3!}\left(\frac{\phi}{2}\right)^3 + \cdots\right] \\ &= \mathbf{1}\cos\left(\frac{\phi}{2}\right) - i\boldsymbol{\sigma}\cdot\hat{\mathbf{n}}\sin\left(\frac{\phi}{2}\right). \end{aligned} \tag{3.2.44}$$

Explicitly, in 2×2 form we have

$$\exp\left(\frac{-i\boldsymbol{\sigma}\cdot\hat{\mathbf{n}}\phi}{2}\right) = \begin{pmatrix} \cos\left(\frac{\phi}{2}\right) - in_z\sin\left(\frac{\phi}{2}\right) & (-in_x - n_y)\sin\left(\frac{\phi}{2}\right) \\ (-in_x + n_y)\sin\left(\frac{\phi}{2}\right) & \cos\left(\frac{\phi}{2}\right) + in_z\sin\left(\frac{\phi}{2}\right) \end{pmatrix} \tag{3.2.45}$$

Just as the operator $\exp(-i\mathbf{S}\cdot\hat{\mathbf{n}}\phi/\hbar)$ acts on a state ket $|\alpha\rangle$, the 2×2 matrix $\exp(-i\boldsymbol{\sigma}\cdot\hat{\mathbf{n}}\phi/2)$ acts on a two-component spinor χ. Under rotations we change χ as follows:

$$\chi \to \exp\left(\frac{-i\boldsymbol{\sigma}\cdot\hat{\mathbf{n}}\phi}{2}\right)\chi. \tag{3.2.46}$$

On the other hand, the σ_k's themselves are to remain *unchanged* under rotations. So strictly speaking, despite its appearance, $\boldsymbol{\sigma}$ is not to be regarded as a vector; rather, it is $\chi^\dagger\boldsymbol{\sigma}\chi$ which obeys the transformation property of a vector:

$$\chi^\dagger\sigma_k\chi \to \sum_l R_{kl}\chi^\dagger\sigma_l\chi. \tag{3.2.47}$$

An explicit proof of this may be given using

$$\exp\left(\frac{i\sigma_3\phi}{2}\right)\sigma_1\exp\left(\frac{-i\sigma_3\phi}{2}\right) = \sigma_1\cos\phi - \sigma_2\sin\phi \tag{3.2.48}$$

and so on, which is the 2×2 matrix analogue of (3.2.6).

In discussing a 2π rotation using the ket formalism, we have seen that a spin $\frac{1}{2}$ ket $|\alpha\rangle$ goes into $-|\alpha\rangle$. The 2×2 analogue of this statement is

$$\exp\left(\frac{-i\boldsymbol{\sigma}\cdot\hat{\mathbf{n}}\phi}{2}\right)\bigg|_{\phi=2\pi} = -1, \quad \text{for any } \hat{\mathbf{n}}, \tag{3.2.49}$$

which is evident from (3.2.44).

As an instructive application of rotation matrix (3.2.45), let us see how we can construct an eigenspinor of $\boldsymbol{\sigma}\cdot\hat{\mathbf{n}}$ with eigenvalue $+1$, where $\hat{\mathbf{n}}$ is a unit vector in some specified direction. Our object is to construct χ satisfying

$$\boldsymbol{\sigma}\cdot\hat{\mathbf{n}}\chi = \chi. \tag{3.2.50}$$

In other words, we look for the two-component column matrix representation of $|\mathbf{S}\cdot\hat{\mathbf{n}};+\rangle$ defined by

$$\mathbf{S}\cdot\hat{\mathbf{n}}|\mathbf{S}\cdot\hat{\mathbf{n}};+\rangle = \left(\frac{\hbar}{2}\right)|\mathbf{S}\cdot\hat{\mathbf{n}};+\rangle. \tag{3.2.51}$$

Actually this can be solved as a straightforward eigenvalue problem (see Problem 9 in Chapter 1), but here we present an alternative method based on rotation matrix (3.2.45).

Let the polar and the azimuthal angles that characterize $\hat{\mathbf{n}}$ be β and α, respectively. We start with $\begin{pmatrix}1\\0\end{pmatrix}$, the two-component spinor that represents the spin-up state. Given this, we first rotate about the y-axis by angle β; we subsequently rotate by angle α about the z-axis. We see that the desired

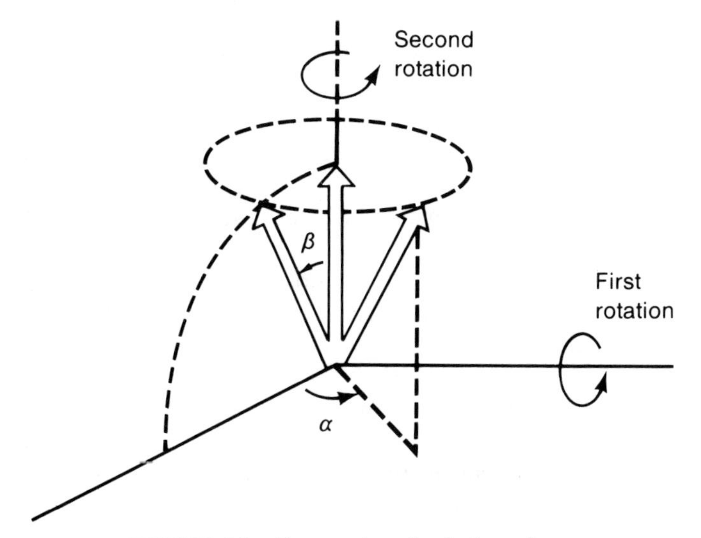

FIGURE 3.3. Construction of $\boldsymbol{\sigma}\cdot\hat{\mathbf{n}}$ eigenspinor.

spin state is then obtained; see Figure 3.3. In the Pauli spinor language this sequence of operations is equivalent to applying $\exp(-i\sigma_2\beta/2)$ to $\begin{pmatrix}1\\0\end{pmatrix}$ followed by an application of $\exp(-i\sigma_3\alpha/2)$. The net result is

$$\chi=\left[\cos\left(\frac{\alpha}{2}\right)-i\sigma_3\sin\left(\frac{\alpha}{2}\right)\right]\left[\cos\left(\frac{\beta}{2}\right)-i\sigma_2\sin\left(\frac{\beta}{2}\right)\right]\begin{pmatrix}1\\0\end{pmatrix}$$

$$=\begin{pmatrix}\cos\left(\frac{\alpha}{2}\right)-i\sin\left(\frac{\alpha}{2}\right) & 0\\ 0 & \cos\left(\frac{\alpha}{2}\right)+i\sin\left(\frac{\alpha}{2}\right)\end{pmatrix}\begin{pmatrix}\cos\left(\frac{\beta}{2}\right) & -\sin\left(\frac{\beta}{2}\right)\\ \sin\left(\frac{\beta}{2}\right) & \cos\left(\frac{\beta}{2}\right)\end{pmatrix}\begin{pmatrix}1\\0\end{pmatrix}$$

$$=\begin{pmatrix}\cos\left(\frac{\beta}{2}\right)e^{-i\alpha/2}\\ \sin\left(\frac{\beta}{2}\right)e^{i\alpha/2}\end{pmatrix}, \tag{3.2.52}$$

in complete agreement with Problem 9 of Chapter 1 if we realize that a phase common to both the upper and lower components is devoid of physical significance.

3.3. SO(3), SU(2), AND EULER ROTATIONS

Orthogonal Group

We will now study a little more systematically the group properties of the operations with which we have been concerned in the previous two sections.

The most elementary approach to rotations is based on specifying the axis of rotation and the angle of rotation. It is clear that we need three real numbers to characterize a general rotation: the polar and the azimuthal angles of the unit vector $\hat{\mathbf{n}}$ taken in the direction of the rotation axis and the rotation angle ϕ itself. Equivalently, the same rotation can be specified by the three Cartesian components of the vector $\hat{\mathbf{n}}\phi$. However, these ways of characterizing rotation are not so convenient from the point of view of studying the group properties of rotations. For one thing, unless ϕ is infinitesimal or $\hat{\mathbf{n}}$ is always in the same direction, we cannot add vectors of the form $\hat{\mathbf{n}}\phi$ to characterize a succession of rotations. It is much easier to work with a 3×3 orthogonal matrix R because the effect of successive rotations can be obtained just by multiplying the appropriate orthogonal matrices.

How many independent parameters are there in a 3×3 orthogonal matrix? A real 3×3 matrix has 9 entries, but we have the orthogonality constraint

$$RR^T=1. \tag{3.3.1}$$

which corresponds to 6 independent equations because the product RR^T, being the same as R^TR, is a symmetrical matrix with 6 independent entries. As a result, there are 3 (that is, $9-6$) independent numbers in R, the same number previously obtained by a more elementary method.

The set of all multiplication operations with orthogonal matrices forms a group. By this we mean that the following four requirements are satisfied:

1. The product of any two orthogonal matrices is another orthogonal matrix, which is satisfied because
$$(R_1R_2)(R_1R_2)^T = R_1R_2R_2^TR_1^T = 1. \tag{3.3.2}$$
2. The associative law holds:
$$R_1(R_2R_3) = (R_1R_2)R_3. \tag{3.3.3}$$
3. The identity matrix 1—physically corresponding to no rotation —defined by
$$R1 = 1R = R \tag{3.3.4}$$
is a member of the class of all orthogonal matrices.
4. The inverse matrix R^{-1}—physically corresponding to rotation in the opposite sense—defined by
$$RR^{-1} = R^{-1}R = 1 \tag{3.3.5}$$
is also a member.

This group has the name SO(3), where S stands for special, O stands for orthogonal, 3 for three dimensions. Note only rotational operations are considered here, hence we have SO(3) rather than O(3) (which can include the inversion operation of Chapter 4 later).

Unitary unimodular group

In the previous section we learned yet another way to characterize an arbitrary rotation—that is, to look at the 2×2 matrix (3.2.45) that acts on the two-component spinor χ. Clearly, (3.2.45) is unitary. As a result, for the c_+ and c_-, defined in (3.2.28),
$$|c_+|^2 + |c_-|^2 = 1 \tag{3.3.6}$$
is left invariant. Furthermore, matrix (3.2.45) is unimodular; that is, its determinant is 1, as will be shown explicitly below.

We can write the most general unitary unimodular matrix as
$$U(a,b) = \begin{pmatrix} a & b \\ -b^* & a^* \end{pmatrix}, \tag{3.3.7}$$
where a and b are *complex* numbers satisfying the unimodular condition
$$|a|^2 + |b|^2 = 1. \tag{3.3.8}$$

We can easily establish the unitary property of (3.3.7) as follows:

$$U(a,b)^{\dagger}U(a,b)=\begin{pmatrix} a^* & -b \\ b^* & a \end{pmatrix}\begin{pmatrix} a & b \\ -b^* & a^* \end{pmatrix}=1, \qquad (3.3.9)$$

where we have used (3.3.8). Notice that the number of independent real parameters in (3.3.7) is again three.

We can readily see that the 2×2 matrix (3.2.45) that characterizes a rotation of a spin $\frac{1}{2}$ system can be written as $U(a,b)$. Comparing (3.2.45) with (3.3.7), we identify

$$\begin{aligned} \mathrm{Re}(a) &= \cos\left(\frac{\phi}{2}\right), & \mathrm{Im}(a) &= -n_z\sin\left(\frac{\phi}{2}\right), \\ \mathrm{Re}(b) &= -n_y\sin\left(\frac{\phi}{2}\right), & \mathrm{Im}(b) &= -n_x\sin\left(\frac{\phi}{2}\right), \end{aligned} \qquad (3.3.10)$$

from which the unimodular property of (3.3.8) is immediate. Conversely, it is clear that the most general unitary unimodular matrix of form (3.3.7) can be interpreted as representing a rotation.

The two complex numbers a and b are known as **Cayley-Klein parameters**. Historically the connection between a unitary unimodular matrix and a rotation was known long before the birth of quantum mechanics. In fact, the Cayley-Klein parameters were used to characterize complicated motions of gyroscopes in rigid-body kinematics.

Without appealing to the interpretations of unitary unimodular matrices in terms of rotations, we can directly check the group properties of multiplication operations with unitary unimodular matrices. Note in particular that

$$U(a_1,b_1)U(a_2,b_2)=U(a_1a_2-b_1b_2^*,a_1b_2+a_2^*b_1), \qquad (3.3.11)$$

where the unimodular condition for the product matrix is

$$|a_1a_2-b_1b_2^*|^2+|a_1b_2+a_2^*b_1|^2=1. \qquad (3.3.12)$$

For the inverse of U we have

$$U^{-1}(a,b)=U(a^*,-b). \qquad (3.3.13)$$

This group is known as SU(2), where S stands for special, U for unitary, and 2 for dimensionality 2. In contrast, the group defined by multiplication operations with general 2×2 unitary matrices (not necessarily constrained to be unimodular) is known as U(2). The most general unitary matrix in two dimensions has four independent parameters and can be written as $e^{i\gamma}$ (with γ real) times a unitary unimodular matrix:

$$U=e^{i\gamma}\begin{pmatrix} a & b \\ -b^* & a^* \end{pmatrix}, \qquad |a|^2+|b|^2=1, \qquad \gamma^*=\gamma. \qquad (3.3.14)$$

SU(2) is called a **subgroup** of U(2).

Because we can characterize rotations using both the SO(3) language and the SU(2) language, we may be tempted to conclude that the *groups* SO(3) and SU(2) are isomorphic—that is, that there is a one-to-one correspondence between an element of SO(3) and an element of SU(2). This inference is not correct. Consider a rotation by 2π and another one by 4π. In the SO(3) language, the matrices representing a 2π rotation and a 4π rotation are both 3×3 identity matrices; however, in the SU(2) language the corresponding matrices are -1 times the 2×2 identity matrix and the identity matrix itself, respectively. More generally, $U(a,b)$ and $U(-a,-b)$ both correspond to a *single* 3×3 matrix in the SO(3) language. The correspondence therefore is two-to-one; for a given R, the corresponding U is double-valued. One can say, however, that the two groups are *locally* isomorphic.

Euler Rotations

From classical mechanics the reader may be familiar with the fact that an arbitrary rotation of a rigid body can be accomplished in three steps, known as **Euler rotations**. The Euler rotation language, specified by three Euler angles, provides yet another way to characterize the most general rotation in three dimensions.

The three steps of Euler rotations are as follows. First, rotate the rigid body counterclockwise (as seen from the positive z-side) about the z-axis by angle α. Imagine now that there is a body y-axis embedded, so to speak, in the rigid body such that before the z-axis rotation is carried out, the body y-axis coincides with the usual y-axis, referred to as the **space-fixed y-axis**. Obviously, after the rotation about the z-axis, the body y-axis no longer coincides with the space-fixed y-axis; let us call the former the y'-axis. To see how all this may appear for a thin disk, refer to Figure 3.4a. We now perform a second rotation, this time about the y'-axis by angle β. As a result, the body z-axis no longer points in the space-fixed z-axis direction. We call the body-fixed z-axis after the second rotation the z'-axis; see Figure 3.4b. The third and final rotation is about the z'-axis by angle γ. The body y-axis now becomes the y''-axis of Figure 3.4c. In terms of 3×3 orthogonal matrices the product of the three operations can be written as

$$R(\alpha,\beta,\gamma)\equiv R_{z'}(\gamma)R_{y'}(\beta)R_z(\alpha). \qquad (3.3.15)$$

A cautionary remark is in order here. Most textbooks in classical mechanics prefer to perform the second rotation (the middle rotation) about the body x-axis rather than about the body y-axis (see, for example, Goldstein 1980). This convention is to be avoided in quantum mechanics for a reason that will become apparent in a moment.

In (3.3.15) there appear $R_{y'}$ and $R_{z'}$, which are matrices for rotations about body axes. This approach to Euler rotations is rather inconvenient in

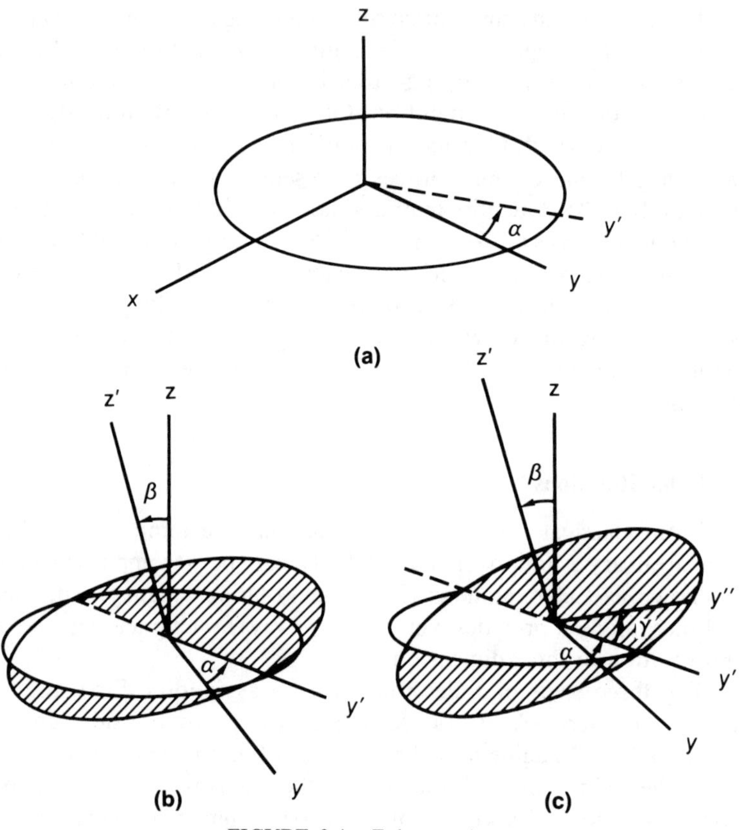

FIGURE 3.4. Euler rotations.

quantum mechanics because we earlier obtained simple expressions for the space-fixed (unprimed) axis components of the **S** operator, but not for the body-axis components. It is therefore desirable to express the body-axis rotations we considered in terms of space-fixed axis rotations. Fortunately there is a very simple relation, namely,

$$R_{y'}(\beta) = R_z(\alpha) R_y(\beta) R_z^{-1}(\alpha). \tag{3.3.16}$$

The meaning of the right-hand side is as follows. First, bring the body y-axis of Figure 3.4a (that is, the y'-axis) back to the original fixed-space y-direction by rotating *clockwise* (as seen from the positive z-side) about the z-axis by angle α; then rotate about the y-axis by angle β. Finally, return the body y-axis to the direction of the y'-axis by rotating about the fixed-space z-axis (*not* about the z'-axis!) by angle α. Equation (3.3.16) tells us that the net effect of these rotations is a single rotation about the y'-axis by angle β.

To prove this assertion, let us look more closely at the effect of both sides of (3.3.16) on the circular disc of Figure 3.4a. Clearly, the orientation of the body y-axis is unchanged in both cases, namely, in the y'-direction. Furthermore, the orientation of the final body z-axis is the same whether we apply $R_{y'}(\beta)$ or $R_z(\alpha)R_y(\beta)R_z^{-1}(\alpha)$. In both cases the final body z-axis makes a polar angle β with the fixed z-axis (the same as the initial z-axis), and its azimuthal angle, as measured in the fixed-coordinate system, is just α. In other words, the final body z-axis is the same as the z'-axis of Figure 3.4b. Similarly, we can prove

$$R_{z'}(\gamma) = R_{y'}(\beta)R_z(\gamma)R_{y'}^{-1}(\beta). \tag{3.3.17}$$

Using (3.3.16) and (3.3.17), we can now rewrite (3.3.15). We obtain

$$\begin{aligned} R_{z'}(\gamma)R_{y'}(\beta)R_z(\alpha) &= R_{y'}(\beta)R_z(\gamma)R_{y'}^{-1}(\beta)R_{y'}(\beta)R_z(\alpha) \\ &= R_z(\alpha)R_y(\beta)R_z^{-1}(\alpha)R_z(\gamma)R_z(\alpha) \\ &= R_z(\alpha)R_y(\beta)R_z(\gamma), \end{aligned} \tag{3.3.18}$$

where in the final step we used the fact that $R_z(\gamma)$ and $R_z(\alpha)$ commute. To summarize,

$$R(\alpha,\beta,\gamma) = R_z(\alpha)R_y(\beta)R_z(\gamma), \tag{3.3.19}$$

where all three matrices on the right-hand side refer to *fixed*-axis rotations.

Now let us apply this set of operations to spin $\frac{1}{2}$ systems in quantum mechanics. Corresponding to the product of orthogonal matrices in (3.3.19) there exists a product of rotation operators in the ket space of the spin $\frac{1}{2}$ system under consideration:

$$\mathscr{D}(\alpha,\beta,\gamma) = \mathscr{D}_z(\alpha)\mathscr{D}_y(\beta)\mathscr{D}_z(\gamma). \tag{3.3.20}$$

The 2×2 matrix representation of this product is

$$\begin{aligned} &\exp\left(\frac{-i\sigma_3\alpha}{2}\right)\exp\left(\frac{-i\sigma_2\beta}{2}\right)\exp\left(\frac{-i\sigma_3\gamma}{2}\right) \\ &= \begin{pmatrix} e^{-i\alpha/2} & 0 \\ 0 & e^{i\alpha/2} \end{pmatrix} \begin{pmatrix} \cos(\beta/2) & -\sin(\beta/2) \\ \sin(\beta/2) & \cos(\beta/2) \end{pmatrix} \begin{pmatrix} e^{-i\gamma/2} & 0 \\ 0 & e^{i\gamma/2} \end{pmatrix} \\ &= \begin{pmatrix} e^{-i(\alpha+\gamma)/2}\cos(\beta/2) & -e^{-i(\alpha-\gamma)/2}\sin(\beta/2) \\ e^{i(\alpha-\gamma)/2}\sin(\beta/2) & e^{i(\alpha+\gamma)/2}\cos(\beta/2) \end{pmatrix}, \end{aligned} \tag{3.3.21}$$

where (3.2.44) was used. This matrix is clearly of the unitary unimodular form. Conversely, the most general 2×2 unitary unimodular matrix can be written in this Euler angle form.

Notice that the matrix elements of the second (middle) rotation $\exp(-i\sigma_y\phi/2)$ are purely real. This would not have been the case had we chosen to rotate about the x-axis rather than the y-axis, as done in most textbooks in classical mechanics. In quantum mechanics it pays to stick to our convention because we prefer the matrix elements of the second rotation, which is the only rotation matrix containing off-diagonal elements, to be purely real.*

The 2×2 matrix in (3.3.21) is called the $j=\frac{1}{2}$ irreducible representation of the rotation operator $\mathscr{D}(\alpha,\beta,\gamma)$. Its matrix elements are denoted by $\mathscr{D}^{(1/2)}_{m'm}(\alpha,\beta,\gamma)$. In terms of the angular-momentum operators we have

$$\mathscr{D}^{(1/2)}_{m'm}(\alpha,\beta,\gamma)=\left\langle j=\frac{1}{2},m'\right|\exp\left(\frac{-iJ_z\alpha}{\hbar}\right)$$

$$\times\exp\left(\frac{-iJ_y\beta}{\hbar}\right)\exp\left(\frac{-iJ_z\gamma}{\hbar}\right)\left|j=\frac{1}{2},m\right\rangle. \qquad (3.3.22)$$

In Section 3.5 we will extensively study higher j-analogues of (3.3.21).

3.4. DENSITY OPERATORS AND PURE VERSUS MIXED ENSEMBLES

Polarized Versus Unpolarized Beams

The formalism of quantum mechanics developed so far makes statistical predictions on an *ensemble*, that is, a collection, of identically prepared physical systems. More precisely, in such an ensemble all members are supposed to be characterized by the same state ket $|\alpha\rangle$. A good example of this is a beam of silver atoms coming out of an SG filtering apparatus. Every atom in the beam has its spin pointing in the same direction, namely, the direction determined by the inhomogeneity of the magnetic field of the filtering apparatus. We have not yet discussed how to describe quantum mechanically an ensemble of physical systems for which some, say 60%, are characterized by $|\alpha\rangle$, and the remaining 40% are characterized by some other ket $|\beta\rangle$.

To illustrate vividly the incompleteness of the formalism developed so far, let us consider silver atoms coming directly out of a hot oven, yet to be subjected to a filtering apparatus of the Stern-Gerlach type. On symmetry grounds we expect that such atoms have *random* spin orientations; in other words, there should be no preferred direction associated with such an ensemble of atoms. According to the formalism developed so far, the most general state ket of a spin $\frac{1}{2}$ system is given by

$$|\alpha\rangle=c_+|+\rangle+c_-|-\rangle. \qquad (3.4.1)$$

*This, of course, depends on our convention that the matrix elements of S_y (or, more generally, J_y) are taken to be purely imaginary.

Is this equation capable of describing a collection of atoms with random spin orientations? The answer is clearly no; (3.4.1) characterizes a state ket whose spin is pointing in *some definite direction*, namely, in the direction of $\hat{\mathbf{n}}$, whose polar and azimuthal angles, β and α, respectively, are obtained by solving

$$\frac{c_+}{c_-} = \frac{\cos(\beta/2)}{e^{i\alpha}\sin(\beta/2)}; \tag{3.4.2}$$

see (3.2.52).

To cope with a situation of this kind we introduce the concept of **fractional population**, or probability weight. An ensemble of silver atoms with completely random spin orientation can be viewed as a collection of silver atoms in which 50% of the members of the ensemble are characterized by $|+\rangle$ and the remaining 50% by $|-\rangle$. We specify such an ensemble by assigning

$$w_+ = 0.5, \qquad w_- = 0.5, \tag{3.4.3}$$

where w_+ and w_- are the fractional population for spin-up and -down, respectively. Because there is no preferred direction for such a beam, it is reasonable to expect that this *same* ensemble can be regarded equally well as a 50-50 mixture of $|S_x;+\rangle$ and $|S_x;-\rangle$. The mathematical formalism needed to accomplish this will appear shortly.

It is very important to note that we are simply introducing here two *real* numbers w_+ and w_-. There is no information on the relative phase between the spin-up and the spin-down ket. Quite often we refer to such a situation as an **incoherent mixture** of spin-up and spin-down states. What we are doing here is to be clearly distinguished from what we did with a coherent linear superposition, for example,

$$\left(\frac{1}{\sqrt{2}}\right)|+\rangle + \left(\frac{1}{\sqrt{2}}\right)|-\rangle, \tag{3.4.4}$$

where the phase relation between $|+\rangle$ and $|-\rangle$ contains vital information on the spin orientation in the xy-plane, in this case in the positive x-direction. In general, we should not confuse w_+ and w_- with $|c_+|^2$ and $|c_-|^2$. The probability concept associated with w_+ and w_- is much closer to that encountered in classical probability theory. The situation encountered in dealing with silver atoms directly from the hot oven may be compared with that of a graduating class in which 50% of the graduating seniors are male, the remaining 50% female. When we pick a student at random, the probability that the particular student is male (or female) is 0.5. Whoever heard of a student referred to as a coherent linear superposition of male and female with a particular phase relation?

The beam of silver atoms coming directly out of the oven is an example of a **completely random ensemble**; the beam is said to be **un-**

polarized because there is no preferred direction for spin orientation. In contrast, the beam that has gone through a selective Stern-Gerlach–type measurement is an example of a **pure ensemble**; the beam is said to be **polarized** because all members of the ensemble are characterized by a single common ket that describes a state with spin pointing in some definite direction. To appreciate the difference between a completely random ensemble and a pure ensemble, let us consider a rotatable SG apparatus where we can vary the direction of the inhomogeneous **B** just by rotating the apparatus. When a completely unpolarized beam directly out of the oven is subjected to such an apparatus, we *always* obtain two emerging beams of *equal* intensity *no matter what the orientation of the apparatus may be*. In contrast, if a polarized beam is subjected to such an apparatus, the relative intensities of the two emerging beams vary as the apparatus is rotated. For some *particular* orientation the ratio of the intensities actually becomes one to zero. In fact, the formalism we developed in Chapter 1 tells us that the relative intensities are simply $\cos^2(\beta/2)$ and $\sin^2(\beta/2)$, where β is the angle between the spin direction of the atoms and the direction of the inhomogeneous magnetic field in the SG apparatus.

A complete random ensemble and a pure ensemble can be regarded as the extremes of what is known as a **mixed ensemble**. In a mixed ensemble a certain fraction—for example, 70%—of the members are characterized by a state ket $|\alpha\rangle$, the remaining 30% by $|\beta\rangle$. In such a case the beam is said to be **partially polarized**. Here $|\alpha\rangle$ and $|\beta\rangle$ need not even be orthogonal; we can, for example, have 70% with spin in the positive x-direction and 30% with spin in the negative z-direction.*

Ensemble Averages and Density Operator

We now present the density operator formalism, pioneered by J. von Neumann in 1927, that quantitatively describes physical situations with mixed as well as pure ensembles. Our general discussion here is not restricted to spin $\frac{1}{2}$ systems, but for illustrative purposes we return repeatedly to spin $\frac{1}{2}$ systems.

A pure ensemble by definition is a collection of physical systems such that every member is characterized by the same ket $|\alpha\rangle$. In contrast, in a mixed ensemble, a fraction of the members with relative population w_1 are characterized by $|\alpha^{(1)}\rangle$, some other fraction with relative population w_2, by $|\alpha^{(2)}\rangle$, and so on. Roughly speaking, a mixed ensemble can be viewed as a mixture of pure ensembles, just as the name suggests. The fractional

*In the literature what we call pure and mixed ensembles are often referred to as pure and mixed states. In this book, however, we use *state* to mean a physical system described by a definite state ket $|\alpha\rangle$.

populations are constrained to satisfy the normalization condition

$$\sum_i w_i = 1. \tag{3.4.5}$$

As we mentioned previously, $|\alpha^{(1)}\rangle$ and $|\alpha^{(2)}\rangle$ need not be orthogonal. Furthermore, the number of terms in the i sum of (3.4.5) need not coincide with the dimensionality N of the ket space; it can easily exceed N. For example, for spin $\frac{1}{2}$ systems with $N = 2$, we may consider 40% with spin in the positive z-direction, 30% with spin in the positive x-direction, and the remaining 30% with spin in the negative y-direction.

Suppose we make a measurement on a mixed ensemble of some observable A. We may ask what is the average measured value of A when a large number of measurements are carried out. The answer is given by the **ensemble average** of A, which is defined by

$$\begin{aligned}[A] &\equiv \sum_i w_i \langle\alpha^{(i)}|A|\alpha^{(i)}\rangle \\ &= \sum_i \sum_{a'} w_i |\langle a'|\alpha^{(i)}\rangle|^2 a',\end{aligned} \tag{3.4.6}$$

where $|a'\rangle$ is an eigenket of A. Recall that $\langle\alpha^{(i)}|A|\alpha^{(i)}\rangle$ is the usual quantum mechanical expectation value of A taken with respect to state $|\alpha^{(i)}\rangle$. Equation (3.4.6) tells us that these expectation values must further be weighted by the corresponding fractional populations w_i. Notice how probabilistic concepts enter twice; first in $|\langle a'|\alpha^{(i)}\rangle|^2$ for the quantum-mechanical probability for state $|\alpha^{(i)}\rangle$ to be found in an A eigenstate $|a'\rangle$; second, in the probability factor w_i for finding in the ensemble a quantum-mechanical state characterized by $|\alpha^{(i)}\rangle$.*

We can now rewrite ensemble average (3.4.6) using a more general basis, $\{|b'\rangle\}$:

$$\begin{aligned}[A] &= \sum_i w_i \sum_{b'} \sum_{b''} \langle\alpha^{(i)}|b'\rangle\langle b'|A|b''\rangle\langle b''|\alpha^{(i)}\rangle \\ &= \sum_{b'} \sum_{b''} \left(\sum_i w_i \langle b''|\alpha^{(i)}\rangle\langle\alpha^{(i)}|b'\rangle \right) \langle b'|A|b''\rangle.\end{aligned} \tag{3.4.7}$$

The number of terms in the sum of the b' (b'') is just the dimensionality of the ket space, while the number of terms in the sum of the i depends on how the mixed ensemble is viewed as a mixture of pure ensembles. Notice that in this form the basic property of the ensemble which does not depend on the particular observable A is factored out. This motivates us to define the **density operator** ρ as follows:

$$\rho \equiv \sum_i w_i |\alpha^{(i)}\rangle\langle\alpha^{(i)}|. \tag{3.4.8}$$

*Quite often in the literature the ensemble average is also called the *expectation value*. However, in this book, the term expectation value is reserved for the average measured value when measurements are carried on a pure ensemble.

The elements of the corresponding **density matrix** have the following form:

$$\langle b''|\rho|b'\rangle = \sum_i w_i \langle b''|\alpha^{(i)}\rangle\langle\alpha^{(i)}|b'\rangle. \tag{3.4.9}$$

The density operator contains all the physically significant information we can possibly obtain about the ensemble in question. Returning to (3.4.7), we see that the ensemble average can be written as

$$\begin{aligned}[A] &= \sum_{b'}\sum_{b''}\langle b''|\rho|b'\rangle\langle b'|A|b''\rangle \\ &= \mathrm{tr}(\rho A).\end{aligned} \tag{3.4.10}$$

Because the trace is independent of representations, $\mathrm{tr}(\rho A)$ can be evaluated using any convenient basis. As a result, (3.4.10) is an extremely powerful relation.

There are two properties of the density operator worth recording. First, the density operator is Hermitian, as is evident from (3.4.8). Second, the density operator satisfies the normalization condition

$$\begin{aligned}\mathrm{tr}(\rho) &= \sum_i\sum_{b'} w_i\langle b'|\alpha^{(i)}\rangle\langle\alpha^{(i)}|b'\rangle \\ &= \sum_i w_i\langle\alpha^{(i)}|\alpha^{(i)}\rangle \\ &= 1.\end{aligned} \tag{3.4.11}$$

Because of the Hermiticity and the normalization condition, for spin $\frac{1}{2}$ systems with dimensionality 2 the density operator, or the corresponding density matrix, is characterized by three independent real parameters. Four real numbers characterize a 2×2 Hermitian matrix. However, only three are independent because of the normalization condition. The three numbers needed are $[S_x]$, $[S_y]$, and $[S_z]$; the reader may verify that knowledge of these three ensemble averages is sufficient to reconstruct the density operator. The manner in which a mixed ensemble is formed can be rather involved. We may mix pure ensembles characterized by all kinds of $|\alpha^{(i)}\rangle$'s with appropriate w_i's; yet for spin $\frac{1}{2}$ systems three real numbers completely characterize the ensemble in question. This strongly suggests that a mixed ensemble can be decomposed into pure ensembles in many different ways. A problem to illustrate this point appears at the end of this chapter.

A pure ensemble is specified by $w_i = 1$ for some $|\alpha^{(i)}\rangle$—with $i = n$, for instance—and $w_i = 0$ for all other conceivable state kets, so the corresponding density operator is written as

$$\rho = |\alpha^{(n)}\rangle\langle\alpha^{(n)}| \tag{3.4.12}$$

with no summation. Clearly, the density operator for a pure ensemble is idempotent, that is,

$$\rho^2 = \rho \tag{3.4.13}$$

or, equivalently,

$$\rho(\rho-1)=0. \tag{3.4.14}$$

Thus, for a pure ensemble only, we have

$$\operatorname{tr}(\rho^2)=1. \tag{3.4.15}$$

in addition to (3.4.11). The eigenvalues of the density operator for a pure ensemble are zero or one, as can be seen by inserting a complete set of base kets that diagonalize the Hermitian operator ρ between ρ and $(\rho-1)$ of (3.4.14). When diagonalized, the density matrix for a pure ensemble must therefore look like

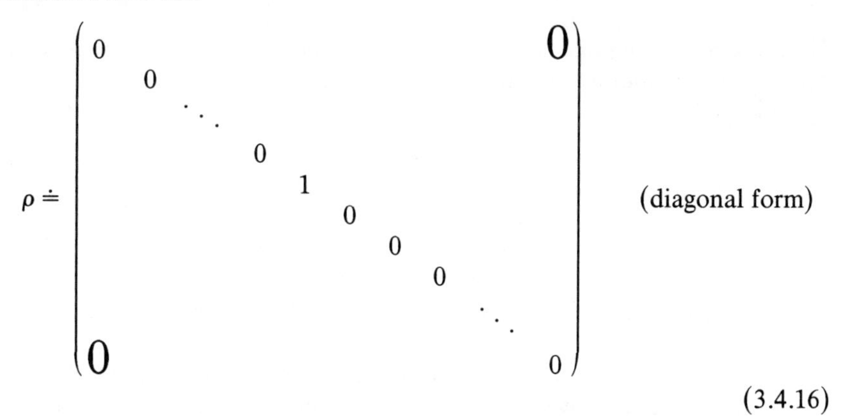

(3.4.16)

It can be shown that $\operatorname{tr}(\rho^2)$ is maximal when the ensemble is pure; for a mixed ensemble $\operatorname{tr}(\rho^2)$ is a positive number less than one.

Given a density operator, let us see how we can construct the corresponding density matrix in some specified basis. To this end we first recall that

$$|\alpha\rangle\langle\alpha| = \sum_{b'}\sum_{b''}|b'\rangle\langle b'|\alpha\rangle\langle\alpha|b''\rangle\langle b''|. \tag{3.4.17}$$

this shows that we can form the square matrix corresponding to $|\alpha^{(i)}\rangle\langle\alpha^{(i)}|$ by combining, in the sense of outer product, the column matrix formed by $\langle b'|\alpha^{(i)}\rangle$ with the row matrix formed by $\langle\alpha^{(i)}|b''\rangle$, which, of course, is equal to $\langle b''|\alpha^{(i)}\rangle^*$. The final step is to sum such square matrices with weighting factors w_i, as indicated in (3.4.8). The final form agrees with (3.4.9), as expected.

It is instructive to study several examples, all referring to spin $\frac{1}{2}$ systems.

Example 1. A completely polarized beam with S_z+.

$$\rho = |+\rangle\langle+| \doteq \begin{pmatrix}1\\0\end{pmatrix}(1,0)$$

$$= \begin{pmatrix}1 & 0\\0 & 0\end{pmatrix}. \tag{3.4.18}$$

Example 2. A completely polarized beam with $S_x \pm$.

$$\rho = |S_x; \pm\rangle\langle S_x; \pm| = \left(\frac{1}{\sqrt{2}}\right)(|+\rangle \pm |-\rangle)\left(\frac{1}{\sqrt{2}}\right)(\langle +| \pm \langle -|)$$

$$\doteq \begin{pmatrix} \frac{1}{2} & \pm\frac{1}{2} \\ \pm\frac{1}{2} & \frac{1}{2} \end{pmatrix}. \tag{3.4.19}$$

The ensembles of Examples 1 and 2 are both pure.

Example 3. An unpolarized beam. This can be regarded as an incoherent mixture of a spin-up ensemble and a spin-down ensemble with equal weights (50% each):

$$\rho = (\tfrac{1}{2})|+\rangle\langle +| + (\tfrac{1}{2})|-\rangle\langle -|$$

$$\doteq \begin{pmatrix} \frac{1}{2} & 0 \\ 0 & \frac{1}{2} \end{pmatrix}, \tag{3.4.20}$$

which is just the identity matrix divided by 2. As we remarked earlier, the same ensemble can also be regarded as an incoherent mixture of an $S_x +$ ensemble and an $S_x -$ ensemble with equal weights. It is gratifying that our formalism automatically satisfies the expectation

$$\begin{pmatrix} \frac{1}{2} & 0 \\ 0 & \frac{1}{2} \end{pmatrix} = \frac{1}{2}\begin{pmatrix} \frac{1}{2} & \frac{1}{2} \\ \frac{1}{2} & \frac{1}{2} \end{pmatrix} + \frac{1}{2}\begin{pmatrix} \frac{1}{2} & -\frac{1}{2} \\ -\frac{1}{2} & \frac{1}{2} \end{pmatrix}, \tag{3.4.21}$$

where we see from Example 2 that the two terms on the right-hand side are the density matrices for pure ensemble with $S_x +$ and $S_x -$. Because ρ in this case is just the identity operator divided by 2 (the dimensionality), we have

$$\mathrm{tr}(\rho S_x) = \mathrm{tr}(\rho S_y) = \mathrm{tr}(\rho S_z) = 0, \tag{3.4.22}$$

where we used the fact that S_k is traceless. Thus for the ensemble average of $\mathbf{S}$ we have

$$[\mathbf{S}] = 0. \tag{3.4.23}$$

This is reasonable because there should be no preferred spin direction in a completely random ensemble of spin $\frac{1}{2}$ systems.

Example 4. As an example of a partially polarized beam, let us consider a 75-25 mixture of two pure ensembles, one with $S_z +$ and the other with $S_x +$:

$$w(S_z +) = 0.75, \qquad w(S_x +) = 0.25. \tag{3.4.24}$$

The corresponding ρ can be represented by

$$\rho \doteq \tfrac{3}{4}\begin{pmatrix} 1 & 0 \\ 0 & 0 \end{pmatrix} + \tfrac{1}{4}\begin{pmatrix} \frac{1}{2} & \frac{1}{2} \\ \frac{1}{2} & \frac{1}{2} \end{pmatrix}$$
$$= \begin{pmatrix} \frac{7}{8} & \frac{1}{8} \\ \frac{1}{8} & \frac{1}{8} \end{pmatrix}, \tag{3.4.25}$$

from which follows

$$[S_x] = \frac{\hbar}{8}, \qquad [S_y] = 0, \qquad [S_z] = \frac{3\hbar}{8}. \tag{3.4.26}$$

We leave as an exercise for the reader the task of showing that this ensemble can be decomposed in ways other than (3.4.24).

Time Evolution of Ensembles

How does the density operator ρ change as a function of time? Let us suppose that at some time t_0 the density operator is given by

$$\rho(t_0) = \sum_i w_i |\alpha^{(i)}\rangle\langle\alpha^{(i)}|. \tag{3.4.27}$$

If the ensemble is to be left undisturbed, we cannot change the fractional population w_i. So the change in ρ is governed solely by the time evolution of state ket $|\alpha^{(i)}\rangle$:

$$|\alpha^{(i)}\rangle \quad \text{at} \quad t_0 \to |\alpha^{(i)}, t_0; t\rangle. \tag{3.4.28}$$

From the fact that $|\alpha^{(i)}, t_0; t\rangle$ satisfies the Schrödinger equation we obtain

$$i\hbar\frac{\partial\rho}{\partial t} = \sum_i w_i\left(H|\alpha^{(i)}, t_0; t\rangle\langle\alpha^{(i)}, t_0; t| - |\alpha^{(i)}, t_0; t\rangle\langle\alpha^{(i)}, t_0; t|H\right)$$
$$= -[\rho, H]. \tag{3.4.29}$$

This looks like the Heisenberg equation of motion except that the sign is wrong! This is not disturbing because ρ is not a dynamic observable in the Heisenberg picture. On the contrary, ρ is built up of Schrödinger-picture state kets and state bras which evolve in time according to the Schrödinger equation.

It is amusing that (3.4.29) can be regarded as the quantum-mechanical analogue of Liouville's theorem in classical statistical mechanics,

$$\frac{\partial\rho_{\text{classical}}}{\partial t} = -[\rho_{\text{classical}}, H]_{\text{classical}}, \tag{3.4.30}$$

where $\rho_{\text{classical}}$ stands for the density of representative points in phase space.* Thus the name *density operator* for the ρ appearing in (3.4.29) is

*Remember, a pure classical state is one represented by a single moving point in phase space $(q_1, \ldots, q_f, p_1, \ldots, p_f)$ at each instant of time. A classical statistical state, on the other hand, is described by our nonnegative density function $\rho_{\text{classical}}(q_1, \ldots, q_f, p_1, \ldots, p_f, t)$ such that the probability that a system is found in the interval $dq_1, \ldots, dp_f$ at time t is $\rho_{\text{classical}}\, dq_1, \ldots, dp_f$.

indeed appropriate. The classical analogue of (3.4.10) for the ensemble average of some observable A is given by

$$A_{\text{average}} = \frac{\int \rho_{\text{classical}} A(q,p)\, d\Gamma_{q,p}}{\int \rho_{\text{classical}}\, d\Gamma_{q,p}}, \tag{3.4.31}$$

where $d\Gamma_{q,p}$ stands for a volume element in phase space.

Continuum Generalizations

So far we have considered density operators in ket space where the base kets are labeled by the discrete-eigenvalues of some observable. The concept of density matrix can be generalized to cases where the base kets used are labeled by continuous eigenvalues. In particular, let us consider the ket space spanned by the position eigenkets $|\mathbf{x}'\rangle$. The analogue of (3.4.10) is given by

$$[A] = \int d^3x' \int d^3x'' \langle \mathbf{x}''|\rho|\mathbf{x}'\rangle\langle \mathbf{x}'|A|\mathbf{x}''\rangle. \tag{3.4.32}$$

The density matrix here is actually a function of $\mathbf{x}'$ and $\mathbf{x}''$, namely,

$$\begin{aligned}\langle \mathbf{x}''|\rho|\mathbf{x}'\rangle &= \langle \mathbf{x}''|\left(\sum_i w_i|\alpha^{(i)}\rangle\langle\alpha^{(i)}|\right)|\mathbf{x}'\rangle \\ &= \sum_i w_i \psi_i(\mathbf{x}'')\psi_i^*(\mathbf{x}'),\end{aligned} \tag{3.4.33}$$

where ψ_i is the wave function corresponding to the state ket $|\alpha^{(i)}\rangle$. Notice that the diagonal element (that is, $\mathbf{x}' = \mathbf{x}''$) of this is just the weighted sum of the probability densities. Once again, the term *density matrix* is indeed appropriate.

In continuum cases, too, it is important to keep in mind that the same mixed ensemble can be decomposed in different ways into pure ensembles. For instance, it is possible to regard a "realistic" beam of particles either as a mixture of plane-wave states (monoenergetic free-particle states) or as a mixture of wave-packet states.

Quantum Statistical Mechanics

We conclude this section with a brief discussion on the connection between the density operator formalism and statistical mechanics. Let us first record some properties of completely random and of pure ensembles.

The density matrix of a completely random ensemble looks like

$$\rho \doteq \frac{1}{N}\begin{pmatrix} 1 & & & & & & 0 \\ & 1 & & & & & \\ & & 1 & & & & \\ & & & \ddots & & & \\ & & & & 1 & & \\ & & & & & 1 & \\ 0 & & & & & & 1 \end{pmatrix} \tag{3.4.34}$$

in any representation [compare Example 3 with (3.4.20)]. This follows from the fact that all states corresponding to the base kets with respect to which the density matrix is written are equally populated. In contrast, in the basis where ρ is diagonalized, we have (3.4.16) for the matrix representation of the density operator for a pure ensemble. The two diagonal matrices (3.4.34) and (3.4.16), both satisfying the normalization requirement (3.4.11), cannot look more different. It would be desirable if we could somehow construct a quantity that characterizes this dramatic difference.

Thus we define a quantity called σ by

$$\sigma = -\operatorname{tr}(\rho \ln \rho). \tag{3.4.35}$$

The logarithm of the operator ρ may appear rather formidable, but the meaning of (3.4.35) is quite unambiguous if we use the basis in which ρ is diagonal:

$$\sigma = -\sum_k \rho_{kk}^{(\text{diag})} \ln \rho_{kk}^{(\text{diag})}. \tag{3.4.36}$$

Because each element $\rho_{kk}^{(\text{diag})}$ is a real number between 0 and 1, σ is necessarily positive semidefinite. For a completely random ensemble (3.4.34), we have

$$\sigma = -\sum_{k=1}^{N} \frac{1}{N} \ln\left(\frac{1}{N}\right) = \ln N. \tag{3.4.37}$$

In contrast, for a pure ensemble (3.4.16) we have

$$\sigma = 0 \tag{3.4.38}$$

where we have used

$$\rho_{kk}^{(\text{diag})} = 0 \quad \textit{or} \quad \ln \rho_{kk}^{(\text{diag})} = 0 \tag{3.4.39}$$

for each term in (3.4.36).

We now argue that physically σ can be regarded as a quantitative measure of disorder. A pure ensemble is an ensemble with a maximum amount of order because all members are characterized by the same quantum-mechanical state ket; it may be likened to marching soldiers in a well-regimented army. According to (3.4.38), σ vanishes for such an ensemble. On the other extreme, a completely random ensemble, in which all quantum-mechanical states are equally likely, may be likened to drunken soldiers wandering around in random directions. According to (3.4.37), σ is large; indeed, we will show later that $\ln N$ is the maximum possible value for σ subject to the normalization condition

$$\sum_k \rho_{kk} = 1. \tag{3.4.40}$$

In thermodynamics we learn that a quantity called **entropy** measures disorder. It turns out that our σ is related to the entropy per constituent member, denoted by S, of the ensemble via

$$S = k\sigma, \tag{3.4.41}$$

where k is a universal constant identifiable with the Boltzmann constant. In fact, (3.4.41) may be taken as the *definition* of entropy in quantum statistical mechanics.

We now show how the density operator ρ can be obtained for an ensemble in thermal equilibrium. The basic assumption we make is that nature tends to maximize σ subject to the constraint that the ensemble average of the Hamiltonian has a certain prescribed value. To justify this assumption would involve us in a delicate discussion of how equilibrium is established as a result of interactions with the environment, which is beyond the scope of this book. In any case, once thermal equilibrium is established, we expect

$$\frac{\partial \rho}{\partial t} = 0. \tag{3.4.42}$$

Because of (3.4.29), this means that ρ and H can be simultaneously diagonalized. So the kets used in writing (3.4.36) may be taken to be energy eigenkets. With this choice ρ_{kk} stands for the fractional population for an energy eigenstate with energy eigenvalue E_k.

Let us maximize σ by requiring that

$$\delta\sigma = 0. \tag{3.4.43}$$

However, we must take into account the constraint that the ensemble average of H has a certain prescribed value. In the language of statistical mechanics, $[H]$ is identified with the internal energy per constituent denoted by U:

$$[H] = \mathrm{tr}(\rho H) = U. \tag{3.4.44}$$

In addition, we should not forget the normalization constraint (3.4.40). So our basic task is to require (3.4.43) subject to the constraints

$$\delta[H] = \sum_k \delta\rho_{kk} E_k = 0 \tag{3.4.45a}$$

and

$$\delta(\operatorname{tr}\rho) = \sum_k \delta\rho_{kk} = 0. \tag{3.4.45b}$$

We can most readily accomplish this by using Lagrange multipliers. We obtain

$$\sum_k \delta\rho_{kk}\left[(\ln \rho_{kk} + 1) + \beta E_k + \gamma\right] = 0, \tag{3.4.46}$$

which for an arbitrary variation is possibly only if

$$\rho_{kk} = \exp(-\beta E_k - \gamma - 1). \tag{3.4.47}$$

The constant γ can be eliminated using the normalization condition (3.4.40), and our final result is

$$\rho_{kk} = \frac{\exp(-\beta E_k)}{\sum_l^N \exp(-\beta E_l)}, \tag{3.4.48}$$

which directly gives the fractional population for an energy eigenstate with eigenvalue E_k. It is to be understood throughout that the sum is over distinct energy eigenstates; if there is degeneracy we must sum over states with the same energy eigenvalue.

The density matrix element (3.4.48) is appropriate for what is known in statistical mechanics as a **canonical ensemble**. Had we attempted to maximize σ without the internal-energy constraint (3.4.45a), we would have obtained instead

$$\rho_{kk} = \frac{1}{N}, \quad \text{(independent of } k\text{)}, \tag{3.4.49}$$

which is the density matrix element appropriate for a completely random ensemble. Comparing (3.4.48) with (3.4.49), we infer that a completely random ensemble can be regarded as the $\beta \to 0$ limit (physically the high-temperature limit) of a canonical ensemble.

We recognize the denominator of (3.4.48) as the partition function

$$Z = \sum_k^N \exp(-\beta E_k) \tag{3.4.50}$$

in statistical mechanics. It can also be written as

$$Z = \operatorname{tr}(e^{-\beta H}). \tag{3.4.51}$$

Knowing ρ_{kk} given in the energy basis, we can write the density operator as

$$\rho = \frac{e^{-\beta H}}{Z}. \tag{3.4.52}$$

This is the most basic equation from which everything follows. We can immediately evaluate the ensemble average of any observable A:

$$[A] = \frac{\mathrm{tr}(e^{-\beta H}A)}{Z}$$

$$= \frac{\left[\sum_k^N \langle A\rangle_k \exp(-\beta E_k)\right]}{\sum_k^N \exp(-\beta E_k)}. \tag{3.4.53}$$

In particular, for the internal energy per constituent we obtain

$$U = \frac{\left[\sum_k^N E_k \exp(-\beta E_k)\right]}{\sum_k^N \exp(-\beta E_k)}$$

$$= -\frac{\partial}{\partial\beta}(\ln Z), \tag{3.4.54}$$

a formula well known to every student of statistical mechanics.

The parameter β is related to the temperature T as follows:

$$\beta = \frac{1}{kT} \tag{3.4.55}$$

where k is the Boltzmann constant. It is instructive to convince ourselves of this identification by comparing the ensemble average $[H]$ of simple harmonic oscillators with the kT expected for the internal energy in the classical limit, which is left as an exercise. We have already commented that in the high-temperature limit, a canonical ensemble becomes a completely random ensemble in which all energy eigenstates are equally populated. In the opposite low-temperature limit ($\beta \to \infty$), (3.4.48) tells us that a canonical ensemble becomes a pure ensemble where only the ground state is populated.

As a simple illustrative example, consider a canonical ensemble made up of spin $\frac{1}{2}$ systems, each with a magnetic moment $e\hbar/2m_e c$ subjected to a uniform magnetic field in the z-direction. The Hamiltonian relevant to this problem has already been given [see (3.2.16)]. Because H and S_z commute, the density matrix for this canonical ensemble is diagonal in the S_z basis. Thus

$$\rho \doteq \frac{\begin{pmatrix} e^{-\beta\hbar\omega/2} & 0 \\ 0 & e^{\beta\hbar\omega/2} \end{pmatrix}}{Z}, \tag{3.4.56}$$

where the partition function is just

$$Z = e^{-\beta\hbar\omega/2} + e^{\beta\hbar\omega/2}. \tag{3.4.57}$$

From this we compute

$$[S_x] = [S_y] = 0, \quad [S_z] = -\left(\frac{\hbar}{2}\right) \tanh\left(\frac{\beta\hbar\omega}{2}\right). \tag{3.4.58}$$

The ensemble average of the magnetic moment component is just e/m_ec times $[S_z]$. The paramagnetic susceptibility χ may be computed from

$$\left(\frac{e}{m_ec}\right) [S_z] = \chi B. \tag{3.4.59}$$

In this way we arrive at Brillouin's formula for χ:

$$\chi = \left(\frac{|e|\hbar}{2m_ecB}\right)\tanh\left(\frac{\beta\hbar\omega}{2}\right). \tag{3.4.60}$$

3.5. EIGENVALUES AND EIGENSTATES OF ANGULAR MOMENTUM

Up to now our discussion of angular momentum has been confined exclusively to spin $\frac{1}{2}$ systems with dimensionality $N = 2$. In this and subsequent sections we study more-general angular momentum states. To this end we first work out the eigenvalues and eigenkets of $\mathbf{J}^2$ and J_z and derive the expressions for matrix elements of angular momentum operators, first obtained in a 1926 paper by M. Born, W. Heisenberg, and P. Jordan.

Commutation Relations and the Ladder Operators

Everything we will do follows from the angular-momentum commutation relations (3.1.20), where we may recall that J_i is defined as the generator of infinitesimal rotation. The first important property we derive from the basic commutation relations is the existence of a new operator $\mathbf{J}^2$, defined by

$$\mathbf{J}^2 \equiv J_xJ_x + J_yJ_y + J_zJ_z, \tag{3.5.1}$$

that commutes with every one of J_k:

$$\left[\mathbf{J}^2, J_k\right] = 0, \quad (k = 1, 2, 3). \tag{3.5.2}$$

To prove this let us look at the $k = 3$ case:

$$\begin{aligned}\left[J_xJ_x + J_yJ_y + J_zJ_z, J_z\right] &= J_x[J_x, J_z] + [J_x, J_z]J_x + J_y\left[J_y, J_z\right] + \left[J_y, J_z\right]J_y \\ &= J_x(-i\hbar J_y) + (-i\hbar J_y)J_x + J_y(i\hbar J_x) + (i\hbar J_x)J_y \\ &= 0. \end{aligned} \tag{3.5.3}$$

The proofs for the cases where $k = 1$ and 2 follow by cyclic permutation $(1 \to 2 \to 3 \to 1)$ of the indices. Because J_x, J_y, and J_z do not commute with each other, we can choose only one of them to be the observable to be diagonalized simultaneously with J^2. By convention we choose J_z for this purpose.

We now look for the simultaneous eigenkets of $\mathbf{J}^2$ and J_z. We denote the eigenvalues of $\mathbf{J}^2$ and J_z by a and b, respectively:

$$\mathbf{J}^2|a,b\rangle = a|a,b\rangle \tag{3.5.4a}$$

$$J_z|a,b\rangle = b|a,b\rangle. \tag{3.5.4b}$$

To determine the allowed values for a and b, it is convenient to work with the non-Hermitian operators

$$J_\pm \equiv J_x \pm iJ_y, \tag{3.5.5}$$

called the **ladder operators**, rather than with J_x and J_y. They satisfy the commutation relations

$$[J_+, J_-] = 2\hbar J_z \tag{3.5.6a}$$

and

$$[J_z, J_\pm] = \pm\hbar J_\pm, \tag{3.5.6b}$$

which can easily be obtained from (3.1.20). Note also that

$$[\mathbf{J}^2, J_\pm] = 0, \tag{3.5.7}$$

which is an obvious consequence of (3.5.2).

What is the physical meaning of $J_\pm$? To answer this we examine how J_z acts on $J_\pm|a,b\rangle$:

$$\begin{aligned} J_z(J_\pm|a,b\rangle) &= ([J_z, J_\pm] + J_\pm J_z)|a,b\rangle \\ &= (b \pm \hbar)(J_\pm|a,b\rangle) \end{aligned} \tag{3.5.8}$$

where we have used (3.5.6b). In other words, if we apply $J_+(J_-)$ to a J_z eigenket, the resulting ket is still a J_z eigenket except that its eigenvalue is now increased (decreased) by one unit of $\hbar$. So now we see why $J_\pm$, which step one step up (down) on the "ladder" of J_z eigenvalues, are known as the ladder operators.

We now digress to recall that the commutation relations in (3.5.6b) are reminiscent of some commutation relations we encountered in the earlier chapters. In discussing the translation operator $\mathscr{T}(\mathbf{l})$ we had

$$[x_i, \mathscr{T}(\mathbf{l})] = l_i \mathscr{T}(\mathbf{l}); \tag{3.5.9}$$

also, in discussing the simple harmonic oscillator we had

$$[N, a^\dagger] = a^\dagger, \qquad [N, a] = -a. \tag{3.5.10}$$

We see that both (3.5.9) and (3.5.10) have a structure similar to (3.5.6b). The physical interpretation of the translation operator is that it changes the eigenvalue of the position operator $\mathbf{x}$ by $\mathbf{l}$ in much the same way as the ladder operator J_+ changes the eigenvalue of J_z by one unit of $\hbar$. Likewise, the oscillator creation operator $a^\dagger$ increases the eigenvalue of the number operator N by unity.

Even though $J_\pm$ changes the eigenvalue of J_z by one unit of $\hbar$, it does not change the eigenvalue of $\mathbf{J}^2$:

$$\begin{aligned}\mathbf{J}^2(J_\pm|a,b\rangle) &= J_\pm \mathbf{J}^2|a,b\rangle \\ &= a(J_\pm|a,b\rangle),\end{aligned} \tag{3.5.11}$$

where we have used (3.5.7). To summarize, $J_\pm|a,b\rangle$ are simultaneous eigenkets of $\mathbf{J}^2$ and J_z with eigenvalues a and $b \pm \hbar$. We may write

$$J_\pm|a,b\rangle = c_\pm|a,b\pm\hbar\rangle, \tag{3.5.12}$$

where the proportionality constant $c_\pm$ will be determined later from the normalization requirement of the angular-momentum eigenkets.

Eigenvalues of $\mathbf{J}^2$ and J_z

We now have the machinery needed to construct angular-momentum eigenkets and to study their eigenvalue spectrum. Suppose we apply J_+ successively, say n times, to a simultaneous eigenket of $\mathbf{J}^2$ and J_z. We then obtain another eigenket of $\mathbf{J}^2$ and J_z with the J_z eigenvalue increased by $n\hbar$, while its $\mathbf{J}^2$ eigenvalue is unchanged. However, this process cannot go on indefinitely. It turns out that there exists an upper limit to b (the J_z eigenvalue) for a given a (the $\mathbf{J}^2$ eigenvalue):

$$a \geq b^2. \tag{3.5.13}$$

To prove this assertion we first note that

$$\begin{aligned}\mathbf{J}^2 - J_z^2 &= \tfrac{1}{2}(J_+J_- + J_-J_+) \\ &= \tfrac{1}{2}(J_+J_+^\dagger + J_+^\dagger J_+).\end{aligned} \tag{3.5.14}$$

Now $J_+J_+^\dagger$ and $J_+^\dagger J_+$ must have nonnegative expectation values because

$$J_+^\dagger|a,b\rangle \overset{\text{DC}}{\leftrightarrow} \langle a,b|J_+, \qquad J_+|a,b\rangle \overset{\text{DC}}{\leftrightarrow} \langle a,b|J_+^\dagger; \tag{3.5.15}$$

thus

$$\langle a,b|(\mathbf{J}^2 - J_z^2)|a,b\rangle \geq 0, \tag{3.5.16}$$

which, in turn, implies (3.5.13). It therefore follows that there must be a $b_{\max}$ such that

$$J_+|a,b_{\max}\rangle = 0. \tag{3.5.17}$$

Stated another way, the eigenvalue of b cannot be increased beyond $b_{\max}$. Now (3.5.17) also implies

$$J_-J_+|a,b_{\max}\rangle = 0. \tag{3.5.18}$$

But

$$J_- J_+ = J_x^2 + J_y^2 - i(J_y J_x - J_x J_y)$$
$$= \mathbf{J}^2 - J_z^2 - \hbar J_z. \tag{3.5.19}$$

So

$$(\mathbf{J}^2 - J_z^2 - \hbar J_z)|a, b_{\max}\rangle = 0. \tag{3.5.20}$$

Because $|a, b_{\max}\rangle$ itself is not a null ket, this relationship is possible only if

$$a - b_{\max}^2 - b_{\max}\hbar = 0 \tag{3.5.21}$$

or

$$a = b_{\max}(b_{\max} + \hbar). \tag{3.5.22}$$

In a similar manner we argue from (3.5.13) that there must also exist a $b_{\min}$ such that

$$J_-|a, b_{\min}\rangle = 0. \tag{3.5.23}$$

By writing $J_+ J_-$ as

$$J_+ J_- = \mathbf{J}^2 - J_z^2 + \hbar J_z \tag{3.5.24}$$

in analogy with (3.5.19), we conclude that

$$a = b_{\min}(b_{\min} - \hbar). \tag{3.5.25}$$

By comparing (3.5.22) with (3.5.25) we infer that

$$b_{\max} = -b_{\min}, \tag{3.5.26}$$

with $b_{\max}$ positive, and that the allowed values of b lie within

$$-b_{\max} \leq b \leq b_{\max}. \tag{3.5.27}$$

Clearly, we must be able to reach $|a, b_{\max}\rangle$ by applying J_+ successively to $|a, b_{\min}\rangle$ a finite number of times. We must therefore have

$$b_{\max} = b_{\min} + n\hbar, \tag{3.5.28}$$

where n is some integer. As a result, we get

$$b_{\max} = \frac{n\hbar}{2}. \tag{3.5.29}$$

It is more conventional to work with j, defined to be $b_{\max}/\hbar$, instead of with $b_{\max}$ so that

$$j = \frac{n}{2}. \tag{3.5.30}$$

The maximum value of the J_z eigenvalue is $j\hbar$, where j is either an integer or a half-integer. Equation (3.5.22) implies that the eigenvalue of $\mathbf{J}^2$ is given by

$$a = \hbar^2 j(j+1). \tag{3.5.31}$$

Let us also define m so that

$$b \equiv m\hbar. \tag{3.5.32}$$

If j is an integer, all m values are integers; if j is a half-integer, all m values are half-integers. The allowed m-values for a given j are

$$m = \underbrace{-j, -j+1, \ldots, j-1, j}_{2j+1 \text{ states}}. \tag{3.5.33}$$

Instead of $|a, b\rangle$ it is more convenient to denote a simultaneous eigenket of $\mathbf{J}^2$ and J_z by $|j, m\rangle$. The basic eigenvalue equations now read

$$\mathbf{J}^2|j, m\rangle = j(j+1)\hbar^2|j, m\rangle \tag{3.5.34a}$$

and

$$J_z|j, m\rangle = m\hbar|j, m\rangle, \tag{3.5.34b}$$

with j either an integer or a half-integer and m given by (3.5.33). It is very important to recall here that we have used only the commutation relations (3.1.20) to obtain these results. The quantization of angular momentum, manifested in (3.5.34), is a direct consequence of the angular-momentum commutation relations, which, in turn, follow from the properties of rotations together with the definition of J_k as the generator of rotation.

Matrix Elements of Angular-Momentum Operators

Let us work out the matrix elements of the various angular-momentum operators. Assuming $|j, m\rangle$ to be normalized, we obviously have from (3.5.34)

$$\langle j', m'|\mathbf{J}^2|j, m\rangle = j(j+1)\hbar^2\delta_{j'j}\delta_{m'm} \tag{3.5.35a}$$

and

$$\langle j', m'|J_z|j, m\rangle = m\hbar\delta_{j'j}\delta_{m'm}. \tag{3.5.35b}$$

To obtain the matrix elements of $J_\pm$, we first consider

$$\begin{aligned}\langle j, m|J_+^\dagger J_+|j, m\rangle &= \langle j, m|(\mathbf{J}^2 - J_z^2 - \hbar J_z)|j, m\rangle \\ &= \hbar^2[j(j+1) - m^2 - m].\end{aligned} \tag{3.5.36}$$

Now $J_+|j, m\rangle$ must be the same as $|j, m+1\rangle$ (normalized) up to a multiplicative constant [see (3.5.12)]. Thus

$$J_+|j, m\rangle = c_{jm}^+|j, m+1\rangle. \tag{3.5.37}$$

Comparison with (3.5.36) leads to

$$\begin{aligned}|c_{jm}^+|^2 &= \hbar^2[j(j+1) - m(m+1)] \\ &= \hbar^2(j-m)(j+m+1).\end{aligned} \tag{3.5.38}$$

So we have determined c_{jm}^{+} up to an arbitrary phase factor. It is customary to choose c_{jm}^{+} to be real and positive by convention. So

$$J_{+}|j,m\rangle = \sqrt{(j-m)(j+m+1)}\,\hbar|j,m+1\rangle. \tag{3.5.39}$$

Similarly, we can derive

$$J_{-}|j,m\rangle = \sqrt{(j+m)(j-m+1)}\,\hbar|j,m-1\rangle. \tag{3.5.40}$$

Finally, we determine the matrix elements of $J_{\pm}$ to be

$$\langle j',m'|J_{\pm}|j,m\rangle = \sqrt{(j\mp m)(j\pm m+1)}\,\hbar\delta_{j'j}\delta_{m',m\pm 1}. \tag{3.5.41}$$

Representations of the Rotation Operator

Having obtained the matrix elements of J_z and $J_{\pm}$, we are now in a position to study the matrix elements of the rotation operator $\mathscr{D}(R)$. If a rotation R is specified by $\hat{\mathbf{n}}$ and ϕ, we can define its matrix elements by

$$\mathscr{D}_{m'm}^{(j)}(R) = \langle j,m'|\exp\left(\frac{-i\mathbf{J}\cdot\hat{\mathbf{n}}\phi}{\hbar}\right)|j,m\rangle. \tag{3.5.42}$$

These matrix elements are sometimes called **Wigner functions** after E. P. Wigner, who made pioneering contributions to the group-theoretical properties of rotations in quantum mechanics. Notice here that the same j-value appears in the ket and bra of (3.5.42); we need not consider matrix elements of $\mathscr{D}(R)$ between states with different j-values because they all vanish trivially. This is because $\mathscr{D}(R)|j,m\rangle$ is still an eigenket of $\mathbf{J}^2$ with the same eigenvalue $j(j+1)\hbar^2$:

$$\begin{aligned}\mathbf{J}^2\mathscr{D}(R)|j,m\rangle &= \mathscr{D}(R)\mathbf{J}^2|j,m\rangle \\ &= j(j+1)\hbar^2[\mathscr{D}(R)|j,m\rangle],\end{aligned} \tag{3.5.43}$$

which follows directly from the fact that $\mathbf{J}^2$ commutes with J_k (hence with any function of J_k). Simply stated, rotations cannot change the j-value, which is an eminently sensible result.

Often in the literature the $(2j+1)\times(2j+1)$ matrix formed by $\mathscr{D}_{m'm}^{(j)}(R)$ is referred to as the $(2j+1)$-*dimensional irreducible representation* of the rotation operator $\mathscr{D}(R)$. This means that the matrix which corresponds to an arbitrary rotation operator in ket space *not* necessarily characterized by a single j-value can, with a suitable choice of basis, be brought to block-diagonal form:

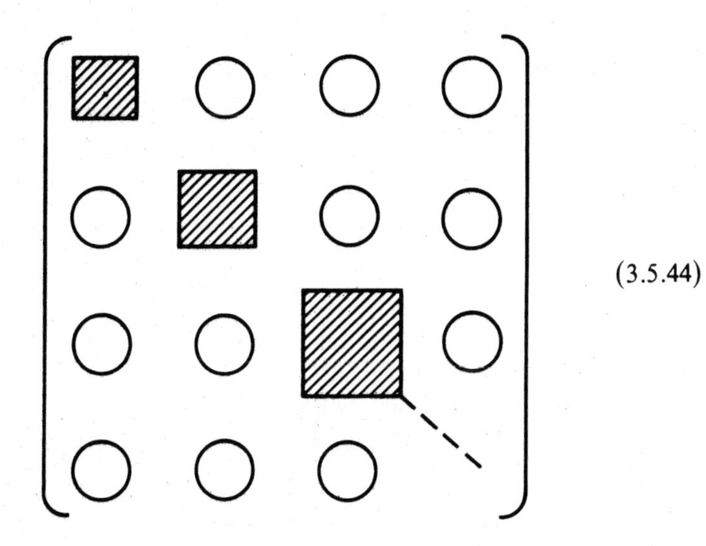

(3.5.44)

where each shaded square is a $(2j+1) \times (2j+1)$ square matrix formed by $\mathscr{D}_{m'm}^{(j)}$ with some definite value of j. Furthermore, each square matrix itself cannot be broken into smaller blocks

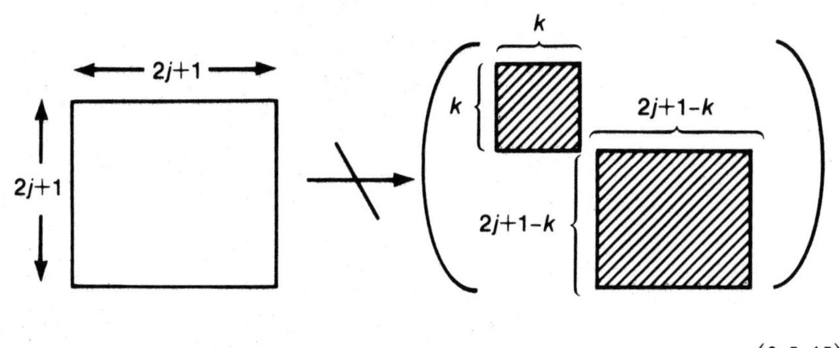

(3.5.45)

with any choice of basis.

The rotation matrices characterized by definite j form a group. First, the identity is a member because the rotation matrix corresponding to no rotation ($\phi = 0$) is the $(2j+1) \times (2j+1)$ identity matrix. Second, the inverse is also a member; we simply reverse the rotation angle ($\phi \to -\phi$) without changing the rotation axis $\hat{\mathbf{n}}$. Third, the product of any two members is also a member; explicitly we have

$$\sum_{m'} \mathscr{D}^{(j)}_{m''m'}(R_1)\mathscr{D}^{(j)}_{m'm}(R_2) = \mathscr{D}^{(j)}_{m''m}(R_1 R_2), \tag{3.5.46}$$

where the product $R_1 R_2$ represents a single rotation. We also note that the rotation matrix is unitary because the corresponding rotation operator is unitary; explicitly we have

$$\mathscr{D}_{m'm}(R^{-1}) = \mathscr{D}^*_{mm'}(R). \tag{3.5.47}$$

To appreciate the physical significance of the rotation matrix let us start with a state represented by $|j, m\rangle$. We now rotate it:

$$|j, m\rangle \to \mathscr{D}(R)|j, m\rangle. \tag{3.5.48}$$

Even though this rotation operation does not change j, we generally obtain states with m-values other than the original m. To find the amplitude for being found in $|j, m'\rangle$, we simply expand the rotated state as follows:

$$\begin{aligned}\mathscr{D}(R)|j, m\rangle &= \sum_{m'} |j, m'\rangle\langle j, m'|\mathscr{D}(R)|j, m\rangle \\ &= \sum_{m'} |j, m'\rangle \mathscr{D}^{(j)}_{m'm}(R),\end{aligned} \tag{3.5.49}$$

where, in using the completeness relation, we took advantage of the fact that $\mathscr{D}(R)$ connects only states with the same j. So the matrix element $\mathscr{D}^{(j)}_{m'm}(R)$ is simply the amplitude for the rotated state to be found in $|j, m'\rangle$ when the original unrotated state is given by $|j, m\rangle$.

In Section 3.3 we saw how Euler angles may be used to characterize the most general rotation. We now consider the matrix realization of (3.3.20) for an arbitrary j (not necessarily $\frac{1}{2}$):

$$\begin{aligned}\mathscr{D}^{(j)}_{m'm}(\alpha, \beta, \gamma) &= \langle j, m'|\exp\left(\frac{-iJ_z\alpha}{\hbar}\right)\exp\left(\frac{-iJ_y\beta}{\hbar}\right)\exp\left(\frac{-iJ_z\gamma}{\hbar}\right)|j, m\rangle \\ &= e^{-i(m'\alpha + m\gamma)}\langle j, m'|\exp\left(\frac{-iJ_y\beta}{\hbar}\right)|j, m\rangle.\end{aligned} \tag{3.5.50}$$

Notice that the only nontrivial part is the middle rotation about the y-axis, which mixes different m-values. It is convenient to define a new matrix $d^{(j)}(\beta)$ as

$$d^{(j)}_{m'm}(\beta) \equiv \langle j, m'|\exp\left(\frac{-iJ_y\beta}{\hbar}\right)|j, m\rangle. \tag{3.5.51}$$

Finally, let us turn to some examples. The $j = \frac{1}{2}$ case has already been worked out in Section 3.3. See the middle matrix of (3.3.21),

$$d^{(1/2)} = \begin{pmatrix} \cos\left(\frac{\beta}{2}\right) & -\sin\left(\frac{\beta}{2}\right) \\ \sin\left(\frac{\beta}{2}\right) & \cos\left(\frac{\beta}{2}\right) \end{pmatrix}. \tag{3.5.52}$$

The next simplest case is $j=1$, which we consider in some detail. Clearly, we must first obtain the 3×3 matrix representation of J_y. Because

$$J_y = \frac{(J_+ - J_-)}{2i} \tag{3.5.53}$$

from the defining equation (3.5.5) for $J_\pm$, we can use (3.5.41) to obtain

$$J_y^{(j=1)} = \left(\frac{\hbar}{2}\right) \begin{matrix} m=1 & m=0 & m=-1 \\ \end{matrix} \begin{pmatrix} 0 & -\sqrt{2}\,i & 0 \\ \sqrt{2}\,i & 0 & -\sqrt{2}\,i \\ 0 & \sqrt{2}\,i & 0 \end{pmatrix} \begin{matrix} m'=1 \\ m'=0. \\ m'=-1 \end{matrix} \tag{3.5.54}$$

Our next task is to work out the Taylor expansion of $\exp(-iJ_y\beta/\hbar)$. Unlike the case $j=\frac{1}{2}$, $[J_y^{(j=1)}]^2$ is *independent* of 1 and $J_y^{(j=1)}$. However, it is easy to work out

$$\left(\frac{J_y^{(j=1)}}{\hbar}\right)^3 = \frac{J_y^{(j=1)}}{\hbar}. \tag{3.5.55}$$

Consequently, for $j=1$ *only*, it is legitimate to replace

$$\exp\left(\frac{-iJ_y\beta}{\hbar}\right) \to 1 - \left(\frac{J_y}{\hbar}\right)^2 (1-\cos\beta) - i\left(\frac{J_y}{\hbar}\right)\sin\beta, \tag{3.5.56}$$

as the reader may verify in detail. Explicitly we have

$$d^{(1)}(\beta) = \begin{pmatrix} \left(\frac{1}{2}\right)(1+\cos\beta) & -\left(\frac{1}{\sqrt{2}}\right)\sin\beta & \left(\frac{1}{2}\right)(1-\cos\beta) \\ \left(\frac{1}{\sqrt{2}}\right)\sin\beta & \cos\beta & -\left(\frac{1}{\sqrt{2}}\right)\sin\beta \\ \left(\frac{1}{2}\right)(1-\cos\beta) & \left(\frac{1}{\sqrt{2}}\right)\sin\beta & \left(\frac{1}{2}\right)(1+\cos\beta) \end{pmatrix}. \tag{3.5.57}$$

Clearly, this method becomes time-consuming for large j. In Section 3.8 we will learn a much easier method for obtaining $d_{mm'}^{(j)}(\beta)$ for *any* j.

3.6. ORBITAL ANGULAR MOMENTUM

We introduced the concept of angular momentum by defining it to be the generator of an infinitesimal rotation. There is another way to approach the subject of angular momentum when spin-angular momentum is zero or can be ignored. The angular momentum **J** for a single particle is then the same as orbital angular momentum, which is defined as

$$\mathbf{L} = \mathbf{x} \times \mathbf{p}. \tag{3.6.1}$$

In this section we explore the connection between the two approaches.

Orbital Angular Momentum as Rotation Generator

We first note that the orbital angular-momentum operator defined as (3.6.1) satisfies the angular-momentum commutation relations

$$[L_i, L_j] = i\varepsilon_{ijk}\hbar L_k \tag{3.6.2}$$

by virtue of the commutation relations among the components of $\mathbf{x}$ and $\mathbf{p}$. This can easily be proved as follows:

$$\begin{aligned}
[L_x, L_y] &= [yp_z - zp_y, zp_x - xp_z] \\
&= [yp_z, zp_x] + [zp_y, xp_z] \\
&= yp_x[p_z, z] + p_y x[z, p_z] \\
&= i\hbar(xp_y - yp_x) \\
&= i\hbar L_z \\
&\ \vdots
\end{aligned} \tag{3.6.3}$$

Next we let

$$1 - i\left(\frac{\delta\phi}{\hbar}\right)L_z = 1 - i\left(\frac{\delta\phi}{\hbar}\right)(xp_y - yp_x) \tag{3.6.4}$$

act on an arbitrary position eigenket $|x', y', z'\rangle$ to examine whether it can be interpreted as the infinitesimal rotation operator about the z-axis by angle $\delta\phi$. Using the fact that momentum is the generator of translation, we obtain [see (1.6.32)]

$$\begin{aligned}
\left[1 - i\left(\frac{\delta\phi}{\hbar}\right)L_z\right]|x', y', z'\rangle &= \left[1 - i\left(\frac{p_y}{\hbar}\right)(\delta\phi x') + i\left(\frac{p_x}{\hbar}\right)(\delta\phi y')\right]|x', y', z'\rangle \\
&= |x' - y'\delta\phi, y' + x'\delta\phi, z'\rangle.
\end{aligned} \tag{3.6.5}$$

This is precisely what we expect if L_z generates an infinitesimal rotation about the z-axis. So we have demonstrated that if $\mathbf{p}$ generates translation, then $\mathbf{L}$ generates rotation.

Suppose the wave function for an arbitrary physical state of a spinless particle is given by $\langle x', y', z'|\alpha\rangle$. After an infinitesimal rotation about the z-axis is performed, the wave function for the rotated state is

$$\langle x', y', z'|\left[1 - i\left(\frac{\delta\phi}{\hbar}\right)L_z\right]|\alpha\rangle = \langle x' + y'\delta\phi, y' - x'\delta\phi, z'|\alpha\rangle. \tag{3.6.6}$$

It is actually more transparent to change the coordinate basis

$$\langle x', y', z'|\alpha\rangle \to \langle r, \theta, \phi|\alpha\rangle. \tag{3.6.7}$$

For the rotated state we have, according to (3.6.6),

$$\begin{aligned}
\langle r, \theta, \phi|\left[1 - i\left(\frac{\delta\phi}{\hbar}\right)L_z\right]|\alpha\rangle &= \langle r, \theta, \phi - \delta\phi|\alpha\rangle \\
&= \langle r, \theta, \phi|\alpha\rangle - \delta\phi\frac{\partial}{\partial\phi}\langle r, \theta, \phi|\alpha\rangle.
\end{aligned} \tag{3.6.8}$$

Because $\langle r, \theta, \phi|$ is an arbitrary position eigenket, we can identify

$$\langle \mathbf{x}'|L_z|\alpha\rangle = -i\hbar \frac{\partial}{\partial\phi}\langle \mathbf{x}'|\alpha\rangle, \tag{3.6.9}$$

which is a well-known result from wave mechanics. Even though this relation can also be obtained just as easily using the position representation of the momentum operator, the derivation given here emphasizes the role of L_z as the generator of rotation.

We next consider a rotation about the x-axis by angle $\delta\phi_x$. In analogy with (3.6.6) we have

$$\langle x', y', z'|\left[1 - i\left(\frac{\delta\phi_x}{\hbar}\right)L_x\right]|\alpha\rangle = \langle x', y' + z'\delta\phi_x, z' - y'\delta\phi_x|\alpha\rangle. \tag{3.6.10}$$

By expressing x', y', and z' in spherical coordinates, we can show that

$$\langle \mathbf{x}'|L_x|\alpha\rangle = -i\hbar\left(-\sin\phi\frac{\partial}{\partial\theta} - \cot\theta\cos\phi\frac{\partial}{\partial\phi}\right)\langle \mathbf{x}'|\alpha\rangle. \tag{3.6.11}$$

Likewise,

$$\langle \mathbf{x}'|L_y|\alpha\rangle = -i\hbar\left(\cos\phi\frac{\partial}{\partial\theta} - \cot\theta\sin\phi\frac{\partial}{\partial\phi}\right)\langle \mathbf{x}'|\alpha\rangle. \tag{3.6.12}$$

Using (3.6.11) and (3.6.12), for the ladder operator $L_\pm$ defined as in (3.5.5), we have

$$\langle \mathbf{x}'|L_\pm|\alpha\rangle = -i\hbar e^{\pm i\phi}\left(\pm i\frac{\partial}{\partial\theta} - \cot\theta\frac{\partial}{\partial\phi}\right)\langle \mathbf{x}'|\alpha\rangle. \tag{3.6.13}$$

Finally, it is possible to write $\langle \mathbf{x}'|\mathbf{L}^2|\alpha\rangle$ using

$$\mathbf{L}^2 = L_z^2 + \left(\tfrac{1}{2}\right)(L_+L_- + L_-L_+), \tag{3.6.14}$$

(3.6.9), and (3.6.13), as follows:

$$\langle \mathbf{x}'|\mathbf{L}^2|\alpha\rangle = -\hbar^2\left[\frac{1}{\sin^2\theta}\frac{\partial^2}{\partial\phi^2} + \frac{1}{\sin\theta}\frac{\partial}{\partial\theta}\left(\sin\theta\frac{\partial}{\partial\theta}\right)\right]\langle \mathbf{x}'|\alpha\rangle. \tag{3.6.15}$$

Apart from $1/r^2$, we recognize the differential operator that appears here to be just the angular part of the Laplacian in spherical coordinates.

It is instructive to establish this connection between the $\mathbf{L}^2$ operator and the angular part of the Laplacian in another way by looking directly at the kinetic-energy operator. We first record an important operator identity,

$$\mathbf{L}^2 = \mathbf{x}^2\mathbf{p}^2 - (\mathbf{x}\cdot\mathbf{p})^2 + i\hbar\mathbf{x}\cdot\mathbf{p}, \tag{3.6.16}$$

where $\mathbf{x}^2$ is understood to be the operator $\mathbf{x}\cdot\mathbf{x}$, just as $\mathbf{p}^2$ stands for the operator $\mathbf{p}\cdot\mathbf{p}$. The proof of this is straightforward:

$$\begin{aligned}
\mathbf{L}^2 &= \sum_{ijlmk} \varepsilon_{ijk} x_i p_j \varepsilon_{lmk} x_l p_m \\
&= \sum_{ijlm} (\delta_{il}\delta_{jm} - \delta_{im}\delta_{jl}) x_i p_j x_l p_m \\
&= \sum_{ijlm} \left[\delta_{il}\delta_{jm} x_i (x_l p_j - i\hbar\delta_{jl}) p_m - \delta_{im}\delta_{jl} x_i p_j (p_m x_l + i\hbar\delta_{lm})\right] \\
&= \mathbf{x}^2\mathbf{p}^2 - i\hbar\mathbf{x}\cdot\mathbf{p} - \sum_{ijlm} \delta_{im}\delta_{jl}\left[x_i p_m (x_l p_j - i\hbar\delta_{jl}) + i\hbar\delta_{lm} x_i p_j\right] \\
&= \mathbf{x}^2\mathbf{p}^2 - (\mathbf{x}\cdot\mathbf{p})^2 + i\hbar\mathbf{x}\cdot\mathbf{p}.
\end{aligned} \tag{3.6.17}$$

Before taking the preceding expression between $\langle\mathbf{x}'|$ and $|\alpha\rangle$, first note that

$$\begin{aligned}
\langle\mathbf{x}'|\mathbf{x}\cdot\mathbf{p}|\alpha\rangle &= \mathbf{x}'\cdot(-i\hbar\nabla'\langle\mathbf{x}'|\alpha\rangle) \\
&= -i\hbar r\frac{\partial}{\partial r}\langle\mathbf{x}'|\alpha\rangle.
\end{aligned} \tag{3.6.18}$$

Likewise,

$$\begin{aligned}
\langle\mathbf{x}'|(\mathbf{x}\cdot\mathbf{p})^2|\alpha\rangle &= -\hbar^2 r\frac{\partial}{\partial r}\left(r\frac{\partial}{\partial r}\langle\mathbf{x}'|\alpha\rangle\right) \\
&= -\hbar^2\left(r^2\frac{\partial^2}{\partial r^2}\langle\mathbf{x}'|\alpha\rangle + r\frac{\partial}{\partial r}\langle\mathbf{x}'|\alpha\rangle\right).
\end{aligned} \tag{3.6.19}$$

Thus

$$\langle\mathbf{x}'|\mathbf{L}^2|\alpha\rangle = r^2\langle\mathbf{x}'|\mathbf{p}^2|\alpha\rangle + \hbar^2\left(r^2\frac{\partial^2}{\partial r^2}\langle\mathbf{x}'|\alpha\rangle + 2r\frac{\partial}{\partial r}\langle\mathbf{x}'|\alpha\rangle\right). \tag{3.6.20}$$

In terms of the kinetic energy $\mathbf{p}^2/2m$, we have

$$\begin{aligned}
\frac{1}{2m}\langle\mathbf{x}'|\mathbf{p}^2|\alpha\rangle &= -\left(\frac{\hbar^2}{2m}\right)\nabla'^2\langle\mathbf{x}'|\alpha\rangle \\
&= -\left(\frac{\hbar^2}{2m}\right)\left(\frac{\partial^2}{\partial r^2}\langle\mathbf{x}'|\alpha\rangle + \frac{2}{r}\frac{\partial}{\partial r}\langle\mathbf{x}'|\alpha\rangle - \frac{1}{\hbar^2 r^2}\langle\mathbf{x}'|\mathbf{L}^2|\alpha\rangle\right).
\end{aligned} \tag{3.6.21}$$

The first two terms in the last line are just the radial part of the Laplacian acting on $\langle\mathbf{x}'|\alpha\rangle$. The last term must then be the angular part of the Laplacian acting on $\langle\mathbf{x}'|\alpha\rangle$, in complete agreement with (3.6.15).

Spherical Harmonics

Consider a spinless particle subjected to a spherical symmetrical potential. The wave equation is known to be separable in spherical coordinates and the energy eigenfunctions can be written as

$$\langle \mathbf{x}'|n,l,m\rangle = R_{nl}(r)Y_l^m(\theta,\phi), \tag{3.6.22}$$

where the position vector $\mathbf{x}'$ is specified by the spherical coordinates r, θ, and ϕ, and n stands for some quantum number other than l and m, for example, the radial quantum number for bound-state problems or the energy for a free-particle spherical wave. As will be made clearer in Section 3.10, this form can be regarded as a direct consequence of the rotational invariance of the problem. When the Hamiltonian is spherically symmetrical, H commutes with L_z and $\mathbf{L}^2$, and the energy eigenkets are expected to be eigenkets of $\mathbf{L}^2$ and L_z also. Because L_k with $k=1,2,3$ satisfy the angular-momentum commutation relations, the eigenvalues of $\mathbf{L}^2$ and L_z are expected to be $l(l+1)\hbar^2$, and $m\hbar = [-l\hbar, (-l+1)\hbar, \ldots, (l-1)\hbar, l\hbar]$.

Because the angular dependence is common to all problems with spherical symmetry, we can isolate it and consider

$$\langle \hat{\mathbf{n}}|l,m\rangle = Y_l^m(\theta,\phi) = Y_l^m(\hat{\mathbf{n}}), \tag{3.6.23}$$

where we have defined a **direction eigenket** $|\hat{\mathbf{n}}\rangle$. From this point of view, $Y_l^m(\theta,\phi)$ is the amplitude for a state characterized by l, m to be found in the direction $\hat{\mathbf{n}}$ specified by θ and ϕ.

Suppose we have relations involving orbital angular-momentum eigenkets. We can immediately write the corresponding relations involving the spherical harmonics. For example, take the eigenvalue equation

$$L_z|l,m\rangle = m\hbar|l,m\rangle. \tag{3.6.24}$$

Multiplying $\langle \hat{\mathbf{n}}|$ on the left and using (3.6.9), we obtain

$$-i\hbar\frac{\partial}{\partial\phi}\langle \hat{\mathbf{n}}|l,m\rangle = m\hbar\langle \hat{\mathbf{n}}|l,m\rangle. \tag{3.6.25}$$

We recognize this equation to be

$$-i\hbar\frac{\partial}{\partial\phi}Y_l^m(\theta,\phi) = m\hbar Y_l^m(\theta,\phi), \tag{3.6.26}$$

which implies that the ϕ-dependence $Y_l^m(\theta,\phi)$ must behave like $e^{im\phi}$. Likewise, corresponding to

$$\mathbf{L}^2|l,m\rangle = l(l+1)\hbar^2|l,m\rangle, \tag{3.6.27}$$

we have [see (3.6.15)]

$$\left[\frac{1}{\sin\theta}\frac{\partial}{\partial\theta}\left(\sin\theta\frac{\partial}{\partial\theta}\right) + \frac{1}{\sin^2\theta}\frac{\partial^2}{\partial\phi^2} + l(l+1)\right]Y_l^m = 0, \tag{3.6.28}$$

which is simply the partial differential equation satisfied by Y_l^m itself. The orthogonality relation

$$\langle l',m'|l,m\rangle = \delta_{ll'}\delta_{mm'} \tag{3.6.29}$$

leads to

$$\int_0^{2\pi} d\phi \int_{-1}^{1} d(\cos\theta) Y_{l'}^{m'*}(\theta,\phi) Y_l^m(\theta,\phi) = \delta_{ll'}\delta_{mm'}, \tag{3.6.30}$$

where we have used the completeness relation for the direction eigenkets,

$$\int d\Omega_{\hat{\mathbf{n}}} |\hat{\mathbf{n}}\rangle\langle\hat{\mathbf{n}}| = 1. \tag{3.6.31}$$

To obtain the Y_l^m themselves, we may start with the $m = l$ case. We have

$$L_+ |l, l\rangle = 0, \tag{3.6.32}$$

which, because of (3.6.13), leads to

$$-i\hbar e^{i\phi}\left[i\frac{\partial}{\partial\theta} - \cot\theta\frac{\partial}{\partial\phi}\right]\langle\hat{\mathbf{n}}|l, l\rangle = 0. \tag{3.6.33}$$

Remembering that the ϕ-dependence must behave like $e^{il\phi}$, we can easily show that this partial differential equation is satisfied by

$$\langle\hat{\mathbf{n}}|l, l\rangle = Y_l^l(\theta,\phi) = c_l e^{il\phi}\sin^l\theta, \tag{3.6.34}$$

where c_l is the normalization constant determined from (3.6.30) to be*

$$c_l = \left[\frac{(-1)^l}{2^l l!}\right]\sqrt{\frac{[(2l+1)(2l)!]}{4\pi}}. \tag{3.6.35}$$

Starting with (3.6.34) we can use

$$\begin{aligned}\langle\hat{\mathbf{n}}|l, m-1\rangle &= \frac{\langle\hat{\mathbf{n}}|L_-|l, m\rangle}{\sqrt{(l+m)(l-m+1)}\,\hbar} \\ &= \frac{1}{\sqrt{(l+m)(l-m+1)}} e^{-i\phi}\left(-\frac{\partial}{\partial\theta} + i\cot\theta\frac{\partial}{\partial\phi}\right)\langle\hat{\mathbf{n}}|l, m\rangle\end{aligned} \tag{3.6.36}$$

successively to obtain all Y_l^m with l fixed. Because this is done in many textbooks on elementary quantum mechanics, we will not work out the details here. The result for $m \geq 0$ is

$$Y_l^m(\theta,\phi) = \frac{(-1)^l}{2^l l!}\sqrt{\frac{(2l+1)}{4\pi}\frac{(l+m)!}{(l-m)!}}\, e^{im\phi}\frac{1}{\sin^m\theta}\frac{d^{l-m}}{d(\cos\theta)^{l-m}}(\sin\theta)^{2l}, \tag{3.6.37}$$

*Normalization condition (3.6.30), of course, does not determine the phase of c_l. The factor $(-1)^l$ is inserted so that when we use the L_- operator successively to reach the state $m = 0$, we obtain Y_l^0 with the same sign as the Legendre polynomial $P_l(\cos\theta)$ whose phase is fixed by $P_l(1) = 1$ [see (3.6.39)].

and we define Y_l^{-m} by

$$Y_l^{-m}(\theta,\phi) = (-1)^m [Y_l^m(\theta,\phi)]^*. \tag{3.6.38}$$

Regardless of whether m is positive or negative, the θ-dependent part of $Y_l^m(\theta,\phi)$ is $[\sin\theta]^{|m|}$ times a polynomial in $\cos\theta$ with a highest power of $l-|m|$. For $m=0$, we obtain

$$Y_l^0(\theta,\phi) = \sqrt{\frac{2l+1}{4\pi}}\, P_l(\cos\theta). \tag{3.6.39}$$

From the point of view of the angular-momentum commutation relations alone, it might not appear obvious why l cannot be a half-integer. It turns out that several arguments can be advanced against half-integer l-values. First, for half-integer l, and hence for half-integer m, the wave function would acquire a minus sign,

$$e^{im(2\pi)} = -1, \tag{3.6.40}$$

under a 2π rotation. As a result, the wave function would not be single-valued; we pointed out in Section 2.4 that the wave function must be single-valued because of the requirement that the expansion of a state ket in terms of position eigenkets be unique. We can prove that if $\mathbf{L}$, defined to be $\mathbf{x}\times\mathbf{p}$, is to be identified as the generator of rotation, then the wave function must acquire a plus sign under a 2π rotation. This follows from the fact that the wave function for a 2π-rotated state is the original wave function itself with no sign change:

$$\begin{aligned}\langle \mathbf{x}' | \exp\left(\frac{-iL_z 2\pi}{\hbar}\right) |\alpha\rangle &= \langle x'\cos 2\pi + y'\sin 2\pi,\ y'\cos 2\pi - x'\sin 2\pi,\ z' |\alpha\rangle \\ &= \langle \mathbf{x}'|\alpha\rangle,\end{aligned} \tag{3.6.41}$$

where we have used the finite-angle version of (3.6.6). Next, let us suppose $Y_l^m(\theta,\phi)$ with a half-integer l were possible. To be specific, we choose the simplest case, $l=m=\frac{1}{2}$. According to (3.6.34) we would have

$$Y_{1/2}^{1/2}(\theta,\phi) = c_{1/2}e^{i\phi/2}\sqrt{\sin\theta}. \tag{3.6.42}$$

From the property of L_- [see (3.6.36)] we would then obtain

$$\begin{aligned}Y_{1/2}^{-1/2}(\theta,\phi) &= e^{-i\phi}\left(-\frac{\partial}{\partial\theta} + i\cot\theta\frac{\partial}{\partial\phi}\right)\left(c_{1/2}e^{i\phi/2}\sqrt{\sin\theta}\right) \\ &= -c_{1/2}e^{-i\phi/2}\cot\theta\sqrt{\sin\theta}.\end{aligned} \tag{3.6.43}$$

This expression is not permissible because it is singular at $\theta = 0, \pi$. What is worse, from the partial differential equation

$$\begin{aligned}\left\langle \hat{\mathbf{n}}|L_-|\tfrac{1}{2},-\tfrac{1}{2}\right\rangle &= -i\hbar e^{-i\phi}\left(-i\frac{\partial}{\partial\theta} - \cot\theta\frac{\partial}{\partial\phi}\right)\left\langle \hat{\mathbf{n}}|\tfrac{1}{2},-\tfrac{1}{2}\right\rangle \\ &= 0\end{aligned} \tag{3.6.44}$$

we directly obtain

$$Y_{1/2}^{-1/2} = c'_{1/2} e^{-i\phi/2} \sqrt{\sin\theta}\,, \tag{3.6.45}$$

in sharp contradiction with (3.6.43). Finally, we know from the Sturm-Liouville theory of differential equations that the solutions of (3.6.28) with l integer form a complete set. An arbitrary function of θ and ϕ can be expanded in terms of Y_l^m with integer l and m only. For all these reasons it is futile to contemplate orbital angular momentum with half-integer l-values.

Spherical Harmonics as Rotation Matrices

We conclude this section on orbital angular momentum by discussing the spherical harmonics from the point of view of the rotation matrices introduced in the last section. We can readily establish the desired connection between the two approaches by constructing the most general direction eigenket $|\hat{\mathbf{n}}\rangle$ by applying appropriate rotation operators to $|\hat{\mathbf{z}}\rangle$, the direction eigenket in the positive z-direction. We wish to find $\mathscr{D}(R)$ such that

$$|\hat{\mathbf{n}}\rangle = \mathscr{D}(R)|\hat{\mathbf{z}}\rangle. \tag{3.6.46}$$

We can rely on the technique used in constructing the eigenspinor of $\boldsymbol{\sigma}\cdot\hat{\mathbf{n}}$ in Section 3.2. We first rotate about the y-axis by angle θ, then around the z-axis by angle ϕ; see Figure 3.3 with $\beta \to \theta$, $\alpha \to \phi$. In the notation of Euler angles we have

$$\mathscr{D}(R) = \mathscr{D}(\alpha = \phi, \beta = \theta, \gamma = 0). \tag{3.6.47}$$

Writing (3.6.46) as

$$|\hat{\mathbf{n}}\rangle = \sum_l \sum_m \mathscr{D}(R)|l, m\rangle\langle l, m|\hat{\mathbf{z}}\rangle, \tag{3.6.48}$$

we see that $|\hat{\mathbf{n}}\rangle$, when expanded in terms of $|l, m\rangle$, contains all possible l-values. However, when this equation is multiplied by $\langle l, m'|$ on the left, only one term in the l-sum contributes, namely,

$$\langle l, m'|\hat{\mathbf{n}}\rangle = \sum_m \mathscr{D}^{(l)}_{m'm}(\alpha = \phi, \beta = \theta, \gamma = 0)\langle l, m|\hat{\mathbf{z}}\rangle. \tag{3.6.49}$$

Now $\langle l, m|\hat{\mathbf{z}}\rangle$ is just a number; in fact, it is precisely $Y_l^{m*}(\theta, \phi)$ evaluated at $\theta = 0$ with ϕ undetermined. At $\theta = 0$, Y_l^m is known to vanish for $m \neq 0$, which can also be seen directly from the fact that $|\hat{\mathbf{z}}\rangle$ is an eigenket of L_z (which equals $xp_y - yp_x$) with eigenvalue zero. So we can write

$$\begin{aligned}
\langle l, m|\hat{\mathbf{z}}\rangle &= Y_l^{m*}(\theta = 0, \phi \text{ undetermined})\,\delta_{m0} \\
&= \sqrt{\frac{(2l+1)}{4\pi}}\, P_l(\cos\theta)\Big|_{\cos\theta = 1}\,\delta_{m0} \\
&= \sqrt{\frac{(2l+1)}{4\pi}}\,\delta_{m0}.
\end{aligned} \tag{3.6.50}$$

Returning to (3.6.49), we have

$$Y_l^{m'*}(\theta,\phi)=\sqrt{\frac{(2l+1)}{4\pi}}\,\mathscr{D}_{m'0}^{(l)}(\alpha=\phi,\beta=\theta,\gamma=0) \qquad (3.6.51)$$

or

$$\mathscr{D}_{m0}^{(l)}(\alpha,\beta,\gamma=0)=\sqrt{\frac{4\pi}{(2l+1)}}\,Y_l^{m*}(\theta,\phi)\bigg|_{\theta=\beta,\,\phi=\alpha}. \qquad (3.6.52)$$

Notice the $m=0$ case, which is of particular importance:

$$d_{00}^{(l)}(\beta)\big|_{\beta=\theta} = P_l(\cos\theta). \qquad (3.6.53)$$

3.7. ADDITION OF ANGULAR MOMENTA

Angular-momentum addition has important applications in all areas of modern physics—from atomic spectroscopy to nuclear and particle collisions. Furthermore, a study of angular-momentum addition provides an excellent opportunity to illustrate the concept of change of basis, which we discussed extensively in Chapter 1.

Simple Examples of Angular-Momentum Addition

Before studying a formal theory of angular-momentum addition, it is worth looking at two simple examples with which the reader may be familiar: (1) how to add orbital angular momentum and spin-angular momentum and (2) how to add the spin-angular momenta of two spin $\frac{1}{2}$ particles.

Previously we studied both spin $\frac{1}{2}$ systems with all quantum-mechanical degrees of freedom other than spin—such as position and momentum—ignored and quantum-mechanical particles with the space degrees of freedom (such as position or momentum) taken into account but the internal degrees of freedom (such as spin) ignored. A realistic description of a particle with spin must of course take into account both the space degree of freedom and the internal degrees of freedom. The base ket for a spin $\frac{1}{2}$ particle may be visualized to be in the direct-product space of the infinite-dimensional ket space spanned by the position eigenkets $\{|\mathbf{x}'\rangle\}$ and the two-dimensional spin space spanned by $|+\rangle$ and $|-\rangle$. Explicitly, we have for the base ket

$$|\mathbf{x}',\pm\rangle=|\mathbf{x}'\rangle\otimes|\pm\rangle, \qquad (3.7.1)$$

where any operator in the space spanned by $\{|\mathbf{x}'\rangle\}$ commutes with any operator in the two-dimensional space spanned by $|\pm\rangle$.

The rotation operator still takes the form $\exp(-i\mathbf{J}\cdot\hat{\mathbf{n}}\phi/\hbar)$ but $\mathbf{J}$, the generator of rotations, is now made up of two parts, namely,

$$\mathbf{J} = \mathbf{L} + \mathbf{S}. \tag{3.7.2}$$

It is actually more obvious to write (3.7.2) as

$$\mathbf{J} = \mathbf{L} \otimes 1 + 1 \otimes \mathbf{S}, \tag{3.7.3}$$

where the 1 in $\mathbf{L} \otimes 1$ stands for the identity operator in the spin space, while the 1 in $1 \otimes \mathbf{S}$ stands for the identity operator in the infinite-dimensional ket space spanned by the position eigenkets. Because $\mathbf{L}$ and $\mathbf{S}$ commute, we can write

$$\begin{aligned}\mathscr{D}(R) &= \mathscr{D}^{(\text{orb})}(R) \otimes \mathscr{D}^{(\text{spin})}(R)\\ &= \exp\left(\frac{-i\mathbf{L}\cdot\hat{\mathbf{n}}\phi}{\hbar}\right) \otimes \exp\left(\frac{-i\mathbf{S}\cdot\hat{\mathbf{n}}\phi}{\hbar}\right).\end{aligned} \tag{3.7.4}$$

The wave function for a particle with spin is written as

$$\langle \mathbf{x}', \pm | \alpha \rangle = \psi_{\pm}(\mathbf{x}'). \tag{3.7.5}$$

The two components $\psi_{\pm}$ are often arranged in column matrix form as follows:

$$\begin{pmatrix} \psi_{+}(\mathbf{x}') \\ \psi_{-}(\mathbf{x}') \end{pmatrix}, \tag{3.7.6}$$

where $|\psi_{\pm}(\mathbf{x}')|^2$ stands for the probability density for the particle to be found at $\mathbf{x}'$ with spin up and down, respectively. Instead of $|\mathbf{x}'\rangle$ as the base kets for the space part, we may use $|n, l, m\rangle$, which are eigenkets of $\mathbf{L}^2$ and L_z with eigenvalues $\hbar^2 l(l+1)$ and $m_l\hbar$, respectively. For the spin part, $|\pm\rangle$ are eigenkets of $\mathbf{S}^2$ and S_z with eigenvalues $3\hbar^2/4$ and $\pm\hbar/2$, respectively. However, as we will show later, we can also use base kets which are eigenkets of $\mathbf{J}^2$, J_z, $\mathbf{L}^2$, and $\mathbf{S}^2$. In other words, we can expand a state ket of a particle with spin in terms of simultaneous eigenkets of $\mathbf{L}^2$, $\mathbf{S}^2$, L_z, and S_z *or* in terms of simultaneous eigenkets of $\mathbf{J}^2$, J_z, $\mathbf{L}^2$, and $\mathbf{S}^2$. We will study in detail how the two descriptions are related.

As a second example, we study two spin $\frac{1}{2}$ particles—say two electrons—with the orbital degree of freedom suppressed. The total spin operator is usually written as

$$\mathbf{S} = \mathbf{S}_1 + \mathbf{S}_2, \tag{3.7.7}$$

but again it is to be understood as

$$\mathbf{S}_1 \otimes 1 + 1 \otimes \mathbf{S}_2, \tag{3.7.8}$$

where the 1 in the first (second) term stands for the identity operator in the spin space of electron 2 (1). We, of course, have

$$[S_{1x}, S_{2y}] = 0 \tag{3.7.9}$$

and so forth. Within the space of electron 1(2) we have the usual commutation relations

$$\left[S_{1x}, S_{1y}\right] = i\hbar S_{1z}, \left[S_{2x}, S_{2y}\right] = i\hbar S_{2z}, \ldots. \tag{3.7.10}$$

As a direct consequence of (3.7.9) and (3.7.10), we have

$$\left[S_x, S_y\right] = i\hbar S_z \tag{3.7.11}$$

and so on for the *total* spin operator.

The eigenvalues of the various spin operators are denoted as follows:

$$\begin{aligned}
&\mathbf{S}^2 = (\mathbf{S}_1 + \mathbf{S}_2)^2 \!: s(s+1)\hbar^2 \\
&S_z = S_{1z} + S_{2z} \quad : m\hbar \\
&S_{1z} \qquad\qquad\quad : m_1\hbar \\
&S_{2z} \qquad\qquad\quad : m_2\hbar
\end{aligned} \tag{3.7.12}$$

Again, we can expand the ket corresponding to an arbitrary spin state of two electrons in terms of either the eigenkets of $\mathbf{S}^2$ and S_z *or* the eigenkets of S_{1z} and S_{2z}. The two possibilities are as follows:

1. The $\{m_1, m_2\}$ representation based on the eigenkets of S_{1z} and S_{2z}:
$$|++\rangle, |+-\rangle, |-+\rangle, \quad \text{and} \quad |--\rangle, \tag{3.7.13}$$
where $|+-\rangle$ stands for $m_1 = \frac{1}{2}$, $m_2 = -\frac{1}{2}$, and so forth.
2. The $\{s, m\}$ representation (or the triplet-singlet representation) based on the eigenkets of $\mathbf{S}^2$ and S_z:
$$|s=1, m=\pm 1, 0\rangle, \; |s=0, m=0\rangle, \tag{3.7.14}$$
where $s = 1$ $(s = 0)$ is referred to as spin triplet (spin singlet).

Notice that in each set there are four base kets. The relationship between the two sets of base kets is as follows:

$$|s=1, m=1\rangle = |++\rangle, \tag{3.7.15a}$$

$$|s=1, m=0\rangle = \left(\frac{1}{\sqrt{2}}\right)(|+-\rangle + |-+\rangle), \tag{3.7.15b}$$

$$|s=1, m=-1\rangle = |--\rangle, \tag{3.7.15c}$$

$$|s=0, m=0\rangle = \left(\frac{1}{\sqrt{2}}\right)(|+-\rangle - |-+\rangle). \tag{3.7.15d}$$

The right-hand side of (3.7.15a) tells us that we have both electrons with spin up; this situation can correspond only to $s = 1$, $m = 1$. We can obtain (3.7.15b) from (3.7.15a) by applying the ladder operator

$$\begin{aligned}
S_- &\equiv S_{1-} + S_{2-} \\
&= (S_{1x} - iS_{1y}) + (S_{2x} - iS_{2y})
\end{aligned} \tag{3.7.16}$$

to both sides of (3.7.15a). In doing so we must remember that an electron 1 operator like S_{1-} affects just the first entry of $|++\rangle$, and so on. We can write

$$S_{-}|s=1, m=1\rangle = (S_{1-} + S_{2-})|++\rangle \tag{3.7.17}$$

as

$$\sqrt{(1+1)(1-1+1)}\,|s=1, m=0\rangle = \sqrt{\left(\tfrac{1}{2}+\tfrac{1}{2}\right)\left(\tfrac{1}{2}-\tfrac{1}{2}+1\right)} \times |-+\rangle + \sqrt{\left(\tfrac{1}{2}+\tfrac{1}{2}\right)\left(\tfrac{1}{2}-\tfrac{1}{2}+1\right)}\,|+-\rangle, \tag{3.7.18}$$

which immediately leads to (3.7.15b). Likewise, we can obtain $|s=1, m=-1\rangle$ by applying (3.7.16) once again to (3.7.15b). Finally, we can obtain (3.7.15d) by requiring it to be orthogonal to the other three kets, in particular to (3.7.15b).

The coefficients that appear on the right-hand side of (3.7.15) are the simplest example of **Clebsch-Gordan coefficients** to be discussed further at a later time. They are simply the elements of the transformation matrix that connects the $\{m_1, m_2\}$ basis to the $\{s, m\}$ basis. It is instructive to derive these coefficients in another way. Suppose we write the 4×4 matrix corresponding to

$$\begin{aligned} \mathbf{S}^2 &= \mathbf{S}_1^2 + \mathbf{S}_2^2 + 2\mathbf{S}_1\cdot\mathbf{S}_2 \\ &= \mathbf{S}_1^2 + \mathbf{S}_2^2 + 2S_{1z}S_{2z} + S_{1+}S_{2-} + S_{1-}S_{2+} \end{aligned} \tag{3.7.19}$$

using the (m_1, m_2) basis. The square matrix is obviously not diagonal because an operator like S_{1+} connects $|-+\rangle$ with $|++\rangle$. The unitary matrix that diagonalizes this matrix carries the $|m_1, m_2\rangle$ base kets into the $|s, m\rangle$ base kets. The elements of this unitary matrix are precisely the Clebsch-Gordan coefficients for this problem. The reader is encouraged to work out all this in detail.

Formal Theory of Angular-Momentum Addition

Having gained some physical insight by considering simple examples, we are now in a position to study more systematically the formal theory of angular-momentum addition. Consider two angular-momentum operators $\mathbf{J}_1$ and $\mathbf{J}_2$ in different subspaces. The components of $\mathbf{J}_1(\mathbf{J}_2)$ satisfy the usual angular-momentum commutation relations:

$$\left[J_{1i}, J_{1j}\right] = i\hbar\varepsilon_{ijk}J_{1k} \tag{3.7.20a}$$

and

$$\left[J_{2i}, J_{2j}\right] = i\hbar\varepsilon_{ijk}J_{2k}. \tag{3.7.20b}$$

However, we have

$$[J_{1k}, J_{2l}] = 0 \tag{3.7.21}$$

between any pair of operators from different subspaces.

The infinitesimal rotation operator that affects both subspace 1 and subspace 2 is written as

$$\left(1 - \frac{i\mathbf{J}_1 \cdot \hat{\mathbf{n}}\delta\phi}{\hbar}\right) \otimes \left(1 - \frac{i\mathbf{J}_2 \cdot \hat{\mathbf{n}}\delta\phi}{\hbar}\right) = 1 - \frac{i(\mathbf{J}_1 \otimes 1 + 1 \otimes \mathbf{J}_2) \cdot \hat{\mathbf{n}}\delta\phi}{\hbar}. \tag{3.7.22}$$

We define the total angular momentum by

$$\mathbf{J} \equiv \mathbf{J}_1 \otimes 1 + 1 \otimes \mathbf{J}_2, \tag{3.7.23}$$

which is more commonly written as

$$\mathbf{J} = \mathbf{J}_1 + \mathbf{J}_2. \tag{3.7.24}$$

The finite-angle version of (3.7.22) is

$$\mathscr{D}_1(R) \otimes \mathscr{D}_2(R) = \exp\left(\frac{-i\mathbf{J}_1 \cdot \hat{\mathbf{n}}\phi}{\hbar}\right) \otimes \exp\left(\frac{-i\mathbf{J}_2 \cdot \hat{\mathbf{n}}\phi}{\hbar}\right). \tag{3.7.25}$$

Notice the appearance of the same axis of rotation and the same angle of rotation.

It is very important to note that the total **J** satisfies the angular-momentum commutation relations

$$[J_i, J_j] = i\hbar\varepsilon_{ijk}J_k \tag{3.7.26}$$

as a direct consequence of (3.7.20) and (3.7.21). In other words, **J** is an angular momentum in the sense of Section 3.1. Physically this is reasonable because **J** is the generator for the *entire* system. Everything we learned in Section 3.5—for example, the eigenvalue spectrum of $\mathbf{J}^2$ and J_z and the matrix elements of the ladder operators—also holds for the total **J**.

As for the choice of base kets we have two options.

Option A: Simultaneous eigenkets of $\mathbf{J}_1^2$, $\mathbf{J}_2^2$, J_{1z}, and J_{2z}, denoted by $|j_1 j_2; m_1 m_2\rangle$. Obviously the four operators commute with each other. The defining equations are

$$\mathbf{J}_1^2 |j_1 j_2; m_1 m_2\rangle = j_1(j_1+1)\hbar^2 |j_1 j_2; m_1 m_2\rangle, \tag{3.7.27a}$$

$$J_{1z} |j_1 j_2; m_1 m_2\rangle = m_1\hbar |j_1 j_2; m_1 m_2\rangle, \tag{3.7.27b}$$

$$\mathbf{J}_2^2 |j_1 j_2; m_1 m_2\rangle = j_2(j_2+1)\hbar^2 |j_1 j_2; m_1 m_2\rangle, \tag{3.7.27c}$$

$$J_{2z} |j_1 j_2; m_1 m_2\rangle = m_2\hbar |j_1 j_2; m_1 m_2\rangle. \tag{3.7.27d}$$

Option B: Simultaneous eigenkets of $\mathbf{J}^2$, $\mathbf{J}_1^2$, $\mathbf{J}_2^2$, and J_z. First, note that this set of operators mutually commute. In particular, we have

$$[\mathbf{J}^2, \mathbf{J}_1^2] = 0, \tag{3.7.28}$$

which can readily be seen by writing $\mathbf{J}^2$ as

$$\mathbf{J}^2 = \mathbf{J}_1^2 + \mathbf{J}_2^2 + 2J_{1z}J_{2z} + J_{1+}J_{2-} + J_{1-}J_{2+}. \tag{3.7.29}$$

We use $|j_1, j_2; jm\rangle$ to denote the base kets of option B:

$$\mathbf{J}_1^2|j_1 j_2; jm\rangle = j_1(j_1+1)\hbar^2|j_1 j_2; jm\rangle, \tag{3.7.30a}$$

$$\mathbf{J}_2^2|j_1 j_2; jm\rangle = j_2(j_2+1)\hbar^2|j_1 j_2; jm\rangle, \tag{3.7.30b}$$

$$\mathbf{J}^2|j_1 j_2; jm\rangle = j(j+1)\hbar^2|j_1 j_2; jm\rangle, \tag{3.7.30c}$$

$$J_z|j_1 j_2; jm\rangle = m\hbar|j_1 j_2; jm\rangle. \tag{3.7.30d}$$

Quite often j_1, j_2 are understood, and the base kets are written simply as $|j, m\rangle$.

It is very important to note that even though

$$[\mathbf{J}^2, J_z] = 0, \tag{3.7.31}$$

we have

$$[\mathbf{J}^2, J_{1z}] \neq 0, \qquad [\mathbf{J}^2, J_{2z}] \neq 0, \tag{3.7.32}$$

as the reader may easily verify using (3.7.29). This means that we cannot add $\mathbf{J}^2$ to the set of operators of option A. Likewise, we cannot add J_{1z} and/or J_{2z} to the set of operators of option B. We have two possible sets of base kets corresponding to the two maximal sets of mutually compatible observables we have constructed.

Let us consider the unitary transformation in the sense of Section 1.5 that connects the two bases:

$$|j_1 j_2; jm\rangle = \sum_{m_1}\sum_{m_2}|j_1 j_2; m_1 m_2\rangle\langle j_1 j_2; m_1 m_2|j_1 j_2; jm\rangle, \tag{3.7.33}$$

where we have used

$$\sum_{m_1}\sum_{m_2}|j_1 j_2; m_1 m_2\rangle\langle j_1 j_2; m_1 m_2| = 1 \tag{3.7.34}$$

and where the right-hand side is the identity operator in the ket space of given j_1 and j_2. The elements of this transformation matrix $\langle j_1 j_2; m_1 m_2|j_1 j_2; jm\rangle$ are Clebsch-Gordan coefficients.

There are many important properties of Clebsch-Gordan coefficients that we are now ready to study. First, the coefficients vanish unless

$$m = m_1 + m_2. \tag{3.7.35}$$

To prove this, first note that

$$(J_z - J_{1z} - J_{2z})|j_1 j_2; jm\rangle = 0. \tag{3.7.36}$$

Multiplying $\langle j_1 j_2; m_1 m_2|$ on the left, we obtain

$$(m - m_1 - m_2)\langle j_1 j_2; m_1 m_2|j_1 j_2; jm\rangle = 0, \tag{3.7.37}$$

which proves our assertion. Admire the power of the Dirac notation! It really pays to write the Clebsch-Gordan coefficients in Dirac's bracket form, as we have done.

Second, the coefficients vanish unless

$$|j_1 - j_2| \le j \le j_1 + j_2. \tag{3.7.38}$$

This property may appear obvious from the vector model of angular-momentum addition, where we visualize $\mathbf{J}$ to be the vectorial sum of $\mathbf{J}_1$ and $\mathbf{J}_2$. However, it is worth checking this point by showing that if (3.7.38) holds, then the dimensionality of the space spanned by $\{|j_1 j_2; m_1 m_2\rangle\}$ is the same as that of the space spanned by $\{|j_1 j_2; jm\rangle\}$. For the (m_1, m_2) way of counting we obtain

$$N = (2j_1 + 1)(2j_2 + 1) \tag{3.7.39}$$

because for given j_1 there are $2j_1 + 1$ possible values of m_1; a similar statement is true for the other angular momentum j_2. As for the (j, m) way of counting, we note that for each j, there are $2j + 1$ states, and according to (3.7.38), j itself runs from $j_1 - j_2$ to $j_1 + j_2$, where we have assumed, without loss of generality, that $j_1 \ge j_2$. We therefore obtain

$$\begin{aligned} N &= \sum_{j = j_1 - j_2}^{j_1 + j_2} (2j + 1) \\ &= \tfrac{1}{2}\left[\{2(j_1 - j_2) + 1\} + \{2(j_1 + j_2) + 1\}\right](2j_2 + 1) \\ &= (2j_1 + 1)(2j_2 + 1). \end{aligned} \tag{3.7.40}$$

Because both ways of counting give the same N-value, we see that (3.7.38) is quite consistent.*

The Clebsch-Gordan coefficients form a unitary matrix. Furthermore, the matrix elements are taken to be real by convention. An immediate consequence of this is that the inverse coefficient $\langle j_1 j_2; jm | j_1 j_2; m_1 m_2\rangle$ is the same as $\langle j_1 j_2; m_1 m_2 | j_1 j_2; jm\rangle$ itself. A real unitary matrix is orthogonal, so we have the orthogonality condition

$$\sum_j \sum_m \langle j_1 j_2; m_1 m_2 | j_1 j_2; jm\rangle \langle j_1 j_2; m_1' m_2' | j_1 j_2; jm\rangle = \delta_{m_1 m_1'} \delta_{m_2 m_2'}, \tag{3.7.41}$$

which is obvious from the orthonormality of $\{|j_1 j_2; m_1 m_2\rangle\}$ together with the reality of the Clebsch-Gordan coefficients. Likewise, we also have

$$\sum_{m_1} \sum_{m_2} \langle j_1 j_2; m_1 m_2 | j_1 j_2; jm\rangle \langle j_1 j_2; m_1 m_2 | j_1 j_2; j'm'\rangle = \delta_{jj'} \delta_{mm'}. \tag{3.7.42}$$

*A complete proof of (3.7.38) is given in Gottfried 1966, 215, and also in Appendix B of this book.

As a special case of this we may set $j' = j$, $m' = m = m_1 + m_2$. We then obtain

$$\sum_{m_1} \sum_{m_2} |\langle j_1 j_2; m_1 m_2 | j_1 j_2; jm \rangle|^2 = 1, \tag{3.7.43}$$

which is just the normalization condition for $|j_1 j_2; jm\rangle$.

Some authors use somewhat different notations for the Clebsch-Gordan coefficients. Instead of $\langle j_1 j_2; m_1 m_2 | j_1 j_2; jm \rangle$ we sometimes see $\langle j_1 m_1 j_2 m_2 | j_1 j_2 jm \rangle$, $C(j_1 j_2 j; m_1 m_2 m)$, $C_{j_1 j_2}(jm; m_1 m_2)$, and so on. They can also be written in terms of **Wigner's 3-j symbol**, which is occasionally found in the literature:

$$\langle j_1 j_2; m_1 m_2 | j_1 j_2; jm \rangle = (-1)^{j_1 - j_2 + m} \sqrt{2j+1} \begin{pmatrix} j_1 & j_2 & j \\ m_1 & m_2 & -m \end{pmatrix}. \tag{3.7.44}$$

Recursion Relations for the Clebsch-Gordan Coefficients*

With j_1, j_2, and j fixed, the coefficients with different m_1 and m_2 are related to each other by **recursion relations**. We start with

$$J_\pm |j_1 j_2; jm\rangle = (J_{1\pm} + J_{2\pm}) \sum_{m_1} \sum_{m_2} |j_1 j_2; m_1 m_2\rangle \langle j_1 j_2; m_1 m_2 | j_1 j_2; jm\rangle. \tag{3.7.45}$$

Using (3.5.39) and (3.5.40) we obtain (with $m_1 \to m_1'$, $m_2 \to m_2'$)

$$\begin{aligned} &\sqrt{(j \mp m)(j \pm m + 1)}\, |j_1 j_2; j, m \pm 1\rangle \\ &= \sum_{m_1'} \sum_{m_2'} \Big(\sqrt{(j_1 \mp m_1')(j_1 \pm m_1' + 1)}\, |j_1 j_2; m_1' \pm 1, m_2'\rangle \\ &\quad + \sqrt{(j_2 \mp m_2')(j_2 \pm m_2' + 1)}\, |j_1 j_2; m_1', m_2' \pm 1\rangle \Big) \\ &\quad \times \langle j_1 j_2; m_1' m_2' | j_1 j_2; jm\rangle. \end{aligned} \tag{3.7.46}$$

Our next step is to multiply by $\langle j_1 j_2; m_1 m_2|$ on the left and use orthonormality, which means that nonvanishing contributions from the right-hand side are possible only with

$$m_1 = m_1' \pm 1, \qquad m_2 = m_2' \tag{3.7.47}$$

for the first term and

$$m_1 = m_1', \qquad m_2 = m_2' \pm 1 \tag{3.7.48}$$

*More-detailed discussion of Clebsch-Gordan and Racah coefficients, recoupling, and the like is given in A. R. Edmonds 1960, for instance.

for the second term. In this manner we obtain the desired recursion relations:

$$\sqrt{(j \mp m)(j \pm m+1)}\,\langle j_1 j_2; m_1 m_2 | j_1 j_2; j, m \pm 1\rangle$$
$$= \sqrt{(j_1 \mp m_1 + 1)(j_1 \pm m_1)}\,\langle j_1 j_2; m_1 \mp 1, m_2 | j_1 j_2; jm\rangle$$
$$+ \sqrt{(j_2 \mp m_2 + 1)(j_2 \pm m_2)}\,\langle j_1 j_2; m_1, m_2 \mp 1 | j_1 j_2; jm\rangle. \tag{3.7.49}$$

It is important to note that because the $J_\pm$ operators have shifted the m-values, the nonvanishing condition (3.7.35) for the Clebsch-Gordan coefficients has now become [when applied to (3.7.49)]

$$m_1 + m_2 = m \pm 1. \tag{3.7.50}$$

We can appreciate the significance of the recursion relations by looking at (3.7.49) in an $m_1 m_2$-plane. The J_+ recursion relation (upper sign) tells us that the coefficient at (m_1, m_2) is related to the coefficients at $(m_1 - 1, m_2)$ and $(m_1, m_2 - 1)$, as shown in Figure 3.5a. Likewise, the J_- recursion relation (lower sign) relates the three coefficients whose m_1, m_2 values are given in Figure 3.5b.

Recursion relations (3.7.49), together with normalization condition (3.7.43), almost uniquely determine all Clebsch-Gordan coefficients. (We say "almost uniquely" because certain sign conventions have yet to be specified.) Our strategy is as follows. We go back to the $m_1 m_2$-plane, again for fixed j_1, j_2, and j, and plot the boundary of the allowed region determined by

$$|m_1| \le j_1, \qquad |m_2| \le j_2, \qquad -j \le m_1 + m_2 \le j; \tag{3.7.51}$$

see Figure 3.6a. We may start with the upper right-hand corner, denoted by A. Because we work near A at the start, a more detailed "map" is in order;

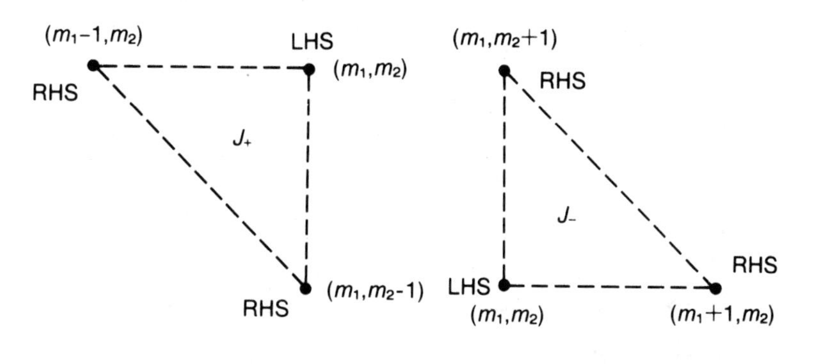

FIGURE 3.5. $m_1 m_2$-plane showing the Clebsch-Gordan coefficients related by the recursion relations (3.7.49).

see Figure 3.6b. We apply the J_- recursion relation (3.7.49) (lower sign), with (m_1, m_2+1) corresponding to A. Observe now that the recursion relation connects A with only B because the site corresponding to (m_1+1, m_2) is forbidden by $m_1 \le j_1$. As a result, we can obtain the Clebsch-Gordan coefficient of B in terms of the coefficient of A. Next, we form a J_+ triangle made up of A, B, and D. This enables us to obtain the coefficient of D once the coefficient of A is specified. We can continue in this fashion: Knowing B and D, we can get to E; knowing B and E we can get to C, and so on. With enough patience we can obtain the Clebsch-Gordan coefficient of every site in terms of the coefficient of starting site, A. For overall normalization we use (3.7.43). The final overall sign is fixed by convention. (See the following example.)

As an important practical example we consider the problem of adding the orbital and spin-angular momenta of a single spin $\frac{1}{2}$ particle. We have

$$j_1 = l \quad \text{(integer)}, \qquad m_1 = m_l,$$
$$j_2 = s = \tfrac{1}{2}, \qquad m_2 = m_s = \pm\tfrac{1}{2}. \tag{3.7.52}$$

The allowed values of j are given by

$$j = l \pm \tfrac{1}{2}, \quad l > 0; \qquad j = \tfrac{1}{2}, \quad l = 0 \tag{3.7.53}$$

So for each l there are two possible j-values; for example, for $l = 1$ (p state) we get, in spectroscopic notation, $p_{3/2}$ and $p_{1/2}$, where the subscript refers to j. The $m_1 m_2$-plane, or better the $m_l m_s$-plane, of this problem is particularly simple. The allowed sites form only two rows: the upper row for $m_s = \frac{1}{2}$ and the lower row for $m_s = -\frac{1}{2}$; see Figure 3.7. Specifically, we work out

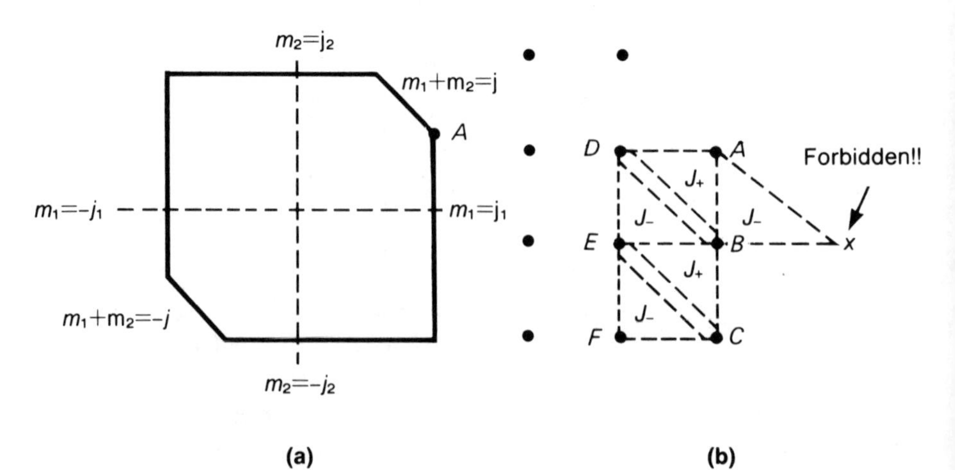

FIGURE 3.6. Use of the recursion relations to obtain the Clebsch-Gordan coefficients.

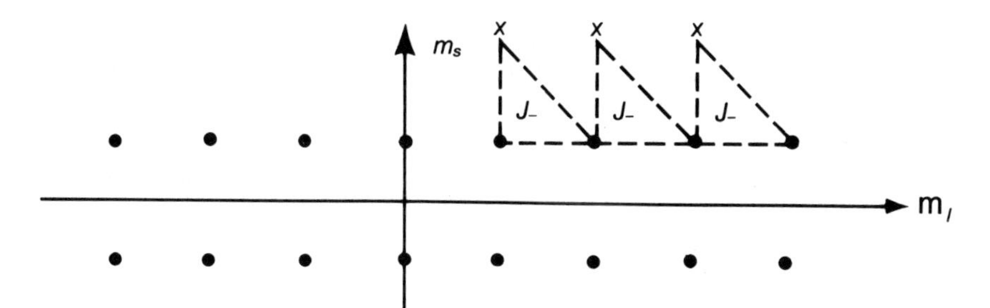

FIGURE 3.7. Recursion relations used to obtain the Clebsch-Gordan coefficients for $j_1 = l$ and $j_2 = s = \frac{1}{2}$.

the case $j = l + \frac{1}{2}$. Because m_s cannot exceed $\frac{1}{2}$, we can use the J_- recursion in such a way that we always stay in the upper row ($m_2 = m_s = \frac{1}{2}$), while the m_l-value changes by one unit each time we consider a new J_- triangle. Suppressing $j_1 = l$, $j_2 = \frac{1}{2}$ in writing the Clebsch-Gordan coefficients, we obtain from (3.7.49) (lower sign)

$$\sqrt{(l+\tfrac{1}{2}+m+1)(l+\tfrac{1}{2}-m)}\,\langle m-\tfrac{1}{2},\tfrac{1}{2}|l+\tfrac{1}{2},m\rangle$$
$$=\sqrt{(l+m+\tfrac{1}{2})(l-m+\tfrac{1}{2})}\,\langle m+\tfrac{1}{2},\tfrac{1}{2}|l+\tfrac{1}{2},m+1\rangle, \tag{3.7.54}$$

where we have used

$$m_1 = m_l = m - \tfrac{1}{2}, \qquad m_2 = m_s = \tfrac{1}{2}. \tag{3.7.55}$$

In this way we can move horizontally by one unit:

$$\left\langle m-\frac{1}{2},\frac{1}{2}\right|l+\frac{1}{2},m\left.\right\rangle = \sqrt{\frac{l+m+\frac{1}{2}}{l+m+\frac{3}{2}}}\left\langle m+\frac{1}{2},\frac{1}{2}\right|l+\frac{1}{2},m+1\left.\right\rangle. \tag{3.7.56}$$

We can in turn express $\langle m+\frac{1}{2},\frac{1}{2}|l+\frac{1}{2},m+1\rangle$ in terms of $\langle m+\frac{3}{2},\frac{1}{2}|l+\frac{1}{2},m+2\rangle$, and so forth. Clearly, this procedure can be continued until m_l reaches l, the maximum possible value:

$$\left\langle m-\frac{1}{2},\frac{1}{2}\right|l+\frac{1}{2},m\left.\right\rangle = \sqrt{\frac{l+m+\frac{1}{2}}{l+m+\frac{3}{2}}}\sqrt{\frac{l+m+\frac{3}{2}}{l+m+\frac{5}{2}}}\left\langle m+\frac{3}{2},\frac{1}{2}\right|l+\frac{1}{2},m+2\left.\right\rangle$$
$$= \sqrt{\frac{l+m+\frac{1}{2}}{l+m+\frac{3}{2}}}\sqrt{\frac{l+m+\frac{3}{2}}{l+m+\frac{5}{2}}}\sqrt{\frac{l+m+\frac{5}{2}}{l+m+\frac{7}{2}}}$$
$$\times\left\langle m+\frac{5}{2},\frac{1}{2}\right|l+\frac{1}{2},m+3\left.\right\rangle$$
$$\vdots$$
$$= \sqrt{\frac{l+m+\frac{1}{2}}{2l+1}}\left\langle l,\frac{1}{2}\right|l+\frac{1}{2},l+\frac{1}{2}\left.\right\rangle. \tag{3.7.57}$$

Consider the angular-momentum configuration in which m_l and m_s are both maximal, that is, l and $\frac{1}{2}$, respectively. The total $m=m_l+m_s$ is $l+\frac{1}{2}$, which is possible only for $j=l+\frac{1}{2}$ and not for $j=l-\frac{1}{2}$. So $|m_l=l, m_s=\frac{1}{2}\rangle$ must be equal to $|j=l+\frac{1}{2}, m=l+\frac{1}{2}\rangle$ up to a phase factor. We take this phase factor to be real and positive by convention. With this choice we have

$$\langle l,\tfrac{1}{2}|l+\tfrac{1}{2},l+\tfrac{1}{2}\rangle=1. \tag{3.7.58}$$

Returning to (3.7.57), we finally obtain

$$\left\langle m-\frac{1}{2},\frac{1}{2}\middle|l+\frac{1}{2},m\right\rangle=\sqrt{\frac{l+m+\frac{1}{2}}{2l+1}}. \tag{3.7.59}$$

But this is only about one-fourth of the story. We must still determine the value of the question marks that appear in the following:

$$\begin{aligned}\left|j=l+\frac{1}{2},m\right\rangle &= \sqrt{\frac{l+m+\frac{1}{2}}{2l+1}}\left|m_l=m-\frac{1}{2},m_s=\frac{1}{2}\right\rangle \\ &+?\left|m_l=m+\frac{1}{2},m_s=-\frac{1}{2}\right\rangle, \\ \left|j=l-\frac{1}{2},m\right\rangle &= ?\left|m_l=m-\frac{1}{2},m_s=\frac{1}{2}\right\rangle+?\left|m_l=m+\frac{1}{2},m_s=-\frac{1}{2}\right\rangle.\end{aligned} \tag{3.7.60}$$

We note that the transformation matrix with fixed m from the (m_l, m_s) basis to the (j, m) basis is, because of orthogonality, expected to have the form

$$\begin{pmatrix}\cos\alpha & \sin\alpha \\ -\sin\alpha & \cos\alpha\end{pmatrix}. \tag{3.7.61}$$

Comparison with (3.7.60) shows that $\cos\alpha$ is (3.7.59) itself; so we can readily determine $\sin\alpha$ up to a sign ambiguity:

$$\sin^2\alpha=1-\frac{(l+m+\frac{1}{2})}{(2l+1)}=\frac{(l-m+\frac{1}{2})}{(2l+1)}. \tag{3.7.62}$$

We claim that $\langle m_l=m+\frac{1}{2}, m_s=-\frac{1}{2}|j=l+\frac{1}{2},m\rangle$ must be positive because all $j=l+\frac{1}{2}$ states are reachable by applying the J_- operator successively to $|j=l+\frac{1}{2}, m=l+\frac{1}{2}\rangle$, and the matrix elements of J_- are always positive by convention. So the 2×2 transformation matrix (3.7.61) can be only

$$\begin{pmatrix}\sqrt{\dfrac{l+m+\frac{1}{2}}{2l+1}} & \sqrt{\dfrac{l-m+\frac{1}{2}}{2l+1}} \\ -\sqrt{\dfrac{l-m+\frac{1}{2}}{2l+1}} & \sqrt{\dfrac{l+m+\frac{1}{2}}{2l+1}}\end{pmatrix}. \tag{3.7.63}$$

We define **spin-angular functions** in two-component form as follows:

$$\begin{aligned}\mathscr{Y}_l^{j=l\pm1/2,m} &= \pm\sqrt{\frac{l\pm m+\frac{1}{2}}{2l+1}}\,Y_l^{m-1/2}(\theta,\phi)\chi_+ \\ &\quad+\sqrt{\frac{l\mp m+\frac{1}{2}}{2l+1}}\,Y_l^{m+1/2}(\theta,\phi)\chi_- \\ &= \frac{1}{\sqrt{2l+1}}\begin{pmatrix}\pm\sqrt{l\pm m+\frac{1}{2}}\,Y_l^{m-1/2}(\theta,\phi)\\ \sqrt{l\mp m+\frac{1}{2}}\,Y_l^{m+1/2}(\theta,\phi)\end{pmatrix}. \end{aligned} \tag{3.7.64}$$

They are, by construction, simultaneous eigenfunctions of $\mathbf{L}^2$, $\mathbf{S}^2$, $\mathbf{J}^2$, and J_z. They are also eigenfunctions of $\mathbf{L}\cdot\mathbf{S}$ but $\mathbf{L}\cdot\mathbf{S}$, being just

$$\mathbf{L}\cdot\mathbf{S} = \left(\tfrac{1}{2}\right)(\mathbf{J}^2-\mathbf{L}^2-\mathbf{S}^2), \tag{3.7.65}$$

is not independent. Indeed, its eigenvalue can easily be computed as follows:

$$\left(\frac{\hbar^2}{2}\right)\left[j(j+1)-l(l+1)-\frac{3}{4}\right] = \begin{cases}\dfrac{l\hbar^2}{2} & \text{for } j=l+\frac{1}{2},\\[2ex] -\dfrac{(l+1)\hbar^2}{2} & \text{for } j=l-\frac{1}{2}.\end{cases} \tag{3.7.66}$$

Clebsch-Gordan Coefficients and Rotation Matrices

Angular-momentum addition may be discussed from the point of view of rotation matrices. Consider the rotation operator $\mathscr{D}^{(j_1)}(R)$ in the ket space spanned by the angular-momentum eigenkets with eigenvalue j_1. Likewise, consider $\mathscr{D}^{(j_2)}(R)$. The product $\mathscr{D}^{(j_1)}\otimes\mathscr{D}^{(j_2)}$ is reducible in the sense that after suitable choice of base kets, its matrix representation can take the following form:

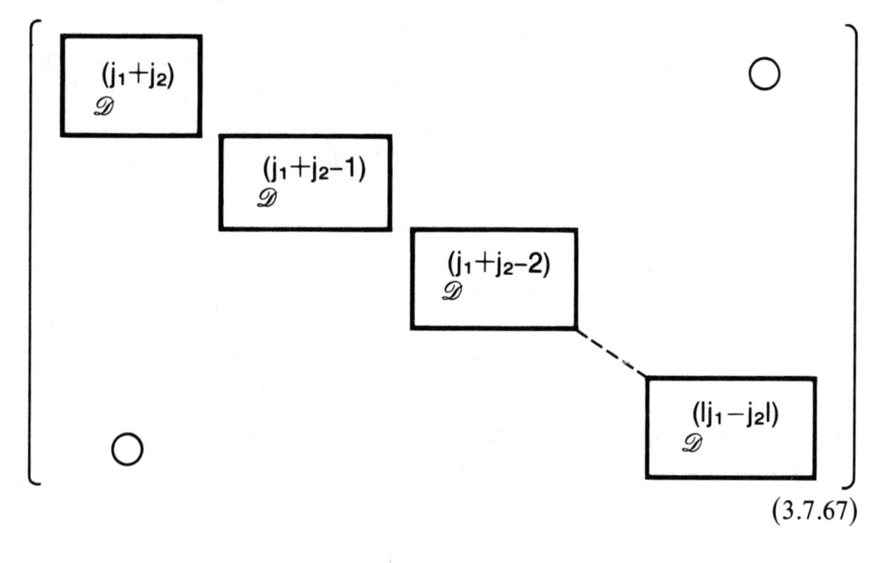

(3.7.67)

In the notation of group theory this is written as

$$\mathscr{D}^{(j_1)} \otimes \mathscr{D}^{(j_2)} = \mathscr{D}^{(j_1+j_2)} \oplus \mathscr{D}^{(j_1+j_2-1)} \oplus \cdots \oplus \mathscr{D}^{(|j_1-j_2|)}. \tag{3.7.68}$$

In terms of the elements of rotation matrices, we have an important expansion known as the **Clebsch-Gordan series**:

$$\mathscr{D}^{(j_1)}_{m_1 m_1'}(R)\mathscr{D}^{(j_2)}_{m_2 m_2'}(R) = \sum_j \sum_m \sum_{m'} \langle j_1 j_2; m_1 m_2 | j_1 j_2; jm\rangle \times \langle j_1 j_2; m_1' m_2' | j_1 j_2; jm'\rangle \mathscr{D}^{(j)}_{mm'}(R), \tag{3.7.69}$$

where the j-sum runs from $|j_1 - j_2|$ to $j_1 + j_2$. The proof of this equation follows. First, note that the left-hand side of (3.7.69) is the same as

$$\begin{aligned}\langle j_1 j_2; m_1 m_2 | \mathscr{D}(R) | j_1 j_2; m_1' m_2'\rangle &= \langle j_1 m_1 | \mathscr{D}(R) | j_1 m_1'\rangle \langle j_2 m_2 | \mathscr{D}(R) | j_2 m_2'\rangle \\ &= \mathscr{D}^{(j_1)}_{m_1 m_1'}(R)\mathscr{D}^{(j_2)}_{m_2 m_2'}(R).\end{aligned} \tag{3.7.70}$$

But the same matrix element is also computable by inserting a complete set of states in the (j, m) basis. Thus

$$\begin{aligned}&\langle j_1 j_2; m_1 m_2 | \mathscr{D}(R) | j_1 j_2; m_1' m_2'\rangle \\ &\quad = \sum_j \sum_m \sum_{j'} \sum_{m'} \langle j_1 j_2; m_1 m_2 | j_1 j_2; jm\rangle \langle j_1 j_2; jm | \mathscr{D}(R) | j_1 j_2; j'm'\rangle \\ &\qquad \times \langle j_1 j_2; j'm' | j_1 j_2; m_1' m_2'\rangle \\ &\quad = \sum_j \sum_m \sum_{j'} \sum_{m'} \langle j_1 j_2; m_1 m_2 | j_1 j_2; jm\rangle \mathscr{D}^{(j)}_{mm'}(R)\delta_{jj'} \\ &\qquad \times \langle j_1 j_2; m_1' m_2' | j_1 j_2; j'm'\rangle,\end{aligned} \tag{3.7.71}$$

which is just the right-hand side of (3.7.69).

As an interesting application of (3.7.69), we derive an important formula for an integral involving three spherical harmonics. First, recall the connection between $\mathscr{D}^{(l)}_{m0}$ and Y_l^{m*} given by (3.6.52). Letting $j_1 \to l_1$, $j_2 \to l_2$, $m_1' \to 0$, $m_2' \to 0$ (hence $m' \to 0$) in (3.7.69), we obtain, after complex conjugation,

$$Y_{l_1}^{m_1}(\theta,\phi) Y_{l_2}^{m_2}(\theta,\phi) = \frac{\sqrt{(2l_1+1)(2l_2+1)}}{4\pi} \sum_{l'} \sum_{m'} \langle l_1 l_2; m_1 m_2 | l_1 l_2; l'm'\rangle \times \langle l_1 l_2; 00 | l_1 l_2; l'0\rangle \sqrt{\frac{4\pi}{2l'+1}}\, Y_{l'}^{m'}(\theta,\phi). \tag{3.7.72}$$

We multiply both sides by $Y_l^{m*}(\theta,\phi)$ and integrate over solid angles. The summations drop out because of the orthogonality of spherical harmonics, and we are left with

$$\int d\Omega\, Y_l^{m*}(\theta,\phi) Y_{l_1}^{m_1}(\theta,\phi) Y_{l_2}^{m_2}(\theta,\phi)$$

$$= \sqrt{\frac{(2l_1+1)(2l_2+1)}{4\pi(2l+1)}} \langle l_1 l_2; 00 | l_1 l_2; l0\rangle \langle l_1 l_2; m_1 m_2 | l_1 l_2; lm\rangle. \tag{3.7.73}$$

The square root factor times the first Clebsch-Gordan coefficient is independent of orientations; that is, of m_1 and m_2. The second Clebsch-Gordan coefficient is the one appropriate for adding l_1 and l_2 to obtain total l. Equation (3.7.73) turns out to be a special case of the Wigner-Eckart theorem to be derived in Section 3.10. This formula is extremely useful in evaluating multipole matrix elements in atomic and nuclear spectroscopy.

3.8. SCHWINGER'S OSCILLATOR MODEL OF ANGULAR MOMENTUM

Angular Momentum and Uncoupled Oscillators

There exists a very interesting connection between the algebra of angular momentum and the algebra of two independent (that is, uncoupled) oscillators, which was worked out in J. Schwinger's notes [see *Quantum Theory of Angular Momentum*, edited by L. C. Biedenharn and H. Van Dam, Academic Press (1965), p. 229]. Let us consider two simple harmonic oscillators, which we call the *plus type* and the *minus type*. We have the annihilation and creation operators, denoted by a_+ and $a_+^\dagger$ for the plus-type oscillator; likewise, we have a_- and $a_-^\dagger$ for the minus-type oscillators. We also define the number operators N_+ and N_- as follows:

$$N_+ \equiv a_+^\dagger a_+, \qquad N_- \equiv a_-^\dagger a_-. \tag{3.8.1}$$

We assume that the usual commutation relations among a, $a^\dagger$, and N hold for oscillators of the same type (see Section 2.3).

$$[a_+, a_+^\dagger] = 1, \qquad [a_-, a_-^\dagger] = 1, \tag{3.8.2a}$$

$$[N_+, a_+] = -a_+, \qquad [N_-, a_-] = -a_-, \tag{3.8.2b}$$

$$[N_+, a_+^\dagger] = a_+^\dagger, \qquad [N_-, a_-^\dagger] = a_-^\dagger. \tag{3.8.2c}$$

However, we assume that any pair of operators between different oscillators commute:

$$[a_+, a_-^\dagger] = [a_-, a_+^\dagger] = 0 \tag{3.8.3}$$

and so forth. So it is in this sense that we say the two oscillators are uncoupled.

Because N_+ and N_- commute by virtue of (3.8.3), we can build up simultaneous eigenkets of N_+ and N_- with eigenvalues n_+ and n_-, respectively. So we have the following eigenvalue equations for $N_\pm$:

$$N_+|n_+,n_-\rangle = n_+|n_+,n_-\rangle, \qquad N_-|n_+,n_-\rangle = n_-|n_+,n_-\rangle. \tag{3.8.4}$$

In complete analogy with (2.3.16) and (2.3.17), the creation and annihilation operators, $a_\pm^\dagger$ and $a_\pm$, act on $|n_+,n_-\rangle$ as follows:

$$a_+^\dagger|n_+,n_-\rangle = \sqrt{n_+ + 1}\,|n_+ + 1,n_-\rangle, \qquad a_-^\dagger|n_+,n_-\rangle = \sqrt{n_- + 1}\,|n_+,n_- + 1\rangle, \tag{3.8.5a}$$

$$a_+|n_+,n_-\rangle = \sqrt{n_+}\,|n_+ - 1,n_-\rangle, \qquad a_-|n_+,n_-\rangle = \sqrt{n_-}\,|n_+,n_- - 1\rangle. \tag{3.8.5b}$$

We can obtain the most general eigenkcts of N_+ and N_- by applying $a_+^\dagger$ and $a_-^\dagger$ successively to the **vacuum ket** defined by

$$a_+|0,0\rangle = 0, \qquad a_-|0,0\rangle = 0. \tag{3.8.6}$$

In this way we obtain

$$|n_+,n_-\rangle = \frac{(a_+^\dagger)^{n_+}(a_-^\dagger)^{n_-}}{\sqrt{n_+!}\sqrt{n_-!}}|0,0\rangle. \tag{3.8.7}$$

Next, we *define*

$$J_+ \equiv \hbar a_+^\dagger a_-, \qquad J_- \equiv \hbar a_-^\dagger a_+, \tag{3.8.8a}$$

and

$$J_z \equiv \left(\frac{\hbar}{2}\right)(a_+^\dagger a_+ - a_-^\dagger a_-) = \left(\frac{\hbar}{2}\right)(N_+ - N_-). \tag{3.8.8b}$$

We can readily prove that these operators satisfy the angular-momentum commutation relations of the usual form

$$[J_z, J_\pm] = \pm\hbar J_\pm, \tag{3.8.9a}$$

$$[J_+, J_-] = 2\hbar J_z. \tag{3.8.9b}$$

For example, we prove (3.8.9b) as follows:

$$\begin{aligned}\hbar^2[a_+^\dagger a_-, a_-^\dagger a_+] &= \hbar^2 a_+^\dagger a_- a_-^\dagger a_+ - \hbar^2 a_-^\dagger a_+ a_+^\dagger a_- \\ &= \hbar^2 a_+^\dagger\,(a_-^\dagger a_- + 1)a_+ - \hbar^2 a_-^\dagger(a_+^\dagger a_+ + 1)a_- \\ &= \hbar^2(a_+^\dagger a_+ - a_-^\dagger a_-) = 2\hbar J_z.\end{aligned} \tag{3.8.10}$$

Defining the *total* N to be

$$N \equiv N_+ + N_- = a_+^\dagger a_+ + a_-^\dagger a_-, \tag{3.8.11}$$

we can also prove

$$\mathbf{J}^2 \equiv J_z^2 + \left(\frac{1}{2}\right)(J_+J_- + J_-J_+)$$
$$= \left(\frac{\hbar^2}{2}\right)N\left(\frac{N}{2}+1\right), \tag{3.8.12}$$

which is left as an exercise.

What are the physical interpretations of all this? We associate spin up ($m = \frac{1}{2}$) with one quantum unit of the plus-type oscillator and spin down ($m = -\frac{1}{2}$) with one quantum unit of the minus-type oscillator. If you like, you may imagine one spin $\frac{1}{2}$ "particle" with spin up (down) with each quantum unit of the plus- (minus-) type oscillator. The eigenvalues n_+ and n_- are just the number of spins up and spins down, respectively. The meaning of J_+ is that it destroys one unit of spin down with the z-component of spin-angular momentum $-\hbar/2$ and creates one unit of spin up with the z-component of spin-angular momentum $+\hbar/2$; the z-component of angular momentum is therefore increased by $\hbar$. Likewise J_- destroys one unit of spin up and creates one unit of spin down; the z-component of angular momentum is therefore decreased by $\hbar$. As for the J_z operator, it simply counts $\hbar/2$ times the difference of n_+ and n_-, just the z-component of the total angular momentum. With (3.8.5) at our disposal we can easily examine how $J_\pm$ and J_z act on $|n_+, n_-\rangle$ as follows:

$$J_+|n_+, n_-\rangle = \hbar a_+^\dagger a_-|n_+, n_-\rangle = \sqrt{n_-(n_+ + 1)}\,\hbar|n_+ + 1, n_- - 1\rangle, \tag{3.8.13a}$$

$$J_-|n_+, n_-\rangle = \hbar a_-^\dagger a_+|n_+, n_-\rangle = \sqrt{n_+(n_- + 1)}\,\hbar|n_+ - 1, n_- + 1\rangle, \tag{3.8.13b}$$

$$J_z|n_+, n_-\rangle = \left(\frac{\hbar}{2}\right)(N_+ - N_-)|n_+, n_-\rangle = \left(\frac{1}{2}\right)(n_+ - n_-)\hbar|n_+, n_-\rangle. \tag{3.8.13c}$$

Notice that in all these operations, the sum $n_+ + n_-$, which corresponds to the total number of spin $\frac{1}{2}$ particles remains unchanged.

Observe now that (3.8.13a), (3.8.13b), and (3.8.13c) reduce to the familiar expressions for the $J_\pm$ and J_z operators we derived in Section 3.5, provided we substitute

$$n_+ \to j + m, \qquad n_- \to j - m. \tag{3.8.14}$$

The square root factors in (3.8.13a) and (3.8.13b) change to

$$\sqrt{n_-(n_+ + 1)} \to \sqrt{(j-m)(j+m+1)},$$
$$\sqrt{n_+(n_- + 1)} \to \sqrt{(j+m)(j-m+1)}, \tag{3.8.15}$$

which are exactly the square root factors appearing in (3.5.39) and (3.5.41).

Notice also that the eigenvalue of the $\mathbf{J}^2$ operator defined by (3.8.12) changes as follows:

$$\left(\frac{\hbar^2}{2}\right)(n_+ + n_-)\left[\frac{(n_+ + n_-)}{2} + 1\right] \to \hbar^2 j(j+1). \qquad (3.8.16)$$

All this may not be too surprising because we have already proved that $J_\pm$ and $\mathbf{J}^2$ operators we constructed out of the oscillator operators satisfy the usual angular-momentum commutation relations. But it is instructive to see in an explicit manner the connection between the oscillator matrix elements and angular-momentum matrix elements. In any case, it is now natural to use

$$j \equiv \frac{(n_+ + n_-)}{2}, \qquad m \equiv \frac{(n_+ - n_-)}{2} \qquad (3.8.17)$$

in place of n_+ and n_- to characterize simultaneous eigenkets of $\mathbf{J}^2$ and J_z. According to (3.8.13a) the action of J_+ changes n_+ into $n_+ + 1$, n_- into $n_- - 1$, which means that j is unchanged and m goes into $m+1$. Likewise, we see that the J_- operator that changes n_+ into $n_+ - 1$; n_- into $n_- + 1$ lowers m by one unit without changing j. We can now write as (3.8.7) for the most general N_+, N_- eigenket

$$|j,m\rangle = \frac{(a_+^\dagger)^{j+m}(a_-^\dagger)^{j-m}}{\sqrt{(j+m)!(j-m)!}}|0\rangle, \qquad (3.8.18)$$

where we have used $|0\rangle$ for the vacuum ket, earlier denoted by $|0,0\rangle$.

A special case of (3.8.18) is of interest. Let us set $m = j$, which physically means that the eigenvalue of J_z is as large as possible for a given j. We have

$$|j,j\rangle = \frac{(a_+^\dagger)^{2j}}{\sqrt{(2j)!}}|0\rangle. \qquad (3.8.19)$$

We can imagine this state to be built up of $2j$ spin $\frac{1}{2}$ particles with their spins all pointing in the positive z-direction.

In general, we note that a complicated object of high j can be visualized as being made up of primitive spin $\frac{1}{2}$ particles, $j+m$ of them with spin up and the remaining $j-m$ of them with spin down. This picture is extremely convenient even though we obviously cannot always regard an object of angular momentum j literally as a composite system of spin $\frac{1}{2}$ particles. All we are saying is that, *as far as the transformation properties under rotations are concerned*, we can visualize any object of angular momentum j as a composite system of $2j$ spin $\frac{1}{2}$ particles formed in the manner indicated by (3.8.18).

From the point of view of angular-momentum addition developed in the previous section, we can add the spins of $2j$ spin $\frac{1}{2}$ particles to obtain states with angular momentum $j, j-1, j-2, \ldots$. As a simple example, we can add the spin-angular momenta of two spin $\frac{1}{2}$ particles to obtain a total angular momentum of zero as well as one. In Schwinger's oscillator scheme, however, we obtain only states with angular momentum j when we start with $2j$ spin $\frac{1}{2}$ particles. In the language of permutation symmetry to be developed in Chapter 6, only totally symmetrical states are constructed by this method. The primitive spin $\frac{1}{2}$ particles appearing here are actually *bosons*! This method is quite adequate if our purpose is to examine the properties under rotations of states characterized by j and m without asking how such states are built up initially.

The reader who is familiar with isospin in nuclear and particle physics may note that what we are doing here provides a new insight into the isospin (or isotopic spin) formalism. The operator J_+ that destroys one unit of the minus type and creates one unit of the plus type is completely analogous to the isospin ladder operator T_+ (sometimes denoted by I_+) that annihilates a neutron (isospin down) and creates a proton (isospin up), thus raising the z-component of isospin by one unit. In contrast, J_z is analogous to T_z, which simply counts the difference between the number of protons and neutrons in nuclei.

Explicit Formula for Rotation Matrices

Schwinger's scheme can be used to derive, in a very simple way, a closed formula for rotation matrices, first obtained by E. P. Wigner using a similar (but not identical) method. We apply the rotation operator $\mathscr{D}(R)$ to $|j, m\rangle$, written as (3.8.18). In the Euler angle notation the only nontrivial rotation is the second one about the y-axis, so we direct our attention to

$$\mathscr{D}(R) = \mathscr{D}(\alpha, \beta, \gamma)|_{\alpha=\gamma=0} = \exp\left(\frac{-iJ_y\beta}{\hbar}\right). \tag{3.8.20}$$

We have

$$\mathscr{D}(R)|j, m\rangle = \frac{[\mathscr{D}(R)a_+^\dagger\mathscr{D}^{-1}(R)]^{j+m}[\mathscr{D}(R)a_-^\dagger\mathscr{D}^{-1}(R)]^{j-m}}{\sqrt{(j+m)!(j-m)!}}\mathscr{D}(R)|0\rangle. \tag{3.8.21}$$

Now, $\mathscr{D}(R)$ acting on $|0\rangle$ just reproduces $|0\rangle$ because, by virtue of (3.8.6), only the leading term, 1, in the expansion of exponential (3.8.20) contributes. So

$$\mathscr{D}(R)a_\pm^\dagger\mathscr{D}^{-1}(R) = \exp\left(\frac{-iJ_y\beta}{\hbar}\right)a_\pm^\dagger \exp\left(\frac{iJ_y\beta}{\hbar}\right). \tag{3.8.22}$$

Thus we may use formula (2.3.47). Letting

$$G \to \frac{-J_y}{\hbar}, \qquad \lambda \to \beta \tag{3.8.23}$$

in (2.3.47), we realize that we must look at various commutators, namely,

$$\begin{gathered}\left[\frac{-J_y}{\hbar}, a_+^\dagger\right] = \left(\frac{1}{2i}\right)\left[a_-^\dagger a_+, a_+^\dagger\right] = \left(\frac{1}{2i}\right)a_-^\dagger, \\ \left[\frac{-J_y}{\hbar}, \left[\frac{-J_y}{\hbar}, a_+^\dagger\right]\right] = \left[\frac{-J_y}{\hbar}, \frac{a_-^\dagger}{2i}\right] = \left(\frac{1}{4}\right)a_+^\dagger\end{gathered} \tag{3.8.24}$$

and so forth. Clearly, we always obtain either $a_+^\dagger$ or $a_-^\dagger$. Collecting terms, we get

$$\mathscr{D}(R) a_+^\dagger \mathscr{D}^{-1}(R) = a_+^\dagger \cos\left(\frac{\beta}{2}\right) + a_-^\dagger \sin\left(\frac{\beta}{2}\right). \tag{3.8.25}$$

Likewise,

$$\mathscr{D}(R) a_-^\dagger \mathscr{D}^{-1}(R) = a_-^\dagger \cos\left(\frac{\beta}{2}\right) - a_+^\dagger \sin\left(\frac{\beta}{2}\right). \tag{3.8.26}$$

Actually this result is not surprising. After all, the basic spin-up state is supposed to transform as

$$a_+^\dagger |0\rangle \to \cos\left(\frac{\beta}{2}\right) a_+^\dagger |0\rangle + \sin\left(\frac{\beta}{2}\right) a_-^\dagger |0\rangle \tag{3.8.27}$$

under a rotation about the y-axis. Substituting (3.8.25) and (3.8.26) into (3.8.21) and recalling the binomial theorem

$$(x+y)^N = \sum_k \frac{N! x^{N-k} y^k}{(N-k)! k!}, \tag{3.8.28}$$

we obtain

$$\begin{aligned}\mathscr{D}(\alpha=0,\beta,\gamma=0|j,m\rangle = &\sum_k \sum_l \frac{(j+m)!(j-m)!}{(j+m-k)!k!(j-m-l)!l!} \\ &\times \frac{\left[a_+^\dagger \cos(\beta/2)\right]^{j+m-k}\left[a_-^\dagger \sin(\beta/2)\right]^k}{\sqrt{(j+m)!(j-m)!}} \\ &\times\left[-a_+^\dagger \sin(\beta/2)\right]^{j-m-l}\left[a_-^\dagger \cos(\beta/2)\right]^l |0\rangle.\end{aligned} \tag{3.8.29}$$

We may compare (3.8.29) with

$$\begin{aligned}\mathscr{D}(\alpha=0,\beta,\gamma=0)|j,m\rangle &= \sum_{m'} |j,m'\rangle d_{m'm}^{(j)}(\beta) \\ &= \sum_{m'} d_{m'm}^{(j)}(\beta) \frac{\left(a_+^\dagger\right)^{j+m'}\left(a_-^\dagger\right)^{j-m'}}{\sqrt{(j+m')!(j-m')!}}|0\rangle.\end{aligned} \tag{3.8.30}$$

We can obtain an explicit form for $d^{(j)}_{m'm}(\beta)$ by equating the coefficients of powers of $a_+^\dagger$ in (3.8.29) and (3.8.30). Specifically, we want to compare $a_+^\dagger$ raised to $j+m'$ in (3.8.30) with $a_+^\dagger$ raised to $2j-k-l$, so we identify

$$l = j - k - m'. \tag{3.8.31}$$

We are seeking $d_{m'm}(\beta)$ with m' fixed. The k-sum and the l-sum in (3.8.29) are not independent of each other; we eliminate l in favor of k by taking advantage of (3.8.31). As for the powers of $a_-^\dagger$, we note that $a_-^\dagger$ raised to $j-m'$ in (3.8.30) automatically matches with $a_-^\dagger$ raised to $k+l$ in (3.8.29) when (3.8.31) is imposed. The last step is to identify the exponents of $\cos(\beta/2)$, $\sin(\beta/2)$, and (-1), which are, respectively,

$$j + m - k + l = 2j - 2k + m - m', \tag{3.8.32a}$$

$$k + j - m - l = 2k - m + m', \tag{3.8.32b}$$

$$j - m - l = k - m + m', \tag{3.8.32c}$$

where we have used (3.8.31) to eliminate l. In this way we obtain **Wigner's formula** for $d^{(j)}_{m'm}(\beta)$:

$$d^{(j)}_{m'm}(\beta) = \sum_k (-1)^{k-m+m'} \frac{\sqrt{(j+m)!(j-m)!(j+m')!(j-m')!}}{(j+m-k)!k!(j-k-m')!(k-m+m')!}$$

$$\times \left(\cos\frac{\beta}{2}\right)^{2j-2k+m-m'} \left(\sin\frac{\beta}{2}\right)^{2k-m+m'}, \tag{3.8.33}$$

where we take the sum over k whenever none of the arguments of factorials in the denominator are negative.

3.9. SPIN CORRELATION MEASUREMENTS AND BELL'S INEQUALITY

Correlations in Spin-Singlet States

The simplest example of angular-momentum addition we encountered in Section 3.7 was concerned with a composite system made up of spin $\frac{1}{2}$ particles. In this section we use such a system to illustrate one of the most astonishing consequences of quantum mechanics.

Consider a two-electron system in a spin-singlet state, that is, with a total spin of zero. We have already seen that the state ket can be written as [see (3.7.15d)]

$$|\text{spin singlet}\rangle = \left(\frac{1}{\sqrt{2}}\right)(|\hat{\mathbf{z}}+;\hat{\mathbf{z}}-\rangle - |\hat{\mathbf{z}}-;\hat{\mathbf{z}}+\rangle), \tag{3.9.1}$$

where we have explicitly indicated the quantization direction. Recall that

$|\hat{\mathbf{z}}+;\hat{\mathbf{z}}-\rangle$ means that electron 1 is in the spin-up state and electron 2 is in the spin-down state. The same is true for $|\hat{\mathbf{z}}-;\hat{\mathbf{z}}+\rangle$.

Suppose we make a measurement on the spin component of one of the electrons. Clearly, there is a 50-50 chance of getting either up or down because the composite system may be in $|\hat{\mathbf{z}}+;\hat{\mathbf{z}}-\rangle$ or $|\hat{\mathbf{z}}-;\hat{\mathbf{z}}+\rangle$ with equal probabilities. But if one of the components is shown to be in the spin-up state, the other is necessarily in the spin-down state, and vice versa. When the spin component of electron 1 is shown to be up, the measurement apparatus has selected the first term, $|\hat{\mathbf{z}}+;\hat{\mathbf{z}}-\rangle$ of (3.9.1); a subsequent measurement of the spin component of electron 2 must ascertain that the state ket of the composite system is given by $|\hat{\mathbf{z}}+;\hat{\mathbf{z}}-\rangle$.

It is remarkable that this kind of correlation can persist even if the two particles are well separated and have ceased to interact provided that as they fly apart, there is no change in their spin states. This is certainly the case for a $J=0$ system disintegrating spontaneously into two spin $\frac{1}{2}$ particles with no relative orbital angular momentum, because angular-momentum conservation must hold in the disintegration process. An example of this would be a rare decay of the η meson (mass 549 MeV$/c^2$) into a muon pair

$$\eta \rightarrow \mu^+ + \mu^- \tag{3.9.2}$$

which, unfortunately, has a branching ratio of only approximately 6×10^{-6}. More realistically, in proton-proton scattering at low kinetic energies, the Pauli principle to be discussed in Chapter 6 forces the interacting protons to be in ${}^1S_0$ (orbital angular momentum 0, spin-singlet state), and the spin states of the scattered protons must be correlated in the manner indicated by (3.9.1) even after they get separated by a *macroscopic distance*.

To be more pictorial we consider a system of two spin $\frac{1}{2}$ particles moving in opposite directions, as in Figure 3.8. Observer A specializes in measuring S_z of particle 1 (flying to the right), while observer B specializes in measuring S_z of particle 2 (flying to the left). To be specific, let us assume that observer A finds S_z to be positive for particle 1. Then he or she can predict, even before B performs any measurement, the outcome of B's measurement with certainty: B must find S_z to be negative for particle 2. On the other hand, if A makes no measurement, B has a 50-50 chance of getting S_z+ or S_z-.

This by itself might not be so peculiar. One may say, "It is just like an urn known to contain one black ball and one white ball. When we blindly pick one of them, there is a 50-50 chance of getting black or white. But if the first ball we pick is black, then we can predict with certainty that the second ball will be white."

It turns out that this analogy is too simple. The actual quantum-mechanical situation is far more sophisticated than that! This is because observers may choose to measure S_x in place of S_z. The *same* pair of

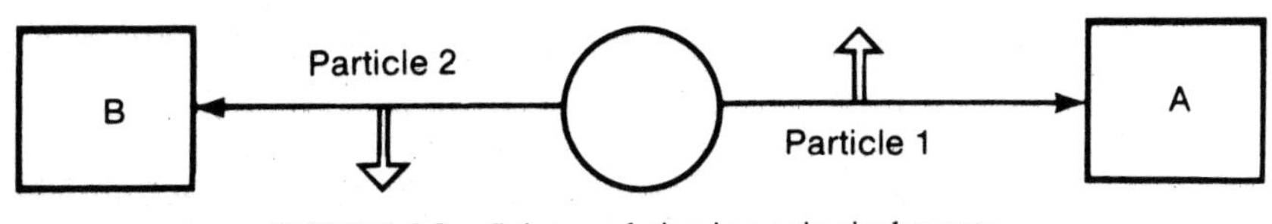

FIGURE 3.8. Spin correlation in a spin-singlet state.

"quantum-mechanical balls" can be analyzed either in terms of black and white *or* in terms of blue and red!

Recall now that for a single spin $\frac{1}{2}$ system the S_x eigenkets and S_z eigenkets are related as follows:

$$|\hat{\mathbf{x}}\pm\rangle = \left(\frac{1}{\sqrt{2}}\right)(|\hat{\mathbf{z}}+\rangle \pm |\hat{\mathbf{z}}-\rangle), \qquad |\hat{\mathbf{z}}\pm\rangle = \left(\frac{1}{\sqrt{2}}\right)(|\hat{\mathbf{x}}+\rangle \pm |\hat{\mathbf{x}}-\rangle). \tag{3.9.3}$$

Returning now to our composite system, we can rewrite spin-singlet ket (3.9.1) by choosing the x-direction as the axis of quantization:

$$|\text{spin singlet}\rangle = \left(\frac{1}{\sqrt{2}}\right)(|\hat{\mathbf{x}}-;\hat{\mathbf{x}}+\rangle - |\hat{\mathbf{x}}+;\hat{\mathbf{x}}-\rangle). \tag{3.9.4}$$

Apart from the overall sign, which in any case is a matter of convention, we could have guessed this form directly from (3.9.1) because spin-singlet states have no preferred direction in space. Let us now suppose that observer A can choose to measure S_z or S_x of particle 1 by changing the orientation of his or her spin analyzer, while observer B always specializes in measuring S_x of particle 2. If A determines S_z of particle 1 to be positive, B clearly has a 50-50 chance for getting S_x+ or S_x-; even though S_z of particle 2 is known to be negative with certainty, its S_x is completely undetermined. On the other hand, let us suppose that A also chooses to measure S_x; if observer A determines S_x of particle 1 to be positive, then without fail, observer B will measure S_x of particle 2 to be negative. Finally, if A chooses to make no measurement, B, of course, will have a 50-50 chance of getting S_x+ or S_x-. To sum up:

1. If A measures S_z and B measures S_x, there is a completely random correlation between the two measurements.
2. If A measures S_x and B measures S_x, there is a 100% (opposite sign) correlation between the two measurements.
3. If A makes no measurement, B's measurements show random results.

Table 3.1 shows all possible results of such measurements when B and A are allowed to choose to measure S_x or S_z.

TABLE 3.1. Spin-correlation Measurements

Spin component measured by A	A's result	Spin component measured by B	B's result
z	$+$	z	$-$
z	$-$	x	$+$
x	$-$	z	$-$
x	$-$	z	$+$
z	$+$	x	$-$
x	$+$	x	$-$
z	$+$	x	$+$
x	$-$	x	$+$
z	$-$	z	$+$
z	$-$	x	$-$
x	$+$	z	$+$
x	$+$	z	$-$

These considerations show that the outcome of B's measurement appears to depend on what kind of measurement A decides to perform: an S_x measurement, an S_z measurement, or no measurement. Notice again that A and B can be miles apart with no possibility of communications or mutual interactions. Observer A can decide how to orient his or her spin-analyzer apparatus long after the two particles have separated. It is as though particle 2 "knows" which spin component of particle 1 is being measured.

The orthodox quantum-mechanical interpretation of this situation is as follows. A measurement is a selection (or filtration) process. When S_z of particle 1 is measured to be positive, then component $|\hat{\mathbf{z}}+;\hat{\mathbf{z}}-\rangle$ is selected. A subsequent measurement of the other particle's S_z merely ascertains that the system is still in $|\hat{\mathbf{z}}+;\hat{\mathbf{z}}-\rangle$. We must accept that a measurement on what appears to be a part of the system is to be regarded as a measurement on the whole system.

Einstein's Locality Principle and Bell's Inequality

Many physicists have felt uncomfortable with the preceding orthodox interpretation of spin-correlation measurements. Their feelings are typified in the following frequently quoted remarks by A. Einstein, which we call **Einstein's locality principle**: "But on one supposition we should, in my opinion, absolutely hold fast: The real factual situation of the system S_2 is independent of what is done with the system S_1, which is spatially separated from the former." Because this problem was first discussed in a

1935 paper of A. Einstein, B. Podolsky, and N. Rosen, it is sometimes known as the Einstein-Podolsky-Rosen paradox.*

Some have argued that the difficulties encountered here are inherent in the probabilistic interpretations of quantum mechanics and that the dynamic behavior at the microscopic level appears probabilistic only because some yet unknown parameters—so-called hidden variables—have not been specified. It is not our purpose here to discuss various alternatives to quantum mechanics based on hidden-variable or other considerations. Rather, let us ask, Do such theories make predictions different from those of quantum mechanics? Until 1964, it could be thought that the alternative theories could be concocted in such a way that they would give no predictions, other than the usual quantum-mechanical predictions, that could be verified experimentally. The whole debate would have belonged to the realm of metaphysics rather than physics. It was then pointed out by J. S. Bell that the alternative theories based on Einstein's locality principle actually predict a *testable inequality relation* among the observables of spin-correlation experiments that *disagrees* with the predictions of quantum mechanics.

We derive Bell's inequality within the framework of a simple model, conceived by E. P. Wigner, that incorporates the essential features of the various alternative theories. Proponents of this model agree that it is impossible to determine S_x and S_z simultaneously. However, when we have a large number of spin $\frac{1}{2}$ particles, we assign a certain fraction of them to have the following property:

If S_z is measured, we obtain a plus sign with certainty.
If S_x is measured, we obtain a minus sign with certainty.

A particle satisfying this property is said to belong to type $(\hat{\mathbf{z}}+, \hat{\mathbf{x}}-)$. Notice that we are not asserting that we can simultaneously measure S_z and S_x to be $+$ and $-$, respectively. When we measure S_z, we do not measure S_x, and vice versa. We are assigning definite values of spin components *in more than one direction* with the understanding that only one or the other of the components can actually be measured. Even though this approach is fundamentally different from that of quantum mechanics, the quantum-mechanical predictions for S_z and S_x measurements performed on the spin-up (S_z+) state are reproduced provided there are as many particles belonging to type $(\hat{\mathbf{z}}+, \hat{\mathbf{x}}+)$ as to type $(\hat{\mathbf{z}}+, \hat{\mathbf{x}}-)$.

Let us now examine how this model can account for the results of spin-correlation measurements made on composite spin-singlet systems.

*To be historically accurate, the original Einstein-Podolsky-Rosen paper dealt with measurements of x and p. The use of composite spin $\frac{1}{2}$ systems to illustrate the Einstein-Podolsky-Rosen paradox started with D. Bohm.

Clearly, for a particular pair, there must be a perfect matching between particle 1 and particle 2 to ensure zero total angular momentum: If particle 1 is of type $(\hat{\mathbf{z}}+,\hat{\mathbf{x}}-)$, then particle 2 must belong to type $(\hat{\mathbf{z}}-,\hat{\mathbf{x}}+)$, and so forth. The results of correlation measurements, such as in Table 3.1, can be reproduced if particle 1 and particle 2 are matched as follows:

$$\begin{array}{cc} \text{particle 1} & \text{particle 2} \\ (\hat{\mathbf{z}}+,\hat{\mathbf{x}}-) \leftrightarrow & (\hat{\mathbf{z}}-,\hat{\mathbf{x}}+), \end{array} \tag{3.9.5a}$$

$$(\hat{\mathbf{z}}+,\hat{\mathbf{x}}+) \leftrightarrow (\hat{\mathbf{z}}-,\hat{\mathbf{x}}-), \tag{3.9.5b}$$

$$(\hat{\mathbf{z}}-,\hat{\mathbf{x}}+) \leftrightarrow (\hat{\mathbf{z}}+,\hat{\mathbf{x}}-), \tag{3.9.5c}$$

$$(\hat{\mathbf{z}}-,\hat{\mathbf{x}}-) \leftrightarrow (\hat{\mathbf{z}}+,\hat{\mathbf{x}}+) \tag{3.9.5d}$$

with equal populations, that is, 25% each. A very important assumption is implied here. Suppose a particular pair belongs to type (3.9.5a) and observer A decides to measure S_z of particle 1; then he or she necessarily obtains a plus sign regardless of whether B decides to measure S_z or S_x. It is in this sense that Einstein's locality principle is incorporated in this model: A's result is predetermined independently of B's choice as to what to measure.

In the examples considered so far, this model has been successful in reproducing the predictions of quantum mechanics. We now consider more-complicated situations where the model leads to predictions different from the usual quantum-mechanical predictions. This time we start with three unit vectors $\hat{\mathbf{a}}$, $\hat{\mathbf{b}}$, and $\hat{\mathbf{c}}$, which are, in general, not mutually orthogonal. We imagine that one of the particles belongs to some definite type, say $(\hat{\mathbf{a}}-,\hat{\mathbf{b}}+,\hat{\mathbf{c}}+)$, which means that if $\mathbf{S}\cdot\hat{\mathbf{a}}$ is measured, we obtain a minus sign with certainty; if $\mathbf{S}\cdot\hat{\mathbf{b}}$ is measured, we obtain a plus sign with certainty; if $\mathbf{S}\cdot\hat{\mathbf{c}}$ is measured, we obtain a plus with certainty. Again there must be a perfect matching in the sense that the other particle necessarily belongs to type $(\hat{\mathbf{a}}+,\hat{\mathbf{b}}-,\hat{\mathbf{c}}-)$ to ensure zero total angular momentum. In any given event, the particle pair in question must be a member of one of the eight types shown in Table 3.2. These eight possibilities are mutually exclusive and disjoint. The population of each type is indicated in the first column.

Let us suppose that observer A finds $\mathbf{S}_1\cdot\hat{\mathbf{a}}$ to be plus and observer B finds $\mathbf{S}_2\cdot\hat{\mathbf{b}}$ to be plus also. It is clear from Table 3.2 that the pair belong to either type 3 or type 4, so the number of particle pairs for which this situation is realized is $N_3 + N_4$. Because N_i is positive semidefinite, we must have inequality relations like

$$N_3 + N_4 \leq (N_2 + N_4) + (N_3 + N_7). \tag{3.9.6}$$

Let $P(\hat{\mathbf{a}}+;\hat{\mathbf{b}}+)$ be the probability that, in a random selection, observer A measures $\mathbf{S}_1\cdot\hat{\mathbf{a}}$ to be + and observer B measures $\mathbf{S}_2\cdot\hat{\mathbf{b}}$ to be +, and so on.

TABLE 3.2. Spin-component Matching in the Alternative Theories

Population	Particle 1	Particle 2
N_1	$(\hat{\mathbf{a}}+, \hat{\mathbf{b}}+, \hat{\mathbf{c}}+)$	$(\hat{\mathbf{a}}-, \hat{\mathbf{b}}-, \hat{\mathbf{c}}-)$
N_2	$(\hat{\mathbf{a}}+, \hat{\mathbf{b}}+, \hat{\mathbf{c}}-)$	$(\hat{\mathbf{a}}-, \hat{\mathbf{b}}-, \hat{\mathbf{c}}+)$
N_3	$(\hat{\mathbf{a}}+, \hat{\mathbf{b}}-, \hat{\mathbf{c}}+)$	$(\hat{\mathbf{a}}-, \hat{\mathbf{b}}+, \hat{\mathbf{c}}-)$
N_4	$(\hat{\mathbf{a}}+, \hat{\mathbf{b}}-, \hat{\mathbf{c}}-)$	$(\hat{\mathbf{a}}-, \hat{\mathbf{b}}+, \hat{\mathbf{c}}+)$
N_5	$(\hat{\mathbf{a}}-, \hat{\mathbf{b}}+, \hat{\mathbf{c}}+)$	$(\hat{\mathbf{a}}+, \hat{\mathbf{b}}-, \hat{\mathbf{c}}-)$
N_6	$(\hat{\mathbf{a}}-, \hat{\mathbf{b}}+, \hat{\mathbf{c}}-)$	$(\hat{\mathbf{a}}+, \hat{\mathbf{b}}-, \hat{\mathbf{c}}+)$
N_7	$(\hat{\mathbf{a}}-, \hat{\mathbf{b}}-, \hat{\mathbf{c}}+)$	$(\hat{\mathbf{a}}+, \hat{\mathbf{b}}+, \hat{\mathbf{c}}-)$
N_8	$(\hat{\mathbf{a}}-, \hat{\mathbf{b}}-, \hat{\mathbf{c}}-)$	$(\hat{\mathbf{a}}+, \hat{\mathbf{b}}+, \hat{\mathbf{c}}+)$

Clearly, we have

$$P(\hat{\mathbf{a}}+;\hat{\mathbf{b}}+) = \frac{(N_3 + N_4)}{\Sigma_i^8 N_i}. \tag{3.9.7}$$

In a similar manner, we obtain

$$P(\hat{\mathbf{a}}+;\hat{\mathbf{c}}+) = \frac{(N_2 + N_4)}{\Sigma_i^8 N_i} \quad \text{and} \quad P(\hat{\mathbf{c}}+;\hat{\mathbf{b}}+) = \frac{(N_3 + N_7)}{\Sigma_i^8 N_i}. \tag{3.9.8}$$

The positivity condition (3.9.6) now becomes

$$P(\hat{\mathbf{a}}+;\hat{\mathbf{b}}+) \leq P(\hat{\mathbf{a}}+;\hat{\mathbf{c}}+) + P(\hat{\mathbf{c}}+;\hat{\mathbf{b}}+). \tag{3.9.9}$$

This is **Bell's inequality**, which follows from Einstein's locality principle.

Quantum Mechanics and Bell's Inequality

We now return to the world of quantum mechanics. In quantum mechanics we do not talk about a certain fraction of particle pairs, say $N_3/\Sigma_i^8 N_i$, belonging to type 3. Instead, we characterize all spin-singlet systems by the same ket (3.9.1); in the language of Section 3.4 we are concerned here with a pure ensemble. Using this ket and the rules of quantum mechanics we have developed, we can unambiguously calculate each of the three terms in inequality (3.9.9).

We first evaluate $P(\hat{\mathbf{a}}+;\hat{\mathbf{b}}+)$. Suppose observer A finds $\mathbf{S}_1 \cdot \hat{\mathbf{a}}$ to be positive; because of the 100% (opposite sign) correlation we discussed earlier, B's measurement of $\mathbf{S}_2 \cdot \hat{\mathbf{a}}$ will yield a minus sign with certainty. But to calculate $P(\hat{\mathbf{a}}+;\hat{\mathbf{b}}+)$ we must consider a new quantization axis $\hat{\mathbf{b}}$ that makes an angle θ_{ab} with $\hat{\mathbf{a}}$; see Figure 3.9. According to the formalism of Section 3.2, the probability that the $\mathbf{S}_2 \cdot \hat{\mathbf{b}}$ measurement yields + when particle 2 is known to be in an eigenket of $\mathbf{S}_2 \cdot \hat{\mathbf{a}}$ with negative eigenvalue is

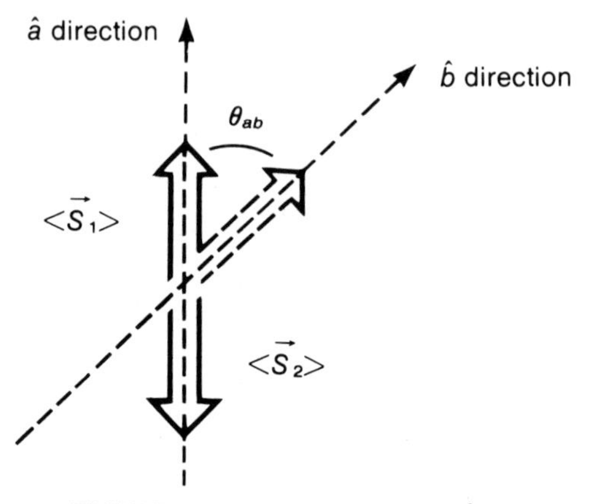

FIGURE 3.9. Evaluation of $P(\hat{\mathbf{a}}+;\hat{\mathbf{b}}+)$.

given by

$$\cos^2\left[\frac{(\pi-\theta_{ab})}{2}\right]=\sin^2\left(\frac{\theta_{ab}}{2}\right). \tag{3.9.10}$$

As a result, we obtain

$$P(\hat{\mathbf{a}}+;\hat{\mathbf{b}}+)=\left(\frac{1}{2}\right)\sin^2\left(\frac{\theta_{ab}}{2}\right), \tag{3.9.11}$$

where the factor $\frac{1}{2}$ arises from the probability of initially obtaining $\mathbf{S}_1\cdot\hat{\mathbf{a}}$ with $+$. Using (3.9.11) and its generalization to the other two terms of (3.9.9), we can write Bell's inequality as

$$\sin^2\left(\frac{\theta_{ab}}{2}\right)\le\sin^2\left(\frac{\theta_{ac}}{2}\right)+\sin^2\left(\frac{\theta_{cb}}{2}\right). \tag{3.9.12}$$

We now show that inequality (3.9.12) is not always possible from a geometric point of view. For simplicity let us choose $\hat{\mathbf{a}}$, $\hat{\mathbf{b}}$, and $\hat{\mathbf{c}}$ to lie in a plane, and let $\hat{\mathbf{c}}$ bisect the two directions defined by $\hat{\mathbf{a}}$ and $\hat{\mathbf{b}}$:

$$\theta_{ab}=2\theta,\qquad \theta_{ac}=\theta_{cb}=\theta. \tag{3.9.13}$$

Inequality (3.9.12) is then violated for

$$0<\theta<\frac{\pi}{2}. \tag{3.9.14}$$

For example, take $\theta=\pi/4$; we then obtain

$$0.500\le 0.292 \quad ?? \tag{3.9.15}$$

So the quantum-mechanical predictions are not compatible with Bell's inequality. There is a real observable—in the sense of being experimentally verifiable—difference between quantum mechanics and the alternative theories satisfying Einstein's locality principle.

Several experiments have been performed to test Bell's inequality. In one of the experiments spin correlations between the final protons in low-energy proton-proton scattering were measured. In all other experiments photon-polarization correlations between a pair of photons in a cascade transition of an excited atom (Ca, Hg, ...),

$$(J=0)\xrightarrow{\gamma}(J=1)\xrightarrow{\gamma}(J=0), \tag{3.9.16}$$

or in the decay of a positronium (an e^+e^- bound state in $^1S_0$) were measured; studying photon-polarization correlations should be just as good in view of the analogy developed in Section 1.1:*

$$\begin{aligned} S_z+ &\rightarrow \hat{\boldsymbol{\varepsilon}} \text{ in } x\text{-direction,}\\ S_z- &\rightarrow \hat{\boldsymbol{\varepsilon}} \text{ in } y\text{-direction,}\\ S_x+ &\rightarrow \hat{\boldsymbol{\varepsilon}} \text{ in } 45^\circ \text{ diagonal direction,}\\ S_x- &\rightarrow \hat{\boldsymbol{\varepsilon}} \text{ in } 135^\circ \text{ diagonal direction.} \end{aligned} \tag{3.9.17}$$

The results of all recent precision experiments have conclusively established that Bell's inequality was violated, in one case by more than nine standard deviations. Furthermore, in all these experiments the inequality relation was violated in such a way that the quantum-mechanical predictions were fulfilled within error limits. In this controversy, quantum mechanics has triumphed with flying colors.

The fact that the quantum-mechanical predictions have been verified does not mean that the whole subject is now a triviality. Despite the experimental verdict we may still feel psychologically uncomfortable about many aspects of measurements of this kind. Consider in particular the following point: Right after observer A performs a measurement on particle 1, how does particle 2—which may, in principle, be many light years away from particle 1—get to "know" how to orient its spin so that the remarkable correlations apparent in Table 3.1 are realized? In one of the experiments to test Bell's inequality (performed by A. Aspect and collaborators) the analyzer settings were changed so rapidly that A's decision as to what to measure could not be made until it was too late for any kind of influence, traveling slower than light, to reach B.

We conclude this section by showing that despite these peculiarities we cannot use spin-correlation measurements to transmit any useful information between two macroscopically separated points. In particular, superluminal (faster than light) communications are impossible.

Suppose A and B both agree in advance to measure S_z; then, without asking A, B knows precisely what A is getting. But this does not mean that

*It should be kept in mind here that by working with photons we are going outside the realm of nonrelativistic quantum mechanics, which is the subject of this book.

A and B are communicating; B just observes a random sequence of positive and negative signs. There is obviously no useful information contained in it. B verifies the remarkable correlations predicted by quantum mechanics only after he or she gets together with A and compares the notes (or computer sheets).

It might be thought that A and B can communicate if one of them suddenly changes the orientation of his or her analyzing apparatus. Let us suppose that A agrees initially to measure S_z, and B, S_x. The results of A's measurements are completely uncorrelated with the results of B's measurements, so there is no information transferred. But then, suppose A suddenly breaks his or her promise and without telling B starts measuring S_x. There are now complete correlations between A's results and B's results. However, B has no way of inferring that A has changed the orientation of his or her analyzer. B continues to see just a random sequence of +'s and −'s by looking at his or her own notebook *only*. So again there is no information transferred.

3.10. TENSOR OPERATORS

Vector Operator

We have been using notations such as **x**, **p**, **S**, and **L**, but as yet we have not systematically discussed their rotational properties. They are vector operators, but what are their properties under rotations? In this section we give a precise quantum-mechanical definition of vector operators based on their commutation relations with the angular-momentum operator. We then generalize to tensor operators with more-complicated transformation properties and derive an important theorem on the matrix elements of vector and tensor operators.

We all know that a **vector** in classical physics is a quantity with three components that transforms by definition like $V_i \to \Sigma_j R_{ij} V_j$ under a rotation. It is reasonable to demand that the expectation value of a vector operator V in quantum mechanics be transformed like a classical vector under rotation. Specifically, as the state ket is changed under rotation according to

$$|\alpha\rangle \to \mathscr{D}(R)|\alpha\rangle, \tag{3.10.1}$$

the expectation value of **V** is assumed to change as follows:

$$\langle\alpha|V_i|\alpha\rangle \to \langle\alpha|\mathscr{D}^\dagger(R)V_i\mathscr{D}(R)|\alpha\rangle = \sum_j R_{ij}\langle\alpha|V_j|\alpha\rangle. \tag{3.10.2}$$

This must be true for an arbitrary ket $|\alpha\rangle$. Therefore,

$$\mathscr{D}^\dagger(R)V_i\mathscr{D}(R) = \sum_j R_{ij}V_j \tag{3.10.3}$$

must hold as an **operator equation**, where R_{ij} is the 3×3 matrix that corresponds to rotation R.

Let us now consider a specific case, an infinitesimal rotation. When the rotation is infinitesimal, we have

$$\mathscr{D}(R) = 1 - \frac{i\varepsilon \mathbf{J} \cdot \hat{\mathbf{n}}}{\hbar}. \tag{3.10.4}$$

We can now write (3.10.3) as

$$V_i + \frac{\varepsilon}{i\hbar}[V_i, \mathbf{J} \cdot \hat{\mathbf{n}}] = \sum_j R_{ij}(\hat{\mathbf{n}}; \varepsilon) V_j. \tag{3.10.5}$$

In particular, for $\hat{\mathbf{n}}$ along the z-axis, we have

$$R(\hat{\mathbf{z}}; \varepsilon) = \begin{pmatrix} 1 & -\varepsilon & 0 \\ \varepsilon & 1 & 0 \\ 0 & 0 & 1 \end{pmatrix}, \tag{3.10.6}$$

so

$$i = 1: \quad V_x + \frac{\varepsilon}{i\hbar}[V_x, J_z] = V_x - \varepsilon V_y \tag{3.10.7a}$$

$$i = 2: \quad V_y + \frac{\varepsilon}{i\hbar}[V_y, J_z] = \varepsilon V_x + V_y \tag{3.10.7b}$$

$$i = 3: \quad V_z + \frac{\varepsilon}{i\hbar}[V_z, J_z] = V_z. \tag{3.10.7c}$$

This means that $\mathbf{V}$ must satisfy the commutation relations

$$[V_i, J_j] = i\varepsilon_{ijk}\hbar V_k. \tag{3.10.8}$$

Clearly, the behavior of $\mathbf{V}$ under a *finite* rotation is completely determined by the preceding commutation relations; we just apply the by-now familiar formula (2.3.47) to

$$\exp\left(\frac{iJ_j\phi}{\hbar}\right) V_i \exp\left(\frac{-iJ_j\phi}{\hbar}\right). \tag{3.10.9}$$

We simply need to calculate

$$[J_j, [J_j, [\cdots [J_j, V_i] \cdots]]]. \tag{3.10.10}$$

Multiple commutators keep on giving back to us V_i or V_k $(k \neq i, j)$ as in spin case (3.2.7).

We can use (3.10.8) as the *defining* property of a vector operator. Notice that the angular-momentum commutation relations are a special case of (3.10.8) in which we let $V_i \to J_i$, $V_k \to J_k$. Other special cases are $[y, L_z] = i\hbar x$, $[x, L_z] = -i\hbar y$, $[p_x, L_z] = -i\hbar p_y$, $[p_y, L_z] = i\hbar p_x$; these can be proved explicitly.

Cartesian Tensors Versus Irreducible Tensors

In classical physics it is customary to define a tensor $T_{ijk\cdots}$ by generalizing $V_i \to \Sigma_j R_{ij} V_j$ as follows:

$$T_{ijk\cdots} \to \sum_{i'} \sum_{j'} \sum_{k'} \cdots R_{ii'} R_{jj'} \ldots T_{i'j'k'\cdots} \tag{3.10.11}$$

under a rotation specified by the 3×3 orthogonal matrix R. The number of indices is called the **rank** of a tensor. Such a tensor is known as a **Cartesian tensor**.

The simplest example of a Cartesian tensor of rank 2 is a **dyadic** formed out of two vectors **U** and **V**. One simply takes a Cartesian component of **U** and a Cartesian component of **V** and puts them together:

$$T_{ij} \equiv U_i V_j. \tag{3.10.12}$$

Notice that we have nine components altogether. They obviously transform like (3.10.11) under rotation.

The trouble with a Cartesian tensor like (3.10.12) is that it is reducible—that is, it can be decomposed into objects that transform differently under rotations. Specifically, for the dyadic in (3.10.12) we have

$$U_i V_j = \frac{\mathbf{U}\cdot\mathbf{V}}{3}\delta_{ij} + \frac{(U_i V_j - U_j V_i)}{2} + \left(\frac{U_i V_j + U_j V_i}{2} - \frac{\mathbf{U}\cdot\mathbf{V}}{3}\delta_{ij}\right). \tag{3.10.13}$$

The first term on the right-hand side, $\mathbf{U}\cdot\mathbf{V}$, is a scalar product invariant under rotation. The second is an antisymmetric tensor which can be written as vector product $\varepsilon_{ijk}(\mathbf{U}\times\mathbf{V})_k$. There are altogether 3 independent components. The last is a 3×3 symmetric traceless tensor with 5 ($=6-1$, where 1 comes from the traceless condition) independent components. The number of independent components checks:

$$3\times 3 = 1+3+5. \tag{3.10.14}$$

We note that the numbers appearing on the right-hand side of (3.10.14) are precisely the multiplicities of objects with angular momentum $l=0$, $l=1$, and $l=2$, respectively. This suggests that the dyadic has been decomposed into tensors that can transform like spherical harmonics with $l=0$, 1, and 2. In fact, (3.10.13) is the simplest nontrivial example to illustrate the reduction of a Cartesian tensor into irreducible **spherical tensors**.

Before presenting the precise definition of a spherical tensor, we first give an example of a spherical tensor of rank k. Suppose we take a spherical harmonic $Y_l^m(\theta,\phi)$. We have already seen that it can be written as $Y_l^m(\hat{\mathbf{n}})$, where the orientation of $\hat{\mathbf{n}}$ is characterized by θ and ϕ. We now replace $\hat{\mathbf{n}}$ by some vector **V**. The result is that we have a spherical tensor of rank k (in place of l) with magnetic quantum number q (in place of m), namely,

$$T_q^{(k)} = Y_{l=k}^{m=q}(\mathbf{V}). \tag{3.10.15}$$

Specifically, in the case $k=1$ we take spherical harmonics with $l=1$ and replace $(z/r) = (\hat{\mathbf{n}})_z$ by V_z, and so on.

$$\begin{aligned} Y_1^0 &= \sqrt{\frac{3}{4\pi}}\cos\theta = \sqrt{\frac{3}{4\pi}}\frac{z}{r} \to T_0^{(1)} = \sqrt{\frac{3}{4\pi}}V_z, \\ Y_1^{\pm 1} &= \mp\sqrt{\frac{3}{4\pi}}\frac{x \pm iy}{\sqrt{2}\,r} \to T_{\pm 1}^{(1)} = \sqrt{\frac{3}{4\pi}}\left(\mp\frac{V_x \pm iV_y}{\sqrt{2}}\right). \end{aligned} \tag{3.10.16}$$

Obviously this can be generalized for higher k, for example,

$$Y_2^{\pm 2} = \sqrt{\frac{15}{32\pi}}\frac{(x \pm iy)^2}{r^2} \to T_{\pm 2}^{(2)} = \sqrt{\frac{15}{32\pi}}\left(V_x \pm iV_y\right)^2. \tag{3.10.17}$$

$T_q^{(k)}$ are irreducible, just as Y_l^m are. For this reason, working with spherical tensors is more satisfactory than working with Cartesian tensors.

To see the transformation of spherical tensors constructed in this manner, let us first review how Y_l^m transform under rotations. First, we have for the direction eigenket;

$$|\hat{\mathbf{n}}\rangle \to \mathscr{D}(R)|\hat{\mathbf{n}}\rangle \equiv |\hat{\mathbf{n}}'\rangle, \tag{3.10.18}$$

which defines the rotated eigenket $|\hat{\mathbf{n}}'\rangle$. We wish to examine how $Y_l^m(\hat{\mathbf{n}}') = \langle\hat{\mathbf{n}}'|l,m\rangle$ would look in terms of $Y_l^m(\hat{\mathbf{n}})$. We can easily see this by starting with

$$\mathscr{D}(R^{-1})|l,m\rangle = \sum_{m'}|l,m'\rangle\mathscr{D}_{m'm}^{(l)}(R^{-1}) \tag{3.10.19}$$

and contracting with $\langle\hat{\mathbf{n}}|$ on the left, using (3.10.18):

$$Y_l^m(\hat{\mathbf{n}}') = \sum_{m'} Y_l^{m'}(\hat{\mathbf{n}})\mathscr{D}_{m'm}^{(l)}(R^{-1}). \tag{3.10.20}$$

If there is an operator that acts like $Y_l^m(\mathbf{V})$, it is then reasonable to expect

$$\mathscr{D}^\dagger(R)Y_l^m(\mathbf{V})\mathscr{D}(R) = \sum_{m'} Y_l^{m'}(\mathbf{V})\mathscr{D}_{mm'}^{(l)*}(R), \tag{3.10.21}$$

where we have used the unitarity of the rotation operator to rewrite $\mathscr{D}_{m'm}^{(l)}(R^{-1})$.

All this work is just to motivate the definition of a spherical tensor. We now consider spherical tensors in quantum mechanics. Motivated by (3.10.21) we define a spherical tensor operator of rank k with $(2k+1)$ components as

$$\mathscr{D}^\dagger(R)T_q^{(k)}\mathscr{D}(R) = \sum_{q'=-k}^{k} \mathscr{D}_{qq'}^{(k)*}(R)T_{q'}^{(k)} \tag{3.10.22a}$$

or, equivalently,

$$\mathscr{D}(R)T_q^{(k)}\mathscr{D}^\dagger(R) = \sum_{q'=-k}^{k} \mathscr{D}_{q'q}^{(k)}(R)T_{q'}^{(k)}. \tag{3.10.22b}$$

This definition holds regardless of whether $T_q^{(k)}$ can be written as $Y_{l=k}^{m=q}(\mathbf{V})$; for example, $(U_x + iU_y)(V_x + iV_y)$ is the $q = +2$ component of a spherical tensor of rank 2 even though, unlike $(V_x + iV_y)^2$, it cannot be written as $Y_k^q(\mathbf{V})$.

A more convenient definition of a spherical tensor is obtained by considering the infinitesimal form of (3.10.22b), namely,

$$\left(1 + \frac{i\mathbf{J}\cdot\hat{\mathbf{n}}\varepsilon}{\hbar}\right)T_q^{(k)}\left(1 - \frac{i\mathbf{J}\cdot\hat{\mathbf{n}}\varepsilon}{\hbar}\right) = \sum_{q'=-k}^{k} T_{q'}^{(k)}\langle kq'|\left(1 + \frac{i\mathbf{J}\cdot\hat{\mathbf{n}}\varepsilon}{\hbar}\right)|kq\rangle \tag{3.10.23}$$

or

$$[\mathbf{J}\cdot\hat{\mathbf{n}}, T_q^{(k)}] = \sum_{q'} T_{q'}^{(k)}\langle kq'|\mathbf{J}\cdot\hat{\mathbf{n}}|kq\rangle. \tag{3.10.24}$$

By taking $\hat{\mathbf{n}}$ in the $\hat{\mathbf{z}}$- and in the $(\hat{\mathbf{x}} \pm i\hat{\mathbf{y}})$ directions and using the nonvanishing matrix elements of J_z and $J_\pm$ [see (3.5.35b) and (3.5.41)], we obtain

$$\left[J_z, T_q^{(k)}\right] = \hbar q T_q^{(k)} \tag{3.10.25a}$$

and

$$\left[J_\pm, T_q^{(k)}\right] = \hbar\sqrt{(k \mp q)(k \pm q + 1)}\, T_{q\pm 1}^{(k)}. \tag{3.10.25b}$$

These commutation relations can be considered as a definition of spherical tensors in place of (3.10.22).

Product of Tensors

We have seen how to form a scalar, vector (or antisymmetric tensor), and a traceless symmetric tensor out of two vectors using the Cartesian tensor language. Of course, spherical tensor language can also be used (Baym 1969, Chapter 17), for example,

$$\begin{aligned}
T_0^{(0)} &= \frac{-\mathbf{U}\cdot\mathbf{V}}{3} = \frac{(U_{+1}V_{-1} + U_{-1}V_{+1} - U_0V_0)}{3},\\
T_q^{(1)} &= \frac{(\mathbf{U}\times\mathbf{V})_q}{i\sqrt{2}},\\
T_{\pm 2}^{(2)} &= U_{\pm 1}V_{\pm 1},\\
T_{\pm 1}^{(2)} &= \frac{U_{\pm 1}V_0 + U_0V_{\pm 1}}{\sqrt{2}},\\
T_0^{(2)} &= \frac{U_{+1}V_{-1} + 2U_0V_0 + U_{-1}V_{+1}}{\sqrt{6}},
\end{aligned} \tag{3.10.26}$$

where $U_q(V_q)$ is the qth component of a spherical tensor of rank 1, corresponding to vector $\mathbf{U}(\mathbf{V})$. The preceding transformation properties can be checked by comparing with Y_l^m and remembering that $U_{+1} = -(U_x + iU_y)/\sqrt{2}$, $U_{-1} = (U_x - iU_y)/\sqrt{2}$, $U_0 = U_z$. A similar check can be made for $V_{\pm 1,0}$. For instance,

$$Y_2^0 = \sqrt{\frac{5}{16\pi}}\,\frac{3z^2 - r^2}{r^2},$$

where $3z^2 - r^2$ can be written as

$$2z^2 + 2\left[-\frac{(x+iy)}{\sqrt{2}}\frac{(x-iy)}{\sqrt{2}}\right];$$

hence, Y_2^0 is just a special case of $T_0^{(2)}$ for $\mathbf{U} = \mathbf{V} = \mathbf{r}$.

A more systematic way of forming tensor products goes as follows. We start by stating a theorem:

Theorem. *Let $X_{q_1}^{(k_1)}$ and $Z_{q_2}^{(k_2)}$ be irreducible spherical tensors of rank k_1 and k_2, respectively. Then*

$$T_q^{(k)} = \sum_{q_1}\sum_{q_2}\langle k_1 k_2; q_1 q_2 | k_1 k_2; kq\rangle X_{q_1}^{(k_1)} Z_{q_2}^{(k_2)} \tag{3.10.27}$$

is a spherical (irreducible) tensor of rank k.

Proof. We must show that under rotation $T_q^{(k)}$ must transform according to (3.10.22)

$$\begin{aligned}
\mathscr{D}^\dagger(R) T_q^{(k)} \mathscr{D}(R) &= \sum_{q_1}\sum_{q_2}\langle k_1 k_2; q_1 q_2 | k_1 k_2; kq\rangle \\
&\quad \times \mathscr{D}^\dagger(R) X_{q_1}^{(k_1)} \mathscr{D}(R) \mathscr{D}^\dagger(R) Z_{q_2}^{(k_2)} \mathscr{D}(R) \\
&= \sum_{q_1}\sum_{q_2}\sum_{q_1'}\sum_{q_2'}\langle k_1 k_2; q_1 q_2 | k_1 k_2; kq\rangle \\
&\quad \times X_{q_1'}^{(k_1)} \mathscr{D}_{q_1' q_1}^{(k_1)}(R^{-1}) Z_{q_2'}^{(k_2)} \mathscr{D}_{q_2' q_2}^{(k_2)}(R^{-1}) \\
&= \sum_{k''}\sum_{q_1}\sum_{q_2}\sum_{q_1'}\sum_{q_2'}\sum_{q''}\sum_{q'}\langle k_1 k_2; q_1 q_2 | k_1 k_2; kq\rangle \\
&\quad \times \langle k_1 k_2; q_1' q_2' | k_1, k_2; k'' q'\rangle \\
&\quad \times \langle k_1 k_2; q_1 q_2 | k_1 k_2; k'' q''\rangle \mathscr{D}_{q' q''}^{(k'')}(R^{-1}) X_{q_1'}^{(k_1)} Z_{q_2'}^{(k_2)},
\end{aligned}$$

where we have used the Clebsch-Gordan series formula (3.7.69). The preceding expression becomes

$$= \sum_{k''}\sum_{q_1'}\sum_{q_2'}\sum_{q''}\sum_{q'}\delta_{kk''}\delta_{qq''}\langle k_1 k_2; q_1' q_2' | k_1 k_2; k'' q'\rangle \mathscr{D}_{q' q''}^{(k'')}(R^{-1}) X_{q_1'}^{(k_1)} Z_{q_2'}^{(k_2)},$$

where we have used the orthogonality of Clebsch-Gordan coefficients (3.7.42). Finally, this expression reduces to

$$= \sum_{q'} \left(\sum_{q_1'} \sum_{q_2'} \langle k_1 k_2; q_1' q_2' | k_1 k_2; k q' \rangle X_{q_1'}^{(k_1)} Z_{q_2'}^{(k_2)} \right) \mathscr{D}_{q'q}^{(k)}(R^{-1})$$

$$= \sum_{q'} T_{q'}^{(k)} \mathscr{D}_{q'q}^{(k)}(R^{-1}) = \sum_{q'} \mathscr{D}_{qq'}^{(k)*}(R) T_{q'}^{(k)} \qquad \square$$

The foregoing shows how we can construct tensor operators of higher or lower ranks by multiplying two tensor operators. Furthermore, the manner in which we construct tensor products out of two tensors is completely analogous to the manner in which we construct an angular-momentum eigenstate by adding two angular momentums; exactly the same Clebsch-Gordan coefficients appear if we let $k_{1,2} \to j_{1,2}, q_{1,2} \to m_{1,2}$.

Matrix Elements of Tensor Operators; the Wigner-Eckart Theorem

In considering the interactions of an electromagnetic field with atoms and nuclei, it is often necessary to evaluate matrix elements of tensor operators with respect to angular-momentum eigenstates. Examples of this will be given in Chapter 5. In general, it is a formidable dynamic task to calculate such matrix elements. However, there are certain properties of these matrix elements that follow purely from kinematic or geometric considerations, which we now discuss.

First, there is a very simple m-selection rule:

m-selection Rule

$$\langle \alpha', j'm' | T_q^{(k)} | \alpha, jm \rangle = 0, \quad \text{unless } m' = q + m. \tag{3.10.28}$$

Proof. Using (3.10.25a), we have

$$\langle \alpha', j'm' | \left(\left[J_z, T_q^{(k)} \right] - \hbar q T_q^{(k)} \right) | \alpha, jm \rangle = \left[(m' - m)\hbar - \hbar q \right] \times \langle \alpha', j'm' | T_q^{(k)} | \alpha, jm \rangle = 0;$$

hence,

$$\langle \alpha', j'm' | T_q^{(k)} | \alpha, jm \rangle = 0 \quad \text{unless } m' = q + m. \qquad \square$$

Another way to see this is to note that transformation property of $T_q^{(k)} | \alpha, jm \rangle$ under rotation, namely,

$$\mathscr{D} T_q^{(k)} | \alpha, jm \rangle = \mathscr{D} T_q^{(k)} \mathscr{D}^\dagger \mathscr{D} | \alpha, jm \rangle. \tag{3.10.29}$$

If we now let $\mathscr{D}$ stand for a rotation operator around the z-axis, we get [see

(3.10.22b) and (3.1.16)]

$$\mathscr{D}(\hat{\mathbf{z}}, \phi) T_q^{(k)} |\alpha, jm\rangle = e^{-iq\phi} e^{-im\phi} T_q^{(k)} |\alpha, jm\rangle, \qquad (3.10.30)$$

which is orthogonal to $|\alpha', j'm'\rangle$ unless $q + m = m'$.

We are going to prove one of the most important theorems in quantum mechanics, the **Wigner-Eckart theorem.**

The Wigner-Eckart Theorem. The matrix elements of tensor operators with respect to angular-momentum eigenstates satisfy

$$\langle \alpha', j'm'|T_q^{(k)}|\alpha, jm\rangle = \langle jk; mq|jk; j'm'\rangle \frac{\langle \alpha' j' \| T^{(k)} \| \alpha j\rangle}{\sqrt{2j+1}}, \qquad (3.10.31)$$

where the **double-bar matrix element** *is independent of m and m', and q.*

Before we present a proof of this theorem, let us look at its significance. First, we see that the matrix element is written as the product of two factors. The first factor is a Clebsch-Gordan coefficient for adding j and k to get j'. It depends only on the geometry, that is, the way the system is oriented with respect to the z-axis. There is no reference whatsoever to the particular nature of the tensor operator. The second factor does depend on the dynamics, for instance, α may stand for the radial quantum number and its evaluation may involve, for example, evaluation of radial integrals. On the other hand, it is completely independent of the magnetic quantum numbers m, m', and q, which specify the orientation of the physical system. To evaluate $\langle \alpha', j'm'|T_q^{(k)}|\alpha, jm\rangle$ with various combinations of m, m', and q' it is sufficient to know just one of them; all others can be related geometrically because they are proportional to Clebsch-Gordan coefficients, which are known. The common proportionality factor is $\langle \alpha' j' \| T^{(k)} \| \alpha j\rangle$, which makes no reference whatsoever to the geometric features.

The selection rules for the tensor operator matrix element can be immediately read off from the selection rules for adding angular momentum. Indeed, from the requirement that the Clebsch-Gordan coefficient be nonvanishing, we immediately obtain the m-selection rule (3.10.28) derived before and also the triangular relation

$$|j-k| \le j' \le j + k. \qquad (3.10.32)$$

Now we prove the theorem.

Proof. Using (3.10.25b) we have

$$\langle \alpha', j'm'|\left[J_\pm, T_q^{(k)}\right]|\alpha, jm\rangle = \hbar\sqrt{(k \mp q)(k \pm q+1)} < \alpha', j'm'|T_{q\pm1}^{(k)}|\alpha, jm\rangle, \qquad (3.10.33)$$

or using (3.5.39) and (3.5.40) we have

$$\begin{aligned}\sqrt{(j'\pm m')(j'\mp m'+1)}\,\langle\alpha', j', m'\mp 1|T_q^{(k)}|\alpha, jm\rangle \\ =\sqrt{(j\mp m)(j\pm m+1)}\,\langle\alpha', j'm'|T_q^{(k)}|\alpha, j, m\pm 1\rangle \\ +\sqrt{(k\mp q)(k\pm q+1)}\,\langle\alpha', j'm'|T_{q\pm 1}^{(k)}|\alpha, jm\rangle. \end{aligned} \tag{3.10.34}$$

Compare this with the recursion relation for the Clebsch-Gordan coefficient (3.7.49). Note the striking similarity if we substitute $j'\to j$, $m'\to m$, $j\to j_1$, $m\to m_1$, $k\to j_2$, and $q\to m_2$. Both recursion relations are of the form $\Sigma_j a_{ij}x_j=0$, that is, first-order linear homogeneous equations with the same coefficients a_{ij}. Whenever we have

$$\sum_j a_{ij}x_j=0, \qquad \sum_j a_{ij}y_j=0, \tag{3.10.35}$$

we cannot solve for the x_j (or y_j) individually but we can solve for the ratios; so

$$\frac{x_j}{x_k}=\frac{y_j}{y_k} \quad \text{or} \quad x_j=cy_j, \tag{3.10.36}$$

where c is a universal proportionality factor. Noting that $\langle j_1j_2;m_1,m_2\pm 1|j_1j_2;jm\rangle$ in the Clebsch-Gordan recursion relation (3.7.49) corresponds to $\langle\alpha', j'm'|T_{q\pm 1}^{(k)}|\alpha, jm\rangle$, we see that

$$\begin{aligned}\langle\alpha', j'm'|T_{q\pm 1}^{(k)}|\alpha, jm\rangle &= (\text{universal proportionality constant independent of} \\ & m,\ q,\ \text{and}\ m')\langle jk; m\,q\pm 1|jk; j'm'\rangle,\end{aligned} \tag{3.10.37}$$

which proves the theorem. □

Let us now look at two simple examples of the Wigner-Eckart theorem.

Example 1. Tensor of rank 0, that is, scalar $T_0^{(0)}=S$. The matrix element of a scalar operator satisfies

$$\langle\alpha', j'm'|S|\alpha, jm\rangle=\delta_{jj'}\delta_{mm'}\frac{\langle\alpha' j'\|S\|\alpha j\rangle}{\sqrt{2j+1}} \tag{3.10.38}$$

because S acting on $|\alpha, jm\rangle$ is like adding an angular momentum of zero. Thus the scalar operator cannot change j, m values.

Example 2. Vector operator which in the spherical tensor language is a rank 1 tensor. The spherical component of $\mathbf{V}$ can be written as $V_{q=\pm 1,0}$, so we have the selection rule

$$\Delta m\equiv m'-m=\pm 1,0 \qquad \Delta j\equiv j'-j=\begin{cases}\pm 1\\ 0.\end{cases} \tag{3.10.39}$$

In addition, the $0 \to 0$ transition is forbidden. This selection rule is of fundamental importance in the theory of radiation; it is the dipole selection rule obtained in the long-wavelength limit of emitted photons.*

For $j = j'$ the Wigner-Eckart theorem—when applied to the vector operator—takes a particularly simple form, often known as the **projection theorem** for obvious reasons.

The Projection Theorem

$$\langle \alpha', jm'|V_q|\alpha, jm\rangle = \frac{\langle \alpha', jm|\mathbf{J}\cdot\mathbf{V}|\alpha, jm\rangle}{\hbar^2 j(j+1)}\langle jm'|J_q|jm\rangle, \quad (3.10.40)$$

where analogous to our discussion after (3.10.26) we choose

$$J_{\pm 1} = \mp\frac{1}{\sqrt{2}}(J_x \pm iJ_y) = \mp\frac{1}{\sqrt{2}}J_{\pm}, \qquad J_0 = J_z. \quad (3.10.41)$$

Proof. Noting (3.10.26) we have

$$\begin{aligned}\langle \alpha', jm|\mathbf{J}\cdot\mathbf{V}|\alpha, jm\rangle &= \langle \alpha', jm|(J_0V_0 - J_{+1}V_{-1} - J_{-1}V_{+1})|\alpha, jm\rangle \\ &= m\hbar\langle \alpha', jm|V_0|\alpha, jm\rangle + \frac{\hbar}{\sqrt{2}}\sqrt{(j+m)(j-m+1)} \\ &\quad \times\langle \alpha', j\,m-1|V_{-1}|\alpha, jm\rangle \\ &\quad - \frac{\hbar}{\sqrt{2}}\sqrt{(j-m)(j+m+1)}\,\langle \alpha', j\,m+1|V_{+1}|\alpha, jm\rangle \\ &= c_{jm}\langle \alpha' j\|\mathbf{V}\|\alpha j\rangle \qquad (3.10.42)\end{aligned}$$

by the Wigner-Eckart theorem (3.10.31), where c_{jm} is independent of α, α', and $\mathbf{V}$, and the matrix elements of $V_{0,\pm 1}$ are all proportional to the double-bar matrix element (sometimes also called the **reduced matrix element**). Furthermore, c_{jm} is independent of m because $\mathbf{J}\cdot\mathbf{V}$ is a scalar operator, so we may as well write it as c_j. Because c_j does not depend on $\mathbf{V}$, (3.10.42) holds even if we let $\mathbf{V} \to \mathbf{J}$ and $\alpha' \to \alpha$, that is,

$$\langle \alpha, jm|\mathbf{J}^2|\alpha, jm\rangle = c_j\langle \alpha j\|\mathbf{J}\|\alpha j\rangle. \quad (3.10.43)$$

Returning to the Wigner-Eckart theorem applied to V_q and J_q, we have

$$\frac{\langle \alpha', jm'|V_q|\alpha, jm\rangle}{\langle \alpha, jm'|J_q|\alpha, jm\rangle} = \frac{\langle \alpha' j\|\mathbf{V}\|\alpha j\rangle}{\langle \alpha j\|\mathbf{J}\|\alpha j\rangle}. \quad (3.10.44)$$

*Additional parity selection rules are discussed in Chapter 4, Section 2. They lead to these $E1$ dipole selection rules.

But the right-hand side of (3.10.44) is the same as $\langle\alpha', jm|\mathbf{J}\cdot\mathbf{V}|\alpha, jm\rangle / \langle\alpha, jm|\mathbf{J}^2|\alpha, jm\rangle$ by (3.10.42) and (3.10.43). Moreover, the left-hand side of (3.10.43) is just $j(j+1)\hbar^2$. So

$$\langle\alpha', jm'|V_q|\alpha, jm\rangle = \frac{\langle\alpha', jm|\mathbf{J}\cdot\mathbf{V}|\alpha, jm\rangle}{\hbar^2 j(j+1)}\langle jm'|J_q|jm\rangle, \quad (3.10.45)$$

which proves the projection theorem. □

We will give applications of the theorem in subsequent sections.

PROBLEMS

1. Find the eigenvalues and eigenvectors of $\sigma_y = \begin{pmatrix} 0 & -i \\ i & 0 \end{pmatrix}$. Suppose an electron is in the spin state $\begin{pmatrix} \alpha \\ \beta \end{pmatrix}$. If s_y is measured, what is the probability of the result $\hbar/2$?
2. Consider the 2×2 matrix defined by

$$U = \frac{a_0 + i\boldsymbol{\sigma}\cdot\mathbf{a}}{a_0 - i\boldsymbol{\sigma}\cdot\mathbf{a}},$$

where a_0 is a real number and $\mathbf{a}$ is a three-dimensional vector with real components.

   a. Prove that U is unitary and unimodular.

   b. In general, a 2×2 unitary unimodular matrix represents a rotation in three dimensions. Find the axis and angle of rotation appropriate for U in terms of a_0, a_1, a_2, and a_3.

3. The spin-dependent Hamiltonian of an electron-positron system in the presence of a uniform magnetic field in the z-direction can be written as

$$H = A\mathbf{S}^{(e^-)}\cdot\mathbf{S}^{(e^+)} + \left(\frac{eB}{mc}\right)\left(S_z^{(e^-)} - S_z^{(e^+)}\right).$$

Suppose the spin function of the system is given by $\chi_+^{(e^-)}\chi_-^{(e^+)}$.

   a. Is this an eigenfunction of H in the limit $A \to 0$, $eB/mc \neq 0$? If it is, what is the energy eigenvalue? If it is not, what is the expectation value of H?

   b. Same problem when $eB/mc \to 0$, $A \neq 0$.

4. Consider a *spin* 1 particle. Evaluate the matrix elements of

$$S_z(S_z + \hbar)(S_z - \hbar) \quad \text{and} \quad S_x(S_x + \hbar)(S_x - \hbar).$$

5. Let the Hamiltonian of a rigid body be

$$H = \frac{1}{2}\left(\frac{K_1^2}{I_1} + \frac{K_2^2}{I_2} + \frac{K_3^2}{I_3}\right),$$

where $\mathbf{K}$ is the angular momentum in the body frame. From this expression obtain the Heisenberg equation of motion for **K** and then find Euler's equation of motion in the correspondence limit.

6. Let $U = e^{iG_3\alpha}e^{iG_2\beta}e^{iG_3\gamma}$, where (α, β, γ) are the Eulerian angles. In order that U represent a rotation (α, β, γ), what are the commutation rules satisfied by the G_k? Relate **G** to the angular momentum operators.
7. What is the meaning of the following equation:

$$U^{-1}A_k U = \sum R_{kl} A_l,$$

where the three components of **A** are matrices? From this equation show that matrix elements $\langle m|A_k|n\rangle$ transform like vectors.

8. Consider a sequence of Euler rotations represented by

$$\mathscr{D}^{(1/2)}(\alpha, \beta, \gamma) = \exp\left(\frac{-i\sigma_3\alpha}{2}\right)\exp\left(\frac{-i\sigma_2\beta}{2}\right)\exp\left(\frac{-i\sigma_3\gamma}{2}\right)$$

$$= \begin{pmatrix} e^{-i(\alpha+\gamma)/2}\cos\dfrac{\beta}{2} & -e^{-i(\alpha-\gamma)/2}\sin\dfrac{\beta}{2} \\ e^{i(\alpha-\gamma)/2}\sin\dfrac{\beta}{2} & e^{i(\alpha+\gamma)/2}\cos\dfrac{\beta}{2} \end{pmatrix}.$$

Because of the group properties of rotations, we expect that this sequence of operations is equivalent to a *single* rotation about some axis by an angle θ. Find θ.

9. a. Consider a pure ensemble of identically prepared spin $\frac{1}{2}$ systems. Suppose the expectation values $\langle S_x\rangle$ and $\langle S_z\rangle$ and the sign of $\langle S_y\rangle$ are known. Show how we may determine the state vector. Why is it unnecessary to know the magnitude of $\langle S_y\rangle$?

   b. Consider a mixed ensemble of spin $\frac{1}{2}$ systems. Suppose the ensemble averages $[S_x]$, $[S_y]$, and $[S_z]$ are all known. Show how we may construct the 2×2 density matrix that characterizes the ensemble.

10. a. Prove that the time evolution of the density operator ρ (in the Schrödinger picture) is given by

$$\rho(t) = \mathscr{U}(t, t_0)\rho(t_0)\mathscr{U}^\dagger(t, t_0).$$

   b. Suppose we have a pure ensemble at $t = 0$. Prove that it cannot evolve into a mixed ensemble as long as the time evolution is governed by the Schrödinger equation.

11. Consider an ensemble of spin 1 systems. The density matrix is now a 3×3 matrix. How many independent (real) parameters are needed to characterize the density matrix? What must we know in addition to $[S_x]$, $[S_y]$, and $[S_z]$ to characterize the ensemble completely?
12. An angular-momentum eigenstate $|j, m = m_{\max} = j\rangle$ is rotated by an infinitesimal angle ε about the y-axis. Without using the explicit form of the $d^{(j)}_{m'm}$ function, obtain an expression for the probability for the new rotated state to be found in the original state up to terms of order ε^2.

13. Show that the 3×3 matrices G_i $(i=1,2,3)$ whose elements are given by

$$(G_i)_{jk} = -i\hbar\varepsilon_{ijk},$$

where j and k are the row and column indices, satisfy the angular momentum commutation relations. What is the physical (or geometric) significance of the transformation matrix that connects G_i to the more usual 3×3 representations of the angular-momentum operator J_i with J_3 taken to be diagonal? Relate your result to

$$\mathbf{V} \to \mathbf{V} + \hat{\mathbf{n}}\delta\phi \times \mathbf{V}$$

under infinitesimal rotations. (*Note*: This problem may be helpful in understanding the photon spin.)

14. a. Let $\mathbf{J}$ be angular momentum. It may stand for orbital $\mathbf{L}$, spin $\mathbf{S}$, or $\mathbf{J}_{\text{total}}$.) Using the fact that $J_x, J_y, J_z (J_\pm \equiv J_x \pm iJ_y)$ satisfy the usual angular-momentum commutation relations, prove

$$\mathbf{J}^2 = J_z^2 + J_+J_- - \hbar J_z.$$

b. Using (a) (or otherwise), derive the "famous" expression for the coefficient c_- that appears in

$$J_-\psi_{jm} = c_-\psi_{j,m-1}.$$

15. The wave function of a particle subjected to a spherically symmetrical potential $V(r)$ is given by

$$\psi(\mathbf{x}) = (x+y+3z)f(r).$$

a. Is ψ an eigenfunction of $\mathbf{L}^2$? If so, what is the l-value? If not, what are the possible values of l we may obtain when $\mathbf{L}^2$ is measured?

b. What are the probabilities for the particle to be found in various m_l states?

c. Suppose it is known somehow that $\psi(\mathbf{x})$ is an energy eigenfunction with eigenvalue E. Indicate how we may find $V(r)$.

16. A particle in a spherically symmetrical potential is known to be in an eigenstate of $\mathbf{L}^2$ and L_z with eigenvalues $\hbar^2 l(l+1)$ and $m\hbar$, respectively. Prove that the expectation values between $|lm\rangle$ states satisfy

$$\langle L_x\rangle = \langle L_y\rangle = 0, \qquad \langle L_x^2\rangle = \langle L_y^2\rangle = \frac{[l(l+1)\hbar^2 - m^2\hbar^2]}{2}.$$

Interpret this result semiclassically.

17. Suppose a half-integer l-value, say $\frac{1}{2}$, were allowed for orbital angular momentum. From

$$L_+Y_{1/2,1/2}(\theta,\phi) = 0,$$

we may deduce, as usual,

$$Y_{1/2,1/2}(\theta,\phi) \propto e^{i\phi/2}\sqrt{\sin\theta}.$$

Now try to construct $Y_{1/2,-1/2}(\theta,\phi)$; by (a) applying L_- to $Y_{1/2,1/2}(\theta,\phi)$; and (b) using $L_- Y_{1/2,-1/2}(\theta,\phi)=0$. Show that the two procedures lead to contradictory results. (This gives an argument against half-integer l-values for orbital angular momentum.)

18. Consider an orbital angular-momentum eigenstate $|l=2, m=0\rangle$. Suppose this state is rotated by an angle β about the y-axis. Find the probability for the new state to be found in $m=0$, ± 1, and ± 2. (The spherical harmonics for $l=0$, 1, and 2 given in Appendix A may be useful.)

19. What is the physical significance of the operators

$$K_+ \equiv a_+^\dagger a_-^\dagger \quad \text{and} \quad K_- \equiv a_+ a_-$$

in Schwinger's scheme for angular momentum? Give the nonvanishing matrix elements of $K_\pm$.

20. We are to add angular momenta $j_1=1$ and $j_2=1$ to form $j=2$, 1, and 0 states. Using either the ladder operator method or the recursion relation, express all (nine) $\{j, m\}$ eigenkets in terms of $|j_1 j_2; m_1 m_2\rangle$. Write your answer as

$$|j=1, m=1\rangle = \frac{1}{\sqrt{2}}|+,0\rangle - \frac{1}{\sqrt{2}}|0,+\rangle, \ldots,$$

where + and 0 stand for $m_{1,2}=1,0$, respectively.

21. a. Evaluate

$$\sum_{m=-j}^{j} |d_{mm'}^{(j)}(\beta)|^2 m$$

for *any* j (integer or half-integer); then check your answer for $j=\frac{1}{2}$.

b. Prove, for any j,

$$\sum_{m=-j}^{j} m^2 |d_{m'm}^{(j)}(\beta)|^2 = \frac{1}{2}j(j+1)\sin^2\beta + m'^2\frac{1}{2}(3\cos^2\beta - 1).$$

[*Hint*: This can be proved in many ways. You may, for instance, examine the rotational properties of J_z^2 using the spherical (irreducible) tensor language.]

22. a. Consider a system with $j=1$. Explicitly write

$$\langle j=1, m'|J_y|j=1, m\rangle$$

in 3×3 matrix form.

b. Show that for $j=1$ only, it is legitimate to replace $e^{-iJ_y\beta/\hbar}$ by

$$1 - i\left(\frac{J_y}{\hbar}\right)\sin\beta - \left(\frac{J_y}{\hbar}\right)^2(1-\cos\beta).$$

c. Using (b), prove

$$d^{(j=1)}(\beta) = \begin{pmatrix} \left(\frac{1}{2}\right)(1+\cos\beta) & -\left(\frac{1}{\sqrt{2}}\right)\sin\beta & \left(\frac{1}{2}\right)(1-\cos\beta) \\ \left(\frac{1}{\sqrt{2}}\right)\sin\beta & \cos\beta & -\left(\frac{1}{\sqrt{2}}\right)\sin\beta \\ \left(\frac{1}{2}\right)(1-\cos\beta) & \left(\frac{1}{\sqrt{2}}\right)\sin\beta & \left(\frac{1}{2}\right)(1+\cos\beta) \end{pmatrix}.$$

23. Express the matrix element $\langle \alpha_2\beta_2\gamma_2 | J_3^2 | \alpha_1\beta_1\gamma_1 \rangle$ in terms of a series in

$$\mathscr{D}^j_{mn}(\alpha\beta\gamma) = \langle \alpha\beta\gamma | jmn \rangle.$$

24. Consider a system made up of two spin $\frac{1}{2}$ particles. Observer A specializes in measuring the spin components of one of the particles (s_{1z}, s_{1x} and so on), while observer B measures the spin components of the other particle. Suppose the system is known to be in a spin-singlet state, that is, $S_{\text{total}} = 0$.
    a. What is the probability for observer A to obtain $s_{1z} = \hbar/2$ when observer B makes no measurement? Same problem for $s_{1x} = \hbar/2$.
    b. Observer B determines the spin of particle 2 to be in the $s_{2z} = \hbar/2$ state with certainty. What can we then conclude about the outcome of observer A's measurement if (i) A measures s_{1z} and (ii) A measures s_{1x}? Justify your answer.
25. Consider a spherical tensor of rank 1 (that is, a vector)

$$V^{(1)}_{\pm 1} = \mp \frac{V_x \pm iV_y}{\sqrt{2}}, \qquad V^{(1)}_0 = V_z.$$

Using the expression for $d^{(j=1)}$ given in Problem 22, evaluate

$$\sum_{q'} d^{(1)}_{qq'}(\beta) V^{(1)}_{q'}$$

and show that your results are just what you expect from the transformation properties of $V_{x,y,z}$ under rotations about the y-axis.
26. a. Construct a spherical tensor of rank 1 out of two different vectors $\mathbf{U} = (U_x, U_y, U_z)$ and $\mathbf{V} = (V_x, V_y, V_z)$. Explicitly write $T^{(1)}_{\pm 1, 0}$ in terms of $U_{x,y,z}$ and $V_{x,y,z}$.
    b. Construct a spherical tensor of rank 2 out of two different vectors $\mathbf{U}$ and $\mathbf{V}$. Write down explicitly $T^{(2)}_{\pm 2, \pm 1, 0}$ in terms of $U_{x,y,z}$ and $V_{x,y,z}$.
27. Consider a spinless particle bound to a fixed center by a central force potential.

a. Relate, as much as possible, the matrix elements

$$\langle n', l', m'|\mp \frac{1}{\sqrt{2}}(x \pm iy)|n, l, m\rangle \quad \text{and} \quad \langle n', l', m'|z|n, l, m\rangle$$

using *only* the Wigner-Eckart theorem. Make sure to state under what conditions the matrix elements are nonvanishing.

b. Do the same problem using wave functions $\psi(\mathbf{x}) = R_{nl}(r)Y_l^m(\theta, \phi)$.

28. a. Write xy, xz, and $(x^2 - y^2)$ as components of a spherical (irreducible) tensor of rank 2.

b. The expectation value

$$Q \equiv e\langle \alpha, j, m = j|(3z^2 - r^2)|\alpha, j, m = j\rangle$$

is known as the *quadrupole moment*. Evaluate

$$e\langle \alpha, j, m'|(x^2 - y^2)|\alpha, j, m = j\rangle,$$

(where $m' = j, j-1, j-2, \ldots$) in terms of Q and appropriate Clebsch-Gordan coefficients.

29. A spin $\frac{3}{2}$ nucleus situated at the origin is subjected to an external inhomogeneous electric field. The basic electric quadrupole interaction may by taken to be

$$H_{\text{int}} = \frac{eQ}{2s(s-1)\hbar^2}\left[\left(\frac{\partial^2\phi}{\partial x^2}\right)_0 S_x^2 + \left(\frac{\partial^2\phi}{\partial y^2}\right)_0 S_y^2 + \left(\frac{\partial^2\phi}{\partial z^2}\right)_0 S_z^2\right],$$

where ϕ is the electrostatic potential satisfying Laplace's equation and the coordinate axes are so chosen that

$$\left(\frac{\partial^2\phi}{\partial x\partial y}\right)_0 = \left(\frac{\partial^2\phi}{\partial y\partial z}\right)_0 = \left(\frac{\partial^2\phi}{\partial x\partial z}\right)_0 = 0.$$

Show that the interaction energy can be written as

$$A(3S_z^2 - \mathbf{S}^2) + B(S_+^2 + S_-^2),$$

and express A and B in terms of $(\partial^2\phi/\partial x^2)_0$ and so on. Determine the energy eigenkets (in terms of $|m\rangle$, where $m = \pm\frac{3}{2}, \pm\frac{1}{2}$) and the corresponding energy eigenvalues. Is there any degeneracy?

CHAPTER 4

Symmetry in Quantum Mechanics

Having studied the theory of rotation in detail, we are in a position to discuss, in more general terms, the connection between symmetries, degeneracies, and conservation laws. We have deliberately postponed this very important topic until now so that we can discuss it using the rotation symmetry of Chapter 3 as an example.

4.1. SYMMETRIES, CONSERVATION LAWS, AND DEGENERACIES

Symmetries in Classical Physics

We begin with an elementary review of the concepts of symmetry and conservation law in classical physics. In the Lagrangian formulation of quantum mechanics, we start with the Lagrangian L, which is a function of a generalized coordinate q_i and the corresponding generalized velocity $\dot{q}_i$. If L is unchanged under displacement,

$$q_i \to q_i + \delta q_i, \tag{4.1.1}$$

then we must have

$$\frac{\partial L}{\partial q_i} = 0. \tag{4.1.2}$$

It then follows, by virtue of the Lagrange equation, $d/dt\ (\partial L/\partial \dot{q}_i) - \partial L/\partial q_i = 0$, that

$$\frac{dp_i}{dt} = 0, \tag{4.1.3}$$

where the canonical momentum is defined as

$$p_i = \frac{\partial L}{\partial \dot{q}_i}. \tag{4.1.4}$$

So if L is unchanged under displacement (4.1.1), then we have a conserved quantity, the canonical momentum conjugate to q_i.

Likewise, in the Hamiltonian formulation based on H regarded as a function of q_i and p_i, we have

$$\frac{dp_i}{dt} = 0 \tag{4.1.5}$$

whenever

$$\frac{\partial H}{\partial q_i} = 0. \tag{4.1.6}$$

So if the Hamiltonian does not explicitly depend on q_i, which is another way of saying H has a symmetry under $q_i \to q_i + \delta q_i$, we have a conserved quantity.

Symmetry in Quantum Mechanics

In quantum mechanics we have learned to associate a **unitary operator**, say $\mathscr{S}$, with an operation like translation or rotation. It has become customary to call $\mathscr{S}$ a **symmetry operator** regardless of whether the physical system itself possesses the symmetry corresponding to $\mathscr{S}$. Further, we have learned that for symmetry operations that differ infinitesimally from the identity transformation, we can write

$$\mathscr{S} = 1 - \frac{i\varepsilon}{\hbar} G, \tag{4.1.7}$$

where G is the Hermitian generator of the symmetry operator in question. Let us now suppose that H is invariant under $\mathscr{S}$. We then have

$$\mathscr{S}^\dagger H \mathscr{S} = H. \tag{4.1.8}$$

But this is equivalent to

$$[G, H] = 0. \tag{4.1.9}$$

By virtue of the Heisenberg equation of motion, we have

$$\frac{dG}{dt} = 0; \tag{4.1.10}$$

hence, G is a constant of the motion. For instance, if H is invariant under translation, then momentum is a constant of the motion; if H is invariant under rotation, then angular momentum is a constant of the motion.

It is instructive to look at the connection between (4.1.9) and conservation of G from the point of view of an eigenket of G when G commutes with H. Suppose at t_0 the system is in an eigenstate of G. Then the ket at a later time obtained by applying the time-evolution operator

$$|g', t_0; t\rangle = U(t, t_0)|g'\rangle \tag{4.1.11}$$

is also an eigenket of G with the same eigenvalue g'. In other words, once a ket is a G eigenket, it is always a G eigenket with the same eigenvalue. The proof of this is extremely simple once we realize that (4.1.9) and (4.1.10) also imply that G commutes with the time-evolution operator, namely,

$$G[U(t, t_0)|g'\rangle] = U(t, t_0)G|g'\rangle = g'[U(t, t_0)|g'\rangle]. \tag{4.1.12}$$

Degeneracies

Let us now turn to the concept of degeneracies. Even though degeneracies may be discussed at the level of classical mechanics—for instance in discussing closed (nonprecessing) orbits in the Kepler problem (Goldstein 1980)—this concept plays a far more important role in quantum mechanics. Let us suppose that

$$[H, \mathscr{S}] = 0 \tag{4.1.13}$$

for some symmetry operator and $|n\rangle$ is an energy eigenket with eigenvalue E_n. Then $\mathscr{S}|n\rangle$ is also an energy eigenket with the same energy, because

$$H(\mathscr{S}|n\rangle) = \mathscr{S}H|n\rangle = E_n(\mathscr{S}|n\rangle). \tag{4.1.14}$$

Suppose $|n\rangle$ and $\mathscr{S}|n\rangle$ represent different states. Then these are two states with the same energy, that is, they are degenerate. Quite often $\mathscr{S}$ is characterized by continuous parameters, say λ, in which case all states of the form $\mathscr{S}(\lambda)|n\rangle$ have the same energy.

We now consider rotation specifically. Suppose the Hamiltonian is rotationally invariant, so

$$[\mathscr{D}(R), H] = 0, \tag{4.1.15}$$

which necessarily implies that

$$[\mathbf{J}, H] = 0, \qquad [\mathbf{J}^2, H] = 0. \tag{4.1.16}$$

We can then form simultaneous eigenkets of H, $\mathbf{J}^2$, and J_z, denoted by $|n; j, m\rangle$. The argument just given implies that all states of the form

$$\mathscr{D}(R)|n; j, m\rangle \tag{4.1.17}$$

have the same energy. We saw in Chapter 3 that under rotation different

m-values get mixed up. In general, $\mathscr{D}(R)|n;\, j, m\rangle$ is a linear combination of $2j+1$ independent states. Explicitly,

$$\mathscr{D}(R)|n;\, j,m\rangle = \sum_{m'} |n;\, j,m'\rangle \mathscr{D}^{(j)}_{m'm}(R), \tag{4.1.18}$$

and by changing the continuous parameter that characterizes the rotation operator $\mathscr{D}(R)$, we can get different linear combinations of $|n;\, j, m'\rangle$. If all states of form $\mathscr{D}(R)|n;\, j, m\rangle$ with arbitrary $\mathscr{D}(R)$ are to have the same energy, it is then essential that each of $|n;\, j, m\rangle$ with different m must have the same energy. So the degeneracy here is $(2j+1)$-fold, just equal to the number of possible m-values. This point is also evident from the fact that all states obtained by successively applying $J_{\pm}$, which commutes with H, to $|n;\, jm\rangle$ have the same energy.

As an application, consider an atomic electron whose potential is written as $V(r)+V_{LS}(r)\mathbf{L}\cdot\mathbf{S}$. Because r and $\mathbf{L}\cdot\mathbf{S}$ are both rotationally invariant, we expect a $(2j+1)$-fold degeneracy for each atomic level. On the other hand, suppose there is an external electric or magnetic field, say in the z-direction. The rotational symmetry is now manifestly broken; as a result, the $(2j+1)$-fold degeneracy is no longer expected and states characterized by different m-values no longer have the same energy. We will examine how this splitting arises in Chapter 5.

4.2. DISCRETE SYMMETRIES, PARITY, OR SPACE INVERSION

So far we have considered continuous symmetry operators—that is, operations that can be obtained by applying successively infinitesimal symmetry operations. All symmetry operations useful in quantum mechanics are not necessarily of this form. In this chapter we consider three symmetry operations that can be considered to be discrete, as opposed to continuous—parity, lattice translation, and time reversal.

The first operation we consider is **parity**, or space inversion. The parity operation, as applied to transformation on the coordinate system, changes a right-handed (RH) system into a left-handed (LH) system, as shown in Figure 4.1. However, in this book we consider a transformation on state kets rather than on the coordinate system. Given $|\alpha\rangle$, we consider a space-inverted state, assumed to be obtained by applying a unitary operator π known as the **parity operator**, as follows:

$$|\alpha\rangle \rightarrow \pi|\alpha\rangle. \tag{4.2.1}$$

We require the expectation value of $\mathbf{x}$ taken with respect to the space inverted state to be opposite in sign.

$$\langle\alpha|\pi^{\dagger}\mathbf{x}\pi|\alpha\rangle = -\langle\alpha|\mathbf{x}|\alpha\rangle, \tag{4.2.2}$$

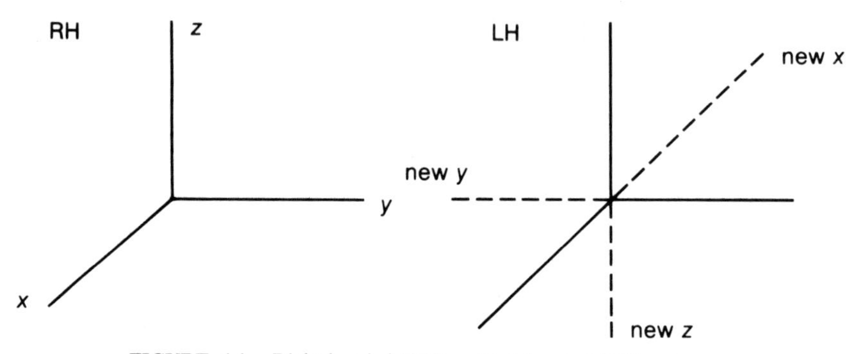

FIGURE 4.1. Right-handed (RH) and left-handed (LH) systems.

a very reasonable requirement. This is accomplished if

$$\pi^{\dagger}\mathbf{x}\pi = -\mathbf{x} \tag{4.2.3}$$

or

$$\mathbf{x}\pi = -\pi\mathbf{x}, \tag{4.2.4}$$

where we have used the fact that π is unitary. In other words, $\mathbf{x}$ and π must *anti*commute.

How does an eigenket of the position operator transform under parity? We claim that

$$\pi|\mathbf{x}'\rangle = e^{i\delta}|-\mathbf{x}'\rangle \tag{4.2.5}$$

where $e^{i\delta}$ is a phase factor (δ real). To prove this assertion let us note that

$$\mathbf{x}\pi|\mathbf{x}'\rangle = -\pi\mathbf{x}|\mathbf{x}'\rangle = (-\mathbf{x}')\pi|\mathbf{x}'\rangle. \tag{4.2.6}$$

This equation says that $\pi|\mathbf{x}'\rangle$ is an eigenket of $\mathbf{x}$ with eigenvalue $-\mathbf{x}'$, so it must be the same as a position eigenket $|-\mathbf{x}'\rangle$ up to a phase factor.

It is customary to take $e^{i\delta}=1$ by convention. Substituting this in (4.2.5), we have $\pi^2|\mathbf{x}'\rangle = |\mathbf{x}'\rangle$; hence, $\pi^2=1$—that is, we come back to the same state by applying π twice. We easily see from (4.2.5) that π is now not only unitary but also Hermitian:

$$\pi^{-1} = \pi^{\dagger} = \pi. \tag{4.2.7}$$

Its eigenvalue can be only $+1$ or -1.

What about the momentum operator? The momentum $\mathbf{p}$ is like $m\,d\mathbf{x}/dt$, so it is natural to expect it to be odd under parity, like $\mathbf{x}$. A more satisfactory argument considers the momentum operator as the generator of translation. Since translation followed by parity is equivalent to parity followed by translation in the *opposite* direction, as can be seen from Figure

4.2, then

$$\pi\mathscr{T}(d\mathbf{x}') = \mathscr{T}(-d\mathbf{x}')\pi \tag{4.2.8}$$

$$\pi\left(1 - \frac{i\mathbf{p}\cdot d\mathbf{x}'}{\hbar}\right)\pi^{\dagger} = 1 + \frac{i\mathbf{p}\cdot d\mathbf{x}'}{\hbar}, \tag{4.2.9}$$

from which follows

$$\{\pi, \mathbf{p}\} = 0 \quad \text{or} \quad \pi^{\dagger}\mathbf{p}\pi = -\mathbf{p}. \tag{4.2.10}$$

We can now discuss the behavior of $\mathbf{J}$ under parity. First, for orbital angular momentum we clearly have

$$[\pi, \mathbf{L}] = 0 \tag{4.2.11}$$

because

$$\mathbf{L} = \mathbf{x} \times \mathbf{p}, \tag{4.2.12}$$

and both $\mathbf{x}$ and $\mathbf{p}$ are odd under parity. However, to show that this property also holds for spin, it is best to use the fact that $\mathbf{J}$ is the generator of rotation. For 3×3 orthogonal matrices, we have

$$R^{(\text{parity})}R^{(\text{rotation})} = R^{(\text{rotation})}R^{(\text{parity})}, \tag{4.2.13}$$

where explicitly

$$R^{(\text{parity})} = \begin{pmatrix} -1 & & 0 \\ & -1 & \\ 0 & & -1 \end{pmatrix}; \tag{4.2.14}$$

that is, the parity and rotation operations commute. In quantum mechanics, it is natural to postulate the corresponding relation for the unitary operators, so

$$\pi\mathscr{D}(R) = \mathscr{D}(R)\pi, \tag{4.2.15}$$

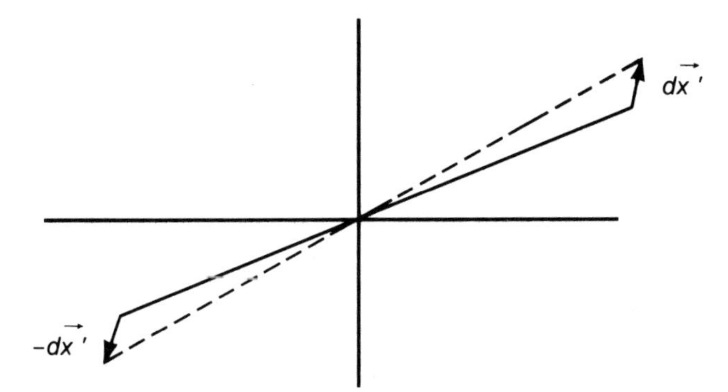

FIGURE 4.2. Translation followed by parity, and vice versa.

where $\mathscr{D}(R) = 1 - i\mathbf{J}\cdot\hat{\mathbf{n}}\varepsilon/\hbar$. From (4.2.15) it follows that

$$[\pi, \mathbf{J}] = 0 \quad \text{or} \quad \pi^\dagger \mathbf{J} \pi = \mathbf{J}. \tag{4.2.16}$$

This together with (4.2.11) means that the spin operator (given by $\mathbf{J} = \mathbf{L} + \mathbf{S}$) also transforms in the same way as $\mathbf{L}$.

Under rotations, $\mathbf{x}$ and $\mathbf{J}$ transform in the same way, so they are both vectors, or spherical tensors, of rank 1. However, $\mathbf{x}$ (or $\mathbf{p}$) is odd under parity [see (4.2.3) and (4.2.10)], while $\mathbf{J}$ is even under parity [see (4.2.16)]. Vectors that are odd under parity are called **polar vectors**, while vectors that are even under parity are called **axial vectors**, or **pseudovectors**.

Let us now consider operators like $\mathbf{S}\cdot\mathbf{x}$. Under rotations they transform like ordinary scalars, such as $\mathbf{S}\cdot\mathbf{L}$ or $\mathbf{x}\cdot\mathbf{p}$. Yet under space inversion we have

$$\pi^{-1}\mathbf{S}\cdot\mathbf{x}\pi = -\mathbf{S}\cdot\mathbf{x}, \tag{4.2.17}$$

while for ordinary scalars we have

$$\pi^{-1}\mathbf{L}\cdot\mathbf{S}\pi = \mathbf{L}\cdot\mathbf{S} \tag{4.2.18}$$

and so on. The operator $\mathbf{S}\cdot\mathbf{x}$ is an example of a **pseudoscalar**.

Wave Functions Under Parity

Let us now look at the parity property of wave functions. First, let ψ be the wave function of a spinless particle whose state ket is $|\alpha\rangle$:

$$\psi(\mathbf{x}') = \langle \mathbf{x}'|\alpha\rangle. \tag{4.2.19}$$

The wave function of the space-inverted state, represented by the state ket $\pi|\alpha\rangle$, is

$$\langle \mathbf{x}'|\pi|\alpha\rangle = \langle -\mathbf{x}'|\alpha\rangle = \psi(-\mathbf{x}'). \tag{4.2.20}$$

Suppose $|\alpha\rangle$ is an eigenket of parity. We have already seen that the eigenvalue of parity must be ± 1, so

$$\pi|\alpha\rangle = \pm|\alpha\rangle. \tag{4.2.21}$$

Let us look at its corresponding wave function,

$$\langle \mathbf{x}'|\pi|\alpha\rangle = \pm\langle \mathbf{x}'|\alpha\rangle. \tag{4.2.22}$$

But we also have

$$\langle \mathbf{x}'|\pi|\alpha\rangle = \langle -\mathbf{x}'|\alpha\rangle, \tag{4.2.23}$$

so the state $|\alpha\rangle$ is even or odd under parity depending on whether the corresponding wave function satisfies

$$\psi(-\mathbf{x}') = \pm\psi(\mathbf{x}')\begin{cases} \text{even parity,} \\ \text{odd parity.} \end{cases} \tag{4.2.24}$$

Not all wave functions of physical interest have definite parities in the sense of (4.2.24). Consider, for instance, the momentum eigenket. The momentum operator anticommutes with the parity operator, so the momentum eigenket is not expected to be a parity eigenket. Indeed, it is easy to see that the plane wave, which is the wave function for a momentum eigenket, does not satisfy (4.2.24).

An eigenket of orbital angular momentum is expected to be a parity eigenket because $\mathbf{L}$ and π commute [see (4.2.11)]. To see how an eigenket of $\mathbf{L}^2$ and L_z behaves under parity, let us examine the properties of its wave function under space inversion,

$$\langle \mathbf{x}'|\alpha, lm\rangle = R_\alpha(r)Y_l^m(\theta,\phi). \tag{4.2.25}$$

The transformation $\mathbf{x}' \to -\mathbf{x}'$ is accomplished by letting

$$\begin{aligned} &r \to r \\ &\theta \to \pi - \theta \qquad (\cos\theta \to -\cos\theta) \\ &\phi \to \phi + \pi \qquad \left(e^{im\phi} \to (-1)^m e^{im\phi}\right). \end{aligned} \tag{4.2.26}$$

Using the explicit form of

$$Y_l^m = (-1)^m \sqrt{\frac{(2l+1)(l-m)!}{4\pi(l+m)!}}\, P_l^m(\cos\theta)e^{im\phi} \tag{4.2.27}$$

for positive m, with (3.6.38), where

$$P_l^{|m|}(\cos\theta) = \frac{(-1)^{m+l}}{2^l l!}\frac{(l+|m|)!}{(l-|m|)!}\sin^{-|m|}\theta\left(\frac{d}{d(\cos\theta)}\right)^{l-|m|}\sin^{2l}\theta, \tag{4.2.28}$$

we can readily show that

$$Y_l^m \to (-1)^l Y_l^m \tag{4.2.29}$$

as θ and ϕ are changed, as in (4.2.26). Therefore, we can conclude that

$$\pi|\alpha, lm\rangle = (-1)^l|\alpha, lm\rangle. \tag{4.2.30}$$

It is actually not necessary to look at Y_l^m; an easier way to obtain the same result is to work with $m=0$ and note that $L_\pm^r|l, m=0\rangle$ $(r=0,1,\ldots,l)$ must have the same parity because π and $(L_\pm)^r$ commute.

Let us now look at the parity properties of energy eigenstates. We begin by stating a very important theorem.

Theorem. *Suppose*

$$[H,\pi]=0 \tag{4.2.31}$$

and $|n\rangle$ *is a nondegenerate eigenket of* H *with eigenvalue* E_n:

$$H|n\rangle = E_n|n\rangle; \tag{4.2.32}$$

then $|n\rangle$ is also a parity eigenket.

Proof. We prove this theorem by first noting that

$$\tfrac{1}{2}(1 \pm \pi)|n\rangle \tag{4.2.33}$$

is a parity eigenket with eigenvalues ± 1 (just use $\pi^2 = 1$). But this is also an energy eigenket with eigenvalue E_n. Furthermore, $|n\rangle$ and (4.2.33) must represent the same state, otherwise there would be two states with the same energy—a contradiction of our nondegenerate assumption. It therefore follows that $|n\rangle$, which is the same as (4.2.33) up to a multiplicative constant, must be a parity eigenket with parity ± 1. □

As an example, let us look at the simple harmonic oscillator (SHO). The ground state $|0\rangle$ has even parity because its wave function, being Gaussian, is even under $\mathbf{x}' \to -\mathbf{x}'$. The first excited state,

$$|1\rangle = a^{\dagger}|0\rangle, \tag{4.2.34}$$

must have an odd parity because $a^{\dagger}$ is linear in x and p, which are both odd [see (2.3.2)]. In general, the parity of the nth excited state of the simple harmonic operator is given by $(-1)^n$.

It is important to note that the nondegenerate assumption is essential here. For instance, consider the hydrogen atom in nonrelativistic quantum mechanics. As is well known, the energy eigenvalues depend only on the principal quantum number n (for example, $2p$ and $2s$ states are degenerate) —the Coulomb potential is obviously invariant under parity—yet an energy eigenket

$$c_p|2p\rangle + c_s|2s\rangle \tag{4.2.35}$$

is obviously not a parity eigenket.

As another example, consider a momentum eigenket. Momentum anticommutes with parity, so—even though free-particle Hamiltonian H is invariant under parity—the momentum eigenket (though obviously an energy eigenket) is not a parity eigenket. Our theorem remains intact because we have here a degeneracy between $|\mathbf{p}'\rangle$ and $|-\mathbf{p}'\rangle$, which have the same energy. In fact, we can easily construct linear combinations $(1/\sqrt{2})(|\mathbf{p}'\rangle \pm |-\mathbf{p}'\rangle)$, which are parity eigenkets with eigenvalues ± 1. In terms of wave-function language, $e^{i\mathbf{p}'\cdot\mathbf{x}'/\hbar}$ does not have a definite parity, but $\cos \mathbf{p}'\cdot\mathbf{x}'/\hbar$ and $\sin \mathbf{p}'\cdot\mathbf{x}'/\hbar$ do.

Symmetrical Double-Well Potential

As an elementary but instructive example, we consider a symmetrical double-well potential; see Figure 4.3. The Hamiltonian is obviously invariant under parity. In fact, the two lowest lying states are as shown in

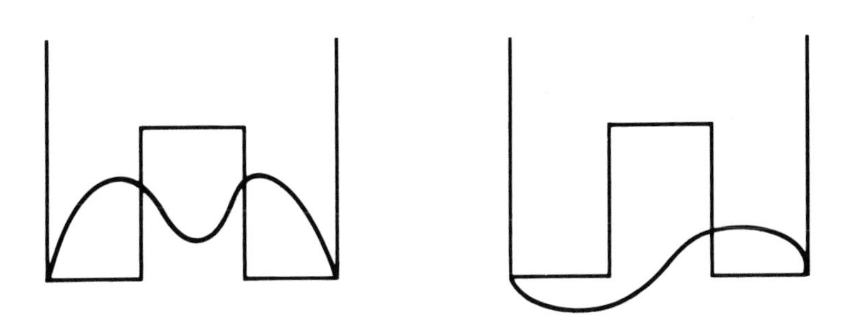

FIGURE 4.3. The symmetrical double well with the two lowest lying states $|S\rangle$ (symmetrical) and $|A\rangle$ (antisymmetrical) shown.

Figure 4.3, we can see by working out the explicit solutions involving sine and cosine in classically allowed regions and sinh and cosh in the classically forbidden region. The solutions are matched where the potential is discontinuous; we call them the **symmetrical state** $|S\rangle$ and the **antisymmetrical state** $|A\rangle$. Of course, they are simultaneous eigenkets of H and π. Calculation also shows that

$$E_A > E_S, \tag{4.2.36}$$

which we can infer from Figure 4.3 by noting that the wave function of the antisymmetrical state has a greater curvature. The energy difference is very tiny if the middle barrier is high, a point which we will discuss later.

We can form

$$|R\rangle = \frac{1}{\sqrt{2}}(|S\rangle + |A\rangle) \tag{4.2.37a}$$

and

$$|L\rangle = \frac{1}{\sqrt{2}}(|S\rangle - |A\rangle). \tag{4.2.37b}$$

The wave functions of (4.2.37a) and (4.2.37b) are largely concentrated in the right-hand side and the left-hand side, respectively. They are obviously not parity eigenstates; in fact, under parity $|R\rangle$ and $|L\rangle$ are interchanged. Note that they are not energy eigenstates either. Indeed, they are classical examples of **nonstationary states**. To be precise, let us assume that the system is represented by $|R\rangle$ at $t=0$. At a later time, we have

$$\begin{aligned}|R, t_0=0; t\rangle &= \frac{1}{\sqrt{2}}\left(e^{-iE_st/\hbar}|S\rangle + e^{-iE_At/\hbar}|A\rangle\right)\\ &= \frac{1}{\sqrt{2}}e^{-iE_st/\hbar}\left(|S\rangle + e^{-i(E_A-E_S)t/\hbar}|A\rangle\right).\end{aligned} \tag{4.2.38}$$

At time $t = T/2 \equiv 2\pi\hbar/2(E_A - E_S)$, the system is found in pure $|L\rangle$. At $t = T$, we are back to pure $|R\rangle$, and so forth. Thus, in general, we have an oscillation between $|R\rangle$ and $|L\rangle$ with angular frequency

$$\omega = \frac{(E_A - E_S)}{\hbar}. \tag{4.2.39}$$

This oscillatory behavior can also be considered from the viewpoint of tunneling in quantum mechanics. A particle initially confined to the right-hand side can tunnel through the classically forbidden region (the middle barrier) into the left-hand side, then back to the right-hand side, and so on. But now let the middle barrier become infinitely high; see Figure 4.4. The $|S\rangle$ and $|A\rangle$ states are now degenerate, so (4.2.37a) and (4.2.37b) are also energy eigenkets even though they are not parity eigenkets. Once the system is found in $|R\rangle$, it remains so forever (oscillation time between $|S\rangle$ and $|A\rangle$ is now ∞). Because the middle barrier is infinitely high, there is no possibility for tunneling. Thus when there is degeneracy, the physically realizable energy eigenkets need not be parity eigenkets. We have a ground state which is asymmetrical despite the fact that the Hamiltonian itself is symmetrical under space inversion, so with degeneracy the symmetry of H is not necessarily obeyed by energy eigenstates $|S\rangle$ and $|A\rangle$.

This is a very simple example of broken symmetry and degeneracy. Nature is full of situations analogous to this. Consider a ferromagnet. The basic Hamiltonian for iron atoms is rotationally invariant, but the ferromagnet clearly has a definite direction in space; hence, the (infinite) number of ground states is *not* rotationally invariant, since the spins are all aligned along some definite (but arbitrary) direction.

A textbook example of a system that illustrates the actual importance of the symmetrical double well is an ammonia molecule, NH_3; see Figure 4.5. We imagine that the three H atoms form the three corners of an equilateral triangle. The N atom can be up or down, where the directions up and down are defined because the molecule is rotating around the axis as

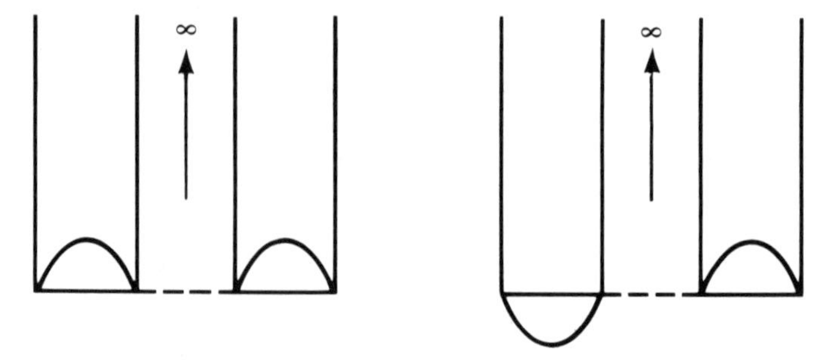

FIGURE 4.4. The symmetrical double well with an infinitely high middle barrier.

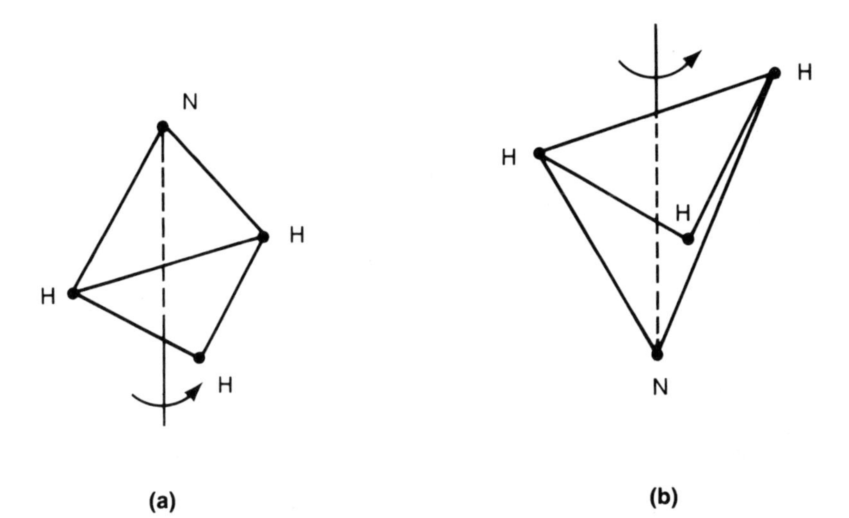

FIGURE 4.5. An ammonia molecule, NH_3, where the three H atoms form the three corners of an equilateral triangle.

shown in Figure 4.5. The up and down positions for the N atom are analogous to R and L of the double-well potential. The parity and energy eigenstates are *superpositions* of Figure 4.5a and Figure 4.5b in the sense of (4.2.37a) and (4.2.37b), respectively, and the energy difference between the simultaneous eigenstates of energy and parity correspond to an oscillation frequency of 24,000 MHz—a wavelength of about 1 cm, which is in the microwave region. In fact, NH_3 is of fundamental importance in maser physics.

There are naturally occurring organic molecules, such as sugar or amino acids, which are of the R-type (or L-type) only. Such molecules which have definite handedness are called **optical isomers**. In many cases the oscillation time is practically infinite—on the order of 10^4–10^6 years—so R-type molecules remain right-handed for all practical purposes. It is amusing that if we attempt to synthesize such organic molecules in the laboratory, we find equal mixtures of R and L. Why we have a preponderance of one type is nature's deepest mystery. Is it due to a genetic accident, like the spiral shell of a snail or the fact that our hearts are on the left-hand side?

Parity-Selection Rule

Suppose $|\alpha\rangle$ and $|\beta\rangle$ are parity eigenstates:

$$\pi|\alpha\rangle = \varepsilon_\alpha|\alpha\rangle \tag{4.2.40a}$$

and

$$\pi|\beta\rangle = \varepsilon_\beta|\beta\rangle, \tag{4.2.40b}$$

where ε_α, ε_β are the parity eigenvalues (± 1). We can show that

$$\langle\beta|\mathbf{x}|\alpha\rangle = 0 \tag{4.2.41}$$

unless $\varepsilon_\alpha = -\varepsilon_\beta$. In other words, the parity-odd operator **x** connects states of opposite parity. The proof of this follows:

$$\langle\beta|\mathbf{x}|\alpha\rangle = \langle\beta|\pi^{-1}\pi\mathbf{x}\pi^{-1}\pi|\alpha\rangle = \varepsilon_\alpha\varepsilon_\beta(-\langle\beta|\mathbf{x}|\alpha\rangle), \tag{4.2.42}$$

which is impossible for a finite nonzero $\langle\beta|\mathbf{x}|\alpha\rangle$ unless ε_α and ε_β are opposite in sign. Perhaps the reader is familiar with this argument from

$$\int \psi_\beta^* \mathbf{x} \psi_\alpha \, d\tau = 0 \tag{4.2.43}$$

if ψ_β and ψ_α have the same parity. This selection rule, due to Wigner, is of importance in discussing radiative transitions between atomic states. As we will discuss in greater detail later, radiative transitions take place between states of opposite parity as a consequence of multipole expansion formalism. This rule was known phenomenologically from analysis of spectral lines, before the birth of quantum mechanics, as **Laporte's rule**. It was Wigner who showed that Laporte's rule is a consequence of the parity-selection rule.

If the basic Hamiltonian H is invariant under parity, nondegenerate energy eigenstates [as a corollary of (4.2.43)] cannot possess a permanent electric dipole moment:

$$\langle n|\mathbf{x}|n\rangle = 0. \tag{4.2.44}$$

This follows trivially from (4.2.43), because with the nondegenerate assumption, energy eigenstates are also parity eigenstates [see (4.2.32) and (4.2.33)]. For a degenerate state, it is perfectly all right to have an electric dipole moment. We will see an example of this when we discuss the linear Stark effect in Chapter 5.

Our considerations can be generalized: Operators that are odd under parity, like **p** or $\mathbf{S}\cdot\mathbf{x}$, have nonvanishing matrix elements only between states of opposite parity. In contrast, operators that are even under parity connect states of the same parity.

Parity Nonconservation

The basic Hamiltonian responsible for the so-called weak interaction of elementary particles is not invariant under parity. In decay processes we can have final states which are superpositions of opposite parity states. Observable quantities like the angular distribution of decay products can depend on pseudoscalars such as $\langle \mathbf{S} \rangle \cdot \mathbf{p}$. It is remarkable that parity conservation was believed to be a sacred principle until 1956, when Lee and Yang speculated that parity is not conserved in weak interactions and proposed crucial experiments to test the validity of parity conservation. Subsequent experiments indeed showed that observable effects do depend on pseudoscalar quantities such as correlation between $\langle \mathbf{S} \rangle$ and $\mathbf{p}$. Because parity is not conserved in weak interactions, previously thought "pure" nuclear and atomic states are, in fact, parity mixtures. These subtle effects have also been found experimentally.

4.3. LATTICE TRANSLATION AS A DISCRETE SYMMETRY

We now consider another kind of discrete symmetry operation, namely, lattice translation. This subject has extremely important applications in solid-state physics.

Consider a periodic potential in one dimension, where $V(x \pm a) = V(x)$, as depicted in Figure 4.6. Realistically, we may consider the motion

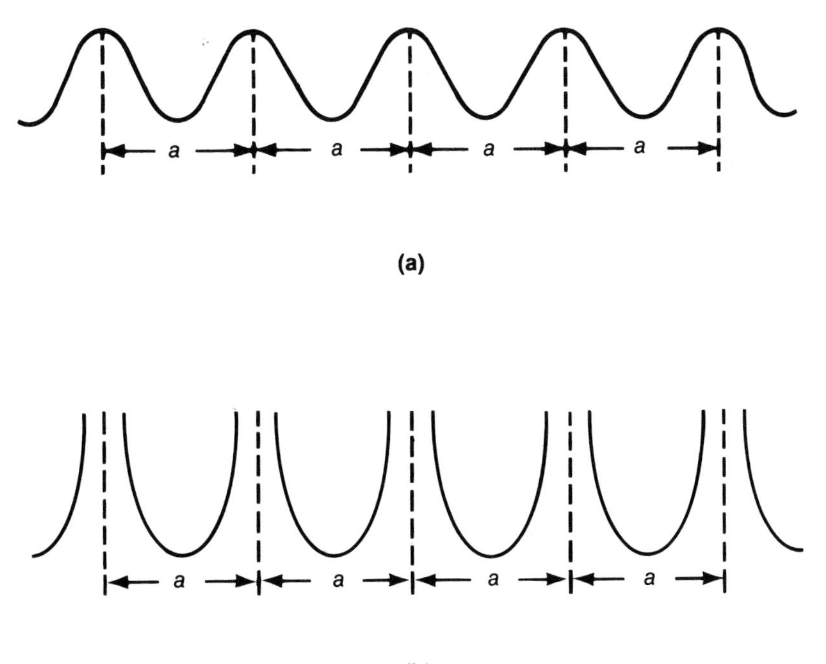

FIGURE 4.6. (a) Periodic potential in one dimension with periodicity a. (b) The periodic potential when the barrier height between two adjacent lattice sites becomes infinite.

of an electron in a chain of regularly spaced positive ions. In general, the Hamiltonian is not invariant under a translation represented by $\tau(l)$ with l arbitrary, where $\tau(l)$ has the property (see Section 1.6)

$$\tau^\dagger(l)x\tau(l) = x + l, \qquad \tau(l)|x'\rangle = |x' + l\rangle. \tag{4.3.1}$$

However, when l coincides with the lattice spacing a, we do have

$$\tau^\dagger(a)V(x)\tau(a) = V(x + a) = V(x). \tag{4.3.2}$$

Because the kinetic-energy part of the Hamiltonian H is invariant under the translation with any displacement, the entire Hamiltonian satisfies

$$\tau^\dagger(a)H\tau(a) = H. \tag{4.3.3}$$

Because $\tau(a)$ is unitary, we have [from (4.3.3)]

$$[H, \tau(a)] = 0, \tag{4.3.4}$$

so the Hamiltonian and $\tau(a)$ can be simultaneously diagonalized. Although $\tau(a)$ is unitary, it is not Hermitian, so we expect the eigenvalue to be a *complex* number of modulus 1.

Before we determine the eigenkets and eigenvalues of $\tau(a)$ and examine their physical significance, it is instructive to look at a special case of periodic potential when the barrier height between two adjacent lattice sites is made to go to infinity, as in Figure 4.6b. What is the ground state for the potential of Figure 4.6b? Clearly, a state in which the particle is completely localized in one of the lattice sites can be a candidate for the ground state. To be specific let us assume that the particle is localized at the nth site and denote the corresponding ket by $|n\rangle$. This is an energy eigenket with energy eigenvalue E_0, namely, $H|n\rangle = E_0|n\rangle$. Its wave function $\langle x'|n\rangle$ is finite only in the nth site. However, we note that a similar state localized at some other site also has the same energy E_0, so actually there are denumerably infinite ground states n, where n runs from $-\infty$ to $+\infty$.

Now $|n\rangle$ is obviously not an eigenket of the lattice-translation operator, because when the lattice-translation operator is applied to it, we obtain $|n+1\rangle$:

$$\tau(a)|n\rangle = |n+1\rangle. \tag{4.3.5}$$

So despite the fact that $\tau(a)$ commutes with H, $|n\rangle$—which is an eigenket of H—is not an eigenket of $\tau(a)$. This is quite consistent with our earlier theorem on symmetry because we have an infinitefold degeneracy. When there is such degeneracy, the symmetry of the world need not be the symmetry of energy eigenkets. Our task is to find a *simultaneous* eigenket of H and $\tau(a)$.

Here we may recall how we handled a somewhat similar situation with the symmetrical double-well potential of the previous section. We noted that even though neither $|R\rangle$ nor $|L\rangle$ is an eigenket of π, we could easily form a symmetrical and an antisymmetrical combination of $|R\rangle$ and $|L\rangle$ that are parity eigenkets. The case is analogous here. Let us specifically form a linear combination

$$|\theta\rangle \equiv \sum_{n=-\infty}^{\infty} e^{in\theta}|n\rangle, \tag{4.3.6}$$

where θ is a real parameter with $-\pi \leq \theta \leq \pi$. We assert that $|\theta\rangle$ is a simultaneous eigenket of H and $\tau(a)$. That it is an H eigenket is obvious because $|n\rangle$ is an energy eigenket with eigenvalue E_0, independent of n. To show that it is also an eigenket of the lattice translation operator we apply $\tau(a)$ as follows:

$$\begin{aligned}\tau(a)|\theta\rangle &= \sum_{n=-\infty}^{\infty} e^{in\theta}|n+1\rangle = \sum_{n=-\infty}^{\infty} e^{i(n-1)\theta}|n\rangle \\ &= e^{-i\theta}|\theta\rangle.\end{aligned} \tag{4.3.7}$$

Note that this simultaneous eigenket of H and $\tau(a)$ is parameterized by a continuous parameter θ. Furthermore, the energy eigenvalue E_0 is independent of θ.

Let us now return to the more realistic situation of Figure 4.6a, where the barrier between two adjacent lattice sites is not infinitely high. We can construct a localized ket $|n\rangle$ just as before with the property $\tau(a)|n\rangle = |n+1\rangle$. However, this time we expect that there is some leakage possible into neighboring lattice sites due to quantum-mechanical tunneling. In other words, the wave function $\langle x'|n\rangle$ has a tail extending to sites other than the nth site. The diagonal elements of H in the $\{|n\rangle\}$ basis are all equal because of translation invariance, that is,

$$\langle n|H|n\rangle = E_0, \tag{4.3.8}$$

independent of n, as before. However we suspect that H is not completely diagonal in the $\{|n\rangle\}$ basis due to leakage. Now, suppose the barriers between adjacent sites are high (but not infinite). We then expect matrix elements of H between distant sites to be completely negligible. Let us assume that the only nondiagonal elements of importance connect immediate neighbors. That is,

$$\langle n'|H|n\rangle \neq 0 \quad \text{only if } n' = n \quad \text{or} \quad n' = n \pm 1. \tag{4.3.9}$$

In solid-state physics this assumption is known as the **tight-binding approximation**. Let us define

$$\langle n \pm 1|H|n\rangle = -\Delta. \tag{4.3.10}$$

Clearly, Δ is again independent of n due to translation invariance of the Hamiltonian. To the extent that $|n\rangle$ and $|n'\rangle$ are orthogonal when $n \neq n'$, we obtain

$$H|n\rangle = E_0|n\rangle - \Delta|n+1\rangle - \Delta|n-1\rangle. \tag{4.3.11}$$

Note that $|n\rangle$ is no longer an energy eigenket.

As we have done with the potential of Figure 4.6b, let us form a linear combination

$$|\theta\rangle = \sum_{n=-\infty}^{\infty} e^{in\theta}|n\rangle. \tag{4.3.12}$$

Clearly, $|\theta\rangle$ is an eigenket of translation operator $\tau(a)$ because the steps in (4.3.7) still hold. A natural question is, is $|\theta\rangle$ an energy eigenket? To answer this question, we apply H:

$$\begin{aligned} H\sum e^{in\theta}|n\rangle &= E_0\sum e^{in\theta}|n\rangle - \Delta\sum e^{in\theta}|n+1\rangle - \Delta\sum e^{in\theta}|n-1\rangle \\ &= E_0\sum e^{in\theta}|n\rangle - \Delta\sum \left(e^{in\theta-i\theta} + e^{in\theta+i\theta}\right)|n\rangle \\ &= (E_0 - 2\Delta\cos\theta)\sum e^{in\theta}|n\rangle. \end{aligned} \tag{4.3.13}$$

The big difference between this and the previous situation is that the energy eigenvalue now depends on the continuous real parameter θ. The degeneracy is lifted as Δ becomes finite, and we have a continuous distribution of energy eigenvalues between $E_0 - 2\Delta$ and $E_0 + 2\Delta$. See Figure 4.7, where we visualize how the energy levels start forming a continuous energy band as Δ is increased from zero.

To see the physical meaning of the parameter θ let us study the wave function $\langle x'|\theta\rangle$. For the wave function of the lattice-translated state $\tau(a)|\theta\rangle$,

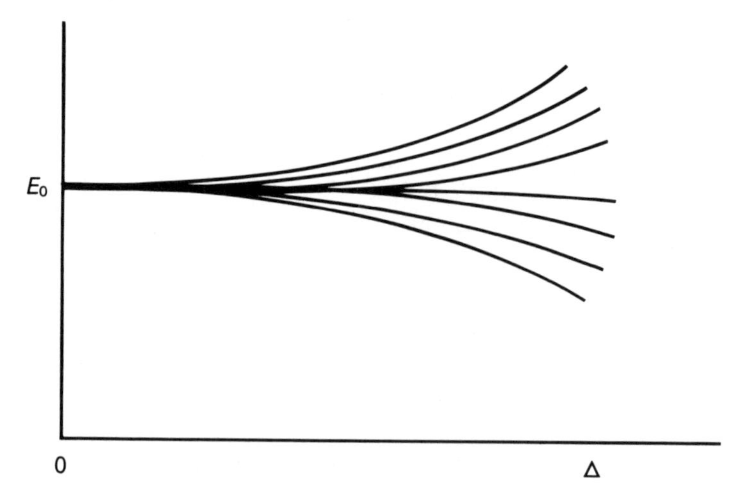

FIGURE 4.7. Energy levels forming a continuous energy band as Δ is increased from zero.

we obtain

$$\langle x'|\tau(a)|\theta\rangle = \langle x'-a|\theta\rangle \tag{4.3.14}$$

by letting $\tau(a)$ act on $\langle x'|$. But we can also let $\tau(a)$ operate on $|\theta\rangle$ and use (4.3.7). Thus

$$\langle x'|\tau(a)|\theta\rangle = e^{-i\theta}\langle x'|\theta\rangle, \tag{4.3.15}$$

so

$$\langle x'-a|\theta\rangle = \langle x'|\theta\rangle e^{-i\theta}. \tag{4.3.16}$$

We solve this equation by setting

$$\langle x'|\theta\rangle = e^{ikx'}u_k(x'), \tag{4.3.17}$$

with $\theta = ka$, where $u_k(x')$ is a periodic function with period a, as we can easily verify by explicit substitutions, namely,

$$e^{ik(x'-a)}u_k(x'-a) = e^{ikx'}u_k(x')e^{-ika}. \tag{4.3.18}$$

Thus we get the important condition known as **Bloch's theorem**: The wave function of $|\theta\rangle$, which is an eigenket of $\tau(a)$, can be written as a plane wave $e^{ikx'}$ times a periodic function with periodicity a. Notice that the only fact we used was that $|\theta\rangle$ is an eigenket of $\tau(a)$ with eigenvalue $e^{-i\theta}$ [see (4.3.7)]. In particular, the theorem holds even if the tight-binding approximation (4.3.9) breaks down.

We are now in a position to interpret our earlier result (4.3.13) for $|\theta\rangle$ given by (4.3.12). We know that the wave function is a plane wave characterized by the propagation wave vector k modulated by a periodic function $u_k(x')$ [see (4.3.17)]. As θ varies from $-\pi$ to π, the wave vector k varies from $-\pi/a$ to π/a. The energy eigenvalue E now depends on k as follows:

$$E(k) = E_0 - 2\Delta\cos ka. \tag{4.3.19}$$

Notice that this energy eigenvalue equation is independent of the detailed shape of the potential as long as the tight-binding approximation is valid. Note also that there is a cutoff in the wave vector k of the Bloch wave function (4.3.17) given by $|k| = \pi/a$. Equation (4.3.19) defines a dispersion curve, as shown in Figure 4.8. As a result of tunneling, the denumerably infinitefold degeneracy is now completely lifted, and the allowed energy values form a continuous band between $E_0 - 2\Delta$ and $E_0 + 2\Delta$, known as the **Brillouin zone**.

So far we have considered only one particle moving in a periodic potential. In a more realistic situation we must look at many electrons moving in such a potential. Actually the electrons satisfy the Pauli exclusion principle, as we will discuss more systematically in Chapter 6, and they start filling the band. In this way, the main qualitative features of metals,

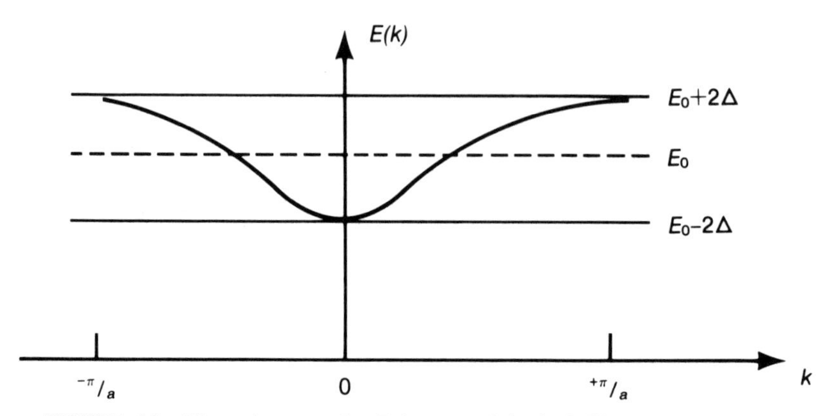

FIGURE 4.8. Dispersion curve for $E(k)$ versus k in the Brillouin zone $|k| \leq \pi/a$.

semiconductors, and the like can be understood as a consequence of translation invariance supplemented by the exclusion principle.

The reader may have noted the similarity between the symmetrical double-well problem of Section 4.2 and the periodic potential of this section. Comparing Figures 4.3 and 4.6, we note that they can be regarded as opposite extremes (two versus infinite) of potentials with a finite number of troughs.

4.4. THE TIME-REVERSAL DISCRETE SYMMETRY

In this section we study another discrete symmetry operator, called **time reversal**. This is a difficult topic for the novice, partly because the term *time reversal* is a misnomer; it reminds us of science fiction. Actually what we do in this section can be more appropriately characterized by the term *reversal of motion*. Indeed, that is the terminology used by E. Wigner, who formulated time reversal in a very fundamental paper written in 1932.

For orientation purposes let us look at classical mechanics. Suppose there is a trajectory of a particle subject to a certain force field; see Figure 4.9. At $t=0$, let the particle stop and reverse its motion: $\mathbf{p}|_{t=0} \rightarrow -\mathbf{p}|_{t=0}$. The particle traverses backward along the same trajectory. If you run the motion picture of trajectory (a) backward as in (b), you may have a hard time telling whether this is the correct sequence.

More formally, if $\mathbf{x}(t)$ is a solution to

$$m\ddot{\mathbf{x}} = -\nabla V(\mathbf{x}), \tag{4.4.1}$$

then $\mathbf{x}(-t)$ is also a possible solution in the same force field derivable from V. It is, of course, important to note that we do not have a dissipative force

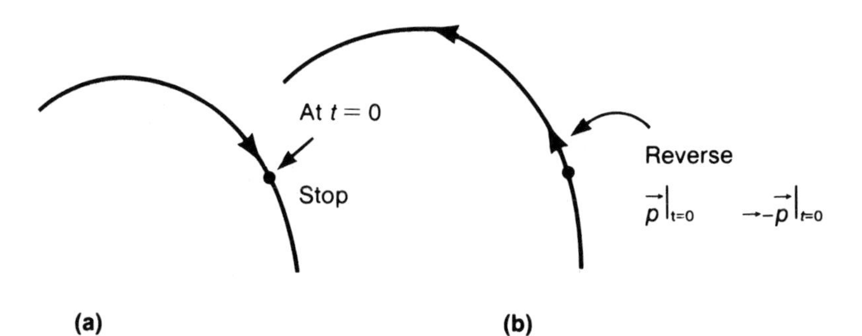

FIGURE 4.9. (a) Classical trajectory which stops at $t = 0$ and (b) reverses its motion $\mathbf{p}|_{t=0} \to -\mathbf{p}|_{t=0}$.

here. A block sliding on a table decelerates (due to friction) and eventually stops. But have you ever seen a block on a table spontaneously start to move and accelerate?

With a magnetic field you may be able to tell the difference. Imagine that you are taking the motion picture of a spiraling electron trajectory in a magnetic field. You may be able to tell whether the motion picture is run forward or backward by comparing the sense of rotation with the magnetic pole labeling N and S. However, from a microscopic point of view, **B** is produced by moving charges via an electric current; if you could reverse the current that causes **B**, then the situation would be quite symmetrical. In terms of the picture shown in Figure 4.10, you may have figured out that N and S are mislabeled! Another more formal way of saying all this is that the Maxwell equations, for example,

$$\nabla\cdot\mathbf{E} = 4\pi\rho, \quad \nabla\times\mathbf{B} - \frac{1}{c}\frac{\partial\mathbf{E}}{\partial t} = \frac{4\pi\mathbf{j}}{c}, \quad \nabla\times\mathbf{E} = -\frac{1}{c}\frac{\partial\mathbf{B}}{\partial t}, \tag{4.4.2}$$

and the Lorentz force equation $\mathbf{F} = e[\mathbf{E} + (1/c)(\mathbf{v}\times\mathbf{B})]$ are invariant under $t \to -t$ provided we also let

$$\mathbf{E} \to \mathbf{E}, \quad \mathbf{B} \to -\mathbf{B}, \quad \rho \to \rho, \quad \mathbf{j} \to -\mathbf{j}, \quad \mathbf{v} \to -\mathbf{v}. \tag{4.4.3}$$

Let us now look at wave mechanics, where the basic equation of the Schrödinger wave equation is

$$i\hbar\frac{\partial\psi}{\partial t} = \left(-\frac{\hbar^2}{2m}\nabla^2 + V\right)\psi. \tag{4.4.4}$$

Suppose $\psi(\mathbf{x}, t)$ is a solution. We can easily verify that $\psi(\mathbf{x}, -t)$ is not a solution, because of the appearance of the first-order time derivative. However, $\psi^*(\mathbf{x}, -t)$ is a solution, as you may verify by complex conjugation of (4.4.4). It is instructive to convince ourselves of this point for an

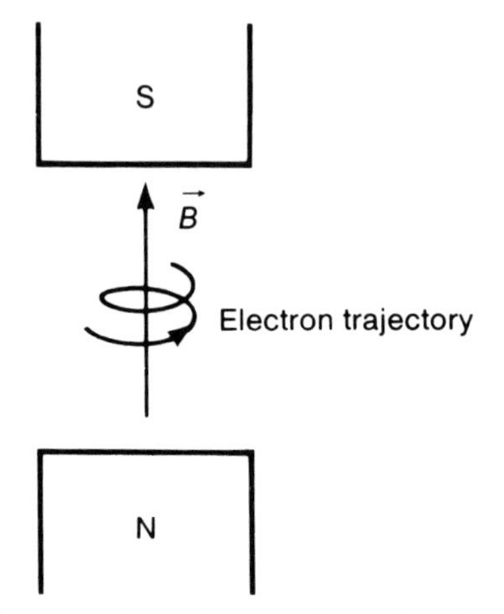

FIGURE 4.10. Electron trajectory between the north and south poles of a magnet.

energy eigenstate, that is, by substituting

$$\psi(\mathbf{x}, t) = u_n(\mathbf{x})e^{-iE_nt/\hbar}, \qquad \psi^*(\mathbf{x}, -t) = u_n^*(\mathbf{x})e^{-iE_nt/\hbar} \tag{4.4.5}$$

into the Schrödinger equation (4.4.4). Thus we conjecture that time reversal must have something to do with complex conjugation. If at $t = 0$ the wave function is given by

$$\psi = \langle \mathbf{x}|\alpha\rangle, \tag{4.4.6}$$

then the wave function for the corresponding time-reversed state is given by $\langle \mathbf{x}|\alpha\rangle^*$. We will later show that this is indeed the case for the wave function of a spinless system. As an example, you may easily check this point for the wave function of a plane wave; see Problem 8 of this chapter.

Digression on Symmetry Operations

Before we begin a systematic treatment of the time-reversal operator, some general remarks on symmetry operations are in order. Consider a symmetry operation

$$|\alpha\rangle \to |\tilde{\alpha}\rangle, \qquad |\beta\rangle \to |\tilde{\beta}\rangle. \tag{4.4.7}$$

One may argue that it is natural to require the inner product $\langle\beta|\alpha\rangle$ to be preserved, that is,

$$\langle\tilde{\beta}|\tilde{\alpha}\rangle = \langle\beta|\alpha\rangle. \tag{4.4.8}$$

Indeed, for symmetry operations such as rotations, translations, and even parity, this is indeed the case. If $|\alpha\rangle$ is rotated and $|\beta\rangle$ is also rotated in the same manner, $\langle \beta | \alpha \rangle$ is unchanged. Formally this arises from the fact that, for the symmetry operations considered in the previous sections, the corresponding symmetry operator is unitary, so

$$\langle \beta | \alpha \rangle \to \langle \beta | U^\dagger U | \alpha \rangle = \langle \beta | \alpha \rangle. \tag{4.4.9}$$

However, in discussing time reversal, we see that requirement (4.4.8) turns out to be too restrictive. Instead, we merely impose the weaker requirement that

$$|\langle \tilde{\beta} | \tilde{\alpha} \rangle| = |\langle \beta | \alpha \rangle|. \tag{4.4.10}$$

Requirement (4.4.8) obviously satisfies (4.4.10). But this is not the only way;

$$\langle \tilde{\beta} | \tilde{\alpha} \rangle = \langle \beta | \alpha \rangle^* = \langle \alpha | \beta \rangle \tag{4.4.11}$$

works equally well. We pursue the latter possibility in this section because from our earlier discussion based on the Schrödinger equation we inferred that time reversal has something to do with complex conjugation.

Definition. *The transformation*

$$|\alpha\rangle \to |\tilde{\alpha}\rangle = \theta |\alpha\rangle, \qquad |\beta\rangle \to |\tilde{\beta}\rangle = \theta |\beta\rangle \tag{4.4.12}$$

is said to be ***antiunitary*** *if*

$$\langle \tilde{\beta} | \tilde{\alpha} \rangle = \langle \beta | \alpha \rangle^*, \tag{4.4.13a}$$

$$\theta(c_1 |\alpha\rangle + c_2 |\beta\rangle) = c_1^* \theta |\alpha\rangle + c_2^* \theta |\beta\rangle. \tag{4.4.13b}$$

In such a case the operator θ is an antiunitary operator. Relation (4.4.13b) alone defines an **antilinear** operator.

We now claim that an antiunitary operator can be written as

$$\theta = UK, \tag{4.4.14}$$

where U is a unitary operator and K is the complex-conjugate operator that forms the complex conjugate of any coefficient that multiplies a ket (and stands on the right of K). Before checking (4.4.13) let us examine the property of the K operator. Suppose we have a ket multiplied by a complex number c. We then have

$$Kc|\alpha\rangle = c^* K |\alpha\rangle. \tag{4.4.15}$$

One may further ask, What happens if $|\alpha\rangle$ is expanded in terms of base kets $\{|a'\rangle\}$? Under the action K we have

$$\begin{aligned} |\alpha\rangle = \sum_{a'} |a'\rangle\langle a'|\alpha\rangle &\overset{K}{\to} |\tilde{\alpha}\rangle = \sum_{a'} \langle a'|\alpha\rangle^* K |a'\rangle \\ &= \sum_{a'} \langle a'|\alpha\rangle^* |a'\rangle. \end{aligned} \tag{4.4.16}$$

Notice that K acting on the base ket does not change the base ket. The explicit representation of $|a'\rangle$ is

$$|a'\rangle = \begin{pmatrix} 0 \\ 0 \\ \vdots \\ 0 \\ 1 \\ 0 \\ \vdots \\ 0 \end{pmatrix}, \tag{4.4.17}$$

and there is nothing to be changed by K. The reader may wonder, for instance, whether the S_y eigenkets for a spin $\frac{1}{2}$ system change under K. The answer is that if the S_z eigenkets are used as base kets, we must change the S_y eigenkets because the S_y eigenkets (1.1.14) undergo under K

$$K\left(\frac{1}{\sqrt{2}}|+\rangle \pm \frac{i}{\sqrt{2}}|-\rangle\right) \to \frac{1}{\sqrt{2}}|+\rangle \mp \frac{i}{\sqrt{2}}|-\rangle. \tag{4.4.18}$$

On the other hand, if the S_y eigenkets themselves are used as the base kets, we do not change the S_y eigenkets under the action of K. Thus the effect of K changes with the basis. As a result, the form of U in (4.4.14) also depends on the particular representation (that is, the choice of base kets) used. Gottfried puts it aptly: "If the basis is changed, the work of U and K has to be reapportioned."

Returning to $\theta = UK$ and (4.4.13), let us first check property (4.4.13b). We have

$$\begin{aligned}\theta(c_1|\alpha\rangle + c_2|\beta\rangle) &= UK(c_1|\alpha\rangle + c_2|\beta\rangle) \\ &= c_1^* UK|\alpha\rangle + c_2^* UK|\beta\rangle \\ &= c_1^*\theta|\alpha\rangle + c_2^*\theta|\beta\rangle,\end{aligned} \tag{4.4.19}$$

so (4.4.13b) indeed holds. Before checking (4.4.13a), we assert that it is always safer to work with the action of θ on kets only. We can figure out how the bras change just by looking at the corresponding kets. In particular, it is not necessary to consider θ acting on bras from the right, nor is it necessary to define $\theta^\dagger$. We have

$$\begin{aligned}|\alpha\rangle \overset{\theta}{\to} |\tilde{\alpha}\rangle &= \sum_{a'} \langle a'|\alpha\rangle^* UK|a'\rangle \\ &= \sum_{a'} \langle a'|\alpha\rangle^* U|a'\rangle \\ &= \sum_{a'} \langle \alpha|a'\rangle U|a'\rangle.\end{aligned} \tag{4.4.20}$$

As for $|\beta\rangle$, we have

$$|\tilde{\beta}\rangle = \sum_{a'} \langle a'|\beta\rangle^* U|a'\rangle \overset{\text{DC}}{\leftrightarrow} \langle\tilde{\beta}| = \sum_{a'} \langle a'|\beta\rangle\langle a'|U^\dagger$$

$$\begin{aligned}\langle\tilde{\beta}|\tilde{\alpha}\rangle &= \sum_{a''}\sum_{a'} \langle a''|\beta\rangle\langle a''|U^\dagger U|a'\rangle\langle\alpha|a'\rangle \\ &= \sum_{a'} \langle\alpha|a'\rangle\langle a'|\beta\rangle = \langle\alpha|\beta\rangle \\ &= \langle\beta|\alpha\rangle^*, \end{aligned} \tag{4.4.21}$$

so this checks.

In order for (4.4.10) to be satisfied, it is of physical interest to consider just two types of transformation—unitary and antiunitary. Other possibilities are related to either of the preceding via trivial phase changes. The proof of this assertion is actually very difficult and will not be discussed further here (Gottfried 1966, 226–28).

Time-Reversal Operator

We are finally in a position to present a formal theory of time reversal. Let us denote the time-reversal operator by Θ, to be distinguished from θ, a general antiunitary operator. Consider

$$|\alpha\rangle \rightarrow \Theta|\alpha\rangle, \tag{4.4.22}$$

where $\Theta|\alpha\rangle$ is the time-reversed state. More appropriately, $\Theta|\alpha\rangle$ should be called the motion-reversed state. If $|\alpha\rangle$ is a momentum eigenstate $|\mathbf{p}'\rangle$, we expect $\Theta|\alpha\rangle$ to be $|-\mathbf{p}'\rangle$ up to a possible phase. Likewise, $\mathbf{J}$ is to be reversed under time reversal.

We now deduce the fundamental property of the time-reversal operator by looking at the time evolution of the time-reversed state. Consider a physical system represented by a ket $|\alpha\rangle$, say at $t=0$. Then at a slightly later time $t = \delta t$, the system is found in

$$|\alpha, t_0 = 0; t = \delta t\rangle = \left(1 - \frac{iH}{\hbar}\delta t\right)|\alpha\rangle, \tag{4.4.23}$$

where H is the Hamiltonian that characterizes the time evolution. Instead of the preceding equation, suppose we first apply Θ, say at $t=0$, and then let the system evolve under the influence of the Hamiltonian H. We then have at δt

$$\left(1 - \frac{iH\delta t}{\hbar}\right)\Theta|\alpha\rangle. \tag{4.4.24a}$$

If motion obeys symmetry under time reversal, we expect the preceding state ket to be the same as

$$\Theta|\alpha, t_0 = 0; t = -\delta t\rangle \tag{4.4.24b}$$

that is, first consider a state ket at *earlier* time $t = -\delta t$, and then reverse $\mathbf{p}$ and $\mathbf{J}$; see Figure 4.11. Mathematically,

$$\left(1 - \frac{iH}{\hbar}\delta t\right)\Theta|\alpha\rangle = \Theta\left(1 - \frac{iH}{\hbar}(-\delta t)\right)|\alpha\rangle. \tag{4.4.25}$$

If the preceding relation is to be true for any ket, we must have

$$-iH\Theta|\rangle = \Theta iH|\rangle, \tag{4.4.26}$$

where the blank ket $|\ \rangle$ emphasizes that (4.4.26) is to be true for any ket.

We now argue that Θ *cannot* be unitary if the motion of time reversal is to make sense. Suppose Θ were unitary. It would then be legitimate to cancel the i's in (4.4.26), and we would have the operator equation

$$-H\Theta = \Theta H. \tag{4.4.27}$$

Consider an energy eigenket $|n\rangle$ with energy eigenvalue E_n. The corresponding time-reversed state would be $\Theta|n\rangle$, and we would have, because of (4.4.27),

$$H\Theta|n\rangle = -\Theta H|n\rangle = (-E_n)\Theta|n\rangle. \tag{4.4.28}$$

This equation says that $\Theta|n\rangle$ is an eigenket of the Hamiltonian with energy eigenvalues $-E_n$. But this is nonsensical even in the very elementary case of a free particle. We know that the energy spectrum of the free particle is positive semidefinite—from 0 to $+\infty$. There is no state lower than a

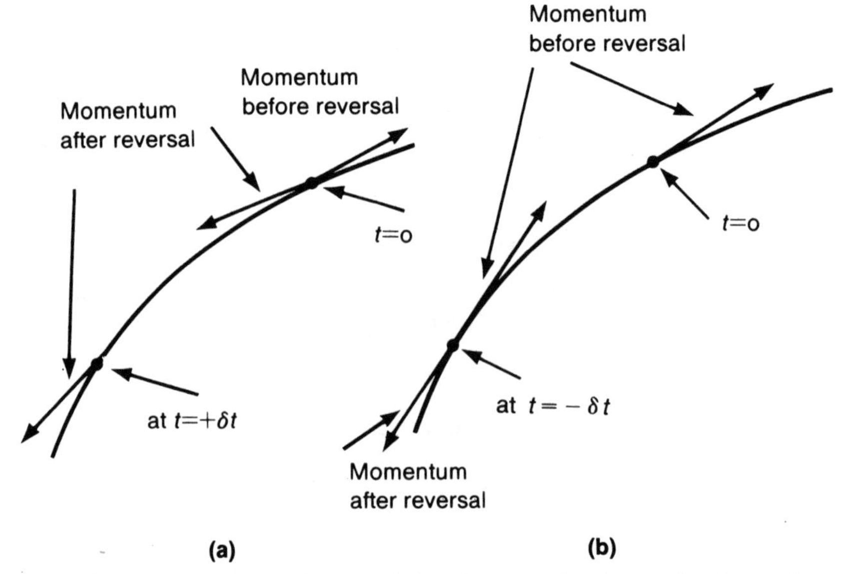

FIGURE 4.11. Momentum before and after time reversal at time $t = 0$ and $t = \pm\delta t$.

particle at rest (momentum eigenstate with momentum eigenvalue zero); the energy spectrum ranging from $-\infty$ to 0 would be completely unacceptable. We can also see this by looking at the structure of the free-particle Hamiltonian. We expect $\mathbf{p}$ to change sign but not $\mathbf{p}^2$; yet (4.4.27) would imply that

$$\Theta^{-1}\frac{\mathbf{p}^2}{2m}\Theta = \frac{-\mathbf{p}^2}{2m}. \tag{4.4.29}$$

All these arguments strongly suggest that if time reversal is to be a useful symmetry at all, we are not allowed to cancel the i's in (4.4.26); hence, Θ had better be antiunitary. In this case the right-hand side of (4.4.26) becomes

$$\Theta iH|\,\rangle = -i\Theta H|\,\rangle \tag{4.4.30}$$

by antilinear property (4.4.13b). Now at last we can cancel the i's in (4.4.26) leading, finally, via (4.4.30) to

$$\Theta H = H\Theta. \tag{4.4.31}$$

Equation (4.4.31) expresses the fundamental property of the Hamiltonian under time reversal. With this equation the difficulties mentioned earlier [see (4.4.27) to (4.4.29)] are absent, and we obtain physically sensible results. From now on we will always take Θ to be antiunitary.

We mentioned earlier that it is best to avoid an antiunitary operator acting on bras from the right. Nevertheless, we may use

$$\langle\beta|\Theta|\alpha\rangle, \tag{4.4.32}$$

which is always to be understood as

$$(\langle\beta|)\cdot(\Theta|\alpha\rangle) \tag{4.4.33}$$

and never as

$$(\langle\beta|\Theta)\cdot|\alpha\rangle. \tag{4.4.34}$$

In fact, we do not even attempt to define $\langle\beta|\Theta$. This is one place where the Dirac bra-ket notation is a little confusing. After all, that notation was invented to handle linear operators, not antilinear operators.

With this cautionary remark, we are in a position to discuss the behavior of operators under time reversal. We continue to take the point of view that the Θ operator is to act on kets

$$|\tilde{\alpha}\rangle = \Theta|\alpha\rangle, \qquad |\tilde{\beta}\rangle = \Theta|\beta\rangle, \tag{4.4.35}$$

yet it is often convenient to talk about operators—in particular, observables—which are odd or even under time reversal. We start with an important identity, namely,

$$\langle\beta|\otimes|\alpha\rangle = \langle\tilde{\alpha}|\Theta\otimes^{\dagger}\Theta^{-1}|\tilde{\beta}\rangle, \tag{4.4.36}$$

where $\otimes$ is a linear operator. This identity follows solely from the antiunitary nature of Θ. To prove this let us define

$$|\gamma\rangle \equiv \otimes^\dagger|\beta\rangle. \tag{4.4.37}$$

By dual correspondence we have

$$|\gamma\rangle \overset{\text{DC}}{\leftrightarrow} \langle\beta|\otimes = \langle\gamma|. \tag{4.4.38}$$

Hence,

$$\begin{aligned}\langle\beta|\otimes|\alpha\rangle &= \langle\gamma|\alpha\rangle = \langle\tilde{\alpha}|\tilde{\gamma}\rangle \\ &= \langle\tilde{\alpha}|\Theta\otimes^\dagger|\beta\rangle = \langle\tilde{\alpha}|\Theta\otimes^\dagger\Theta^{-1}\Theta|\beta\rangle \\ &= \langle\tilde{\alpha}|\Theta\otimes^\dagger\Theta^{-1}|\tilde{\beta}\rangle,\end{aligned} \tag{4.4.39}$$

which proves the identity. In particular, for *Hermitian* observables A we get

$$\langle\beta|A|\alpha\rangle = \langle\tilde{\alpha}|\Theta A\Theta^{-1}|\tilde{\beta}\rangle. \tag{4.4.40}$$

We say that observables are even or odd under time reversal according to whether we have the upper or lower sign in

$$\Theta A\Theta^{-1} = \pm A. \tag{4.4.41}$$

Note that this equation, together with (4.4.40), gives a phase restriction on the matrix elements of A taken with respect to time reversed states as follows:

$$\langle\beta|A|\alpha\rangle = \pm\langle\tilde{\beta}|A|\tilde{\alpha}\rangle^*. \tag{4.4.42}$$

If $|\beta\rangle$ is identical to $|\alpha\rangle$, so that we are talking about expectation values, we have

$$\langle\alpha|A|\alpha\rangle = \pm\langle\tilde{\alpha}|A|\tilde{\alpha}\rangle, \tag{4.4.43}$$

where $\langle\tilde{\alpha}|A|\tilde{\alpha}\rangle$ is the expectation value taken with respect to the time-reversed state.

As an example, let us look at the expectation value of $\mathbf{p}$. It is reasonable to expect that the expectation value of $\mathbf{p}$ taken with respect to the time-reversed state be of opposite sign. Thus

$$\langle\alpha|\mathbf{p}|\alpha\rangle = -\langle\tilde{\alpha}|\mathbf{p}|\tilde{\alpha}\rangle, \tag{4.4.44}$$

so we take $\mathbf{p}$ to be an odd operator, namely,

$$\Theta\mathbf{p}\Theta^{-1} = -\mathbf{p}. \tag{4.4.45}$$

This implies that

$$\begin{aligned}\mathbf{p}\Theta|\mathbf{p}'\rangle &= -\Theta\mathbf{p}\Theta^{-1}\Theta|\mathbf{p}'\rangle \\ &= (-\mathbf{p}')\Theta|\mathbf{p}'\rangle.\end{aligned} \tag{4.4.46}$$

Equation (4.4.46) agrees with our earlier assertion that $\Theta|\mathbf{p}'\rangle$ is a momentum eigenket with eigenvalue $-\mathbf{p}'$. It can be identified with $|-\mathbf{p}'\rangle$ itself with a suitable choice of phase. Likewise, we obtain

$$\Theta \mathbf{x} \Theta^{-1} = \mathbf{x}$$
$$\Theta|\mathbf{x}'\rangle = |\mathbf{x}'\rangle \qquad \text{(up to a phase)} \tag{4.4.47}$$

from the (eminently reasonable) requirement

$$\langle \alpha|\mathbf{x}|\alpha\rangle = \langle \tilde{\alpha}|\mathbf{x}|\tilde{\alpha}\rangle. \tag{4.4.48}$$

We can now check the invariance of the fundamental commutation relation

$$[x_i, p_j]|\ \rangle = i\hbar\delta_{ij}|\ \rangle, \tag{4.4.49}$$

where the blank ket $|\ \rangle$ stands for any ket. Applying Θ to both sides of (4.4.49), we have

$$\Theta[x_i, p_j]\Theta^{-1}\Theta|\ \rangle = \Theta i\hbar\delta_{ij}|\ \rangle, \tag{4.4.50}$$

which leads to, after passing Θ through $i\hbar$,

$$[x_i, (-p_j)]\Theta|\ \rangle = -i\hbar\delta_{ij}\Theta|\ \rangle. \tag{4.4.51}$$

Note that the fundamental commutation relation $[x_i, p_j] = i\hbar\delta_{ij}$ is preserved by virtue of the fact that Θ is antiunitary. This can be given as yet another reason for taking Θ to be antiunitary; otherwise, we will be forced to abandon either (4.4.45) or (4.4.47)! Similarly, to preserve

$$[J_i, J_j] = i\hbar\varepsilon_{ijk}J_k, \tag{4.4.52}$$

the angular-momentum operator must be odd under time reversal, that is,

$$\Theta \mathbf{J} \Theta^{-1} = -\mathbf{J}. \tag{4.4.53}$$

This is consistent for spinless system where $\mathbf{J}$ is just $\mathbf{x} \times \mathbf{p}$. Alternatively, we could have deduced this relation by noting that the rotational operator and the time-reversal operator commute (note the extra i!).

Wave Function

Suppose at some given time, say at $t = 0$, a spinless single-particle system is found in a state represented by $|\alpha\rangle$. Its wave function $\langle \mathbf{x}'|\alpha\rangle$ appears as the expansion coefficient in the position representation

$$|\alpha\rangle = \int d^3x' |\mathbf{x}'\rangle\langle \mathbf{x}'|\alpha\rangle. \tag{4.4.54}$$

Applying the time-reversal operator

$$\begin{aligned}\Theta|\alpha\rangle &= \int d^3x' \Theta|\mathbf{x}'\rangle\langle \mathbf{x}'|\alpha\rangle^* \\ &= \int d^3x' |\mathbf{x}'\rangle\langle \mathbf{x}'|\alpha\rangle^*,\end{aligned} \tag{4.4.55}$$

where we have chosen the phase convention so that $\Theta|\mathbf{x}'\rangle$ is $|\mathbf{x}'\rangle$ itself. We then recover the rule

$$\psi(\mathbf{x}') \to \psi^*(\mathbf{x}') \tag{4.4.56}$$

inferred earlier by looking at the Schrödinger wave equation [see (4.4.5)]. The angular part of the wave function is given by a spherical harmonic Y_l^m. With the usual phase convention we have

$$Y_l^m(\theta,\phi) \to Y_l^{m*}(\theta,\phi) = (-1)^m Y_l^{-m}(\theta,\phi). \tag{4.4.57}$$

Now $Y_l^m(\theta,\phi)$ is the wave function for $|l,m\rangle$ [see (3.6.23)]; therefore, from (4.4.56) we deduce

$$\Theta|l,m\rangle = (-1)^m|l,-m\rangle. \tag{4.4.58}$$

If we study the probability current density (2.4.16) for a wave function of type (3.6.22) going like $R(r)Y_l^m$, we shall conclude that for $m>0$ the current flows in the counterclockwise direction, as seen from the positive z-axis. The wave function for the corresponding time-reversed state has its probability current flowing in the opposite direction because the sign of m is reversed. All this is very reasonable.

As a nontrivial consequence of time-reversal invariance, we state an important theorem on the reality of the energy eigenfunction of a spinless particle.

Theorem. *Suppose the Hamiltonian is invariant under time reversal and the energy eigenket $|n\rangle$ is nondegenerate; then the corresponding energy eigenfunction is real (or, more generally, a real function times a phase factor independent of* **x**).

Proof. To prove this, first note that

$$H\Theta|n\rangle = \Theta H|n\rangle = E_n\Theta|n\rangle, \tag{4.4.59}$$

so $|n\rangle$ and $\Theta|n\rangle$ have the same energy. The nondegeneracy assumption prompts us to conclude that $|n\rangle$ and $\Theta|n\rangle$ must represent the *same* state; otherwise there would be two different states with the same energy E_n, an obvious contradiction! Let us recall that the wave functions for $|n\rangle$ and $\Theta|n\rangle$ are $\langle\mathbf{x}'|n\rangle$ and $\langle\mathbf{x}'|n\rangle^*$, respectively. They must be the same—that is,

$$\langle\mathbf{x}'|n\rangle = \langle\mathbf{x}'|n\rangle^* \tag{4.4.60}$$

for all practical purposes—or, more precisely, they can differ at most by a phase factor independent of **x**. □

Thus if we have, for instance, a nondegenerate bound state, its wave function is always real. On the other hand, in the hydrogen atom with $l \neq 0$, $m \neq 0$, the energy eigenfunction characterized by definite (n,l,m) quantum numbers is complex because Y_l^m is complex; this does not contradict the theorem because $|n,l,m\rangle$ and $|n,l,-m\rangle$ are degenerate. Similarly, the wave function of a plane wave $e^{i\mathbf{p}\cdot\mathbf{x}/\hbar}$ is complex, but it is degenerate with $e^{-i\mathbf{p}\cdot\mathbf{x}/\hbar}$.

We see that for a spinless system, the wave function for the time-reversed state, say at $t=0$, is simply obtained by complex conjugation. In terms of ket $|\alpha\rangle$ written as in (4.4.16) or in (4.4.54), the Θ operator is the complex conjugate operator K itself because K and Θ have the same effect when acting on the base ket $|a'\rangle$ (or $|\mathbf{x}'\rangle$). We may note, however, that the situation is quite different when the ket $|\alpha\rangle$ is expanded in terms of the momentum eigenket because Θ must change $|\mathbf{p}'\rangle$ into $|-\mathbf{p}'\rangle$ as follows:

$$\Theta|\alpha\rangle = \int d^3p'|-\mathbf{p}'\rangle\langle\mathbf{p}'|\alpha\rangle^* = \int d^3p'|\mathbf{p}'\rangle\langle-\mathbf{p}'|\alpha\rangle^*. \tag{4.4.61}$$

It is apparent that the momentum-space wave function of the time-reversed state is not just the complex conjugate of the original momentum-space wave function; rather, we must identify $\phi^*(-\mathbf{p}')$ as the momentum-space wave function for the time-reversed state. This situation once again illustrates the basic point that the particular form of Θ depends on the particular representation used.

Time Reversal for a Spin $\frac{1}{2}$ System

The situation is even more interesting for a particle with spin—spin $\frac{1}{2}$, in particular. We recall from Section 3.2 that the eigenket of $\mathbf{S}\cdot\hat{\mathbf{n}}$ with eigenvalue $\hbar/2$ can be written as

$$|\hat{\mathbf{n}};+\rangle = e^{-iS_z\alpha/\hbar}e^{-iS_y\beta/\hbar}|+\rangle, \tag{4.4.62}$$

where $\hat{\mathbf{n}}$ is characterized by the polar and azimuthal angles β and α, respectively. Noting (4.4.53) we have

$$\Theta|\hat{\mathbf{n}};+\rangle = e^{-iS_z\alpha/\hbar}e^{-iS_y\beta/\hbar}\Theta|+\rangle = \eta|\hat{\mathbf{n}};-\rangle. \tag{4.4.63}$$

On the other hand, we can easily verify that

$$|\hat{\mathbf{n}};-\rangle = e^{-i\alpha S_z/\hbar}e^{-i(\pi+\beta)S_y/\hbar}|+\rangle. \tag{4.4.64}$$

In general, we saw earlier that the product UK is an antiunitary operator. Comparing (4.4.63) and (4.4.64) with Θ set equal to UK, and noting that K acting on the base ket $|+\rangle$ gives just $|+\rangle$, we see that

$$\Theta = \eta e^{-i\pi S_y/\hbar}K = -i\eta\left(\frac{2S_y}{\hbar}\right)K, \tag{4.4.65}$$

where η stands for an arbitrary phase (a complex number of modulus unity). Another way to be convinced of (4.4.65) is to verify that if $\chi(\hat{\mathbf{n}};+)$ is the two-component eigenspinor corresponding to $|\hat{\mathbf{n}};+\rangle$ [in the sense that

$\boldsymbol{\sigma}\cdot\hat{\mathbf{n}}\chi(\hat{\mathbf{n}};\,+) = \chi(\hat{\mathbf{n}};\,+)$], then

$$-i\sigma_y\chi^*(\hat{\mathbf{n}};\,+) \tag{4.4.66}$$

(note the complex conjugation!) is the eigenspinor corresponding to $|\hat{\mathbf{n}};\,-\rangle$, again up to an arbitrary phase, see Problem 7 of this chapter. The appearance of S_y or σ_y can be traced to the fact that we are using the representation in which S_z is diagonal and the nonvanishing matrix elements of S_y are purely imaginary.

Let us now note

$$e^{-i\pi S_y/\hbar}|+\rangle = +|-\rangle, \qquad e^{-i\pi S_y/\hbar}|-\rangle = -|+\rangle. \tag{4.4.67}$$

Using (4.4.67), we are in a position to work out the effect of Θ, written as (4.4.65), on the most general spin $\frac{1}{2}$ ket:

$$\Theta(c_+|+\rangle + c_-|-\rangle) = +\eta c_+^*|-\rangle - \eta c_-^*|+\rangle. \tag{4.4.68}$$

Let us apply Θ once again:

$$\begin{aligned}\Theta^2(c_+|+\rangle + c_-|-\rangle) &= -|\eta|^2 c_+|+\rangle - |\eta|^2 c_-|-\rangle \\ &= -(c_+|+\rangle + c_-|-\rangle)\end{aligned} \tag{4.4.69}$$

or

$$\Theta^2 = -1, \tag{4.4.70}$$

(where -1 is to be understood as -1 times the identity operator) for *any* spin orientation. This is an extraordinary result. It is crucial to note here that our conclusion is completely independent of the choice of phase; (4.4.70) holds no matter what phase convention we may use for η. In contrast, we may note that two successive applications of Θ to a spinless state give

$$\Theta^2 = +1 \tag{4.4.71}$$

as is evident from, say, (4.4.58).

More generally, we now prove

$$\Theta^2|j \text{ half-integer}\rangle = -|j \text{ half-integer}\rangle \tag{4.4.72a}$$

$$\Theta^2|j \text{ integer}\rangle = +|j \text{ integer}\rangle. \tag{4.4.72b}$$

Thus the eigenvalue of Θ^2 is given by $(-1)^{2j}$. We first note that (4.4.65) generalizes for arbitrary j to

$$\Theta = \eta e^{-i\pi J_y/\hbar}K. \tag{4.4.73}$$

For a ket $|\alpha\rangle$ expanded in terms of $|j,m\rangle$ base eigenkets, we have

$$\begin{aligned}\Theta\left(\Theta\sum|jm\rangle\langle jm|\alpha\rangle\right) &= \Theta\left(\eta\sum e^{-i\pi J_y/\hbar}|jm\rangle\langle jm|\alpha\rangle^*\right) \\ &= |\eta|^2 e^{-2i\pi J_y/\hbar}\sum|jm\rangle\langle jm|\alpha\rangle.\end{aligned} \tag{4.4.74}$$

But

$$e^{-2i\pi J_y/\hbar}|jm\rangle = (-1)^{2j}|jm\rangle, \tag{4.4.75}$$

as is evident from the properties of angular-momentum eigenstates under rotation by 2π.

In (4.4.72b), $|j \text{ integer}\rangle$ may stand for the spin state

$$\frac{1}{\sqrt{2}}(|+-\rangle \pm |-+\rangle) \tag{4.4.76}$$

of a two-electron system or the orbital state $|l, m\rangle$ of a spinless particle. It is important only that j is an integer. Likewise, $|j \text{ half-integer}\rangle$ may stand, for example, for a three-electron system in any configuration. Actually, for a system made up exclusively of electrons, any system with an odd (even) number of electrons—regardless of their spatial orientation (for example, relative orbital angular momentum)—is odd (even) under Θ^2; they need not even be $\mathbf{J}^2$ eigenstates!

We make a parenthetical remark on the phase convention. In our earlier discussion based on the position representation, we saw that with the usual convention for spherical harmonics it is natural to choose the arbitrary phase for $|l, m\rangle$ under time reversal so that

$$\Theta|l, m\rangle = (-1)^m|l, -m\rangle. \tag{4.4.77}$$

Some authors find it attractive to generalize this to obtain

$$\Theta|j, m\rangle = (-1)^m|j, -m\rangle \quad (j \text{ an integer}) \tag{4.4.78}$$

regardless of whether j refers to l or s (for an integer spin system). We may naturally ask, is this compatible with (4.4.72a) for a spin $\frac{1}{2}$ system when we visualize $|j, m\rangle$ as being built up of "primitive" spin $\frac{1}{2}$ objects according to Wigner and Schwinger. It is easy to see that (4.4.72a) is indeed consistent provided we choose η in (4.4.73) to be $+i$. In fact, in general, we can take

$$\Theta|j, m\rangle = i^{2m}|j, -m\rangle \tag{4.4.79}$$

for any j—either a half-integer j or an integer j; see Problem 10 of this chapter. The reader should be warned, however, that this is not the only convention found in the literature. (See, for instance, Frauenfelder and Henley 1974.) For some physical applications, it is more convenient to use other choices; for instance, the phase convention that makes the $\mathbf{J}_\pm$ operator matrix elements simple is *not* the phase convention that makes the time-reversal operator properties simple. We emphasize once again that (4.4.70) is completely independent of phase convention.

Having worked out the behavior of angular-momentum eigenstates under time reversal, we are in a position to study once again the expectation values of a Hermitian operator. Recalling (4.4.43), we obtain under time reversal (canceling the i^{2m} factors)

$$\langle \alpha, j, m|A|\alpha, j, m\rangle = \pm\langle \alpha, j, -m|A|\alpha, j, -m\rangle. \tag{4.4.80}$$

Now suppose A is a component of a spherical tensor $T_q^{(k)}$. Because of the Wigner-Eckart theorem, it is sufficient to examine just the matrix element of the $q=0$ component. In general, $T^{(k)}$ (assumed to be Hermitian) is said to be even or odd under time reversal depending on how its $q=0$ component satisfies the upper or lower sign in

$$\Theta T_{q=0}^{(k)}\Theta^{-1} = \pm T_{q=0}^{(k)}. \tag{4.4.81}$$

Equation (4.4.80) for $A = T_0^{(k)}$ becomes

$$\langle \alpha, j, m|T_0^{(k)}|\alpha, j, m\rangle = \pm\langle \alpha, j, -m|T_0^{(k)}|\alpha, j, -m\rangle. \tag{4.4.82}$$

Due to (3.6.46)–(3.6.49), we expect $|\alpha, j, -m\rangle = \mathscr{D}(0,\pi,0)|\alpha, j, m\rangle$ up to a phase. We next use (3.10.22) for $T_0^{(k)}$, which leads to

$$\mathscr{D}^\dagger(0,\pi,0)T_0^{(k)}\mathscr{D}(0,\pi,0) = (-1)^k T_0^{(k)} + (q \neq 0 \text{ components}), \tag{4.4.83}$$

where we have used $\mathscr{D}_{00}^{(k)}(0,\pi,0) = P_k(\cos\pi) = (-1)^k$, and the $q \neq 0$ components give vanishing contributions when sandwiched between $\langle \alpha, j, m|$ and $|\alpha, j, m\rangle$. The net result is

$$\langle \alpha, j, m|T_0^{(k)}|\alpha, j, m\rangle = \pm(-1)^k\langle \alpha, j, m|T_0^{(k)}|\alpha, j, m\rangle. \tag{4.4.84}$$

As an example, taking $k=1$, the expectation value $\langle \mathbf{x}\rangle$ taken with respect to eigenstates of j, m vanishes. We may argue that we already know $\langle \mathbf{x}\rangle = 0$ from parity inversion if the expectation value is taken with respect to parity eigenstates [see (4.2.41)]. But note that here $|\alpha, j, m\rangle$ need not be parity eigenkets! For example, the $|j, m\rangle$ for spin $\frac{1}{2}$ particles could be $c_s|s_{1/2}\rangle + c_p|p_{1/2}\rangle$.

Interactions with Electric and Magnetic Fields; Kramers Degeneracy

Consider charged particles in an external electric or magnetic field. If we have only a static electric field interacting with the electric charge, the interaction part of the Hamiltonian is just

$$V(\mathbf{x}) = e\phi(\mathbf{x}), \tag{4.4.85}$$

where $\phi(\mathbf{x})$ is the electrostatic potential. Because $\phi(\mathbf{x})$ is a real function of the time-reversal even operator $\mathbf{x}$, we have

$$[\Theta, H] = 0. \tag{4.4.86}$$

Unlike the parity case, (4.4.86) does not lead to an interesting conservation law. The reason is that

$$\Theta U(t,t_0) \neq U(t,t_0)\Theta \tag{4.4.87}$$

even if (4.4.86) holds, so our discussion following (4.1.9) of Section 4.1 breaks down. As a result, there is no such thing as the "conservation of time-reversal quantum number." As we already mentioned, requirement (4.4.86) does, however, lead to a nontrivial phase restriction—the reality of a nondegenerate wave function for a spinless system [see (4.4.59) and (4.4.60)].

Another far-reaching consequence of time-reversal invariance is the **Kramers degeneracy**. Suppose H and Θ commute, and let $|n\rangle$ and $\Theta|n\rangle$ be the energy eigenket and its time-reversed state, respectively. It is evident from (4.4.86) that $|n\rangle$ and $\Theta|n\rangle$ belong to the same energy eigenvalue $E_n(H\Theta|n\rangle = \Theta H|n\rangle = E_n\Theta|n\rangle)$. The question is, Does $|n\rangle$ represent the same state as $\Theta|n\rangle$? If it does, $|n\rangle$ and $\Theta|n\rangle$ can differ at most by a phase factor. Hence,

$$\Theta|n\rangle = e^{i\delta}|n\rangle. \tag{4.4.88}$$

Applying Θ again to (4.4.88), we have $\Theta^2|n\rangle = \Theta e^{i\delta}|n\rangle = e^{-i\delta}\Theta|n\rangle = e^{-i\delta}e^{+i\delta}|n\rangle$; hence,

$$\Theta^2|n\rangle = +|n\rangle. \tag{4.4.89}$$

But this relation is impossible for half-integer j systems, for which Θ^2 is always -1, so we are led to conclude that $|n\rangle$ and $\Theta|n\rangle$, which have the same energy, must correspond to distinct states—that is, there must be a degeneracy. This means, for instance, that for a system composed of an odd number of electrons in an external electric field **E**, each energy level must be at least twofold degenerate no matter how complicated **E** may be. Considerations along this line have interesting applications to electrons in crystals where odd-electron and even-electron systems exhibit very different behaviors. Historically, Kramers inferred degeneracy of this kind by looking at explicit solutions of the Schrödinger equation; subsequently, Wigner pointed out that Kramers degeneracy is a consequence of time-reversal invariance.

Let us now turn to interactions with an external magnetic field. The Hamiltonian H may then contain terms like

$$\mathbf{S}\cdot\mathbf{B}, \qquad \mathbf{p}\cdot\mathbf{A}+\mathbf{A}\cdot\mathbf{p}, \qquad (\mathbf{B}=\nabla\times\mathbf{A}), \tag{4.4.90}$$

where the magnetic field is to be regarded as external. The operators **S** and **p** are odd under time reversal; these interaction terms therefore do lead to

$$\Theta H \neq H\Theta. \tag{4.4.91}$$

As a trivial example, for a spin $\frac{1}{2}$ system the spin-up state $|+\rangle$ and its time-reversed state $|-\rangle$ no longer have the same energy in the presence of an external magnetic field. In general, Kramers degeneracy in a system containing an odd number of electrons can be lifted by applying an external magnetic field.

Notice that when we treat **B** as external, we do not change **B** under time reversal; this is because the atomic electron is viewed as a closed quantum-mechanical system to which we apply the time-reversal operator.

This should not be confused with our earlier remarks concerning the invariance of the Maxwell equations (4.4.2) and the Lorentz force equation under $t \to -t$ and (4.4.3). There we were to apply time reversal to the *whole world*, for example, even to the currents in the wire that produces the **B** field!

PROBLEMS

1. Calculate the *three lowest* energy levels, together with their degeneracies, for the following systems (assume equal mass *distinguishable* particles):
   a. Three noninteracting spin $\frac{1}{2}$ particles in a box of length L.
   b. Four noninteracting spin $\frac{1}{2}$ particles in a box of length L.
2. Let $\mathscr{T}_{\mathbf{d}}$ denote the translation operator (displacement vector $\mathbf{d}$); $\mathscr{D}(\hat{\mathbf{n}}, \phi)$, the rotation operator ($\hat{\mathbf{n}}$ and ϕ are the axis and angle of rotation, respectively); and π the parity operator. Which, if any, of the following pairs commute? Why?
   a. $\mathscr{T}_{\mathbf{d}}$ and $\mathscr{T}_{\mathbf{d}'}$ ($\mathbf{d}$ and $\mathbf{d}'$ in different directions).
   b. $\mathscr{D}(\hat{\mathbf{n}}, \phi)$ and $\mathscr{D}(\hat{\mathbf{n}}', \phi')$ ($\hat{\mathbf{n}}$ and $\hat{\mathbf{n}}'$ in different directions).
   c. $\mathscr{T}_{\mathbf{d}}$ and π.
   d. $\mathscr{D}(\hat{\mathbf{n}}, \phi)$ and π.
3. A quantum-mechanical state Ψ is known to be a simultaneous eigenstate of two Hermitian operators A and B which *anticommute*,

$$AB + BA = 0.$$

   What can you say about the eigenvalues of A and B for state Ψ? Illustrate your point using the parity operator (which can be chosen to satisfy $\pi = \pi^{-1} = \pi^{\dagger}$) and the momentum operator.
4. A spin $\frac{1}{2}$ particle is bound to a fixed center by a spherically symmetrical potential.
   a. Write down the spin angular function $\mathscr{Y}_{l=0}^{j=1/2,\, m=1/2}$.
   b. Express $(\boldsymbol{\sigma}\cdot\mathbf{x})\, \mathscr{Y}_{l=0}^{j=1/2,\, m=1/2}$ in terms of some other $\mathscr{Y}_{l}^{j,\, m}$.
   c. Show that your result in (b) is understandable in view of the transformation properties of the operator $\mathbf{S}\cdot\mathbf{x}$ under rotations and under space inversion (parity).
5. Because of weak (neutral-current) interactions there is a parity-violating potential between the atomic electron and the nucleus as follows:

$$V = \lambda\left[\delta^{(3)}(\mathbf{x})\mathbf{S}\cdot\mathbf{p} + \mathbf{S}\cdot\mathbf{p}\delta^{(3)}(\mathbf{x})\right],$$

   where $\mathbf{S}$ and $\mathbf{p}$ are the spin and momentum operators of the electron, and the nucleus is assumed to be situated at the origin. As a result, the ground state of an alkali atom, usually characterized by $|n, l, j, m\rangle$ actually contains very tiny contributions from other eigenstates as follows:

$$|n, l, j, m\rangle \to |n, l, j, m\rangle + \sum_{n'l'j'm'} C_{n'l'j'm'}|n', l', j', m'\rangle.$$

On the basis of symmetry considerations *alone*, what can you say about (n', l', j', m'), which give rise to nonvanishing contributions? Suppose the radial wave functions and the energy levels are all known. Indicate how you may calculate $C_{n'l'j'm'}$. Do we get further restrictions on (n', l', j', m')?

6. Consider a symmetric rectangular double-well potential:

$$V = \begin{cases} \infty & \text{for } |x| > a + b; \\ 0 & \text{for } a < |x| < a + b; \\ V_0 > 0 & \text{for } |x| < a. \end{cases}$$

Assuming that V_0 is very high compared to the quantized energies of low-lying states, obtain an approximate expression for the energy splitting between the two lowest-lying states.

7. a. Let $\psi(\mathbf{x}, t)$ be the wave function of a spinless particle corresponding to a plane wave in three dimensions. Show that $\psi^*(\mathbf{x}, -t)$ is the wave function for the plane wave with the momentum direction reversed.
   b. Let $\chi(\hat{\mathbf{n}})$ be the two-component eigenspinor of $\boldsymbol{\sigma}\cdot\hat{\mathbf{n}}$ with eigenvalue $+1$. *Using the explicit form of* $\chi(\hat{\mathbf{n}})$ (in terms of the polar and azimuthal angles β and γ that characterize $\hat{\mathbf{n}}$) verify that $-i\sigma_2\chi^*(\hat{\mathbf{n}})$ is the two-component eigenspinor with the spin direction reversed.

8. a. Assuming that the Hamiltonian is invariant under time reversal, prove that the wave function for a spinless nondegenerate system at any given instant of time can always be chosen to be real.
   b. The wave function for a plane-wave state at $t = 0$ is given by a complex function $e^{i\mathbf{p}\cdot\mathbf{x}/\hbar}$. Why does this not violate time-reversal invariance?

9. Let $\phi(\mathbf{p}')$ be the momentum-space wave function for state $|\alpha\rangle$, that is, $\phi(\mathbf{p}') = \langle\mathbf{p}'|\alpha\rangle$. Is the momentum-space wave function for the time-reversed state $\Theta|\alpha\rangle$ given by $\phi(\mathbf{p}')$, $\phi(-\mathbf{p}')$, $\phi^*(\mathbf{p}')$, or $\phi^*(-\mathbf{p}')$? Justify your answer.

10. a. What is the time-reversed state corresponding to $\mathscr{D}(R)|j, m\rangle$?
    b. Using the properties of time reversal and rotations, prove

$$\mathscr{D}^{(j)*}_{m'm}(R) = (-1)^{m-m'}\mathscr{D}^{(j)}_{-m',-m}(R).$$

    c. Prove $\Theta|j, m\rangle = i^{2m}|j, -m\rangle$.

11. Suppose a spinless particle is bound to a fixed center by a potential $V(\mathbf{x})$ so asymmetrical that no energy level is degenerate. Using time-

reversal invariance prove

$$\langle \mathbf{L} \rangle = 0$$

for any energy eigenstate. (This is known as **quenching** of orbital angular momentum.) If the wave function of such a nondegenerate eigenstate is expanded as

$$\sum_l \sum_m F_{lm}(r) Y_l^m(\theta, \phi),$$

what kind of phase restrictions do we obtain on $F_{lm}(r)$?

12. The Hamiltonian for a spin 1 system is given by

$$H = AS_z^2 + B\left(S_x^2 - S_y^2\right).$$

Solve this problem *exactly* to find the normalized energy eigenstates and eigenvalues. (A spin-dependent Hamiltonian of this kind actually appears in crystal physics.) Is this Hamiltonian invariant under time reversal? How do the normalized eigenstates you obtained transform under time reversal?

CHAPTER 5

Approximation Methods

Few problems in quantum mechanics—with either time-independent or time-dependent Hamiltonians—can be solved exactly. Inevitably we are forced to resort to some form of approximation methods. One may argue that with the advent of high-speed computers it is always possible to obtain the desired solution numerically to the requisite degree of accuracy; nevertheless, it remains important to understand the basic physics of the approximate solutions even before we embark on ambitious computer calculations. This chapter is devoted to a fairly systematic discussion of approximate solutions to bound-state problems.

5.1. TIME-INDEPENDENT PERTURBATION THEORY: NONDEGENERATE CASE

Statement of the Problem

The approximation method we consider here is time-independent perturbation theory—sometimes known as the Rayleigh-Schrödinger perturbation theory. We consider a time-independent Hamiltonian H such that it can be split into two parts, namely,

$$H = H_0 + V, \tag{5.1.1}$$

where the $V = 0$ problem is assumed to have been solved in the sense that both the exact energy eigenkets $|n^{(0)}\rangle$ and the exact energy eigenvalues $E_n^{(0)}$ are known:

$$H_0|n^{(0)}\rangle = E_n^{(0)}|n^{(0)}\rangle. \tag{5.1.2}$$

We are required to find approximate eigenkets and eigenvalues for the *full* Hamiltonian problem

$$(H_0 + V)|n\rangle = E_n|n\rangle, \tag{5.1.3}$$

where V is known as the **perturbation**; it is not, in general, the full-potential operator. For example, suppose we consider the hydrogen atom in an external electric or magnetic field. The unperturbed Hamiltonian H_0 is taken to be the kinetic energy $\mathbf{p}^2/2m$ *and* the Coulomb potential due to the presence of the proton nucleus $-e^2/r$. Only that part of the potential due to the interaction with the external $\mathbf{E}$ or $\mathbf{B}$ field is represented by the perturbation V.

Instead of (5.1.3) it is customary to solve

$$(H_0 + \lambda V)|n\rangle = E_n|n\rangle, \tag{5.1.4}$$

where λ is a continuous real parameter. This parameter is introduced to keep track of the number of times the perturbation enters. At the end of the calculation we may set $\lambda \to 1$ to get back to the full-strength case. In other words, we assume that the strength of the perturbation can be controlled. The parameter λ can be visualized to vary continuously from 0 to 1, the $\lambda = 0$ case corresponding to the unperturbed problem and $\lambda = 1$ corresponding to the full-strength problem of (5.1.3). In physical situations where this approximation method is applicable, we expect to see a smooth transition of $|n^0\rangle$ into $|n\rangle$ and $E_n^{(0)}$ into E_n as λ is "dialed" from 0 to 1.

The method rests on the expansion of the energy eigenvalues and energy eigenkets in power of λ. This means that we implicitly assume the analyticity of the energy eigenvalues and eigenkets in a complex λ-plane around $\lambda = 0$. Of course, if our method is to be of practical interest, good approximations can better be obtained by taking only one or two terms in the expansion.

The Two-State Problem

Before we embark on a systematic presentation of the basic method, let us see how the expansion in λ might indeed be valid in the exactly soluble two-state problem we have encountered many times already. Suppose we have a Hamiltonian that can be written as

$$H = E_1^{(0)}|1^{(0)}\rangle\langle 1^{(0)}| + E_2^{(0)}|2^{(0)}\rangle\langle 2^{(0)}| + \lambda V_{12}|1^{(0)}\rangle\langle 2^{(0)}| + \lambda V_{21}|2^{(0)}\rangle\langle 1^{(0)}|, \tag{5.1.5}$$

where $|1^{(0)}\rangle$ and $|2^{(0)}\rangle$ are the energy eigenkets for the $\lambda = 0$ problem, and we consider the case $V_{11} = V_{22} = 0$. In this representation the H may be represented by a square matrix as follows:

$$H = \begin{pmatrix} E_1^{(0)} & \lambda V_{12} \\ \lambda V_{21} & E_2^{(0)} \end{pmatrix}, \tag{5.1.6}$$

where we have used the basis formed by the unperturbed energy eigenkets. The V matrix must, of course, be Hermitian; let us solve the case when V_{12} and V_{21} are real:

$$V_{12} = V_{12}^*, \qquad V_{21} = V_{21}^*; \tag{5.1.7}$$

hence, by Hermiticity

$$V_{12} = V_{21}. \tag{5.1.8}$$

This can always be done by adjusting the phase of $|2^{(0)}\rangle$ relative to that of $|1^{(0)}\rangle$. The problem of obtaining the energy eigenvalues here is completely analogous to that of solving the spin-orientation problem, where the analogue of (5.1.6) is

$$H = a_0 + \boldsymbol{\sigma}\cdot\mathbf{a} = \begin{pmatrix} a_0 + a_3 & a_1 \\ a_1 & a_0 - a_3 \end{pmatrix}, \tag{5.1.9}$$

where we assume $\mathbf{a} = (a_1, 0, a_3)$ is small and a_0, a_1, a_3 are all real. The eigenvalues for this problem are known to be just

$$E = a_0 \pm \sqrt{a_1^2 + a_3^2}\,. \tag{5.1.10}$$

By analogy the corresponding eigenvalues for (5.1.6) are

$$\begin{Bmatrix} E_1 \\ E_2 \end{Bmatrix} = \frac{\left(E_1^{(0)} + E_2^{(0)}\right)}{2} \pm \sqrt{\left[\frac{\left(E_1^{(0)} - E_2^{(0)}\right)^2}{4} + \lambda^2|V_{12}|^2\right]}. \tag{5.1.11}$$

Let us suppose $\lambda|V_{12}|$ is small compared with the relevant energy scale, the difference of the energy eigenvalues of the unperturbed problem:

$$\lambda|V_{12}| \ll |E_1^{(0)} - E_2^{(0)}|. \tag{5.1.12}$$

We can then use

$$\sqrt{1+\varepsilon} = 1 + \frac{1}{2}\varepsilon - \frac{\varepsilon^2}{8} + \cdots \tag{5.1.13}$$

to obtain the expansion of the energy eigenvalues in the presence of perturbation $\lambda|V_{12}|$, namely,

$$E_1 = E_1^{(0)} + \frac{\lambda^2|V_{12}|^2}{\left(E_1^{(0)} - E_2^{(0)}\right)} + \cdots$$

$$E_2 = E_2^{(0)} + \frac{\lambda^2|V_{12}|^2}{\left(E_2^{(0)} - E_1^{(0)}\right)} + \cdots. \tag{5.1.14}$$

These are expressions that we can readily obtain using the general formalism to be developed shortly. It is also possible to write down the energy eigenkets in analogy with the spin-orientation problem.

The reader might be led to believe that a perturbation expansion always exists for a sufficiently weak perturbation. Unfortunately this is not necessarily the case. As an elementary example, consider a one-dimensional problem involving a particle of mass m in a very weak square-well potential of depth V_0 ($V = -V_0$ for $-a < x < a$, $V = 0$ for $|x| > a$). This problem admits one bound state of energy,

$$E = -(2ma^2/\hbar^2)|\lambda V|^2, \quad \lambda > 0 \text{ for attraction.} \tag{5.1.15}$$

We might regard the square well as a very weak perturbation to be added to the free-particle Hamiltonian and interpret result (5.1.15) as the energy shift in the ground state from zero to $|\lambda V|^2$. Specifically, because (5.1.15) is quadratic in V, we might be tempted to associate this as the energy shift of the ground state computed according to second-order perturbation theory. However, this view is false because if this were the case, the system would also admit an $E < 0$ state for a repulsive potential case with λ negative, which would be sheer nonsense.

Let us now examine the radius of convergence of series expansion (5.1.14). If we go back to the exact expression of (5.1.11) and regard it as a function of a *complex* variable λ, we see that as $|\lambda|$ is increased from zero, branch points are encountered at

$$\lambda|V_{12}| = \frac{\pm i\left(E_1^{(0)} - E_2^{(0)}\right)}{2}. \tag{5.1.16}$$

The condition for the convergence of the series expansion for the $\lambda = 1$ full-strength case is

$$|V_{12}| < \frac{|E_1^{(0)} - E_2^{(0)}|}{2}. \tag{5.1.17}$$

If this condition is not met, perturbation expansion (5.1.14) is meaningless.*

*See the discussion on convergence following (5.1.44), under general remarks.

Formal Development of Perturbation Expansion

We now state in more precise terms the basic problem we wish to solve. Suppose we know completely and exactly the energy eigenkets and energy eigenvalues of

$$H_0|n^{(0)}\rangle = E_n^{(0)}|n^{(0)}\rangle. \tag{5.1.18}$$

The set $\{|n^{(0)}\rangle\}$ is complete in the sense that the closure relation $1 = \sum_n |n^{(0)}\rangle\langle n^{(0)}|$ holds. Furthermore, we assume here that the energy spectrum is nondegenerate; in the next section we will relax this assumption. We are interested in obtaining the energy eigenvalues and eigenkets for the problem defined by (5.1.4). To be consistent with (5.1.18) we should write (5.1.4) as

$$(H_0 + \lambda V)|n\rangle_\lambda = E_n^{(\lambda)}|n\rangle_\lambda \tag{5.1.19}$$

to denote the fact that the energy eigenvalues $E_n^{(\lambda)}$ and energy eigenkets $|n\rangle_\lambda$ are functions of the continuous parameter λ; however, we will usually dispense with this correct but more cumbersome notation.

As the continuous parameter λ is increased from zero, we expect the energy eigenvalue E_n for the nth eigenket to depart from its unperturbed value $E_n^{(0)}$, so we define the energy shift for the nth level as follows:

$$\Delta_n \equiv E_n - E_n^{(0)}. \tag{5.1.20}$$

The basic Schrödinger equation to be solved (approximately) is

$$\left(E_n^{(0)} - H_0\right)|n\rangle = (\lambda V - \Delta_n)|n\rangle. \tag{5.1.21}$$

We may be tempted to invert the operator $E_n^{(0)} - H_0$; however, in general, the inverse operator $1/(E_n^{(0)} - H_0)$ is ill defined because it may act on $|n^{(0)}\rangle$. Fortunately in our case $(\lambda V - \Delta_n)|n\rangle$ has no component along $|n^{(0)}\rangle$, as can easily be seen by multiplying both sides of (5.1.21) by $\langle n^{(0)}|$ on the left:

$$\langle n^{(0)}|(\lambda V - \Delta_n)|n\rangle = 0. \tag{5.1.22}$$

Suppose we define the complementary projection operator

$$\phi_n \equiv 1 - |n^{(0)}\rangle\langle n^{(0)}| = \sum_{k \neq n} |k^{(0)}\rangle\langle k^{(0)}|. \tag{5.1.23}$$

The inverse operator $1/(E_n^{(0)} - H_0)$ is well defined when it multiplies ϕ_n on the right. Explicitly,

$$\frac{1}{E_n^{(0)} - H_0}\phi_n = \sum_{k \neq n} \frac{1}{E_n^{(0)} - E_k^{(0)}}|k^{(0)}\rangle\langle k^{(0)}|. \tag{5.1.24}$$

Also from (5.1.22) and (5.1.23), it is evident that

$$(\lambda V - \Delta_n)|n\rangle = \phi_n(\lambda V - \Delta_n)|n\rangle. \tag{5.1.25}$$

We may therefore be tempted to rewrite (5.1.21) as

$$|n\rangle \stackrel{?}{=} \frac{1}{E_n^{(0)} - H_0}\phi_n(\lambda V - \Delta_n)|n\rangle. \tag{5.1.26}$$

However, this cannot be correct because as $\lambda \to 0$, we must have $|n\rangle \to |n^{(0)}\rangle$ and $\Delta_n \to 0$. Nevertheless, even for $\lambda \neq 0$, we can always add to $|n\rangle$ a solution to the homogeneous equation (5.1.18), namely, $c_n|n^{(0)}\rangle$, so a suitable final form is

$$|n\rangle = c_n(\lambda)|n^{(0)}\rangle + \frac{1}{E_n^{(0)} - H_0}\phi_n(\lambda V - \Delta_n)|n\rangle, \tag{5.1.27}$$

where

$$\lim_{\lambda \to 0} c_n(\lambda) = 1. \tag{5.1.28}$$

Note that

$$c_n(\lambda) = \langle n^{(0)}|n\rangle. \tag{5.1.29}$$

For reasons we will see later, it is convenient to depart from the usual normalization convention

$$\langle n|n\rangle = 1. \tag{5.1.30}$$

Rather, we set

$$\langle n^{(0)}|n\rangle = c_n(\lambda) = 1, \tag{5.1.31}$$

even for $\lambda \neq 0$. We can always do this if we are not worried about the overall normalization because the only effect of setting $c_n \neq 1$ is to introduce a common multiplicative factor. Thus, if desired, we can always normalize the ket at the very end of the calculation. It is also customary to write

$$\frac{1}{E_n^{(0)} - H_0}\phi_n \to \frac{\phi_n}{E_n^{(0)} - H_0} \tag{5.1.32}$$

and similarly

$$\frac{1}{E_n^{(0)} - H_0}\phi_n = \phi_n\frac{1}{E_n^{(0)} - H_0} = \phi_n\frac{1}{E_n^{(0)} - H_0}\phi_n, \tag{5.1.33}$$

so we have

$$|n\rangle = |n^{(0)}\rangle + \frac{\phi_n}{E_n^{(0)} - H_0}(\lambda V - \Delta_n)|n\rangle. \tag{5.1.34}$$

We also note from (5.1.22) and (5.1.31) that

$$\Delta_n = \lambda\langle n^{(0)}|V|n\rangle. \tag{5.1.35}$$

Everything depends on the two equations in (5.1.34) and (5.1.35). Our basic strategy is to expand $|n\rangle$ and Δ_n in the powers of λ and then match the appropriate coefficients. This is justified because (5.1.34) and (5.1.35) are identities which hold for all values of λ between 0 and 1. We begin by writing

$$\begin{aligned} |n\rangle &= |n^{(0)}\rangle + \lambda|n^{(1)}\rangle + \lambda^2|n^{(2)}\rangle + \cdots \\ \Delta_n &= \lambda\Delta_n^{(1)} + \lambda^2\Delta_n^{(2)} + \cdots . \end{aligned} \tag{5.1.36}$$

Substituting (5.1.36) into (5.1.35) and equating the coefficient of various powers of λ, we obtain

$$\begin{aligned} 0(\lambda^1):&\quad \Delta_n^{(1)} = \langle n^{(0)}|V|n^{(0)}\rangle \\ 0(\lambda^2):&\quad \Delta_n^{(2)} = \langle n^{(0)}|V|n^{(1)}\rangle \\ \vdots\quad&\quad\quad \vdots \\ 0(\lambda^N):&\quad \Delta_n^{(N)} = \langle n^{(0)}|V|n^{(N-1)}\rangle, \\ \vdots\quad&\quad\quad \vdots \end{aligned} \tag{5.1.37}$$

so to evaluate the energy shift up to order λ^N it is sufficient to know $|n\rangle$ only up to order λ^{N-1}. We now look at (5.1.34); when it is expanded using (5.1.36), we get

$$\begin{aligned} |n^{(0)}\rangle &+ \lambda|n^{(1)}\rangle + \lambda^2|n^{(2)}\rangle + \cdots \\ &= |n^{(0)}\rangle + \frac{\phi_n}{E_n^{(0)} - H_0}\left(\lambda V - \lambda\Delta_n^{(1)} - \lambda^2\Delta_n^{(2)} - \cdots\right) \\ &\quad \times \left(|n^{(0)}\rangle + \lambda|n^{(1)}\rangle + \cdots\right). \end{aligned} \tag{5.1.38}$$

Equating the coefficient of powers of λ, we have

$$0(\lambda):\quad |n^{(1)}\rangle = \frac{\phi_n}{E_n^{(0)} - H_0} V|n^{(0)}\rangle, \tag{5.1.39}$$

where we have used $\phi_n\Delta_n^{(1)}|n^{(0)}\rangle = 0$. Armed with $|n^{(1)}\rangle$, it is now profitable for us to go back to our earlier expression for $\Delta_n^{(2)}$ [see (5.1.37)]:

$$\Delta_n^{(2)} = \langle n^{(0)}|V\frac{\phi_n}{E_n^{(0)} - H_0} V|n^{(0)}\rangle. \tag{5.1.40}$$

Knowing $\Delta_n^{(2)}$, we can work out the λ^2-term in ket equation (5.1.38) also using (5.1.39) as follows:

$$\begin{aligned} 0(\lambda^2):\quad |n^{(2)}\rangle &= \frac{\phi_n}{E_n^{(0)} - H_0} V \frac{\phi_n}{E_n^{(0)} - H_0} V|n^{(0)}\rangle \\ &\quad - \frac{\phi_n}{E_n^{(0)} - H_0}\langle n^{(0)}|V|n^{(0)}\rangle \frac{\phi_n}{E_n^{(0)} - H_0} V|n^{(0)}\rangle. \end{aligned} \tag{5.1.41}$$

Clearly, we can continue in this fashion as long as we wish. Our operator method is very compact; it is not necessary to write down the indices each time. Of course, to do practical calculations we must use at the end the explicit form of ϕ_n as given by (5.1.23).

To see how all this works, we write down the explicit expansion for the energy shift

$$\begin{aligned}\Delta_n &\equiv E_n - E_n^{(0)} \\ &= \lambda V_{nn} + \lambda^2 \sum_{k \neq n} \frac{|V_{nk}|^2}{E_n^{(0)} - E_k^{(0)}} + \cdots, \end{aligned} \tag{5.1.42}$$

where

$$V_{nk} \equiv \langle n^{(0)}|V|k^{(0)}\rangle \neq \langle n|V|k\rangle, \tag{5.1.43}$$

that is, the matrix elements are taken with respect to *unperturbed* kets. Notice that when we apply the expansion to the two-state problem we recover the earlier expression (5.1.14). The expansion for the perturbed ket goes as follows:

$$\begin{aligned}|n\rangle &= |n^{(0)}\rangle + \lambda \sum_{k \neq n} |k^{(0)}\rangle \frac{V_{kn}}{E_n^{(0)} - E_k^{(0)}} \\ &+ \lambda^2 \left(\sum_{k \neq n} \sum_{l \neq n} \frac{|k^{(0)}\rangle V_{kl} V_{ln}}{\left(E_n^{(0)} - E_k^{(0)}\right)\left(E_n^{(0)} - E_l^{(0)}\right)} - \sum_{k \neq n} \frac{|k^{(0)}\rangle V_{nn} V_{kn}}{\left(E_n^{(0)} - E_k^{(0)}\right)^2} \right) \\ &+ \cdots . \end{aligned} \tag{5.1.44}$$

Equation (5.1.44) says that the nth level is no longer proportional to the unperturbed ket $|n^{(0)}\rangle$ but acquires components along other unperturbed energy kets; stated another way, the perturbation V mixes various unperturbed energy eigenkets.

A few general remarks are in order. First, to obtain the first-order energy shift it is sufficient to evaluate the expectation value of V with respect to the unperturbed kets. Second, it is evident from the expression of the second-order energy shift (5.1.42) that two energy levels, say the ith level and the jth level, when connected by V_{ij} tend to repel each other; the lower one, say the ith level, tends to get depressed as a result of mixing with the higher jth level by $|V_{ij}|^2/(E_j^{(0)} - E_i^{(0)})$, while the energy of the jth level goes up by the same amount. This is a special case of the **no-level crossing theorem**, which states that a pair of energy levels connected by perturbation do not cross as the strength of the perturbation is varied.

Suppose there is more than one pair of levels with appreciable matrix elements but the ket $|n\rangle$, whose energy we are concerned with, refers to the ground state; then each term in (5.1.42) for the second-order energy shift is negative. This means that the second-order energy shift is always negative for the ground state; the lowest state tends to get even lower as a result of mixing.

It is clear that perturbation expansions (5.1.42) and (5.1.44) will converge if $|V_{il}/(E_i^{(0)} - E_l^{(0)})|$ is sufficiently "small." A more specific criterion can be given for the case in which H_0 is simply the kinetic-energy operator (then this Rayleigh-Schrödinger perturbation expansion is just the Born series): At an energy $E_0 < 0$, the Born series converges if and only if neither $H_0 + V$ nor $H_0 - V$ has bound states of energy $E \leq E_0$ (R. G. Newton 1982, p. 233).

Wave-function Renormalization

We are in a position to look at the normalization of the perturbed ket. Recalling the normalization convention we use, (5.1.31), we see that the perturbed ket $|n\rangle$ is not normalized in the usual manner. We can renormalize the perturbed ket by defining

$$|n\rangle_N = Z_n^{1/2}|n\rangle, \tag{5.1.45}$$

where Z_n is simply a constant with ${}_N\langle n|n\rangle_N = 1$. Multiplying $\langle n^{(0)}|$ on the left we obtain [because of (5.1.31)]

$$Z_n^{1/2} = \langle n^{(0)}|n\rangle_N. \tag{5.1.46}$$

What is the physical meaning of Z_n? Because $|n\rangle_N$ satisfies the usual normalization requirement (5.1.30), Z_n can be regarded as the probability for the perturbed energy eigenstate to be found in the corresponding unperturbed energy eigenstate. Noting

$${}_N\langle n|n\rangle_N = Z_n\langle n|n\rangle = 1, \tag{5.1.47}$$

we have

$$\begin{aligned} Z_n^{-1} = \langle n|n\rangle &= \left(\langle n^{(0)}| + \lambda\langle n^{(1)}| + \lambda^2\langle n^{(2)}| + \cdots\right) \\ &\quad \times \left(|n^{(0)}\rangle + \lambda|n^{(1)}\rangle + \lambda^2|n^{(2)}\rangle + \cdots\right) \\ &= 1 + \lambda^2\langle n^{(1)}|n^{(1)}\rangle + 0(\lambda^3) \\ &= 1 + \lambda^2 \sum_{k \neq n} \frac{|V_{kn}|^2}{\left(E_n^{(0)} - E_k^{(0)}\right)^2} + 0(\lambda^3), \end{aligned} \tag{5.1.48a}$$

so up to order λ^2, we get for the probability of the perturbed state to be found in the corresponding unperturbed state

$$Z_n \simeq 1 - \lambda^2 \sum_{k \neq n} \frac{|V_{kn}|^2}{\left(E_n^0 - E_k^0\right)^2}. \tag{5.1.48b}$$

The second term in (5.1.48b) is to be understood as the probability for "leakage" to states other than $|n^{(0)}\rangle$. Notice that Z_n is less than 1, as expected on the basis of the probability interpretation for Z.

It is also amusing to note from (5.1.42) that to order λ^2, Z is related to the derivative of E_n with respect to $E_n^{(0)}$ as follows:

$$Z_n = \frac{\partial E_n}{\partial E_n^{(0)}}. \tag{5.1.49}$$

We understand, of course, that in taking the partial derivative of E_n with respect to $E_n^{(0)}$, we must regard the matrix elements of V as fixed quantities. Result (5.1.49) is actually quite general and not restricted to second-order perturbation theory.

Elementary Examples

To illustrate the perturbation method we have developed, let us look at two examples. The first one concerns a simple harmonic oscillator whose unperturbed Hamiltonian is the usual one:

$$H_0 = \frac{p^2}{2m} + \frac{1}{2}m\omega^2x^2. \tag{5.1.50}$$

Suppose the spring constant $k = m\omega^2$ is changed slightly. We may represent the modification by adding an extra potential

$$V = \tfrac{1}{2}\varepsilon m\omega^2 x^2, \tag{5.1.51}$$

where ε is a dimensionless parameter such that $\varepsilon \ll 1$. From a certain point of view this is the silliest problem in the world to which to apply perturbation theory; the exact solution is immediately obtained just by changing ω as follows:

$$\omega \to \sqrt{1+\varepsilon}\,\omega, \tag{5.1.52}$$

yet this is an instructive example because it affords a comparison between the perturbation approximation and the exact approach.

We are concerned here with the new ground-state ket $|0\rangle$ in the presence of V and the ground-state energy shift Δ_0:

$$|0\rangle = |0^{(0)}\rangle + \sum_{k\neq 0} |k^{(0)}\rangle \frac{V_{k0}}{E_0^{(0)} - E_k^{(0)}} + \cdots \tag{5.1.53a}$$

and

$$\Delta_0 = V_{00} + \sum_{k\neq 0} \frac{|V_{k0}|^2}{E_0^{(0)} - E_k^{(0)}} + \cdots. \tag{5.1.53b}$$

The relevant matrix elements are (see Problem 5 in this chapter)

$$\begin{aligned} V_{00} &= \left(\frac{\varepsilon m\omega^2}{2}\right)\langle 0^{(0)}|x^2|0^{(0)}\rangle = \frac{\varepsilon\hbar\omega}{4} \\ V_{20} &= \left(\frac{\varepsilon m\omega^2}{2}\right)\langle 2^{(0)}|x^2|0^{(0)}\rangle = \frac{\varepsilon\hbar\omega}{2\sqrt{2}}. \end{aligned} \tag{5.1.54}$$

All other matrix elements of form V_{k0} vanish. Noting that the nonvanishing energy denominators in (5.1.53a) and (5.1.53b) are $-2\hbar\omega$, we can combine everything to obtain

$$|0\rangle = |0^{(0)}\rangle - \frac{\varepsilon}{4\sqrt{2}}|2^{(0)}\rangle + 0(\varepsilon^2) \tag{5.1.55a}$$

and

$$\Delta_0 = E_0 - E_0^{(0)} = \hbar\omega\left[\frac{\varepsilon}{4} - \frac{\varepsilon^2}{16} + 0(\varepsilon^3)\right]. \tag{5.1.55b}$$

Notice that as a result of perturbation, the ground-state ket, when expanded in terms of original unperturbed energy eigenkets $\{|n^{(0)}\rangle\}$, acquires a component along the second excited state. The absence of a component along the first excited state is not surprising because our *total* H is invariant under parity; hence, an energy eigenstate is expected to be a parity eigenstate.

A comparison with the exact method can easily be made for the energy shift as follows:

$$\frac{\hbar\omega}{2} \to \left(\frac{\hbar\omega}{2}\right)\sqrt{1+\varepsilon} = \left(\frac{\hbar\omega}{2}\right)\left[1 + \frac{\varepsilon}{2} - \frac{\varepsilon^2}{8} + \cdots\right], \tag{5.1.56}$$

in complete agreement with (5.1.55b). As for the perturbed ket, we look at the change in the wave function. In the absence of V the ground-state wave function is

$$\langle x|0^{(0)}\rangle = \frac{1}{\pi^{1/4}}\frac{1}{\sqrt{x_0}}e^{-x^2/2x_0^2}, \tag{5.1.57}$$

where

$$x_0 \equiv \sqrt{\frac{\hbar}{m\omega}}. \tag{5.1.58}$$

Substitution (5.1.52) leads to

$$x_0 \to \frac{x_0}{(1+\varepsilon)^{1/4}}; \tag{5.1.59}$$

hence,

$$\begin{aligned}\langle x|0^{(0)}\rangle &\to \frac{1}{\pi^{1/4}\sqrt{x_0}}(1+\varepsilon)^{1/8}\exp\left[-\left(\frac{x^2}{2x_0^2}\right)(1+\varepsilon)^{1/2}\right]\\ &\simeq \frac{1}{\pi^{1/4}}\frac{1}{\sqrt{x_0}}e^{-x^2/2x_0^2} + \frac{\varepsilon}{\pi^{1/4}\sqrt{x_0}}e^{-x^2/2x_0^2}\left[\frac{1}{8} - \frac{1}{4}\frac{x^2}{x_0^2}\right]\\ &= \langle x|0^{(0)}\rangle - \frac{\varepsilon}{4\sqrt{2}}\langle x|2^{(0)}\rangle,\end{aligned} \tag{5.1.60}$$

where we have used

$$\langle x|2^{(0)}\rangle = \frac{1}{2\sqrt{2}}\langle x|0^{(0)}\rangle H_2\left(\frac{x}{x_0}\right)$$

$$= \frac{1}{2\sqrt{2}}\frac{1}{\pi^{1/4}}\frac{1}{\sqrt{x_0}}e^{-x^2/2x_0^2}\left[-2 + 4\left(\frac{x}{x_0}\right)^2\right], \tag{5.1.61}$$

and $H_2(x/x_0)$ is a Hermite polynomial of order 2.

As another illustration of nondegenerate perturbation theory, we discuss the **quadratic Stark effect**. A one-electron atom—the hydrogen atom or a hydrogenlike atom with one valence electron outside the closed (spherically symmetrical) shell—is subjected to a uniform electric field in the positive z-direction. The Hamiltonian H is split into two parts,

$$H_0 = \frac{\mathbf{p}^2}{2m} + V_0(r) \quad \text{and} \quad V = -e|\mathbf{E}|z \quad (e < 0 \text{ for the electron}). \tag{5.1.62}$$

[*Editor's Note:* Since the perturbation $V \to -\infty$ as $z \to -\infty$, particles bound by H_0 can, of course, escape now, and all formerly bound states acquire a finite lifetime. However, we can still formally use perturbation theory to calculate the shift in the energy. (The imaginary part of this shift, which we shall ignore here, would give us the lifetime of the state or the width of the corresponding resonance.)]

It is assumed that the energy eigenkets and the energy spectrum for the unperturbed problem (H_0 only) are completely known. The electron spin turns out to be irrelevant in this problem, and we assume that with spin degrees of freedom ignored, no energy level is degenerate. This assumption does not hold for $n \neq 1$ levels of the hydrogen atoms, where V_0 is the pure Coulomb potential; we will treat such cases later. The energy shift is given by

$$\Delta_k = -e|\mathbf{E}|z_{kk} + e^2|\mathbf{E}|^2\sum_{j\neq k}\frac{|z_{kj}|^2}{E_k^{(0)} - E_j^{(0)}} + \cdots, \tag{5.1.63}$$

where we have used k rather than n to avoid confusion with the principal quantum number n. With no degeneracy, $|k^{(0)}\rangle$ is espected to be a parity eigenstate; hence,

$$z_{kk} = 0, \tag{5.1.64}$$

as we saw in Section 4.2. Physically speaking, there can be no linear Stark effect, that is, there is no term in the energy shift proportional to $|\mathbf{E}|$ because the atom possesses a vanishing permanent electric dipole, so the energy shift is *quadratic* in $|\mathbf{E}|$ if terms of order $e^3|\mathbf{E}|^3$ or higher are ignored.

Let us now look at z_{kj}, which appears in (5.1.63), where k (or j) is the **collective index** that stands for (n, l, m) and (n', l', m'). First, we recall the selection rule [see (3.10.39)]

$$\langle n', l'm'|z|n, lm\rangle = 0 \quad \text{unless} \begin{cases} l' = l \pm 1 \\ m' = m \end{cases} \tag{5.1.65}$$

that follows from angular momentum (the Wigner-Eckart theorem with $T^{(1)}_{q=0}$) and parity considerations.

There is another way to look at the m-selection rule. In the presence of V, the full spherical symmetry of the Hamiltonian is destroyed by the external electric field that selects the positive z-direction, but V (hence the total H) is still invariant under rotation around the z-axis; in other words, we still have a cylindrical symmetry. Formally this is reflected by the fact that

$$[V, L_z] = 0. \tag{5.1.66}$$

This means that L_z is still a good quantum number even in the presence of V. As a result, the perturbation can be written as a superposition of eigenkets of L_z with the same m—$m = 0$ in our case. This statement is true for all orders, in particular, for the first-order ket. Also, because the second-order energy shift is obtained from the first-order ket [see (5.1.40)], we can understand why only the $m = 0$ terms contribute to the sum.

The polarizability α of an atom is defined in terms of the energy shift of the atomic state as follows:

$$\Delta = -\tfrac{1}{2}\alpha|\mathbf{E}|^2. \tag{5.1.67}$$

Let us consider the special case of the ground state of the hydrogen atom. Even though the spectrum of the hydrogen atom is degenerate for excited states, the ground state (with spin ignored) is nondegenerate, so the formalism of nondegenerate perturbation theory can be applied. The ground state $|0^{(0)}\rangle$ is denoted in the (n, l, m) notation by $(1,0,0)$, so

$$\alpha = -2e^2 \sum_{k \neq 0}^{\infty} \frac{|\langle k^{(0)}|z|1,0,0\rangle|^2}{\left[E_0^{(0)} - E_k^{(0)}\right]}, \tag{5.1.68}$$

where the sum over k includes not only all bound states $|n, l, m\rangle$ (for $n > 1$) but also the positive-energy continuum states of hydrogen.

There are many ways to estimate approximately or evaluate exactly the sum in (5.1.68) with various degrees of sophistication. We present here the simplest of all the approaches. Suppose the denominator in (5.1.68) were constant. Then we could obtain the sum by considering

$$\begin{aligned}\sum_{k \neq 0} |\langle k^{(0)}|z|1,0,0\rangle|^2 &= \sum_{\text{all } k} |\langle k^{(0)}|z|1,0,0\rangle|^2 \\ &= \langle 1,0,0|z^2|1,0,0\rangle,\end{aligned} \tag{5.1.69}$$

where we have used the completeness relation in the last step. But we can easily evaluate $\langle z^2\rangle$ for the ground state as follows:

$$\langle z^2\rangle = \langle x^2\rangle = \langle y^2\rangle = \tfrac{1}{3}\langle r^2\rangle, \tag{5.1.70}$$

and using the explicit form for the wave function we obtain

$$\langle z^2\rangle = a_0^2,$$

where a_0 stands for the Bohr radius. Unfortunately the expression for polarizability α involves the energy denominator that depends on $E_k^{(0)}$, but we know that the inequality

$$-E_0^{(0)}+E_k^{(0)} \geq -E_0^{(0)}+E_1^{(0)}=\frac{e^2}{2a_0}\left[1-\frac{1}{4}\right] \tag{5.1.71}$$

holds for every energy denominator in (5.1.68). As a result, we can obtain an upper limit for the polarizability of the ground state of the hydrogen atom, namely,

$$\alpha < \frac{16a_0^3}{3} \simeq 5.3a_0^3. \tag{5.1.72}$$

It turns out that we can evaluate exactly the sum in (5.1.68) using a method due to A. Dalgarno and J. T. Lewis (Merzbacher 1970, 424, for example), which also agrees with the experimentally measured value. This gives

$$\alpha = \frac{9a_0^3}{2} = 4.5a_0^3. \tag{5.1.73}$$

We obtain the same result (without using perturbation theory) by solving the Schrödinger equation exactly using parabolic coordinates.

5.2. TIME-INDEPENDENT PERTURBATION THEORY: THE DEGENERATE CASE

The perturbation method we developed in the previous section fails when the unperturbed energy eigenkets are degenerate. The method of the previous section assumes that there is a unique and well-defined unperturbed ket of energy $E_n^{(0)}$ which the perturbed ket approaches as $\lambda \to 0$. With degeneracy present, however, any linear combination of unperturbed kets has the same unperturbed energy; in such a case it is not a priori obvious to what linear combination of the unperturbed kets the perturbed ket is reduced in the limit $\lambda \to 0$. Here specifying just the energy eigenvalue is not enough; some other observable is needed to complete the picture. To be more specific, with degeneracy we can take as our base kets simultaneous eigenkets of H_0 and some other observable A, and we can continue labeling the unperturbed energy eigenket by $|k^{(0)}\rangle$, where k now symbolizes a collective index that stands for both the energy eigenvalue *and* the A eigenvalue. When the perturbation operator V does not commute with A, the **zeroth-order** eigenkets for H (including the perturbation) are in fact *not* A eigenkets.

From a more practical point of view, a blind application of formulas like (5.1.42) and (5.1.44) obviously runs into difficulty because

$$\frac{V_{nk}}{E_n^{(0)}-E_k^{(0)}} \tag{5.2.1}$$

becomes singular if V_{nk} is nonvanishing and $E_n^{(0)}$ and $E_k^{(0)}$ are equal. We must modify the method of the previous section to accommodate such a situation.

Whenever there is degeneracy we are free to choose our base set of unperturbed kets. We should, by all means, exploit this freedom. Intuitively we suspect that the catastrophe of vanishing denominators may be avoided by choosing our base kets in such a way that V has no off-diagonal matrix elements (such as $V_{nk} = 0$ in (5.2.1)). In other words, we should use the linear combinations of the degenerate unperturbed kets that diagonalize H in the subspace spanned by the degenerate unperturbed kets. This is indeed the correct procedure to use.

Suppose there is a g-fold degeneracy before the perturbation V is switched on. This means that there are g different eigenkets all with the same unperturbed energy $E_D^{(0)}$. Let us denote these kets by $\{|m^{(0)}\rangle\}$. In general, the perturbation removes the degeneracy in the sense that there will be g perturbed eigenkets all with different energies. Let them form a set $\{|l\rangle\}$. As λ goes to zero $|l\rangle \to |l^{(0)}\rangle$, and various $|l^{(0)}\rangle$ are eigenkets of H_0 all with the same energy $E_m^{(0)}$. However, the set $|l^{(0)}\rangle$ need not coincide with $\{|m^{(0)}\rangle\}$ even though the two sets of unperturbed eigenkets span the same degenerate subspace, which we call D. We can write

$$|l^{(0)}\rangle = \sum_{m \in D} \langle m^{(0)}|l^{(0)}\rangle |m^{(0)}\rangle,$$

where the sum is over the energy eigenkets in the degenerate subspace.

Before expanding in λ, there is a rearrangement of the Schrödinger equation that will make it much easier to carry out the expansion. Let P_0 be a projection operator onto the space defined by $\{|m^{(0)}\rangle\}$. We define $P_1 = 1 - P_0$ to be the projection onto the remaining states. We shall then write the Schrödinger equation for the states $|l\rangle$ as

$$\begin{aligned} 0 &= (E - H_0 - \lambda V)|l\rangle \\ &= (E - E_D^{(0)} - \lambda V)P_0|l\rangle + (E - H_0 - \lambda V)P_1|l\rangle. \end{aligned} \tag{5.2.2}$$

We next separate (5.2.2) into two equations by projecting from the left on (5.2.2) with P_0 and P_1,

$$(E - E_D^{(0)} - \lambda P_0 V)P_0|l\rangle - \lambda P_0 V P_1|l\rangle = 0 \tag{5.2.3}$$

$$-\lambda P_1 V P_0|l\rangle + (E - H_0 - \lambda P_1 V)P_1|l\rangle = 0. \tag{5.2.4}$$

We can solve (5.2.4) in the P_1 subspace because $P_1(E - H_0 - \lambda P_1 V P_1)$ is not singular in this subspace since E is close to $E_D^{(0)}$ and the eigenvalues of $P_1 H_0 P_1$ are all different from $E_D^{(0)}$. Hence we can write

$$P_1|l\rangle = P_1 \frac{\lambda}{E - H_0 - \lambda P_1 V P_1} P_1 V P_0|l\rangle \tag{5.2.5}$$

or written out explicitly to order λ when $|l\rangle$ is expanded as $|l\rangle = |l^{(0)}\rangle + \lambda|l^{(1)}\rangle + \cdots$,

$$P_1|l^{(1)}\rangle = \sum_{k \notin D} \frac{|k^{(0)}\rangle V_{kl}}{E_D^{(0)} - E_k^{(0)}}. \tag{5.2.6}$$

To calculate $P_0|l\rangle$, we substitute (5.2.5) into (5.2.3) to obtain

$$\left(E - E_D^{(0)} - \lambda P_0 V P_0 - \lambda^2 P_0 V P_1 \frac{1}{E - H_0 - \lambda V} P_1 V P_0\right) P_0 |l\rangle = 0. \quad (5.2.7)$$

Although there is a term of order λ^2 in (5.2.7) that results from the substitution, we shall find that it produces a term of order λ in the state $P_0|l\rangle$. To order λ we obtain the equation for the energies to order λ and eigenfunctions to order zero,

$$(E - E_D^{(0)} - \lambda P_0 V P_0)(P_0|l^{(0)}\rangle) = 0. \quad (5.2.8)$$

This is an equation in the g dimensional degenerate subspace and clearly means that the eigenvectors are just the eigenvectors of the $g \times g$ matrix $P_0 V P_0$ and the eigenvalues $E^{(1)}$ are just the roots of the secular equation

$$\det[V - (E - E_D^{(0)})] = 0 \quad (5.2.9)$$

where $V =$ matrix of $P_0 V P_0$ with matrix elements $\langle m^{(0)}|V|m'^{(0)}\rangle$. Explicitly in matrix form we have

$$\begin{pmatrix} V_{11} & V_{12} & \cdots \\ V_{21} & V_{22} & \cdots \\ \vdots & \vdots & \ddots \end{pmatrix} \begin{pmatrix} \langle 1^{(0)}|l^{(0)}\rangle \\ \langle 2^{(0)}|l^{(0)}\rangle \\ \vdots \end{pmatrix} = \Delta_l^{(1)} \begin{pmatrix} \langle 1^{(0)}|l^{(0)}\rangle \\ \langle 2^{(0)}|l^{(0)}\rangle \\ \vdots \end{pmatrix}. \quad (5.2.10)$$

The roots determine the eigenvalues $\Delta_l^{(1)}$—there are g altogether—and by substituting them into (5.2.10), we can solve for $\langle m^{(0)}|l^{(0)}\rangle$ for each l up to an overall normalization constant. Thus by solving the eigenvalue problem, we obtain in one stroke both the first-order energy shifts and the correct zeroth-order eigenkets. Notice that the zeroth-order kets we obtain as $\lambda \to 0$ are just the linear combinations of the various $|m^{(0)}\rangle$'s that diagonalize the perturbation V, the diagonal elements immediately giving the first-order shift

$$\Delta_l^{(1)} = \langle l^{(0)}|V|l^{(0)}\rangle. \quad (5.2.11)$$

Note also that if the degenerate subspace were the whole space, we would have solved the problem exactly in this manner. The presence of unperturbed "distant" eigenkets not belonging to the degenerate subspace will show up only in higher orders—first order and higher for the energy eigenkets and second order and higher for the energy eigenvalues.

Expression (5.2.11) looks just like the first-order energy shift [see (5.1.37)] in the nondegenerate case except that here we have to make sure that the base kets used are such that V does not have nonvanishing off-diagonal matrix elements in the subspace spanned by the degenerate unperturbed eigenkets. If the V operator is already diagonal in the base ket representation we are using, we can immediately write down the first-order shift by taking the expectation value of V, just as in the nondegenerate case.

Let us now look at (5.2.7). To be safe we keep all terms in the $g \times g$ effective Hamiltonian that appears in (5.2.7) to order λ^2 although we want $P_0|l\rangle$ only to order λ. We find

$$\left(E - E_D^{(0)} - \lambda P_0 V P_0 - \lambda^2 P_0 V P_1 \frac{1}{E_D^{(0)} - H_0} P_1 V P_0\right) P_0|l\rangle = 0. \quad (5.2.12)$$

Let us call the eigenvalues of the $g \times g$ matrix $P_0 V P_0 v_i$ and the eigenvectors $P_0|l_i^{(0)}\rangle$. The eigen energies to first order are $E_i^{(1)} = E_D^{(0)} + \lambda v_i$. We assume that the degeneracy is completely resolved so that $E_i^{(1)} - E_j^{(1)} = \lambda(v_i - v_j)$ are all non-zero. We can now apply non-degenerate perturbation theory (5.1.39) to the $g \times g$ dimensional Hamiltonian that appears in (5.2.12). The resulting correction to the eigenvectors $P_0|l_i^{(0)}\rangle$ is

$$P_0|l_i^{(1)}\rangle = \sum_{j \neq i} \lambda \frac{P_0|l_j^{(0)}\rangle}{v_j - v_i} \langle l_j^{(0)}|VP_1 \frac{1}{E_D^{(0)} - H_0} P_1 V|l_i^{(0)}\rangle$$

or more explicitly

$$P_0|l_i^{(1)}\rangle = \sum_{j \neq i} \lambda \frac{P_0|l_j^{(0)}\rangle}{v_j - v_i} \sum_{k \notin D} \langle l_j^{(0)}|V|k\rangle \frac{1}{E_D^{(0)} - E_k^{(0)}} \langle k|V|l_i^{(0)}\rangle. \quad (5.2.14)$$

Thus, although the third term in the effective Hamiltonian that appears in (5.2.12) is of order λ^2, it is divided by energy denominators of order λ in forming the correction to the eigenvector, which then gives terms of order λ in the vector. If we add together (5.2.6) and (5.2.14), we get the eigenvector accurate to order λ.

As in the nondegenerate case, it is convenient to adopt the normalization convention $\langle l^{(0)}|l\rangle = 1$. We then have, from (5.2.3) and (5.2.4), $\lambda\langle l^{(0)}|V|l\rangle = \Delta_l = \lambda\Delta_l^{(1)} + \lambda^2\Delta_l^{(2)} + \cdots$. The λ-term just reproduces (5.2.11). As for the λ^2-term, we obtain $\Delta_l^{(2)} = \langle l^{(0)}|V|l^{(1)}\rangle = \langle l^{(0)}|V|P_1 l^{(1)}\rangle + \langle l^{(0)}|V|P_0 l_i^{(1)}\rangle$. Since the vectors $P_0|l_j^{(0)}\rangle$ are eigenvectors of V, the correction to the vector, (5.2.14), gives no contribution to the second-order energy shift, so we find using (5.2.6)

$$\Delta_l^{(2)} = \sum_{k \in D} \frac{|V_{kl}|^2}{E_D^{(0)} - E_k^{(0)}} \quad (5.2.15)$$

Our procedure works provided that there is no degeneracy in the roots of secular equation (5.2.9). Otherwise we still have an ambiguity as to which linear contribution of the degenerate unperturbed kets the perturbed kets are reduced in the limit $\lambda \rightarrow 0$. Put in another way, if our method is to work, the degeneracy should be removed completely in first order. A challenge for the experts: How must we proceed if the degeneracy is not removed in first order, that is, if some of the roots of the secular equation are equal? (See Problem 12 of this chapter.)

Let us now summarize the basic procedure of degenerate perturbation theory:

1. Identify degenerate unperturbed eigenkets and construct the perturbation matrix V, a $g \times g$ matrix if the degeneracy is g-fold.
2. Diagonalize the perturbation matrix by solving, as usual, the appropriate secular equation.
3. Identify the roots of the secular equation with the first-order energy shifts; the base kets that diagonalize the V matrix are the correct zeroth-order kets to which the perturbed kets approach in the limit $\lambda \rightarrow 0$.
4. For higher orders use the formulas of the corresponding nondegenerate perturbation theory except in the summations, where we exclude all contributions from the unperturbed kets in the degenerate subspace D.

Linear Stark Effect

As an example of degenerate perturbation theory, let us study the effect of a uniform electric field on excited states of the hydrogen atom. As is well known, in the Schrödinger theory with a pure Coulomb potential with no spin dependence, the bound state energy of the hydrogen atom depends only on the principal quantum number n. This leads to degeneracy for all but the ground state because the allowed values of l for a given n satisfy

$$0 \leq l < n. \tag{5.2.16}$$

To be specific, for the $n = 2$ level, there is an $l = 0$ state called $2s$ and three $l = 1$ $(m = \pm 1, 0)$ states called $2p$, all with the same energy, $-e^2/8a_0$. As we apply a uniform electric field in the z-direction, the appropriate perturbation operator is given by

$$V = -ez|\mathbf{E}|, \tag{5.2.17}$$

which we must now diagonalize. Before we evaluate the matrix elements in detail using the usual (nlm) basis, let us note that the perturbation (5.2.17) has nonvanishing matrix elements only between states of opposite parity, that is, between $l = 1$ and $l = 0$ in our case. Furthermore, in order for the matrix element to be nonvanishing, the m-values must be the same because z behaves like a spherical tensor of rank one with spherical component (*magnetic quantum number*) zero. So the only nonvanishing matrix elements are between $2s$ ($m = 0$ necessarily) and $2p$ with $m = 0$. Thus

$$V = \begin{array}{c} \begin{matrix} 2s & 2p \;\; m=0 & 2p \;\; m=1 & 2p \;\; m=-1 \end{matrix} \\ \begin{pmatrix} 0 & \langle 2s|V|2p, m=0\rangle & 0 & 0 \\ \langle 2p, m=0|V|2s\rangle & 0 & 0 & 0 \\ 0 & 0 & 0 & 0 \\ 0 & 0 & 0 & 0 \end{pmatrix} \end{array} \tag{5.2.18}$$

Explicitly,

$$\begin{aligned}\langle 2s|V|2p, m=0\rangle &= \langle 2p, m=0|V|2s\rangle \\ &= 3ea_0|\mathbf{E}|.\end{aligned} \tag{5.2.19}$$

It is sufficient to concentrate our attention on the upper left-hand corner of the square matrix. It then looks very much like the σ_x matrix, and we can immediately write down the answer—for the energy shifts we get

$$\nabla_{\pm}^{(1)} = \pm\ 3ea_0|\mathbf{E}|, \tag{5.2.20}$$

where the subscripts $\pm$ refer to the zeroth-order kets that diagonalize V:

$$|\pm\rangle = \frac{1}{\sqrt{2}}(|2s,m=0\rangle \pm |2p,m=0\rangle). \tag{5.2.21}$$

Schematically the energy levels are as shown in Figure 5.1.

Notice that the shift is *linear* in the applied electric field strength, hence the term the **linear Stark effect**. One way we can visualize the existence of this effect is to note that the energy eigenkets (5.2.21) are not parity eigenstates and are therefore allowed to have nonvanishing electric permanent dipole moments, as we can easily see by explicitly evaluating $\langle z\rangle$. Quite generally, for an energy state that we can write as a superposition of opposite parity states, it is permissible to have a nonvanishing permanent electric dipole moment, which gives rise to the linear Stark effect.

An interesting question can now be asked. If we look at the "real" hydrogen atom, the $2s$ level and $2p$ level are not really degenerate. Due to the spin orbit force, $2p_{3/2}$ is separated from $2p_{1/2}$, as we will show in the next section, and even the degeneracy between the $2s_{1/2}$ and $2p_{1/2}$ levels that persists in the single particle Dirac theory is removed by quantum electrodynamics effects (the *Lamb shift*). We might therefore ask, Is it realistic to apply degenerate perturbation theory to this problem? A comparison with the exact result shows that if the perturbation matrix elements

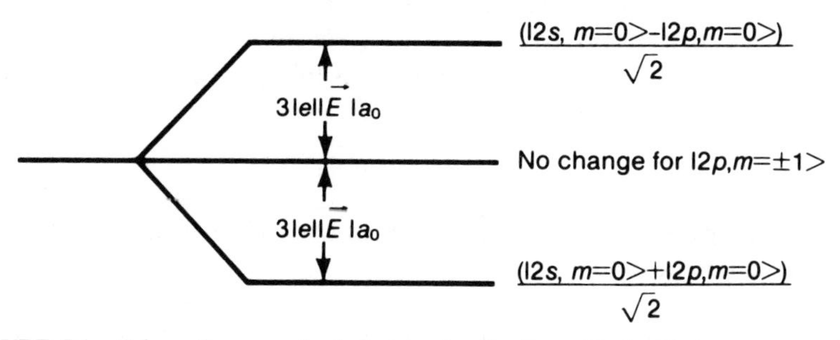

FIGURE 5.1. Schematic energy-level diagram for the linear Stark effect as an example of degenerate perturbation theory.

are much larger when compared to the Lamb shift splitting, then the energy shift is linear in $|\mathbf{E}|$ for all practical purposes and the formalism of degenerate perturbation theory is applicable. On the opposite extreme, if the perturbation matrix elements are small compared to the Lamb shift splitting, then the energy shift is quadratic and we can apply nondegenerate perturbation theory; see Problem 13 of this chapter. This incidentally shows that the formalism of degenerate perturbation theory is still useful when the energy levels are almost degenerate compared to the energy scale defined by the perturbation matrix element. In intermediate cases we must work harder; it is safer to attempt to diagonalize the Hamiltonian exactly in the space spanned by all the nearby levels.

5.3. HYDROGENLIKE ATOMS: FINE STRUCTURE AND THE ZEEMAN EFFECT

Spin-Orbit Interaction and Fine Structure

In this section we study the atomic levels of hydrogenlike atoms, that is, atoms with one valence electron outside the closed shell. Alkali atoms such as sodium (Na) and potassium (K) belong to this category.

The central (spin-independent) potential $V_c(r)$ appropriate for the valence electron is no longer of the pure Coulomb form. This is because the electrostatic potential $\phi(r)$ that appears in

$$V_c(r) = e\phi(r) \tag{5.3.1}$$

is no longer due just to the nucleus of electric charge $|e|Z$; we must take into account the cloud of negatively charged electrons in the inner shells. A precise form of $\phi(r)$ does not concern us here. We simply remark that the degeneracy characteristics of the pure Coulomb potential are now removed in such a way that the higher l states lie higher for a given n. Physically this arises from the fact that the higher l states are more susceptible to the repulsion due to the electron cloud.

Instead of studying the details of $V_c(r)$, which determines the gross structure of hydrogenlike atoms, we discuss the effect of the spin-orbit ($\mathbf{L}\cdot\mathbf{S}$) interaction that gives rise to *fine structure*. We can understand the existence of this interaction in a qualitative fashion as follows. Because of the central force part (5.3.1), the valence electron experiences the electric field

$$\mathbf{E} = -\left(\frac{1}{e}\right)\nabla V_c(r). \tag{5.3.2}$$

But whenever a moving charge is subjected to an electric field, it "feels" an effective magnetic field given by

$$\mathbf{B}_{\text{eff}} = -\left(\frac{\mathbf{v}}{c}\right)\times\mathbf{E}. \tag{5.3.3}$$

Because the electron has a magnetic moment μ given by

$$\mu = \frac{e\mathbf{S}}{m_e c}, \tag{5.3.4}$$

we suspect a spin-orbit potential V_{LS} contribution to H as follows:

$$\begin{aligned} H_{LS} &\stackrel{?}{=} -\mu\cdot\mathbf{B}_{\text{eff}} \\ &= \mu\cdot\left(\frac{\mathbf{v}}{c}\times\mathbf{E}\right) \\ &= \left(\frac{e\mathbf{S}}{m_e c}\right)\cdot\left[\frac{\mathbf{p}}{m_e c}\times\left(\frac{\mathbf{x}}{r}\right)\frac{1}{(-e)}\frac{dV_c}{dr}\right] \\ &= \frac{1}{m_e^2 c^2}\frac{1}{r}\frac{dV_c}{dr}(\mathbf{L}\cdot\mathbf{S}). \end{aligned} \tag{5.3.5}$$

When this expression is compared with the observed spin-orbit interaction, it is seen to have the correct sign, but the magnitude turns out to be too large by a factor of two. There is a classical explanation for this due to spin precession (*Thomas precession* after L. H. Thomas), but we shall not bother with that (Jackson 1975, for example). We simply treat the spin-orbit interaction phenomenologically and take V_{LS} to be one-half of (5.3.5). The correct quantum-mechanical explanation for this discrepancy must await the Dirac (relativistic) theory of the electron (Sakurai 1967, for instance).

We are now in a position to apply perturbation theory to hydrogenic atoms using V_{LS} as the perturbation (V of Sections 5.1 and 5.2). The unperturbed Hamiltonian H_0 is taken to be

$$H_0 = \frac{\mathbf{p}^2}{2m} + V_c(r), \tag{5.3.6}$$

where the central potential V_c is no longer of the pure Coulomb form for alkali atoms. With just H_0 we have freedom in choosing the base kets:

$$\begin{aligned} &\text{Set 1:} \quad \text{The eigenkets of } \mathbf{L}^2, L_z, \mathbf{S}^2, S_z. \\ &\text{Set 2:} \quad \text{The eigenkets of } \mathbf{L}^2, \mathbf{S}^2, \mathbf{J}^2, J_z. \end{aligned} \tag{5.3.7}$$

Without V_{LS} (or H_{LS}) either set is satisfactory in the sense that the base kets are also energy eigenkets. With H_{LS} added it is far superior to use set 2 of (5.3.7) because $\mathbf{L}\cdot\mathbf{S}$ does not commute with L_z and S_z, while it does commute with $\mathbf{J}^2$ and J_z. Remember the cardinal rule: Choose unperturbed kets that diagonalize the perturbation. You have to be either a fool or a masochist to use the L_z, S_z eigenkets [set 1 of (5.3.7)] as the base kets for this problem; if we proceeded to apply blindly the method of degenerate perturbation theory starting with set 1 as our base kets, we would be forced to diagonalize the $V_{LS}(H_{LS})$ matrix written in the L_z, S_z representation. The results of this, after a lot of hard algebra, give us just the $\mathbf{J}^2$, J_z eigenkets as the zeroth-order unperturbed kets to be used!

In degenerate perturbation theory, if the perturbation is already diagonal in the representation we are using, all we need to do for the first-order energy shift is to take the expectation value. The wave function in the two-component form is explicitly written as

$$\psi_{nlm} = R_{nl}(r)\,\mathcal{Y}_l^{j=l\pm 1/2,m} \tag{5.3.8}$$

where $\mathcal{Y}_l^{j=l\pm 1/2,m}$ is the spin-angular function of Section 3.7 [see (3.7.64)]. For the first-order shift, we obtain

$$\Delta_{nlj} = \frac{1}{2m_e^2c^2}\left\langle \frac{1}{r}\frac{dV_c}{dr}\right\rangle_{nl}\frac{\hbar^2}{2}\left\{\begin{matrix} l \\ [-(l+1)] \end{matrix}\right\} \quad \begin{matrix} j=l+\frac{1}{2} \\ j=l-\frac{1}{2} \end{matrix}$$

$$\left\langle \frac{1}{r}\frac{dV_c}{dr}\right\rangle_{nl} \equiv \int_0^\infty R_{nl}\frac{1}{r}\frac{dV_c}{dr}R_{nl}r^2\,dr \tag{5.3.9}$$

where we have used the m-independent identity [see (3.7.65)]

$$\int \mathcal{Y}^\dagger \mathbf{S}\cdot\mathbf{L}\mathcal{Y}\,d\Omega = \frac{1}{2}\left[j(j+1)-l(l+1)-\frac{3}{4}\right]\hbar^2 = \frac{\hbar^2}{2}\left\{\begin{matrix} l \\ -(l+1) \end{matrix}\right\} \quad \begin{matrix} j=l+\frac{1}{2} \\ j=l-\frac{1}{2} \end{matrix} \tag{5.3.10}$$

Equation (5.3.9) is known as **Lande's interval rule**.

To be specific, consider a sodium atom. From standard atomic spectroscopy notation, the ground-state configuration is

$$(1s)^2(2s)^2(2p)^6(3s). \tag{5.3.11}$$

The inner 10 electrons can be visualized to form a spherically symmetrical electron cloud. We are interested in the excitation of the eleventh electron from $3s$ to a possible higher state. The nearest possibility is excitation to $3p$. Because the central potential is no longer of the pure Coulomb form, $3s$ and $3p$ are now split. The fine structure brought about by V_{LS} refers to even a finer split within $3p$, between $3p_{1/2}$ and $3p_{3/2}$, where the subscript refers to j. Experimentally, we observe two closely separated yellow lines—known as the **sodium *D* lines**—one at 5896 Å, the other at 5890 Å; see Figure 5.2. Notice that $3p_{3/2}$ lies higher because the radial integral in (5.3.9) is positive.

To appreciate the order of magnitude of the fine-structure splitting, let us note that for $Z \simeq 1$

$$\left\langle \frac{1}{r}\frac{dV_c}{dr}\right\rangle_{nl} \sim \frac{e^2}{a_0^3} \tag{5.3.12}$$

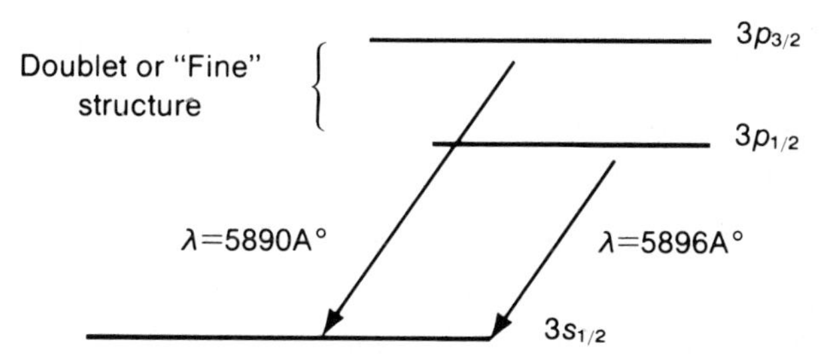

FIGURE 5.2. Schematic diagram of $3s$ and $3p$ lines. The $3s$ and $3p$ degeneracy is lifted because $V_c(r)$ is now the screened Coulomb potential due to core electrons rather than pure Coulombic; V_{LS} then removes the $3p_{1/2}$ and $3p_{3/2}$ degeneracy.

just on the basis of dimensional considerations. So the fine-structure splitting is of order $(e^2/a_0^3)(\hbar/m_e c)^2$, which is to be compared with Balmer splittings of order e^2/a_0. It is useful to recall here that the classical radius of the electron, the Compton wavelength of the electron, and the Bohr radius are related in the following way:

$$\frac{e^2}{m_e c^2} : \frac{\hbar}{m_e c} : a_0 :: 1 : 137 : (137)^2, \tag{5.3.13}$$

where we have used

$$\frac{e^2}{\hbar c} = \frac{1}{137}. \tag{5.3.14}$$

Typically, fine-structure splittings are then related to typical Balmer splittings via

$$\left(\frac{e^2}{a_0^3}\frac{\hbar^2}{m_e^2 c^2}\right) : \left(\frac{e^2}{a_0}\right) :: \left(\frac{1}{137}\right)^2 : 1, \tag{5.3.15}$$

which explains the origin of the term *fine structure*. There are other effects of similar orders of magnitude. Specifically, the relativistic mass correction arising from the expansion

$$\sqrt{m_e^2 c^4 + p^2 c^2} - m_e c^2 \simeq \frac{\mathbf{p}^2}{2m_e} - \frac{\mathbf{p}^4}{8m_e^3 c^2} \tag{5.3.16}$$

is of the same order.

The Zeeman Effect

We now discuss hydrogen or hydrogenlike (one-electron) atoms in a uniform magnetic field—the **Zeeman effect**, sometimes called the *anomalous Zeeman effect* with the electron spin taken into account. Recall that a uniform magnetic field B is derivable from a vector potential

$$\mathbf{A} = \tfrac{1}{2}(\mathbf{B}\times\mathbf{r}). \tag{5.3.17}$$

For $\mathbf{B}$ in the positive z-direction ($\mathbf{B} = B\hat{\mathbf{z}}$),

$$\mathbf{A} = -\tfrac{1}{2}(By\hat{\mathbf{x}} - Bx\hat{\mathbf{y}}) \tag{5.3.18}$$

suffices where B stands for $|\mathbf{B}|$. Apart from the spin term, the interaction Hamiltonian is generated by the substitution

$$\mathbf{p} \to \mathbf{p} - \frac{e\mathbf{A}}{c}. \tag{5.3.19}$$

We therefore have

$$H = \frac{\mathbf{p}^2}{2m_e} + V_c(r) - \frac{e}{2m_e c}(\mathbf{p}\cdot\mathbf{A} + \mathbf{A}\cdot\mathbf{p}) + \frac{e^2\mathbf{A}^2}{2m_e c^2}. \tag{5.3.20}$$

Because

$$\begin{aligned}\langle\mathbf{x}'|\mathbf{p}\cdot\mathbf{A}(\mathbf{x})|\rangle &= -i\hbar\nabla'\cdot[\mathbf{A}(\mathbf{x}')\langle\mathbf{x}'|\rangle] \\ &= \langle\mathbf{x}'|\mathbf{A}(\vec{\mathbf{x}})\cdot\mathbf{p}|\rangle + \langle\mathbf{x}'|\rangle[-i\hbar\nabla'\cdot\mathbf{A}(\mathbf{x}')],\end{aligned} \tag{5.3.21}$$

it is legitimate to replace $\mathbf{p}\cdot\mathbf{A}$ by $\mathbf{A}\cdot\mathbf{p}$ whenever

$$\nabla\cdot\mathbf{A}(\mathbf{x}) = 0, \tag{5.3.22}$$

which is the case for the vector potential of (5.3.18). Noting

$$\begin{aligned}\mathbf{A}\cdot\mathbf{p} &= |\mathbf{B}|\left(-\tfrac{1}{2}yp_x + \tfrac{1}{2}xp_y\right) \\ &= \tfrac{1}{2}|\mathbf{B}|L_z\end{aligned} \tag{5.3.23}$$

and

$$\mathbf{A}^2 = \tfrac{1}{4}|\mathbf{B}|^2(x^2 + y^2), \tag{5.3.24}$$

we obtain for (5.3.20)

$$H = \frac{\mathbf{p}^2}{2m_e} + V_c(r) - \frac{e}{2m_e c}|\mathbf{B}|L_z + \frac{e^2}{8m_e c^2}|\mathbf{B}|^2(x^2 + y^2). \tag{5.3.25}$$

To this we may add the spin magnetic-moment interaction

$$-\boldsymbol{\mu}\cdot\mathbf{B} = \frac{-e}{m_e c}\mathbf{S}\cdot\mathbf{B} = \frac{-e}{m_e c}|\mathbf{B}|S_z. \tag{5.3.26}$$

The quadratic $|\mathbf{B}|^2(x^2 + y^2)$ is unimportant for a one-electron atom; the analogous term is important for the ground state of the helium atom where $L_z^{(\text{tot})}$ and $S_z^{(\text{tot})}$ both vanish. The reader may come back to this problem when he or she computes diamagnetic susceptibilities in Problems 18 and 19 of this chapter.

To summarize, omitting the quadratic term, the total Hamiltonian is made up of the following three terms:

$$H_0 = \frac{\mathbf{p}^2}{2m_e} + V_c(r) \tag{5.3.27a}$$

$$H_{LS} = \frac{1}{2m_e^2c^2}\frac{1}{r}\frac{dV_c(r)}{dr}\mathbf{L}\cdot\mathbf{S} \tag{5.3.27b}$$

$$H_B = \frac{-e|\mathbf{B}|}{2m_ec}(L_z + 2S_z). \tag{5.3.27c}$$

Notice the factor 2 in front of S_z; this reflects the fact that the g-factor of the electron is 2.

Suppose H_B is treated as a small perturbation. We can study the effect of H_B using the eigenkets of $H_0 + H_{LS}$—the $\mathbf{J}^2$, J_z eigenkets—as our base kets. Noting

$$L_z + 2S_z = J_z + S_z, \tag{5.3.28}$$

the first-order shift can be written as

$$\frac{-e|\mathbf{B}|}{2m_ec}\langle J_z + S_z\rangle_{j=l\pm 1/2,\, m}. \tag{5.3.29}$$

The expectation value of J_z immediately gives $m\hbar$. As for $\langle S_z\rangle$, we first recall

$$\left|j = l \pm \frac{1}{2}, m\right\rangle = \pm\sqrt{\frac{l \pm m + \frac{1}{2}}{2l+1}}$$

$$\times\left|m_l = m - \frac{1}{2},\ m_s = \frac{1}{2}\right\rangle + \sqrt{\frac{l \mp m + \frac{1}{2}}{2l+1}}\left|m_l = m + \frac{1}{2}, m_s = -\frac{1}{2}\right\rangle. \tag{5.3.30}$$

The expectation value of S_z can then easily be computed:

$$\begin{aligned}\langle S_z\rangle_{j=l\pm 1/2,\, m} &= \frac{\hbar}{2}\left(|c_+|^2 - |c_-|^2\right)\\ &= \frac{\hbar}{2}\frac{1}{(2l+1)}\left[\left(l \pm m + \frac{1}{2}\right) - \left(l \mp m + \frac{1}{2}\right)\right] = \pm\frac{m\hbar}{(2l+1)}.\end{aligned} \tag{5.3.31}$$

In this manner we obtain Lande's formula for the energy shift (due to $\mathbf{B}$ field),

$$\Delta E_B = \frac{-e\hbar B}{2m_ec}m\left[1 \pm \frac{1}{(2l+1)}\right]. \tag{5.3.32}$$

We see that the energy shift of (5.3.32) is proportional to m. To understand the physical origin for this, we present another method for deriving (5.3.31). We recall that the expectation value of S_z can also be obtained using the projection theorem of Section 3.10. We get [see (3.10.45)]

$$\begin{aligned}
\langle S_z \rangle_{j=l\pm 1/2,m} &= \left[\langle \mathbf{S}\cdot\mathbf{J} \rangle_{j=l\pm 1/2} \right] \frac{m\hbar}{\hbar^2 j(j+1)} \\
&= \frac{m\langle \mathbf{J}^2 + \mathbf{S}^2 - \mathbf{L}^2 \rangle_{j=l\pm 1/2}}{2\hbar j(j+1)} \\
&= m\hbar \left[\frac{(l\pm\frac{1}{2})(l\pm\frac{1}{2}+1)+\frac{3}{4}-l(l+1)}{2(l\pm\frac{1}{2})(l\pm\frac{1}{2}+1)} \right] \\
&= \pm\frac{m\hbar}{(2l+1)},
\end{aligned} \tag{5.3.33}$$

which is in complete agreement with (5.3.31).

In the foregoing discussion the magnetic field is treated as a small perturbation. We now consider the opposite extreme—**the Paschen-Back limit**—with a magnetic field so intense that the effect of H_B is far more important than that of H_{LS}, which we later add as a small perturbation. With $H_0 + H_B$ only, the good quantum numbers are L_z and S_z. Even $\mathbf{J}^2$ is no good because spherical symmetry is completely destroyed by the strong **B** field that selects a particular direction in space, the z-direction.We are left with cylindrical symmetry only—that is, invariance under rotation around the z-axis. So the L_z, S_z eigenkets $|l, s=\frac{1}{2}, m_l, m_s\rangle$ are to be used as our base kets. The effect of the main term H_B can easily be computed:

$$\langle H_B \rangle_{m_l m_s} = \frac{-e|\mathbf{B}|\hbar}{2m_e c}(m_l + 2m_s). \tag{5.3.34}$$

The $2(2l+1)$ degeneracy in m_l and m_s we originally had with H_0 [see (5.3.27a)] is now reduced by H_B to states with the same $(m_l)+(2m_s)$, namely, $(m_l)+(1)$ and $(m_l+2)+(-1)$. Clearly we must evaluate the expectation value of $\mathbf{L}\cdot\mathbf{S}$ with respect to $|m_l, m_s\rangle$:

$$\begin{aligned}
\langle \mathbf{L}\cdot\mathbf{S} \rangle &= \langle L_z S_z + \tfrac{1}{2}(L_+ S_- + L_- S_+) \rangle_{m_l m_s} \\
&= \hbar^2 m_l m_s,
\end{aligned} \tag{5.3.35}$$

where we have used

$$\langle L_\pm \rangle_{m_l} = 0, \qquad \langle S_\pm \rangle_{m_s} = 0. \tag{5.3.36}$$

Hence,

$$\langle H_{LS} \rangle_{m_l m_s} = \frac{\hbar^2 m_l m_s}{2m_e^2 c^2} \left\langle \frac{1}{r}\frac{dV_c}{dr} \right\rangle. \tag{5.3.37}$$

TABLE 5-1

	Dominant interaction	Almost good	No good	Always good
Weak **B**	H_{LS}	$\mathbf{J}^2$ (or $\mathbf{L}\cdot\mathbf{S}$)	L_z, S_z*	$\mathbf{L}^2, \mathbf{S}^2, J_z$
Strong **B**	H_B	L_z, S_z	$\mathbf{J}^2$ (or $\mathbf{L}\cdot\mathbf{S}$)	

*The exception is the stretched configuration, for example, $p_{3/2}$ with $m = \pm\frac{3}{2}$. Here L_z and S_z are both good; this is because magnetic quantum number J_z, $m = m_l + m_s$ can be satisfied in only one way.

In many elementary books there are pictorial interpretations of the weak-field result (5.3.32) and the strong-field result (5.3.34), but we do not bother with them here. We simply summarize our results in Table 5-1, where weak and strong **B** fields are "calibrated" by comparing their magnitudes $e\hbar B/2m_e c$ with $(1/137)^2 e^2/a_0$. In this table *almost good* simply means good to the extent that the less dominant interaction could be ignored.

Specifically, let us look at the level scheme of a p electron $l=1$ ($p_{3/2}, p_{1/2}$). In the weak **B** case the energy shifts are linear in **B**, with slopes determined by

$$m\left[1 \pm \left(\frac{1}{2l+1}\right)\right].$$

As we now increase **B**, mixing becomes possible between states with the same m-value—for example, $p_{3/2}$ with $m = \pm\frac{1}{2}$ and $p_{1/2}$ with $m = \pm\frac{1}{2}$; in this connection note the operator $L_z + 2S_z$ that appears in H_B [(5.3.27c)] is a rank 1 tensor operator $T_{q=0}^{(k=1)}$ with spherical component $q = 0$. In the intermediate **B** region simple formulas like (5.3.32) and (5.3.34) for the expectation values are not possible; it is really necessary to diagonalize the appropriate 2×2 matrix (Gottfried 1966, 371–73). In the strong **B** limit the energy shifts are again proportional to $|\mathbf{B}|$; as we see in (5.3.34), the slopes are determined by $m_l + 2m_s$.

Van der Waals' Interaction

An important, nice application of the Rayleigh-Schrödinger perturbation theory is to calculate the long-range interaction, or **van der Waals' force**, between two hydrogen atoms in their ground states. It is easy to show that the energy between the two atoms for large separation r is attractive and varies as r^{-6}.

Consider the two protons of the hydrogen atoms to be *fixed* at a distance r (along the z-axis) with $\mathbf{r}_1$ the vector from the first proton to its electron and $\mathbf{r}_2$ the vector from the second proton to its electron; see Figure

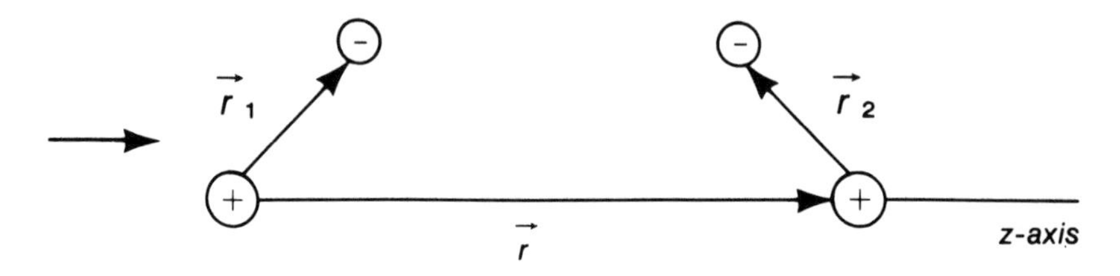

FIGURE 5.3. Two hydrogen atoms with their protons (+) separated by a fixed distance r and their electrons (−) at displacements $\mathbf{r}_i$ from them.

5.3. Then the Hamiltonian H can be written as

$$H = H_0 + V$$

$$H_0 = -\frac{\hbar^2}{2m}\left(\nabla_1^2 + \nabla_2^2\right) - \frac{e^2}{r_1} - \frac{e^2}{r_2} \tag{5.3.38}$$

$$V = \frac{e^2}{r} + \frac{e^2}{|\mathbf{r}+\mathbf{r}_2-\mathbf{r}_1|} - \frac{e^2}{|\mathbf{r}+\mathbf{r}_2|} - \frac{e^2}{|\mathbf{r}-\mathbf{r}_1|}$$

The lowest-energy solution of H_0 is simply the product of the ground-state wave functions of the noninteracting hydrogen atoms

$$U_0^{(0)} = U_{100}^{(0)}(\mathbf{r}_1)U_{100}^{(0)}(\mathbf{r}_2) \tag{5.3.39}$$

Now for large r ($\gg$ the Bohr radius a_0) expand the perturbation V in powers of $\mathbf{r}_i/\mathbf{r}$ to obtain

$$V = \frac{e^2}{r^3}(x_1x_2 + y_1y_2 - 2z_1z_2) + 0\left(\frac{1}{r^4}\right) + \cdots \tag{5.3.40}$$

The lowest-order r^{-3}-term in (5.3.40) corresponds to the interaction of two electric dipoles $e\mathbf{r}_1$ and $e\mathbf{r}_2$ separated by $\mathbf{r}$. The higher-order terms represent higher-order multipole interactions, and thus every term in V involves spherical harmonics Y_l^m with $l_i > 0$ for each hydrogen atom. Hence, for each term in (5.3.40) the first-order perturbation energy matrix element $V_{00} \simeq 0$, since the ground state $U_0^{(0)}$ wave function (5.3.39) has $l_i = 0$ (and $\int d\Omega\, Y_l^m(\Omega) = 0$ for l and $m \neq 0$). The second-order perturbation

$$E^{(2)}(r) = \frac{e^4}{r^6}\sum_{k\neq 0}\frac{|\langle k^{(0)}|x_1x_2 + y_1y_2 - 2z_1z_2|0^{(0)}\rangle|^2}{E_0^{(0)} - E_k^{(0)}} \tag{5.3.41}$$

will be nonvanishing. We immediately see that this interaction varies as $1/r^6$; since $E_k^{(0)} > E_0^{(0)}$, it is negative. This $1/r^6$ long-range attractive van der Waals' potential is a general property of the interaction between two atoms in their ground state.*

*See the treatment in Schiff (1968), pages 261–263, which gives a lower and upper bound on the magnitude of the van der Waals' potential from (5.3.41) and from a variational calculation. Also note the first footnote on page 263 of Schiff concerning retardation effects.

5.4. VARIATIONAL METHODS

The perturbation theory developed in the previous section is, of course, of no help unless we already know exact solutions to a problem whose Hamiltonian is sufficiently similar. The variational method we now discuss is very useful for estimating the ground state energy E_0 when such exact solutions are not available.

We attempt to guess the ground-state energy E_0 by considering a "trial ket" $|\tilde{0}\rangle$, which tries to imitate the true ground-state ket $|0\rangle$. To this end we first obtain a theorem of great practical importance. We define $\overline{H}$ such that

$$\overline{H} \equiv \frac{\langle\tilde{0}|H|\tilde{0}\rangle}{\langle\tilde{0}|\tilde{0}\rangle}, \tag{5.4.1}$$

where we have accommodated the possibility that $|\tilde{0}\rangle$ might not be normalized. We can then prove the following.

Theorem.

$$\overline{H} \geq E_0. \tag{5.4.2}$$

This means that we can obtain an *upper bound* to E_0 by considering various kinds of $|\tilde{0}\rangle$. The proof of this is very straightforward.

Proof. Even though we do not know the energy eigenket of the Hamiltonian H, we can imagine that $|\tilde{0}\rangle$ can be expanded as

$$|\tilde{0}\rangle = \sum_{k=0}^{\infty} |k\rangle\langle k|\tilde{0}\rangle \tag{5.4.3}$$

where $|k\rangle$ is an *exact* energy eigenket of H:

$$H|k\rangle = E_k|k\rangle. \tag{5.4.4}$$

The theorem (5.4.2) follows when we use $E_k = E_k - E_0 + E_0$ to evaluate $\overline{H}$ in (5.4.1). We have

$$\overline{H} = \frac{\sum_{k=0} |\langle k|\tilde{0}\rangle|^2 E_k}{\sum_{k=0} |\langle k|\tilde{0}\rangle|^2} \tag{5.4.5a}$$

$$= \frac{\sum_{k=1}^{\infty} |\langle k|\tilde{0}\rangle|^2 (E_k - E_0)}{\sum_{k=0} |\langle k|\tilde{0}\rangle|^2} + E_0 \tag{5.4.5b}$$

$$\geq E_0, \tag{5.4.5c}$$

where we have used the fact that $E_k - E_0$ in the first sum of (5.4.5b) is necessarily positive. It is also obvious from this proof that the equality sign in (5.4.2) holds only if $|\tilde{0}\rangle$ coincides exactly with $|0\rangle$, that is, if the coefficients $\langle k|\tilde{0}\rangle$ all vanish for $k \neq 0$. □

The theorem (5.4.2) is quite powerful because $\overline{H}$ provides an upper bound to the true ground-state energy. Furthermore, a relatively poor trial ket can give a fairly good energy estimate for the ground state because if

$$\langle k|\tilde{0}\rangle \sim 0(\varepsilon) \quad \text{for } k \neq 0, \tag{5.4.6}$$

then from (5.4.5) we have

$$\overline{H} - E_0 \sim 0(\varepsilon^2). \tag{5.4.7}$$

We see an example of this in a moment. Of course, the method does not say anything about the discrepancy between $\overline{H}$ and E_0; all we know is that $\overline{H}$ is larger than (or equal to) E_0.

Another way to state the theorem is to assert that $\overline{H}$ is stationary with respect to the variation

$$|\tilde{0}\rangle \to |\tilde{0}\rangle + \delta|\tilde{0}\rangle; \tag{5.4.8}$$

that is, $\delta\overline{H} = 0$ when $|\tilde{0}\rangle$ coincides with $|0\rangle$. By this we mean that if $|0\rangle + \delta|\tilde{0}\rangle$ is used in place of $|\tilde{0}\rangle$ in (5.4.5) and we calculate $\overline{H}$, then the error we commit in estimating the true ground-state energy involves $|\tilde{0}\rangle$ to order $(\delta|\tilde{0}\rangle)^2$.

The variational method per se does not tell us what kind of trial kets are to be used to estimate the ground-state energy. Quite often we must appeal to physical intuition—for example, the asymptotic behavior of wave function at large distances. What we do in practice is to characterize trial kets by one or more parameters $\lambda_1, \lambda_2, \ldots$, and compute $\overline{H}$ as a function of $\lambda_1, \lambda_2, \ldots$. We then minimize $\overline{H}$ by (1) setting the derivative with respect to the parameters all zero, namely,

$$\frac{\partial \overline{H}}{\partial \lambda_1} = 0, \qquad \frac{\partial \overline{H}}{\partial \lambda_2} = 0, \ldots, \tag{5.4.9}$$

(2) determining the optimum values of $\lambda_1, \lambda_2, \ldots$, and (3) substituting them back to the expression for $\overline{H}$.

If the wave function for the trial ket already has a functional form of the exact ground-state energy eigenfunction, we of course obtain the true ground-state energy function by this method. For example, suppose somebody has the foresight to guess that the wave function for the ground state of the hydrogen atom must be of the form

$$\langle \mathbf{x}|0\rangle \propto e^{-r/a}, \tag{5.4.10}$$

where a is regarded as a parameter to be varied. We then find, upon minimizing $\overline{H}$ with (5.4.10), the correct ground-state energy $-e^2/2a_0$. Not surprisingly, the minimum is achieved when a coincides with the Bohr radius a_0.

As a second example, we attempt to estimate the ground state of the infinite-well (one-dimensional box) problem defined by

$$V = \begin{cases} 0, & \text{for } |x| < a \\ \infty, & \text{for } |x| > a. \end{cases} \tag{5.4.11}$$

The exact solutions are, of course, well known:

$$\begin{aligned} \langle x|0\rangle &= \frac{1}{\sqrt{a}} \cos\left(\frac{\pi x}{2a}\right), \\ E_0 &= \left(\frac{\hbar^2}{2m}\right)\left(\frac{\pi^2}{4a^2}\right). \end{aligned} \tag{5.4.12}$$

But suppose we did not know these. Evidently the wave function must vanish at $x = \pm a$; furthermore, for the ground state the wave function cannot have any wiggles. The simplest analytic function that satisfies both requirements is just a parabola going through $x = \pm a$:

$$\langle x|\tilde{0}\rangle = a^2 - x^2, \tag{5.4.13}$$

where we have not bothered to normalize $|\tilde{0}\rangle$. Here there is no variational parameter. We can compute $\overline{H}$ as follows:

$$\begin{aligned} \overline{H} &= \frac{\left(\frac{-\hbar^2}{2m}\right) \int_{-a}^{a} (a^2 - x^2) \frac{d^2}{dx^2} (a^2 - x^2) dx}{\int_{-a}^{a} (a^2 - x^2)^2 \, dx} \\ &= \left(\frac{10}{\pi^2}\right) \left(\frac{\pi^2 \hbar^2}{8a^2 m}\right) \simeq 1.0132 \, E_0. \end{aligned} \tag{5.4.14}$$

It is remarkable that with such a simple trial function we can come within 1.3% of the true ground-state energy.

A much better result can be obtained if we use a more sophisticated trial function. We try

$$\langle x|\tilde{0}\rangle = |a|^\lambda - |x|^\lambda, \tag{5.4.15}$$

where λ is now regarded as a variational parameter. Straightforward algebra gives

$$\overline{H} = \left[\frac{(\lambda + 1)(2\lambda + 1)}{(2\lambda - 1)}\right] \left(\frac{\hbar^2}{4ma^2}\right), \tag{5.4.16}$$

which has a minimum at

$$\lambda = \frac{(1 + \sqrt{6})}{2} \simeq 1.72, \tag{5.4.17}$$

not far from $\lambda = 2$ (a parabola) considered earlier. This gives

$$\overline{H}_{\min} = \left(\frac{5+2\sqrt{6}}{\pi^2} \right) E_0 \simeq 1.00298 E_0. \tag{5.4.18}$$

So the variational method with (5.4.15) gives the correct ground-state energy within 0.3%—a fantastic result considering the simplicity of the trial function used.

How well does this trial function imitate the true ground-state wave function? It is amusing that we can answer this question without explicitly evaluating the overlap integral $\langle 0|\tilde{0}\rangle$. Assuming that $|\tilde{0}\rangle$ is normalized, we have [from (5.4.1)–(5.4.4)]

$$\begin{aligned} \overline{H}_{\min} &= \sum_{k=0}^{\infty} |\langle k|\tilde{0}\rangle|^2 E_k \\ &\geq |\langle 0|\tilde{0}\rangle|^2 E_0 + 9E_0\left(1 - |\langle 0|\tilde{0}\rangle|^2\right) \end{aligned} \tag{5.4.19}$$

where $9E_0$ is the energy of the second excited state; the first excited state $(k=1)$ gives no contribution by parity conservation. Solving for $|\langle 0|\tilde{0}\rangle|$ and using (5.4.18), we have

$$|\langle 0|\tilde{0}\rangle|^2 \geq \frac{9E_0 - \overline{H}_{\min}}{8E_0} = 0.99963. \tag{5.4.20}$$

Departure from unity characterizes a component of $|\tilde{0}\rangle$ in a direction orthogonal to $|0\rangle$. If we are talking about "angle" θ defined by

$$\langle 0|\tilde{0}\rangle = \cos\theta, \tag{5.4.21}$$

then (5.4.20) corresponds to

$$\theta \lesssim 1.1°, \tag{5.4.22}$$

so $|0\rangle$ and $|\tilde{0}\rangle$ are nearly "parallel."

One of the earliest applications of the variational method involved the ground-state energy of the helium atom, which we will discuss in Section 6.1. We can also use the variational method to estimate the energies of first excited states; all we need to do is work with a trial ket orthogonal to the ground-state wave function—either exact, if known, or an approximate one obtained by the variational method.

5.5. TIME-DEPENDENT POTENTIALS: THE INTERACTION PICTURE

Statement of the Problem

So far in this book we have been concerned with Hamiltonians that do not contain time explicitly. In nature, however, there are many quantum-mechanical systems of importance with time dependence. In the remaining part of this chapter we show how to deal with situations with time-dependent potentials.

We consider a Hamiltonian H such that it can be split into two parts,

$$H = H_0 + V(t), \tag{5.5.1}$$

where H_0 does not contain time explicitly. The problem $V(t) = 0$ is assumed to be solved in the sense that the energy eigenkets $|n\rangle$ and the energy eigenvalues E_n defined by

$$H_0|n\rangle = E_n|n\rangle \tag{5.5.2}$$

are completely known.* We may be interested in situations where initially only one of the energy eigenstates of H_0—for example, $|i\rangle$—is populated. As time goes on, however, states other than $|i\rangle$ are populated because with $V(t) \neq 0$ we are no longer dealing with "stationary" problems; the time-evolution operator is no longer as simple as $e^{-iHt/\hbar}$ when H itself involves time. Quite generally the time-dependent potential $V(t)$ can cause transitions to states other than $|i\rangle$. The basic question we address is, What is the probability as a function of time for the system to be found in $|n\rangle$, with $n \neq i$?

More generally, we may be interested in how an arbitrary state ket changes as time goes on, where the total Hamiltonian is the sum of H_0 and $V(t)$. Suppose at $t = 0$, the state ket of a physical system is given by

$$|\alpha\rangle = \sum_n c_n(0)|n\rangle. \tag{5.5.3}$$

We wish to find $c_n(t)$ for $t > 0$ such that

$$|\alpha, t_0 = 0; t\rangle = \sum_n c_n(t) e^{-iE_n t/\hbar}|n\rangle \tag{5.5.4}$$

where the ket on the left side stands for the state ket in the Schrödinger picture at t of a physical system whose state ket at $t = 0$ was found to be $|\alpha\rangle$.

The astute reader may have noticed the manner in which we have separated the time dependence of the coefficient of $|n\rangle$ in (5.5.4). The factor $e^{-iE_n t/\hbar}$ is present even if V is absent. This way of writing the time dependence makes it clear that the time evolution of $c_n(t)$ is due solely to the presence of $V(t)$; $c_n(t)$ would be identically equal to $c_n(0)$ and hence independent of t if V were zero. As we shall see in a moment, this separation is convenient because $c_n(t)$ satisfies a relatively simple differential equation. The probability of finding $|n\rangle$ is found by evaluating $|c_n(t)|^2$.

*In (5.5.2) we no longer use the notation $|n^{(0)}\rangle$, $E_n^{(0)}$.

The Interaction Picture

Before we discuss the differential equation for $c_n(t)$, we discuss the interaction picture. Suppose we have a physical system such that its state ket coincides with $|\alpha\rangle$ at $t = t_0$, where t_0 is often taken to be zero. At a later time, we denote the state ket in the Schrödinger picture by $|\alpha, t_0; t\rangle_S$, where the subscript S reminds us that we are dealing with the state ket of the Schrödinger picture.

We now *define*

$$|\alpha, t_0; t\rangle_I = e^{iH_0t/\hbar}|\alpha, t_0; t\rangle_S, \tag{5.5.5}$$

where $|\ \rangle_I$ stands for a state ket that represents the same physical situation in the *interaction picture*. At $t = 0$, $|\ \rangle_I$ evidently coincides with $|\ \rangle_S$. For operators (representing observables) we define observables in the interaction picture as

$$A_I \equiv e^{iH_0t/\hbar}A_S e^{-iH_0t/\hbar}. \tag{5.5.6}$$

In particular,

$$V_I = e^{iH_0t/\hbar}Ve^{-iH_0t/\hbar} \tag{5.5.7}$$

where V without a subscript is understood to be the time-dependent potential in the Schrödinger picture. The reader may recall here the connection between the Schrödinger picture and the Heisenberg picture:

$$|\alpha\rangle_H = e^{+iHt/\hbar}|\alpha, t_0 = 0; t\rangle_S \tag{5.5.8}$$

$$A_H = e^{iHt/\hbar}A_S e^{-iHt/\hbar}. \tag{5.5.9}$$

The basic difference between (5.5.8) and (5.5.9) on the one hand and (5.5.6) and (5.5.7) on the other is that H rather than H_0 appears in the exponential.

We now derive the fundamental differential equation that characterizes the time evolution of a state ket in the interaction picture. Let us take the time derivative of (5.5.5) with the full H given by (5.5.1):

$$\begin{aligned} i\hbar\frac{\partial}{\partial t}|\alpha, t_0; t\rangle_I &= i\hbar\frac{\partial}{\partial t}\left(e^{iH_0t/\hbar}|\alpha, t_0; t\rangle_S\right) \\ &= -H_0 e^{iH_0t/\hbar}|\alpha, t_0; t\rangle_S + e^{iH_0t/\hbar}(H_0 + V)|\alpha, t_0; t\rangle_S \\ &= e^{iH_0t/\hbar}Ve^{-iH_0t/\hbar}e^{iH_0t/\hbar}|\alpha, t_0; t\rangle_S. \end{aligned} \tag{5.5.10}$$

We thus see

$$i\hbar\frac{\partial}{\partial t}|\alpha, t_0; t\rangle_I = V_I|\alpha, t_0; t\rangle_I, \tag{5.5.11}$$

which is a Schrödinger-like equation with the total H replaced by V_I. In other words $|\alpha, t_0; t\rangle_I$ would be a ket fixed in time if V_I were absent. We can also show for an observable A (that does not contain time t explicitly in the Schrödinger picture) that

$$\frac{dA_I}{dt} = \frac{1}{i\hbar}[A_I, H_0], \tag{5.5.12}$$

which is a Heisenberg-like equation with H replaced by H_0.

TABLE 5.2

	Heisenberg picture	Interaction picture	Schrödinger picture
State ket	No change	Evolution determined by V_I	Evolution determined by H
Observable	Evolution determined by H	Evolution determined by H_0	No change

In many respects, the interaction picture, or *Dirac picture*, is intermediate between the Schrödinger picture and the Heisenberg picture; This should be evident from Table 5.2.

In the interaction picture we continue using $|n\rangle$ as our base kets. Thus we expand $|\ \rangle_I$ as follows:

$$|\alpha, t_0; t\rangle_I = \sum_n c_n(t)|n\rangle. \tag{5.5.13}$$

With t_0 set equal to 0, we see that the $c_n(t)$ appearing here are the same as the $c_n(t)$ introduced earlier in (5.5.4), as can easily be verified by multiplying both sides of (5.5.4) by $e^{iH_0t/\hbar}$ using (5.5.2).

We are finally in a position to write the differential equation for $c_n(t)$. Multiplying both sides of (5.5.11) by $\langle n|$ from the left, we obtain

$$i\hbar\frac{\partial}{\partial t}\langle n|\alpha, t_0; t\rangle_I = \sum_m \langle n|V_I|m\rangle\langle m|\alpha, t_0; t\rangle_I. \tag{5.5.14}$$

This can also be written using

$$\langle n|e^{iH_0t/\hbar}V(t)e^{-iH_0t/\hbar}|m\rangle = V_{nm}(t)e^{i(E_n - E_m)t/\hbar}$$

and

$$c_n(t) = \langle n|\alpha, t_0; t\rangle_I$$

[from (5.5.13)] as

$$i\hbar\frac{d}{dt}c_n(t) = \sum_m V_{nm}e^{i\omega_{nm}t}c_m(t), \tag{5.5.15}$$

where

$$\omega_{nm} \equiv \frac{(E_n - E_m)}{\hbar} = -\omega_{mn}. \tag{5.5.16}$$

Explicitly,

$$i\hbar\begin{pmatrix}\dot{c}_1\\ \dot{c}_2\\ \dot{c}_3\\ \vdots\end{pmatrix} = \begin{pmatrix} V_{11} & V_{12}e^{i\omega_{12}t} & & \cdots\\ V_{21}e^{i\omega_{21}t} & V_{22} & & \cdots\\ & & V_{33} & \cdots\\ \vdots & \vdots & & \ddots\end{pmatrix}\begin{pmatrix}c_1\\ c_2\\ c_3\\ \vdots\end{pmatrix}. \tag{5.5.17}$$

This is the basic coupled differential equation that must be solved to obtain the probability of finding $|n\rangle$ as a function of t.

Time-Dependent Two-State Problems: Nuclear Magnetic Resonance, Masers, and So Forth

Exact soluble problems with time-dependent potentials are rather rare. In most cases we have to resort to perturbation expansion to solve the coupled differential equations (5.5.17), as we will discuss in the next section. There is, however, a problem of enormous practical importance, which can be solved exactly—a two-state problem with a sinusoidal oscillating potential.

The problem is defined by

$$\begin{aligned} H_0 &= E_1|1\rangle\langle 1| + E_2|2\rangle\langle 2| \quad (E_2 > E_1) \\ V(t) &= \gamma e^{i\omega t}|1\rangle\langle 2| + \gamma e^{-i\omega t}|2\rangle\langle 1|, \end{aligned} \tag{5.5.18}$$

where γ and ω are real and positive. In the language of (5.5.14) and (5.5.15), we have

$$\begin{aligned} V_{12} &= V_{21}^* = \gamma e^{i\omega t} \\ V_{11} &= V_{22} = 0. \end{aligned} \tag{5.5.19}$$

We thus have a time-dependent potential that connects the two energy eigenstates of H_0. In other words, we can have a transition between the two states $|1\rangle \rightleftarrows |2\rangle$.

An exact solution to this problem is available. If initially—at $t = 0$ —only the lower level is populated so that [see (5.5.3)]

$$c_1(0) = 1, \qquad c_2(0) = 0, \tag{5.5.20}$$

then the probability for being found in each of the two states is given by (**Rabi's formula**, after I. I. Rabi, who is the father of molecular beam techniques)

$$|c_2(t)|^2 = \frac{\gamma^2/\hbar^2}{\gamma^2/\hbar^2 + (\omega - \omega_{21})^2/4} \sin^2\left\{\left[\frac{\gamma^2}{\hbar^2} + \frac{(\omega - \omega_{21})^2}{4}\right]^{1/2} t\right\} \tag{5.5.21a}$$

$$|c_1(t)|^2 = 1 - |c_2(t)|^2, \tag{5.5.21b}$$

where

$$\omega_{21} \equiv \frac{(E_2 - E_1)}{\hbar}, \tag{5.5.22}$$

as the reader may verify by working out Problem 30 of this chapter.

Let us now look at $|c_2|^2$ a little more closely. We see that the probability for finding the upper state E_2 exhibits an oscillatory time dependence with angular frequency, two times that of

$$\Omega = \sqrt{\left(\frac{\gamma^2}{\hbar^2}\right) + \frac{(\omega - \omega_{21})^2}{4}}\,. \tag{5.5.23}$$

The amplitude of oscillation is very large when

$$\omega \simeq \omega_{21} = \frac{(E_2 - E_1)}{\hbar}, \tag{5.5.24}$$

that is, when the angular frequency of the potential—usually due to an externally applied electric or magnetic field—is nearly equal to the angular frequency characteristic of the two-state system. Equation (5.5.24) is therefore known as the **resonance condition.**

It is instructive to look at (5.5.21a) and (5.5.21b) a little closely exactly at resonance:

$$\omega = \omega_{21}, \qquad \Omega = \frac{\gamma}{\hbar}. \tag{5.5.25}$$

We can plot $|c_1(t)|^2$ and $|c_2(t)|^2$ as a function of t; see Figure 5.4. From $t = 0$ to $t = \pi\hbar/2\gamma$, the two-level system absorbs energy from the time-dependent potential $V(t)$; $|c_1(t)|^2$ decreases from unity as $|c_2(t)|^2$ grows. At $t = \pi\hbar/2\gamma$, only the upper state is populated. From $t = \pi\hbar/2\gamma$ to $t = \pi\hbar/\gamma$, the system gives up its excess energy [of the excited (upper) state] to $V(t)$; $|c_2|^2$ decreases and $|c_1|^2$ increases. This *absorption-emission cycle* is repeated indefinitely, as is also shown in Figure 5.4, so $V(t)$ can be regarded as a source or sink of energy; put in another way, $V(t)$ can cause a transition from $|1\rangle$ to $|2\rangle$ (absorption) or from $|2\rangle$ to $|1\rangle$ (emission). We will come back to this point of view when we discuss emission and absorption of radiation.

The absorption-emission cycle takes place even away from resonance. However, the amplitude of oscillation for $|2\rangle$ is now reduced; $|c_2(t)|^2_{\max}$ is no longer 1 and $|c_1(t)|^2$ does not go down all the way to 0. In Figure 5.5 we plot $|c_2(t)|^2_{\max}$ as a function of ω. This curve has a resonance peak centered around $\omega = \omega_{21}$, and the full width at half maxima is given by

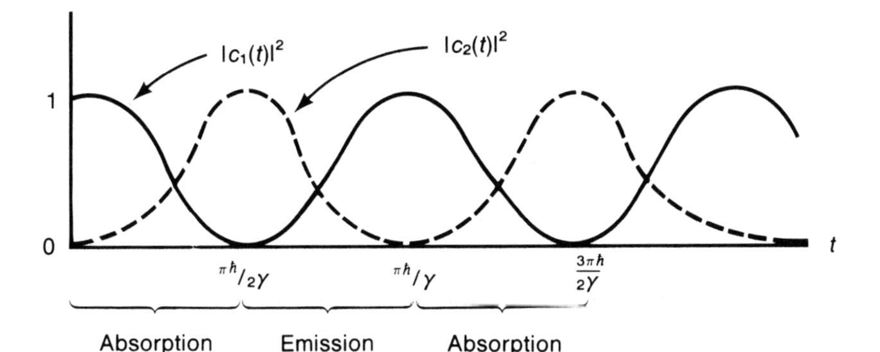

FIGURE 5.4. Plot of $|c_1(t)|^2$ and $|c_2(t)|^2$ against time t exactly at resonance $\omega = \omega_{21}$ and $\Omega = \gamma/\hbar$. The graph also illustrates the back-and-forth behavior between $|1\rangle$ and $|2\rangle$.

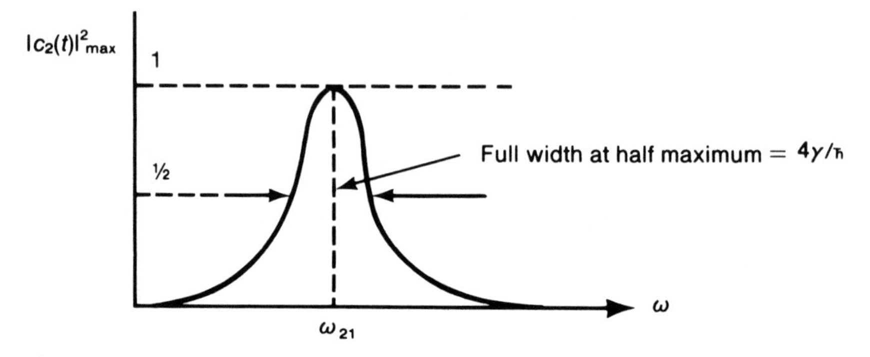

FIGURE 5.5. Graph of $|c_2(t)|^2_{\max}$ as a function of ω, where $\omega = \omega_{21}$ corresponds to the resonant frequency.

$4\gamma/\hbar$. It is worth noting that the weaker the time-dependent potential (γ small), the narrower the resonance peak.

Spin Magnetic Resonance

The two-state problem defined by (5.5.18) has many physical applications. As a first example, consider a spin $\frac{1}{2}$ system—say a bound electron —subjected to a t-independent uniform magnetic field in the z-direction *and*, in addition, a t-dependent magnetic field rotating in the xy-plane:

$$\mathbf{B} = B_0\hat{\mathbf{z}} + B_1(\hat{\mathbf{x}}\cos\omega t + \hat{\mathbf{y}}\sin\omega t) \tag{5.5.26}$$

with B_0 and B_1 constant. We can treat the effect of the uniform t-independent field as H_0 and the effect of the rotating field as V. For

$$\boldsymbol{\mu} = \frac{e}{m_e c}\mathbf{S} \tag{5.5.27}$$

we have

$$H_0 = -\left(\frac{e\hbar B_0}{2m_e c}\right)(|+\rangle\langle+| - |-\rangle\langle-|)$$

$$V(t) = -\left(\frac{e\hbar B_1}{2m_e c}\right)[\cos\omega t(|+\rangle\langle-| + |-\rangle\langle+|)$$

$$+\sin\omega t(-i|+\rangle\langle-| + i|-\rangle\langle+|)], \tag{5.5.28}$$

where we have used the ket-bra forms of $2S_j/\hbar$ [see (3.2.1)]. With $e<0, E_+$ has a higher energy than E_-, and we can identify

$$\begin{aligned} |+\rangle &\to |2\rangle \quad \text{(upper level)} \\ |-\rangle &\to |1\rangle \quad \text{(lower level)} \end{aligned} \tag{5.5.29}$$

to make correspondence with the notation of (5.5.18). The angular frequency characteristic of the two-state system is

$$\omega_{21} = \frac{|e|B_0}{m_e c}, \tag{5.5.30}$$

which is just the spin-precession frequency for the $B_0 \neq 0$, $B_1 = 0$ problem already treated in Section 2.1. Even though the expectation values of $\langle S_{x,y} \rangle$ change due to spin precession in the counterclockwise direction (seen from the positive z-side), $|c_+|^2$ and $|c_-|^2$ remain unchanged in the absence of the rotating field. We now add a new feature as a result of the rotating field: $|c_+|^2$ and $|c_-|^2$ do change as a function of time. This can be seen by identifying

$$\frac{-e\hbar B_1}{2m_e c} \to \gamma, \qquad \omega \to \omega \tag{5.5.31}$$

to make correspondence to the notation of (5.5.18); our time-dependent interaction (5.5.28) is precisely of form (5.5.18). The fact that $|c_+(t)|^2$ and $|c_-(t)|^2$ vary in the manner indicated by Figure 5.4 for $\omega = \omega_{21}$ and the correspondence (5.5.29), for example, implies that the spin $\frac{1}{2}$ system undergoes a succession of spin-flops, $|+\rangle \rightleftarrows |-\rangle$, in addition to spin precession. Semiclassically, spin-flops of this kind can be interpreted as being due to the driving torque exerted by rotating **B**.

The resonance condition is satisfied whenever the frequency of the rotating magnetic field coincides with the frequency of spin precession determined by the strength of the uniform magnetic field. We see that the probability of spin-flops is particularly large.

In practice, a rotating magnetic field may be difficult to produce experimentally. Fortunately, a horizontally oscillating magnetic field—for instance, in the x-direction— is just as good. To see this, we first note that such an oscillating field can be decomposed into a counterclockwise component and a clockwise component as follows:

$$2B_1\hat{\mathbf{x}}\cos\omega t = B_1(\hat{\mathbf{x}}\cos\omega t + \hat{\mathbf{y}}\sin\omega t) + B_1(\hat{\mathbf{x}}\cos\omega t - \hat{\mathbf{y}}\sin\omega t). \tag{5.5.32}$$

We can obtain the effect of the counterclockwise component simply by reversing the sign of ω. Suppose the resonance condition is met for the counterclockwise component

$$\omega \simeq \omega_{21}. \tag{5.5.33}$$

Under a typical experimental condition,

$$\frac{B_1}{B_0} \ll 1, \tag{5.5.34}$$

which implies from (5.5.30) and (5.5.31) that

$$\frac{\gamma}{\hbar} \ll \omega_{21}; \tag{5.5.35}$$

as a result, whenever the resonance condition is met for the counterclockwise component, the effect of the clockwise component becomes completely negligible, since it amounts to $\omega \to -\omega$, and the amplitude becomes small in magnitude as well as very rapidly oscillating.

The resonance problem we have solved is of fundamental importance in interpreting atomic molecular beam and nuclear magnetic resonance experiments. By varying the frequency of oscillating field, it is possible to make a very precise measurement of magnetic moment. We have based our discussion on the solution to differential equations (5.5.17); this problem can also be solved, perhaps more elegantly, by introducing the *rotating axis representation* of Rabi, Schwinger, and Van Vleck.

Maser

As another application of the time-dependent two-state problem, let us consider a **maser**. Specifically, we consider an ammonia molecule NH_3, which—as we may recall from Section 4.2—has two parity eigenstates $|S\rangle$ and $|A\rangle$ lying close together such that $|A\rangle$ is slightly higher. Let $\boldsymbol{\mu}_{el}$ be the electric dipole operator of the molecule. From symmetry considerations we expect that $\boldsymbol{\mu}_{el}$ is proportional to $\mathbf{x}$, the position operator for the N atom. The basic interaction is like $-\boldsymbol{\mu}_{el}\cdot\mathbf{E}$, where for a maser $\mathbf{E}$ is a time-dependent electric field in a microwave cavity:

$$\mathbf{E} = |\mathbf{E}|_{\max}\hat{\mathbf{z}}\cos\omega t. \tag{5.5.36}$$

It is legitimate to ignore the spatial variation of $\mathbf{E}$ because the wavelength in the microwave region is far larger than molecular dimension. The frequency ω is tuned to the energy difference between $|A\rangle$ and $|S\rangle$:

$$\omega \simeq \frac{(E_A - E_S)}{\hbar}. \tag{5.5.37}$$

The diagonal matrix elements of the dipole operator vanish by parity:

$$\langle A|\boldsymbol{\mu}_{el}|A\rangle = \langle S|\boldsymbol{\mu}_{el}|S\rangle = 0, \tag{5.5.38}$$

but the off-diagonal elements are, in general, nonvanishing:

$$\langle S|\mathbf{x}|A\rangle = \langle A|\mathbf{x}|S\rangle \neq 0. \tag{5.5.39}$$

This means that there is a time-dependent potential that connects $|S\rangle$ and $|A\rangle$, and the general two-state problem we discussed earlier is now applicable.

We are now in a position to discuss how masers work. Given a molecular beam of NH_3 containing both $|S\rangle$ and $|A\rangle$, we first eliminate the $|S\rangle$ component by letting the beam go through a region of time-independent inhomogeneous electric field. Such an electric field separates $|S\rangle$ from $|A\rangle$ in much the same way as the inhomogeneous magnetic field in the Stern-Gerlach experiment separates $|+\rangle$ from $|-\rangle$. A pure beam of $|A\rangle$ then enters a microwave cavity tuned to the energy difference $E_A - E_S$. The dimension of the cavity is such that the time spent by the molecule is just $(\pi/2)\hbar/\gamma$. As a result we stay in the first emission phase of Figure 5.4; we have $|A\rangle$ in and $|S\rangle$ out. The excess energy of $|A\rangle$ is given up to the time-dependent potential as $|A\rangle$ turns into $|S\rangle$ and the radiation (microwave) field gains energy. In this way we obtain Microwave Amplification by Stimulated Emission of Radiation, or MASER.

There are many other applications of the general time-dependent two-state problem, such as the atomic clock and optical pumping. In fact, it is amusing to see that as many as four Nobel Prizes in physics have been awarded to those who exploited time-dependent two-state systems of some form.*

5.6. TIME-DEPENDENT PERTURBATION THEORY

Dyson Series

With the exception of a few problems like the two-level time-dependent problem of the previous section, exact solutions to the differential equation for $c_n(t)$ are usually not available. We must be content with approximate solutions to (5.5.17) obtained by perturbation expansion:

$$c_n(t) = c_n^{(0)} + c_n^{(1)} + c_n^{(2)} + \cdots, \tag{5.6.1}$$

where $c_n^{(1)}, c_n^{(2)}, \ldots$ signify amplitudes of first order, second order, and so on in the strength parameter of the time-dependent potential. The iteration method used to solve this problem is similar to what we did in time-independent perturbation theory. If initially only the state i is populated, we approximate c_n on the right-hand side of differential equation (5.5.17) by $c_n^{(0)} = \delta_{ni}$ (independent of t) and relate it to the time derivative of $c_n^{(1)}$, integrate the differential equation to obtain $c_n^{(1)}$, plug $c_n^{(1)}$ into the right-hand side [of (5.5.17)] again to obtain the differential equation for $c_n^{(2)}$, and so on. This is how Dirac developed time-dependent perturbation theory in 1927.

Instead of working with $c_n(t)$, we propose to look at the time evolution operator $U_I(t, t_0)$ in the interaction picture, which we will define later. We obtain a perturbation expansion for $U_I(t, t_0)$, and at the very end

---

Nobel Prize winners* who took advantage of resonance in the two-level systems are Rabi (1944) on molecular beams and nuclear magnetic resonance; Bloch and Purcell (1952) on **B field in atomic nuclei and nuclear magnetic moments; Townes, Basov, and Prochorov (1964) on masers, lasers, and quantum optics; and Kastler (1966) on optical pumping.

we relate the matrix elements of U_I to $c_n(t)$. If we are interested only in solving simple problems in nonrelativistic quantum mechanics, all this might look superfluous; however, the operator formalism we develop is very powerful because it can immediately be applied to more-advanced problems, such as relativistic quantum field theory and many-body theory.

The time-evolution operator in the interaction picture is defined by

$$|\alpha, t_0; t\rangle_I = U_I(t, t_0)|\alpha, t_0; t_0\rangle_I. \tag{5.6.2}$$

Differential equation (5.5.11) for the state ket of the interaction picture is equivalent to

$$i\hbar \frac{d}{dt} U_I(t, t_0) = V_I(t) U_I(t, t_0). \tag{5.6.3}$$

We must solve this operator differential equation subject to the initial condition

$$U_I(t, t_0)|_{t=t_0} = 1. \tag{5.6.4}$$

First, let us note that the differential equation together with the initial condition is equivalent to the following integral equation:

$$U_I(t, t_0) = 1 - \frac{i}{\hbar}\int_{t_0}^{t} V_I(t') U_I(t', t_0)\, dt'. \tag{5.6.5}$$

We can obtain an approximate solution to this equation by iteration:

$$\begin{aligned}
U_I(t, t_0) &= 1 - \frac{i}{\hbar}\int_{t_0}^{t} V_I(t')\left[1 - \frac{i}{\hbar}\int_{t_0}^{t'} V_I(t'') U_I(t'', t_0)\, dt''\right] dt' \\
&= 1 - \frac{i}{\hbar}\int_{t_0}^{t} dt'\, V_I(t') + \left(\frac{-i}{\hbar}\right)^2 \int_{t_0}^{t} dt' \int_{t_0}^{t'} dt''\, V_I(t') V_I(t'') \\
&\quad + \cdots + \left(\frac{-i}{\hbar}\right)^n \int_{t_0}^{t} dt' \int_{t_0}^{t'} dt'' \cdots \\
&\quad \times \int_{t_0}^{t^{(n-1)}} dt^{(n)} V_I(t') V_I(t'') \cdots V_I(t^{(n)}) \\
&\quad + \cdots .
\end{aligned} \tag{5.6.6}$$

This series is known as the **Dyson series** after Freeman J. Dyson, who applied this method to covariant quantum electrodynamics (QED).* Setting aside the difficult question of convergence, we can compute $U_I(t, t_0)$ to any finite order of perturbation theory.

*Note that in QED, the time-ordered product ($t' > t'' > \cdots$) is introduced, and then this perturbation series can be summed into an exponential form. This exponential form immediately gives $U(t, t_0) = U(t, t_1)U(t_1, t_0)$ (Bjorken and Drell 1965, 175–78).

Transition Probability

Once $U_I(t, t_0)$ is given, we can predict the time development of any state ket. For example, if the initial state at $t = 0$ is one of the energy eigenstates of H_0, then to obtain the initial state ket at a later time, all we need to do is multiply by $U_I(t, 0)$:

$$\begin{aligned} |i, t_0 = 0; t\rangle_I &= U_I(t, 0)|i\rangle \\ &= \sum_n |n\rangle\langle n|U_I(t, 0)|i\rangle. \end{aligned} \tag{5.6.7}$$

In fact, $\langle n|U_I(t, 0)|i\rangle$ is nothing more than what we called $c_n(t)$ earlier [see (5.5.13)]. We will say more about this later.

We earlier introduced the time-evolution operator $U(t, t_0)$ in the Schrödinger picture (see Section 2.2). Let us now explore the connection between $U(t, t_0)$ and $U_I(t, t_0)$. We note from (2.2.13) and (5.5.5) that

$$\begin{aligned} |\alpha, t_0; t\rangle_I &= e^{iH_0t/\hbar}|\alpha, t_0; t\rangle_S \\ &= e^{iH_0t/\hbar}U(t, t_0)|\alpha, t_0; t_0\rangle_S \\ &= e^{iH_0t/\hbar}U(t, t_0)e^{-iH_0t_0/\hbar}|\alpha, t_0; t_0\rangle_I. \end{aligned} \tag{5.6.8}$$

So we have

$$U_I(t, t_0) = e^{iH_0t/\hbar}U(t, t_0)e^{-iH_0t_0/\hbar}. \tag{5.6.9}$$

Let us now look at the matrix element of $U_I(t, t_0)$ between energy eigenstates of H_0:

$$\langle n|U_I(t, t_0)|i\rangle = e^{i(E_nt - E_it_0)/\hbar}\langle n|U(t, t_0)|i\rangle. \tag{5.6.10}$$

We recall from Section 2.2 that $\langle n|U(t, t_0)|i\rangle$ is defined to be the transition amplitude. Hence our $\langle n|U_I(t, t_0)|i\rangle$ here is not quite the same as the transition amplitude defined earlier. However, the transition *probability* defined as the square of the modulus of $\langle n|U(t, t_0)|i\rangle$ is the same as the analogous quantity in the interaction picture

$$|\langle n|U_I(t, t_0)|i\rangle|^2 = |\langle n|U(t, t_0)|i\rangle|^2. \tag{5.6.11}$$

Parenthetically, we may remark that if the matrix elements of U_I are taken between initial and final states that are not energy eigenstates—for example, between $|a'\rangle$ and $|b'\rangle$ (eigenkets of A and B, respectively), where $[H_0, A] \neq 0$ and/or $[H_0, B] \neq 0$—we have, in general,

$$|\langle b'|U_I(t, t_0)|a'\rangle| \neq |\langle b'|U(t, t_0)|a'\rangle|,$$

as the reader may easily verify. Fortunately, in problems where the interaction picture is found to be useful, the initial and final states are usually taken to be H_0 eigenstates. Otherwise, all that is needed is to expand $|a'\rangle$, $|b'\rangle$, and so on in terms of the energy eigenkets of H_0.

Coming back to $\langle n|U_I(t,t_0)|i\rangle$, we illustrate by considering a physical situation where at $t=t_0$, the system is known to be in state $|i\rangle$. The state ket in the Schrödinger picture $|i,t_0;t\rangle_S$ is then equal to $|i\rangle$ up to a phase factor. In applying the interaction picture, it is convenient to choose the phase factor at $t=t_0$ so that

$$|i,t_0;t_0\rangle_S = e^{-iE_it_0/\hbar}|i\rangle, \tag{5.6.12}$$

which means that in the interaction picture we have the simple equation

$$|i,t_0;t_0\rangle_I = |i\rangle. \tag{5.6.13}$$

At a later time we have

$$|i,t_0;t\rangle_I = U_I(t,t_0)|i\rangle. \tag{5.6.14}$$

Comparing this with the expansion

$$|i,t_0;t\rangle_I = \sum_n c_n(t)|n\rangle, \tag{5.6.15}$$

we see

$$c_n(t) = \langle n|U_I(t,t_0)|i\rangle. \tag{5.6.16}$$

We now go back to the perturbation expansion for $U_I(t,t_0)$ [see (5.6.6)]. We can also expand $c_n(t)$ as in (5.6.1), where $c_n^{(1)}$ is first order in $V_I(t)$, $c_n^{(2)}$ is second order in $V_I(t)$, and so on. Comparing the expansion of both sides of (5.6.16), we obtain [using (5.5.7)]

$$\begin{aligned}
c_n^{(0)}(t) &= \delta_{ni} \qquad \text{(independent of } t\text{)}\\
c_n^{(1)}(t) &= \frac{-i}{\hbar}\int_{t_0}^{t}\langle n|V_I(t')|i\rangle\, dt'\\
&= \frac{-i}{\hbar}\int_{t_0}^{t} e^{i\omega_{ni}t'}V_{ni}(t')\, dt' \qquad\qquad (5.6.17)\\
c_n^{(2)}(t) &= \left(\frac{-i}{\hbar}\right)^2 \sum_m \int_{t_0}^{t} dt' \int_{t_0}^{t'} dt''\, e^{i\omega_{nm}t'}V_{nm}(t')e^{i\omega_{mi}t''}V_{mi}(t''),
\end{aligned}$$

where we have used

$$e^{i(E_n-E_i)t/\hbar} = e^{i\omega_{ni}t}. \tag{5.6.18}$$

The transition probability for $|i\rangle \to |n\rangle$ with $n \neq i$ is obtained by

$$P(i\to n) = |c_n^{(1)}(t) + c_n^{(2)}(t) + \cdots|^2. \tag{5.6.19}$$

Constant Perturbation

As an application of (5.6.17), let us consider a constant perturbation turned on at $t=0$:

$$V(t) = \begin{cases} 0, & \text{for } t<0\\ V \quad \text{(independent of } t\text{)}, & \text{for } t\ge 0. \end{cases} \tag{5.6.20}$$

Even though the operator V has no explicit dependence on time, it is, in general, made up of operators like $\mathbf{x}$, $\mathbf{p}$, and $\mathbf{s}$. Now suppose at $t=0$, we have only $|i\rangle$. With t_0 taken to be zero, we obtain

$$c_n^{(0)} = c_n^{(0)}(0) = \delta_{in},$$

$$c_n^{(1)} = \frac{-i}{\hbar} V_{ni} \int_0^t e^{i\omega_{ni}t'} dt'$$

$$= \frac{V_{ni}}{E_n - E_i}\left(1 - e^{i\omega_{ni}t}\right), \tag{5.6.21}$$

or

$$|c_n^{(1)}|^2 = \frac{|V_{ni}|^2}{|E_n - E_i|^2}(2 - 2\cos\omega_{ni}t)$$

$$= \frac{4|V_{ni}|^2}{|E_n - E_i|^2} \sin^2\left[\frac{(E_n - E_i)t}{2\hbar}\right]. \tag{5.6.22}$$

The probability of finding $|n\rangle$ depends not only on $|V_{ni}|^2$ but also on the energy difference $E_n - E_i$, so let us try to see how (5.6.22) looks as a function of E_n. In practice, we are interested in this way of looking at (5.6.22) when there are many states with $E \sim E_n$ so that we can talk about a continuum of final states with nearly the same energy. To this end we define

$$\omega \equiv \frac{E_n - E_i}{\hbar} \tag{5.6.23}$$

and plot $4\sin^2(\omega t/2)/\omega^2$ as a function of ω for fixed t, the time interval during which the perturbation has been on; see Figure 5.6. We see that the height of the middle peak, centered around $\omega = 0$, is t^2 and the width is proportional to $1/t$. As t becomes large, $|c_n^{(1)}(t)|^2$ is appreciable only for those final states that satisfy

$$t \sim \frac{2\pi}{|\omega|} = \frac{2\pi\hbar}{|E_n - E_i|}. \tag{5.6.24}$$

If we call Δt the time interval during which the perturbation has been turned on, a transition with appreciable probability is possible only if

$$\Delta t \Delta E \sim \hbar, \tag{5.6.25}$$

where by ΔE we mean the energy change involved in a transition with appreciable probability. If Δt is small, we have a broader peak in Figure 5.6, and as a result we can tolerate a fair amount of energy *non*conservation. On the other hand, if the perturbation has been on for a very long time, we have a very narrow peak, and approximate energy conservation is required for a transition with appreciable probability. Note that this "uncertainty relation" is fundamentally different from the $x - p$ uncertainty relation of

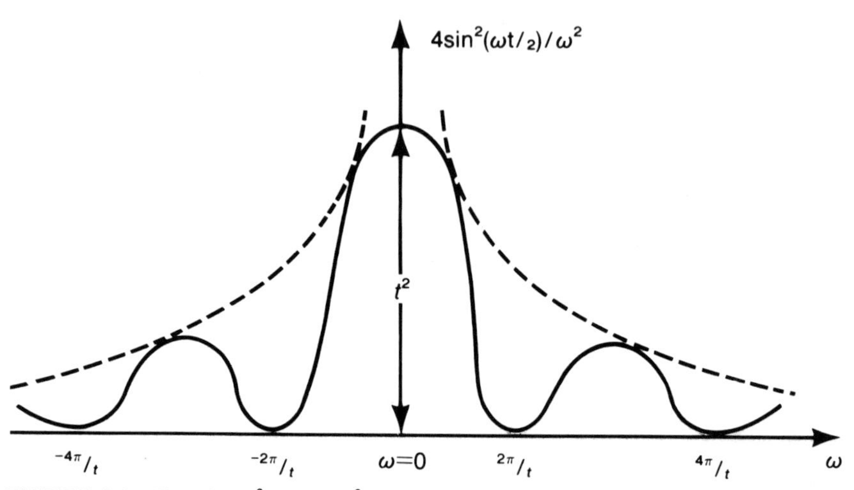

FIGURE 5.6. Plot of $4\sin^2(\omega t/2)/\omega^2$ versus ω for a fixed t, where in $\omega = (E_n - E_i)/\hbar$ we have regarded E_n as a continuous variable.

Section 1.6. There x and p are both observables. In contrast, time in nonrelativistic quantum mechanics is a parameter, not an observable.

For those transitions with exact energy conservation $E_n = E_i$, we have

$$|c_n^{(1)}(t)|^2 = \frac{1}{\hbar^2}|V_{ni}|^2 t^2. \tag{5.6.26}$$

The probability of finding $|n\rangle$ after a time interval t is quadratic, *not* linear, in the time interval during which V has been on. This may appear intuitively unreasonable. There is no cause for alarm, however. In a realistic situation where our formalism is applicable, there is usually a group of final states, all with nearly the same energy as the energy of the initial state $|i\rangle$. In other words, a final state forms a continuous energy spectrum in the neighborhood of E_i. We give two examples along this line. Consider for instance, elastic scattering by some finite range potential (see Figure 5.7), which we will consider in detail in Chapter 7. The initial state is taken to be a plane wave state with its propagation direction oriented in the positive z-direction; the final state may also be a plane wave state of the same energy but with its propagation direction, in general, in a direction other than the positive z-direction. Another example of interest is the de-excitation of an excited atomic state via the emission of an Auger electron. The simplest example is a helium atom. The initial state may be $(2s)^2$, where both the electrons are excited; the final state may be $(1s)$ (that is, one of the electrons still bound) of the He^+ ion, while the second electron escapes with a positive energy E; see Figure 5.8. In such a case we are interested in the

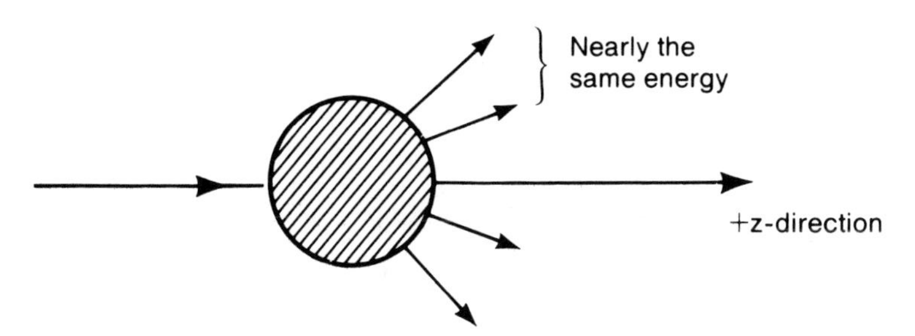

FIGURE 5.7. Elastic scattering of plane wave by some finite range potential.

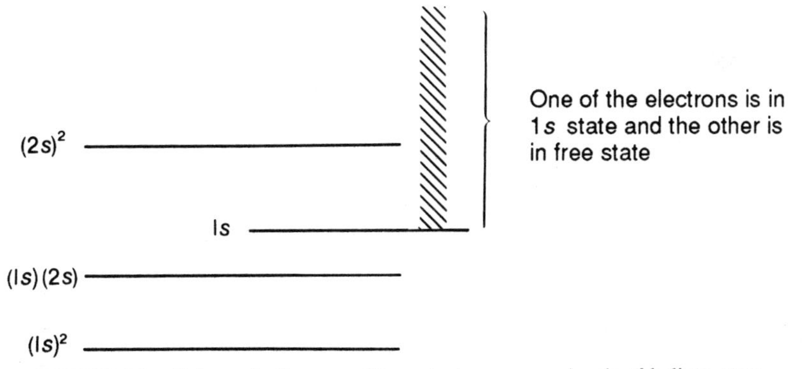

FIGURE 5.8. Schematic diagram of two electron energy levels of helium atom.

total probability—that is, the transition probabilities summed over final states with $E_n \simeq E_i$:

$$\sum_{n,\, E_n \simeq E_i} |c_n^{(1)}|^2. \tag{5.6.27}$$

It is customary to define the density of final states as the number of states within energy interval $(E, E + dE)$ as

$$\rho(E)\, dE. \tag{5.6.28}$$

We can then write (5.6.27) as

$$\sum_{n,\, E_n \simeq E_i} |c_n^{(1)}|^2 \Rightarrow \int dE_n \rho(E_n) |c_n^{(1)}|^2$$

$$= 4 \int \sin^2 \left[\frac{(E_n - E_i)t}{2\hbar} \right] \frac{|V_{ni}|^2}{|E_n - E_i|^2} \rho(E_n)\, dE_n. \tag{5.6.29}$$

As $t \to \infty$, we can take advantage of

$$\lim_{t \to \infty} \frac{1}{|E_n - E_i|^2} \sin^2 \left[\frac{(E_n - E_i)t}{2\hbar} \right] = \frac{\pi t}{2\hbar} \delta(E_n - E_i), \tag{5.6.30}$$

which follows from

$$\lim_{\alpha\to\infty}\frac{1}{\pi}\frac{\sin^2\alpha x}{\alpha x^2}=\delta(x). \tag{5.6.31}$$

It is now possible to take the average of $|V_{ni}|^2$ outside the integral sign and perform the integration with the δ-function:

$$\lim_{t\to\infty}\int dE_n\rho(E_n)|c_n^{(1)}(t)|^2=\left(\frac{2\pi}{\hbar}\right)\overline{|V_{ni}|^2}\rho(E_n)t\bigg|_{E_n\simeq E_i}. \tag{5.6.32}$$

Thus the total transition probability *is* proportional to t for large values of t, which is quite reasonable. Notice that this linearity in t is a consequence of the fact that the total transition probability is proportional to the area under the peak of Figure 5.6, where the height varies as t^2 and the width varies as $1/t$.

It is conventional to consider the **transition rate**—that is, the transition probability per unit time. Expression (5.6.32) tells us that the total transition rate, defined by

$$\frac{d}{dt}\left(\sum_n|c_n^{(1)}|^2\right), \tag{5.6.33}$$

is constant in t for large t. Calling (5.6.33) $w_{i\to[n]}$, where $[n]$ stands for a group of final states with energy similar to i, we obtain

$$w_{i\to[n]}=\frac{2\pi}{\hbar}\overline{|V_{ni}|^2}\rho(E_n)_{E_n\simeq E_i} \tag{5.6.34}$$

independent of t, provided the first-order time-dependent perturbation theory is valid. This formula is of great practical importance; it is called **Fermi's golden rule** even though the basic formalism of t-dependent perturbation theory is due to Dirac.* We sometimes write (5.6.34) as

$$w_{i\to n}=\left(\frac{2\pi}{\hbar}\right)|V_{ni}|^2\delta(E_n-E_i), \tag{5.6.35}$$

where it must be understood that this expression is integrated with $\int dE_n\rho(E_n)$.

We should also understand what is meant by $\overline{|V_{ni}|^2}$. If the final states $|n\rangle$ form a quasi-continuum, the matrix elements V_{ni} are often similar if $|n\rangle$ are similar. However, it may happen that all energy eigenstates with the same E_n do not necessarily have similar matrix elements. Consider, for example, elastic scattering. The $|V_{ni}|^2$ that determines the scattering cross section may depend on the final momentum direction. In such a case the

**Editorial Note.* Dr. Edward Teller, in teaching a quantum mechanics class at Berkeley in 1955 (in which the editor was in attendance), mentioned that this was actually Fermi's golden rule 2. Since Teller did not subscribe to golden rule 1, the class was not informed what rule 1 was. However, golden rule 1 is given in E. Fermi 1950, 136 and 148.

group of final states we should consider must have not only approximately the same energy, but they must also have approximately the same momentum direction. This point becomes clearer when we discuss the photoelectric effect.

Let us now look at the second-order term, still with the constant perturbation of (5.6.20). From (5.6.17) we have

$$c_n^{(2)} = \left(\frac{-i}{\hbar}\right)^2 \sum_m V_{nm} V_{mi} \int_0^t dt' \, e^{i\omega_{nm}t'} \int_0^{t'} dt'' \, e^{i\omega_{mi}t''}$$

$$= \frac{i}{\hbar} \sum_m \frac{V_{nm} V_{mi}}{E_m - E_i} \int_0^t \left(e^{i\omega_{ni}t'} - e^{i\omega_{nm}t'}\right) dt'. \tag{5.6.36}$$

The first term on the right-hand side has the same t-dependence as $c_n^{(1)}$ [see (5.6.21)]. If this were the only term, we could then repeat the same argument as before and conclude that as $t \to \infty$, the only important contribution arises from $E_n \simeq E_i$. Indeed, when E_m differs from E_n and E_i, the second contribution gives rise to a rapid oscillation, which does not give a contribution to the transition probability that grows with t.

With $c^{(1)}$ and $c^{(2)}$ together, we have

$$w_{i \to [n]} = \frac{2\pi}{\hbar} \left. \overline{\left| V_{ni} + \sum_m \frac{V_{nm} V_{mi}}{E_i - E_m} \right|^2} \rho(E_n) \right|_{E_n \simeq E_i}. \tag{5.6.37}$$

The formula has the following physical interpretation. We visualize that the transition due to the second-order term takes place in two steps. First, $|i\rangle$ makes an energy nonconserving transition to $|m\rangle$; subsequently, $|m\rangle$ makes an energy nonconserving transition to $|n\rangle$, where between $|n\rangle$ and $|i\rangle$ there is overall energy conservation. Such energy nonconserving transitions are often called *virtual transitions*. Energy need not be conserved for those virtual transitions into (or from) virtual intermediate states. In contrast, the first-order term V_{ni} is often said to represent a direct energy-conserving "real" transition. A special treatment is needed if $V_{nm}V_{mi} \neq 0$ with $E_m \simeq E_i$. The best way to treat this is to use the slow-turn-on method $V \to e^{\eta t} V$, which we will discuss in Section 5.8 and Problem 31 of this chapter. The net result is to change the energy denominator in (5.6.37) as follows:

$$E_i - E_m \to E_i - E_m + i\varepsilon. \tag{5.6.38}$$

Harmonic Perturbation

We now consider a sinusoidally varying time-dependent potential, commonly referred to as **harmonic perturbation**:

$$V(t) = \mathscr{V} e^{i\omega t} + \mathscr{V}^\dagger e^{-i\omega t}, \tag{5.6.39}$$

where $\mathscr{V}$ may still depend on $\mathbf{x}$, $\mathbf{p}$, $\mathbf{s}$, and so on. Actually, we have already encountered a time-dependent potential of this kind in Section 5.5 in discussing t-dependent two-level problems.

Again assume that only one of the eigenstates of H_0 is populated initially. Perturbation (5.6.39) is assumed to be turned on at $t=0$, so

$$\begin{aligned} c_n^{(1)} &= \frac{-i}{\hbar}\int_0^t \left(\mathscr{V}_{ni}e^{i\omega t'} + \mathscr{V}_{ni}^{\dagger}e^{-i\omega t'}\right)e^{i\omega_{ni}t'}\,dt' \\ &= \frac{1}{\hbar}\left[\frac{1-e^{i(\omega+\omega_{ni})t}}{\omega+\omega_{ni}}\mathscr{V}_{ni} + \frac{1-e^{i(\omega_{ni}-\omega)t}}{-\omega+\omega_{ni}}\mathscr{V}_{ni}^{\dagger}\right] \end{aligned} \tag{5.6.40}$$

where $\mathscr{V}_{ni}^{\dagger}$ actually stands for $(\mathscr{V}^{\dagger})_{ni}$. We see that this formula is similar to the constant perturbation case. The only change needed is

$$\omega_{ni} = \frac{E_n - E_i}{\hbar} \to \omega_{ni} \pm \omega. \tag{5.6.41}$$

So as $t\to\infty$, $|c_n^{(1)}|^2$ is appreciable only if

$$\omega_{ni} + \omega \simeq 0 \quad \text{or} \quad E_n \simeq E_i - \hbar\omega \tag{5.6.42a}$$

$$\omega_{ni} - \omega \simeq 0 \quad \text{or} \quad E_n \simeq E_i + \hbar\omega. \tag{5.6.42b}$$

Clearly, whenever the first term is important because of (5.6.42a), the second term is unimportant, and vice versa. We see that we have no energy-conservation condition satisfied by the quantum-mechanical system alone; rather the apparent lack of energy conservation is compensated by the energy given out to—or energy taken away from—the "external" potential $V(t)$. Pictorially, we have Figure 5.9. In the first case (*stimulated emission*), the quantum-mechanical system gives up energy $\hbar\omega$ to V; this is clearly possible only if the initial state is excited. In the second case (*absorption*), the quantum-mechanical system receives energy $\hbar\omega$ from V and ends up as an excited state. Thus a time-dependent perturbation can be regarded as an inexhaustible source or sink of energy.

In complete analogy with (5.6.34), we have

$$\begin{aligned} w_{i\to[n]} &= \frac{2\pi}{\hbar}\overline{|\mathscr{V}_{ni}|^2}\rho(E_n)\Big|_{E_n\cong E_i-\hbar\omega} \\ w_{i\to[n]} &= \frac{2\pi}{\hbar}\overline{|\mathscr{V}_{ni}^{\dagger}|^2}\rho(E_n)\Big|_{E_n\cong E_i+\hbar\omega} \end{aligned} \tag{5.6.43}$$

or, more commonly,

$$w_{i\to n} = \frac{2\pi}{\hbar}\left\{\begin{matrix}|\mathscr{V}_{ni}|^2 \\ |\mathscr{V}_{ni}^{\dagger}|^2\end{matrix}\right\}\delta(E_n - E_i \pm \hbar\omega). \tag{5.6.44}$$

Note also that

$$|\mathscr{V}_{ni}|^2 = |\mathscr{V}_{in}^{\dagger}|^2, \tag{5.6.45}$$

FIGURE 5.9. (i) Stimulated emission: Quantum-mechanical system gives up $\hbar\omega$ to V (possible only if initial state is excited). (ii) Absorption: Quantum mechanical system receives $\hbar\omega$ from V and ends up as an excited state.

which is a consequence of

$$\langle i|\mathscr{V}^\dagger|n\rangle = \langle n|\mathscr{V}|i\rangle^* \tag{5.6.46}$$

(remember $\mathscr{V}^\dagger|n\rangle \overset{\text{DC}}{\leftrightarrow} \langle n|\mathscr{V}$). Combining (5.6.43) and (5.6.45), we have

$$\frac{\text{emission rate for } i \to [n]}{\text{density of final states for } [n]} = \frac{\text{absorption rate for } n \to [i]}{\text{density of final states for } [i]}, \tag{5.6.47}$$

where in the absorption case we let i stand for final states. Equation (5.6.47), which expresses symmetry between emission and absorption, is known as **detailed balancing.**

To summarize, for constant perturbation, we obtain appreciable transition probability for $|i\rangle \to |n\rangle$ only if $E_n \simeq E_i$. In contrast, for harmonic perturbation we have appreciable transition probability only if $E_n \simeq E_i - \hbar\omega$ (stimulated emission) or $E_n \simeq E_i + \hbar\omega$ (absorption).

5.7. APPLICATIONS TO INTERACTIONS WITH THE CLASSICAL RADIATION FIELD

Absorption and Stimulated Emission

We apply the formalism of time-dependent perturbation theory to the interactions of atomic electron with the classical radiation field. By a **classical radiation field** we mean the electric or magnetic field derivable from a classical (as opposed to quantized) radiation field.

The basic Hamiltonian, with $|\mathbf{A}|^2$ omitted, is

$$H = \frac{\mathbf{p}^2}{2m_e} + e\phi(\mathbf{x}) - \frac{e}{m_e c}\mathbf{A}\cdot\mathbf{p}, \tag{5.7.1}$$

which is justified if

$$\nabla\cdot\mathbf{A} = 0; \tag{5.7.2}$$

specifically, we work with a monochromatic field of the plane wave for

$$\mathbf{A} = 2A_0\hat{\boldsymbol{\varepsilon}}\cos\left(\frac{\omega}{c}\hat{\mathbf{n}}\cdot\mathbf{x} - \omega t\right) \tag{5.7.3}$$

where $\hat{\boldsymbol{\varepsilon}}$ and $\hat{\mathbf{n}}$ are the (linear) polarization and propagation direction. Equation (5.7.3) obviously satisfies (5.7.2) because $\hat{\boldsymbol{\varepsilon}}$ is perpendicular to the propagation direction $\hat{\mathbf{n}}$. We write

$$\cos\left(\frac{\omega}{c}\hat{\mathbf{n}}\cdot\mathbf{x} - \omega t\right) = \frac{1}{2}\left[e^{i(\omega/c)\hat{\mathbf{n}}\cdot\mathbf{x} - i\omega t} + e^{-i(\omega/c)\hat{\mathbf{n}}\cdot\mathbf{x} + i\omega t}\right] \tag{5.7.4}$$

and treat $-(e/m_e c)\mathbf{A}\cdot\mathbf{p}$ as time-dependent potential, where we express $\mathbf{A}$ in (5.7.3) as

$$\mathbf{A} = A_0\hat{\boldsymbol{\varepsilon}}\left[e^{i(\omega/c)\hat{\mathbf{n}}\cdot\mathbf{x} - i\omega t} + e^{-i(\omega/c)\hat{\mathbf{n}}\cdot\mathbf{x} + i\omega t}\right]. \tag{5.7.5}$$

Comparing this result with (5.6.39), we see that the $e^{-i\omega t}$-term in

$$-\left(\frac{e}{m_e c}\right)\mathbf{A}\cdot\mathbf{p} = -\left(\frac{e}{m_e c}\right)A_0\hat{\boldsymbol{\varepsilon}}\cdot\mathbf{p}\left[e^{i(\omega/c)\hat{\mathbf{n}}\cdot\mathbf{x} - i\omega t} + e^{-i(\omega/c)\hat{\mathbf{n}}\cdot\mathbf{x} + i\omega t}\right] \tag{5.7.6}$$

is responsible for absorption, while the $e^{+i\omega t}$-term is responsible for stimulated emission.

Let us now treat the absorption case in detail. We have

$$\mathscr{V}_{ni}^{\dagger} = -\frac{eA_0}{m_e c}\left(e^{i(\omega/c)(\hat{\mathbf{n}}\cdot\mathbf{x})}\hat{\boldsymbol{\varepsilon}}\cdot\mathbf{p}\right)_{ni} \tag{5.7.7}$$

and

$$w_{i\to n} = \frac{2\pi}{\hbar}\frac{e^2}{m_e^2c^2}|A_0|^2|\langle n|e^{i(\omega/c)(\hat{\mathbf{n}}\cdot\mathbf{x})}\hat{\boldsymbol{\varepsilon}}\cdot\mathbf{p}|i\rangle|^2\delta(E_n - E_i - \hbar\omega). \tag{5.7.8}$$

The meaning of the δ-function is clear. If $|n\rangle$ forms a continuum, we simply integrate with $\rho(E_n)$. But even if $|n\rangle$ is discrete, because $|n\rangle$ cannot be a ground state (albeit a bound-state energy level), its energy is not infinitely sharp; there may be a natural broadening due to a finite lifetime (see Section 5.8); there can also be a mechanism for broadening due to collisions. In such cases, we regard $\delta(\omega - \omega_{ni})$ as

$$\delta(\omega - \omega_{ni}) = \lim_{\gamma\to 0}\left(\frac{\gamma}{2\pi}\right)\frac{1}{\left[(\omega - \omega_{ni})^2 + \gamma^2/4\right]}. \tag{5.7.9}$$

Finally, the incident electromagnetic wave itself is not perfectly monochromatic; in fact, there is always a finite frequency width.

We derive an absorption cross section as

$$\frac{(\text{Energy/unit time) absorbed by the atom }(i\to n)}{\text{Energy flux of the radiation field}}. \tag{5.7.10}$$

For the energy flux (energy per area per unit time), classical electromagnetic theory gives us

$$c\mathscr{U} = \frac{1}{2\pi}\frac{\omega^2}{c}|A_0|^2, \tag{5.7.11}$$

where we have used

$$\mathscr{U} = \frac{1}{2}\left(\frac{E_{\max}^2}{8\pi} + \frac{B_{\max}^2}{8\pi}\right) \tag{5.7.12}$$

for energy density (energy per unit volume) with

$$\mathbf{E} = -\frac{1}{c}\frac{\partial}{\partial t}\mathbf{A}, \qquad \mathbf{B} = \nabla \times \mathbf{A}. \tag{5.7.13}$$

Putting everything together, we get (remembering that $\hbar\omega$ = energy absorbed by atom for each absorption process)

$$\begin{aligned}\sigma_{abs} &= \frac{\hbar\omega(2\pi/\hbar)(e^2/m_e^2c^2)|A_0|^2|\langle n|e^{i(\omega/c)(\hat{\mathbf{n}}\cdot\mathbf{x})}\hat{\boldsymbol{\varepsilon}}\cdot\mathbf{p}|i\rangle|^2\delta(E_n - E_i - \hbar\omega)}{(1/2\pi)(\omega^2/c)|A_0|^2}\\ &= \frac{4\pi^2\hbar}{m_e^2\omega}\left(\frac{e^2}{\hbar c}\right)|\langle n|e^{i(\omega/c)(\hat{\mathbf{n}}\cdot\mathbf{x})}\hat{\boldsymbol{\varepsilon}}\cdot\mathbf{p}|i\rangle|^2\delta(E_n - E_i - \hbar\omega).\end{aligned} \tag{5.7.14}$$

Equation (5.7.14) has the correct dimension $[1/(M^2/T)](M^2L^2/T^2)T = L^2$ if we recognize that $\alpha = e^2/\hbar c \simeq 1/137$ (dimensionless) and $\delta(E_n - E_i - \hbar\omega) = (1/\hbar)\delta(\omega_{ni} - \omega)$, where $\delta(\omega_{ni} - \omega)$ has time dimension T.

Electric Dipole Approximation

The *electric dipole approximation* (E1 approximation) is based on the fact that the wavelength of the radiation field is far longer than the atomic dimension, so that the series (remember $\omega/c = 1/\lambda\!\!\!^{-}$)

$$e^{i(\omega/c)\hat{\mathbf{n}}\cdot\mathbf{x}} = 1 + i\frac{\omega}{c}\hat{\mathbf{n}}\cdot\mathbf{x} + \cdots \tag{5.7.15}$$

can be approximated by its leading term, 1. The validity of this for a light atom is as follows: First, the $\hbar\omega$ of the radiation field must be of order of atomic level spacing, so

$$\hbar\omega \sim \frac{Ze^2}{(a_0/Z)} \simeq \frac{Ze^2}{R_{\text{atom}}}. \tag{5.7.16}$$

This leads to

$$\frac{c}{\omega} = \lambda\!\!\!^{-} \sim \frac{c\hbar R_{\text{atom}}}{Ze^2} \simeq \frac{137 R_{\text{atom}}}{Z}. \tag{5.7.17}$$

In other words,

$$\frac{1}{\lambda\!\!\!^{-}}R_{\text{atom}} \sim \frac{Z}{137} \ll 1 \tag{5.7.18}$$

for light atoms (small Z). Because the matrix element of x is of order R_{atom}, that of x^2, of order R^2_{atom}, and so on, we see that the approximation of replacing (5.7.15) by its leading term is an excellent one.

Now we have

$$\langle n|e^{i(\omega/c)(\hat{\mathbf{n}}\cdot\mathbf{x})}\hat{\boldsymbol{\varepsilon}}\cdot\mathbf{p}|i\rangle \to \hat{\boldsymbol{\varepsilon}}\cdot\langle n|\mathbf{p}|i\rangle. \tag{5.7.19}$$

In particular, we take $\hat{\boldsymbol{\varepsilon}}$ along the x-axis (and $\hat{\mathbf{n}}$ along the z-axis). We must calculate $\langle n|p_x|i\rangle$. Using

$$[x, H_0] = \frac{i\hbar p_x}{m}, \tag{5.7.20}$$

we have

$$\begin{aligned}\langle n|p_x|i\rangle &= \frac{m}{i\hbar}\langle n|[x, H_0]|i\rangle \\ &= im\omega_{ni}\langle n|x|i\rangle.\end{aligned} \tag{5.7.21}$$

Because of the approximation of the dipole operator, this approximation scheme is called the **electric dipole approximation**. We may here recall [see (3.10.39)] the selection rule for the dipole matrix element. Since $\mathbf{x}$ is a spherical tensor of rank 1 with $q = \pm 1$, we must have $m' - m = \pm 1$, $|j' - j| = 0,1$ (no $0 \to 0$ transition). If $\hat{\boldsymbol{\varepsilon}}$ is along the y-axis, the same selection rule applies. On the other hand, if $\hat{\boldsymbol{\varepsilon}}$ is in the z-direction, $q = 0$; hence, $m' = m$.

With the electric dipole approximation, the absorption cross section (5.7.14) now takes a simpler form upon using (5.7.19) and (5.7.21) as

$$\sigma_{abs} = 4\pi^2\alpha\omega_{ni}|\langle n|x|i\rangle|^2\delta(\omega - \omega_{ni}). \tag{5.7.22}$$

In other words, σ_{abs} treated as a function of ω exhibits a sharp δ-function-like peak whenever $\hbar\omega$ corresponds to the energy-level spacing at $\omega \simeq (E_n - E_i)/\hbar$. Suppose $|i\rangle$ is the ground state, then ω_{ni} is necessarily positive; integrating (5.7.22), we get

$$\int \sigma_{abs}(\omega)\,d\omega = \sum_n 4\pi^2\alpha\omega_{ni}|\langle n|x|i\rangle|^2. \tag{5.7.23}$$

In atomic physics we define **oscillator strength**, f_{ni}, as

$$f_{ni} \equiv \frac{2m\omega_{ni}}{\hbar}|\langle n|x|i\rangle|^2. \tag{5.7.24}$$

It is then straightforward (consider $[x,[x, H_0]]$) to establish the **Thomas-Reiche-Kuhn sum rule**,

$$\sum_n f_{ni} = 1. \tag{5.7.25}$$

In terms of the integration over the absorption cross section, we have

$$\int \sigma_{abs}(\omega)\,d\omega = \frac{4\pi^2\alpha\hbar}{2m_e} = 2\pi^2 c\left(\frac{e^2}{m_e c^2}\right). \tag{5.7.26}$$

Notice how $\hbar$ has disappeared. Indeed, this is just the oscillation sum rule already known in classical electrodynamics (Jackson 1975, for instance). Historically, this was one of the first examples of how "new quantum mechanics" led to the correct classical result. This sum rule is quite remarkable because we did not specify in detail the form of the Hamiltonian.

Photoelectric Effect

We now consider the **photoelectric effect**—that is, the ejection of an electron when an atom is placed in the radiation field. The basic process is considered to be the transition from an atomic (bound) state to a continuum state $E > 0$. Therefore, $|i\rangle$ is the ket for an atomic state, while $|n\rangle$ is the ket for a continuum state, which can be taken to be a plane-wave state $|\mathbf{k}_f\rangle$, an approximation that is valid if the final electron is not too slow. Our earlier formula for $\sigma_{abs}(\omega)$ can still be used, except that we must now integrate $\delta(\omega_{ni} - \omega)$ together with the density of final states $\rho(E_n)$.

Our basic task is to calculate the number of final states per unit energy interval. As we will see in a moment, this is an example where the matrix element depends not only on the final state energy but also on the momentum *direction*. We must therefore consider a group of final states with both similar momentum directions and similar energies.

To count the number of states it is convenient to use the box normalization convention for plane-wave states. We consider a plane-wave state normalized if when we integrate the square modulus of its wave function for a cubic box of side L, we obtain unity. Furthermore, the state is assumed to satisfy the periodic boundary condition with periodicity of the side of the box. The wave function must then be of form

$$\langle \mathbf{x}|\mathbf{k}_f\rangle = \frac{e^{i\mathbf{k}_f\cdot\mathbf{x}}}{L^{3/2}}, \tag{5.7.27}$$

where the allowed values of k_x must satisfy

$$k_x = \frac{2\pi n_x}{L}, \ldots, \tag{5.7.28}$$

with n_x a positive or negative integer. Similar restrictions hold for k_y and k_z. Notice that as $L \to \infty$, k_x, k_y, and k_z become continuous variables.

The problem of counting the number of states is reduced to that of counting the number of dots in three-dimensional lattice space. We define n such that

$$n^2 = n_x^2 + n_y^2 + n_z^2. \tag{5.7.29}$$

As $L \to \infty$, it is a good approximation to treat n as a continuous variable; in fact it is just the magnitude of the radial vector in the lattice space. Let us consider a small-volume element such that the radial vector falls within n and $n + dn$ and the solid angle element $d\Omega$; clearly, it is of volume $n^2\,dn\,d\Omega$. The energy of the final-state plane wave is related to k_f and hence to n; we have

$$E = \frac{\hbar^2 k_f^2}{2m_e} = \frac{\hbar^2}{2m_e}\frac{n^2(2\pi)^2}{L^2}. \tag{5.7.30}$$

Furthermore, the direction of the radial vector in the lattice space is just the momentum direction of the final state, so the number of states in the interval between E and $E + dE$ with direction into $d\Omega$ being $\mathbf{k}_f$ is (remember $dE = (\hbar^2 k_f/m_e)\,dk_f$) given by*

$$\begin{aligned} n^2\,d\Omega\frac{dn}{dE}\,dE &= \left(\frac{L}{2\pi}\right)^3 \left(\mathbf{k}_f^2\right)\frac{dk_f}{dE}\,d\Omega\,dE \\ &= \left(\frac{L}{2\pi}\right)^3 \frac{m_e}{\hbar^2}k_f\,dE\,d\Omega. \end{aligned} \tag{5.7.31}$$

We can now put everything together to obtain an expression for the differential cross section for the photoelectric effect:

$$\frac{d\sigma}{d\Omega} = \frac{4\pi^2\alpha\hbar}{m_e^2\omega}|\langle\mathbf{k}_f|e^{i(\omega/c)(\hat{\mathbf{n}}\cdot\mathbf{x})}\hat{\boldsymbol{\varepsilon}}\cdot\mathbf{p}|i\rangle|^2\frac{m_e k_f L^3}{\hbar^2(2\pi)^3}. \tag{5.7.32}$$

To be specific, let us consider the ejection of a K shell (the innermost shell) electron caused by absorption of light. The initial-state wave function is essentially the same as the ground-state hydrogen atom wave function except that the Bohr radius a_0 is replaced by a_0/Z. Thus

$$\begin{aligned} \langle\mathbf{k}_f|e^{i(\omega/c)(\hat{\mathbf{n}}\cdot\mathbf{x})}\hat{\boldsymbol{\varepsilon}}\cdot\mathbf{p}|i\rangle &= \hat{\boldsymbol{\varepsilon}}\cdot\int d^3x\frac{e^{-i\mathbf{k}_f\cdot\mathbf{x}}}{L^{3/2}}e^{i(\omega/c)(\hat{\mathbf{n}}\cdot\mathbf{x})} \\ &\quad\times(-i\hbar\nabla)\left[e^{-Zr/a_0}\left(\frac{Z}{a_0}\right)^{3/2}\right]. \end{aligned} \tag{5.7.33}$$

Integrating by parts, we can pass ∇ to the left side. Furthermore,

$$\hat{\boldsymbol{\varepsilon}}\cdot\left[\nabla e^{i(\omega/c)(\hat{\mathbf{n}}\cdot\mathbf{x})}\right] = 0 \tag{5.7.34}$$

because $\hat{\boldsymbol{\varepsilon}}$ is perpendicular to $\hat{\mathbf{n}}$. On the other hand, ∇ acting on $e^{-i\mathbf{k}_f\cdot\mathbf{x}}$ brings down $-i\mathbf{k}_f$, which can be taken outside the integral. Thus to evaluate (5.7.33), all we need to do is take the Fourier transform of the atomic wave function with respect to

$$\mathbf{q} \equiv \mathbf{k}_f - \left(\frac{\omega}{c}\right)\hat{\mathbf{n}}. \tag{5.7.35}$$

---

*This is equivalent to taking one state per cube $d^3x\,d^3p/(2\pi\hbar)^3$ in phase space.

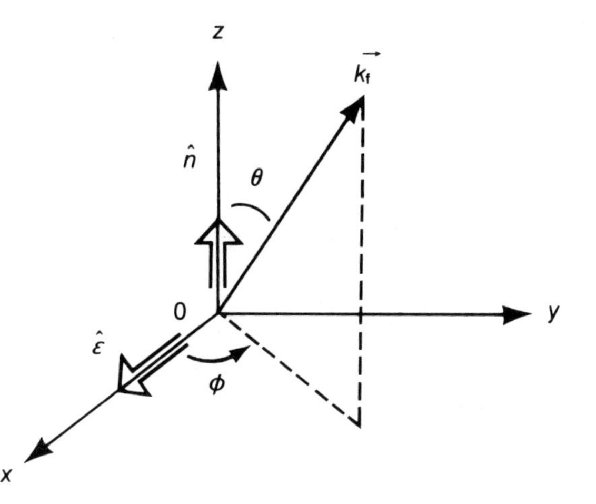

FIGURE 5.10. Polar coordinate system with $\boldsymbol{\varepsilon}$ and $\hat{\mathbf{n}}$ along $0x$ and $0z$, respectively, and $\mathbf{k}_f = (k_f \sin\theta\cos\phi,\ k_f \sin\theta\sin\phi,\ k_f\cos\theta)$.

The final answer is (see Problem 39 of this chapter for the Fourier transform of the hydrogen atom wave function)

$$\frac{d\sigma}{d\Omega} = 32e^2 k_f \frac{(\hat{\boldsymbol{\varepsilon}}\cdot\mathbf{k}_f)^2}{mc\omega}\frac{Z^5}{a_0^5}\frac{1}{\left[(Z^2/a_0^2)+q^2\right]^4}. \tag{5.7.36}$$

If we introduce the coordinate system shown in Figure 5.10, we can write the differential cross section in terms of θ and ϕ using

$$(\hat{\boldsymbol{\varepsilon}}\cdot\mathbf{k}_f)^2 = k_f^2 \sin^2\theta\cos^2\phi$$

$$q^2 = k_f^2 - 2k_f\frac{\omega}{c}\cos\theta + \left(\frac{\omega}{c}\right)^2. \tag{5.7.37}$$

5.8. ENERGY SHIFT AND DECAY WIDTH

Our considerations so far have been restricted to the question of how states other than the initial state get populated. In other words, we have been concerned with the time development of the coefficient $c_n(t)$ with $n \neq i$. The question naturally arises, What happens to $c_i(t)$ itself?

To avoid the effect of a sudden change in the Hamiltonian, we propose to increase the perturbation very slowly. In the remote past $(t \to -\infty)$ the time-dependent potential is assumed to be zero. We then *gradually* turn on the perturbation to its full value; specifically,

$$V(t) = e^{\eta t} V \tag{5.8.1}$$

where V is assumed to be constant and η is small and positive. At the end of the calculation, we let $\eta \to 0$ (see Figure 5.11), and the potential then becomes constant at all times.

In the remote past, we take this time to be $-\infty$, so the state ket in the interaction picture is assumed to be $|i\rangle$. Our basic aim is to evaluate $c_i(t)$. However, before we do that, let us make sure that the old formula of the golden rule (see Section 5.6) can be reproduced using this slow-turn-on method. For $c_n(t)$ with $n \neq i$, we have [using (5.6.17)]

$$\begin{aligned} c_n^{(0)}(t) &= 0 \\ c_n^{(1)}(t) &= \frac{-i}{\hbar} V_{ni} \lim_{t_0 \to -\infty} \int_{t_0}^{t} e^{\eta t'} e^{i\omega_{ni} t'} dt' \\ &= \frac{-i}{\hbar} V_{ni} \frac{e^{\eta t + i\omega_{ni} t}}{\eta + i\omega_{ni}}. \end{aligned} \tag{5.8.2}$$

To lowest nonvanishing order, the transition probability is therefore given by

$$|c_n(t)|^2 \simeq \frac{|V_{ni}|^2}{\hbar^2} \frac{e^{2\eta t}}{\eta^2 + \omega_{ni}^2}, \tag{5.8.3}$$

or

$$\frac{d}{dt} |c_n(t)|^2 \simeq \frac{2|V_{ni}|^2}{\hbar^2} \left(\frac{\eta e^{2\eta t}}{\eta^2 + \omega_{ni}^2} \right). \tag{5.8.4}$$

We now let $\eta \to 0$. Clearly, it is all right to replace $e^{\eta t}$ by unity, but note

$$\lim_{\eta \to 0} \frac{\eta}{\eta^2 + \omega_{ni}^2} = \pi \delta(\omega_{ni}) = \pi \hbar \delta(E_n - E_i). \tag{5.8.5}$$

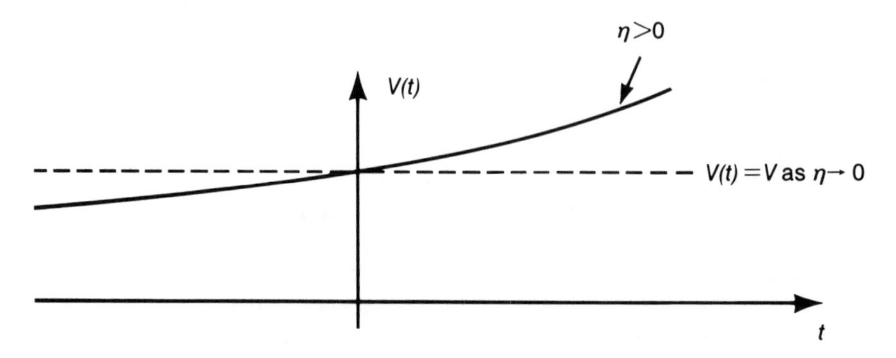

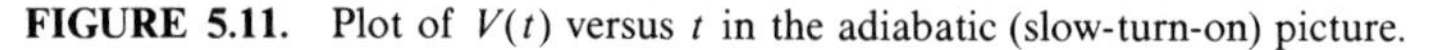

FIGURE 5.11. Plot of $V(t)$ versus t in the adiabatic (slow-turn-on) picture.

This leads to the golden rule,

$$w_{i\to n} \simeq \left(\frac{2\pi}{\hbar}\right)|V_{ni}|^2\delta(E_n - E_i). \tag{5.8.6}$$

Encouraged by this result, let us calculate $c_i^{(0)}$, $c_i^{(1)}$, and $c_i^{(2)}$, again using (5.6.17). We have

$$c_i^{(0)} = 1$$

$$c_i^{(1)} = \frac{-i}{\hbar}V_{ii}\lim_{t_0\to-\infty}\int_{t_0}^{t} e^{\eta t'}\,dt' = \frac{-i}{\hbar\eta}V_{ii}e^{\eta t}$$

$$c_i^{(2)} = \left(\frac{-i}{\hbar}\right)^2\sum_m |V_{mi}|^2\lim_{t_0\to-\infty}\int_{t_0}^{t} dt'\,e^{i\omega_{im}t'+\eta t'}\frac{e^{i\omega_{mi}t'+\eta t'}}{i(\omega_{mi}-i\eta)}$$

$$= \left(\frac{-i}{\hbar}\right)^2|V_{ii}|^2\frac{e^{2\eta t}}{2\eta^2} + \left(\frac{-i}{\hbar}\right)\sum_{m\neq i}\frac{|V_{mi}|^2 e^{2\eta t}}{2\eta(E_i - E_m + i\hbar\eta)}. \tag{5.8.7}$$

Thus up to second order we have

$$c_i(t) \simeq 1 - \frac{i}{\hbar\eta}V_{ii}e^{\eta t} + \left(\frac{-i}{\hbar}\right)^2|V_{ii}|^2\frac{e^{2\eta t}}{2\eta^2} + \left(\frac{-i}{\hbar}\right)\sum_{m\neq i}\frac{|V_{mi}|^2 e^{2\eta t}}{2\eta(E_i - E_m + i\hbar\eta)}. \tag{5.8.8}$$

Now consider the time derivative of c_i $[dc_i(t)/dt \equiv \dot{c}_i]$, which we have from (5.8.8). Upon dividing by c_i and letting $\eta \to 0$ (thus replacing $e^{\eta t}$ and $e^{2\eta t}$ by unity), we get

$$\frac{\dot{c}_i}{c_i} \simeq \frac{\dfrac{-i}{\hbar}V_{ii} + \left(\dfrac{-i}{\hbar}\right)^2\dfrac{|V_{ii}|^2}{\eta} + \left(\dfrac{-i}{\hbar}\right)\displaystyle\sum_{m\neq i}\frac{|V_{mi}|^2}{(E_i - E_m + i\hbar\eta)}}{1 - \dfrac{i}{\hbar}\dfrac{V_{ii}}{\eta}}$$

$$\simeq \frac{-i}{\hbar}V_{ii} + \left(\frac{-i}{\hbar}\right)\sum_{m\neq i}\frac{|V_{mi}|^2}{E_i - E_m + i\hbar\eta}. \tag{5.8.9}$$

Expansion (5.8.9) is formally correct up to second order in V. Note here that $\dot{c}_i(t)/c_i(t)$ is now independent of t. Equation (5.8.9) is a differential equation that is to hold *at all times*. Having obtained this, it is convenient to renormalize c_i so that $c_i(0) = 1$. We now try the ansatz

$$c_i(t) = e^{-i\Delta_i t/\hbar}, \qquad \frac{\dot{c}_i(t)}{c_i(t)} = \frac{-i}{\hbar}\Delta_i \tag{5.8.10}$$

with Δ_i constant (in time) but not necessarily real. Clearly (5.8.10) is consistent with (5.8.9) because the right-hand side of (5.8.10) is constant. We can see the physical meaning of Δ_i by noting that $e^{-i\Delta_i t/\hbar}|i\rangle$ in the interaction picture implies $e^{-i\Delta_i t/\hbar - iE_i t/\hbar}|i\rangle$ in the Schrödinger picture. In other words,

$$E_i \to E_i + \Delta_i \tag{5.8.11}$$

as a result of perturbation. That is, we have calculated the *level shift* using time-*dependent* perturbation theory. Now expand, as usual,

$$\Delta_i = \Delta_i^{(1)} + \Delta_i^{(2)} + \cdots, \tag{5.8.12}$$

and compare (5.8.10) with (5.8.9); we get to first order

$$\Delta_i^{(1)} = V_{ii}. \tag{5.8.13}$$

But this is just what we expect from *t-independent perturbation theory.* Before we look at $\Delta_i^{(2)}$, recall

$$\lim_{\varepsilon \to 0} \frac{1}{x + i\varepsilon} = \text{Pr.}\frac{1}{x} - i\pi\delta(x). \tag{5.8.14}$$

Thus

$$\text{Re}\left(\Delta_i^{(2)}\right) = \text{Pr.} \sum_{m \neq i} \frac{|V_{mi}|^2}{E_i - E_m} \tag{5.8.15a}$$

$$\text{Im}\left(\Delta_i^{(2)}\right) = -\pi \sum_{m \neq i} |V_{mi}|^2 \delta(E_i - E_m). \tag{5.8.15b}$$

But the right-hand side of (5.8.15b) is familiar from the golden rule, so we can identify

$$\sum_{m \neq i} w_{i \to m} = \frac{2\pi}{\hbar} \sum_{m \neq i} |V_{mi}|^2 \delta(E_i - E_m) = -\frac{2}{\hbar}\text{Im}\left[\Delta_i^{(2)}\right]. \tag{5.8.16}$$

Coming back to $c_i(t)$, we can write (5.8.10) as

$$c_i(t) = e^{-(i/\hbar)[\text{Re}(\Delta_i)t] + (1/\hbar)[\text{Im}(\Delta_i)t]}. \tag{5.8.17}$$

If we define

$$\frac{\Gamma_i}{\hbar} \equiv -\frac{2}{\hbar}\text{Im}(\Delta_i), \tag{5.8.18}$$

then

$$|c_i|^2 = e^{2\,\text{Im}(\Delta_i)t/\hbar} = e^{-\Gamma_i t/\hbar}. \tag{5.8.19}$$

Therefore, Γ_i characterizes the rate at which state $|i\rangle$ disappears.

It is worth checking the probability conservation up to second order in V for small t:

$$|c_i|^2 + \sum_{m \neq i} |c_m|^2 = (1 - \Gamma_i t/\hbar) + \sum_{m \neq i} w_{i \to m} t = 1, \tag{5.8.20}$$

where (5.8.16) has been used. Thus the probabilities for finding the initial state and all other states add up to 1. Put in another way, the depletion of state $|i\rangle$ is compensated by the growth of states other than $|i\rangle$.

To summarize, the real part of the energy shift is what we usually associate with the level shift. The imaginary part of the energy shift is, apart from -2 [see (5.8.18)], the **decay width**. Note also

$$\frac{\hbar}{\Gamma_i} = \tau_i \tag{5.8.21}$$

where τ_i is the mean lifetime of state $|i\rangle$ because

$$|c_i|^2 = e^{-t/\tau_i}. \tag{5.8.22}$$

To see why Γ_i is called *width*, we look at the Fourier decomposition

$$\int f(E) e^{-iEt/\hbar} dE = e^{-i[E_i + \mathrm{Re}(\Delta_i)]t/\hbar - \Gamma_i t/2\hbar}. \tag{5.8.23}$$

Using the Fourier inversion formula, we get

$$|f(E)|^2 \propto \frac{1}{\{E - [E_i + \mathrm{Re}(\Delta_i)]\}^2 + \Gamma_i^2/4} \tag{5.8.24}$$

Therefore, Γ_i has the usual meaning of full width at half maximum. Notice that we get the time-energy uncertainty relation from (5.8.21)

$$\Delta t \Delta E \sim \hbar, \tag{5.8.25}$$

where we identify the uncertainty in the energy with Γ_i and the mean lifetime with Δt.

Even though we discussed the subject of energy shift and decay width using the constant perturbation V obtained as the limit of (5.8.1) when $\eta \to 0$, we can easily generalize our considerations to the harmonic perturbation case discussed in Section 5.6. All we must do is to let

$$E_{n(m)} - E_i \to E_{n(m)} - E_i \pm \hbar\omega \tag{5.8.26}$$

in (5.8.2), (5.8.8), and (5.8.15), and so on. The quantum-mechanical description of unstable states we have developed here is originally due to Wigner and Weisskopf in 1930.

PROBLEMS

1. A simple harmonic oscillator (in one dimension) is subjected to a perturbation

$$\lambda H_1 = bx$$

where b is a real constant.

a. Calculate the energy shift of the ground state to *lowest nonvanishing order*.

b. Solve this problem *exactly* and compare with your result obtained in (a).
[You may assume without proof

$$\langle u_{n'}|x|u_n\rangle = \sqrt{\frac{\hbar}{2m\omega}}\left(\sqrt{n+1}\,\delta_{n',n+1} + \sqrt{n}\,\delta_{n',n-1}\right).\Bigg]$$

2. In nondegenerate time-independent perturbation theory, what is the probability of finding in a perturbed energy eigenstate ($|k\rangle$) the corresponding unperturbed eigenstate ($|k^{(0)}\rangle$)? Solve this up to terms of order g^2.
3. Consider a particle in a two-dimensional potential

$$V_0 = \begin{cases} 0 & \text{for } 0 \le x \le L, 0 \le y \le L, \\ \infty & \text{otherwise.} \end{cases}$$

Write the energy eigenfunctions for the ground and first excited states. We now add a time-independent perturbation of the form

$$V_1 = \begin{cases} \lambda xy & \text{for } 0 \le x \le L, 0 \le y \le L, \\ 0 & \text{otherwise.} \end{cases}$$

Obtain the zeroth-order energy eigenfunctions and the first-order energy shifts for the ground and first excited states.
4. Consider an isotropic harmonic oscillator in *two* dimensions. The Hamiltonian is given by

$$H_0 = \frac{p_x^2}{2m} + \frac{p_y^2}{2m} + \frac{m\omega^2}{2}(x^2 + y^2).$$

a. What are the energies of the three lowest-lying states? Is there any degeneracy?
b. We now apply a perturbation

$$V = \delta m\omega^2 xy,$$

where δ is a dimensionless real number much smaller than unity. Find the zeroth-order energy eigenket and the corresponding energy to first order [that is, the unperturbed energy obtained in (a) plus the first-order energy shift] for each of the three lowest-lying states.
c. Solve the $H_0 + V$ problem *exactly*. Compare with the perturbation results obtained in (b).
[You may use $\langle n'|x|n\rangle = \sqrt{\hbar/2m\omega}\,(\sqrt{n+1}\,\delta_{n',n+1} + \sqrt{n}\,\delta_{n',n-1})$.]
5. Establish (5.1.54) for the one-dimensional harmonic oscillator given by (5.1.50) with an additional perturbation $V = \frac{1}{2}\varepsilon m\omega^2 x^2$. Show that all other matrix elements V_{k0} vanish.
6. A slightly anisotropic three-dimensional harmonic oscillator has $\omega_z \approx \omega_x = \omega_y$. A charged particle moves in the field of this oscillator and is at the same time exposed to a uniform magnetic field in the x-direction. Assuming that the Zeeman splitting is comparable to the splitting produced by the anisotropy, but small compared to $\hbar\omega$, cal-

culate to first order the energies of the components of the first excited state. Discuss various limiting cases. (From Merzbacher, *Quantum Mechanics, 2/e*, © 1970. Reprinted by permission of Ellis Horwood, Ltd.)

7. A one-electron atom whose ground state is nondegenerate is placed in a uniform electric field in the z-direction. Obtain an approximate expression for the induced electric dipole moment of the ground state by considering the expectation value of ez with respect to the perturbed state vector computed to first order. Show that the same expression can also be obtained from the energy shift $\Delta = -\alpha|\mathbf{E}|^2/2$ of the ground state computed to second order. (*Note*: α stands for the polarizability.) Ignore spin.
8. Evaluate the matrix elements (or expectation values) given below. If any vanishes, explain why it vanishes using simple symmetry (or other) arguments.
   a. $\langle n=2, l=1, m=0|x|\, n=2, l=0, m=0\rangle$.
   b. $\langle n=2, l=1, m=0|p_z|\, n=2, l=0, m=0\rangle$.
   [In (a) and (b), $|nlm\rangle$ stands for the energy eigenket of a nonrelativistic hydrogen atom with spin ignored.]
   c. $\langle L_z\rangle$ for an electron in a central field with $j=\frac{9}{2}$, $m=\frac{7}{2}$, $l=4$.
   d. $\langle \text{singlet}, m_s=0|S_z^{(e-)} - S_z^{(e+)}|\text{triplet}, m_s=0\rangle$ for an s-state positronium.
   e. $\langle \mathbf{S}^{(1)}\cdot\mathbf{S}^{(2)}\rangle$ for the ground state of a hydrogen *molecule*.
9. A p-orbital electron characterized by $|n, l=1, m=\pm 1, 0\rangle$ (ignore spin) is subjected to a potential

$$V = \lambda(x^2 - y^2) \quad (\lambda = \text{constant}).$$

   a. Obtain the "correct" zeroth-order energy eigenstates that diagonalize the perturbation. You need not evaluate the energy shifts in detail, but show that the original threefold degeneracy is now completely removed.
   b. Because V is invariant under time reversal and because there is no longer any degeneracy, we expect each of the energy eigenstates obtained in (a) to go into itself (up to a phase factor or sign) under time reversal. Check this point explicitly.
10. Consider a spinless particle in a two-dimensional infinite square well:

$$V = \begin{cases} 0 & \text{for } 0 \le x \le a, 0 \le y \le a, \\ \infty & \text{otherwise.} \end{cases}$$

   a. What are the energy eigenvalues for the three lowest states? Is there any degeneracy?
   b. We now add a potential

$$V_1 = \lambda xy, \quad 0 \le x \le a, 0 \le y \le a.$$

Taking this as a weak perturbation, answer the following:

(i) Is the energy shift due to the perturbation linear or quadratic in λ for each of the three states?

(ii) Obtain expressions for the energy shifts of the three lowest states accurate to order λ. (You need not evaluate integrals that may appear.)

(iii) Draw an energy diagram with and without the perturbation for the three energy states. Make sure to specify which unperturbed state is connected to which perturbed state.

11. The Hamiltonian matrix for a two-state system can be written as

$$\mathscr{H} = \begin{pmatrix} E_1^0 & \lambda\Delta \\ \lambda\Delta & E_2^0 \end{pmatrix}.$$

Clearly the energy eigenfunctions for the unperturbed problems ($\lambda = 0$) are given by

$$\phi_1^{(0)} = \begin{pmatrix} 1 \\ 0 \end{pmatrix}, \qquad \phi_2^{(0)} = \begin{pmatrix} 0 \\ 1 \end{pmatrix}.$$

a. Solve this problem *exactly* to find the energy eigenfunctions ψ_1 and ψ_2 and the energy eigenvalues E_1 and E_2.

b. Assuming that $\lambda|\Delta| \ll |E_1^0 - E_2^0|$, solve the same problem using time-independent perturbation theory up to first order in the energy eigenfunctions and up to second order in the energy eigenvalues. Compare with the exact results obtained in (a).

c. Suppose the two unperturbed energies are "almost degenerate," that is,

$$|E_1^0 - E_2^0| \ll \lambda|\Delta|.$$

Show that the exact results obtained in (a) closely resemble what you would expect by applying *degenerate* perturbation theory to this problem with E_1^0 set exactly equal to E_2^0.

12. (This is a tricky problem because the degeneracy between the first and the second state is not removed in first order. See also Gottfried 1966, 397, Problem 1.) This problem is from Schiff 1968, 295, Problem 4. A system that has three unperturbed states can be represented by the perturbed Hamiltonian matrix

$$\begin{pmatrix} E_1 & 0 & a \\ 0 & E_1 & b \\ a^* & b^* & E_2 \end{pmatrix}$$

where $E_2 > E_1$. The quantities a and b are to be regarded as perturbations that are of the same order and are small compared with $E_2 - E_1$. Use the second-order nondegenerate perturbation theory to calculate the perturbed eigenvalues. (Is this procedure correct?) Then diagonalize the matrix to find the exact eigenvalues. Finally, use the second-order degenerate perturbation theory. Compare the three results obtained.

13. Compute the Stark effect for the $2S_{1/2}$ and $2P_{1/2}$ levels of hydrogen for a field ε sufficiently weak so that $e\varepsilon a_0$ is small compared to the fine structure, but take the Lamb shift δ ($\delta = 1057$ MHz) into account (that is, ignore $2P_{3/2}$ in this calculation). Show that for $e\varepsilon a_0 \ll \delta$, the energy shifts are quadratic in ε, whereas for $e\varepsilon a_0 \gg \delta$ they are linear in ε. (The radial integral you need is $\langle 2s|r|2p\rangle = 3\sqrt{3}\, a_0$.) Briefly discuss the consequences (if any) of time reversal for this problem. This problem is from Gottfried 1966, Problem 7-3.
14. Work out the Stark effect to lowest nonvanishing order for the $n = 3$ level of the hydrogen atom. Ignoring the spin-orbit force and relativistic correction (Lamb shift), obtain not only the energy shifts to lowest nonvanishing order but also the corresponding zeroth-order eigenket.
15. Suppose the electron had a very small intrinsic *electric* dipole moment analogous to the spin magnetic moment (that is, $\boldsymbol{\mu}_{el}$ proportional to $\boldsymbol{\sigma}$). Treating the hypothetical $-\boldsymbol{\mu}_{el}\cdot\mathbf{E}$ interaction as a small perturbation, discuss qualitatively how the energy levels of the Na atom ($Z = 11$) would be altered in the absence of any external electromagnetic field. Are the level shifts first order or second order? State explicitly which states get mixed with each other. Obtain an expression for the energy shift of the lowest level that is affected by the perturbation. Assume throughout that only the valence electron is subjected to the hypothetical interaction.
16. Consider a particle bound to a fixed center by a spherically symmetric potential $V(r)$.
    a. Prove
    $$|\psi(0)|^2 = \left(\frac{m}{2\pi\hbar^2}\right)\left\langle \frac{dV}{dr} \right\rangle$$
    for all s states, ground and excited.
    b. Check this relation for the ground state of a three-dimensional isotropic oscillator, the hydrogen atom, and so on.
    (*Note*: This relation has actually been found to be useful in guessing the form of the potential between a quark and an antiquark.)
17. a. Suppose the Hamiltonian of a rigid rotator in a magnetic field perpendicular to the axis is of the form (Merzbacher 1970, Problem 17-1)
    $$A\mathbf{L}^2 + BL_z + CL_y$$
    if terms quadratic in the field are neglected. Assuming $B \gg C$, use perturbation theory to lowest nonvanishing order to get approximate energy eigenvalues.

b. Consider the matrix elements

$$\langle n'l'm_l'm_s'|(3z^2 - r^2)|nlm_lm_s\rangle,$$

$$\langle n'l'm_l'm_s'|xy|nlm_lm_s\rangle$$

of a one-electron (for example, alkali) atom. Write the selection rules for Δl, Δm_l, and Δm_s. Justify your answer.

18. Work out the *quadratic* Zeeman effect for the ground-state hydrogen atom $[\langle\mathbf{x}|0\rangle = (1/\sqrt{\pi a_0^3})e^{-r/a_0}]$ due to the usually neglected $e^2\mathbf{A}^2/2m_ec^2$-term in the Hamiltonian taken to first order. Write the energy shift as

$$\Delta = -\tfrac{1}{2}\chi\mathbf{B}^2$$

and obtain an expression for *diamagnetic susceptibility*, χ. (The following definite integral may be useful:

$$\int_0^\infty e^{-\alpha r} r^n \, dr = \frac{n!}{\alpha^{n+1}}.)$$

19. (Merzbacher 1970, 448, Problem 11.) For the He wave function, use

$$\psi(\mathbf{x}_1, \mathbf{x}_2) = (Z_{\text{eff}}^3/\pi a_0^3)\exp\left[\frac{-Z_{\text{eff}}(r_1 + r_2)}{a_0}\right]$$

with $Z_{\text{eff}} = 2 - \frac{5}{16}$, as obtained by the variational method. The measured value of the diamagnetic susceptibility is 1.88×10^{-6} cm^3/mole.

Using the Hamiltonian for an atomic electron in a magnetic field, determine, for a state of zero angular momentum, the energy change to order B^2 if the system is in a uniform magnetic field represented by the vector potential $\mathbf{A} = \frac{1}{2}\mathbf{B}\times\mathbf{r}$.

Defining the atomic diamagnetic susceptibility χ by $E = -\frac{1}{2}\chi B^2$, calculate χ for a helium atom in the ground state and compare the result with the measured value.

20. Estimate the ground-state energy of a one-dimensional simple harmonic oscillator using

$$\langle x|\tilde{0}\rangle = e^{-\beta|x|}$$

as a trial function with β to be varied. (You may use

$$\int_0^\infty e^{-\alpha x} x^n \, dx = \frac{n!}{\alpha^{n+1}}.)$$

21. Estimate the lowest eigenvalue (λ) of the differential equation

$$\frac{d^2\psi}{dx^2} + (\lambda - |x|)\psi = 0, \qquad \psi \to 0 \text{ for } |x| \to \infty$$

using the variational method with

$$\psi = \begin{cases} c(\alpha - |x|), & \text{for } |x| < \alpha \\ 0, & \text{for } |x| > \alpha \end{cases} \qquad (\alpha \text{ to be varied})$$

as a trial function. (*Caution*: $d\psi/dx$ is discontinuous at $x = 0$.) Numerical data that may be useful for this problem are:

$$3^{1/3} = 1.442, \qquad 5^{1/3} = 1.710, \qquad 3^{2/3} = 2.080, \qquad \pi^{2/3} = 2.145.$$

The *exact* value of the lowest eigenvalue can be shown to be 1.019.

22. Consider a one-dimensional simple harmonic oscillator whose classical angular frequency is ω_0. For $t < 0$ it is known to be in the ground state. For $t > 0$ there is also a time-dependent potential

$$V(t) = F_0 x \cos \omega t$$

where F_0 is constant in both space and time. Obtain an expression for the expectation value $\langle x \rangle$ as a function of time using time-dependent perturbation theory to lowest nonvanishing order. Is this procedure valid for $\omega \simeq \omega_0$? [You may use $\langle n'|x|n\rangle = \sqrt{\hbar/2m\omega_0}\,(\sqrt{n+1}\,\delta_{n',n+1} + \sqrt{n}\,\delta_{n',n-1})$.]

23. A one-dimensional harmonic oscillator is in its ground state for $t < 0$. For $t \geq 0$ it is subjected to a time-dependent but spatially uniform *force* (not potential!) in the x-direction,

$$F(t) = F_0 e^{-t/\tau}.$$

a. Using time-dependent perturbation theory to first order, obtain the probability of finding the oscillator in its first excited state for $t > 0$. Show that the $t \to \infty$ (τ finite) limit of your expression is independent of time. Is this reasonable or surprising?

b. Can we find higher excited states?

[You may use

$$\langle n'|x|n\rangle = \sqrt{\hbar/2m\omega}\left(\sqrt{n}\,\delta_{n',n-1} + \sqrt{n+1}\,\delta_{n',n+1}\right).]$$

24. Consider a particle bound in a simple harmonic oscillator potential. Initially ($t < 0$), it is in the ground state. At $t = 0$ a perturbation of the form

$$H'(x,t) = Ax^2 e^{-t/\tau}$$

is switched on. Using time-dependent perturbation theory, calculate the probability that, after a sufficiently long time ($t \gg \tau$), the system will have made a transition to a given excited state. Consider all final states.

25. The unperturbed Hamiltonian of a two-state system is represented by

$$H_0 = \begin{pmatrix} E_1^0 & 0 \\ 0 & E_2^0 \end{pmatrix}.$$

There is, in addition, a time-dependent perturbation

$$V(t) = \begin{pmatrix} 0 & \lambda \cos \omega t \\ \lambda \cos \omega t & 0 \end{pmatrix} \quad (\lambda \text{ real}).$$

a. At $t = 0$ the system is known to be in the first state, represented by

$$\begin{pmatrix} 1 \\ 0 \end{pmatrix}.$$

Using time-dependent perturbation theory and assuming that $E_1^0 - E_2^0$ is *not* close to $\pm \hbar\omega$, derive an expression for the probability that the system be found in the second state represented by

$$\begin{pmatrix} 0 \\ 1 \end{pmatrix}$$

as a function of t $(t > 0)$.

b. Why is this procedure not valid when $E_1^0 - E_2^0$ is close to $\pm \hbar\omega$?

26. A one-dimensional simple harmonic oscillator of angular frequency ω is acted upon by a spatially uniform but time-dependent force (*not* potential)

$$F(t) = \frac{(F_0\tau/\omega)}{(\tau^2 + t^2)}, \qquad -\infty < t < \infty.$$

At $t = -\infty$, the oscillator is known to be in the ground state. Using the time-dependent perturbation theory to first order, calculate the probability that the oscillator is found in the first excited state at $t = +\infty$.

Challenge for experts: $F(t)$ is so normalized that the impulse

$$\int F(t)\,dt$$

imparted to the oscillator is always the same—that is, independent of τ; yet for $\tau \gg 1/\omega$, the probability for excitation is essentially negligible. Is this reasonable? [Matrix element of x: $\langle n'|x|n\rangle = (\hbar/2m\omega)^{1/2}(\sqrt{n}\,\delta_{n',n-1} + \sqrt{n+1}\,\delta_{n',n+1})$.]

27. Consider a particle in one dimension moving under the influence of some time-independent potential. The energy levels and the corresponding eigenfunctions for this problem are assumed to be known. We now subject the particle to a traveling pulse represented by a time-dependent potential,

$$V(t) = A\delta(x - ct).$$

a. Suppose at $t = -\infty$ the particle is known to be in the ground state whose energy eigenfunction is $\langle x|i\rangle = u_i(x)$. Obtain the probability for finding the system in some excited state with energy eigenfunction $\langle x|f\rangle = u_f(x)$ at $t = +\infty$.

b. Interpret your result in (a) physically by regarding the δ-function pulse as a superposition of harmonic perturbations; recall

$$\delta(x - ct) = \frac{1}{2\pi c}\int_{-\infty}^{\infty} d\omega\, e^{i\omega[(x/c) - t]}.$$

Emphasize the role played by energy conservation, which holds even quantum mechanically as long as the perturbation has been on for a very long time.

28. A hydrogen atom in its ground state $[(n, l, m) = (1,0,0)]$ is placed between the plates of a capacitor. A time-dependent but spatial uniform electric field (not potential!) is applied as follows:

$$\mathbf{E} = \begin{cases} 0 & \text{for } t < 0, \\ \mathbf{E}_0 e^{-t/\tau} & \text{for } t > 0 \end{cases} \qquad (\mathbf{E}_0 \text{ in the positive } z\text{-direction}).$$

Using first-order time-dependent perturbation theory, compute the probability for the atom to be found at $t \gg \tau$ in each of the three $2p$ states: $(n, l, m) = (2, 1, \pm 1$ or $0)$. Repeat the problem for the $2s$ state: $(n, l, m) = (2,0,0)$. You need not attempt to evaluate radial integrals, but perform all other integrations (with respect to angles and time).

29. Consider a composite system made up of two spin $\frac{1}{2}$ objects. For $t < 0$, the Hamiltonian does not depend on spin and can be taken to be zero by suitably adjusting the energy scale. For $t > 0$, the Hamiltonian is given by

$$H = \left(\frac{4\Delta}{\hbar^2}\right)\mathbf{S}_1 \cdot \mathbf{S}_2.$$

Suppose the system is in $|+-\rangle$ for $t \leq 0$. Find, as a function of time, the probability for being found in each of the following states $|++\rangle$, $|+-\rangle$, $|-+\rangle$, and $|--\rangle$:

a. By solving the problem exactly.

b. By solving the problem assuming the validity of first-order time-dependent perturbation theory with H as a perturbation switched on at $t = 0$. Under what condition does (b) give the correct results?

30. Consider a two-level system with $E_1 < E_2$. There is a time-dependent potential that connects the two levels as follows:

$$V_{11} = V_{22} = 0, \qquad V_{12} = \gamma e^{i\omega t}, \qquad V_{21} = \gamma e^{-i\omega t} \quad (\gamma \text{ real}).$$

At $t = 0$, it is known that only the lower level is populated—that is, $c_1(0) = 1$, $c_2(0) = 0$.

a. Find $|c_1(t)|^2$ and $|c_2(t)|^2$ for $t > 0$ by *exactly* solving the coupled differential equation

$$i\hbar \dot{c}_k = \sum_{n=1}^{2} V_{kn}(t) e^{i\omega_{kn}t} c_n, \quad (k = 1, 2).$$

b. Do the same problem using time-dependent perturbation theory to lowest nonvanishing order. Compare the two approaches for small values of γ. Treat the following two cases separately: (i) ω very different from ω_{21} and (ii) ω close to ω_{21}.

Answer for (a): (Rabi's formula)

$$|c_2(t)|^2 = \frac{\gamma^2/\hbar^2}{\gamma^2/\hbar^2 + (\omega - \omega_{21})^2/4} \sin^2\left\{\left[\frac{\gamma^2}{\hbar^2} + \frac{(\omega - \omega_{21})^2}{4}\right]^{1/2} t\right\},$$

$$|c_1(t)|^2 = 1 - |c_2(t)|^2.$$

31. Show that the slow-turn-on of perturbation $V \rightarrow Ve^{\eta t}$ (see Baym 1969, 257) can generate contribution from the second term in (5.6.36).
32. a. Consider the positronium problem you solved in Chapter 3, Problem 3. In the presence of a uniform and static magnetic field B along the z-axis, the Hamiltonian is given by

$$H = A\mathbf{S}_1 \cdot \mathbf{S}_2 + \left(\frac{eB}{m_e c}\right)(S_{1z} - S_{2z}).$$

Solve this problem to obtain the energy levels of all four states *using degenerate time-independent perturbation theory* (instead of diagonalizing the Hamiltonian matrix). Regard the first and the second terms in the expression for H as H_0 and V, respectively. Compare your results with the exact expressions

$$E = -\left(\frac{\hbar^2 A}{4}\right)\left[1 \pm 2\sqrt{1 + 4\left(\frac{eB}{m_e c\hbar A}\right)^2}\,\right] \quad \text{for} \begin{cases} \text{singlet } m = 0 \\ \text{triplet } m = 0 \end{cases}$$

$$E = \frac{\hbar^2 A}{4} \qquad \text{for triplet } m = \pm 1,$$

where *triplet* (*singlet*) $m = 0$ stands for the state that becomes a pure triplet (singlet) with $m = 0$ as $B \rightarrow 0$.

b. We now attempt to cause transitions (via stimulated emission and absorption) between the two $m = 0$ states by introducing an oscillating magnetic field of the "right" frequency. Should we orient the magnetic field along the z-axis or along the x- (or y-) axis? Justify your choice. (The original static field is assumed to be along the z-axis throughout.)

c. Calculate the eigenvectors to first order.

32′. Repeat Problem 32 above, but with the atomic hydrogen Hamiltonian

$$H = A\mathbf{S}_1 \cdot \mathbf{S}_2 + \left(\frac{eB}{m_e c}\right)\mathbf{S}_1 \cdot \mathbf{B}$$

where in the hyperfine term $A\mathbf{S}_1 \cdot \mathbf{S}_2$, $\mathbf{S}_1$ is the electron spin, while $\mathbf{S}_2$ is the proton spin. [Note the problem here has less symmetry than that of the positronium case].

33. Consider the spontaneous emission of a photon by an excited atom.

The process is known to be an $E1$ transition. Suppose the magnetic quantum number of the atom decreases by one unit. What is the angular distribution of the emitted photon? Also discuss the polarization of the photon with attention to angular-momentum conservation for the whole (atom plus photon) system.

34. Consider an atom made up of an electron and a singly charged ($Z=1$) triton (^{3}H). Initially the system is in its ground state ($n=1$, $l=0$). Suppose the nuclear charge *suddenly increases* by one unit (realistically by emitting an electron and an antineutrino). This means that the triton nucleus turns into a helium ($Z=2$) nucleus of mass 3 (^{3}He). Obtain the probability for the system to be found in the ground state of the resulting helium ion. The hydrogenic wave function is given by

$$\psi_{n=1,l=0}(\mathbf{x}) = \frac{1}{\sqrt{\pi}}\left(\frac{Z}{a_0}\right)^{3/2} e^{-Zr/a_0}.$$

35. The ground state of a hydrogen atom ($n=1, l=0$) is subjected to a time-dependent potential as follows:

$$V(\mathbf{x},t) = V_0 \cos(kz - \omega t).$$

Using time-dependent perturbation theory, obtain an expression for the transition rate at which the electron is emitted with momentum $\mathbf{p}$. Show, in particular, how you may compute the angular distribution of the ejected electron (in terms of θ and ϕ defined with respect to the z-axis). Discuss *briefly* the similarities and the differences between this problem and the (more realistic) photoelectric effect. (*Note:* For the initial wave function see Problem 34. If you have a normalization problem, the final wave function may be taken to be

$$\psi_f(\mathbf{x}) = \left(\frac{1}{L^{3/2}}\right) e^{i\mathbf{p}\cdot\mathbf{x}/\hbar}$$

with L very large, but you should be able to show that the observable effects are independent of L.)

36. Derive an expression for the density of free particle states in *two* dimensions, that is, the two-dimensional analog of

$$\rho(E)\, dE\, d\Omega = \left(\frac{L}{2\pi}\right)^3 \left(\frac{mk}{\hbar^2}\right) dE\, d\Omega, \quad \left(\mathbf{k} \equiv \frac{\mathbf{p}}{\hbar},\ E = \frac{\mathbf{p}^2}{2m}\right).$$

Your answer should be written as a function of k (or E) times $dE\, d\phi$, where ϕ is the polar angle that characterizes the momentum direction in two dimensions.

37. A particle of mass m constrained to move in one dimension is confined

within $0 < x < L$ by an infinite-wall potential

$$V = \infty \quad \text{for } x < 0, x > L,$$
$$V = 0 \quad \text{for } 0 \leq x \leq L.$$

Obtain an expression for the density of states (that is, the number of states per unit energy interval) for *high* energies as a function of E. (Check your dimension!)

38. Linearly polarized light of angular frequency ω is incident on a one-electron "atom" whose wave function can be approximated by the ground state of a three-dimensional isotropic harmonic oscillator of angular frequency ω_0. Show that the differential cross section for the ejection of a photoelectron is given by

$$\frac{d\sigma}{d\Omega} = \frac{4\alpha\hbar^2 k_f^3}{m^2\omega\omega_0}\sqrt{\frac{\pi\hbar}{m\omega_0}}\exp\left\{-\frac{\hbar}{m\omega_0}\left[k_f^2 + \left(\frac{\omega}{c}\right)^2\right]\right\}$$
$$\times \sin^2\theta\cos^2\phi\exp\left[\left(\frac{2\hbar k_f\omega}{m\omega_0 c}\right)\cos\theta\right]$$

provided the ejected electron of momentum $\hbar k_f$ can be regarded as being in a plane-wave state. (The coordinate system used is shown in Figure 5.10.)

39. Find the probability $|\phi(\mathbf{p}')|^2 d^3p'$ of the particular momentum $\mathbf{p}'$ for the ground-state hydrogen atom. (This is a nice exercise in three-dimensional Fourier transforms. To perform the angular integration choose the z-axis in the direction of $\mathbf{p}$.)

40. Obtain an expression for $\tau(2p \to 1s)$ for the hydrogen atom. Verify that it is equal to 1.6×10^{-9} s.

CHAPTER 6

Identical Particles

This short chapter is devoted to a discussion of some striking quantum-mechanical effects arising from the identity of particles. We also consider some applications to atoms more complex than hydrogen or hydrogenlike atoms.

6.1. PERMUTATION SYMMETRY

In classical physics it is possible to keep track of individual particles even though they may look alike. When we have particle 1 and particle 2 considered as a system, we can, in principle, follow the trajectory of 1 and that of 2 separately at each instant of time. For bookkeeping purposes, you may color one of them blue and the other red and then examine how the red particle moves and how the blue particle moves as time passes.

In quantum mechanics, however, identical particles are truly indistinguishable. This is because we cannot specify more than a complete set of commuting observables for each of the particles; in particular, we cannot label the particle by coloring it blue. Nor can we follow the trajectory because that would entail a position measurement at each instant of time, which necessarily disturbs the system; in particular the two situations (a) and (b) shown in Figure 6.1 cannot be distinguished—not even in principle.

For simplicity consider just two particles. Suppose one of the particles, which we call particle 1, is characterized by $|k'\rangle$, where k' is a

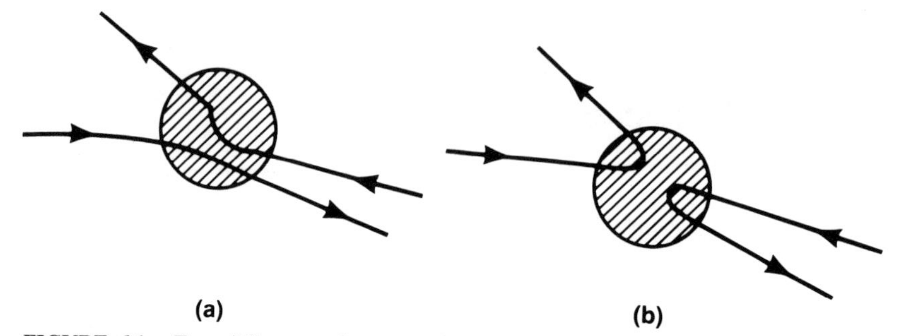

FIGURE 6.1. Two different paths, (a) and (b), of a two-electron system, for example, in which we cannot assert even in principle through which of the paths the electrons pass.

collective index for a complete set of observables. Likewise, we call the ket of the remaining particle $|k''\rangle$. The state ket for the two particles can be written in product form,

$$|k'\rangle|k''\rangle, \tag{6.1.1}$$

where it is understood that the first ket refers to particle 1 and the second ket to particle 2. We can also consider

$$|k''\rangle|k'\rangle, \tag{6.1.2}$$

where particle 1 is characterized by $|k''\rangle$ and particle 2 by $|k'\rangle$. Even though the two particles are indistinguishable, it is worth noting that mathematically (6.1.1) and (6.1.2) are *distinct* kets for $k' \neq k''$. In fact, with $k' \neq k''$, they are orthogonal to each other.

Suppose we make a measurement on the two-particle system. We may obtain k' for one particle and k'' for the other. However, we do not know a priori whether the state ket is $|k'\rangle|k''\rangle$, $|k''\rangle|k'\rangle$, or—for that matter—any linear combination of the two. Put in another way, all kets of form

$$c_1|k'\rangle|k''\rangle + c_2|k''\rangle|k'\rangle \tag{6.1.3}$$

lead to an identical set of eigenvalues when measurement is performed. This is known as **exchange degeneracy**. Exchange degeneracy presents a difficulty because, unlike the single particle case, a specification of the eigenvalue of a complete set of observables does not completely determine the state ket. The way nature avoids this difficulty is quite ingenious. But before proceeding further, let us develop the mathematics of permutation symmetry.

We define the permutation operator P_{12} by

$$P_{12}|k'\rangle|k''\rangle = |k''\rangle|k'\rangle. \tag{6.1.4}$$

Clearly,

$$P_{21} = P_{12} \quad \text{and} \quad P_{12}^2 = 1. \tag{6.1.5}$$

Under P_{12}, particle 1 having k' becomes particle 1 having k''; particle 2 having k'' becomes particle 2 having k'. In other words, it has the effect of interchanging 1 and 2.

In practice we often encounter an observable that has particle labels. For example in $\mathbf{S}_1 \cdot \mathbf{S}_2$ for a two-electron system, $\mathbf{S}_1$ ($\mathbf{S}_2$) stands for the spin operator of particle 1 (2). For simplicity we consider a specific case where the two-particle state ket is completely specified by the eigenvalues of a single observable A for each of the particles:

$$A_1|a'\rangle|a''\rangle = a'|a'\rangle|a''\rangle \tag{6.1.6a}$$

and

$$A_2|a'\rangle|a''\rangle = a''|a'\rangle|a''\rangle, \tag{6.1.6b}$$

where the subscripts on A denote the particle labels, and A_1 and A_2 are thus the observables A for particles 1 and 2, respectively. Applying P_{12} to both sides of (6.1.6a), we have

$$\begin{aligned} P_{12}A_1P_{12}^{-1}P_{12}|a'\rangle|a''\rangle &= a'P_{12}|a'\rangle|a''\rangle \\ &= P_{12}A_1P_{12}^{-1}|a''\rangle|a'\rangle = a'|a''\rangle|a'\rangle. \end{aligned} \tag{6.1.7}$$

This is consistent with (6.1.6b) only if

$$P_{12}A_1P_{12}^{-1} = A_2. \tag{6.1.8}$$

It follows that P_{12} must change the particle labels of observables.

Let us now consider the Hamiltonian of a system of two identical particles. The observables, such as momentum and position operators, must necessarily appear symmetrically in the Hamiltonian—for example,

$$H = \frac{\mathbf{p}_1^2}{2m} + \frac{\mathbf{p}_2^2}{2m} + V_{\text{pair}}(|\mathbf{x}_1 - \mathbf{x}_2|) + V_{\text{ext}}(\mathbf{x}_1) + V_{\text{ext}}(\mathbf{x}_2). \tag{6.1.9}$$

Here we have separated the mutual interaction between the two particles from their interaction with some other external potential. Clearly, we have

$$P_{12}HP_{12}^{-1} = H \tag{6.1.10}$$

for H made up of observables for two identical particles. Because P_{12} commutes with H, we can say that P_{12} is a constant of the motion. The eigenvalues of P_{12} allowed are $+1$ and -1 because of (6.1.5). It therefore follows that if the two-particle state ket is symmetric (antisymmetric) to start with, it remains so at all times.

If we insist on eigenkets of P_{12}, two particular linear combinations are selected:

$$|k'k''\rangle_+ \equiv \frac{1}{\sqrt{2}}\left(|k'\rangle|k''\rangle + |k''\rangle|k'\rangle\right), \tag{6.1.11a}$$

and

$$|k'k''\rangle_{-} \equiv \frac{1}{\sqrt{2}}\left(|k'\rangle|k''\rangle - |k''\rangle|k'\rangle\right). \tag{6.1.11b}$$

We can define the symmetrizer and antisymmetrizer as follows:

$$S_{12} \equiv \tfrac{1}{2}(1+P_{12}), \qquad A_{12} \equiv \tfrac{1}{2}(1-P_{12}). \tag{6.1.12}$$

If we apply $S_{12}(A_{12})$ to an arbitrary linear combination of $|k'\rangle|k''\rangle$ and $|k''\rangle|k'\rangle$, the resulting ket is necessarily symmetric (antisymmetric). This can easily be seen as follows:

$$\begin{aligned}\begin{Bmatrix} S_{12} \\ A_{12} \end{Bmatrix}&\left[c_1|k'\rangle|k''\rangle + c_2|k''\rangle|k'\rangle\right] \\ &= \tfrac{1}{2}\left(c_1|k'\rangle|k''\rangle + c_2|k''\rangle|k'\rangle\right) \pm \tfrac{1}{2}\left(c_1|k''\rangle|k'\rangle + c_2|k'\rangle|k''\rangle\right) \\ &= \frac{c_1 \pm c_2}{2}\left(|k'\rangle|k''\rangle \pm |k''\rangle|k'\rangle\right). \end{aligned} \tag{6.1.13}$$

Our consideration can be extended to a system made up of many identical particles. We define

$$\begin{aligned} P_{ij}|k'\rangle|k''\rangle \cdots |k^{i}\rangle|k^{i+1}\rangle \cdots |k^{j}\rangle \cdots \\ = |k'\rangle|k''\rangle \cdots |k^{j}\rangle|k^{i+1}\rangle \cdots |k^{i}\rangle \ldots . \end{aligned} \tag{6.1.14}$$

Clearly,

$$P_{ij}^2 = 1 \tag{6.1.15}$$

just as before, and the allowed eigenvalues of P_{ij} are $+1$ and -1. It is important to note, however, that in general

$$\left[P_{ij}, P_{kl}\right] \neq 0. \tag{6.1.16}$$

It is worth explicitly working out a system of three identical particles. First, there are $3! = 6$ possible kets of form

$$|k'\rangle|k''\rangle|k'''\rangle \tag{6.1.17}$$

where k', k'', and k''' are all different. Thus there is sixfold exchange degeneracy. Yet if we insist that the state be *totally* symmetrical or *totally* antisymmetrical, we can form only one linear combination each. Explicitly, we have

$$\begin{aligned} |k'k''k'''\rangle_{\pm} \equiv \frac{1}{\sqrt{6}}\{&|k'\rangle|k''\rangle|k'''\rangle \pm |k''\rangle|k'\rangle|k'''\rangle \\ &+ |k''\rangle|k'''\rangle|k'\rangle \pm |k'''\rangle|k''\rangle|k'\rangle \\ &+ |k'''\rangle|k'\rangle|k''\rangle \pm |k'\rangle|k'''\rangle|k''\rangle\}. \end{aligned} \tag{6.1.18}$$

These are both simultaneous eigenkets of P_{12}, P_{23}, and P_{13}. We remarked that there are altogether six independent state kets. It therefore follows that there are four independent kets that are neither totally symmetrical nor totally antisymmetrical. We could also introduce the operator P_{123} by defining

$$P_{123}\left(|k'\rangle|k''\rangle|k'''\rangle\right)=|k''\rangle|k'''\rangle|k'\rangle. \tag{6.1.19}$$

Note that $P_{123}=P_{12}P_{13}$ because

$$P_{12}P_{13}\left(|k'\rangle|k''\rangle|k'''\rangle\right)=P_{12}\left(|k'''\rangle|k''\rangle|k'\rangle\right)=|k''\rangle|k'''\rangle|k'\rangle. \tag{6.1.20}$$

In writing (6.1.18) we assumed that k', k'', and k''' are all different. If two of the three indices coincide, it is impossible to have a totally antisymmetrical state. The totally symmetrical state is given by

$$|k'k'k''\rangle_{+}=\frac{1}{\sqrt{3}}\left(|k'\rangle|k'\rangle|k''\rangle+|k'\rangle|k''\rangle|k'\rangle+|k''\rangle|k'\rangle|k'\rangle\right), \tag{6.1.21}$$

where the normalization factor is understood to be $\sqrt{2!/3!}$. For more-general cases we have a normalization factor

$$\sqrt{\frac{N_1!N_2!\cdots N_n!}{N!}}, \tag{6.1.22}$$

where N is the total number of particles and N_i the number of times $|k^{(i)}\rangle$ occurs.

6.2. SYMMETRIZATION POSTULATE

So far we have not discussed whether nature takes advantage of totally symmetrical or totally antisymmetrical states. It turns out that systems containing N identical particles are either totally symmetrical under the interchange of any pair, in which case the particles are said to satisfy **Bose-Einstein** (B-E) **statistics**, hence known as **bosons**, or totally antisymmetrical, in which case the particles are said to satisfy **Fermi-Dirac** (F-D) **statistics**, hence known as **fermions**. Thus

$$P_{ij}|N\text{ identical bosons}\rangle=+\,|N\text{ identical bosons}\rangle \tag{6.2.1a}$$

$$P_{ij}|N\text{ identical fermions}\rangle=-\,|N\text{ identical fermions}\rangle, \tag{6.2.1b}$$

where P_{ij} is the permutation operator that interchanges the ith and the jth particle, with i and j arbitrary. It is an empirical fact that a mixed symmetry does not occur.

Even more remarkable is that there is a connection between the spin of a particle and the statistics obeyed by it:

Half-integer spin particles are fermions; (6.2.2a)

Integer spin particles are bosons. (6.2.2b)

Here particles can be composite; for example, a ^{3}He nucleus is a fermion just as the e^- or the proton; a ^{4}He nucleus is a boson just as the $\pi^{\pm}, \pi^0$ meson.

This spin-statistics connection is, as far as we know, an exact law of nature with no known exceptions. In the framework of nonrelativistic quantum mechanics, this principle must be accepted as an empirical postulate. In the relativistic quantum theory, however, it can be proved that half-integer spin particles cannot be bosons and integer spin particles cannot be fermions.

An immediate consequence of the electron being a fermion is that the electron must satisfy the **Pauli exclusion principle**, which states that no two electrons can occupy the same state. This follows because a state like $|k'\rangle|k'\rangle$ is necessarily symmetrical, which is not possible for a fermion. As is well known, the Pauli exclusion principle is the cornerstone of atomic and molecular physics as well as the whole of chemistry.

To illustrate the dramatic differences between fermions and bosons, let us consider two particles, each of which can occupy only two states, characterized by k' and k''. For a system of two fermions, we have no choice; there is only one possibility:

$$\frac{1}{\sqrt{2}}\left(|k'\rangle|k''\rangle - |k''\rangle|k'\rangle\right). \tag{6.2.3}$$

For bosons there are three states possible:

$$|k'\rangle|k'\rangle, \qquad |k''\rangle|k''\rangle, \qquad \frac{1}{\sqrt{2}}\left(|k'\rangle|k''\rangle + |k''\rangle|k'\rangle\right). \tag{6.2.4}$$

In contrast, for "classical" particles satisfying **Maxwell-Boltzmann** (M-B) **statistics** with no restriction on symmetry, we have altogether four independent states:

$$|k'\rangle|k''\rangle, \qquad |k''\rangle|k'\rangle, \qquad |k'\rangle|k'\rangle, \qquad |k''\rangle|k''\rangle. \tag{6.2.5}$$

We see that in the fermion case it is impossible for both particles to occupy the same state. In the boson case, for two out of the three allowed kets, both particles occupy the same state. In the classical (M-B) statistics case, both particles occupy the same state for two out of the four allowed kets. In this sense fermions are the least sociable; they avoid each other to make sure that they are not in the same state; in contrast, bosons are the most sociable, they really love to be in the same state, even more so than classical particles obeying M-B statistics.

The difference between fermions and bosons show up most dramatically at low temperatures; a system made up of bosons, such as liquid ^{4}He, exhibits a tendency for all particles to get down to the same ground state at extremely low temperatures. This is known as **Bose-Einstein condensation**, a feature not shared by a system made up of fermions.

6.3. TWO-ELECTRON SYSTEM

Let us now consider specifically a two-electron system. The eigenvalue of the permutation operator is necessarily -1. Suppose the base kets we use may be specified by $\mathbf{x}_1$, $\mathbf{x}_2$, m_{s1}, and m_{s2}, where m_{s1} and m_{s2} stand for the spin-magnetic quantum numbers of electron 1 and electron 2, respectively.

We can express the wave function for a two-electron system as a linear combination of the state ket with eigenbras of $\mathbf{x}_1$, $\mathbf{x}_2$, m_{s1}, and m_{s2} as follows:

$$\psi = \sum_{m_{s1}} \sum_{m_{s2}} C(m_{s1}, m_{s2}) \langle \mathbf{x}_1, m_{s1}; \mathbf{x}_2, m_{s2} | \alpha \rangle. \tag{6.3.1}$$

If the Hamiltonian commutes with $\mathbf{S}^2_{\text{tot}}$,

$$\left[\mathbf{S}^2_{\text{tot}}, H\right] = 0, \tag{6.3.2}$$

then the energy eigenfunction is expected to be an eigenfunction of $\mathbf{S}^2_{\text{tot}}$, and if ψ is written as

$$\psi = \phi(\mathbf{x}_1, \mathbf{x}_2)\chi, \tag{6.3.3}$$

then the spin function χ is expected to be one of the following:

$$\chi(m_{s1}, m_{s2}) = \begin{cases} \left.\begin{array}{l} \chi_{++} \\ \dfrac{1}{\sqrt{2}}(\chi_{+-} + \chi_{-+}) \\ \chi_{--} \end{array}\right\} & \text{triplet (symmetrical)} \\ \dfrac{1}{\sqrt{2}}(\chi_{+-} - \chi_{-+}) & \text{singlet (antisymmetrical),} \end{cases} \tag{6.3.4}$$

where χ_{+-} corresponds to $\chi(m_{s1} = \frac{1}{2}, m_{s2} = -\frac{1}{2})$. Notice that the triplet spin functions are all symmetrical; this is reasonable because the ladder operator $S_{1-} + S_{2-}$ commutes with P_{12} and the $|+\rangle|+\rangle$ state is even under P_{12}.

We note

$$\langle \mathbf{x}_1, m_{s1}; \mathbf{x}_2, m_{s2} | P_{12} | \alpha \rangle = \langle \mathbf{x}_2, m_{s2}; \mathbf{x}_1, m_{s1} | \alpha \rangle. \tag{6.3.5}$$

Fermi-Dirac statistics thus requires

$$\langle \mathbf{x}_1, m_{s1}; \mathbf{x}_2, m_{s2} | \alpha \rangle = -\langle \mathbf{x}_2, m_{s2}; \mathbf{x}_1, m_{s1} | \alpha \rangle. \tag{6.3.6}$$

Clearly, P_{12} can be written as

$$P_{12} = P_{12}^{(\text{space})} P_{12}^{(\text{spin})} \tag{6.3.7}$$

where $P_{12}^{(\text{space})}$ just interchanges the position coordinate, while $P_{12}^{(\text{spin})}$ just interchanges the spin states. It is amusing that we can express $P_{12}^{(\text{spin})}$ as

$$P_{12}^{(\text{spin})} = \frac{1}{2}\left(1 + \frac{4}{\hbar^2}\mathbf{S}_1 \cdot \mathbf{S}_2\right), \tag{6.3.8}$$

which follows because

$$\mathbf{S}_1 \cdot \mathbf{S}_2 = \begin{cases} \dfrac{\hbar^2}{4} & \text{(triplet)} \\ \dfrac{-3\hbar^2}{4} & \text{(singlet)} \end{cases} \tag{6.3.9}$$

It follows from (6.3.3) that letting

$$|\alpha\rangle \to P_{12}|\alpha\rangle \tag{6.3.10}$$

amounts to

$$\phi(\mathbf{x}_1, \mathbf{x}_2) \to \phi(\mathbf{x}_2, \mathbf{x}_1), \qquad \chi(m_{s1}, m_{s2}) \to \chi(m_{s2}, m_{s1}). \tag{6.3.11}$$

This together with (6.3.6) implies that if the space part of the wave function is symmetrical (antisymmetrical) the spin part must be antisymmetrical (symmetrical). As a result, the spin triplet state has to be combined with an antisymmetrical space function and the spin singlet state has to be combined with a space symmetrical function.

The space part of the wave function $\phi(\mathbf{x}_1, \mathbf{x}_2)$ provides the usual probabilistic interpretation. The probability for finding electron 1 in a volume element d^3x_1 centered around $\mathbf{x}_1$ and electron 2 in a volume element d^3x_2 is

$$|\phi(\mathbf{x}_1, \mathbf{x}_2)|^2 d^3x_1 d^3x_2. \tag{6.3.12}$$

To see the meaning of this more closely, let us consider the specific case where the mutual interaction between the two electrons [for example, $V_{\text{pair}}(|\mathbf{x}_1 - \mathbf{x}_2|), \mathbf{S}_1 \cdot \mathbf{S}_2$] can be ignored. If there is no spin dependence, the wave equation for the energy eigenfunction ψ [see (6.1.9)],

$$\left[\frac{-\hbar^2}{2m}\nabla_1^2 - \frac{\hbar^2}{2m}\nabla_2^2 + V_{\text{ext}}(\mathbf{x}_1) + V_{\text{ext}}(\mathbf{x}_2)\right]\psi = E\psi, \tag{6.3.13}$$

is now separable. We have a solution of the form $\omega_A(\mathbf{x}_1)\omega_B(\mathbf{x}_2)$ times the spin function. With no spin dependence $\mathbf{S}_{\text{tot}}^2$ necessarily (and trivially) commutes with H, so the spin part must be a triplet or a singlet, which have definite symmetry properties under $P_{12}^{(\text{spin})}$. The space part must then be written as a symmetrical and antisymmetrical combination of $\omega_A(\mathbf{x}_1)\omega_B(\mathbf{x}_2)$

and $\omega_A(\mathbf{x}_2)\omega_B(\mathbf{x}_1)$:

$$\phi(\mathbf{x}_1,\mathbf{x}_2)=\frac{1}{\sqrt{2}}[\omega_A(\mathbf{x}_1)\omega_B(\mathbf{x}_2)\pm\omega_A(\mathbf{x}_2)\omega_B(\mathbf{x}_1)] \qquad (6.3.14)$$

where the upper sign is for a spin singlet and the lower is for a spin triplet. The probability of observing electron 1 in d^3x_1 around $\mathbf{x}_1$ and electron 2 in d^3x_2 around $\mathbf{x}_2$ is given by

$$\tfrac{1}{2}\{|\omega_A(\mathbf{x}_1)|^2|\omega_B(\mathbf{x}_2)|^2+|\omega_A(\mathbf{x}_2)|^2|\omega_B(\mathbf{x}_1)|^2$$
$$\pm 2\,\mathrm{Re}[\omega_A(\mathbf{x}_1)\omega_B(\mathbf{x}_2)\omega_A^*(\mathbf{x}_2)\omega_B^*(\mathbf{x}_1)]\}\,d^3x_1d^3x_2. \qquad (6.3.15)$$

The last term in the curly bracket is known as the **exchange density**.

We immediately see that when the electrons are in a spin-triplet state, the probability of finding the second electron at the same point in space vanishes. Put another way, the electrons tend to avoid each other when their spins are in a triplet state. In contrast, when their spins are in a singlet state, there is enhanced probability of finding them at the same point in space because of the presence of the exchange density.

Clearly, the question of identity is important only when the exchange density is nonnegligible or when there is substantial overlap between function ω_A and function ω_B. To see this point clearly, let us take the extreme case where $|\omega_A(\mathbf{x})|^2$ (where $\mathbf{x}$ may refer to $\mathbf{x}_1$ or $\mathbf{x}_2$) is big only in region A and $|\omega_B(\mathbf{x})|^2$ is big only in region B such that the two regions are widely separated. Now choose d^3x_1 in region A and d^3x_2 in region B; see Figure 6.2. The only important term then is just the first term in (6.3.15),

$$|\omega_A(\mathbf{x}_1)|^2|\omega_B(\mathbf{x}_2)|^2, \qquad (6.3.16)$$

which is nothing more than the joint probability density expected for classical particles. In this connection, recall that classical particles are necessarily well localized and the question of identity simply does not arise. Thus the exchange-density term is unimportant if regions A and B do not overlap. There is no need to antisymmetrize if the electrons are far apart

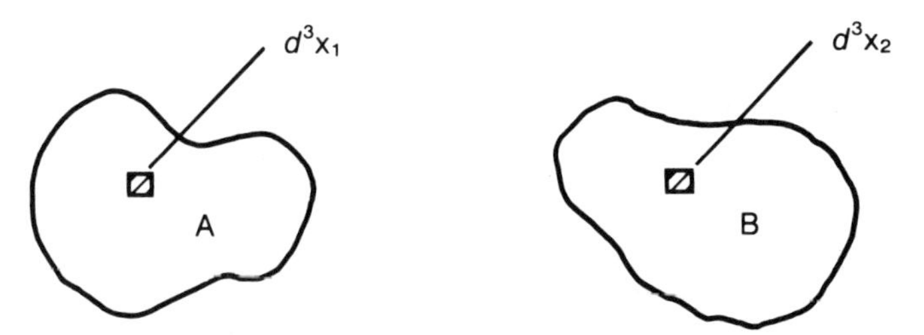

FIGURE 6.2. Two widely separated regions A and B; $|\omega_A(\mathbf{x})|^2$ is large in region A while $|\omega_B(\mathbf{x})|^2$ is large in region B.

and the overlap is negligible. This is quite gratifying. We never have to worry about the question of antisymmetrization with 10 billion electrons, nor is it necessary to take into account the antisymmetrization requirement between an electron in Los Angeles and an electron in Beijing.

6.4. THE HELIUM ATOM

A study of the helium atom is rewarding for several reasons. First of all, it is the simplest realistic problem where the question of identity—which we encountered in Section 6.3—plays an important role. Second, even though it is a simple system, the two-particle Schrödinger equation cannot be solved analytically; therefore, this is a nice place to illustrate the use of perturbation theory and also the use of the variational method.

The basic Hamiltonian is given by

$$H = \frac{\mathbf{p}_1^2}{2m} + \frac{\mathbf{p}_2^2}{2m} - \frac{2e^2}{r_1} - \frac{2e^2}{r_2} + \frac{e^2}{r_{12}}, \tag{6.4.1}$$

where $r_1 \equiv |\mathbf{x}_1|$, $r_2 \equiv |\mathbf{x}_2|$, and $r_{12} \equiv |\mathbf{x}_1 - \mathbf{x}_2|$; see Figure 6.3. Suppose the e^2/r_{12}-term were absent. Then, with the identity question ignored, the wave function would be just the product of two hydrogen atom wave functions with $Z = 1$ changed into $Z = 2$. The total spin is a constant of the motion, so the spin state is either singlet or triplet. The space part of the wave function for the important case where one of the electrons is in the ground state and the other in an excited state characterized by (nlm) is

$$\phi(\mathbf{x}_1, \mathbf{x}_2) = \frac{1}{\sqrt{2}} [\psi_{100}(\mathbf{x}_1)\psi_{nlm}(\mathbf{x}_2) \pm \psi_{100}(\mathbf{x}_2)\psi_{nlm}(\mathbf{x}_1)] \tag{6.4.2}$$

where the upper (lower) sign is for the spin singlet (triplet). We will come back to this general form for an excited state later.

For the ground state, we need a special treatment. Here the configuration is characterized by $(1s)^2$, that is, both electrons in $n = 1$, $l = 0$.

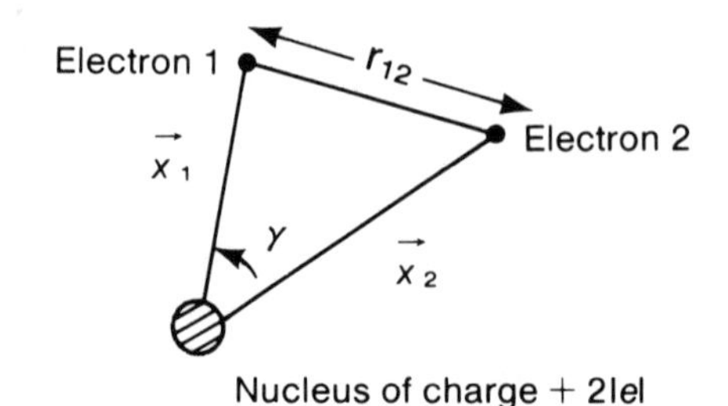

FIGURE 6.3. Schematic diagram of the helium atom.

The space function must then necessarily be symmetric and only the spin singlet function is allowed. So we have

$$\psi_{100}(\mathbf{x}_1)\psi_{100}(\mathbf{x}_2)\chi_{\text{singlet}} = \frac{Z^3}{\pi a_0^3} e^{-Z(r_1+r_2)/a_0}\chi \tag{6.4.3}$$

with $Z = 2$. Not surprisingly, this "unperturbed" wave function gives

$$E = 2\times 4\left(-\frac{e^2}{2a_0}\right) \tag{6.4.4}$$

for the ground-state energy, which is about 30% bigger than the experimental value.

This is just the starting point of our investigation because in obtaining the above form (6.4.3), we have completely ignored the last term in (6.4.1) that describes the interaction between the two electrons. One way to approach the problem of obtaining a better energy value is to apply first-order perturbation theory using (6.4.3) as the unperturbed wave function and e^2/r_{12} as the perturbation. We obtain

$$\Delta_{(1s)^2} = \left\langle \frac{e^2}{r_{12}} \right\rangle_{(1s)^2} = \int\int \frac{Z^6}{\pi^2 a_0^6} e^{-2Z(r_1+r_2)/a_0} \frac{e^2}{r_{12}} d^3x_1\, d^3x_2. \tag{6.4.5}$$

To carry out the indicated integration we first note

$$\frac{1}{r_{12}} = \frac{1}{\sqrt{r_1^2 + r_2^2 - 2r_1 r_2 \cos\gamma}} = \sum_{l=0}^{\infty} \frac{r_<^l}{r_>^{l+1}} P_l(\cos\gamma), \tag{6.4.6}$$

where $r_>$ ($r_<$) is the larger (smaller) of r_1 and r_2 and γ is the angle between $\mathbf{x}_1$ and $\mathbf{x}_2$. The angular integration is easily performed by expressing $P_l(\cos\gamma)$ in terms of $Y_l^m(\theta_1, \phi_1)$ and $Y_l^m(\theta_2, \phi_2)$ using the addition theorem of spherical harmonics:

$$P_l(\cos\gamma) = \frac{4\pi}{2l+1} \sum_{m=-l}^{l} Y_l^{m*}(\theta_1, \phi_1) Y_l^m(\theta_2, \phi_2). \tag{6.4.7}$$

The angular integration is now trivial:

$$\int Y_l^m(\theta_i, \phi_i)\, d\Omega_i = \frac{1}{\sqrt{4\pi}}(4\pi)\delta_{l0}\delta_{m0}. \tag{6.4.8}$$

The radial integration is elementary (but involves tedious algebra!); it leads to

$$\int_0^\infty \left[\int_0^{r_1} \frac{1}{r_1} e^{-(2Z/a_0)(r_1+r_2)} r_2^2\, dr_2 + \int_{r_1}^\infty \frac{1}{r_2} e^{-(2Z/a_0)(r_1+r_2)} r_2^2\, dr_2\right] r_1^2\, dr_1$$

$$= \frac{5}{128}\frac{a_0^5}{Z^5}. \tag{6.4.9}$$

Combining everything, we have (for $Z = 2$)

$$\Delta_{(1s)^2} = \left(\frac{Z^6 e^2}{\pi^2 a_0^6}\right) 4\pi (\sqrt{4\pi})^2 \left(\frac{5}{128}\right)\left(\frac{a_0^5}{Z^5}\right)$$

$$= \left(\frac{5}{2}\right)\left(\frac{e^2}{2a_0}\right). \tag{6.4.10}$$

Adding this energy shift to (6.4.4), we have

$$E_{\text{cal}} = \left(-8 + \frac{5}{2}\right)\left(\frac{e^2}{2a_0}\right) \simeq -74.8 \text{ eV}. \tag{6.4.11}$$

Compare this with the experimental value,

$$E_{\text{exp}} = -78.8 \text{ eV}. \tag{6.4.12}$$

This is not bad, but we can do better! We propose to use the variational method with Z, which we call Z_{eff}, as a variational parameter. The physical reason for this choice is that the effective Z seen by one of the electrons is smaller than 2 because the positive charge of 2 units at the origin (see Figure 6.3) is "screened" by the negatively charged cloud of the other electron; in other words, the other electron tends to neutralize the positive charge due to the helium nucleus at the center. For the normalized trial function we use

$$\langle \mathbf{x}_1, \mathbf{x}_2 | \tilde{0} \rangle = \left(\frac{Z_{\text{eff}}^3}{\pi a_0^3}\right) e^{-Z_{\text{eff}}(r_1 + r_2)/a_0}. \tag{6.4.13}$$

From this we obtain

$$\overline{H} = \left\langle \tilde{0} \left| \frac{\mathbf{p}_1^2}{2m} + \frac{\mathbf{p}_2^2}{2m} \right| \tilde{0} \right\rangle - \left\langle \tilde{0} \left| \frac{Ze^2}{r_1} + \frac{Ze^2}{r_2} \right| \tilde{0} \right\rangle + \left\langle \tilde{0} \left| \frac{e^2}{r_{12}} \right| \tilde{0} \right\rangle$$

$$= \left(2\frac{Z_{\text{eff}}^2}{2} - 2ZZ_{\text{eff}} + \frac{5}{8}Z_{\text{eff}}\right)\left(\frac{e^2}{a_0}\right). \tag{6.4.14}$$

We easily see that the minimization of $\overline{H}$ is at

$$Z_{\text{eff}} = 2 - \tfrac{5}{16} = 1.6875. \tag{6.4.15}$$

This is smaller than 2, as anticipated. Using this value for Z_{eff} we get

$$E_{\text{cal}} = -77.5 \text{ eV}, \tag{6.4.16}$$

which is already very close considering the crudeness of the trial wave function.

Historically, this achievement was considered to be one of the earliest signs that Schrödinger's wave mechanics was on the right track. We

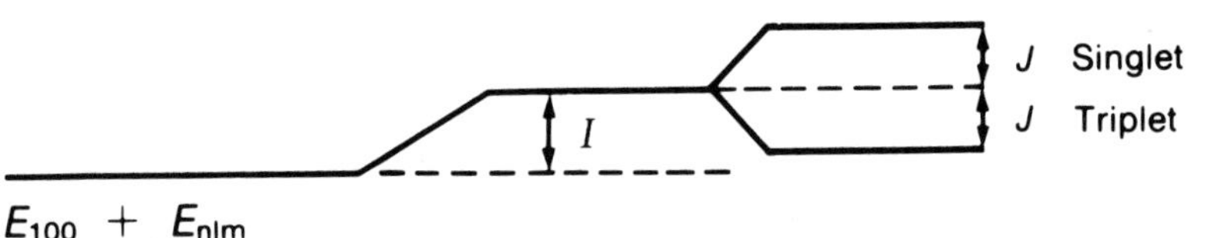

FIGURE 6.4. Schematic diagram for the energy-level splittings of $(1s)(nl)$ for the helium atom.

cannot get this kind of number by the purely algebraic (operator) method. The helium calculation was first done by A. Unsöld in 1927.*

Let us briefly consider excited states. This is more interesting from the point of view of illustrating quantum-mechanical effects due to identity. We consider just $(1s)(nl)$. We write the energy of this state as

$$E = E_{100} + E_{nlm} + \Delta E. \tag{6.4.17}$$

In first-order perturbation theory, ΔE is obtained by evaluating the expectation value of e^2/r_{12}. We can write

$$\left\langle \frac{e^2}{r_{12}} \right\rangle = I \pm J, \tag{6.4.18}$$

where I and J, known respectively as the direct integral and the exchange integral, are given by

$$I = \int d^3x_1 \int d^3x_2 |\psi_{100}(\mathbf{x}_1)|^2 |\psi_{nlm}(\mathbf{x}_2)|^2 \frac{e^2}{r_{12}}, \tag{6.4.19a}$$

$$J = \int d^3x_1 \int d^3x_2 \psi_{100}(\mathbf{x}_1) \psi_{nlm}(\mathbf{x}_2) \frac{e^2}{r_{12}} \psi_{100}^*(\mathbf{x}_2) \psi_{nlm}^*(\mathbf{x}_1). \tag{6.4.19b}$$

The upper (lower) sign goes with the spin singlet (triplet) state. Obviously, I is positive; we can also show that J is positive. So the net result is such that for the same configuration, the spin singlet state lies higher, as shown in Figure 6.4.

The physical interpretation for this is as follows: In the singlet case the space function is symmetric and the electrons have a tendency to come close to each other. Therefore, the effect of the electrostatic repulsion is more serious; hence, a higher energy results. In the triplet case, the space function is antisymmetric and the electrons tend to avoid each other. Helium in spin-singlet states is known as **parahelium**, while helium in spin-triplet states is known as **orthohelium**. Each configuration splits into

*A. Unsöld, Ann. Phys. **82** (1927) 355.

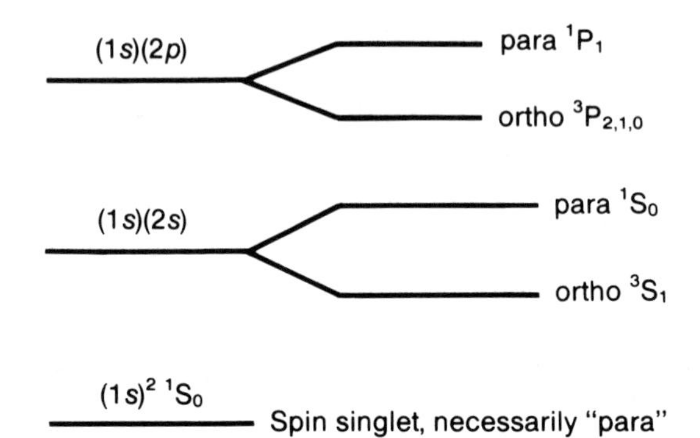

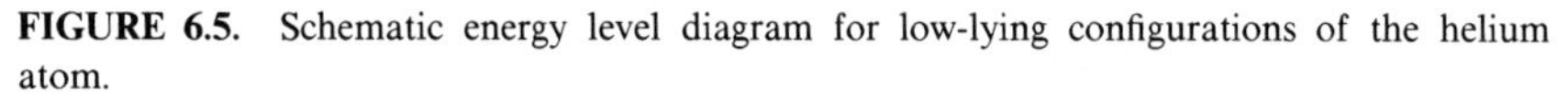

FIGURE 6.5. Schematic energy level diagram for low-lying configurations of the helium atom.

the para state and the ortho state, the para state lying higher. For the ground state only parahelium is possible. See Figure 6.5 for a schematic energy level diagram of the helium atom.

It is very important to recall that the original Hamiltonian is spin-independent because the potential is made up of just three Coulomb terms. There was no $\mathbf{S}_1 \cdot \mathbf{S}_2$-term whatsoever. Yet there is a spin-dependent effect—the electrons with parallel spins have a lower energy—that arises from Fermi-Dirac statistics.

This explanation of the apparent spin dependence of the helium atom energy levels is due to Heisenberg. The physical origin of ferromagnetism—alignment of the electron spins extended over microscopic distances—is also believed to be essentially the same, but the properties of ferromagnets are much harder to calculate quantitatively from first principles.

6.5. PERMUTATION SYMMETRY AND YOUNG TABLEAUX

In keeping track of the requirement imposed by permutation symmetry, there is an extremely convenient bookkeeping technique known as **Young tableaux**, after A. Young, an English clergyman who published a fundamental paper on this subject in 1901. This section is meant to be an introduction to Young tableaux for the practical-minded reader. We do not necessarily present all derivations; this is one of those cases where the rules are simpler than the derivations.

To illustrate the basic techniques involved, we consider once again the spin states of a two-electron system. We have three symmetric states corresponding to the three possible orientations of the spin-triplet state and an antisymmetric state corresponding to the spin singlet. The spin state of an *individual* electron is to be represented by a box. We use $\boxed{1}$ for spin up and $\boxed{2}$ for spin down. These boxes are the basic *primitive objects* of SU(2). A single box represents a doublet.

We define a symmetric tableau, $\begin{array}{|c|c|}\hline \ & \ \\\hline\end{array}$, and an antisymmetric tableau, $\begin{array}{|c|}\hline \ \\\hline \ \\\hline\end{array}$ When applied to the spin states of a two-electron system, $\begin{array}{|c|c|}\hline \ & \ \\\hline\end{array}$ is the Young tableau for a spin triplet, while $\begin{array}{|c|}\hline \ \\\hline \ \\\hline\end{array}$ is the Young tableau for a spin singlet. Returning now to $\boxed{1}$ and $\boxed{2}$, we can build up the three-triplet states as follows:

$$\begin{array}{|c|c|}\hline \ & \ \\\hline\end{array} \left\{ \begin{array}{l} \begin{array}{|c|c|}\hline 1 & 1 \\\hline\end{array} \\ \begin{array}{|c|c|}\hline 1 & 2 \\\hline\end{array} \\ \begin{array}{|c|c|}\hline 2 & 2 \\\hline\end{array} \end{array} \right. . \tag{6.5.1}$$

We do not consider $\begin{array}{|c|c|}\hline 2 & 1 \\\hline\end{array}$ because when we put boxes horizontally, symmetry is understood. So we deduce an important rule: Double counting is avoided if we require that the number (label) *not decrease* going from the left to the right.

As for the antisymmetric spin-singlet state,

$$\begin{array}{|c|}\hline 1 \\\hline 2 \\\hline\end{array} \tag{6.5.2}$$

is the only possibility. Clearly $\begin{array}{|c|}\hline 1 \\\hline 1 \\\hline\end{array}$ and $\begin{array}{|c|}\hline 2 \\\hline 2 \\\hline\end{array}$ are impossible because of the requirement of antisymmetry; for a vertical tableau we cannot have a symmetric state. Furthermore, $\begin{array}{|c|}\hline 2 \\\hline 1 \\\hline\end{array}$ is discarded to avoid double counting. To eliminate the unwanted symmetry states, we therefore require the number (label) to *increase* as we go down.

We take the following to be the general rule. In drawing Young tableau, going from left to right the number cannot decrease; going down the number must increase. We deduced this rule by considering the spin states of two electrons, but we can show this rule to be applicable to the construction of any tableau.

Consider now three electrons. We can construct totally symmetric spin states by the following rule:

$$\begin{array}{|c|c|c|}\hline \ & \ & \ \\\hline\end{array} \left\{ \begin{array}{l} \begin{array}{|c|c|c|}\hline 1 & 1 & 1 \\\hline\end{array} \\ \begin{array}{|c|c|c|}\hline 1 & 1 & 2 \\\hline\end{array} \\ \begin{array}{|c|c|c|}\hline 1 & 2 & 2 \\\hline\end{array} \\ \begin{array}{|c|c|c|}\hline 2 & 2 & 2 \\\hline\end{array} \end{array} \right. . \tag{6.5.3}$$

This method gives four states altogether. This is just the multiplicity of the $j = \frac{3}{2}$ state, which is obviously symmetric as seen from the $m = \frac{3}{2}$ case, where all three spins are aligned in the positive z-direction.

What about the totally antisymmetric states? We may try vertical tableau like

$$\begin{array}{|c|}\hline 1 \\\hline 1 \\\hline 1 \\\hline\end{array} \quad \text{or} \quad \begin{array}{|c|}\hline 1 \\\hline 2 \\\hline 2 \\\hline\end{array} .$$

But these are illegal, because the numbers must increase as we go down. This is not surprising because total antisymmetry is impossible for spin states of three electrons; quite generally, a necessary (but, of course, not sufficient) condition for total antisymmetry is that every state must be different. In fact, in SU(2) we cannot have three boxes in a vertical column.

We now define a mixed symmetry tableau that looks like this: $\yng(2,1)$. Such a tableau can be visualized as either a single box attached to a symmetric tableau [as in Equation (6.5.4a)] or a single box attached to an antisymmetric tableau [as in Equation (6.5.4b)].

If the spin function for three electrons is symmetric in two of the indices, neither of them can be antisymmetric with respect to a third index. For example,

$$(|+\rangle_1|-\rangle_2+|-\rangle_1|+\rangle_2)|-\rangle_3 \tag{6.5.5}$$

satisfies symmetry under $1 \leftrightarrow 2$, but it is neither symmetric nor antisymmetric with respect to $1 \leftrightarrow 3$ (or $2 \leftrightarrow 3$). We may try to enforce antisymmetry under $1 \leftrightarrow 3$ by subtracting the same thing from (6.5.5) with 1 and 3 interchanged:

$$\begin{aligned} &|+\rangle_1|-\rangle_2|-\rangle_3-|+\rangle_3|-\rangle_2|-\rangle_1+|-\rangle_1|+\rangle_2|-\rangle_3-|-\rangle_3|+\rangle_2|-\rangle_1 \\ &\quad = (|+\rangle_1|-\rangle_3-|-\rangle_1|+\rangle_3)|-\rangle_2. \end{aligned} \tag{6.5.6}$$

This satisfies antisymmetry under $1 \leftrightarrow 3$, but we no longer have the original symmetry under $1 \leftrightarrow 2$. In any case the dimensionality of $\yng(2,1)$ is 2; that is, it represents a doublet ($j=\frac{1}{2}$). We can see this by noting that such a tableau corresponds to the angular momentum addition of a spin doublet ($\square$) and a spin triplet ($\yng(2)$). We have used up the totally symmetric quartet $\yng(3)$, so the remainder must be a doublet. Alternatively, we may attach a doublet ($\square$) to a singlet ($\yng(1,1)$) as in (6.5.4b); the net product is obviously a doublet. So no matter how we consider it, $\yng(2,1)$ must represent a doublet. But this is precisely what the rule gives. If the number cannot decrease in the horizontal direction and must increase in the vertical direction, the only possibilities are

$$\begin{ytableau} 1 & 1 \\ 2 \end{ytableau} \quad \text{and} \quad \begin{ytableau} 1 & 2 \\ 2 \end{ytableau}. \tag{6.5.7}$$

Because there are only two possibilities, $\yng(2,1)$ must correspond to a doublet. Notice that we do not consider $\begin{ytableau} \none & \\ & \end{ytableau}$.

We can consider Clebsch-Gordan series or angular-momentum addition as follows:

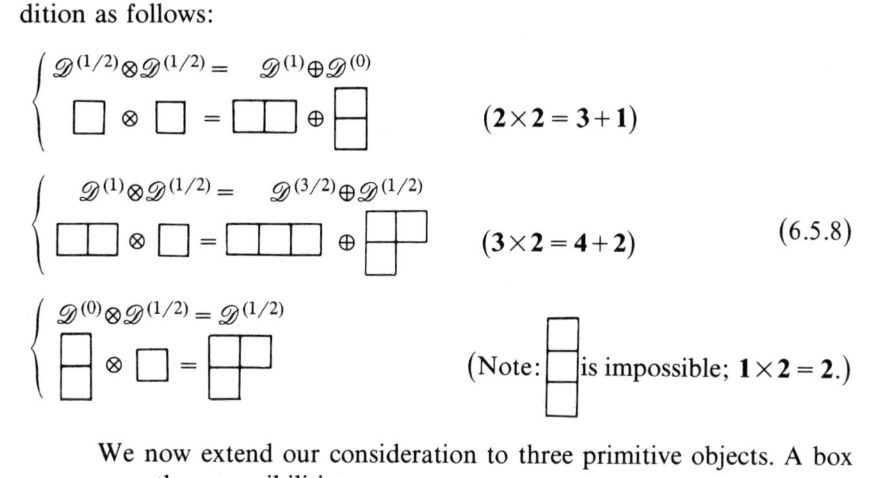

We now extend our consideration to three primitive objects. A box can assume three possibilities:

$$\square : \boxed{1}, \boxed{2}, \boxed{3}. \tag{6.5.9}$$

The labels 1, 2, and 3 may stand for the magnetic quantum numbers of p-orbitals in atomic physics or charge states of the pion π^+, π^0, π^-, or the u, d, and s quarks in the SU(3) classification of elementary particles. We *assume* that the rule we inferred using two primitive objects can be generalized and work out such concepts as the dimensionality; we then check to see whether everything is reasonable.

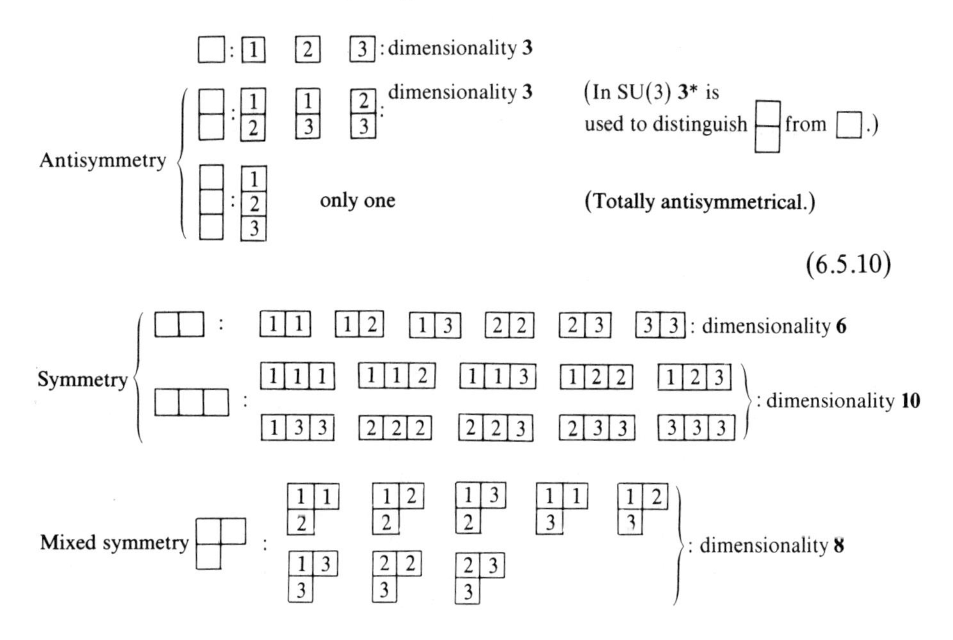

These tableaux correspond to representations of SU(3).

λ_1 boxes in first row
λ_2 boxes in second row
λ_3 boxes in third row

(6.5.11)

For the purpose of figuring the dimensionality, it is legitimate to strike out

(6.5.12)

a "singlet" at the left-hand side of (6.5.11). We can show the dimensionality to be

$$d(\lambda_1, \lambda_2, \lambda_3) = \frac{(p+1)(q+1)(p+q+2)}{2} \tag{6.5.13}$$

where $p = \lambda_1 - \lambda_2$ and $q = \lambda_2 - \lambda_3$.

In the SU(3) language, a tableau corresponds to a definite irreducible representation of SU(3). On the other hand, when each box is interpreted to mean a $j = 1$ object, then a tableau does not correspond to an irreducible representation of the rotation group. Let us discuss this point more specifically. We start with putting together two boxes, where each box represents a $j = 1$ state.

$$(\mathbf{3} \times \mathbf{3} = \mathbf{6} + \mathbf{3}) \tag{6.5.14}$$

The horizontal tableau has six states; the tableau is to be broken down into $j = 2$ (multiplicity 5) and $j = 0$ (multiplicity 1), both of which are symmetric. In fact this is just how we construct a symmetric second-rank tensor out of two vectors. The vertical tableau corresponds to an antisymmetric $j = 1$ state behaving like $\mathbf{a} \times \mathbf{b}$.

To work out the addition of three angular momenta with $j_1 = j_2 = j_3 = 1$, let us first figure out

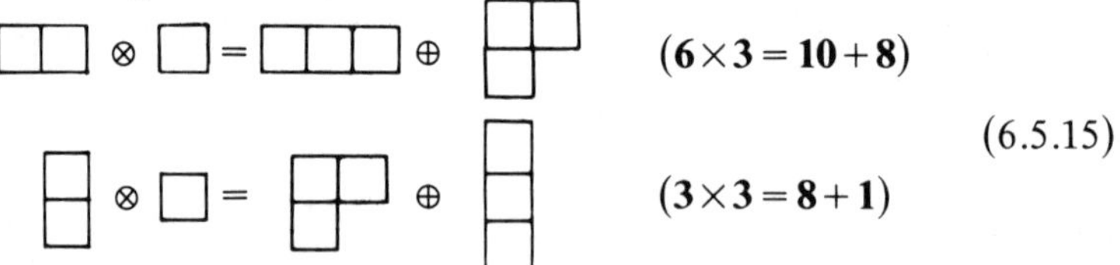

(6.5.15)

To see how breaks down, we note that it must contain $j = 3$. However, this is not the only totally symmetric state; $\mathbf{a}(\mathbf{b}\cdot\mathbf{c}) + \mathbf{b}(\mathbf{c}\cdot\mathbf{a}) + \mathbf{c}(\mathbf{a}\cdot\mathbf{b})$ is also totally symmetric, and this has the transformation property of $j = 1$. So contains both $j = 3$ (seven states) and $j = 1$ (three states). As for with eight possibilities altogether, the argument is more involved, but we note that this **8** cannot be broken into **7**+**1** because **7** is totally symmetric, while **1** is totally antisymmetric when we know that **8** is of *mixed* symmetry. So the only other possibility is **8** = **5** + **3**—in other words $j = 2$ and $j = 1$.

Finally, therefore, we must have

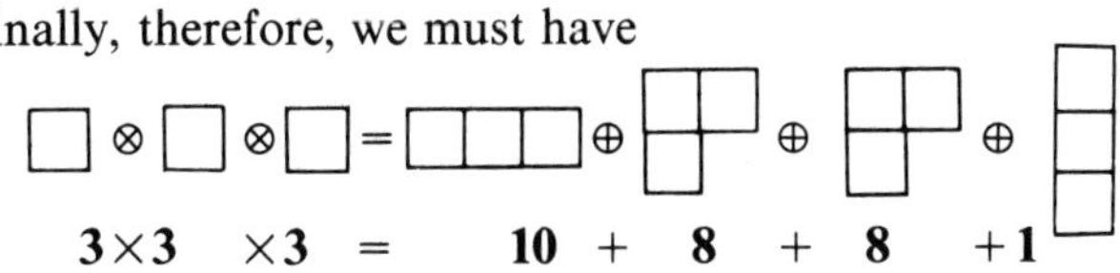

$$\begin{array}{ccccccccccccccc} \square & \otimes & \square & \otimes & \square & = & \square\square\square & \oplus & \begin{smallmatrix}\square\square\\ \square\end{smallmatrix} & \oplus & \begin{smallmatrix}\square\square\\ \square\end{smallmatrix} & \oplus & \begin{smallmatrix}\square\\ \square\\ \square\end{smallmatrix} \\ \mathbf{3} & \times & \mathbf{3} & \times & \mathbf{3} & = & \mathbf{10} & + & \mathbf{8} & + & \mathbf{8} & + & \mathbf{1} \\ & & & & & & \uparrow & & \uparrow & & \uparrow & & \\ & & & & & & 7+3 & & 5+3 & & 5+3 & & \end{array} \tag{6.5.16}$$

In terms of angular-momentum states, we have

$j=3$ (dimension 7)	once	(totally symmetric)
$j=2$ (dimension 5)	twice	(both mixed symmetry)
$j=1$ (dimension 3)	three times	(one totally symmetric, two mixed symmetry)
$j=0$ (dimension 1)	once	(totally antisymmetric)

(6.5.17)

That the $j=0$ state is unique corresponds to the fact that the only product of **a**, **b**, and **c** invariant under rotation is $\mathbf{a}\cdot(\mathbf{b}\times\mathbf{c})$, which is necessarily antisymmetric. We can also look at the $j=1$ states using the tensor approach. We have three independent vectors constructed from **a**, **b**, and **c**: $\mathbf{a}(\mathbf{b}\cdot\mathbf{c})$, $\mathbf{b}(\mathbf{c}\cdot\mathbf{a})$, and $\mathbf{c}(\mathbf{a}\cdot\mathbf{b})$. From these we can construct only one totally symmetric combination,

$$\mathbf{a}(\mathbf{b}\cdot\mathbf{c})+\mathbf{b}(\mathbf{c}\cdot\mathbf{a})+\mathbf{c}(\mathbf{a}\cdot\mathbf{b}). \tag{6.5.18}$$

We can apply this consideration to the $(2p)^3$ configuration of the nitrogen atom, N ($Z=7$, shell structure $(1s)^2(2s)^2(2p)^3$). There are altogether

$$\frac{6!}{3!3!}=\frac{6\cdot5\cdot4}{3\cdot2}=20 \text{ states}. \tag{6.5.19}$$

The space function behaving like $\square\square\square$, which would have $l_{\text{tot}}=3$, is actually impossible because it must be combined with a totally antisymmetric spin function. Remember

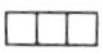

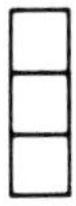

$$\begin{smallmatrix}\square\\ \square\\ \square\end{smallmatrix},$$

where each box stands for spin up or down, is impossible for spin states of three electrons. In contrast the totally antisymmetric space function

$$\begin{smallmatrix}\square\\ \square\\ \square\end{smallmatrix}$$

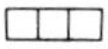

with $l_{\text{tot}}=0$ is perfectly allowable if combined with a totally symmetric spin function $\square\square\square$, which has total spin angular momentum $\frac{3}{2}$; we have ${}^4S_{3/2}$ for this configuration. The mixed-symmetry space function $\begin{smallmatrix}\square\square\\ \square\end{smallmatrix}$ ($l_{\text{tot}}=2,1$) must be combined with a mixed-symmetry spin function, necessarily a spin doublet $\begin{smallmatrix}\square\square\\ \square\end{smallmatrix}$. So we have ${}^2D_{5/2,3/2}$ and ${}^2P_{3/2,1/2}$. The counting of states leads to the following:

$$\begin{array}{ll} j=\frac{5}{2} & 6 \text{ states} \\ j=\frac{3}{2} & 4\times 3 \text{ states} \\ j=\frac{1}{2} & \underline{2 \text{ states}} \\ & 20 \text{ states} \end{array} \tag{6.5.20}$$

This agrees with the other way of counting.

Finally, we consider applications to elementary particle physics. Here we apply SU(3) considerations to the nonrelativistic quark model, where the primitive objects are u, d, s (up, down, and strange, where *up* and *down* refer to *isospin up* and *isospin down*). A box can now stand for u, d, or s. We look at the decuplet representation $\square\square\square$ (**10**):

$$\begin{array}{lll} \Delta^{++,+,0,-} & ddd\,udd\,uud\,uuu & I=\frac{3}{2}, \\ \Sigma^{+,0,-} & dds\,uds\,uus & I=1, \\ \Xi^{0,-} & dss\,uss & I=\frac{1}{2}, \\ \Omega^{-} & sss & I=0, \end{array} \tag{6.5.21}$$

where I stands for isospin.

Now all ten states are known to be spin $\frac{3}{2}$ objects. It is safe to assume that the space part is in a relative S-state for low-lying states of three quarks. We expect total symmetry in the spin degree of freedom. For instance, the $j=\frac{3}{2}$, $m=\frac{3}{2}$ state of Δ can be visualized to have quark spins all aligned.

But the quarks are spin $\frac{1}{2}$ objects; we expect total antisymmetry because of Fermi-Dirac statistics. Yet

$$\begin{array}{ll} \text{Quark label (now called flavor):} & \text{symmetric} \\ \text{Spin:} & \text{symmetric} \\ \text{Space:} & \text{symmetric} \end{array} \tag{6.5.22}$$

for the $j=\frac{3}{2}$ decuplet. But as is evident from (6.5.22), the total symmetry in this case is even! This led to the "statistics paradox," which was especially embarrassing because other aspects of the nonrelativistic quark model were so successful.

The way out of this dilemma is to postulate that there is actually an additional degree of freedom called *color* (red, blue, or yellow) and postulate that the observed hadrons (strongly interacting particles, including $J = \frac{3}{2}^+$ states considered here) are **color singlets**

$$\frac{1}{\sqrt{6}}(|\mathrm{RBY}\rangle - |\mathrm{BRY}\rangle + |\mathrm{BYR}\rangle - |\mathrm{YBR}\rangle + |\mathrm{YRB}\rangle - |\mathrm{RYB}\rangle). \quad (6.5.23)$$

This is in complete analogy with the unique

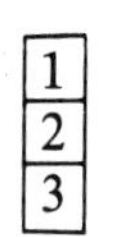

of (6.5.10), a totally antisymmetric combination in color space. The statistics problem is now solved because

$$\underset{(-)}{P_{ij}} = \underset{(+)}{P_{ij}^{(\text{flavor})}} \underset{(+)}{P_{ij}^{(\text{spin})}} \underset{(+)}{P_{ij}^{(\text{space})}} \underset{(-)}{P_{ij}^{(\text{color})}} \quad (6.5.24)$$

One might argue that this is a cheap way to get out of the dilemma. Fortunately, there were also other pieces of evidence in favor of color, such as the decay rate of π^0 and the cross section for electron-positron annihilation into hadrons. In fact, this is a very good example of how attempts to overcome difficulties lead to nontrivial prediction of color.

PROBLEMS

1. a. N identical spin $\frac{1}{2}$ particles are subjected to a one-dimensional simple harmonic oscillator potential. What is the ground-state energy? What is the Fermi energy?
   b. What are the ground state and Fermi energies if we ignore the mutual interactions and assume N to be very large?
2. It is obvious that two nonidentical spin1 particles with no orbital angular momenta (that is, s-states for both) can form $j = 0$, $j = 1$, and $j = 2$. Suppose, however, that the two particles are *identical*. What restrictions do we get?
3. Discuss what would happen to the energy levels of a helium atom if the electron were a spinless boson. Be as quantitative as you can.
4. Three spin 0 particles are situated at the corners of an equilateral triangle. Let us define the z-axis to go through the center and in the direction normal to the plane of the triangle. The whole system is free to rotate about the z-axis. Using statistics considerations, obtain restrictions on the magnetic quantum numbers corresponding to J_z.

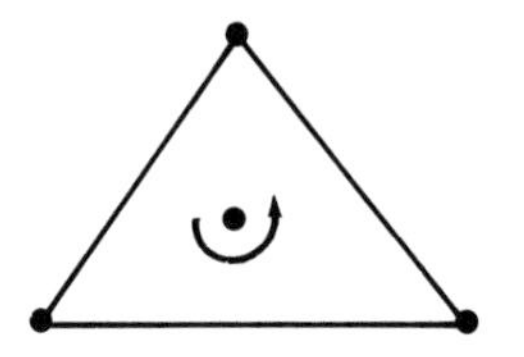

5. Consider three weakly interacting, identical spin 1 particles.
   a. Suppose the space part of the state vector is known to be symmetric under interchange of *any* pair. Using notation $|+\rangle|0\rangle|+\rangle$ for particle 1 in $m_s = +1$, particle 2 in $m_s = 0$, particle 3 in $m_s = +1$, and so on, construct the normalized spin states in the following three cases:
      (i) All three of them in $|+\rangle$.
      (ii) Two of them in $|+\rangle$, one in $|0\rangle$.
      (iii) All three in different spin states.
      What is the total spin in each case?
   b. Attempt to do the same problem when the space part is antisymmetric under interchange of any pair.
6. Suppose the electron were a spin $\frac{3}{2}$ particle obeying Fermi-Dirac statistics. Write the configuration of a hypothetical Ne ($Z = 10$) atom made up of such "electrons" [that is, the analog of $(1s)^2(2s)^2(2p)^6$]. Show that the configuration is highly degenerate. What is the ground state (the lowest term) of the hypothetical Ne atom in spectroscopic notation ($^{2S+1}L_J$, where S, L, and J stand for the total spin, the total orbital angular momentum, and the total angular momentum, respectively) when exchange splitting and spin-orbit splitting are taken into account?
7. Two identical spin $\frac{1}{2}$ fermions move in one dimension under the influence of the infinite-wall potential $V = \infty$ for $x < 0$, $x > L$, and $V = 0$ for $0 \le x \le L$.
   a. Write the ground-state wave function and the ground-state energy when the two particles are constrained to a triplet spin state (*ortho* state).
   b. Repeat (a) when they are in a singlet spin state (*para* state).
   c. Let us now suppose that the two particles interact mutually via a very short-range attractive potential that can be approximated by

$$V = -\lambda\delta(x_1 - x_2) \quad (\lambda > 0).$$

   Assuming that perturbation theory is valid even with such a singular potential, discuss semiquantitatively what happens to the energy levels obtained in (a) and (b).

CHAPTER 7

Scattering Theory

This last chapter of the book is devoted to the theory of scattering and, more generally, collision processes. It is impossible to overemphasize the importance of this subject.

7.1. THE LIPPMANN-SCHWINGER EQUATION

We begin with a time-independent formulation of scattering processes. We assume that the Hamiltonian can be written as

$$H = H_0 + V \tag{7.1.1}$$

where H_0 stands for the kinetic-energy operator

$$H_0 = \frac{\mathbf{p}^2}{2m}. \tag{7.1.2}$$

In the absence of a scatterer, V would be zero, and an energy eigenstate would be just a free particle state $|\mathbf{p}\rangle$. The presence of V causes the energy eigenstate to be different from a free-particle state. However, if the scattering process is to be elastic—that is, no change in energy—we are interested in obtaining a solution to the full-Hamiltonian Schrödinger equation with the same energy eigenvalue. More specifically, let $|\phi\rangle$ be the energy eigenket of H_0:

$$H_0|\phi\rangle = E|\phi\rangle. \tag{7.1.3}$$

(We use $|\phi\rangle$ here rather than $|\mathbf{p}\rangle$ because we may later be interested in free-spherical wave rather than plane-wave states; $|\phi\rangle$ may stand for either.) The basic Schrödinger equation we wish to solve is

$$(H_0 + V)|\psi\rangle = E|\psi\rangle. \tag{7.1.4}$$

Both H_0 and $H_0 + V$ exhibit *continuous* energy spectra. We look for a solution to (7.1.4) such that, as $V \to 0$, we have $|\psi\rangle \to |\phi\rangle$, where $|\phi\rangle$ is the solution to the free-particle Schrödinger equation [(7.1.3)] with the *same* energy eigenvalue.

It may be argued that the desired solution is

$$|\psi\rangle = \frac{1}{E - H_0} V|\psi\rangle + |\phi\rangle, \tag{7.1.5}$$

apart from complications arising from the singular nature of the operator $1/(E - H_0)$. We can see this by noting that $E - H_0$ applied to (7.1.5) immediately gives the correct equation, (7.1.4). The presence of $|\phi\rangle$ is reasonable because $|\psi\rangle$ must reduce to $|\phi\rangle$ as V vanishes. However, without prescriptions for dealing with a singular operator, an equation of type (7.1.5) has no meaning. The trick we used in time-independent perturbation theory—inserting the complimentary projection operator, and so on (see Section 5.1)—does not work well here because both $|\phi\rangle$ and $|\psi\rangle$ exhibit continuous eigenvalues. Instead, this time the solution is specified by making E slightly complex:

$$|\psi^{(\pm)}\rangle = |\phi\rangle + \frac{1}{E - H_0 \pm i\varepsilon} V|\psi^{(\pm)}\rangle. \tag{7.1.6}$$

This is known as the **Lippmann-Schwinger equation**. The physical meaning of $\pm$ is to be discussed in a moment by looking at $\langle\mathbf{x}|\psi^{(\pm)}\rangle$ at large distances.

The Lippmann-Schwinger equation is a *ket equation* independent of particular representations. We now confine ourselves to the position basis by multiplying $\langle\mathbf{x}|$ from the left. Thus

$$\langle\mathbf{x}|\psi^{(\pm)}\rangle = \langle\mathbf{x}|\phi\rangle + \int d^3x' \left\langle \mathbf{x} \left| \frac{1}{E - H_0 \pm i\varepsilon} \right| \mathbf{x}' \right\rangle \langle\mathbf{x}'|V|\psi^{(\pm)}\rangle. \tag{7.1.7}$$

This is an **integral equation** for scattering because the unknown ket $|\psi^{(\pm)}\rangle$ appears under an integral sign. If $|\phi\rangle$ stands for a plane-wave state with momentum $\mathbf{p}$, we can write

$$\langle\mathbf{x}|\phi\rangle = \frac{e^{i\mathbf{p}\cdot\mathbf{x}/\hbar}}{(2\pi\hbar)^{3/2}} \tag{7.1.8}$$

[*Editor's Note:* In contrast to bound states, the plane-wave state (7.1.8) is, of course, not normalizable and not really a vector in Hilbert space. Dealing with such states is one of the inconveniences of time-independent scattering theory. The "normalization" in (7.1.8) is such that

$$\int d^3x \langle \mathbf{p}'|\mathbf{x}\rangle\langle\mathbf{x}|\mathbf{p}\rangle = \delta^{(3)}(\mathbf{p} - \mathbf{p}'). \tag{7.1.9)]}$$

If, on the other hand, the Lippman-Schwinger equation is written using the momentum basis, we obtain

$$\langle\mathbf{p}|\psi^{(\pm)}\rangle = \langle\mathbf{p}|\phi\rangle + \frac{1}{E - (p^2/2m) \pm i\varepsilon}\langle\mathbf{p}|V|\psi^{(\pm)}\rangle. \tag{7.1.10}$$

We shall come back to this equation in Section 7.2.

Let us consider specifically the position basis and work with (7.1.7). To make any progress we must first evaluate the kernel of the integral equation defined by

$$G_{\pm}(\mathbf{x}, \mathbf{x}') \equiv \frac{\hbar^2}{2m}\left\langle\mathbf{x}\left|\frac{1}{E - H_0 \pm i\varepsilon}\right|\mathbf{x}'\right\rangle. \tag{7.1.11}$$

We claim that $G_{\pm}(\mathbf{x}, \mathbf{x}')$ is given by

$$G_{\pm}(\mathbf{x}, \mathbf{x}') = -\frac{1}{4\pi}\frac{e^{\pm ik|\mathbf{x}-\mathbf{x}'|}}{|\mathbf{x} - \mathbf{x}'|} \tag{7.1.12}$$

where $E \equiv \hbar^2k^2/2m$. To show this, we evaluate (7.1.11) as follows:

$$\frac{\hbar^2}{2m}\left\langle\mathbf{x}\left|\frac{1}{E - H_0 \pm i\varepsilon}\right|\mathbf{x}'\right\rangle = \frac{\hbar^2}{2m}\int d^3p' \int d^3p'' \langle\mathbf{x}|\mathbf{p}'\rangle \times \left\langle\mathbf{p}'\left|\frac{1}{E - (\mathbf{p}'^2/2m) \pm i\varepsilon}\right|\mathbf{p}''\right\rangle\langle\mathbf{p}''|\mathbf{x}'\rangle, \tag{7.1.13}$$

where H_0 acts on $\langle\mathbf{p}'|$. Now use

$$\left\langle\mathbf{p}'\left|\frac{1}{E - (\mathbf{p}'^2/2m) \pm i\varepsilon}\right|\mathbf{p}''\right\rangle = \frac{\delta^{(3)}(\mathbf{p}' - \mathbf{p}'')}{E - (\mathbf{p}'^2/2m) \pm i\varepsilon}$$

$$\langle\mathbf{x}|\mathbf{p}'\rangle = \frac{e^{i\mathbf{p}'\cdot\mathbf{x}/\hbar}}{(2\pi\hbar)^{3/2}}, \qquad \langle\mathbf{p}''|\mathbf{x}'\rangle = \frac{e^{-i\mathbf{p}'\cdot\mathbf{x}'/\hbar}}{(2\pi\hbar)^{3/2}}. \tag{7.1.14}$$

The right-hand side of (7.1.13) becomes

$$\frac{\hbar^2}{2m}\int \frac{d^3p'}{(2\pi\hbar)^3}\frac{e^{i\mathbf{p}'\cdot(\mathbf{x}-\mathbf{x}')/\hbar}}{[E - (\mathbf{p}'^2/2m) \pm i\varepsilon]}. \tag{7.1.15}$$

Now write $E = \hbar^2 k^2/2m$ and set $\mathbf{p}' \equiv \hbar \mathbf{q}$. Equation (7.1.15) becomes

$$\frac{1}{(2\pi)^3}\int_0^\infty q^2\,dq\int_0^{2\pi}d\phi\int_{-1}^{+1}\frac{d(\cos\theta)\,e^{i|\mathbf{q}||\mathbf{x}-\mathbf{x}'|\cos\theta}}{k^2-q^2\pm i\varepsilon}$$

$$=\frac{-1}{8\pi^2}\frac{1}{i|\mathbf{x}-\mathbf{x}'|}\int_{-\infty}^{+\infty}\frac{dqq\left(e^{iq|\mathbf{x}-\mathbf{x}'|}-e^{-iq|\mathbf{x}-\mathbf{x}'|}\right)}{q^2-k^2\mp i\varepsilon}$$

$$=-\frac{1}{4\pi}\frac{e^{\pm ik|\mathbf{x}-\mathbf{x}'|}}{|\mathbf{x}-\mathbf{x}'|}. \tag{7.1.16}$$

In the last step we used the method of residues, noting that the integrand has poles in the complex q-plane at

$$q = \pm k\sqrt{1 \pm \left(\frac{i\varepsilon}{k^2}\right)} \simeq \pm k \pm i\varepsilon'. \tag{7.1.17}$$

The reader may recognize that $G_\pm$ is nothing more than Green's function for the Helmholtz equation,

$$(\nabla^2 + k^2)G_\pm(\mathbf{x},\mathbf{x}') = \delta^{(3)}(\mathbf{x}-\mathbf{x}'). \tag{7.1.18}$$

Armed with the explicit form of $G_\pm$ as in Equation (7.1.12), we can use (7.1.11) to write Equation (7.1.7) as

$$\langle \mathbf{x}|\psi^{(\pm)}\rangle = \langle \mathbf{x}|\phi\rangle - \frac{2m}{\hbar^2}\int d^3x'\frac{e^{\pm ik|\mathbf{x}-\mathbf{x}'|}}{4\pi|\mathbf{x}-\mathbf{x}'|}\langle \mathbf{x}'|V|\psi^{(\pm)}\rangle. \tag{7.1.19}$$

Notice that the wave function $\langle \mathbf{x}|\psi^{(\pm)}\rangle$ in the presence of the scatterer is written as the sum of the wave function for the incident wave $\langle \mathbf{x}|\phi\rangle$ and a term that represents the effect of scattering. As we will see explicitly later, at sufficiently large distances the spatial dependence of the second term is $e^{\pm ikr}/r$ provided that the potential is of finite range. This means that the positive solution (negative solution) corresponds to the plane wave plus an outgoing (incoming) spherical wave. In most physical problems we are interested in the positive solution because it is difficult to prepare a system satisfying the boundary condition appropriate for the negative solution.

To see the behavior of $\langle \mathbf{x}|\psi^{(\pm)}\rangle$ more explicitly, let us consider the specific case where V is a local potential—that is, a potential diagonal in the $\mathbf{x}$-representation. Potentials that are functions only of the position operator $\mathbf{x}$ belong to this category. In precise terms V is said to be **local** if it can be written as

$$\langle \mathbf{x}'|V|\mathbf{x}''\rangle = V(\mathbf{x}')\,\delta^{(3)}(\mathbf{x}'-\mathbf{x}''). \tag{7.1.20}$$

As a result, we obtain

$$\begin{aligned}\langle\mathbf{x}'|V|\psi^{(\pm)}\rangle &= \int d^3x''\langle\mathbf{x}'|V|\mathbf{x}''\rangle\langle\mathbf{x}''|\psi^{(\pm)}\rangle \\ &= V(\mathbf{x}')\langle\mathbf{x}'|\psi^{(\pm)}\rangle. \end{aligned} \tag{7.1.21}$$

The integral equation (7.1.19) now simplifies as

$$\langle\mathbf{x}|\psi^{(\pm)}\rangle = \langle\mathbf{x}|\phi\rangle - \frac{2m}{\hbar^2}\int d^3x' \frac{e^{\pm ik|\mathbf{x}-\mathbf{x}'|}}{4\pi|\mathbf{x}-\mathbf{x}'|}V(\vec{\mathbf{x}}')\langle\mathbf{x}'|\psi^{(\pm)}\rangle. \tag{7.1.22}$$

Let us attempt to understand the physics contained in this equation. The vector $\mathbf{x}$ is understood to be directed towards the observation point at which the wave function is evaluated. For a finite range potential, the region that gives rise to a nonvanishing contribution is limited in space. In scattering processes we are interested in studying the effect of the scatterer (that is, the finite range potential) at a point far outside the range of the potential. This is quite relevant from a practical point of view because we cannot put a detector at short distance near the scattering center. Observation is always made by a detector placed very far away from the scatterer at r greatly larger than the range of the potential. In other words, we can safely set

$$|\mathbf{x}| \gg |\mathbf{x}'|, \tag{7.1.23}$$

as depicted in Figure 7.1.

Introducing

$$\begin{aligned} r &= |\mathbf{x}| \\ r' &= |\mathbf{x}'| \end{aligned} \tag{7.1.24}$$

and

$$\alpha = \angle(\mathbf{x}, \mathbf{x}'), \tag{7.1.25}$$

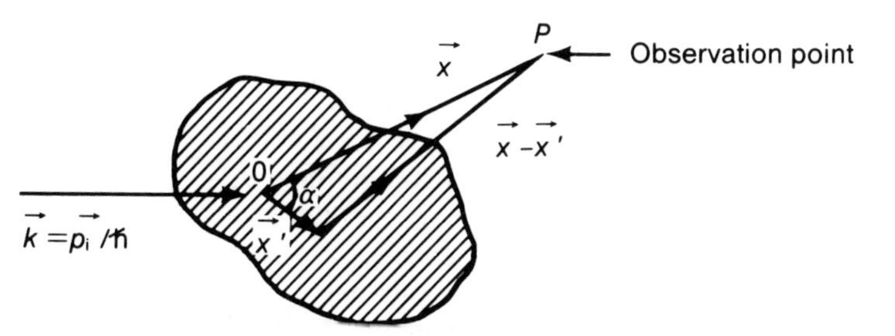

FIGURE 7.1. Finite-range scattering potential. The *observation point P* is where the wave function $\langle\mathbf{x}|\psi^{(\pm)}\rangle$ is to be evaluated, while the contribution to the integral in Equation (7.1.22) is for $|\mathbf{x}'|$ less than the range of the potential, as depicted by the shaded region of the figure.

we have for $r \gg r'$,

$$|\mathbf{x}-\mathbf{x}'| = \sqrt{r^2 - 2rr'\cos\alpha + r'^2}$$
$$= r\left(1 - \frac{2r'}{r}\cos\alpha + \frac{r'^2}{r^2}\right)^{1/2} \tag{7.1.26}$$
$$\simeq r - \hat{\mathbf{r}}\cdot\mathbf{x}'$$

where

$$\hat{\mathbf{r}} \equiv \frac{\mathbf{x}}{|\mathbf{x}|}. \tag{7.1.27}$$

Also define

$$\mathbf{k}' \equiv k\hat{\mathbf{r}}. \tag{7.1.28}$$

The motivation for this definition is that $\mathbf{k}'$ represents the propagation vector for waves reaching observation point $\mathbf{x}$. We then obtain

$$e^{\pm ik|\mathbf{x}-\mathbf{x}'|} \simeq e^{\pm ikr}e^{\mp i\mathbf{k}'\cdot\mathbf{x}'} \tag{7.1.29}$$

for large r. It is legitimate to replace $1/|\mathbf{x}-\mathbf{x}'|$ by just $1/r$. Furthermore, to get rid of the $\hbar$'s in expressions such as $1/(2\pi\hbar)^{3/2}$, it is convenient to use $|\mathbf{k}\rangle$ rather than $|\mathbf{p}_i\rangle$, where

$$\mathbf{k} \equiv \frac{\mathbf{p}_i}{\hbar}. \tag{7.1.30}$$

Because $|\mathbf{k}\rangle$ is normalized as

$$\langle\mathbf{k}|\mathbf{k}'\rangle = \delta^{(3)}(\mathbf{k}-\mathbf{k}') \tag{7.1.31}$$

we have

$$\langle\mathbf{x}|\mathbf{k}\rangle = \frac{e^{i\mathbf{k}\cdot\mathbf{x}}}{(2\pi)^{3/2}}. \tag{7.1.32}$$

So, finally,

$$\langle\mathbf{x}|\psi^{(+)}\rangle \xrightarrow{r \text{ large}} \langle\mathbf{x}|\mathbf{k}\rangle - \frac{1}{4\pi}\frac{2m}{\hbar^2}\frac{e^{ikr}}{r}\int d^3x'\, e^{-i\mathbf{k}'\cdot\mathbf{x}'}V(\mathbf{x}')\langle\mathbf{x}'|\psi^{(+)}\rangle$$
$$= \frac{1}{(2\pi)^{3/2}}\left[e^{i\mathbf{k}\cdot\mathbf{x}} + \frac{e^{ikr}}{r}f(\mathbf{k}',\mathbf{k})\right]. \tag{7.1.33}$$

This form makes it very clear that we have the original plane wave in propagation direction $\mathbf{k}$ plus an outgoing spherical wave with amplitude $f(\mathbf{k}',\mathbf{k})$ given by

$$f(\mathbf{k}',\mathbf{k}) \equiv -\frac{1}{4\pi}\frac{2m}{\hbar^2}(2\pi)^3\int d^3x'\,\frac{e^{-i\mathbf{k}'\cdot\mathbf{x}'}}{(2\pi)^{3/2}}V(\mathbf{x}')\langle\mathbf{x}'|\psi^{(+)}\rangle$$
$$= -\frac{1}{4\pi}(2\pi)^3\frac{2m}{\hbar^2}\langle\mathbf{k}'|V|\psi^{(+)}\rangle. \tag{7.1.34}$$

Similarly, we can readily show from (7.1.22) and (7.1.29) that $\langle \mathbf{x}|\psi^{(-)}\rangle$ corresponds to the original plane wave in propagation direction $\mathbf{k}$ plus an incoming spherical wave with spatial dependence e^{-ikr}/r and amplitude $-(1/4\pi)(2\pi)^3(2m/\hbar^2)\langle -\mathbf{k}'|V|\psi^{(-)}\rangle$.

To obtain the **differential cross section** $d\sigma/d\Omega$, we may consider a large number of identically prepared particles all characterized by the wave function (7.1.32). We can then ask, what is the number of incident particles crossing a plane perpendicular to the incident direction per unit area per unit time? This is just proportional to the probability flux due to the first term on the right-hand side in (7.1.33). Likewise we may ask, what is the number of scattered particles going into a small area $d\sigma$ subtending a differential solid-angle element $d\Omega$? Clearly,

$$\frac{d\sigma}{d\Omega}d\Omega = \frac{\text{number of particles scattered into } d\Omega \text{ per unit time}}{\text{number of incident particles crossing unit area per unit time}}$$

$$= \frac{r^2|\mathbf{j}_{\text{scatt}}|d\Omega}{|\mathbf{j}_{\text{incid}}|} = |f(\mathbf{k}',\mathbf{k})|^2 d\Omega. \tag{7.1.35}$$

Hence the differential cross section is

$$\frac{d\sigma}{d\Omega} = |f(\mathbf{k}',\mathbf{k})|^2. \tag{7.1.36}$$

Wave-Packet Description

The reader may wonder here whether our time-independent formulation of scattering has anything to do with the motion of a particle being

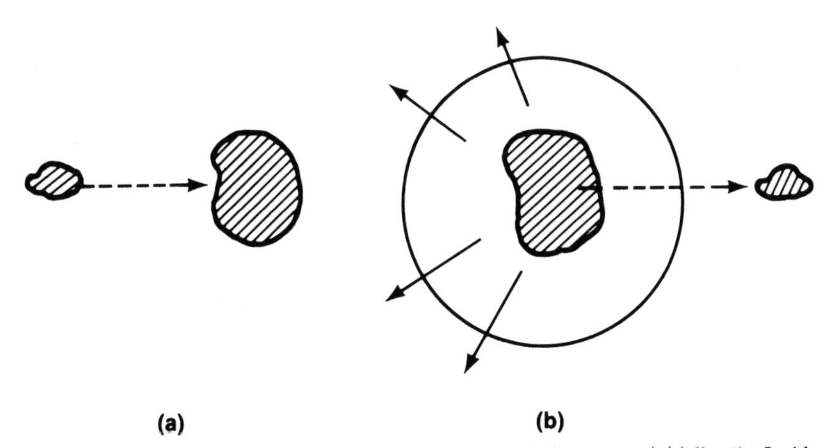

FIGURE 7.2. (a) Incident wave packet approaching scattering center initially. (b) Incident wave packet continuing to move in the original direction plus spherical outgoing wave front (after a long time duration).

bounced by a scattering center. The incident plane wave we have used is infinite in extent in both space and time. In a more realistic situation, we consider a wave packet (a difficult subject!) that approaches the scattering center.* After a long time we have both the original wave packet moving in the original direction plus a spherical wave front that moves outward, as in Figure 7.2. Actually the use of a plane wave is satisfactory as long as the dimension of the wave packet is much larger than the size of the scatterer (or range of V).

7.2. THE BORN APPROXIMATION

Equation (7.1.34) is still not directly useful in computing the differential cross section because in the expression for $f(\mathbf{k}',\mathbf{k})$ the unknown ket $|\psi^{(+)}\rangle$ appears. If the effect of the scatterer is not very strong, we may infer that it is not such a bad approximation to replace $\langle\mathbf{x}'|\psi^{(+)}\rangle$ (which appears under the integral sign) by $\langle\mathbf{x}'|\phi\rangle$—that is,

$$\langle\mathbf{x}'|\psi^{(+)}\rangle \to \langle\mathbf{x}'|\phi\rangle = \frac{e^{i\mathbf{k}\cdot\mathbf{x}'}}{(2\pi)^{3/2}}. \tag{7.2.1}$$

We then obtain an approximate expression for $f(\mathbf{k}',\mathbf{k})$. Because we treat the potential V to first order, the approximate amplitude so obtained is known as the **first-order Born amplitude** and is denoted by $f^{(1)}$:

$$f^{(1)}(\mathbf{k}',\mathbf{k}) = -\frac{1}{4\pi}\frac{2m}{\hbar^2}\int d^3x'\, e^{i(\mathbf{k}-\mathbf{k}')\cdot\mathbf{x}'}V(\mathbf{x}'). \tag{7.2.2}$$

In other words, apart from $-(2m/4\pi\hbar^2)$, the first-order amplitude is just the three-dimensional Fourier transform of the potential V with respect to $\mathbf{q} \equiv \mathbf{k}-\mathbf{k}'$.

For a spherically symmetric potential, $f^{(1)}(\mathbf{k}',\mathbf{k})$ is a function of $|\mathbf{k}-\mathbf{k}'|$, given by (remember $|\mathbf{k}'| = k$ by energy conservation)

$$|\mathbf{k}-\mathbf{k}'| \equiv q = 2k\sin\frac{\theta}{2}; \tag{7.2.3}$$

see Figure 7.3. We can perform the angular integration explicitly to obtain

$$\begin{aligned} f^{(1)}(\theta) &= -\frac{1}{2}\frac{2m}{\hbar^2}\frac{1}{iq}\int_0^\infty \frac{r^2}{r}V(r)\left(e^{iqr}-e^{-iqr}\right)dr \\ &= -\frac{2m}{\hbar^2}\frac{1}{q}\int_0^\infty rV(r)\sin qr\,dr. \end{aligned} \tag{7.2.4}$$

*For a fuller account of the wave-packet approach, see M. L. Goldberger and K. M. Watson, *Collision Theory*, Chapter 3 (New York: John Wiley, 1964); R. G. Newton, *Scattering Theory of Waves and Particles*, Chapter 6 (New York: McGraw-Hill, 1966).

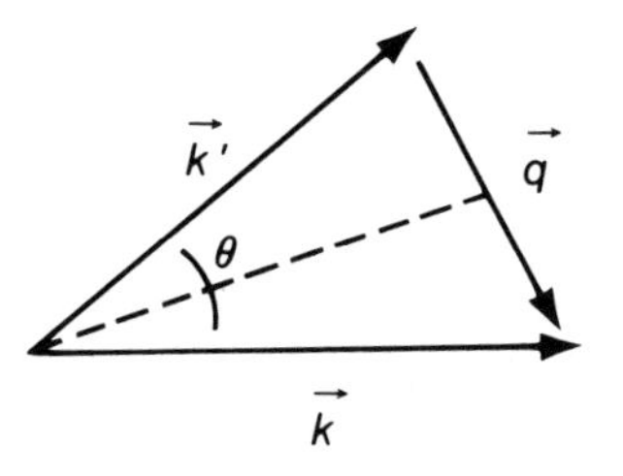

FIGURE 7.3. Scattering through angle θ, where $\mathbf{q} = \mathbf{k} - \mathbf{k}'$.

An example is now in order. Consider scattering by a Yukawa potential

$$V(r) = \frac{V_0 e^{-\mu r}}{\mu r}, \tag{7.2.5}$$

where V_0 is independent of r and $1/\mu$ corresponds, in a certain sense, to the range of the potential. Notice that V goes to zero very rapidly for $r \gg 1/\mu$. For this potential we obtain [from (7.2.4)]

$$f^{(1)}(\theta) = -\left(\frac{2mV_0}{\mu\hbar^2}\right)\frac{1}{q^2+\mu^2}, \tag{7.2.6}$$

where we have used

$$\begin{aligned} \operatorname{Im}\left[\int_0^\infty e^{-\mu r} e^{iqr}\, dr\right] &= -\operatorname{Im}\left(\frac{1}{-\mu + iq}\right) \\ &= \frac{q}{\mu^2 + q^2}. \end{aligned} \tag{7.2.7}$$

Notice also that

$$q^2 = 4k^2 \sin^2\frac{\theta}{2} = 2k^2(1-\cos\theta). \tag{7.2.8}$$

So, in the first Born approximation, the differential cross section for scattering by a Yukawa potential is given by

$$\left(\frac{d\sigma}{d\Omega}\right) \simeq \left(\frac{2mV_0}{\mu\hbar^2}\right)^2 \frac{1}{\left[2k^2(1-\cos\theta)+\mu^2\right]^2}. \tag{7.2.9}$$

It is amusing to observe here that as $\mu \to 0$, the Yukawa potential is reduced to the Coulomb potential, provided the ratio V_0/μ is fixed—for example, to be $ZZ'e^2$—in the limiting process. We see that the first Born differential cross section obtained in this manner becomes

$$\left(\frac{d\sigma}{d\Omega}\right) \simeq \frac{(2m)^2(ZZ'e^2)^2}{\hbar^4} \frac{1}{16k^4\sin^4(\theta/2)}. \tag{7.2.10}$$

Even the $\hbar$ disappears if $\hbar k$ is identified as $|\mathbf{p}|$, so

$$\left(\frac{d\sigma}{d\Omega}\right)=\frac{1}{16}\left(\frac{ZZ'e^2}{E_{KE}}\right)^2\frac{1}{\sin^4(\theta/2)}, \tag{7.2.11}$$

where $E_{KE}=|\mathbf{p}|^2/2m$; this is precisely the Rutherford scattering cross section that can be obtained *classically*.

Coming back to (7.2.4), the Born amplitude with a spherically symmetric potential, there are several general remarks we can make if $f(\mathbf{k}',\mathbf{k})$ can be approximated by the corresponding first Born amplitude, $f^{(1)}$.

1. $d\sigma/d\Omega$, or $f(\theta)$, is a function of q only; that is, $f(\theta)$ depends on the energy $(\hbar^2k^2/2m)$ and θ only through the combination $2k^2(1-\cos\theta)$.
2. $f(\theta)$ is always real.
3. $d\sigma/d\Omega$ is independent of the sign of V.
4. For small k (q necessarily small),
$$f^{(1)}(\theta)=-\frac{1}{4\pi}\frac{2m}{\hbar^2}\int V(r)\,d^3x$$
involving a volume integral independent of θ.
5. $f(\theta)$ is small for large q due to rapid oscillation of the integrand.

Let us now discuss the validity of the first-order Born approximation. It is clear from the derivation that if the Born approximation is to be applicable, $\langle\mathbf{x}|\psi^{(+)}\rangle$ should not be too different from $\langle\mathbf{x}|\phi\rangle$ inside the range of potential—that is, in the region where $V(\mathbf{x})$ is appreciable. Otherwise, it is not legitimate to replace $|\psi^{(+)}\rangle$ by $|\phi\rangle$. In other words, the distortion of the incident wave must be small. Going back to the exact expression (7.1.22), we note that the condition that $\langle\mathbf{x}|\psi^{(+)}\rangle$ be not too different from $\langle\mathbf{x}|\phi\rangle$ at the center of the scattering potential $\mathbf{x}\simeq 0$ is seen to be

$$\left|\frac{2m}{\hbar^2}\frac{1}{4\pi}\int d^3x'\,\frac{e^{ikr'}}{r'}\,V(\mathbf{x}')\,e^{i\mathbf{k}\cdot\mathbf{x}'}\right|\ll 1. \tag{7.2.12}$$

We now consider what the special case of the Yukawa potential in (7.2.5) may imply. At low energies—that is, for small k ($k\ll\mu$)—it is legitimate to replace $e^{ikr'}$ by 1. So we must have

$$\frac{2m}{\hbar^2}\frac{|V_0|}{\mu^2}\ll 1. \tag{7.2.13}$$

This requirement may be compared with the condition for the Yukawa potential to develop a bound state, which we can show to be

$$\frac{2m}{\hbar^2\mu^2}|V_0|\geq 2.7 \tag{7.2.14}$$

with V_0 negative. In other words, if the potential is strong enough to develop a bound state, the Born approximation will probably give a misleading result. In the opposite high k-limit, the condition that the second term in (7.1.22) is small can be shown to imply

$$\frac{2m}{\hbar^2}\frac{|V_0|}{\mu k}\ln\left(\frac{k}{\mu}\right) \ll 1. \tag{7.2.15}$$

As k becomes larger, this inequality is more easily satisfied. Quite generally, the Born approximation tends to get better at higher energies.

The Higher-Order Born Approximation

Let us now consider the higher-order Born approximation. Here, it is more compact to use the symbolic approach. For the transition operator T, we define T such that

$$V|\psi^{(+)}\rangle = T|\phi\rangle. \tag{7.2.16}$$

Multiplying the Lippmann-Schwinger equation (7.1.6) by V, we obtain

$$T|\phi\rangle = V|\phi\rangle + V\frac{1}{E-H_0+i\varepsilon}T|\phi\rangle. \tag{7.2.17}$$

This is supposed to hold for $|\phi\rangle$ taken to be any plane-wave state; furthermore, we know that these momentum eigenkets are complete. Therefore, we must have the following operator equation satisfied:

$$T = V + V\frac{1}{E-H_0+i\varepsilon}T. \tag{7.2.18}$$

The scattering amplitude of (7.1.34) can now be written as [using (7.2.16) with the $|\phi\rangle$ as momentum eigenkets]

$$f(\mathbf{k}',\mathbf{k}) = -\frac{1}{4\pi}\frac{2m}{\hbar^2}(2\pi)^3\langle\mathbf{k}'|T|\mathbf{k}\rangle. \tag{7.2.19}$$

Thus to determine $f(\mathbf{k}',\mathbf{k})$, it is sufficient to know the transition operator T.

We can obtain an iterative solution for T as follows:

$$T = V + V\frac{1}{E-H_0+i\varepsilon}V + V\frac{1}{E-H_0+i\varepsilon}V\frac{1}{E-H_0+i\varepsilon}V + \cdots. \tag{7.2.20}$$

Correspondingly, we can expand f as follows:

$$f(\mathbf{k}',\mathbf{k}) = \sum_{n=1}^{\infty} f^{(n)}(\mathbf{k}',\mathbf{k}), \tag{7.2.21}$$

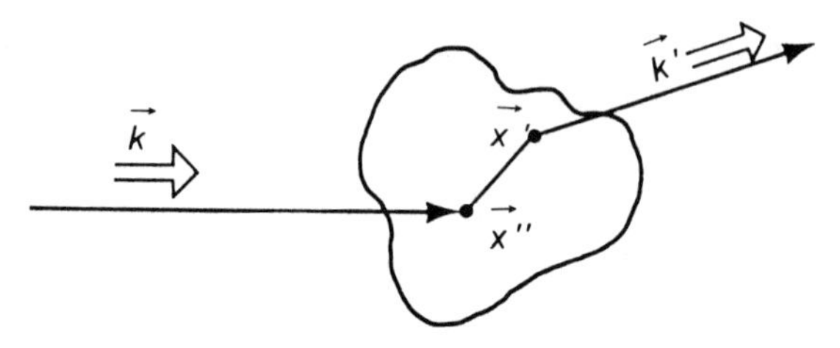

FIGURE 7.4. Physical interpretation of the higher-order Born term $f^{(2)}(\mathbf{k}',\mathbf{k})$.

where n is the number of times the V operator enters. We have

$$f^{(1)}(\mathbf{k}',\mathbf{k}) = -\frac{1}{4\pi}\frac{2m}{\hbar^2}(2\pi)^3\langle\mathbf{k}'|V|\mathbf{k}\rangle$$

$$f^{(2)}(\mathbf{k}',\mathbf{k}) = -\frac{1}{4\pi}\frac{2m}{\hbar^2}(2\pi)^3\langle\mathbf{k}'|V\frac{1}{E-H_0+i\varepsilon}V|\mathbf{k}\rangle \tag{7.2.22}$$

$$\vdots$$

If an explicit form is required for $f^{(2)}$, we can write it as

$$\begin{aligned} f^{(2)} &= -\frac{1}{4\pi}\frac{2m}{\hbar^2}(2\pi)^3\int d^3x'\int d^3x''\langle\mathbf{k}'|\mathbf{x}'\rangle V(\mathbf{x}') \\ &\quad\times\left\langle\mathbf{x}'\left|\frac{1}{E-H_0+i\varepsilon}\right|\mathbf{x}''\right\rangle V(\mathbf{x}'')\langle\mathbf{x}''|\mathbf{k}\rangle \\ &= -\frac{1}{4\pi}\frac{2m}{\hbar^2}\int d^3x'\int d^3x''\, e^{-i\mathbf{k}'\cdot\mathbf{x}'}V(\mathbf{x}') \\ &\quad\times\left[\frac{2m}{\hbar^2}G_+(\mathbf{x}',\mathbf{x}'')\right]V(\mathbf{x}'')e^{i\mathbf{k}\cdot\mathbf{x}''}. \end{aligned} \tag{7.2.23}$$

A physical interpretation of (7.2.23) is given in Figure 7.4, where the incident wave interacts at $\mathbf{x}''$—which explains the appearance of $V(\mathbf{x}'')$—and then propagates from $\mathbf{x}''$ to $\mathbf{x}'$ via Green's function for the Helmholtz equation (7.1.18); subsequently, a second interaction occurs at $\mathbf{x}'$—thus the appearance of $V(\mathbf{x}')$—and, finally, the wave is scattered into the direction $\mathbf{k}'$. In other words, $f^{(2)}$ corresponds to scattering viewed as a two-step process; likewise, $f^{(3)}$ is viewed as a three-step process, and so on.

7.3 OPTICAL THEOREM

There is a very famous relationship popularly attributed to Bohr, Peierls, and Placzek [*Editor's Note:* This relationship is in fact due to Eugene Feenberg* (*Phys. Rev.* **40**, 1932); see R. G. Newton (*Am. J. Phys.* **44**,

---

*As pointed out by Newton in his Review (c.f. Ref. 8 therein), Feenberg's paper is also remarkable for the fact that it was published on 1 April 1932, with a receipt date stated as 8 September 1932—thus violating causality!

639, 1976) for the historical background.] called the **optical theorem**, which relates the imaginary part of the forward scattering amplitude $f(\theta = 0)$ to the total cross section σ_{tot}, as follows:

Optical Theorem

$$\operatorname{Im} f(\theta=0)=\frac{k\sigma_{\text{tot}}}{4\pi} \tag{7.3.1}$$

where

$$f(\theta=0)\equiv f(\mathbf{k},\mathbf{k}), \tag{7.3.2}$$

the setting of $\mathbf{k}'\equiv\mathbf{k}$ *imposes scattering in the forward direction, and*

$$\sigma_{\text{tot}}\equiv\int\frac{d\sigma}{d\Omega}\,d\Omega. \tag{7.3.3}$$

Proof. From (7.2.19) we have

$$f(\theta=0)=f(\mathbf{k},\mathbf{k})=-\frac{1}{4\pi}\frac{2m}{\hbar^2}(2\pi)^3\langle\mathbf{k}|T|\mathbf{k}\rangle. \tag{7.3.4}$$

We next evaluate $\operatorname{Im}\langle\mathbf{k}|T|\mathbf{k}\rangle$ using (7.2.16), (7.1.6), and the hermiticity of V:

$$\begin{aligned}\operatorname{Im}\langle\mathbf{k}|T|\mathbf{k}\rangle&=\operatorname{Im}\langle\mathbf{k}|V|\psi^{(+)}\rangle\\&=\operatorname{Im}\left[\left(\langle\psi^{(+)}|-\langle\psi^{(+)}|V\frac{1}{E-H_0-i\varepsilon}\right)V|\psi^{(+)}\rangle\right].\end{aligned} \tag{7.3.5}$$

Now we use the well-known relation

$$\frac{1}{E-H_0-i\varepsilon}=\text{Pr.}\left(\frac{1}{E-H_0}\right)+i\pi\delta(E-H_0)$$

to reduce the right-hand side of (7.3.5) to the form

$$\begin{aligned}&\operatorname{Im}(\langle\psi^{(+)}|V|\psi^{(+)}\rangle)-\operatorname{Im}\langle\psi^{(+)}|V\,\text{Pr.}\frac{1}{E-H_0}V|\psi^{(+)}\rangle\\&\quad-\operatorname{Im}\langle\psi^{(+)}|Vi\pi\delta(E-H_0)V|\psi^{(+)}\rangle.\end{aligned} \tag{7.3.6}$$

The first two terms of (7.3.6) vanish because of the hermiticity of V and $V\,\text{Pr.}[1/(E-H_0)]V$; hence (7.3.6) reduces to

$$-\pi\langle\psi^{(+)}|V\delta(E-H_0)V|\psi^{(+)}\rangle. \tag{7.3.7}$$

Again, we can recast (7.3.7) using (7.2.16) and $|\phi\rangle=|\mathbf{k}\rangle$ as

$$\begin{aligned}\operatorname{Im}\langle\mathbf{k}|T|\mathbf{k}\rangle&=-\pi\langle\mathbf{k}|T^{\dagger}\delta(E-H_0)T|\mathbf{k}\rangle\\&=-\pi\int d^3k'\,\langle\mathbf{k}|T^{\dagger}|\mathbf{k}'\rangle\langle\mathbf{k}'|T|\mathbf{k}\rangle\delta\left(E-\frac{\hbar^2k'^2}{2m}\right)\\&=-\pi\int d\Omega'\,\frac{mk}{\hbar^2}|\langle\mathbf{k}'|T|\mathbf{k}\rangle|^2,\end{aligned} \tag{7.3.8}$$

where we have used $d^3k' = k'^2 dE(dk'/dE)\,d\Omega'$, the δ-function constraint $E = \hbar^2 k'^2/2m$ [hence $dE = (\hbar^2 k'/m)\,dk'$], and—finally—$k' = k$. From (7.3.4) and (7.3.8), we have

$$\operatorname{Im} f(0) = -\frac{1}{4\pi}\frac{2m}{\hbar^2}(2\pi)^3\left(-\frac{\pi m k}{\hbar^2}\int d\Omega' |\langle \mathbf{k}'|T|\mathbf{k}\rangle|^2\right)$$

$$= \frac{k\sigma_{\text{tot}}}{4\pi}, \tag{7.3.9}$$

where in the last step we have used (7.1.36), (7.2.19), and (7.3.3). □

We can appreciate the physical significance of the optical theorem after we discuss shadow scattering.

7.4. EIKONAL APPROXIMATION

This approximation covers a situation in which $V(\mathbf{x})$ varies very little over a distance of order of wavelength $\lambda\!\!\!^{-}$ (which can be regarded as "small"). Note that V itself need not be weak as long as $E \gg |V|$; hence the domain of validity here is different from the Born approximation. Under these conditions, the semiclassical path concept becomes applicable, and we replace the exact wave function $\psi^{(+)}$ by the semiclassical wave function [see (2.4.18) and (2.4.22)], namely,

$$\psi^{(+)} \sim e^{iS(\mathbf{x})/\hbar}. \tag{7.4.1}$$

This leads to the Hamilton-Jacobi equation for S,

$$\frac{(\nabla S)^2}{2m} + V = E = \frac{\hbar^2 k^2}{2m}, \tag{7.4.2}$$

as discussed in Section 2.4. We propose to compute S from (7.4.2) by making the further approximation that the classical trajectory is a straight-line path, which should be satisfactory for small deflection at high energy.* Consider the situation depicted in Figure 7.5, where the straight-line trajectory is along the z-direction. Integrating (7.4.2) we have

$$\frac{S}{\hbar} = \int_{-\infty}^{z}\left[k^2 - \frac{2m}{\hbar^2}V\left(\sqrt{b^2 + z'^2}\right)\right]^{1/2} dz' + \text{constant}. \tag{7.4.3}$$

The additive constant is to be chosen in such a way that

$$\frac{S}{\hbar} \to kz \quad \text{as} \quad V \to 0 \tag{7.4.4}$$

so that the plane-wave form for (7.4.1) is reproduced in this zero-potential

*Needless to say, solution of (7.4.2) to *determine* the classical trajectory would be a forbidding task in general.

limit. We can then write Equation (7.4.3) as

$$\frac{S}{\hbar} = kz + \int_{-\infty}^{z} \left[\sqrt{k^2 - \frac{2m}{\hbar^2} V\left(\sqrt{b^2 + z'^2}\right)} - k \right] dz'$$

$$\cong kz - \frac{m}{\hbar^2 k} \int_{-\infty}^{z} V\left(\sqrt{b^2 + z'^2}\right) dz' \tag{7.4.5}$$

where for $E \gg V$, we have used

$$\sqrt{k^2 - \frac{2m}{\hbar^2} V\left(\sqrt{b^2 + z'^2}\right)} \sim k - \frac{mV}{\hbar^2 k}$$

at high $E = \hbar^2 k^2 / 2m$. So

$$\psi^{(+)}(\mathbf{x}) = \psi^{(+)}(\mathbf{b} + z\hat{\mathbf{z}}) \simeq \frac{1}{(2\pi)^{3/2}} e^{ikz} \exp\left[\frac{-im}{\hbar^2 k} \int_{-\infty}^{z} V\left(\sqrt{b^2 + z'^2}\right) dz' \right]. \tag{7.4.6}$$

Though (7.4.6) does not have the correct asymptotic form appropriate for an incident plus spherical outgoing wave (that is, it is not of form $e^{i\vec{\mathbf{k}}\cdot\vec{\mathbf{x}}} + f(\theta)(e^{ikr}/r)$ and indeed refers only to motion along the original direction), it can nevertheless still be used in (7.1.34) to obtain an approximate expression for $f(\mathbf{k}', \mathbf{k})$, to wit,

$$f(\mathbf{k}', \mathbf{k}) = -\frac{1}{4\pi} \frac{2m}{\hbar^2} \int d^3x' e^{-i\mathbf{k}'\cdot\mathbf{x}'} V\left(\sqrt{b^2 + z'^2}\right) e^{i\mathbf{k}\cdot\mathbf{x}'}$$

$$\times \exp\left[-\frac{im}{\hbar^2 k} \int_{-\infty}^{z'} V\left(\sqrt{b^2 + z''^2}\right) dz'' \right]. \tag{7.4.7}$$

Note that without the last factor, exp [...], (7.4.7) is just like the first-order Born amplitude in (7.2.2). We perform the three-dimensional (d^3x') integration in (7.4.7) by introducing cylindrical coordinates $d^3x' = b\, db\, d\phi_b\, dz'$ (see

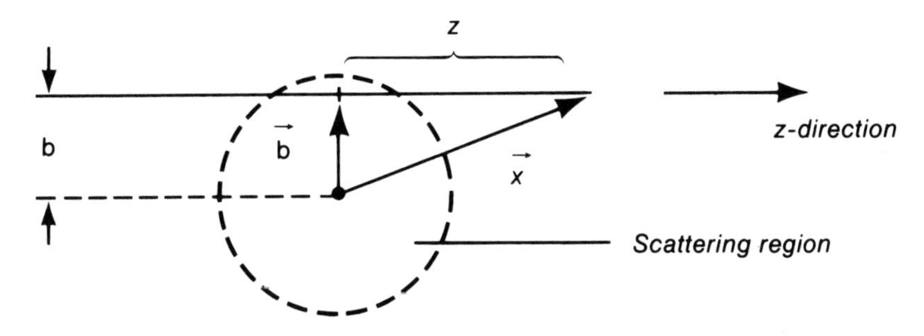

FIGURE 7.5. Schematic diagram of eikonal approximation scattering where the classical straight line trajectory is along the z-direction, $|\mathbf{x}| = r$, and $b = |\mathbf{b}|$ is the impact parameter.

Figure 7.5) and noting that

$$(\mathbf{k}-\mathbf{k}')\cdot\mathbf{x}' = (\mathbf{k}-\mathbf{k}')\cdot(\mathbf{b}+z'\hat{\mathbf{z}}) \simeq -\mathbf{k}'\cdot\mathbf{b}, \tag{7.4.8}$$

where we have used $\mathbf{k}\perp\mathbf{b}$ and $(\mathbf{k}-\mathbf{k}')\cdot\hat{\mathbf{z}} \sim 0(\theta^2)$, which can be ignored for small deflection θ. Without loss of generality we choose scattering to be in the xz-plane and write

$$\mathbf{k}'\cdot\mathbf{b} = (k\sin\theta\hat{\mathbf{x}} + k\cos\theta\hat{\mathbf{z}})\cdot(b\cos\phi_b\hat{\mathbf{x}} + b\sin\phi_b\hat{\mathbf{y}}) \simeq kb\theta\cos\phi_b. \tag{7.4.9}$$

The expression for $f(\mathbf{k}',\mathbf{k})$ becomes

$$f(\mathbf{k}',\mathbf{k}) = -\frac{1}{4\pi}\frac{2m}{\hbar^2}\int_0^\infty b\,db\int_0^{2\pi} d\phi_b\, e^{-ikb\theta\cos\phi_b}$$

$$\times\int_{-\infty}^{+\infty} dz\, V\exp\left[\frac{-im}{\hbar^2 k}\int_{-\infty}^{z} V dz'\right]. \tag{7.4.10}$$

We next use the following identities:

$$\int_0^{2\pi} d\phi_b\, e^{-ikb\theta\cos\phi_b} = 2\pi J_0(kb\theta) \tag{7.4.11}$$

and

$$\int_{-\infty}^{+\infty} dz\, V\exp\left[\frac{-im}{\hbar^2 k}\int_{-\infty}^{z} V dz'\right] = \frac{i\hbar^2 k}{m}\exp\left[\frac{-im}{\hbar^2 k}\int_{-\infty}^{z} V dz'\right]\Bigg|_{z=-\infty}^{z=+\infty}, \tag{7.4.12}$$

where, of course, the contribution from $z = -\infty$ on the right-hand side of (7.4.12) vanishes in the exponent. So, finally

$$f(\mathbf{k}',\mathbf{k}) = -ik\int_0^\infty db\, bJ_0(kb\theta)[e^{2i\Delta(b)}-1], \tag{7.4.13}$$

where

$$\Delta(b) \equiv \frac{-m}{2k\hbar^2}\int_{-\infty}^{+\infty} V(\sqrt{b^2+z^2})\, dz. \tag{7.4.14}$$

In (7.4.14) we fix the impact parameter b and integrate along the straight-line path z, shown in Figure 7.5. There is no contribution from $[e^{2i\Delta(b)}-1]$ in (7.4.13) if b is greater than the range of V.

It can be shown in a straightforward manner that the eikonal approximation satisfies the optical theorem (7.3.1). This proof plus some interesting applications—for example, when V is a Gaussian potential $\Delta(b)$ becomes Gaussian in b-space—are discussed in the literature (Gottfried 1966). For the case where V is a Yukawa potential, see Problem 7 in this chapter.

7.5. FREE-PARTICLE STATES: PLANE WAVES VERSUS SPHERICAL WAVES

In considering scattering by a spherically symmetric potential, we often examine how states with definite angular momenta are affected by the scatterer. Such considerations lead to the method of partial waves, to be discussed in detail in Section 7.6. However, before discussing the angular momentum decomposition of scattering states, let us first talk about free-particle states, which are also eigenstates of angular momentum.

For a free particle the Hamiltonian is just the kinetic-energy operator, which obviously commutes with the momentum operator. It is for this reason that in Section 7.1 we also took $|\phi\rangle$, an eigenket of the free-particle Hamiltonian, to be a momentum eigenket or a plane-wave state $|\mathbf{k}\rangle$, where the eigenvalue of the momentum operator is $\hbar\mathbf{k}$. We note, however, that the free-particle Hamiltonian also commutes with $\mathbf{L}^2$ and L_z. Thus it is possible to consider a simultaneous eigenket of H_0, $\mathbf{L}^2$, and L_z. Ignoring spin, such a state is denoted by $|E, l, m\rangle$, often called a **spherical-wave state**.

More generally, the most general free-particle state can be regarded as a superposition of $|E, l, m\rangle$ with various E, l, and m in much the same way as the most general free-particle state can be regarded as a superposition of $|\mathbf{k}\rangle$ with different $\mathbf{k}$, different in both magnitude and direction. Put in another way, a free-particle state can be analysed using either the plane-wave basis $\{|\mathbf{k}\rangle\}$ or the spherical-wave basis $\{|E, l, m\rangle\}$.

We now derive the transformation function $\langle\mathbf{k}|E, l, m\rangle$ that connects the plane-wave basis with the spherical-wave basis. We can also regard this quantity as the momentum-space wave function for the spherical wave characterized by E, l, and m. We adopt the normalization convention for the spherical-wave eigenket as follows:

$$\langle E', l', m'|E, l, m\rangle = \delta_{ll'}\delta_{mm'}\delta(E - E'). \tag{7.5.1}$$

In analogy with the position-space wave function, we may guess the angular dependence:

$$\langle\mathbf{k}|E, l, m\rangle = g_{lE}(k)Y_l^m(\hat{\mathbf{k}}). \tag{7.5.2}$$

To prove this rigorously, we proceed as follows. First, consider the momentum eigenket $|k\hat{\mathbf{z}}\rangle$—that is, a plane-wave state whose propagation direction is along the positive z-axis. An important property of this state is that it has no orbital angular-momentum component in the z-direction:

$$L_z|k\hat{\mathbf{z}}\rangle = (xp_y - yp_x)|k_x = 0, k_y = 0, k_z = k\rangle = 0. \tag{7.5.3}$$

Actually this is plausible from classical considerations: The angular-momentum component must vanish in the direction of propagation because $\mathbf{L}\cdot\mathbf{p} = (\mathbf{x}\times\mathbf{p})\cdot\mathbf{p} = 0$. Because of (7.5.3)—and since $\langle E', l', m'|k\hat{\mathbf{z}}\rangle = 0$ for

$m' \neq 0$—we must be able to expand $|k\hat{\mathbf{z}}\rangle$ as follows:

$$|k\hat{\mathbf{z}}\rangle = \sum_{l'} \int dE' |E', l', m' = 0\rangle \langle E', l', m' = 0 | k\hat{\mathbf{z}}\rangle. \tag{7.5.4}$$

Notice that there is no m' sum; m' is always zero. We can obtain the most general momentum eigenket, with the direction of $\mathbf{k}$ specified by θ and ϕ, from $|k\hat{\mathbf{z}}\rangle$ by just applying the appropriate rotation operator as follows [see Figure 3.3 and (3.6.47)]:

$$|\mathbf{k}\rangle = \mathscr{D}(\alpha = \phi, \beta = \theta, \gamma = 0)|k\hat{\mathbf{z}}\rangle. \tag{7.5.5}$$

Multiplying (7.5.5) by $\langle E, l, m|$ on the left, we obtain

$$\begin{aligned}
\langle E, l, m|\mathbf{k}\rangle &= \sum_{l'} \int dE' \langle E, l, m|\mathscr{D}(\alpha = \phi, \beta = \theta, \gamma = 0)|E', l', m' = 0\rangle \\
&\quad \times \langle E', l', m' = 0|k\hat{\mathbf{z}}\rangle \\
&= \sum_{l'} \int dE' \mathscr{D}_{m0}^{(l')}(\alpha = \phi, \beta = \theta, \gamma = 0) \\
&\quad \times \delta_{ll'} \delta(E - E') \langle E', l', m' = 0|k\hat{\mathbf{z}}\rangle \\
&= \mathscr{D}_{m0}^{(l)}(\alpha = \phi, \beta = \theta, \gamma = 0) \langle E, l, m = 0|k\hat{\mathbf{z}}\rangle.
\end{aligned} \tag{7.5.6}$$

Now $\langle E, l, m = 0|k\hat{\mathbf{z}}\rangle$ is independent of the orientation of $\mathbf{k}$—that is, independent of θ and ϕ—and we may as well call it $\sqrt{\frac{2l+1}{4\pi}}\, g_{lE}^*(k)$. So we can write, using (3.6.51),

$$\langle \mathbf{k}|E, l, m\rangle = g_{lE}(k)\, Y_l^m(\hat{\mathbf{k}}). \tag{7.5.7}$$

Let us determine $g_{lE}(k)$. First, we note that

$$(H_0 - E)|E, l, m\rangle = 0. \tag{7.5.8}$$

But we also let $H_0 - E$ operate on a momentum eigenbra $\langle \mathbf{k}|$ as follows:

$$\langle \mathbf{k}|(H_0 - E) = \left(\frac{\hbar^2 k^2}{2m} - E\right)\langle \mathbf{k}|. \tag{7.5.9}$$

Multiplying (7.5.9) with $|E, l, m\rangle$ on the right, we obtain

$$\left(\frac{\hbar^2 k^2}{2m} - E\right)\langle \mathbf{k}|E, l, m\rangle = 0. \tag{7.5.10}$$

This means that $\langle \mathbf{k}|E, l, m\rangle$ can be nonvanishing only if $E = \hbar^2 k^2/2m$; so we must be able to write $g_{lE}(k)$ as

$$g_{lE}(k) = N\delta\left(\frac{\hbar^2 k^2}{2m} - E\right). \tag{7.5.11}$$

To determine N we go back to our normalization convention (7.5.1). We obtain

$$\begin{aligned}\langle E', l', m'|E, l, m\rangle &= \int d^3k''\langle E', l', m'|\mathbf{k}''\rangle\langle\mathbf{k}''|E, l, m\rangle \\ &= \int k''^2\, dk'' \int d\Omega_{\mathbf{k}''}|N|^2\delta\left(\frac{\hbar^2 k''^2}{2m} - E'\right) \\ &\quad\times \delta\left(\frac{\hbar^2 k''^2}{2m} - E\right) Y_{l'}^{m'*}(\hat{\mathbf{k}}'') Y_l^m(\hat{\mathbf{k}}'') \\ &= \int \frac{k''^2\, dE''}{dE''/dk''}\int d\Omega_{\mathbf{k}''}|N|^2\delta\left(\frac{\hbar^2 k''^2}{2m} - E'\right)\delta\left(\frac{\hbar^2 k''^2}{2m} - E\right) \\ &\quad\times Y_{l'}^{m'*}(\hat{\mathbf{k}}'') Y_l^m(\hat{\mathbf{k}}'') \\ &= |N|^2\frac{mk'}{\hbar^2}\delta(E - E')\delta_{ll'}\delta_{mm'},\end{aligned} \tag{7.5.12}$$

where we have defined $E'' = \hbar^2 k''^2/2m$ to change k''-integration into E''-integration. Comparing this with (7.5.1), we see that $N = \hbar/\sqrt{mk}$ will suffice. Therefore, we can finally write

$$g_{lE}(k) = \frac{\hbar}{\sqrt{mk}}\delta\left(\frac{\hbar^2 k^2}{2m} - E\right); \tag{7.5.13}$$

hence

$$\langle\mathbf{k}|E, l, m\rangle = \frac{\hbar}{\sqrt{mk}}\delta\left(\frac{\hbar^2 k^2}{2m} - E\right) Y_l^m(\hat{\mathbf{k}}). \tag{7.5.14}$$

From (7.5.14) we infer that the plane-wave state $|\mathbf{k}\rangle$ can be expressed as a superposition of free spherical-wave states with all possible l-values; in particular,

$$\begin{aligned}|\mathbf{k}\rangle &= \sum_l \sum_m \int dE|E, l, m\rangle\langle E, l, m|\mathbf{k}\rangle \\ &= \sum_{l=0}^{\infty}\sum_{m=-l}^{l} |E, l, m\rangle\bigg|_{E=\hbar^2k^2/2m}\left(\frac{\hbar}{\sqrt{mk}} Y_l^{m*}(\hat{\mathbf{k}})\right).\end{aligned} \tag{7.5.15}$$

Because the transverse dimension of the plane wave is infinite, we expect that the plane wave must contain all possible values of impact parameter b (semiclassically, the impact parameter $b \simeq l\hbar/p$). From this point of view it is no surprise that the momentum eigenstates $|\mathbf{k}\rangle$, when analyzed in terms of spherical-wave states, contain all possible values of l.

We have derived the wave function for $|E, l, m\rangle$ in momentum space. Next, we consider the corresponding wave function in position space.

From wave mechanics, the reader should be familiar with the fact that the wave function for a free spherical wave is $j_l(kr)Y_l^m(\hat{\mathbf{r}})$, where $j_l(kr)$ is the spherical Bessel function of order l (see Appendix A). The second solution $n_l(kr)$, although it satisfies the appropriate differential equation, is inadmissible because it is singular at the origin. So we can write

$$\langle \mathbf{x}|E,l,m\rangle = c_l j_l(kr)Y_l^m(\hat{\mathbf{r}}). \tag{7.5.16}$$

To determine c_l, all we have to do is compare

$$\begin{aligned}\langle \mathbf{x}|\mathbf{k}\rangle = \frac{e^{i\mathbf{k}\cdot\mathbf{x}}}{(2\pi)^{3/2}} &= \sum_l \sum_m \int dE \langle \mathbf{x}|E,l,m\rangle\langle E,l,m|\mathbf{k}\rangle \\ &= \sum_l \sum_m \int dE\, c_l j_l(kr)Y_l^m(\hat{\mathbf{r}}) \frac{\hbar}{\sqrt{mk}}\delta\left(E - \frac{\hbar^2k^2}{2m}\right)Y_l^{m*}(\hat{\mathbf{k}}) \\ &= \sum_l \frac{(2l+1)}{4\pi}P_l(\hat{\mathbf{k}}\cdot\hat{\mathbf{r}})\frac{\hbar}{\sqrt{mk}}c_l j_l(kr),\end{aligned} \tag{7.5.17}$$

where we have used the addition theorem

$$\Sigma_m Y_l^m(\hat{\mathbf{r}})Y_l^{m*}(\hat{\mathbf{k}}) = [(2l+1)/4\pi]P_l(\hat{\mathbf{k}}\cdot\hat{\mathbf{r}})$$

in the last step. Now $\langle \mathbf{x}|\mathbf{k}\rangle = e^{i\mathbf{k}\cdot\mathbf{x}}/(2\pi)^{3/2}$ can also be written as

$$\frac{e^{i\mathbf{k}\cdot\mathbf{x}}}{(2\pi)^{3/2}} = \frac{1}{(2\pi)^{3/2}}\sum_l (2l+1)i^l j_l(kr)P_l(\hat{\mathbf{k}}\cdot\hat{\mathbf{r}}), \tag{7.5.18}$$

which can be proved by using the following integral representation for $j_l(kr)$:

$$j_l(kr) = \frac{1}{2i^l}\int_{-1}^{+1} e^{ikr\cos\theta}P_l(\cos\theta)\,d(\cos\theta). \tag{7.5.19}$$

Comparing (7.5.17) with (7.5.18), we have

$$c_l = \frac{i^l}{\hbar}\sqrt{\frac{2mk}{\pi}}. \tag{7.5.20}$$

To summarize, we have

$$\langle \mathbf{k}|E,l,m\rangle = \frac{\hbar}{\sqrt{mk}}\delta\left(E - \frac{\hbar^2k^2}{2m}\right)Y_l^m(\hat{\mathbf{k}}) \tag{7.5.21a}$$

$$\langle \mathbf{x}|E,l,m\rangle = \frac{i^l}{\hbar}\sqrt{\frac{2mk}{\pi}}\, j_l(kr)Y_l^m(\hat{\mathbf{r}}). \tag{7.5.21b}$$

These expressions are extremely useful in developing the partial-wave expansion discussed in the next section.

We conclude this section by applying (7.5.21a) to a decay process. Suppose a parent particle of spin j disintegrates into two spin-zero particles A (spin j) $\to B$ (spin 0)$+\,C$ (spin 0). The basic Hamiltonian responsible for such a decay process is, in general, very complicated. However, we do know that angular momentum is conserved because the basic Hamiltonian must be rotationally invariant. So the momentum-space wave function for the final state must be of the form (7.5.21a), with l identified with the spin of the parent particle. This immediately enables us to compute the angular distribution of the decay product because the momentum-space wave function is nothing more than the probability amplitude for finding the decay product with relative momentum direction **k**.

As a concrete example from nuclear physics, let us consider the decay of an excited nucleus, Ne^{20*}:

$$Ne^{20*} \to O^{16} + He^4. \tag{7.5.22}$$

Both O^{16} and He^4 are known to be spinless particles. Suppose the magnetic quantum number of the parent nucleus is ± 1, relative to some direction z. Then the angular distribution of the decay product is proportional to $|Y_1^{\pm 1}(\theta, \phi)|^2 = (3/8\pi)\sin^2\theta$, where (θ, ϕ) are the polar angles defining the relative direction **k** of the decay product. On the other hand, if the magnetic quantum number is 0 for a parent nucleus with spin 1, the decay angular distribution varies as $|Y_1^0(\theta, \phi)|^2 = (3/4\pi)\cos^2\theta$.

For a general spin orientation we obtain

$$\sum_{m=-1}^{1} w(m)|Y_{l=1}^m|^2. \tag{7.5.23}$$

For an unpolarized nucleus the various $w(m)$ are all equal, and we obtain an isotropic distribution; this is not surprising because there is no preferred direction if the parent particle is unpolarized.

For a higher spin object, the angular distribution of the decay is more involved; the higher the spin of the parent decaying system, the greater the complexity of the angular distribution of the decay products. Quite generally, through a study of the angular distribution of the decay products, it is possible to determine the spin of the parent nucleus.

7.6. METHOD OF PARTIAL WAVES

Partial-Wave Expansion

Let us now come back to the case $V \neq 0$. We assume that the potential is spherically symmetric, that is, invariant under rotations in three dimensions. It then follows that the transition operator T, which is given by (7.2.20), commutes with $\mathbf{L}^2$ and $\mathbf{L}$. In other words, T is a scalar operator.

It is now useful to use the spherical-wave basis because the Wigner-Eckart theorem [see (3.10.38)], applied to a scalar operator, immediately gives

$$\langle E', l', m'|T|E, l, m\rangle = T_l(E)\delta_{ll'}\delta_{mm'}. \tag{7.6.1}$$

In other words, T is diagonal both in l and in m; furthermore, the (nonvanishing) diagonal element depends on E and l but not on m. This leads to an enormous simplification, as we will see shortly.

Let us now look at the scattering amplitude (7.2.19):

$$\begin{aligned}
f(\mathbf{k}',\mathbf{k}) &= -\frac{1}{4\pi}\frac{2m}{\hbar^2}(2\pi)^3\langle\mathbf{k}'|T|\mathbf{k}\rangle \\
&= -\frac{1}{4\pi}\frac{2m}{\hbar^2}(2\pi)^3\sum_l\sum_m\sum_{l'}\sum_{m'}\int dE\int dE'\langle\mathbf{k}'|E'l'm'\rangle \\
&\quad\times\langle E'l'm'|T|Elm\rangle\langle Elm|\mathbf{k}\rangle \\
&= -\frac{1}{4\pi}\frac{2m}{\hbar^2}(2\pi)^3\frac{\hbar^2}{mk}\sum_l\sum_m T_l(E)\bigg|_{E=\hbar^2k^2/2m} Y_l^m(\hat{\mathbf{k}}')Y_l^{m*}(\hat{\mathbf{k}}) \\
&= -\frac{4\pi^2}{k}\sum_l\sum_m T_l(E)\bigg|_{E=\hbar^2k^2/2m} Y_l^m(\hat{\mathbf{k}}')Y_l^{m*}(\hat{\mathbf{k}}).
\end{aligned} \tag{7.6.2}$$

To obtain the angular dependence of the scattering amplitude, let us choose the coordinate system in such a way that $\mathbf{k}$, as usual, is in the positive z-direction. We then have [see (3.6.50)]

$$Y_l^m(\hat{\mathbf{k}}) = \sqrt{\frac{2l+1}{4\pi}}\,\delta_{m0}, \tag{7.6.3}$$

where we have used $P_l(1)=1$; hence only the terms $m=0$ contribute. Taking θ to be the angle between $\mathbf{k}'$ and $\mathbf{k}$, we can write

$$Y_l^0(\hat{\mathbf{k}}') = \sqrt{\frac{2l+1}{4\pi}}\,P_l(\cos\theta). \tag{7.6.4}$$

It is customary here to define the **partial-wave amplitude** $f_l(k)$ as follows:

$$f_l(k) \equiv -\frac{\pi T_l(E)}{k}. \tag{7.6.5}$$

For (7.6.2) we then have

$$f(\mathbf{k}',\mathbf{k}) = f(\theta) = \sum_{l=0}^{\infty}(2l+1)f_l(k)P_l(\cos\theta), \tag{7.6.6}$$

where $f(\theta)$ still depends on k (or the incident energy) even though k is suppressed.

To appreciate the physical significance of $f_l(k)$, let us study the large-distance behavior of the wave function $\langle\mathbf{x}|\psi^{(+)}\rangle$ given by (7.1.33). Using the expansion of a plane wave in terms of spherical waves [(7.5.18)] and noting that (Appendix A)

$$j_l(kr) \xrightarrow{\text{large } r} \frac{e^{i(kr-(l\pi/2))} - e^{-i(kr-(l\pi/2))}}{2ikr}, \qquad (i^l = e^{i(\pi/2)l}) \qquad (7.6.7)$$

and that $f(\theta)$ is given by (7.6.6), we have

$$\begin{aligned}\langle \mathbf{x}|\psi^{(+)}\rangle &\xrightarrow{\text{large } r} \frac{1}{(2\pi)^{3/2}}\left[e^{ikz} + f(\theta)\frac{e^{ikr}}{r}\right]\\ &= \frac{1}{(2\pi)^{3/2}}\left[\sum_l (2l+1)P_l(\cos\theta)\left(\frac{e^{ikr} - e^{-i(kr-l\pi)}}{2ikr}\right)\right.\\ &\quad \left. + \sum_l (2l+1)f_l(k)P_l(\cos\theta)\frac{e^{ikr}}{r}\right]\\ &= \frac{1}{(2\pi)^{3/2}}\sum_l (2l+1)\frac{P_l}{2ik}\left[[1+2ikf_l(k)]\frac{e^{ikr}}{r} - \frac{e^{-i(kr-l\pi)}}{r}\right].\end{aligned} \qquad (7.6.8)$$

The physics of scattering is now clear. When the scatterer is absent, we can analyze the plane wave as the sum of a spherically outgoing wave behaving like e^{ikr}/r and a spherically incoming wave behaving like $-e^{-i(kr-l\pi)}/r$ for each l. The presence of the scatterer changes only the coefficient of the outgoing wave, as follows:

$$1 \to 1 + 2ikf_l(k). \qquad (7.6.9)$$

The incoming wave is completely unaffected.

Unitarity and Phase Shifts

We now examine the consequences of probability conservation, or unitarity. In a time-independent formulation, the flux current density $\mathbf{j}$ must satisfy

$$\nabla \cdot \mathbf{j} = -\frac{\partial |\psi|^2}{\partial t} = 0. \qquad (7.6.10)$$

Let us now consider a spherical surface of very large radius. By Gauss's theorem, we must have

$$\int_{\substack{\text{spherical}\\ \text{surface}}} \mathbf{j} \cdot d\mathbf{S} = 0. \qquad (7.6.11)$$

Physically (7.6.10) and (7.6.11) mean that there is no source or sink of particles. The outgoing flux must equal the incoming flux. Furthermore, because of angular-momentum conservation, this must hold for each partial wave separately. In other words, the coefficient of e^{ikr}/r must be the same in magnitude as the coefficient of e^{-ikr}/r. Defining $S_l(k)$ to be

$$S_l(k) \equiv 1 + 2ikf_l(k), \tag{7.6.12}$$

this means [from (7.6.9)] that

$$|S_l(k)| = 1, \tag{7.6.13}$$

that is, the most that can happen is a change in the phase of the outgoing wave. Equation (7.6.13) is known as the **unitarity relation** for the lth partial wave. In a more advanced treatment of scattering, $S_l(k)$ can be regarded as the lth diagonal element of the S operator, which is required to be unitary as a consequence of probability conservation.

We thus see that the only change in the wave function at a large distance as a result of scattering is to change the *phase* of the outgoing wave. Calling this phase $2\delta_l$ (the factor of 2 here is conventional), we can write

$$S_l = e^{2i\delta_l}, \tag{7.6.14}$$

with δ_l real. It is understood here that δ_l is a function of k even though we do not explicitly write δ_l as $\delta_l(k)$. Returning to f_l, we can write [from (7.6.12)]

$$f_l = \frac{(S_l - 1)}{2ik} \tag{7.6.15}$$

or, explicitly in terms of δ_l,

$$f_l = \frac{e^{2i\delta_l} - 1}{2ik} = \frac{e^{i\delta_l}\sin\delta_l}{k} = \frac{1}{k\cot\delta_l - ik}, \tag{7.6.16}$$

whichever is convenient. For the full scattering amplitude we have

$$\begin{aligned} f(\theta) &= \sum_{l=0} (2l+1)\left(\frac{e^{2i\delta_l} - 1}{2ik}\right) P_l(\cos\theta) \\ &= \frac{1}{k} \sum_{l=0} (2l+1) e^{i\delta_l} \sin\delta_l P_l(\cos\theta) \end{aligned} \tag{7.6.17}$$

with δ_l real. This expression for $f(\theta)$ rests on the twin principles of **rotational invariance** and **probability conservation**. In many books on wave mechanics, (7.6.17) is obtained by explicitly solving the Schrödinger equation with a real, spherically symmetric potential; our derivation of (7.6.17) may be of interest because it can be generalized to situations when the potential described in the context of nonrelativistic quantum mechanics may fail.

The differential cross section $d\sigma/d\Omega$ can be obtained [see (7.1.36)] by just taking the modulus squared of (7.6.17). To obtain the total cross section we have

$$\begin{aligned}\sigma_{\text{tot}} &= \int |f(\theta)|^2 \, d\Omega \\ &= \frac{1}{k^2} \int_0^{2\pi} d\phi \int_{-1}^{+1} d(\cos\theta) \sum_l \sum_{l'} (2l+1)(2l'+1) \\ &\quad \times e^{i\delta_l} \sin\delta_l e^{-i\delta_{l'}} \sin\delta_{l'} P_l P_{l'} \\ &= \frac{4\pi}{k^2} \sum_l (2l+1) \sin^2\delta_l. \end{aligned} \tag{7.6.18}$$

We can check the optical theorem (7.3.1), which we obtained earlier using a more general argument. All we need to do is note from (7.6.17) that

$$\begin{aligned}\operatorname{Im} f(\theta = 0) &= \sum_l \frac{(2l+1)\operatorname{Im}\left[e^{i\delta_l}\sin\delta_l\right]}{k} P_l(\cos\theta)\Big|_{\theta=0} \\ &= \sum_l \frac{(2l+1)}{k} \sin^2\delta_l, \end{aligned} \tag{7.6.19}$$

which is the same as (7.6.18) except for $4\pi/k$.

As a function of energy, δ_l changes; hence $f_l(k)$ changes also. The unitarity relation of (7.6.13) is a restriction on the manner in which f_l can vary. This can be most conveniently seen by drawing an Argand diagram for kf_l. We plot kf_l in a complex plane, as shown in Figure 7.6, which is

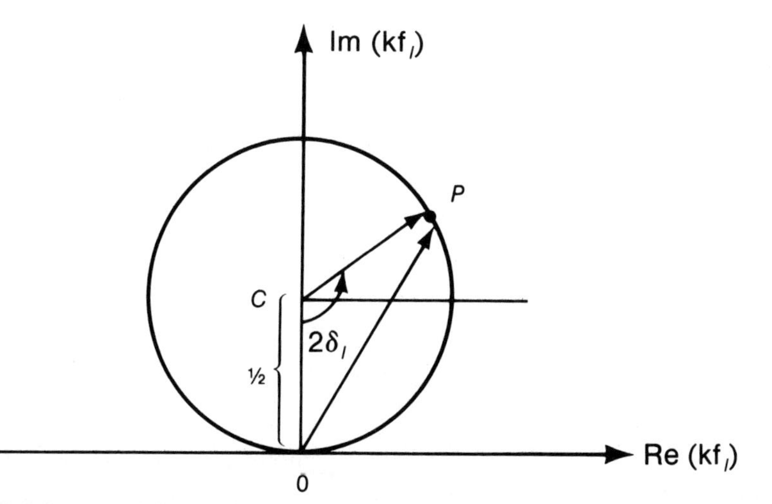

FIGURE 7.6. Argand diagram for kf_l. OP is the magnitude of kf_l, while CO and CP are each radii of length $\frac{1}{2}$ on the unitary circle; angle $OCP = 2\delta_l$.

self-explanatory if we note from (7.6.16) that

$$kf_l = \frac{i}{2} + \frac{1}{2} e^{-(i\pi/2) + 2i\delta_l}. \tag{7.6.20}$$

Notice that there is a circle of radius $\frac{1}{2}$, known as the **unitary circle**, on which kf_l must lie.

We can see many important features from Figure 7.6. Suppose δ_l is small. Then f_l must stay near the bottom of the circle. It may be positive or negative, but f_l is almost purely real:

$$f_l = \frac{e^{i\delta_l} \sin\delta_l}{k} \simeq \frac{(1+i\delta_l)\delta_l}{k} \simeq \frac{\delta_l}{k}. \tag{7.6.21}$$

On the other hand, if δ_l is near $\pi/2$, kf_l is almost purely imaginary, and the magnitude of kf_l is maximal. Under such a condition the lth partial wave may be in resonance, a concept to be discussed in some detail in Section 7.8. Note that the maximum partial cross section

$$\sigma_{\max}^{(l)} = 4\pi\lambdabar^2(2l+1) \tag{7.6.22}$$

is achieved [see (7.6.18)] when $\sin^2\delta_l = 1$.

Connection with the Eikonal Approximation

The eikonal approximation discussed in 7.4 is valid at high energies ($\lambdabar \ll$ range R); hence many partial waves contribute. We may regard l as a continuous variable. As an aside we note the semiclassical argument that $l = bk$ (because angular momentum $l\hbar = bp$, where b is the impact parameter and momentum $p = \hbar k$). We take

$$l_{\max} = kR; \tag{7.6.23}$$

then we make the following substitutions in expression (7.6.17):

$$\sum_l^{l_{\max}=kR} \to k\int db, \qquad P_l(\cos\theta) \overset{\text{large } l,\ \text{small } \theta}{\simeq} J_0(l\theta) = J_0(kb\theta),$$
$$\delta_l \to \Delta(b)|_{b=l/k}, \tag{7.6.24}$$

where $l_{\max} = kR$ implies that

$$e^{2i\delta_l} - 1 = e^{2i\Delta(b)} - 1 = 0 \quad \text{for } l > l_{\max}. \tag{7.6.25}$$

We have

$$f(\theta) \to k\int db \frac{2kb}{2ik}(e^{2i\Delta(b)} - 1) J_0(kb\theta)$$
$$= -ik\int db\, bJ_0(kb\theta)[e^{2i\Delta(b)} - 1]. \tag{7.6.26}$$

The computation of δ_l can be done by using the explicit form for $\Delta(b)$ given by (7.4.14) (see Problem 7 in this chapter).

Determination of Phase Shifts

Let us now consider how we may actually determine the phase shifts given a potential V. We assume that V vanishes for $r > R$, R being the range of the potential. Outside (that is, for $r > R$) the wave function must be that of a free spherical wave. This time, however, there is no reason to exclude $n_l(r)$ because the origin is excluded from our consideration. The wave function is therefore a linear combination of $j_l(kr)P_l(\cos\theta)$ and $n_l(kr)P_l(\cos\theta)$ or, equivalently, $h_l^{(1)}P_l$ and $h_l^{(2)}P_l$, where $h_l^{(1)}$ and $h_l^{(2)}$ are the spherical Hankel functions defined by

$$h_l^{(1)} = j_l + in_l, \qquad h_l^{(2)} = j_l - in_l; \tag{7.6.27}$$

these have the asymptotic behavior (see Appendix A)

$$h_l^{(1)} \xrightarrow{r \text{ large}} \frac{e^{i(kr-(l\pi/2))}}{ikr}, \qquad h_l^{(2)} \xrightarrow{r \text{ large}} -\frac{e^{-i(kr-(l\pi/2))}}{ikr}. \tag{7.6.28}$$

The full-wave function at any r can then be written as:

$$\langle \mathbf{x}|\psi^{(+)}\rangle = \frac{1}{(2\pi)^{3/2}} \sum i^l(2l+1)A_l(r)P_l(\cos\theta) \quad (r > R). \tag{7.6.29}$$

For $r > R$ we have (for the radial-wave function)

$$A_l = c_l^{(1)}h_l^{(1)}(kr) + c_l^{(2)}h_l^{(2)}(kr), \tag{7.6.30}$$

where the coefficient that multiplies A_l in (7.6.29) is chosen so that, for $V = 0$, $A_l(r)$ coincides with $j_l(kr)$ everywhere. Using (7.6.28), we can compare the behavior of the wave function for large r given by (7.6.29) and (7.6.30) with

$$\frac{1}{(2\pi)^{3/2}} \sum_l (2l+1)P_l\left[\frac{e^{2i\delta_l}e^{ikr}}{2ikr} - \frac{e^{-i(kr-l\pi)}}{2ikr}\right]. \tag{7.6.31}$$

Clearly, we must have

$$c_l^{(1)} = \tfrac{1}{2}e^{2i\delta_l}, \qquad c_l^{(2)} = \tfrac{1}{2}. \tag{7.6.32}$$

So the radial-wave function for $r > R$ is now written as

$$A_l(r) = e^{i\delta_l}\left[\cos\delta_l j_l(kr) - \sin\delta_l n_l(kr)\right]. \tag{7.6.33}$$

Using this, we can evaluate the logarithmic derivative at $r = R$—that is, just outside the range of the potential—as follows:

$$\begin{aligned}\beta_l &\equiv \left(\frac{r}{A_l}\frac{dA_l}{dr}\right)_{r=R} \\ &= kR\left[\frac{j_l'(kR)\cos\delta_l - n_l'(kR)\sin\delta_l}{j_l(kR)\cos\delta_l - n_l(kR)\sin\delta_l}\right],\end{aligned} \tag{7.6.34}$$

where $j_l'(kR)$ stands for the derivative of j_l with respect to kr evaluated at

$kr = kR$. Conversely, knowing the logarithmic derivative at R, we can obtain the phase shift as follows:

$$\tan \delta_l = \frac{kRj_l'(kR) - \beta_l j_l(kR)}{kRn_l'(kR) - \beta_l n_l(kR)}. \tag{7.6.35}$$

The problem of determining the phase shift is thus reduced to that of obtaining β_l.

We now look at the solution to the Schrödinger equation for $r < R$ —that is, inside the range of the potential. For a spherically symmetric potential, we can solve the Schrödinger equation in three dimensions by looking at the equivalent one-dimensional equation

$$\frac{d^2u_l}{dr^2} + \left(k^2 - \frac{2m}{\hbar^2}V - \frac{l(l+1)}{r^2}\right)u_l = 0, \tag{7.6.36}$$

where

$$u_l = rA_l(r) \tag{7.6.37}$$

subject to the boundary condition

$$u_l|_{r=0} = 0. \tag{7.6.38}$$

We integrate this one-dimensional Schrödinger equation—if necessary, numerically—up to $r = R$, starting at $r = 0$. In this way we obtain the logarithmic derivative at R. By continuity we must be able to match the logarithmic derivative for the inside and outside solutions at $r = R$:

$$\beta_l|_{\text{inside solution}} = \beta_l|_{\text{outside solution}}, \tag{7.6.39}$$

where the left-hand side is obtained by integrating the Schrödinger equation up to $r = R$, while the right-hand side is expressible in terms of the phase shifts that characterize the large-distance behavior of the wave function. This means that the phase shifts are obtained simply by substituting β_l for the inside solution into $\tan\delta_l$ [(7.6.35)]. For an alternative approach it is possible to derive an integral equation for $A_l(r)$, from which we can obtain phase shifts (see Problem 8 of this chapter).

Hard-Sphere Scattering

Let us work out a specific example. We consider scattering by a hard, or rigid, sphere

$$V = \begin{cases} \infty & \text{for } r < R \\ 0 & \text{for } r > R. \end{cases} \tag{7.6.40}$$

In this problem we need not even evaluate β_l (which is actually ∞). All we need to know is that the wave function must vanish at $r = R$ because the sphere is impenetrable. Therefore,

$$A_l(r)|_{r=R}=0 \tag{7.6.41}$$

or, from (7.6.33),

$$j_l(kR)\cos\delta_l - n_l(kR)\sin\delta_l = 0 \tag{7.6.42}$$

or

$$\tan\delta_l = \frac{j_l(kR)}{n_l(kR)}. \tag{7.6.43}$$

Thus the phase shifts are now known for any l. Notice that no approximations have been made so far.

To appreciate the physical significance of the phase shifts, let us consider the $l=0$ case (S-wave scattering) specifically. Equation (7.6.43) becomes, for $l=0$,

$$\tan\delta_0 = \frac{\sin kR/kR}{-\cos kR/kR} = -\tan kR, \tag{7.6.44}$$

or $\delta_0 = -kR$. The radial-wave function (7.6.33) with $e^{i\delta_0}$ omitted varies as

$$A_{l=0}(r) \propto \frac{\sin kr}{kr}\cos\delta_0 + \frac{\cos kr}{kr}\sin\delta_0 = \frac{1}{kr}\sin(kr+\delta_0). \tag{7.6.45}$$

Therefore, if we plot $rA_{l=0}(r)$ as a function of distance r, we obtain a sinusoidal wave, which is shifted when compared to the free sinusoidal wave by amount R; see Figure 7.7.

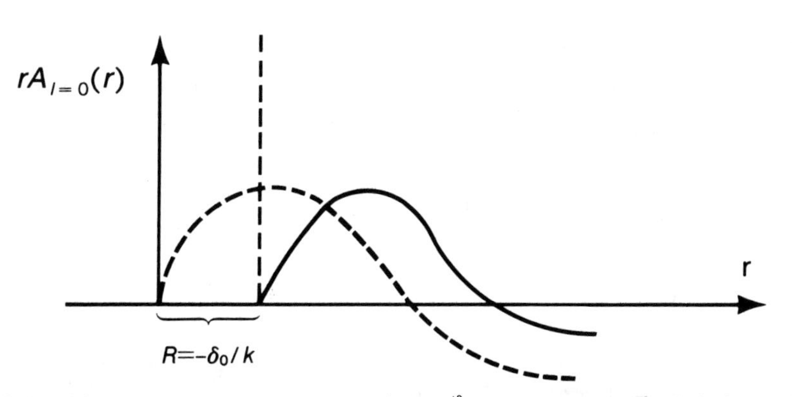

FIGURE 7.7. Plot of $rA_{l=0}(r)$ versus r (with the $e^{i\delta_0}$ factor removed). The dashed curve for $V=0$ behaves like $\sin kr$, while the solid curve is for S-wave hard-sphere scattering, shifted by $R=-\delta_0/k$ from the case $V=0$.

Let us now study the low and high energy limits of $\tan\delta_l$. Low energy means kR small, $kR \ll 1$. We can then use (see Appendix A)

$$\begin{aligned} j_l(kr) &\simeq \frac{(kr)^l}{(2l+1)!!} \\ n_l(kr) &\simeq -\frac{(2l-1)!!}{(kr)^{l+1}} \end{aligned} \tag{7.6.46}$$

to obtain

$$\tan\delta_l = \frac{-(kR)^{2l+1}}{\{(2l+1)[(2l-1)!!]^2\}}. \tag{7.6.47}$$

It is therefore all right to ignore δ_l with $l \neq 0$. In other words, we have S-wave scattering only, which is actually expected for almost any finite-range potential at low energy. Because $\delta_0 = -kR$ regardless of whether k is large or small, we obtain

$$\frac{d\sigma}{d\Omega} = \frac{\sin^2\delta_0}{k^2} \simeq R^2 \quad \text{for } kR \ll 1. \tag{7.6.48}$$

It is interesting that the total cross section, given by

$$\sigma_{\text{tot}} = \int \frac{d\sigma}{d\Omega}\, d\Omega = 4\pi R^2, \tag{7.6.49}$$

is *four* times the *geometric cross section* πR^2. By geometric cross section we mean the area of the disc of radius R that blocks the propagation of the plane wave (and has the same cross section area as that of a hard sphere). Low-energy scattering, of course, means a very large wavelength scattering, and we do not necessarily expect a classically reasonable result.

One might conjecture that the geometric cross section is reasonable to expect for high-energy scattering because at high energies the situation might look similar to the semiclassical situation. At high energies many l-values contribute, up to $l_{\max} \simeq kR$, a reasonable assumption. The total cross section is therefore given by

$$\sigma_{\text{tot}} = \frac{4\pi}{k^2} \sum_{l=0}^{l \simeq kR} (2l+1)\sin^2\delta_l. \tag{7.6.50}$$

But using (7.6.43), we have

$$\sin^2\delta_l = \frac{\tan^2\delta_l}{1+\tan^2\delta_l} = \frac{[j_l(kR)]^2}{[j_l(kR)]^2 + [n_l(kR)]^2} \simeq \sin^2\left(kR - \frac{\pi l}{2}\right), \tag{7.6.51}$$

where we have used

$$j_l(kr) \sim \frac{1}{kr}\sin\left(kr - \frac{l\pi}{2}\right)$$
$$n_l(kr) \sim -\frac{1}{kr}\cos\left(kr - \frac{l\pi}{2}\right). \tag{7.6.52}$$

We see that δ_l decreases by 90° each time l increases by one unit. Thus, for an adjacent pair of partial waves, $\sin^2\delta_l + \sin^2\delta_{l+1} = \sin^2\delta_l + \sin^2(\delta_l - \pi/2) = \sin^2\delta_l + \cos^2\delta_l = 1$, and with so many l-values contributing to (7.6.50), it is legitimate to replace $\sin^2\delta_l$ by its average value, $\frac{1}{2}$. The number of terms in the l-sum is roughly kR, as is the average of $2l+1$. Putting all the ingredients together, (7.6.50) becomes

$$\sigma_{\text{tot}} = \frac{4\pi}{k^2}(kR)^2\frac{1}{2} = 2\pi R^2, \tag{7.6.53}$$

which is not the geometric cross section πR^2 either! To see the origin of the factor of 2, we may split (7.6.17) into two parts:

$$f(\theta) = \frac{1}{2ik}\sum_{l=0}^{kR}(2l+1)e^{2i\delta_l}P_l(\cos\theta) + \frac{i}{2k}\sum_{l=0}^{kR}(2l+1)P_l(\cos\theta)$$
$$= f_{\text{reflection}} + f_{\text{shadow}}. \tag{7.6.54}$$

In evaluating $\int|f_{\text{refl}}|^2 d\Omega$, the orthogonality of the $P_l(\cos\theta)$'s ensures that there is no interference amongst contributions from different l, and we obtain the sum of the square of partial-wave contributions:

$$\int|f_{\text{refl}}|^2 d\Omega = \frac{2\pi}{4k^2}\sum_{l=0}^{l_{\max}}\int_{-1}^{+1}(2l+1)^2[P_l(\cos\theta)]^2 d(\cos\theta)$$
$$= \frac{\pi l_{\max}^2}{k^2} = \pi R^2. \tag{7.6.55}$$

Turning our attention to f_{shad}, we note that it is pure imaginary. It is particularly strong in the forward direction because $P_l(\cos\theta) = 1$ for $\theta = 0$, and the contributions from various l-values all add up coherently—that is, with the same phase, pure imaginary and positive in our case. We can use the small-angle approximation for P_l [see (7.6.24)] to obtain

$$f_{\text{shad}} \simeq \frac{i}{2k}\sum(2l+1)J_0(l\theta)$$
$$\simeq ik\int_0^R bdbJ_0(kb\theta)$$
$$= \frac{iRJ_1(kR\theta)}{\theta}. \tag{7.6.56}$$

But this is just the formula for Fraunhofer diffraction in optics with a strong peaking near $\theta \simeq 0$. Letting $\xi = kR\theta$ and $d\xi/\xi = d\theta/\theta$, we can evaluate

$$\begin{aligned}\int |f_{\text{shad}}|^2 \, d\Omega &= 2\pi \int_{-1}^{+1} \frac{R^2 [J_1(kR\theta)]^2}{\theta^2} d(\cos\theta) \\ &\simeq 2\pi R^2 \int_0^{\infty} \frac{[J_1(\xi)]^2}{\xi} d\xi \\ &\simeq \pi R^2. \end{aligned} \tag{7.6.57}$$

Finally, the interference between f_{shad} and f_{refl} vanishes:

$$\text{Re}\left(f_{\text{shad}}^* f_{\text{refl}}\right) \simeq 0 \tag{7.6.58}$$

because the phase of f_{refl} oscillates ($2\delta_{l+1} = 2\delta_l - \pi$), approximately averaging to zero, while f_{shad} is pure imaginary. Thus

$$\sigma_{\text{tot}} = \underset{\substack{\uparrow \\ \sigma_{\text{refl}}}}{\pi R^2} + \underset{\substack{\uparrow \\ \sigma_{\text{shad}}}}{\pi R^2}. \tag{7.6.59}$$

The second term (coherent contribution in the forward direction) is called a *shadow* because for hard-sphere scattering at high energies, waves with impact parameter less than R must be deflected. So, just *behind* the scatterer there must be zero probability for finding the particle and a shadow must be created. In terms of wave mechanics, this shadow is due to destructive interference between the original wave (which would be there even if the scatterer were absent) and the newly scattered wave. Thus we need scattering in order to create a shadow. That this shadow amplitude must be pure imaginary may be seen by recalling from (7.6.8) that the coefficient of $e^{ikr}/2ikr$ for the lth partial wave behaves like $1 + 2ikf_l(k)$, where the 1 would be present even without the scatterer; hence there must be a positive imaginary term in f_l to get cancellation. In fact, this gives a physical interpretation of the optical theorem, which can be checked explicitly. First note that

$$\frac{4\pi}{k} \text{Im} f(0) \simeq \frac{4\pi}{k} \text{Im}[f_{\text{shad}}(0)] \tag{7.6.60}$$

because $\text{Im}[f_{\text{refl}}(0)]$ averages to zero due to oscillating phase. Using (7.6.54), we obtain

$$\frac{4\pi}{k} \text{Im} f_{\text{shad}}(0) = \frac{4\pi}{k} \text{Im}\left[\frac{i}{2k} \sum_{l=0}^{kR} (2l+1) P_l(1)\right] = 2\pi R^2 \tag{7.6.61}$$

which is indeed equal to σ_{tot}.

7.7. LOW-ENERGY SCATTERING AND BOUND STATES

At low energies—or, more precisely, when $\lambda\!\!\!^{-} = 1/k$ is comparable to or larger than the range R—partial waves for higher l are, in general, unimportant. This point may be obvious classically because the particle cannot penetrate the centrifugal barrier; as a result the potential inside has no effect. In terms of quantum mechanics, the effective potential for the lth partial wave is given by

$$V_{\text{eff}} = V(r) + \frac{\hbar^2}{2m}\frac{l(l+1)}{r^2}; \tag{7.7.1}$$

unless the potential is strong enough to accommodate $l \neq 0$ bound states near $E \simeq 0$, the behavior of the radial-wave function is largely determined by the centrifugal barrier term, which means that it must resemble $j_l(kr)$. More quantitatively, it is possible to estimate the behavior of the phase shift using the integral equation for the partial wave (see Problem 8 of this chapter):

$$\frac{e^{i\delta_l}\sin\delta_l}{k} = -\frac{2m}{\hbar^2}\int_0^\infty j_l(kr)V(r)A_l(r)r^2 dr. \tag{7.7.2}$$

If $A_l(r)$ is not too different from $j_l(kr)$ and $1/k$ is much larger than the range of the potential, the right-hand side would vary as k^{2l}; for small δ_l, the left-hand side must vary as δ_l/k. Hence, the phase shift k goes to zero as

$$\delta_l \sim k^{2l+1} \tag{7.7.3}$$

for small k. This is known as **threshold behavior**.

It is therefore clear that at low energies with a finite range potential, S-wave scattering is important.

Rectangular Well or Barrier

To be specific let us consider S-wave scattering by

$$V = \begin{cases} V_0 = \text{constant} & \text{for } r < R \\ 0 & \text{otherwise} \end{cases} \qquad \begin{cases} V_0 > 0 & \text{repulsive} \\ V_0 < 0 & \text{attractive} \end{cases} \tag{7.7.4}$$

Many of the features we obtain here are common to more-complicated finite range potentials.

We have already seen that the outside-wave function [see (7.6.33) and (7.6.45)] must behave like

$$e^{i\delta_0}[j_0(kr)\cos\delta_0 - n_0(kr)\sin\delta_0] \simeq \frac{e^{i\delta_0}\sin(kr+\delta_0)}{kr}. \tag{7.7.5}$$

The inside solution can also easily be obtained for V_0 a constant:

$$u \equiv rA_{l=0}(r) \propto \sin k'r, \tag{7.7.6}$$

with k' determined by

$$E - V_0 = \frac{\hbar^2 k'^2}{2m}, \tag{7.7.7}$$

where we have used the boundary condition $u = 0$ at $r = 0$. In other words, the inside wave function is also sinusoidal as long as $E > V_0$. The curvature of the sinusoidal wave is different than in the free-particle case; as a result the wave function can be pushed in ($\delta_0 > 0$) or pulled out ($\delta_0 < 0$) depending on whether $V_0 < 0$ (attractive) or $V_0 > 0$ (repulsive), as shown in Figure 7.8. Notice also that (7.7.6) and (7.7.7) hold even if $V_0 > E$, provided we understand sin to mean sinh—that is, the wave function behaves like

$$u(r) \propto \sinh[\kappa r], \tag{7.7.6$'$}$$

where

$$\frac{\hbar^2\kappa^2}{2m} = (V_0 - E). \tag{7.7.7$'$}$$

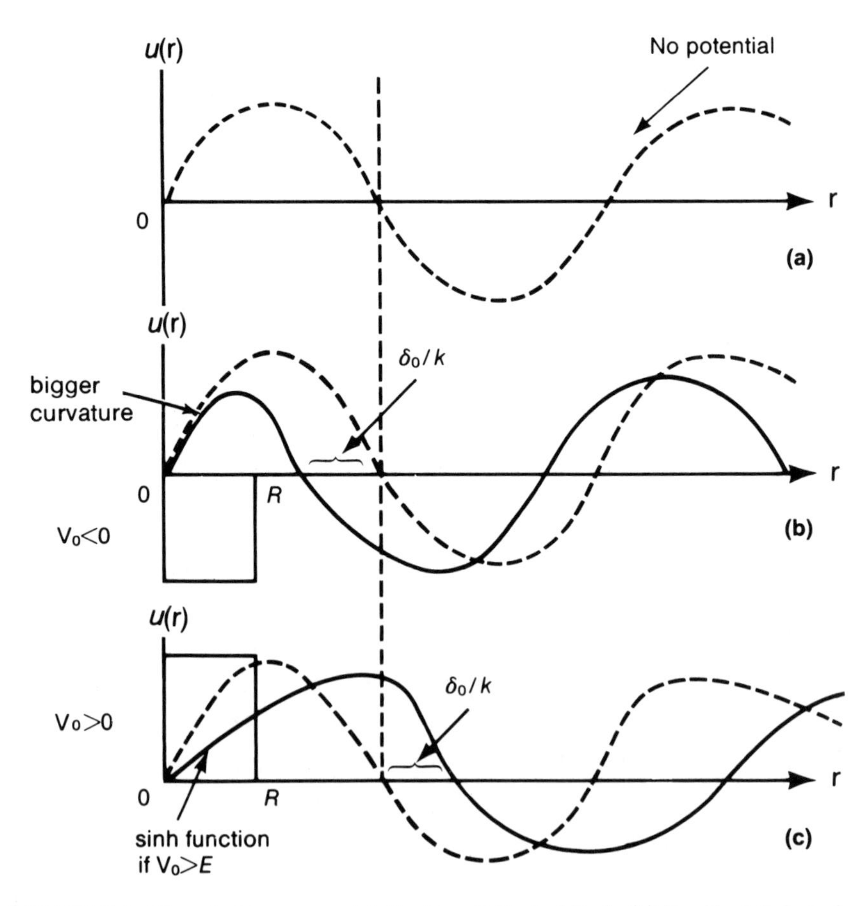

FIGURE 7.8 Plot of $u(r)$ versus r. (a) For $V = 0$ (dashed line). (b) For $V_0 < 0$, $\delta_0 > 0$ with the wave function (solid line) pushed in. (c) For $V_0 > 0$, $\delta_0 < 0$ with the wave function (solid line) pulled out.

We now concentrate on the attractive case and imagine that the magnitude of V_0 is increased. Increased attraction will result in a wave function with a larger curvature. Suppose the attraction is such that the interval $[0, R]$ just accommodates one-fourth cycle of the sinusoidal wave. Working in the low energy $kR \ll 1$ limit, the phase shift is now $\delta_0 = \pi/2$, and this results in a maximal S-wave cross section for a given k because $\sin^2\delta_0$ is unity. Now increase the well depth V_0 even further. Eventually the attraction is so strong that one-half cycle of the sinusoidal wave can be fitted within the range of the potential. The phase shift δ_0 is now π; in other words, the wave function outside R is 180° out of phase compared to the free-particle-wave function. What is remarkable is that the partial cross section vanishes ($\sin^2\delta_0 = 0$),

$$\sigma_{l=0} = 0, \tag{7.7.8}$$

despite the very strong attraction of the potential. In addition, if the energy is low enough for $l \neq 0$ waves still to be unimportant, we then have an almost-perfect transmission of the incident wave. This kind of situation, known as the **Ramsauer-Townsend effect**, is actually observed experimentally for scattering of electrons by such rare gases as argon, krypton, and xenon. This effect was first observed in 1923 prior to the birth of wave mechanics and was considered to be a great mystery. Note the typical parameters here are $R \sim 2 \times 10^{-8}$ cm for electron kinetic energy of order 0.1eV, leading to $kR \sim 0.324$.

Zero-Energy Scattering and Bound States

Let us consider scattering at extremely low energies ($k \simeq 0$). For $r > R$ and for $l = 0$, the outside radial-wave function satisfies

$$\frac{d^2u}{dr^2} = 0. \tag{7.7.9}$$

The obvious solution to (7.7.9) is

$$u(r) = \text{constant}(r - a), \tag{7.7.10}$$

just a straight line! This can be understood as an infinitely long wavelength limit of the usual expression for the outside-wave function [see (7.6.37) and (7.6.45)],

$$\lim_{k\to 0} \sin(kr + \delta_0) = \lim_{k\to 0} \sin\left[k\left(r + \frac{\delta_0}{k}\right)\right], \tag{7.7.11}$$

which looks like (7.7.10). We have

$$\frac{u'}{u} = k\cot\left[k\left(r + \frac{\delta_0}{k}\right)\right] \overset{k\to 0}{\to} \frac{1}{r-a}. \tag{7.7.12}$$

Setting $r = 0$ [even though at $r = 0$, (7.7.10) is not the true wave function], we obtain

$$\lim_{k\to 0} k\cot\delta_0 \overset{k\to 0}{\to} -\frac{1}{a}. \tag{7.7.13}$$

The quantity a is known as the **scattering length**. The limit of the total cross section as $k \to 0$ is given by [see (7.6.16)]

$$\sigma_{\text{tot}} = \sigma_{l=0} = 4\pi \lim_{k\to 0} \left|\frac{1}{k \cot \delta_0 - ik}\right|^2 = 4\pi a^2. \tag{7.7.14}$$

Even though a has the same dimension as the range of the potential R, a and R can differ by orders of magnitude. In particular, for an attractive potential, it is possible for the magnitude of the scattering length to be far greater than the range of the potential. To see the physical meaning of a, we note that a is nothing more than the intercept of the outside-wave function. For a repulsive potential, $a > 0$ and is roughly of order of R, as seen in Figure 7.9a. However, for an attractive potential, the intercept is on the negative side (Figure 7.9b). If we *increase* the attraction, the outside-wave function can again cross the r-axis on the positive side (Figure 7.9c).

The sign change resulting from increased attraction is related to the development of a bound state. To see this point quantitatively, we note from Figure 7.9c that for a very large and positive, the wave function is essentially flat for $r > R$. But (7.7.10) with a very large is not too different from $e^{-\kappa r}$ with κ essentially zero. Now $e^{-\kappa r}$ with $\kappa \simeq 0$ is just a bound-state-wave function for $r > R$ with energy E infinitesimally negative. The inside-wave function ($r < R$) for the $E = 0+$ case (scattering with zero kinetic energy) and the $E = 0-$ case (bound state with infinitesimally small binding energy) are essentially the same because in both cases k' in $\sin k'r$ [(7.7.6)] is determined by

$$\frac{\hbar^2 k'^2}{2m} = E - V_0 \simeq |V_0| \tag{7.7.15}$$

with E infinitesimal (positive or negative).

Because the inside-wave functions are the same for the two physical situations ($E = 0+$ and $E = 0-$), we can equate the logarithmic derivative of the bound-state-wave function with that of the solution involving zero kinetic-energy scattering,

$$-\left.\frac{\kappa e^{-\kappa r}}{e^{-\kappa r}}\right|_{r=R} = \left.\left(\frac{1}{r-a}\right)\right|_{r=R}, \tag{7.7.16}$$

or, if $R \ll a$,

$$\kappa \simeq \frac{1}{a}. \tag{7.7.17}$$

The binding energy satisfies

$$E_{\text{BE}} = -E_{\text{bound state}} = \frac{\hbar^2\kappa^2}{2m} \simeq \frac{\hbar^2}{2ma^2}, \tag{7.7.18}$$

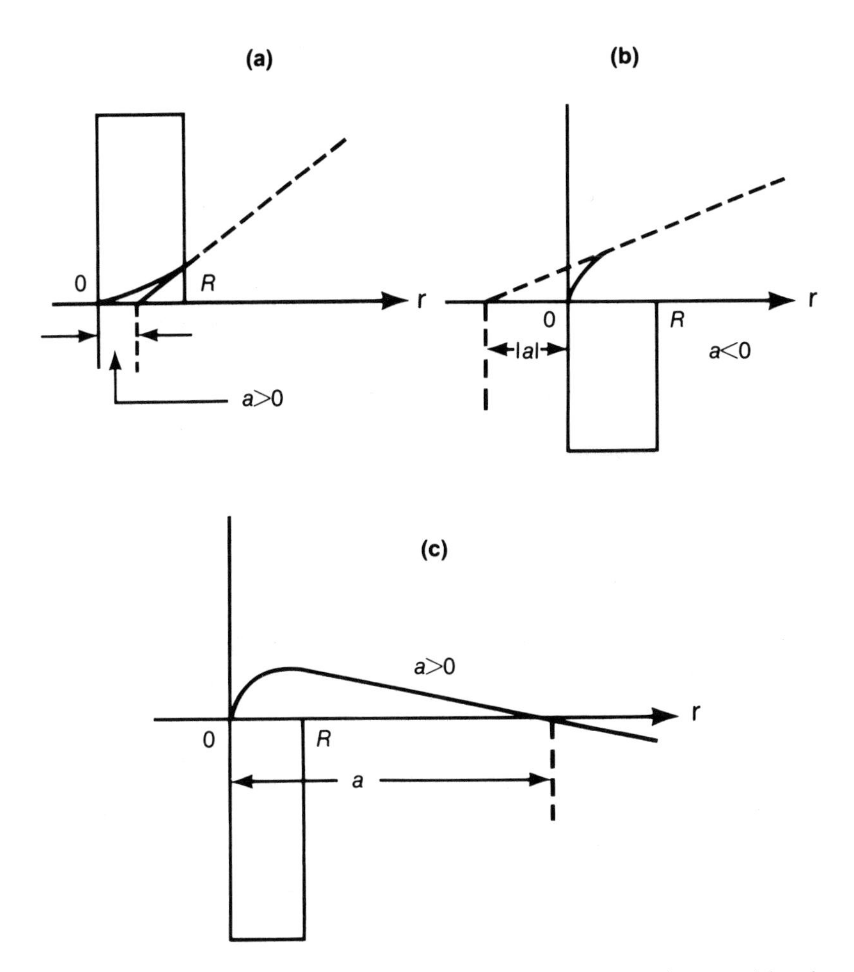

FIGURE 7.9. Plot of $u(r)$ versus r for (a) repulsive potential, (b) attractive potential, and (c) deeper attraction. The intercept a of the zero-energy outside-wave function with the r-axis is shown for each of three cases.

and we have a relation between scattering length and bound-state energy. This is a remarkable result. To wit, if there is a loosely bound state, we can infer its binding energy by performing scattering experiments near zero kinetic energy, provided a is measured to be large compared with the range R of the potential. This connection between the scattering length and the bound-state energy was first pointed out by Wigner, who attempted to apply (7.7.18) to np-scattering.

Experimentally, the ${}^3S_1$-state of the np-system has a bound state, that is, the deuteron with

$$E_{\mathrm{BE}} = 2.22 \text{ MeV}. \tag{7.7.19}$$

The scattering length is measured to be

$$a_{\text{triplet}} = 5.4 \times 10^{-13} \text{ cm}, \tag{7.7.20}$$

leading to the binding-energy prediction

$$\frac{\hbar^2}{2\mu a^2} = \frac{\hbar^2}{m_N a^2} = m_N c^2 \left(\frac{\hbar}{m_N c a} \right)^2$$

$$= (938 \text{ MeV}) \left(\frac{2.1 \times 10^{-14} \text{ cm}}{5.4 \times 10^{-13} \text{ cm}} \right)^2 = 1.4 \text{ MeV} \tag{7.7.21}$$

where μ is the reduced mass approximated by $m_{n,p}/2$. The agreement between (7.7.19) and (7.7.21) is not too satisfactory. The discrepancy is due to the fact that the inside-wave functions are not exactly the same and that $a_{\text{triplet}} \gg R$ is not really such a good approximation for the deuteron. A better result can be obtained by keeping the next term in the expansion of $k \cot \delta$ as a function of k,

$$k \cot \delta_0 = -\frac{1}{a} + \frac{1}{2} r_0 k^2, \tag{7.7.22}$$

where r_0 is known as the effective range (see, for example, Preston 1962, 23).

Bound States as Poles of $S_l(k)$

We conclude this section by studying the analytic properties of the amplitude $S_l(k)$ for $l = 0$. Let us go back to (7.6.8) and (7.6.12), where the radial wave function for $l = 0$ at large distance was found to be proportional to

$$S_{l=0}(k) \frac{e^{ikr}}{r} - \frac{e^{-ikr}}{r}. \tag{7.7.23}$$

Compare this with the wave function for a bound state at large distance,

$$\frac{e^{-\kappa r}}{r}. \tag{7.7.24}$$

The existence of a bound state implies that a nontrivial solution to the Schrödinger equation with $E < 0$ exists only for a particular (discrete) value of κ. We may argue that $e^{-\kappa r}/r$ is like e^{ikr}/r, except that k is now purely imaginary. Apart from k being imaginary, the important difference between (7.7.23) and (7.7.24) is that in the bound-state case, $e^{-\kappa r}/r$ is present even without the analogue of the incident wave. Quite generally only the ratio of the coefficient of e^{ikr}/r to that of e^{-ikr}/r is of physical interest, and this is given by $S_l(k)$. In the bound-state case we can sustain the outgoing wave (with imaginary k) even without an incident wave. So the ratio is ∞, which means that $S_{l=0}(k)$, regarded as a function of a complex variable k, has a

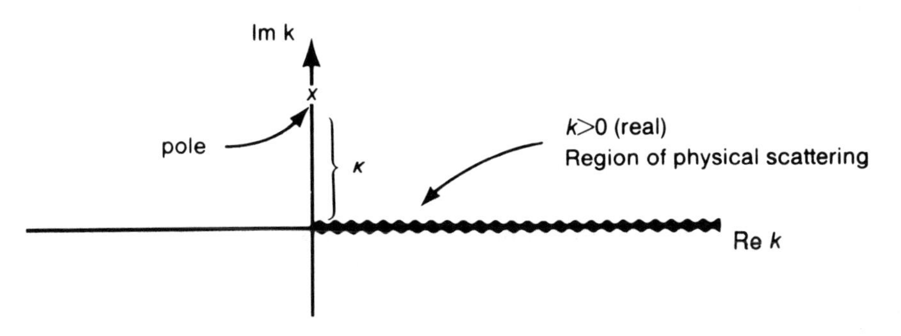

FIGURE 7.10. The complex k-plane with bound-state pole at $k = +i\kappa$.

pole at $k = i\kappa$. Thus a bound state implies a pole (which can be shown to be a simple pole) on the positive imaginary axis of the complex k-plane; see Figure 7.10. For k real and positive, we have the region of physical scattering. Here we must require [compare with (7.6.14)]

$$S_{l=0} = e^{2i\delta_0} \tag{7.7.25}$$

with δ_0 real. Furthermore, as $k \to 0$, $k \cot \delta_0$ has a limiting value $-1/a$ [(7.7.13)], which is finite, so δ_0 must behave as follows:

$$\delta_0 \to 0, \pm \pi, \ldots. \tag{7.7.26}$$

Hence $S_{l=0} = e^{2i\delta_0} \to 1$ as $k \to 0$.

Now let us attempt to construct a simple function satisfying:

1. Pole at $k = i\kappa$ (existence of bound state).
2. $|S_{l=0}| = 1$ for $k > 0$ real (unitarity). (7.7.27)
3. $S_{l=0} = 1$ at $k = 0$ (threshold behavior).

The simplext function that satisfies all three conditions of (7.7.27) is

$$S_{l=0}(k) = \frac{-k - i\kappa}{k - i\kappa}. \tag{7.7.28}$$

[*Editor's Note:* Equation (7.7.28) is chosen for simplicity rather than as a physically realistic example. For reasonable potentials (not hard spheres!) the phase shift vanishes as $k \to \infty$.]

An assumption implicit in choosing this form is that there is no other singularity that is important apart from the bound-state pole. We can then use (7.6.15) to obtain, for $f_{l=0}(k)$,

$$f_{l=0} = \frac{S_{l=0} - 1}{2ik} = \frac{1}{-\kappa - ik}. \tag{7.7.29}$$

Comparing this with (7.6.16),

$$f_{l=0} = \frac{1}{k \cot \delta_0 - ik}, \tag{7.7.30}$$

we see that

$$\lim_{k \to 0} k \cot \delta_0 = -\frac{1}{a} = -\kappa, \tag{7.7.31}$$

precisely the relation between bound state and scattering length [(7.7.17)].

It thus appears that by exploiting unitarity and analyticity of $S_l(k)$ in the k-plane, we may obtain the kind of information that can be secured by solving the Schrödinger equation explicitly. This kind of technique can be very useful in problems where the details of the potential are not known.

7.8. RESONANCE SCATTERING

In atomic, nuclear, and particle physics, we often encounter a situation where the scattering cross section for a given partial wave exhibits a pronounced peak. This section is concerned with the dynamics of such a **resonance.**

We continue to consider a finite-ranged potential $V(r)$. The *effective* potential appropriate for the radial wave function of the lth partial wave is $V(r)$ plus the centrifugal barrier term as given by (7.7.1). Suppose $V(r)$ itself is attractive. Because the second term,

$$\frac{\hbar^2}{2m}\frac{l(l+1)}{r^2}$$

is repulsive, we have a situation where the effective potential has an attractive well followed by a repulsive barrier at larger distances, as shown in Figure 7.11.

Suppose the barrier were infinitely high. It would then be possible for particles to be trapped inside, which is another way of saying that we expect bound states, with energy $E > 0$. They are *genuine* bound states in the sense that they are eigenstates of the Hamiltonian with definite values of E. In other words, they are *stationary* states with infinite lifetime.

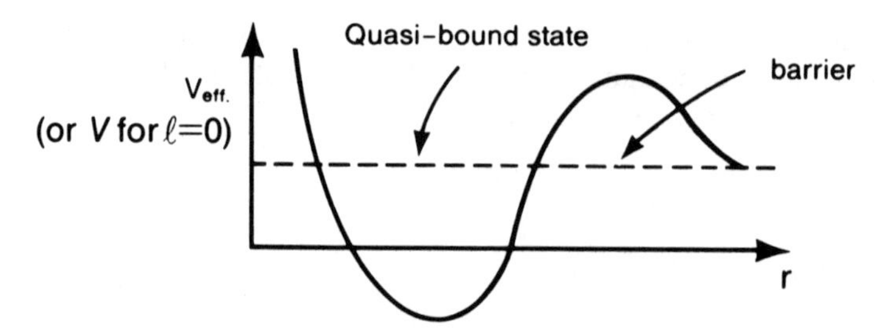

FIGURE 7.11. $V_{\text{eff}} = V(r) + (\hbar^2/2m)[l(l+1)/r^2]$ versus r. For $l \neq 0$ the barrier can be due to $(\hbar^2/2m)[l(l+1)/r^2]$; for $l = 0$ barrier must be due to V itself.

In the more realistic case of a finite barrier, the particle can be trapped inside, but it cannot be trapped forever. Such a trapped state has a finite lifetime due to quantum-mechanical tunneling. In other words, a particle leaks through the barrier to the outside region. Let us call such a state **quasi-bound state** because it would be an honest bound state if the barrier were infinitely high.

The corresponding scattering phase shift δ_l rises through the value $\pi/2$ as the incident energy rises through that of the quasi-bound state, and at the same time the corresponding partial-wave cross section passes through its maximum possible value $4\pi(2l + 1)/k^2$. [*Editor's Note:* Such a sharp rise in the phase shift is, in the time-dependent Schrödinger equation, associated with a delay of the emergence of the trapped particles, rather than an unphysical advance, as would be the case for a sharp decrease through $\pi/2$.]

It is instructive to verify this point with explicit calculations for some known potential. The result of a numerical calculation shows that a resonance behavior is in fact possible for $l \neq 0$ with a spherical-well potential.

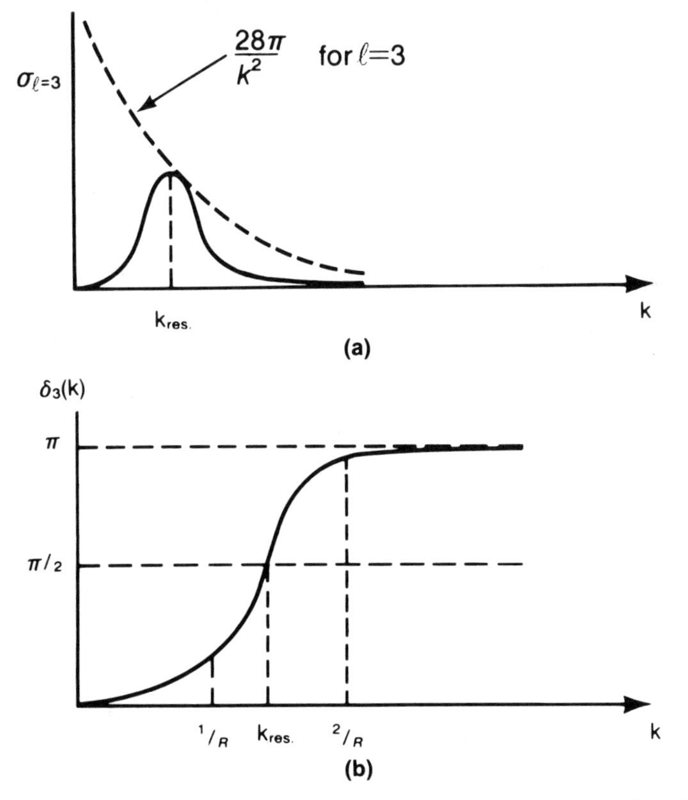

FIGURE 7.12. Plots of (a) $\sigma_{l=3}$ versus k, where at resonance $\delta_3(k_{\text{res}}) = \pi/2$ and $\sigma_{l=3} = (4\pi/k_{\text{res}}^2)\times 7 = 28\pi/k_{\text{res}}^2$ and (b) $\delta_3(k)$ versus k. The curves are for a spherical well with $2mV_0R^2/\hbar^2 = 5.5$.

To be specific we show the results for a spherical well with $2mV_0R^2/\hbar^2 = 5.5$ and $l = 3$ in Figure 7.12. The phase shift (Figure 7.12b), which is small at extremely low energies, starts increasing rapidly past $k = 1/R$, and goes through $\pi/2$ around $k = 1.3/R$.

Another very instructive example is provided by a repulsive δ-shell potential that is exactly soluble (see Problem 9 in this chapter):

$$\frac{2m}{\hbar^2}V(r) = \gamma\delta(r - R). \tag{7.8.1}$$

Here resonances are possible for $l = 0$ because the δ-shell potential itself can trap the particle in the region $0 < r < R$. For the case $\gamma = \infty$, we expect a series of bound states in the region $r < R$ with

$$kR = \pi, 2\pi, \ldots; \tag{7.8.2}$$

this is because the radial wave function for $l = 0$ must vanish not only at $r = 0$ but also at $r = R-$ in this case. For the region $r > R$, we simply have hard-sphere scattering with the S-wave phase shift, given by

$$\delta_0 = -kR. \tag{7.8.3}$$

With $\gamma = \infty$, there is no connection between the two problems because the wall at $r = R$ cannot be penetrated.

The situation is more interesting with a finite barrier, as we can show explicitly. The scattering phase shift exhibits a resonance behavior whenever

$$E_{\text{incident}} \simeq E_{\text{quasi-bound state}}. \tag{7.8.4}$$

Moreover, the larger the γ, the sharper the resonance peak. However, away from the resonance δ_0 looks very much like the hard-sphere phase shift. Thus we have a situation in which a resonance behavior is superimposed on a smoothly behaving background scattering. This serves as a model for neutron-nucleus scattering, where a series of sharp resonance peaks are observed on top of a smoothly varying cross section.

Coming back to our general discussion of resonance scattering, we ask how the scattering amplitudes vary in the vicinity of the resonance energy. If we are to have any connection between σ_l being large and the quasi-bound states, δ_l *must go through* $\pi/2$ (*or* $3\pi/2, \ldots$) *from below*, as discussed above. In other words $\cot\delta_l$ must go through zero from above. Assuming that $\cot\delta_l$ is smoothly varying near the vicinity of resonance, that is,

$$E \simeq E_r, \tag{7.8.5}$$

we may attempt to expand $\cot\delta_l$ as follows:

$$\cot\delta_l = \underbrace{\cot\delta_l|_{E=E_r}}_{0} - c(E - E_r) + 0\left[(E - E_r)^2\right]. \tag{7.8.6}$$

This leads to

$$f_l(k) = \frac{1}{k\cot\delta_l - ik} = \frac{1}{k}\frac{1}{[-c(E-E_r)-i]}$$
$$= -\frac{\Gamma/2}{k\left[(E-E_r)+\dfrac{i\Gamma}{2}\right]}, \tag{7.8.7}$$

where we have defined the *width* Γ by

$$\left.\frac{d(\cot\delta_l)}{dE}\right|_{E=E_r} = -c \equiv -\frac{2}{\Gamma} \tag{7.8.8}$$

Notice that Γ is very small if $\cot\delta_l$ varies rapidly. If a simple resonance dominates the lth partial-wave cross section, we obtain a one-level resonance formula (the Breit-Wigner formula):

$$\sigma_l = \frac{4\pi}{k^2}\frac{(2l+1)(\Gamma/2)^2}{(E-E_r)^2+\Gamma^2/4}. \tag{7.8.9}$$

So it is legitimate to regard Γ as the full width at half-maximum, provided the resonance is reasonably narrow so that variation in $1/k^2$ can be ignored.

7.9. IDENTICAL PARTICLES AND SCATTERING

As an example to illustrate the phase of the scattering amplitude, let us consider the scattering of two identical spinless charged particles via the Coulomb potential (which we discuss further in Section 7.13). The space-wave function must now be symmetric, so the asymptotic wave function must look like

$$e^{i\mathbf{k}\cdot\mathbf{x}} + e^{-i\mathbf{k}\cdot\mathbf{x}} + [f(\theta)+f(\pi-\theta)]\frac{e^{ikr}}{r}, \tag{7.9.1}$$

where $\mathbf{x} = \mathbf{x}_1 - \mathbf{x}_2$ is the relative position vector between the two particles 1 and 2. This results in a differential cross section,

$$\frac{d\sigma}{d\Omega} = |f(\theta)+f(\pi-\theta)|^2$$
$$= |f(\theta)|^2 + |f(\pi-\theta)|^2 + 2\,\mathrm{Re}[f(\theta)f^*(\pi-\theta)]. \tag{7.9.2}$$

The cross section is enhanced through constructive interference at $\theta \simeq \pi/2$.

In contrast, for spin $\frac{1}{2}$ – spin $\frac{1}{2}$ scattering with unpolarized beam and V independent of spin, we have the spin-singlet scattering going with space-symmetrical function and the spin triplet going with space-antisymmetrical wave function (see Section 6.3). If the initial beam is unpolarized, we have the statistical contribution $\frac{1}{4}$ for spin singlet and $\frac{3}{4}$ for spin triplet; hence

$$\begin{aligned}\frac{d\sigma}{d\Omega} &= \frac{1}{4}|f(\theta)+f(\pi-\theta)|^2 + \frac{3}{4}|f(\theta)-f(\pi-\theta)|^2 \\ &= |f(\theta)|^2 + |f(\pi-\theta)|^2 - \mathrm{Re}\left[f(\theta)f^*(\pi-\theta)\right]. \end{aligned} \tag{7.9.3}$$

In other words, we expect destructive interference at $\theta \simeq \pi/2$. This has, in fact, been observed.

7.10. SYMMETRY CONSIDERATIONS IN SCATTERING

Suppose V and H_0 are both invariant under some symmetry operation. We may ask what this implies for the matrix element of T or for the scattering amplitude $f(\mathbf{k}',\mathbf{k})$.

If the symmetry operator is unitary (for example, rotation and parity), everything is quite straightforward. Using the explicit form of T as given by (7.2.20), we see that

$$UH_0U^\dagger = H_0, \qquad UVU^\dagger = V \tag{7.10.1}$$

implies that T is also invariant under U—that is,

$$UTU^\dagger = T. \tag{7.10.2}$$

We define

$$|\tilde{\mathbf{k}}\rangle \equiv U|\mathbf{k}\rangle, \qquad |\tilde{\mathbf{k}}'\rangle \equiv U|\mathbf{k}'\rangle. \tag{7.10.3}$$

Then

$$\begin{aligned}\langle\tilde{\mathbf{k}}'|T|\tilde{\mathbf{k}}\rangle &= \langle\mathbf{k}'|U^\dagger UTU^\dagger U|\mathbf{k}\rangle \\ &= \langle\mathbf{k}'|T|\mathbf{k}\rangle. \end{aligned} \tag{7.10.4}$$

As an example, we consider the specific case where U stands for the parity operator

$$\pi|\mathbf{k}\rangle = |-\mathbf{k}\rangle, \qquad \pi|-\mathbf{k}\rangle = |\mathbf{k}\rangle. \tag{7.10.5}$$

Thus invariance of H_0 and V under parity would mean

$$\langle-\mathbf{k}'|T|-\mathbf{k}\rangle = \langle\mathbf{k}'|T|\mathbf{k}\rangle. \tag{7.10.6}$$

Pictorially, we have the situation illustrated in Figure 7.13a.

We exploited the consequence of angular-momentum conservation when we developed the method of partial waves. The fact that T is diagonal in the $|Elm\rangle$ representation is a direct consequence of T being invariant under rotation. Notice also that $\langle\mathbf{k}'|T|\mathbf{k}\rangle$ depends only on the relative orientation of $\mathbf{k}$ and $\mathbf{k}'$, as depicted in Figure 7.13b.

When the symmetry operation is antiunitary (as in time reversal), we must be more careful. First, we note that the requirement that V as well as

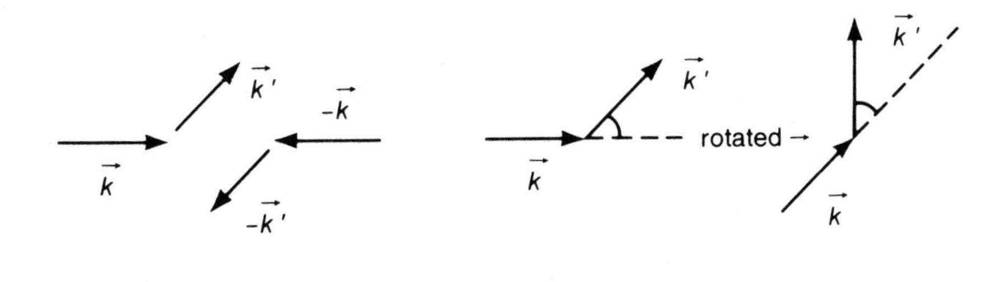

FIGURE 7.13. (a) Equality of T matrix elements between $\mathbf{k} \to \mathbf{k}'$ and $-\mathbf{k} \to -\mathbf{k}'$. (b) Equality of T matrix elements under rotation.

H_0 be invariant under time reversal invariance requires that

$$\Theta T \Theta^{-1} = T^{\dagger}. \tag{7.10.7}$$

This is because the antiunitary operator changes

$$\frac{1}{E - H_0 + i\varepsilon} \quad \text{into} \quad \frac{1}{E - H_0 - i\varepsilon} \tag{7.10.8}$$

in (7.2.20). We also recall that for an antiunitary operator [see (4.4.11)],

$$\langle \beta | \alpha \rangle = \langle \tilde{\alpha} | \tilde{\beta} \rangle, \tag{7.10.9}$$

where

$$|\tilde{\alpha}\rangle \equiv \Theta |\alpha\rangle \quad \text{and} \quad |\tilde{\beta}\rangle \equiv \Theta |\beta\rangle. \tag{7.10.10}$$

Let us consider

$$|\alpha\rangle = T|\mathbf{k}\rangle, \qquad \langle \beta | = \langle \mathbf{k}' |; \tag{7.10.11}$$

then

$$\begin{aligned} |\tilde{\alpha}\rangle &= \Theta T |\mathbf{k}\rangle = \Theta T \Theta^{-1} \Theta |\mathbf{k}\rangle = T^{\dagger} | -\mathbf{k}\rangle \\ |\tilde{\beta}\rangle &= \Theta |\mathbf{k}\rangle = |-\mathbf{k}'\rangle. \end{aligned} \tag{7.10.12}$$

As a result (7.10.9) becomes

$$\langle \mathbf{k}' | T | \mathbf{k} \rangle = \langle -\mathbf{k} | T | -\mathbf{k}' \rangle. \tag{7.10.13}$$

Notice that the initial and final momenta are interchanged, in addition to the fact that the directions of the momenta have been reversed.

It is also interesting to combine the requirements of time reversal [(7.10.13)] and parity [(7.10.6)]:

$$\langle \mathbf{k}' | T | \mathbf{k} \rangle \overset{\text{under } \Theta}{=} \langle -\mathbf{k} | T | -\mathbf{k}' \rangle \overset{\text{under } \pi}{=} \langle \mathbf{k} | T | \mathbf{k}' \rangle; \tag{7.10.14}$$

that is, from (7.2.19) and (7.2.22) we have

$$f(\mathbf{k}, \mathbf{k}') = f(\mathbf{k}', \mathbf{k}), \tag{7.10.15}$$

which results in

$$\frac{d\sigma}{d\Omega}(\mathbf{k} \to \mathbf{k}') = \frac{d\sigma}{d\Omega}(\mathbf{k}' \to \mathbf{k}). \tag{7.10.16}$$

Equation (7.10.16) is known as **detailed balance**.

It is more interesting to look at the analogue of (7.10.14) when we have spin. Here we may characterize the initial free-particle ket by $|\mathbf{k},m_s\rangle$, and we exploit (4.4.79) for the time-reversal portion:

$$\begin{aligned}\langle \mathbf{k}',m'_s|T|\mathbf{k},m_s\rangle &= i^{-2m_s+2m_{s'}}\langle -\mathbf{k},\, -m_s|T|-\mathbf{k}',\, -m'_s\rangle \\ &= i^{-2m_s+2m_{s'}}\langle \mathbf{k},\, -m_s|T|\mathbf{k}',\, -m'_s\rangle. \end{aligned} \tag{7.10.17}$$

For unpolarized initial states, we sum over the initial spin states and divide by $(2s+1)$; if the final polarization is not observed, we must sum over final states. We then obtain detailed balance in the form

$$\overline{\frac{d\sigma}{d\Omega}}(\mathbf{k} \to \mathbf{k}') = \overline{\frac{d\sigma}{d\Omega}}(\mathbf{k}' \to \mathbf{k}), \tag{7.10.18}$$

where we understand the bar on the top of $d\sigma/d\Omega$ in (7.10.18) to mean that we average over the initial spin states and sum over the final spin states.

7.11. TIME-DEPENDENT FORMULATION OF SCATTERING

Our discussion of scattering so far has been based on the time-independent formulation. It is also possible to develop a formalism of scattering based on the time-dependent Schrödinger equation. We show that this formalism leads to the same Lippmann-Schwinger equation, (7.1.6).

In the time-dependent formulation we conceive of a scattering process as a change in the state ket from a free-particle ket to a state ket influenced by the presence of the potential V. The basic equation of motion is

$$\left(i\hbar\frac{\partial}{\partial t} - H_0\right)|\psi;t\rangle = V|\psi;t\rangle, \tag{7.11.1}$$

where $|\psi;t\rangle$ is the time-dependent Schrödinger ket in the presence of V. The boundary condition appropriate for the scattering problem is that in the remote past $(t \to -\infty)$, the particle was free. This requirement is automatically accomplished if we turn on the potential adiabatically—that is, very slowly—as in Section 5.8:

$$V \to \lim_{\eta \to 0} Ve^{\eta t}. \tag{7.11.2}$$

Just as the partial differential equation with an inhomogeneous term is solved by introducing a Green's function [see (7.1.18)], the operator

Schrödinger equation (7.11.1) is solved by introducing the Green's operator $G_+(t, t')$ satisfying

$$\left(i\hbar\frac{\partial}{\partial t} - H_0\right)G_+(t, t') = \delta(t - t'). \tag{7.11.3}$$

The *causality* requirement we impose on G_+ is that the interaction of a particle at t' has an effect only for $t > t'$. We therefore impose the *retarded* boundary condition

$$G_+(t, t') = 0 \quad \text{for } t < t'. \tag{7.11.4}$$

We claim that the solution to (7.11.3) and (7.11.4) is

$$G_+(t, t') = -\frac{i}{\hbar}\theta(t - t')e^{-iH_0(t-t')/\hbar}. \tag{7.11.5}$$

For $t > t'$, a constant times $e^{-iH_0(t-t')/\hbar}$ clearly satisfies the differential equation (7.11.3) because the δ-function on the right-hand side is inoperative. For $t < t'$, we obviously have the desired condition (7.11.4) because $G_+(t,t')$ here is identically zero. At $t = t'$, we note that there is an extra contribution due to the discontinuity of the θ-function that just balances the right-hand side of (7.11.3):

$$i\hbar\frac{\partial}{\partial t}\left[-\frac{i}{\hbar}\theta(t - t')\right] = \delta(t - t'). \tag{7.11.6}$$

The solution to the full problem can now be written as follows:

$$|\psi^{(+)}; t\rangle = |\phi; t\rangle + \int_{-\infty}^{+\infty} G_+(t, t')V|\psi^{(+)}; t'\rangle\, dt', \tag{7.11.7}$$

where

$$\left(i\hbar\frac{\partial}{\partial t} - H_0\right)|\phi; t\rangle = 0 \tag{7.11.8}$$

and the upper limit of the integration in (7.11.7) may as well be t because G_+ vanishes for $t' > t$. As $t \to -\infty$, $|\psi^{(+)}; t\rangle$ coincides with $|\phi; t\rangle$ just as required; this is because $G_+(t, t')$ vanishes as $t \to -\infty$ for any finite value of t'. To see that (7.11.7) satisfies the t-dependent Schrödinger equation (7.11.1), we merely apply the operator $[i\hbar(\partial/\partial t) - H_0]$ to $|\psi^{(+)}; t\rangle$. The $|\phi; t\rangle$ ket makes no contribution; as for its effect on the second term, $[i\hbar(\partial/\partial t) - H_0]$ acting on G_+ just gives $\delta(t - t')$, by means of which the integration is immediately evaluated to be $V|\psi^{(+)}; t\rangle$.

So far we have not even required $|\psi^{(+)}; t\rangle$ to be an energy eigenket. If it is, we can separate the time dependence as usual:

$$\begin{aligned} |\phi; t\rangle &= |\phi\rangle e^{-iEt/\hbar} \\ |\psi; t\rangle &= |\psi\rangle e^{-iEt/\hbar}. \end{aligned} \tag{7.11.9}$$

Here there is the implicit assumption that E does not change if V is switched on adiabatically according to (7.11.2). Equations (7.11.5) and (7.11.7), evaluated at $t = 0$, now yield

$$|\psi^{(+)}\rangle = |\phi\rangle - \frac{i}{\hbar}\int_{-\infty}^{0} dt' e^{iH_0 t'/\hbar} e^{-iEt'/\hbar} V|\psi^{(+)}\rangle. \qquad (7.11.10)$$

The integral may appear to oscillate indefinitely. However, we recall that V is really to be understood as $Ve^{\eta t}$. As a result, the time integration is straightforward:

$$\begin{aligned} |\psi^{(+)}\rangle &= |\phi\rangle - \frac{i}{\hbar} \lim_{t'' \to -\infty} \int_{t''}^{0} dt' e^{i(H_0 - E - i\eta\hbar)t'/\hbar} V|\psi^{(+)}\rangle \\ &= |\phi\rangle - \frac{1}{H_0 - E - i\eta\hbar}\left[1 - \lim_{t'' \to -\infty} e^{[i(H_0 - E)/\hbar + \eta]t''}\right] V|\psi^{(+)}\rangle. \end{aligned} \qquad (7.11.11)$$

But as $t'' \to -\infty$, this is just the Lippmann-Schwinger equation (7.1.6).

The reader should note carefully how the $i\varepsilon$ prescription appears in the two formalisms. Earlier, using the time-independent formalism we saw that the choice of positive sign of the $i\varepsilon$-term (see Section 7.1) corresponds to the statement that the scatterer affects only outgoing spherical waves. In the t-dependent formulation presented here, the presence of $i\varepsilon$ ($\varepsilon = \eta\hbar$) in (7.11.11) arises from the requirement that the particle was free in the remote past.

One may argue that the slow switch-on of the potential (7.11.2) on which we have relied is somewhat artificial. But let us suppose that we are using a wave-packet formalism to describe scattering. When the wave packet is well outside the range of the potential, it does not matter whether the potential is zero or finite. In particular, V may be zero in the remote past, which then leads to no difficulty.

Connection With Time-Dependent Perturbation Theory

Inasmuch as scattering can be discussed using the time-dependent formulation, which is based on the time-dependent Schrödinger equation (7.11.1), we should be able to apply an approximation scheme based on (7.11.1). In particular we should be able to deploy the methods of time-dependent perturbation theory developed earlier in Section 5.6 to the scattering problem—provided, of course, the potential can be regarded as weak in some sense.

First we will show how the golden rule can be applied to compute $d\sigma/d\Omega$, leading to the results of the Born approximation of Section 7.2.

We assume in the remote past that the state ket is represented by a momentum eigenket $|\mathbf{k}\rangle$. When the interaction is slowly switched on, as in (7.11.2), momentum eigenkets other than $|\mathbf{k}\rangle$—for example, $|\mathbf{k}'\rangle$—are populated. As we have seen (see Section 5.6), the transition probability is (to first order) evaluated in the following manner. First we write

$$\langle\mathbf{k}'|U_I^{(1)}(t,-\infty)|\mathbf{k}\rangle = -\frac{i}{\hbar}\int_{-\infty}^{t}\langle\mathbf{k}'|V_I(t')|\mathbf{k}\rangle\,dt', \quad (7.11.12)$$

where $V_I(t') = e^{iH_0t'/\hbar}Ve^{-iH_0t'/\hbar}e^{\eta t'}$ and $E_\mathbf{k} = \hbar^2k^2/2m$. The transition probability of finding $|\mathbf{k}'\rangle$ at time t is then

$$|\langle\mathbf{k}'|U_I^{(1)}(t,-\infty)|\mathbf{k}\rangle|^2 = \frac{|\langle\mathbf{k}'|V|\mathbf{k}\rangle|^2}{\hbar^2}\frac{e^{2\eta t}}{\left[(E_\mathbf{k}-E_{\mathbf{k}'})^2/\hbar^2+\eta^2\right]}. \quad (7.11.13)$$

As $\eta \to 0$ (t finite), the transition rate is just what we expect from the golden rule:

$$\frac{d}{dt}|\langle\mathbf{k}'|U_I^{(1)}(t,-\infty)|\mathbf{k}\rangle|^2 = \frac{2\pi}{\hbar}|\langle\mathbf{k}'|V|\mathbf{k}\rangle|^2\delta(E_k-E_{k'}), \quad (7.11.14)$$

where we have used (5.8.5). Note that the right-hand side of (7.11.14) is independent of t.

It is important here to review the normalization conventions used for plane-wave states. In Section 1.7 we used the δ-function normalization

$$\langle\mathbf{k}'|\mathbf{k}\rangle = \delta^{(3)}(\mathbf{k}-\mathbf{k}'). \quad (7.11.15)$$

The completeness relation is then written as

$$1 = \int d^3k|\mathbf{k}\rangle\langle\mathbf{k}|. \quad (7.11.16)$$

In applying t-dependent perturbation theory to scattering processes, it is more helpful to use box normalization,

$$\langle\mathbf{k}'|\mathbf{k}\rangle = \delta_{\mathbf{k}',\mathbf{k}}, \quad (7.11.17)$$

where the allowed values for $\mathbf{k}$ are given by

$$k_{x,y,z} = \frac{2\pi n_{x,y,z}}{L}. \quad (7.11.18)$$

The wave function is given by

$$\langle\mathbf{x}|\mathbf{k}\rangle = \frac{1}{L^{3/2}}e^{i\mathbf{k}\cdot\mathbf{x}} \quad (7.11.19)$$

instead of $[1/(2\pi)^{3/2}]e^{i\mathbf{k}\cdot\mathbf{x}}$. The completeness relation is

$$1 = \sum_\mathbf{k}|\mathbf{k}\rangle\langle\mathbf{k}|, \quad (7.11.20)$$

but for a very large box we might as well treat the variables $k_{x,y,z}$ as continuous, so

$$1 = \left(\frac{L}{2\pi}\right)^3 \int d^3k |\mathbf{k}\rangle\langle\mathbf{k}|. \tag{7.11.21}$$

This looks like (7.11.16) except for the presence of $(L/2\pi)^3$. A very useful relation that enables us to go from the δ-function normalization convention to that of box normalization is

$$(2\pi)^3 \underset{\text{normalization}}{\langle\mathbf{k}'|V|\mathbf{k}\rangle_{\delta-fn}} = L^3 \underset{\text{normalization}}{\langle\mathbf{k}'|V|\mathbf{k}\rangle_{\text{box}}} \tag{7.11.22}$$

as the reader may easily verify by inserting a complete set of states in the position representation.

The box normalization is very convenient for evaluating the density of states. For this reason it is used more often in a treatment of scattering based on time-dependent perturbation theory. Thus if we are interested in scattering into the solid angle $d\Omega$, the relevant formula is

$$n^2\, dn\, d\Omega = \left(\frac{L}{2\pi}\right)^3 \frac{km}{\hbar^2}\, dE\, d\Omega \tag{7.11.23}$$

[see (5.7.31)].

Coming back to scattering, which we view as a transition rate w from $|\mathbf{k}\rangle$ into a group of states $|\mathbf{k}'\rangle$ subtending the solid-angle element $d\Omega$, we have w given by (for elastic scattering $k' = k$)

$$w = \frac{2\pi}{\hbar} |\langle\mathbf{k}'|V|\mathbf{k}\rangle|^2 \left(\frac{L}{2\pi}\right)^3 \frac{km}{\hbar^2}\, d\Omega \tag{7.11.24}$$

from the golden rule (5.6.34) and (7.11.14). This must be equated to

$$(\text{Incident flux}) \times \frac{d\sigma}{d\Omega}\, d\Omega. \tag{7.11.25}$$

As for the incident flux, we obtain [from $\mathbf{j} = (\hbar/m)\mathrm{Im}(\psi^* \nabla\psi)$]

$$|\mathbf{j}| = \frac{\hbar}{m}\left|\mathrm{Im}\left(\frac{e^{-i\mathbf{k}\cdot\mathbf{x}}}{L^{3/2}} \nabla \frac{e^{i\mathbf{k}\cdot\mathbf{x}}}{L^{3/2}}\right)\right| = \frac{\hbar k}{mL^3}. \tag{7.11.26}$$

Alternatively, incident flux = velocity/volume = $\hbar k/mL^3$. Putting everything together

$$\begin{aligned}\frac{d\sigma}{d\Omega} &= \left(\frac{mL^3}{\hbar k}\right)\left(\frac{2\pi}{\hbar}\right)\left(\frac{L}{2\pi}\right)^3\left(\frac{km}{\hbar^2}\right)\left|\frac{1}{L^3}\int d^3x\, V(\mathbf{x})\, e^{i(\mathbf{k}-\mathbf{k}')\cdot\mathbf{x}}\right|^2 \\ &= \left|\frac{1}{4\pi}\frac{2m}{\hbar^2}\int d^3x\, V(\mathbf{x})\, e^{i(\mathbf{k}-\mathbf{k}')\cdot\mathbf{x}}\right|^2.\end{aligned} \tag{7.11.27}$$

But this is precisely the result of the first-order Born approximation [(7.2.22)]. Higher-order Born terms can be obtained in a similar manner.

To sum up, the time-dependent formulation based on the time-dependent Schrödinger equation [(7.11.1)] enables us to derive easily the results we obtained earlier using the time-independent formalism—the Lippmann-Schwinger equation, the Born approximation, and so on. Furthermore, it turns out that the time-dependent formalism is more suitable for discussing more-general reaction processes other than elastic scattering. As a concrete example to illustrate this point, we now turn to a discussion of the inelastic scattering of electrons by atoms.

7.12. INELASTIC ELECTRON-ATOM SCATTERING

Let us consider the interactions of electron beams with atoms assumed to be in their ground states. The incident electron may get scattered elastically with final atoms unexcited:

$$e^- + \text{atom (ground state)} \to e^- + \text{atom (ground state)}. \quad (7.12.1)$$

This is an example of *elastic scattering*. To the extent that the atom can be regarded as infinitely heavy, the kinetic energy of the electron does not change. It is also possible for the target atom to get excited:

$$e^- + \text{atom (ground state)} \to e^- + \text{atom (excited state)}. \quad (7.12.2)$$

In this case we talk about **inelastic scattering** because the kinetic energy of the final outgoing electron is now less than that of the initial incoming electron, the difference being used to excite the target atom.

The initial ket of the electron plus the atomic system is written as

$$|\mathbf{k}, 0\rangle \quad (7.12.3)$$

where $\mathbf{k}$ refers to the wave vector of the incident electron and 0 stands for the atomic ground state. Strictly speaking (7.12.3) should be understood as the direct product of the incident electron ket $|\mathbf{k}\rangle$ and the ground-state atomic ket $|0\rangle$. The corresponding wave function is

$$\frac{1}{L^{3/2}} e^{i\mathbf{k}\cdot\mathbf{x}} \psi_0(\mathbf{x}_1, \mathbf{x}_2, \ldots, \mathbf{x}_z) \quad (7.12.4)$$

where we use the box normalization for the plane wave.

We may be interested in a final-state electron with a definite wave vector $\mathbf{k}'$. The final-state ket and the corresponding wave function are

$$|\mathbf{k}', n\rangle \quad \text{and} \quad \frac{1}{L^{3/2}} e^{i\mathbf{k}'\cdot\mathbf{x}} \psi_n(\mathbf{x}_1, \ldots, \mathbf{x}_z), \quad (7.12.5)$$

where $n = 0$ for elastic scattering and $n \neq 0$ for inelastic scattering.

Assuming that time-dependent perturbation theory is applicable, we can immediately write the differential cross section, as in the previous section:

$$\frac{d\sigma}{d\Omega}(0 \to n) = \frac{1}{(\hbar k/m_e L^3)} \frac{2\pi}{\hbar} |\langle \mathbf{k}' n|V|\mathbf{k}0\rangle|^2 \left(\frac{L}{2\pi}\right)^3 \left(\frac{k' m_e}{\hbar^2}\right)$$

$$= \left(\frac{k'}{k}\right) L^6 \left|\frac{1}{4\pi}\frac{2m_e}{\hbar^2}\langle \mathbf{k}', n|V|\mathbf{k},0\rangle\right|^2. \tag{7.12.6}$$

Everything is similar, including the cancellation of terms such as L^3, with one important exception: $k' \equiv |\mathbf{k}'|$ is not, in general, equal to $k \equiv |\mathbf{k}|$ for inelastic scattering.

The next question is, What V is appropriate for this problem? The incident electron can interact with the nucleus, assumed to be situated at the origin; it can also interact with each of the atomic electrons. So V is to be written as

$$V = -\frac{Ze^2}{r} + \sum_i \frac{e^2}{|\mathbf{x}-\mathbf{x}_i|}. \tag{7.12.7}$$

Here complications may arise because of the identity of the incident electron with one of the atomic electrons; to treat this rigorously is a nontrivial task. Fortunately, for a relatively fast electron we can legitimately ignore the question of identity; this is because there is little overlap between the bound-state electron and the incident electron in *momentum* space. We must evaluate the matrix element $\langle \mathbf{k}'n|V|\mathbf{k}0\rangle$, which, when explicitly written, is

$$\langle \mathbf{k}'n|V|\mathbf{k}0\rangle = \frac{1}{L^3}\int d^3x e^{i\mathbf{q}\cdot\mathbf{x}}\langle n|-\frac{Ze^2}{r} + \sum_i \frac{e^2}{|\mathbf{x}-\mathbf{x}_i|}|0\rangle$$

$$= \frac{1}{L^3}\int d^3x e^{i\mathbf{q}\cdot\mathbf{x}} \prod_i^z \int d^3x_i \psi_n^*(\mathbf{x}_1,\ldots,\mathbf{x}_z)\left[-\frac{Ze^2}{r} + \sum_i \frac{e^2}{|\mathbf{x}-\mathbf{x}_i|}\right]$$

$$\times \psi_0(\mathbf{x}_1,\ldots,\mathbf{x}_z) \tag{7.12.8}$$

with $\mathbf{q} \equiv \mathbf{k} - \mathbf{k}'$.

Let us see how to evaluate the matrix element of the first term, $-Ze^2/r$, where r actually means $|\mathbf{x}|$. First we note that this is a potential between the incident electron and the nucleus, which is independent of the atomic electron coordinates. So it can be taken outside the integration

$$\prod_i^z \int d^3x_i$$

in (7.12.8); we simply obtain

$$\langle n|0\rangle = \delta_{n0} \tag{7.12.9}$$

for the remainder. In other words, this term contributes only to the elastic-scattering case, where the target atom remains unexcited. In the elastic case we must still integrate $e^{i\mathbf{q}\cdot\mathbf{x}}/r$ with respect to $\mathbf{x}$, which amounts to taking the Fourier transform of the Coulomb potential; this can readily be done because we already evaluated in Section 7.2 the Fourier transform of the Yukawa potential [see (7.2.6) in conjunction with (7.2.2)]. Hence

$$\int d^3x \frac{e^{i\mathbf{q}\cdot\mathbf{x}}}{r} = \lim_{\mu \to 0} \int \frac{d^3x e^{i\mathbf{q}\cdot\mathbf{x} - \mu r}}{r} = \frac{4\pi}{q^2}. \tag{7.12.10}$$

As for the second term in (7.12.8), we can evaluate the Fourier transform of $1/|\mathbf{x}-\mathbf{x}_i|$. We can accomplish this by shifting the coordinate variables $\mathbf{x} \to \mathbf{x}+\mathbf{x}_i$:

$$\sum_i \int \frac{d^3x e^{i\mathbf{q}\cdot\mathbf{x}}}{|\mathbf{x}-\mathbf{x}_i|} = \sum_i \int \frac{d^3x e^{i\mathbf{q}\cdot(\mathbf{x}+\mathbf{x}_i)}}{|\mathbf{x}|} = \frac{4\pi}{q^2} \sum_i e^{i\mathbf{q}\cdot\mathbf{x}_i}. \tag{7.12.11}$$

Notice that this is just the Fourier transform of the Coulomb potential multiplied by the Fourier transform of the electron density due to the atomic electrons situated at $\mathbf{x}_i$:

$$\rho_{\text{atom}}(\mathbf{x}) = \sum_i \delta^{(3)}(\mathbf{x}-\mathbf{x}_i). \tag{7.12.12}$$

We customarily define the **form factor** $F_n(\mathbf{q})$ for excitation $|0\rangle$ to $|n\rangle$ as follows:

$$ZF_n(\mathbf{q}) \equiv \langle n|\sum_i e^{i\mathbf{q}\cdot\mathbf{x}_i}|0\rangle, \tag{7.12.13}$$

which is made of coherent—in the sense of definite phase relationships—contributions from the various electrons. Notice that as $q \to 0$, we have

$$\frac{1}{Z}\langle n|\sum_i e^{i\mathbf{q}\cdot\mathbf{x}_i}|0\rangle \to 1$$

for $n=0$; hence the form factor approaches unity in the elastic-scattering case. For $n \neq 0$ (inelastic scattering), $F_n(\mathbf{q}) \to 0$ as $\mathbf{q} \to 0$ by orthogonality between $|n\rangle$ and $|0\rangle$. We can then write the matrix element in (7.12.8) as

$$\int d^3x e^{i\mathbf{q}\cdot\mathbf{x}} \langle n| \left(-\frac{Ze^2}{r} + \sum_i \frac{e^2}{|\mathbf{x}-\mathbf{x}_i|} \right) |0\rangle = \frac{4\pi Ze^2}{q^2}[-\delta_{n0} + F_n(\mathbf{q})] \tag{7.12.14}$$

We are finally in a position to write the differential cross section for inelastic (or elastic) scattering of electrons by atoms:

$$\begin{aligned} \frac{d\sigma}{d\Omega}(0 \to n) &= \left(\frac{k'}{k}\right) \left| \frac{1}{4\pi} \frac{2m_e}{\hbar^2} \frac{4\pi Ze^2}{q^2} [-\delta_{n0} + F_n(\mathbf{q})] \right|^2 \\ &= \frac{4m_e^2}{\hbar^4} \frac{(Ze^2)^2}{q^4} \left(\frac{k'}{k}\right) |-\delta_{n0} + F_n(\mathbf{q})|^2. \end{aligned} \tag{7.12.15}$$

For inelastic scattering the δ_{n0}-term does not contribute, and it is customary to write the differential cross section in terms of the Bohr radius,

$$a_0 = \frac{\hbar^2}{e^2 m_e}, \tag{7.12.16}$$

as follows:

$$\frac{d\sigma}{d\Omega}(0 \to n) = 4Z^2 a_0^2 \left(\frac{k'}{k}\right) \frac{1}{(qa_0)^4} |F_n(\mathbf{q})|^2. \tag{7.12.17}$$

Quite often $d\sigma/dq$ is used in place of $d\sigma/d\Omega$; using

$$q^2 = |\mathbf{k} - \mathbf{k}'|^2 = k^2 + k'^2 - 2kk'\cos\theta \tag{7.12.18}$$

and $dq = -\,d(\cos\theta)kk'/q$, we can write

$$\frac{d\sigma}{dq} = \frac{2\pi q}{kk'} \frac{d\sigma}{d\Omega}. \tag{7.12.19}$$

The inelastic cross section we have obtained can be used to discuss *stopping power*—the energy loss of a charged particle as it goes through matter. A number of people, including H. A. Bethe and F. Bloch, have discussed the quantum-mechanical derivation of stopping power from the point of view of the inelastic-scattering cross section. We are interested in the energy loss of a charged particle per unit length traversed by the incident charged particle. The collision rate per unit length is $N\sigma$, where N is the number of atoms per unit volume; at each collision process the energy lost by the charged particle is $E_n - E_0$. So dE/dx is written as

$$\begin{aligned}
\frac{dE}{dx} &= N \sum_n (E_n - E_0) \int \frac{d\sigma}{dq}(0 \to n)\, dq \\
&= N \sum_n (E_n - E_0) \frac{4Z^2}{a_0^2} \int_{q_{\min}}^{q_{\max}} \frac{k'}{k} \frac{1}{q^4} \frac{2\pi q}{kk'} |F_n(\mathbf{q})|^2\, dq \\
&= \frac{8\pi N}{k^2 a_0^2} \sum_n (E_n - E_0) \int_{q_{\min}}^{q_{\max}} \left| \langle n | \sum_{i=1}^{z} e^{i\mathbf{q}\cdot\mathbf{x}_i} | 0 \rangle \right|^2 \frac{dq}{q^3}.
\end{aligned} \tag{7.12.20}$$

There are many papers written on how to evaluate the sum in (7.12.20).* The upshot of all this is to justify quantum mechanically Bohr's 1913 formula for stopping power,

$$\frac{dE}{dx} = \frac{4\pi N Z e^4}{m_e v^2} \ln\left(\frac{2m_e v^2}{I}\right), \tag{7.12.21}$$

where I is a semiempirical parameter related to the average excitation

*For a relatively elementary discussion, see K. Gottfried (1966) and H. A. Bethe and R. W. Jackiw (1968).

energy $\langle E_n - E_0 \rangle$. If the charged particle has electric charge $\pm ze$, we just replace Ze^4 by $z^2 Ze^4$. It is also important to note that even if the projectile is not an electron, the m_e that appears in (7.12.21) is still the electron mass, not the mass of the charged particle. So the energy loss is dependent on the charge and the velocity of the projectile but is independent of the mass of the projectile. This has an important application to the detection of charged particles.

Quantum mechanically, we view the energy loss of a charged particle as a series of inelastic-scattering processes. At each interaction between the charged particle and an atom, we may imagine that a "measurement" of the position of the charged particle is made. We may wonder why particle tracks in such media as cloud chambers and nuclear emulsions are nearly straight. The reason is that the differential cross section (7.12.17) is sharply peaked at small q; in an overwhelming number of collisions, the final direction of momentum is nearly the same as the incident electron due to the rapid falloff of q^{-4} and $F_n(\mathbf{q})$ for large q.

Nuclear Form Factor

The excitation of atoms due to inelastic scattering is important for $q \sim 10^9\ \text{cm}^{-1}$ to $10^{10}\ \text{cm}^{-1}$. If q is too large, the contributions due to $F_0(\mathbf{q})$ or $F_n(\mathbf{q})$ drop off very rapidly. At extremely high q, where q is now of order $1/R_{\text{nucleus}} \sim 10^{12}\ \text{cm}^{-1}$, the structure of the nucleus becomes important. The Coulomb potential due to the point nucleus must now be replaced by a Coulomb potential due to an extended object,

$$-\frac{Ze^2}{r} \to -Ze^2 \int \frac{d^3x' N(r')}{|\mathbf{x}-\mathbf{x}'|}, \tag{7.12.22}$$

where $N(r)$ is a nuclear charge distribution, normalized so that

$$\int d^3x' N(r') = 1. \tag{7.12.23}$$

The pointlike nucleus can now be regarded as a special case, with

$$N(r') = \delta^{(3)}(r'). \tag{7.12.24}$$

We can evaluate the Fourier transform of the right-hand side of (7.12.22) in analogy with (7.12.10) as follows:

$$\begin{aligned} Ze^2 \int d^3x \int \frac{d^3x' e^{i\mathbf{q}\cdot\mathbf{x}} N(r')}{|\mathbf{x}-\mathbf{x}'|} &= Ze^2 \int d^3x' e^{i\mathbf{q}\cdot\mathbf{x}'} N(r') \int \frac{d^3x e^{i\mathbf{q}\cdot\mathbf{x}}}{r} \\ &= Ze^2 \frac{4\pi}{q^2} F_{\text{nucleus}}(\mathbf{q}) \end{aligned} \tag{7.12.25}$$

where we have shifted the coordinates $\mathbf{x} \to \mathbf{x} + \mathbf{x}'$ in the first step and

$$F_{\text{nucleus}} \equiv \int d^3x e^{i\mathbf{q}\cdot\mathbf{x}} N(r). \tag{7.12.26}$$

We thus obtain the deviation from the Rutherford formula due to the finite size of the nucleus,

$$\frac{d\sigma}{d\Omega} = \left(\frac{d\sigma}{d\Omega}\right)_{\text{Rutherford}} |F(\mathbf{q})|^2, \tag{7.12.27}$$

where $(d\sigma/d\Omega)_{\text{Rutherford}}$ is the differential cross section for the electric scattering of electrons by a pointlike nucleus of charge $Z|e|$. For small q we have

$$\begin{aligned} F_{\text{nucleus}}(\mathbf{q}) &= \int d^3x \left(1 + i\mathbf{q}\cdot\mathbf{x} - \frac{1}{2}q^2r^2(\hat{\mathbf{q}}\cdot\hat{\mathbf{r}})^2 + \cdots\right) N(r) \\ &= 1 - \frac{1}{6}q^2\langle r^2\rangle_{\text{nucleus}} + \cdots . \end{aligned} \tag{7.12.28}$$

The $\mathbf{q}\cdot\mathbf{x}$-term vanishes because of spherical symmetry, and in the q^2-term we have used the fact that the angular average of $\cos^2\theta$ (where θ is the angle between $\hat{\mathbf{q}}$ and $\hat{\mathbf{r}}$) is just $\frac{1}{3}$:

$$\frac{1}{2}\int_{-1}^{+1} d(\cos\theta)\cos^2\theta = \frac{1}{3}. \tag{7.12.29}$$

The quantity $\langle r^2\rangle_{\text{nucleus}}$ is known as the mean square radius of the nucleus. In this way it is possible to "measure" the size of the nucleus and also of the proton, as done by R. Hofstadter and coworkers. In the proton case the spin (magnetic moment) effect is also important.

7.13. COULOMB SCATTERING

This last section was not written even in preliminary form by Professor Sakurai, but it was listed as an item he would like to consider for *Modern Quantum Mechanics*. Professor Thomas Fulton of The Johns Hopkins University graciously offered to contribute his own lecture notes on Coulomb scattering for this section in honor of Professor Sakurai's memory. The editor deeply appreciates this gesture of friendship. However, the reader needs to be alerted that there could be stylistic and even notational changes here from the rest of the book.

In this revised edition, the material is somewhat reorganized; the principal results previously obtained are still given in the main text, while the mathematical details are now presented in Appendixes C.1 and C.2.

Much of this material is a slightly expanded version of Gottfried's treatment (Gottfried 1966, pp. 148–153) and uses the notation of this text. The treatment of Coulomb partial waves is original. The present discussion of Coulomb scattering differs from the standard treatment (see, for example, Bohm 1951, and Schiff 1968) by dealing directly with the asymptotic forms. The differential equation (7.13.7) is the standard form, which can also be obtained by separating variables in parabolic coordinates. It can be recognized as the differential equation for the confluent hypergeometric function,* and so the properties of this function (including asymptotic forms) can be applied to analyse this case. However, the present alternative approach can pass directly to the asymptotic limit by using Laplace transforms and complex variable techniques (see Appendixes C.1 and C.2), and thus avoid referring to the detailed properties of the confluent hypergeometric function. Clearly, this simpler approach, which can be understood without any reference to material outside the present textbook, loses information about the scattering wave function at the origin. Thus, for example, we can say nothing about the *Gamow factor*,† which is relevant for the study of many nuclear reaction rates. The interested reader is therefore encouraged to supplement his knowledge here with reading from other texts (e.g., Schiff 1968) that deal with the behavior of the Coulomb wave function at the origin via power series expansion of the associated confluent hypergeometric function.

Scattering Solutions of the Hamiltonian of the Coulomb Interaction

We start with the three-dimensional Schrödinger equation with Coulomb potential

$$V(r) = \frac{-Z_1 Z_2 e^2}{r} \tag{7.13.1}$$

for a collision between particles of charges $Z_1 e$ and $-Z_2 e$, as follows:

$$\left(-\frac{\hbar^2}{2m}\nabla^2 - \frac{Z_1 Z_2 e^2}{r}\right)\psi(\mathbf{r}) = E\psi(\mathbf{r}), \qquad E > 0 \tag{7.13.2a}$$

or

$$\left(\nabla^2 + k^2 + \frac{2\gamma k}{r}\right)\psi(\mathbf{r}) = 0 \tag{7.13.2b}$$

where m is the reduced mass $m = m_1 m_2/(m_1 + m_2)$ for the particles $Z_1 e$

---

*See, for instance, E. T. Whittaker and G. N. Watson, *A Course of Modern Analysis*, 4th ed. (London: Cambridge University Press, 1935), Chapter XVI.

†G. Gamow, *Z. Phys.* Vol. *51*, no. 204 (1928); R. W. Gurney and E. U. Condon, *Phys. Rev.* Vol. *33*, no. 127 (1929).

and $-Z_2e$. The γk in (7.13.2b) is given by

$$\gamma k = \frac{Z_1 Z_2 e^2 m}{\hbar^2} \quad \text{or} \quad \gamma = \frac{Z_1 Z_2 e^2}{\hbar v} = \frac{\alpha Z_1 Z_2 c}{v}, \quad \alpha = \frac{e^2}{\hbar c} = \frac{1}{137.03608}. \tag{7.13.3}$$

Note that $\gamma > 0$ corresponds to attraction. As long as we are interested in a pure Coulomb field, it is possible to write our solution $\psi_{\mathbf{k}}(\mathbf{r})$ to (7.13.2) as

$$\begin{aligned} \psi_{\mathbf{k}}(\mathbf{r}) &= e^{i\mathbf{k}\cdot\mathbf{r}}\chi(u) \\ u &= ikr(1 - \cos\theta) = ik(r - z) = ikw \\ \mathbf{k}\cdot\mathbf{r} &= kz. \end{aligned} \tag{7.13.4}$$

The separation of variables for $\psi_{\mathbf{k}}(\mathbf{r})$ in (7.13.4) is plausible if we recognize (1) that the solution will not involve the azimuthal angle ϕ because of the axial symmetry of the problem and (2) since $\psi_{\mathbf{k}}(\mathbf{r})$ represents the complete Coulomb-wave function (incident plus scattered wave), terms must exist in its dominant asymptotic form that contain $e^{i\mathbf{k}\cdot\mathbf{r}}$ and $r^{-1}e^{ikr}$. We will demonstrate shortly [see Equations (7.13.9a), (7.13.13), and (7.13.14)] that, with the choice of independent variables we are about to make, this will indeed be the case. (Another argument for our particular choice of variables is given by Gottfried 1966, 148.)

Let us choose independent variables (z, w, λ) where $w = r - z$ and λ can be taken to be ϕ, on which the solution $\psi_{\mathbf{k}}(\mathbf{r})$ does *not* depend. In changing our Cartesian coordinates (x, y, z) to (z, w, λ), use—for example—such expressions as

$$\begin{aligned} \frac{\partial}{\partial x} &= \frac{\partial w}{\partial x}\frac{\partial}{\partial w} + \frac{\partial z}{\partial x}\frac{\partial}{\partial z} + \frac{\partial \lambda}{\partial x}\frac{\partial}{\partial \lambda} \\ &= \frac{x}{r}\frac{\partial}{\partial w} + \frac{\partial \lambda}{\partial x}\frac{\partial}{\partial \lambda} \end{aligned} \tag{7.13.5}$$

Note that because $e^{ikz}\chi(u)$ is independent of λ, the operation $\partial/\partial\lambda$ makes no contribution. The reader will readily verify that, operating on $e^{ikz}\chi(u)$,

$$\begin{aligned} \frac{\partial^2}{\partial x^2} &= \frac{x^2}{r^2}\frac{\partial^2}{\partial w^2} + \left(\frac{1}{r} - \frac{x^2}{r^3}\right)\frac{\partial}{\partial w} \\ \frac{\partial^2}{\partial y^2} &= \frac{y^2}{r^2}\frac{\partial^2}{\partial w^2} + \left(\frac{1}{r} - \frac{y^2}{r^3}\right)\frac{\partial}{\partial w} \\ \frac{\partial^2}{\partial z^2} &= \frac{w^2}{r^2}\frac{\partial^2}{\partial w^2} + \left(\frac{w}{r^2} - \frac{w^2}{r^3}\right)\frac{\partial}{\partial w} - \frac{2w}{r}\frac{\partial^2}{\partial z \partial w} + \frac{w}{r^2}\frac{\partial}{\partial w} + \frac{\partial^2}{\partial z^2}. \end{aligned} \tag{7.13.6}$$

Using (7.13.4)–(7.13.6) in (7.13.2), we have

$$\left[u\frac{d^2}{du^2} + (1 - u)\frac{d}{du} - i\gamma\right]\chi(u) = 0. \tag{7.13.7}$$

Our first attempt at the solution of (7.13.7) considers the asymptotic case—that is, solutions for *large* $r - z$ (away from the forward direction, since $r - z = r(1 - \cos\theta) = 0$ for $\theta = 0$). Let us try two types of solutions, (1) $\chi \sim u^{\lambda}$ and (2) $\chi \sim e^{u}$.

The first type of solution requires

$$(-\lambda - i\gamma)u^{\lambda} \simeq 0, \quad \text{or} \quad \lambda = -i\gamma \tag{7.13.8}$$

and, hence,

$$\chi(u) \sim u^{-i\gamma} \sim e^{-i\gamma \ln k(r-z)} \tag{7.13.9a}$$

$$\psi_{\mathbf{k}}(\mathbf{r}) \sim e^{i[\mathbf{k}\cdot\mathbf{r} - \gamma \ln k(r-z)]}. \tag{7.13.9b}$$

Equation (7.13.9b) clearly represents the incoming plane-wave piece. The second type of solution, $\chi \sim e^{u}$, manifestly satisfies $u(d^2\chi/du^2) - u(d\chi/du) = 0$, so heuristically we try for a solution that combines features of both solutions in product form,

$$\chi(u) \sim u^{\lambda}e^{u}, \tag{7.13.10}$$

and substitute into (7.13.7). The coefficient of $u^{\lambda+1}e^{u}$ obviously vanishes, while the coefficient of the $u^{\lambda}e^{u}$-term leads to the following relation:

$$2\lambda + (1 - \lambda) - i\gamma = 0, \quad \text{or} \quad \lambda = -1 + i\gamma. \tag{7.13.11}$$

Hence,

$$\chi(u) \simeq \frac{u^{i\gamma}e^{u}}{u} \simeq \frac{e^{ik(r-z)}}{k(r - z)}e^{i\gamma \ln k(r-z)} \tag{7.13.12}$$

and

$$\psi(\mathbf{r}) \sim \frac{e^{i[kr + \gamma \ln k(r-z)]}}{r - z}, \tag{7.13.13}$$

clearly an outgoing spherical wave. Combining the two solution pieces given by (7.13.9) and (7.13.13) linearly and noting that $r - z = r(1 - \cos\theta) = 2r\sin^2(\theta/2)$ while $\ln[2kr\sin^2(\theta/2)] = \ln(2kr) + 2\ln\sin(\theta/2)$, we write the normalized wave function as

$$\psi_{\mathbf{k}}(\mathbf{r}) \xrightarrow[|u|\to\infty]{} \frac{1}{(2\pi)^{3/2}}\left\{e^{i[\mathbf{k}\cdot\mathbf{r} - \gamma \ln(kr - \mathbf{k}\cdot\mathbf{r})]} + \frac{f_c(k, \theta)e^{i[kr + \gamma \ln 2kr]}}{r}\right\}. \tag{7.13.14}$$

There are extra phase effects at large r, which is due to the long range of the Coulomb interaction.

Now we must find the Coulomb scattering amplitude $f_c(k, \theta)$. Interestingly, aside from a factor of γ/k, the magnitude is just from $(r - z)$: $(1/2k \sin^2 \theta/2)$, which appears in the asymptotic form (7.13.13). But we have to work hard to get it! Because the details here are quite involved mathematically (c.f. Appendix C.1), the first-time reader is encouraged to focus only on the final results summarized below. The physical basis of our results are discussed later on in this section.

For the normalized $\psi_{\mathbf{k}}(\mathbf{r})$ we obtain

$$\psi_{\mathbf{k}}(\mathbf{r}) \sim \frac{1}{(2\pi)^{3/2}}\left[e^{i[kz-\gamma \ln k(r-z)]} + \frac{g_1^*(\gamma)}{g_1(\gamma)}\frac{\gamma}{2k\sin^2(\theta/2)}\frac{e^{i[kr+\gamma \ln k(r-z)]}}{r}\right]. \tag{7.13.15}$$

This exhibits explicitly all the r- and k-dependent phases and leads to the identification from (7.13.14) that

$$f_c(k, \theta) = \frac{\gamma e^{i[\Theta(\gamma)+2\gamma \ln \sin(\theta/2)]}}{2k\sin^2(\theta/2)}$$
$$e^{i\Theta(\gamma)} = \frac{g_1^*(\gamma)}{g_1(\gamma)}. \tag{7.13.16}$$

Thus, as in (7.1.36), the differential cross section is

$$\frac{d\sigma_c(k, \theta)}{d\Omega} = |f_c(k, \theta)|^2 = \frac{\gamma^2}{4k^2\sin^4(\theta/2)}, \tag{7.13.17}$$

where γ is given by (7.13.3)—and we are back to the Rutherford formula of (7.2.11).

For the aficionados, we remind the reader of Hankel's formula

$$\int_{C_1} e^s s^{-z} ds = \frac{2\pi i}{\Gamma(z)} \tag{7.13.18}$$

(which checks for $z = 1$ and $\Gamma(1) = 0! = 1$). Thus

$$g_1(\gamma) = \frac{1}{\Gamma(1 - i\gamma)}, \quad e^{i\Theta(\gamma)} = \frac{\Gamma(1 - i\gamma)}{\Gamma(1 + i\gamma)}. \tag{7.13.19}$$

Partial-Wave Analysis for the Coulomb Case

We can rewrite $\psi(\mathbf{r})$ from (7.13.4) and (C.1.5) as

$$\psi(\mathbf{r}) = e^{i\mathbf{k}\cdot\mathbf{r}}\chi(u)$$
$$= Ae^{i\mathbf{k}\cdot\mathbf{r}}\int_C e^{ut}t^{i\gamma-1}(1 - t)^{-i\gamma}dt$$

$$= A \int_C e^{i\mathbf{k}\cdot\mathbf{r}} e^{ik(r-z)t}(1 - t)d(t, \gamma)dt$$

$$= A \int_C e^{ikrt} e^{i\mathbf{k}\cdot\mathbf{r}(1-t)}(1 - t)d(t, \gamma)dt \qquad (7.13.20)$$

where

$$d(t, \gamma) = t^{i\gamma-1}(1 - t)^{-i\gamma-1}. \qquad (7.13.21)$$

We perform the partial-wave expansion of the plane wave analogous to (7.5.18) as

$$\psi(\mathbf{r}) = \sum_{l=0}^{\infty} (2l + 1)i^l P_l(\cos\theta) A_l(kr), \qquad (7.13.22)$$

where

$$A_l(kr) = A \int_C e^{ikrt} j_l[kr(1 - t)](1 - t)d(t, \gamma)dt. \qquad (7.13.23)$$

Now, $j_l = \frac{1}{2}(h_l^{(1)} + h_l^{(1)*})$, where the $h_l^{(1)}$ are spherical Hankel functions (see Appendix A and Whittaker and Watson 1935), $h_l(\rho) = E_l(\rho)e^{i\rho}/\rho$,

l	0	1	2
$E_l(\rho)$	$-i$	$-\left(1 + \frac{i}{\rho}\right)$	$i\left(1 + \frac{3i}{\rho} - \frac{3}{\rho^2}\right)$,

(7.13.24)

and, asymptotically, $E_l(\rho) = (-i)^{l+1} + 0(\rho^{-1})$. Hence, we write

$$A_l(kr) = A_l^{(1)}(kr) + A_l^{(2)}(kr), \qquad (7.13.25)$$

where $A_l^{(1)}$ and $A_l^{(2)}$ correspond to the contributions from h_l and h_l^*, respectively.

The explicit evaluation of $A_l^{(1)}(kr)$ and $A_l^{(2)}(kr)$ is given in Appendix C.2 and leads to the following results

$$A_l^{(1)}(kr) = 0 \qquad (7.13.26)$$

and

$$A_l^{(2)}(kr) \simeq -\frac{Ae^{\pi\gamma/2}}{2ikr}[2\pi i g_1(\gamma)] \times \{e^{-i[kr-(l\pi/2)+\gamma \ln 2kr]} - e^{2i\eta_l(k)}e^{i[kr-(l\pi/2)+\gamma \ln 2kr]}\}, \qquad (7.13.27)$$

where we have optimistically introduced a real phase $\eta_l(k)$:

$$e^{2i\eta_l(k)} = \frac{g_3(\gamma)}{g_1(\gamma)} = \Gamma(1 - i\gamma)g_3(\gamma). \qquad (7.13.28)$$

Note that we write $\eta_l(k)$ because γ depends on k [see (7.13.3)].

Let us look briefly at some simple examples to demonstrate that $\eta_l(k)$ is indeed a real phase. For $l = 0$,

$$E_0^* = i, \qquad 2\pi i g_3(\gamma) = \int_{C_1} e^s s^{-i\gamma-1} ds = 2\pi i g_1^*(\gamma) = \frac{2\pi i}{\Gamma(1 + i\gamma)},$$

$$e^{2i\eta_0} = \frac{\Gamma(1 - i\gamma)}{\Gamma(1 + i\gamma)}. \tag{7.13.29}$$

For $l = 1$, we have

$$E_1^*(\rho) = -\left(1 - \frac{i}{\rho}\right), \qquad E_1^*\left(\frac{-s}{2i}\right) = -\left(1 - \frac{i}{-s/2i}\right) = -\left(1 - \frac{2}{s}\right)$$

$$\begin{aligned} 2\pi i g_3(\gamma) &= -\int_{C_1} e^s s^{-i\gamma-1}\left(1 - \frac{2}{s}\right) ds = -2\pi i\left[\frac{1}{\Gamma(1 + i\gamma)} - \frac{2}{\Gamma(2 + i\gamma)}\right] \\ &= \frac{-2\pi i}{\Gamma(2 + i\gamma)}\,[(1 + i\gamma) - 2] = \frac{2\pi i(1 - i\gamma)}{\Gamma(2 + i\gamma)} \end{aligned} \tag{7.13.30}$$

$$e^{2i\eta_1} = \frac{g_3(\gamma)}{g_1(\gamma)} = \frac{(1 - i\gamma)\Gamma(1 - i\gamma)}{\Gamma(2 + i\gamma)} = \frac{\Gamma(2 - i\gamma)}{\Gamma(2 + i\gamma)}.$$

The same pattern repeats for $l = 2, 3, \ldots$. Thus for a general l, we conjecture

$$e^{2i\eta_l} = \frac{\Gamma(1 + l - i\gamma)}{\Gamma(1 + l + i\gamma)}. \tag{7.13.31}$$

As a final summary we remark that the Coulomb partial-wave-scattering case differs from the general partial-wave case (see Section 7.6) only through the modification

$$\begin{aligned} -\frac{1}{2}\, h_l^{(1)}(kr) &\cong -\frac{e^{i(kr - (l\pi/2))}}{2ikr} \to -\frac{e^{i(kr - (l\pi/2) + \gamma \ln 2kr)}}{2ikr} \\ e^{2i\delta_l} &\to e^{2i\eta_l} = \frac{\Gamma(1 + l - i\gamma)}{\Gamma(1 + l + i\gamma)}. \end{aligned} \tag{7.13.32}$$

Let us note that the asymptotic form of the Coulomb scattering wave function, Eq. (7.13.15), is different from that which appears in Eq. (7.1.33), written for large r as

$$\psi \to e^{ikz} + f(\theta)\,\frac{e^{ikr}}{r}. \tag{7.13.33}$$

The latter form is valid only for short-range forces, while the Coulomb force is a long-range one. The form (7.1.33′) is modified as follows for the Coulomb interaction: The first term on the right-hand side of (7.13.15) contains not only the factor e^{ikz} as in (7.13.33), but also a modifying, coordinate dependent, multiplicative phase factor $e^{-i\gamma \ln k(r-z)}$. Hence, the incident plane wave is slightly distorted, irrespective of how far away the particle is from the origin. Analogously, the second term on the right-hand side of (7.13.15), corresponding to the outgoing spherical wave, contains the r dependent phase factor $e^{i\gamma \ln 2kr}$. Nevertheless, in spite of these long-range effects leading to phase factors that vary with distance, we can still define a scattering cross section for short-range forces because the distorting terms, although they alter physically observable quantities (e.g., the mean current), lead to alterations that vanish as r goes to infinity.

PROBLEMS

1. The Lippmann-Schwinger formalism can also be applied to a *one*-dimensional transmission-reflection problem with a finite-range potential, $V(x) \neq 0$ for $0 < |x| < a$ only.
   a. Suppose we have an incident wave coming from the left: $\langle x|\phi\rangle = e^{ikx}/\sqrt{2\pi}$. How must we handle the singular $1/(E-H_0)$ operator if we are to have a transmitted wave only for $x > a$ and a reflected wave and the original wave for $x < -a$? Is the $E \to E + i\varepsilon$ prescription still correct? Obtain an expression for the appropriate Green's function and write an integral equation for $\langle x|\psi^{(+)}\rangle$.
   b. Consider the special case of an attractive δ-function potential

$$V = -\left(\frac{\gamma\hbar^2}{2m}\right)\delta(x) \quad (\gamma > 0).$$

   Solve the integral equation to obtain the transmission and reflection amplitudes. Check your results with Gottfried 1966, 52.
   c. The one-dimensional δ-function potential with $\gamma > 0$ admits one (and only one) bound state for any value of γ. Show that the transmission and reflection amplitudes you computed have bound-state poles at the expected positions when k is regarded as a complex variable.
2. Prove

$$\sigma_{\text{tot}} \simeq \frac{m^2}{\pi\hbar^4}\int d^3x \int d^3x' V(r)V(r')\frac{\sin^2 k|\mathbf{x}-\mathbf{x}'|}{k^2|\mathbf{x}-\mathbf{x}'|^2}$$

   in each of the following ways.
   a. By integrating the differential cross section computed using the first-order Born approximation.
   b. By applying the optical theorem to the forward-scattering amplitude in the *second*-order Born approximation. [Note that $f(0)$ is real if the first-order Born approximation is used.]
3. Consider a potential

$$V = 0 \quad \text{for } r > R, \qquad V = V_0 = \text{constant} \quad \text{for } r < R,$$

where V_0 may be positive or negative. Using the method of partial waves, show that for $|V_0| \ll E = \hbar^2 k^2/2m$ and $kR \ll 1$ the differential cross section is isotropic and that the total cross section is given by

$$\sigma_{\text{tot}} = \left(\frac{16\pi}{9}\right)\frac{m^2 V_0^2 R^6}{\hbar^4}.$$

Suppose the energy is raised slightly. Show that the angular distribution can then be written as

$$\frac{d\sigma}{d\Omega} = A + B\cos\theta.$$

Obtain an approximate expression for B/A.

4. A spinless particle is scattered by a weak Yukawa potential

$$V = \frac{V_0 e^{-\mu r}}{\mu r}$$

where $\mu > 0$ but V_0 can be positive or negative. It was shown in the text that the first-order Born amplitude is given by

$$f^{(1)}(\theta) = -\frac{2mV_0}{\hbar^2\mu}\frac{1}{\left[2k^2(1-\cos\theta)+\mu^2\right]}.$$

a. Using $f^{(1)}(\theta)$ and assuming $|\delta_l| \ll 1$, obtain an expression for δ_l in terms of a Legendre function of the second kind,

$$Q_l(\zeta) = \frac{1}{2}\int_{-1}^{1}\frac{P_l(\zeta')}{\zeta - \zeta'}\,d\zeta'.$$

b. Use the expansion formula

$$Q_l(\zeta) = \frac{l!}{1\cdot 3\cdot 5\cdots(2l+1)}$$
$$\times\left\{\frac{1}{\zeta^{l+1}} + \frac{(l+1)(l+2)}{2(2l+3)}\frac{1}{\zeta^{l+3}}\right.$$
$$\left. + \frac{(l+1)(l+2)(l+3)(l+4)}{2\cdot 4\cdot(2l+3)(2l+5)}\frac{1}{\zeta^{l+5}} + \cdots\right\} \quad (|\zeta| > 1)$$

to prove each assertion.

(i) δ_l is negative (positive) when the potential is repulsive (attractive).

(ii) When the de Broglie wavelength is much longer than the range of the potential, δ_l is proportional to k^{2l+1}. Find the proportionality constant.

5. Check explicitly the $x - p_x$ uncertainty relation for the ground state of a particle confined inside a hard sphere: $V = \infty$ for $r > a$, $V = 0$ for $r < a$. (*Hint*: Take advantage of spherical symmetry.)
6. Consider the scattering of a particle by an impenetrable sphere

$$V(r) = \begin{cases} 0 & \text{for } r > a \\ \infty & \text{for } r < a. \end{cases}$$

   a. Derive an expression for the s-wave ($l = 0$) phase shift. (You need not know the detailed properties of the spherical Bessel functions to be able to do this simple problem!)
   b. What is the total cross section σ $[\sigma = \int (d\sigma/d\Omega)\, d\Omega]$ in the extreme low-energy limit $k \to 0$? Compare your answer with the geometric cross section πa^2. You may assume without proof:

$$\frac{d\sigma}{d\Omega} = |f(\theta)|^2,$$

$$f(\theta) = \left(\frac{1}{k}\right) \sum_{l=0}^{\infty} (2l+1) e^{i\delta_l} \sin\delta_l P_l(\cos\theta).$$

7. Use $\delta_l = \Delta(b)|_{b=l/k}$ to obtain the phase shift δ_l for scattering at high energies by (a) the Gaussian potential, $V = V_0 \exp(-r^2/a^2)$, and (b) the Yukawa potential, $V = V_0 \exp(-\mu r)/\mu r$. Verify the assertion that δ_l goes to zero very rapidly with increasing l (k fixed) for $l \gg kR$, where R is the "range" of the potential. [The formula for $\Delta(b)$ is given in (7.4.14)].
8. a. Prove

$$\frac{\hbar^2}{2m}\langle \mathbf{x}| \frac{1}{E - H_0 + i\varepsilon} |\mathbf{x}'\rangle = -ik \sum_l \sum_m Y_l^m(\hat{\mathbf{r}}) Y_l^{m*}(\hat{\mathbf{r}}')\, j_l(kr_<) h_l^{(1)}(kr_>)$$

   where $r_<$ ($r_>$) stands for the smaller (larger) of r and r'.
   b. For spherically symmetric potentials, the Lippmann-Schwinger equation can be written for *spherical* waves:

$$|Elm(+)\rangle = |Elm\rangle + \frac{1}{E - H_0 + i\varepsilon} V |Elm(+)\rangle.$$

   Using (a), show that this equation, written in the $\mathbf{x}$-representation, leads to an equation for the radial function, $A_l(k;r)$, as follows:

$$A_l(k;r) = j_l(kr) - \frac{2mik}{\hbar^2} \times \int_0^\infty j_l(kr_<) h_l^{(1)}(kr_>) V(r') A_l(k;r') r'^2\, dr'.$$

By taking r very large, also obtain

$$f_l(k) = e^{i\delta_l}\frac{\sin\delta_l}{k}$$
$$= -\left(\frac{2m}{\hbar^2}\right)\int_0^\infty j_l(kr)A_l(k;r)V(r)r^2\,dr.$$

9. Consider scattering by a repulsive δ-shell potential:

$$\left(\frac{2m}{\hbar^2}\right)V(r) = \gamma\delta(r-R), \quad (\gamma > 0).$$

a. Set up an equation that determines the s-wave phase shift δ_0 as a function of k $(E = \hbar^2k^2/2m)$.
b. Assume now that γ is very large,

$$\gamma \gg \frac{1}{R}, k.$$

Show that if $\tan kR$ is *not* close to zero, the s-wave phase shift resembles the hard-sphere result discussed in the text. Show also that for $\tan kR$ close to (but not exactly equal to) zero, resonance behavior is possible; that is, $\cot\delta_0$ goes through zero from the positive side as k increases. Determine approximately the positions of the resonances keeping terms of order $1/\gamma$; compare them with the bound-state energies for a particle confined *inside* a spherical wall of the same radius,

$$V = 0, \quad r < R; \qquad V = \infty, \quad r > R.$$

Also obtain an approximate expression for the resonance width Γ defined by

$$\Gamma = \frac{-2}{[d(\cot\delta_0)/dE]|_{E=E_r}}$$

and notice, in particular, that the resonances become extremely sharp as γ becomes large. (*Note*: For a different, more sophisticated approach to this problem see Gottfried 1966, 131–141, who discusses the analytic properties of the D_l-function defined by $A_l = j_l/D_l$.)

10. A spinless particle is scattered by a time-dependent potential

$$\mathscr{V}(\mathbf{r},t) = V(\mathbf{r})\cos\omega t.$$

Show that if the potential is treated to first order in the transition amplitude, the energy of the scattered particle is increased or decreased by $\hbar\omega$. Obtain $d\sigma/d\Omega$. Discuss qualitatively what happens if the higher-order terms are taken into account.

11. Show that the differential cross section for the elastic scattering of a fast

electron by the ground state of the hydrogen atom is given by

$$\frac{d\sigma}{d\Omega}=\left(\frac{4m^2e^4}{\hbar^4q^4}\right)\left\{1-\frac{16}{\left[4+(qa_0)^2\right]^2}\right\}^2.$$

(Ignore the effect of identity.)

12. Let the energy of a particle moving in a central field be $E(J_1J_2J_3)$, where (J_1, J_2, J_3) are the three action variables. How does the functional form of E specialize for the Coulomb potential? Using the recipe of the action-angle method, compare the degeneracy of the central field and the Coulomb problems and relate it to the vector **A**.

    If the Hamiltonian is

$$H=\frac{p^2}{2\mu}+V(r)+F(\mathbf{A}^2),$$

how are these statements changed?

Describe the corresponding degeneracies of the central field and Coulomb problems in quantum theory in terms of the usual quantum numbers (n, l, m) and also in terms of the quantum numbers (k, m, n). Here the second set, (k, m, n), labels the wave functions $\mathscr{D}^k_{mn}(\alpha\beta\gamma)$.

How are the wave functions $\mathscr{D}^k_{mn}(\alpha\beta\gamma)$ related to Laguerre times spherical harmonics?

APPENDIX A

Brief Summary of Elementary Solutions to Schrödinger's Wave Equation

Here we summarize the simple solutions to Schrödinger's wave equation for a variety of soluble potential problems.

A.1. FREE PARTICLES ($V=0$)

The plane-wave, or momentum, eigenfunction is

$$\psi_{\mathbf{k}}(\mathbf{x},t)=\frac{1}{(2\pi)^{3/2}}e^{i\mathbf{k}\cdot\mathbf{x}-i\omega t}, \tag{A.1.1}$$

where

$$\mathbf{k}=\frac{\mathbf{p}}{\hbar},\qquad \omega=\frac{E}{\hbar}=\frac{\mathbf{p}^2}{2m\hbar}=\frac{\hbar\mathbf{k}^2}{2m} \tag{A.1.2}$$

and our normalization is

$$\int\psi_{\mathbf{k}'}^{*}\psi_{\mathbf{k}}\,d^3x=\delta^{(3)}(\mathbf{k}-\mathbf{k}'). \tag{A.1.3}$$

The superposition of plane waves leads to the *wave-packet* description. In the one-dimensional case,

$$\psi(x,t)=\frac{1}{\sqrt{2\pi}}\int_{-\infty}^{\infty}dk\,A(k)e^{i(kx-\omega t)}\quad\left(\omega=\frac{\hbar k^2}{2m}\right). \tag{A.1.4}$$

For $|A(k)|$ sharply peaked near $k\simeq k_0$, the wave packet moves with a group velocity

$$v_g \simeq \left(\frac{d\omega}{dk}\right)_{k_0} = \frac{\hbar k_0}{m}. \tag{A.1.5}$$

The time evolution of a minimum wave packet can be described by

$$\psi(x,t) = \left[\frac{(\Delta x)_0^2}{2\pi^3}\right]^{1/4} \int_{-\infty}^{\infty} e^{-(\Delta x)_0^2(k-k_0)^2 + ikx - i\omega(k)t}\, dk, \quad \omega(k) = \frac{\hbar k^2}{2m}, \tag{A.1.6}$$

where

$$|\psi(x,t)|^2 = \left\{\frac{1}{2\pi(\Delta x)_0^2\left[1 + (\hbar^2 t^2/4m^2)(\Delta x)_0^{-4}\right]}\right\}^{1/2}$$
$$\times \exp\left\{-\frac{(x - \hbar k_0 t/m)^2}{2(\Delta x)_0^2\left[1 + (\hbar^2 t^2/4m^2)(\Delta x)_0^{-4}\right]}\right\}. \tag{A.1.7}$$

So the width of the wave packet expands as

$$(\Delta x)_0 \quad \text{at } t = 0 \to (\Delta x)_0\left[1 + \frac{\hbar^2 t^2}{4m^2}(\Delta x)_0^{-4}\right]^{1/2} \quad \text{at } t > 0. \tag{A.1.8}$$

A.2. PIECEWISE CONSTANT POTENTIALS IN ONE DIMENSION

The basic solutions are:

$$E > V = V_0: \quad \psi_E(x) = c_+ e^{ikx} + c_- e^{-ikx}, \qquad k = \sqrt{\frac{2m(E - V_0)}{\hbar^2}}. \tag{A.2.1}$$

$E < V = V_0$ (classically forbidden region):

$$\psi_E(x) = c_+ e^{\kappa x} + c_- e^{-\kappa x}, \quad \kappa = \sqrt{\frac{2m(V_0 - E)}{\hbar^2}} \tag{A.2.2}$$

($c_\pm$ must be set equal to 0 if $x = \pm\infty$ is included in the domain under discussion).

Rigid-Wall Potential (One-dimensional Box)

Here

$$V = \begin{cases} 0 & \text{for } 0 < x < L, \\ \infty & \text{otherwise.} \end{cases} \tag{A.2.3}$$

The wave functions and energy eigenstates are

$$\psi_E(x)=\sqrt{\frac{2}{L}}\sin\left(\frac{n\pi x}{L}\right),\quad n=1,2,3\ldots,$$

$$E=\frac{\hbar^2n^2\pi^2}{2mL^2}. \tag{A.2.4}$$

Square-Well Potential

The potential V is

$$V=\begin{cases}0 & \text{for } |x|>a\\ -V_0 & \text{for } |x|<a \quad (V_0>0)\end{cases} \tag{A.2.5}$$

The bound-state ($E<0$) solutions are:

$$\psi_E\sim\begin{cases}e^{-\kappa|x|} & \text{for } |x|>a,\\ \cos kx \quad \text{(even parity)} & \\ \sin kx \quad \text{(odd parity)} & \text{for } |x|<a,\end{cases} \tag{A.2.6}$$

where

$$k=\sqrt{\frac{2m(-|E|+V_0)}{\hbar^2}},\qquad \kappa=\sqrt{\frac{2m|E|}{\hbar^2}}. \tag{A.2.7}$$

The allowed discrete values of energy $E=-\hbar^2\kappa^2/2m$ are to be determined by solving

$$\begin{aligned} ka\tan ka &= \kappa a \quad \text{(even parity)}\\ ka\cot ka &= -\kappa a \quad \text{(odd parity)}.\end{aligned} \tag{A.2.8}$$

Note also that κ and k are related by

$$\frac{2mV_0a^2}{\hbar^2}=(k^2+\kappa^2)a^2. \tag{A.2.9}$$

A.3. TRANSMISSION—REFLECTION PROBLEMS

In this discussion we define the transmission coefficient T to be the ratio of the flux of the transmitted wave to that of the incident wave. We consider these simple examples.

Square Well ($V=0$ for $|x|>a$, $V=-V_0$ for $|x|<a$.)

$$\begin{aligned} T&=\frac{1}{\left\{1+\left[(k'^2-k^2)^2/4k^2k'^2\right]\sin^2 2k'a\right\}}\\ &=\frac{1}{\left\{1+\left[V_0^2/4E(E+V_0)\right]\sin^2\left(2a\sqrt{2m(E+V_0)/\hbar^2}\right)\right\}}\end{aligned} \tag{A.3.1}$$

where

$$k = \sqrt{\frac{2mE}{\hbar^2}}, \qquad k' = \sqrt{\frac{2m(E+V_0)}{\hbar^2}}. \tag{A.3.2}$$

Note that resonances occur whenever

$$2a\sqrt{\frac{2m(E+V_0)}{\hbar^2}} = n\pi, \quad n = 1,2,3,\ldots. \tag{A.3.3}$$

Potential Barrier ($V = 0$ for $|x| > a$, $V = V_0 > 0$ for $|x| < a$.)

Case 1: $E < V_0$.

$$\begin{aligned} T &= \frac{1}{\left\{1+\left[(k^2+\kappa^2)^2/4k^2\kappa^2\right]\sinh^2 2\kappa a\right\}} \\ &= \frac{1}{\left\{1+\left[V_0^2/4E(V_0-E)\right]\sinh^2\left(2a\sqrt{2m(V_0-E)/\hbar^2}\right)\right\}}. \end{aligned} \tag{A.3.4}$$

Case 2: $E > V_0$. This case is the same as the square-well case with V_0 replaced by $-V_0$.

Potential Step ($V = 0$ for $x < 0$, $V = V_0$ for $x > 0$, and $E > V_0$.)

$$T = \frac{4kk'}{(k+k')^2} = \frac{4\sqrt{(E-V_0)E}}{\left(\sqrt{E}+\sqrt{E-V_0}\right)^2} \tag{A.3.5}$$

with

$$k = \sqrt{\frac{2mE}{\hbar^2}}, \qquad k' = \sqrt{\frac{2m(E-V_0)}{\hbar^2}}. \tag{A.3.6}$$

More General Potential Barrier [$V(x) > E$ for $a < x < b$, $V(x) < E$ outside range $[a,b]$.]

The approximate JWKB solution for T is

$$T \simeq \exp\left\{-2\int_a^b dx\sqrt{\frac{2m[V(x)-E]}{\hbar^2}}\right\}, \tag{A.3.7}$$

where a and b are the classical turning points.*

*JWKB stand for Jeffreys-Wentzel-Kramers-Brillouin.

A.4. SIMPLE HARMONIC OSCILLATOR

Here the potential is

$$V(x) = \frac{m\omega^2 x^2}{2}, \tag{A.4.1}$$

and we introduce a dimensionless variable

$$\xi = \sqrt{\frac{m\omega}{\hbar}}\, x. \tag{A.4.2}$$

The energy eigenfunctions are

$$\psi_E = (2^n n!)^{-1/2}\left(\frac{m\omega}{\pi\hbar}\right)^{1/4} e^{-\xi^2/2} H_n(\xi) \tag{A.4.3}$$

and the energy levels are

$$E = \hbar\omega\left(n + \frac{1}{2}\right), \quad n = 0,1,2,\ldots. \tag{A.4.4}$$

The Hermite polynomials have the following properties:

$$H_n(\xi) = (-1)^n e^{\xi^2} \frac{\partial^n}{\partial \xi^n} e^{-\xi^2}$$

$$\int_{-\infty}^{\infty} H_{n'}(\xi)\, H_n(\xi)\, e^{-\xi^2}\, d\xi = \pi^{1/2} 2^n n!\, \delta_{nn'}$$

$$\frac{d^2}{d\xi^2} H_n - 2\xi \frac{dH_n}{d\xi} + 2nH_n = 0 \tag{A.4.5}$$

$$H_0(\xi) = 1, \quad \mathrm{H}_1(\xi) = 2\xi,$$

$$H_2(\xi) = 4\xi^2 - 2, \; H_3(\xi) = 8\xi^3 - 12\xi,$$

$$H_4(\xi) = 16\xi^4 - 48\xi^2 + 12.$$

A.5. THE CENTRAL FORCE PROBLEM [SPHERICALLY SYMMETRICAL POTENTIAL $V = V(r)$.]

Here the basic differential equation is

$$-\frac{\hbar^2}{2m}\left[\frac{1}{r^2}\frac{\partial}{\partial r}\left(r^2\frac{\partial\psi_E}{\partial r}\right) + \frac{1}{r^2\sin\theta}\frac{\partial}{\partial\theta}\left(\sin\theta\frac{\partial\psi_E}{\partial\theta}\right) + \frac{1}{r^2\sin^2\theta}\frac{\partial^2\psi_E}{\partial\phi^2}\right] + V(r)\psi_E = E\psi_E \tag{A.5.1}$$

where our spherically symmetrical potential $V(r)$ satisfies

$$\lim_{r\to 0} r^2 V(r) \to 0. \tag{A.5.2}$$

The method of separation of variables,

$$\Psi_E(\mathbf{x}) = R(\tilde{x})Y_l^m(\theta,\phi), \tag{A.5.3}$$

leads to the angular equation

$$-\left[\frac{1}{\sin\theta}\frac{\partial}{\partial\theta}\left(\sin\theta\frac{\partial}{\partial\theta}\right)+\frac{1}{\sin^2\theta}\frac{\partial^2}{\partial\phi^2}\right]Y_l^m = l(l+1)Y_l^m, \tag{A.5.4}$$

where the spherical harmonics

$$Y_l^m(\theta,\phi), \quad l=0,1,2,\ldots, m=-l,-l+1,\ldots,+l \tag{A.5.5}$$

satisfy

$$-i\frac{\partial}{\partial\phi}Y_l^m = mY_l^m \tag{A.5.6}$$

and $Y_l^m(\theta,\phi)$ have the following properties:

$$Y_l^m(\theta,\phi) = (-1)^m\sqrt{\frac{2l+1}{4\pi}\frac{(l-m)!}{(l+m)!}}\,P_l^m(\cos\theta)e^{im\phi} \quad \text{for } m \geq 0,$$

$$Y_l^m(\theta,\phi) = (-1)^{|m|}Y_l^{|m|*}(\theta,\phi) \quad \text{for } m<0,$$

$$P_l^m(\cos\theta) = (1-\cos^2\theta)^{m/2}\frac{d^m}{d(\cos\theta)^m}P_l(\cos\theta) \quad \text{for } m \geq 0,$$

$$P_l(\cos\theta) = \frac{(-1)^l}{2^l l!}\frac{d^l(1-\cos^2\theta)^l}{d(\cos\theta)^l},$$

$$Y_0^0 = \frac{1}{\sqrt{4\pi}}, \qquad Y_1^0 = \sqrt{\frac{3}{4\pi}}\cos\theta, \tag{A.5.7}$$

$$Y_1^{\pm 1} = \mp\sqrt{\frac{3}{8\pi}}(\sin\theta)e^{\pm i\phi},$$

$$Y_2^0 = \sqrt{\frac{5}{16\pi}}(3\cos^2\theta-1),$$

$$Y_2^{\pm 1} = \mp\sqrt{\frac{15}{8\pi}}(\sin\theta\cos\theta)e^{\pm i\phi},$$

$$Y_2^{\pm 2} = \sqrt{\frac{15}{32\pi}}(\sin^2\theta)e^{\pm 2i\phi}$$

$$\int Y_{l'}^{m'*}(\theta,\phi)Y_l^m(\theta,\phi)\,d\Omega = \delta_{ll'}\delta_{mm'}\left[\int d\Omega = \int_0^{2\pi}d\phi\int_{-1}^{+1}d(\cos\theta)\right].$$

For the radial piece of (A.5.3), let us define

$$u_E(r) = rR(r). \tag{A.5.8}$$

Then the radial equation is reduced to an equivalent one-dimensional problem, namely,

$$-\frac{\hbar^2}{2m}\frac{d^2u_E}{dr^2}+\left[V(r)+\frac{l(l+1)\hbar^2}{2mr^2}\right]u_E=Eu_E$$

subject to the boundary condition

$$u_E(r)|_{r=0}=0. \tag{A.5.9}$$

For the case of *free* particles $[V(r)=0]$ in our spherical coordinates:

$$R(r)=c_1 j_l(\rho)+c_2 n_l(\rho) \quad (c_2=0 \text{ if the origin is included.}) \tag{A.5.10}$$

where ρ is a dimensionless variable

$$\rho \equiv kr, \qquad k=\sqrt{\frac{2mE}{\hbar^2}}. \tag{A.5.11}$$

We need to list the commonly used properties of the Bessel functions and spherical Bessel and Hankel functions. The spherical Bessel functions are

$$j_l(\rho)=\left(\frac{\pi}{2\rho}\right)^{1/2}J_{l+1/2}(\rho),$$

$$n_l(\rho)=(-1)^{l+1}\left(\frac{\pi}{2\rho}\right)^{1/2}J_{-l-1/2}(\rho),$$

$$j_0(\rho)=\frac{\sin\rho}{\rho}, \qquad n_0(\rho)=-\frac{\cos\rho}{\rho},$$

$$j_1(\rho)=\frac{\sin\rho}{\rho^2}-\frac{\cos\rho}{\rho}, \qquad n_1(\rho)=-\frac{\cos\rho}{\rho^2}-\frac{\sin\rho}{\rho}, \tag{A.5.12}$$

$$j_2(\rho)=\left(\frac{3}{\rho^3}-\frac{1}{\rho}\right)\sin\rho-\frac{3}{\rho^2}\cos\rho,$$

$$n_2(\rho)=-\left(\frac{3}{\rho^3}-\frac{1}{\rho}\right)\cos\rho-\frac{3}{\rho^2}\sin\rho.$$

For $\rho\to 0$, the leading terms are

$$j_l(\rho)\xrightarrow[\rho\to 0]{}\frac{\rho^l}{(2l+1)!!}, \qquad n_l(\rho)\xrightarrow[\rho\to 0]{}-\frac{(2l-1)!!}{\rho^{l+1}}, \tag{A.5.13}$$

where

$$(2l+1)!!\equiv(2l+1)(2l-1)\cdots 5\cdot 3\cdot 1. \tag{A.5.14}$$

In the large ρ-asymptotic limit, we have

$$\begin{aligned} j_l(\rho)&\xrightarrow[\rho\to\infty]{}\frac{1}{\rho}\cos\left[\rho-\frac{(l+1)\pi}{2}\right],\\ n_l(\rho)&\xrightarrow[\rho\to\infty]{}\frac{1}{\rho}\sin\left[\rho-\frac{(l+1)\pi}{2}\right]. \end{aligned} \tag{A.5.15}$$

Because of constraints (A.5.8) and (A.5.9), $R(r)$ must be finite at $r = 0$; hence, from (A.5.10) and (A.5.13) we see that the $n_l(\rho)$-term must be deleted because of its singular behavior as $\rho \to 0$. Thus $R(r) = c_l j_l(\rho)$ [or, in the notation of Chapter VII, Section 7.6, $A_l(r) = R(r) = c_l j_l(\rho)$]. For a three-dimensional square-well potential, $V = -V_0$ for $r < R$ (with $V_0 > 0$), the desired solution is

$$R(r) = A_l(r) = \text{constant } j_l(\alpha r), \tag{A.5.16}$$

where

$$\alpha = \left[\frac{2m(V_0 - |E|)}{\hbar^2}\right]^{1/2}, \quad r < R. \tag{A.5.17}$$

As discussed in (7.6.30), the exterior solution for $r > R$, where $V = 0$, can be written as a linear combination of spherical Hankel functions. These are defined as follows:

$$\begin{aligned} h_l^{(1)}(\rho) &= j_l(\rho) + i n_l(\rho) \\ h_l^{(1)*}(\rho) &= h_l^{(2)}(\rho) = j_l(\rho) - i n_l(\rho) \end{aligned} \tag{A.5.18}$$

which, from (A.5.15), have the asymptotic forms for $\rho \to \infty$ as follows:

$$\begin{aligned} h_l^{(1)}(\rho) &\xrightarrow[\rho \to \infty]{} \frac{1}{\rho} e^{i[\rho - (l+1)\pi/2]} \\ h_l^{(1)*}(\rho) = h_l^{(2)}(\rho) &\xrightarrow[\rho \to \infty]{} \frac{1}{\rho} e^{-i[\rho - (l+1)\pi/2]}. \end{aligned} \tag{A.5.19}$$

If we are interested in the bound-state energy levels of the three-dimensional square-well potential, where $V(r) = 0$, $r > R$, we have

$$\begin{aligned} u_l(r) &= rA_l(r) = \text{constant } e^{-\kappa r} f\left(\frac{1}{\kappa r}\right) \\ \kappa &= \left(\frac{2m|E|}{\hbar^2}\right)^{1/2}. \end{aligned} \tag{A.5.20}$$

To the extent that the asymptotic expansions, of which (A.5.19) give the leading terms, do not contain terms with exponent of opposite sign to that given, we have—for $r > R$—the desired solution from (A.5.20):

$$A_l(r) = \text{constant } h_l^{(1)}(i\kappa r) = \text{constant}\left[j_l(i\kappa r) + i n_l(i\kappa r)\right], \tag{A.5.21}$$

where the first three of these functions are

$$\begin{aligned} h_0^{(1)}(i\kappa r) &= -\frac{1}{\kappa r} e^{-\kappa r} \\ h_1^{(1)}(i\kappa r) &= i\left(\frac{1}{\kappa r} + \frac{1}{\kappa^2 r^2}\right) e^{-\kappa r} \\ h_2^{(1)}(i\kappa r) &= \left(\frac{1}{\kappa r} + \frac{3}{\kappa^2 r^2} + \frac{3}{\kappa^3 r^3}\right) e^{-\kappa r}. \end{aligned} \tag{A.5.22}$$

Finally, we note that in considering the shift from free particles $[V(r)=0]$ to the case of the constant potential $V(r)=V_0$, we need only to replace the E in the free-particle solution [see (A.5.10) and (A.5.11) by $E-V_0$. Note, though, that if $E<V_0$, $h_l^{(1,2)}(i\kappa r)$ is to be used with $\kappa = \sqrt{2m(V_0-E)/\hbar^2}$.

A.6. HYDROGEN ATOM

Here the potential is

$$V(r) = -\frac{Ze^2}{r} \tag{A.6.1}$$

and we introduce the dimensionless variable

$$\rho = \left(\frac{8m_e|E|}{\hbar^2}\right)^{1/2} r. \tag{A.6.2}$$

The energy eigenfunctions and eigenvalues (energy levels) are

$$\psi_{nlm} = R_{nl}(r)Y_l^m(\theta,\phi)$$

$$R_{nl}(r) = -\left\{\left(\frac{2Z}{na_0}\right)^3 \frac{(n-l-1)!}{2n[(n+l)!]^3}\right\}^{1/2} e^{-\rho/2}\rho^l L_{n+l}^{2l+1}(\rho)$$

$$E_n = \frac{-Z^2e^2}{2n^2a_0} \qquad \text{(independent of } l \text{ and } m) \tag{A.6.3}$$

$$a_0 = \text{Bohr radius} = \frac{\hbar^2}{m_e e^2}$$

$$n \geq l+1, \qquad \rho = \frac{2Zr}{na_0}.$$

The associated Laguerre polynomials are defined as follows:

$$L_p^q(\rho) = \frac{d^q}{d\rho^q}L_p(\rho), \tag{A.6.4}$$

where—in particular—

$$L_p(\rho) = e^{\rho}\frac{d^p}{d\rho^p}(\rho^p e^{-\rho}) \tag{A.6.5}$$

and the normalization integral satisfies

$$\int e^{-\rho}\rho^{2l}\left[L_{n+l}^{2l+1}(\rho)\right]^2\rho^2\,d\rho = \frac{2n[(n+l)!]^3}{(n-l-1)!}. \tag{A.6.6}$$

The radial functions for low n are:

$$R_{10}(r) = \left(\frac{Z}{a_0}\right)^{3/2} 2e^{-Zr/a_0}$$

$$R_{20}(r) = \left(\frac{Z}{2a_0}\right)^{3/2} (2 - Zr/a_0)e^{-Zr/2a_0} \tag{A.6.7}$$

$$R_{21}(r) = \left(\frac{Z}{2a_0}\right)^{3/2} \frac{Zr}{\sqrt{3}\,a_0} e^{-Zr/2a_0}.$$

The radial integrals are

$$\langle r^k \rangle \equiv \int_0^\infty dr\, r^{2+k} \, [R_{nl}(r)]^2,$$

$$\langle r \rangle = \left(\frac{a_0}{2Z}\right) [3n^2 - l(l + 1)]$$

$$\langle r^2 \rangle = \left(\frac{a_0^2 n^2}{2Z^2}\right) [5n^2 + 1 - 3l(l + 1)] \tag{A.6.8}$$

$$\left\langle \frac{1}{r} \right\rangle = \frac{Z}{n^2 a_0},$$

$$\left\langle \frac{1}{r^2} \right\rangle = \frac{Z^2}{[n^3 a_0^2 (l + \frac{1}{2})]}.$$

APPENDIX B

Proof of the Angular-Momentum Addition Rule Given by Equation (3.7.38)

It will be instructive to discuss the angular-momentum addition rule from the quantum-mechanical point of view. Let us, for the moment, label our angular momenta, so that $j_1 \geqq j_2$. This we can always do. From Equation (3.7.35), the maximum value of m, $m^{\max}$ is

$$m^{\max} = m_1^{\max} + m_2^{\max} = j_1 + j_2. \tag{B.1.1}$$

There is only one ket that corresponds to the eigenvalue $m^{\max}$, whether the description is in terms of $|j_1 j_2; m_1 m_2\rangle$ or $|j_1 j_2; jm\rangle$. In other words, choosing the phase factor to be 1, we have

$$|j_1 j_2; j_1 j_2\rangle = |j_1 j_2; j_1 + j_2, j_1 + j_2\rangle. \tag{B.1.2}$$

In the $|j_1 j_2; m_1 m_2\rangle$ basis, there are two kets that correspond to the m eigenvalue $m^{\max} - 1$, namely, one ket with $m_1 = m_1^{\max} - 1$ and $m_2 = m_2^{\max}$ and one ket with $m_1 = m_1^{\max}$ and $m_2 = m_2^{\max} - 1$. There is thus a twofold degeneracy in this basis; therefore, there must be a twofold degeneracy in the $|j_1 j_2; jm\rangle$ basis as well. From where could this come? Clearly, $m^{\max} - 1$ is a possible m-value for $j = j_1 + j_2$. It is also a possible m-value for $j = j_1 + j_2 - 1$—in fact, the maximum m-value for this j. So j_1, j_2 can add to j's of $j_1 + j_2$ and $j_1 + j_2 - 1$.

We can continue in this way, but it is clear that the degeneracy cannot increase indefinitely. Indeed, for $m^{\min} = -j_1 - j_2$, there is once again a single ket. The maximum degeneracy is $(2j_2 + 1)$-fold, as is apparent

TABLE B.1. Special Examples of Values of m, m_1, and m_2 for the Two Cases $j_1 = 2,\ j_2 = 1$ and $j_1 = 2,\ j_2 = \frac{1}{2}$, Respectively

$j_1 = 2,\ j_2 = 1$ m	3	2	1	0	-1	-2	-3
(m_1, m_2)	$(2,1)$	$(1,1)$	$(0,1)$	$(-1,1)$	$(-2,1)$		
		$(2,0)$	$(1,0)$	$(0,0)$	$(-1,0)$	$(-2,0)$	
			$(2,-1)$	$(1,-1)$	$(0,-1)$	$(-1,-1)$	$(-2,-1)$
Numbers of States	1	2	3	3	3	2	1
$j_1 = 2,\ j_2 = \frac{1}{2}$ m	$\frac{5}{2}$	$\frac{3}{2}$	$\frac{1}{2}$	$-\frac{1}{2}$	$-\frac{3}{2}$	$-\frac{5}{2}$	
(m_1, m_2)	$(2,\frac{1}{2})$	$(1,\frac{1}{2})$	$(0,\frac{1}{2})$	$(-1,\frac{1}{2})$	$(-2,\frac{1}{2})$		
		$(2,-\frac{1}{2})$	$(1,-\frac{1}{2})$	$(0,-\frac{1}{2})$	$(-1,-\frac{1}{2})$	$(-2,-\frac{1}{2})$	
Numbers of States	1	2	2	2	2	1	

from Table B.1 constructed for two special examples: for $j_1 = 2,\ j_2 = 1$ and for $j_1 = 2,\ j_2 = \frac{1}{2}$. This $(2j_2 + 1)$-fold degeneracy must be associated with the $2j_2 + 1$ states j:

$$j_1 + j_2,\ j_1 + j_2 - 1, \ldots, j_1 - j_2. \tag{B.1.3}$$

If we lift the restriction $j_1 \geqq j_2$, we obtain (3.7.38).

APPENDIX C

Mathematical Details of Coulomb Scattering Formalism

C.1. THE COULOMB SCATTERING AMPLITUDE $f_c(k, \theta)$

Let us go back to the differential equation (7.13.7). The theory of differential equations tells us that the only singularity in the finite u-plane is at $u = 0$.* (We want the solution that is regular at $u = 0$.) We examine $\chi(u)$ in the Laplace transform space, t:

$$\chi(u) = \int_{t_1}^{t_2} e^{ut} f(t)\, dt, \tag{C.1.1}$$

where the path of integration $t_1 \to t_2$ in the complex t-plane will be decided later. If we substitute $\chi(u)$ given by (C.1.1) into (7.13.7), we obtain

$$\begin{aligned}
&\int_{t_1}^{t_2} [ut^2 + (1-u)t - i\gamma] f e^{ut} dt \\
&\quad = \int_{t_1}^{t_2} f\left[(t - i\gamma) + t(t-1)\frac{d}{dt}\right] e^{ut} dt \\
&\quad = t(t-1)e^{ut} f\Big|_{t_1}^{t_2} + \int_{t_1}^{t_2} \left\{(t-i\gamma)f - \frac{d}{dt}[t(t-1)f(t)]\right\} e^{ut} dt = 0,
\end{aligned} \tag{C.1.2}$$

where we have integrated by parts. If we assume that surface terms vanish, evidently

$$(t - i\gamma) f(t) - \frac{d}{dt}[(t-1)tf(t)] = 0, \tag{C.1.3}$$

and we can easily obtain the solution for $f(t)$ as

$$f = At^{i\gamma - 1}(1 - t)^{-i\gamma}, \tag{C.1.4}$$

where A is some constant. As an interim summary we have [from (C.1.1)]

$$\chi(u) = A \int_C e^{ut} t^{i\gamma - 1}(1 - t)^{-i\gamma} dt, \quad \int_{t_1}^{t_2} \to \int_C. \tag{C.1.5}$$

*For the mathematical background, see E. T. Whittaker and G. N. Watson, *A Course of Modern Analysis*, 4th ed. (Cambridge, London: 1935), Chapter XVI.

Now we must choose C. Note that the integrand in (C.1.5) has branch points at $t = 0, 1$ in the complex t-plane; see Figure C.1. For the surface term in (C.1.2) to vanish, we require that the t at the end points contain an imaginary part that goes to $+\infty$. This is because $u = ik(r - z) = ikr(1 - \cos\theta) \equiv i\kappa$ and $\kappa > 0$. Hence,

$$e^{ut} \rightarrow e^{i\kappa t} \xrightarrow[t \rightarrow a + i\infty]{} 0.$$

and surface terms vanish. But there are lots of ways of reaching $t = a + i\infty$! The choice of C indicated in Figure C.1a will give the correct asymptotic forms, as the actual calculations show [see (C.1.10)], and we will adopt this specific contour.

Let $s = ut$. Then (C.1.5) becomes

$$\begin{aligned} \chi(u) &= A \int_{\bar{C}} \frac{ds}{u} e^s \left(\frac{s}{u}\right)^{i\gamma - 1} \left(\frac{u - s}{u}\right)^{-i\gamma} \\ &= A \int_{\bar{C}} ds\, e^s s^{i\gamma - 1} (u - s)^{-i\gamma}, \end{aligned} \tag{C.1.6}$$

where $\bar{C}$ is the s-plane contour depicted in Figure C.2a. We require that $\chi(u)$ is regular at $u = 0$; in fact, it is because

$$\chi(0) = (-1)^{-i\gamma} A \int_{\bar{C}} \frac{e^s}{s}\, ds = (-1)^{-i\gamma} A 2\pi i. \tag{C.1.7}$$

We now evaluate (C.1.6) for $u \rightarrow \infty$—the asymptotic limit—and show that the correct form is obtained. For the curve C_1 in Figure C.2b and for a fixed s, $s = -(s_0 \pm i\varepsilon)$ and $s/u = -(s_0 \pm i\varepsilon)/i\kappa$, but this is small for $\kappa \rightarrow \infty$. So, we can expand $(u - s)$ in powers of s/u. For the curve C_2 in

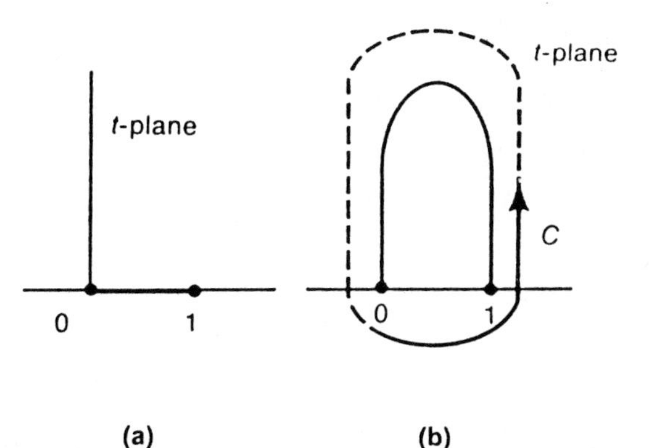

FIGURE C.1. (a) Branch point at $t = 0,1$ and simple straight-line branch cut between (0, 1) in t-plane. (b) "Rubber-band distorted" branch cut between (0, 1) and a possible contour C for (C.1.5).

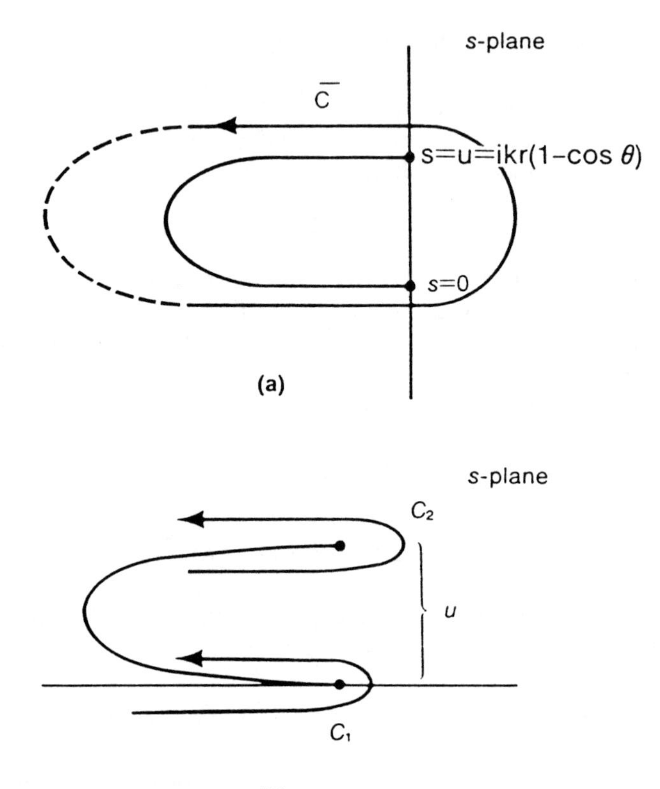

FIGURE C.2. (a) s-plane contour $\overline{C}$ for Equation (C.1.6). (b) Rubber-band distortion of $\overline{C}$ into C_1 plus C_2.

Figure C.2b, $s = -s_0 + i(\kappa \pm \varepsilon)$; hence, $s/u = 1 - (s_0 \pm i\varepsilon)/\kappa \underset{\kappa\to\infty}{\longrightarrow} 1$, so we can not expand in powers of s/u. However, if we change variables $s \to s'$ here, where $s' = s - u$, then expansion about s'/u is possible. Hence, the contour $\overline{C}$ is now replaced by the contour given in Figure C.2b and, further, $s \to s'$ takes $C_2 \to C_1$:

$$\int_{\overline{C}} ds \equiv \int_{C_1} ds + \underbrace{\int_{C_2} ds}_{s = s' + u} = \int_{C_1} ds + \int_{C_1} ds' \tag{C.1.8}$$

and

$$\chi(u) = A \int_{C_1} \{[e^s s^{i\gamma-1}(u - s)^{-i\gamma}]ds + [e^{s'+u}(s' + u)^{i\gamma-1}(-s')^{-i\gamma}]ds'\}. \tag{C.1.9}$$

Here $(s' + u)^{i\gamma - 1}$ is clearly expandable in s'/u, and—since s' is a dummy integration variable for the C_1-contour—we may write (C.1.9) more compactly as

$$\chi(u) = A\left\{u^{-i\gamma}\int_{C_1} e^s s^{i\gamma-1}\left(1 - \left(\frac{s}{u}\right)\right)^{-i\gamma} ds \right.$$
$$\left. + e^u u^{i\gamma-1}(-1)^{-i\gamma}\int_{C_1} e^s\left(1 + \left(\frac{s}{u}\right)\right)^{i\gamma-1} s^{-i\gamma} ds\right\}. \qquad \text{(C.1.10)}$$

Note that the coefficients of the two integrals in (C.1.10), $u^{-i\gamma}$ and $e^u u^{i\gamma-1}(-1)^{-i\gamma}$, do have the correct asymptotic forms [see (7.13.9), (7.13.12), and (7.13.14)] for Coulomb scattering.

The leading terms in (C.1.10) are obtained when we let $s/u \to 0$; then

$$\chi(u) \sim 2\pi i A[u^{-i\gamma}g_1(\gamma) - (-u)^{i\gamma-1}e^u g_2(\gamma)]$$
$$2\pi i g_1(\gamma) = \int_{C_1} e^s s^{i\gamma-1}\,ds \qquad \text{(C.1.11)}$$
$$2\pi i g_2(\gamma) = \int_{C_1} e^s s^{-i\gamma}\,ds.$$

Integrating the expression for $2\pi i g_2(\gamma)$ by parts, we have

$$2\pi i g_2(\gamma) = \int_{C_1} e^s s^{-i\gamma}\,ds = e^s s^{-i\gamma}\Big|_{-\infty - i\varepsilon}^{-\infty + i\varepsilon} - (-i\gamma)\int_{C_1} e^s s^{-i\gamma-1}\,ds$$
$$= +\, i\gamma g_1^*(\gamma)2\pi i. \qquad \text{(C.1.12)}$$

Noting that $-u = u^*$, we find

$$\chi(u) \cong 2\pi i A u^{-i\gamma}g_1(\gamma)\left[1 + \frac{(u^*)^{+i\gamma}}{u^{-i\gamma}}\frac{g_1^*(\gamma)}{g_1(\gamma)}\, i\gamma\,\frac{e^u}{u}\right]$$
$$= 2\pi i A u^{-i\gamma}g_1(\gamma)\left[1 + e^{i\phi(k,u,\gamma)}\, i\gamma\,\frac{e^u}{u}\right], \qquad \text{(C.1.13)}$$

where we have noted that

$$\frac{(u^*)^{+i\gamma}}{u^{-i\gamma}}\frac{g_1^*(\gamma)}{g_1(\gamma)} = \frac{[u^{-i\gamma}g_1(\gamma)]^*}{[u^{-i\gamma}g_1(\gamma)]} = e^{i\phi(k,u,\gamma)}. \qquad \text{(C.1.14)}$$

Using

$$(u^*)^{i\gamma} = (-i)^{i\gamma}[k(r - z)]^{i\gamma} = e^{+\gamma\pi/2}e^{i\gamma \ln k(r-z)} \qquad \text{(C.1.15)}$$
$$(u)^{-i\gamma} = (i)^{-i\gamma}[k(r - z)]^{-i\gamma} = e^{+\gamma\pi/2}e^{-i\gamma \ln k(r-z)}$$

in going from $\chi \to \psi$ [multiplication by e^{ikz} as in (7.13.4)], we obtain the normalized asymptotic $\psi_k(\mathbf{r})$, which is given by Eq. (7.13.15).

C.2. EVALUATION OF $A_l^{(1)}(kr)$ AND $A_l^{(2)}(kr)$ NECESSARY FOR THE PARTIAL WAVE ANALYSIS OF COULOMB SCATTERING

Let us note that

$$\begin{aligned} e^{ikrt} e^{ikr(1-t)} &= e^{ikr}, \\ e^{ikrt} e^{-ikr(1-t)} &= e^{-ikr} e^{2ikrt}, \end{aligned} \tag{C.2.1}$$

so that

$$A_l^{(1)}(kr) = \frac{Ae^{ikr}}{2kr} \int_C E_l[kr(1 - t)] d(t, \gamma) dt. \tag{C.2.2}$$

Since $d(t, \gamma) \sim t^{-2}$ for large $|t|$ and $E_l \sim t^0 \sim 1$ and the only singularities in the t-plane are branch points $t = 0, 1$ and possible poles from E_l when its argument $kr(1 - t)$ vanishes, we close the contour C in the manner shown in Figure C.1b. This leads at once to

$$A_l^{(1)}(kr) = 0, \tag{C.2.3}$$

since—thinking of a stereographic projection of the complex plane on the unit sphere—the contour C can be viewed as surrounding the region exterior to it, where there are no poles.

For $A_l^{(2)}(kr)$, we have

$$A_l^{(2)}(kr) = \frac{Ae^{-ikr}}{2kr} \int_C E_l^* [kr(1 - t)] e^{2ikrt} d(t, \gamma) dt. \tag{C.2.4}$$

We play the same game as we did between (C.1.5) and (C.1.6). Let $\bar{s} = \bar{u}t$, where $\bar{u} = 2ikr$; then $t = \bar{s}/\bar{u}$ and $1 - t = (\bar{u} - \bar{s})/\bar{u}$, so

$$A_l^{(2)}(kr) = \frac{iAe^{-\bar{u}/2}}{\bar{u}} \int_{\bar{C}} e^{\bar{s}} (\bar{s}/\bar{u})^{i\gamma - 1} \left(\frac{\bar{u} - \bar{s}}{\bar{u}}\right)^{-i\gamma - 1} E_l^* \left(\frac{\bar{u} - \bar{s}}{2i}\right) \frac{d\bar{s}}{\bar{u}}, \tag{C.2.5}$$

where $\bar{C}$ is like Figure C.2a except that now the ends of the branch point in the $\bar{s}$-plane are $\bar{s} = 0$ and $\bar{s} = 2ikr$. We recall that $\bar{u}^* = -\bar{u}$, and therefore

$$\begin{aligned} A_l^{(2)}(kr) &= iAe^{-\bar{u}/2} \int_{\bar{C}} e^{\bar{s}} \bar{s}^{i\gamma - 1} (\bar{u} - \bar{s})^{-i\gamma - 1} E_l^* \left(\frac{\bar{u} - \bar{s}}{2i}\right) d\bar{s} \\ &= iAe^{-\bar{u}/2} \int_{C_1} \left[e^s s^{i\gamma - 1} (\bar{u} - s)^{-i\gamma - 1} E_l^* \left(\frac{\bar{u} - s}{2i}\right) \right. \end{aligned}$$

$$+ e^{s+\overline{u}}(s+\overline{u})^{i\gamma-1}(-s)^{-i\gamma-1}E_l^*\left(\frac{-s}{2i}\right)\Bigg]ds$$

$$\cong \frac{iA}{\overline{u}}\left[e^{-\overline{u}/2}\overline{u}^{-i\gamma}\int_{C_1} e^s s^{i\gamma-1}E_l^*\left(\frac{\overline{u}}{2i}\right)ds\right.$$

$$\left.+ e^{\overline{u}/2}\overline{u}^{i\gamma}\int_{C_1} e^s(-s)^{-i\gamma-1}E_l^*\left(\frac{-s}{2i}\right)ds\right]$$

$$\cong \frac{A}{2kr}\left[e^{-\overline{u}/2}\overline{u}^{-i\gamma}\int_{C_1} e^s s^{i\gamma-1}\frac{1}{(-i)^{l+1}}ds\right.$$

$$\left.-(e^{-\overline{u}/2}\overline{u}^{-i\gamma})^*\int_{C_1} e^s s^{-i\gamma-1}E_l^*\left(\frac{-s}{2i}\right)ds\right], \tag{C.2.6}$$

where in the last equation we have used the asymptotic form in $\overline{u}$ for $E_l^*\,(\overline{u}/2i)$ as given in (7.13.24). Note next that $e^{-\overline{u}/2}\,\overline{u}^{-i\gamma} = e^{\pi\gamma/2}e^{-i(kr+\gamma\ln 2kr)}$; hence, (C.2.6) becomes

$$A_l^{(2)}(kr) \simeq -\frac{Ae^{\pi\gamma/2}}{2ikr}\{e^{-i[kr-(l\pi/2)+\gamma\ln 2kr]}2\pi i g_1(\gamma)$$
$$-e^{i[kr-(l\pi/2)+\gamma\ln 2kr]}2\pi i g_3(\gamma)\}, \tag{C.2.7}$$

where

$$2\pi i g_1(\gamma) = \int_{C_1} e^s s^{i\gamma-1}ds = \frac{2\pi i}{\Gamma(1-i\gamma)}$$
$$2\pi i g_3(\gamma) = i^{l-1}\int_{C_1} e^s s^{-i\gamma-1}E_l^*\left(\frac{-s}{2i}\right)ds. \tag{C.2.8}$$

Note $g_1(\gamma)$ and $g_3(\gamma)$ are used in Eq. (7.13.28), and indeed lead to a real phase $\eta_l(k)$.

SUPPLEMENT I

Adiabatic Change and Geometrical Phase

When the author died in 1982, this book was left in manuscript form; subsequently, there have been some new developments in quantum mechanics. The most important development is a definitive formulation of geometrical phases, introduced by M. V. Berry in 1983. The phase factors accompanying adiabatic changes are expressed in concise and elegant forms and have found universal applications in various fields of physics, thus giving a new viewpoint to quantum theory. We review here the physical consequences of these phases, which have in fact been used unconsciously in some cases already, by adding a supplement to the Japanese version of the text. (Here in the new English edition of *Modern Quantum Mechanics* we are providing a translation from Japanese of this supplement, prepared by Professor Akio Sakurai of Kyoto Sangyo University for the Japanese version of the book. The Editor deeply appreciates Professor Akio Sakurai's guidance on an initial translation provided by his student, Yasunaga Suzuki, as a term paper for the graduate quantum mechanics course here at the University of Hawaii—Manoa.)

ADIABATIC THEOREM

Let us first consider a spin with magnitude S in a uniform time-independent magnetic field. Due to the magnetic moment associated with the spin, the energy levels of this quantized system are split into $2S+1$ levels. The state ket $|n\rangle$ with the nth energy eigenvalue E_n undergoes a change in phase factor as (for a Hamiltonian constant in time)

$$|n\rangle \rightarrow e^{-iE_n t/\hbar}\,|n\rangle \tag{S.1}$$

in time evolution, as we learned in Chapter 2. This is the time development of the stationary state. We next change the direction of the magnetic field very slowly. Following the slowly varying magnetic field, the direction of the spin changes also. We keep its component in the direction of the magnetic field constant in time, however. Thus the initial quantum number n does not change at all with the changing magnetic field. This is a typical example of the **adiabatic theorem** in quantum mechanics.

Generally speaking, the adiabatic theorem asserts that there should be mechanical invariants under the slow variation exerted externally on a mechanical system. The entropy in thermodynamic systems and action integral in classical mechanical systems are, like the quantum numbers in quantum mechanical systems, examples of *adiabatic invariants*. Here we simply regard an adiabatic change as a time change that is very slow compared with the time for proper motion in each system (such as periods of oscillation).† We do not discuss how slow change can be regarded as an **adiabatic change** (see A. Messiah, *Quantum Mechanics*, New York: Interscience, 1961, Chapter XVII). Our problem is to discuss how the state ket $|n\rangle$ of a quantum mechanical system changes under adiabatic changes. We find that apart from the simple extension of (S.1), there is another phase factor dependent upon the path of an adiabatic process. The general expression was clearly given by Berry (M. V. Berry, *Proceedings of the Royal Society of London*. Vol. **A392** (1984), p. 45) in his seminal paper.

BERRY'S PHASE

We consider the Hamiltonian of the system with an external time-dependent parameter $\boldsymbol{R}(t)$ by denoting it as $H(\boldsymbol{R}(t))$. The ket $|n(\boldsymbol{R}(t))\rangle$ of the nth energy eigenstate corresponding to $\boldsymbol{R}(t)$ satisfies the eigenvalue equation at time t,

$$H(\boldsymbol{R}(t))\ |n(\boldsymbol{R}(t))\rangle = E_n\ (\boldsymbol{R}(t))\ |n(\boldsymbol{R}(t))\rangle \tag{S.2}$$

where the ket $|n(\boldsymbol{R}(t))\rangle$ has already been normalized. Let us make $\boldsymbol{R}$ evolve in time from $\boldsymbol{R}(0) = \boldsymbol{R}_0$. Suppose that the state ket is $|n(\boldsymbol{R}_0), t_0 = 0;\mathrm{t}\rangle$ at time t. The time-dependent Schrödinger equation that the state ket obeys is

$$H(\boldsymbol{R}(t))\ |n(\boldsymbol{R}_0), t_0; t\rangle = i\hbar \frac{\partial}{\partial t}\ |n(\boldsymbol{R}_0), t_0; t\rangle \tag{S.3}$$

where $t_0 = 0$. When the change of $\boldsymbol{R}(t)$ is slow enough, we expect from the adiabatic theorem that $|n(\boldsymbol{R}_0), t_0; t\rangle$ would be proportional to the nth energy eigenket $|n(\boldsymbol{R}(t))\rangle$ of $H(\boldsymbol{R}(t))$ at time t. Thus we represent it as

$$|n(\boldsymbol{R}_0), t_0; t\rangle = \exp\left\{-\frac{i}{\hbar}\int_0^t E_n\ (\boldsymbol{R}(t'))dt'\right\} \exp(i\gamma_n(t))\ |n(\boldsymbol{R}(t))\rangle. \tag{S.4}$$

The first factor on the r.h.s. is the usual dynamical phase factor summing up all phase changes in stationary states [c.f. (2.6.7)]. On the other hand, by substituting (S.4) into (S.3), the second phase factor $\exp(i\gamma_n(t))$ is shown to be determined by

†For adiabatic invariants in classical periodic systems and Ehrenfest's adiabatic hypothesis, see S. Tomonaga, *Quantum Mechanics I*. (Amsterdam: North-Holland Press, 1962).

$$\frac{d}{dt}\gamma_n(t) = i\langle n(\boldsymbol{R}(t)) \,|\, \boldsymbol{\nabla}_{\boldsymbol{R}} n(\boldsymbol{R}(t)) \rangle \frac{d}{dt}\boldsymbol{R}(t). \tag{S.5}$$

Therefore, $\gamma_n(t)$ is represented by a path integral in the parameter ($\boldsymbol{R}-$) space:

$$\gamma_n(t) = i\int_{\boldsymbol{R}_0\ \mathrm{C}}^{\boldsymbol{R}(t)} \langle n(\boldsymbol{R}(t')) \,|\, \boldsymbol{\nabla}_{\boldsymbol{R}} n(\boldsymbol{R}(t')) \rangle \, d\boldsymbol{R}(t') \tag{S.6}$$

where the path C of integration is that of the adiabatic process as the external parameter $\boldsymbol{R}$ changes from $\boldsymbol{R}_0$ to $\boldsymbol{R}(t)$. We can see that γ_n is a real number by differentiating with respect to $\boldsymbol{R}$ both sides of the normalization condition:

$$\langle n(\boldsymbol{R}(t)) \,|n(\boldsymbol{R}(t))\rangle = 1 \tag{S.7}$$

($\langle n \,|\, \boldsymbol{\nabla}_{\boldsymbol{R}} n\rangle$ is pure imaginary.) Since the integral (S.6) contains the derivative of the eigenstate $|\boldsymbol{\nabla}_{\boldsymbol{R}} n\rangle$, the practical evaluation here might be expected to be complicated. However, if $\boldsymbol{R}$ describes a closed loop in parameter space returning to the starting point after time interval T (the periodicity), as is in the case of $\boldsymbol{R}(T) = \boldsymbol{R}_0$, the path integral becomes greatly simplified. First applying the Stoke's theorem, we transform the line integral along the closed loop C to a surface integral over the surface S(C) enclosed by the closed loop C:

$$\gamma_n\,(\mathrm{C}) = i\oint_{\mathrm{C}} \langle n(\boldsymbol{R}) \,|\, \boldsymbol{\nabla}_{\boldsymbol{R}} n(\boldsymbol{R})\rangle d\boldsymbol{R} = -\iint_{S(\mathrm{C})} \boldsymbol{V}_n(\boldsymbol{R})\cdot d\boldsymbol{S}, \tag{S.8}$$

$$\begin{aligned}\boldsymbol{V}_n\,(\boldsymbol{R}) &= \mathrm{Im}\ \boldsymbol{\nabla}_{\boldsymbol{R}} n \times \langle n(\boldsymbol{R}) \,|\, \boldsymbol{\nabla}_{\boldsymbol{R}} n(\boldsymbol{R})\rangle \\ &= \mathrm{Im}\ \langle \boldsymbol{\nabla}_{\boldsymbol{R}} n(\boldsymbol{R}) \,|\times|\, \boldsymbol{\nabla}_{\boldsymbol{R}} n(\boldsymbol{R})\rangle \\ &= \mathrm{Im} \sum_{m\neq n} \langle \boldsymbol{\nabla}_{\boldsymbol{R}} n(\boldsymbol{R}) \,|m(\boldsymbol{R})\rangle \times \langle m(\boldsymbol{R}) \,|\, \boldsymbol{\nabla}_{\boldsymbol{R}} n(\boldsymbol{R})\rangle,\end{aligned} \tag{S.9}$$

where we have used the vector identity:

$$\boldsymbol{\nabla}\times[f(\boldsymbol{x})\boldsymbol{\nabla} g(\boldsymbol{x})] = (\boldsymbol{\nabla} f(\boldsymbol{x})) \times (\boldsymbol{\nabla} g(\boldsymbol{x})). \tag{S.10}$$

Suppose $|n(\boldsymbol{R})\rangle$ is modified by some phase factor $\exp[i\chi(\boldsymbol{R})]$. Then, even if the integrand of line integral (S.8) changes as

$$\langle n(\boldsymbol{R})\boldsymbol{\nabla}_{\boldsymbol{R}} n(\boldsymbol{R})\rangle \rightarrow \langle n(\boldsymbol{R})\boldsymbol{\nabla}_{\boldsymbol{R}} n(\boldsymbol{R})\rangle + i\boldsymbol{\nabla}_{\boldsymbol{R}}\chi(\boldsymbol{R}), \tag{S.11}$$

$\boldsymbol{V}_n(\boldsymbol{R})$ does not change because $\boldsymbol{\nabla}\times\boldsymbol{\nabla}\chi=0$ (see Section 2.6 on gauge invariance). We have omitted the term $m=n$ in the last line of (S.9) because

$\langle n|\boldsymbol{\nabla}_{\boldsymbol{R}}n\rangle$ is purely imaginary. On the other hand, the off-diagonal matrix element $\langle m|\boldsymbol{\nabla}_{\boldsymbol{R}}n\rangle$ is represented as follows. Multiplying $\langle m|$ from the left to the equation

$$(\boldsymbol{\nabla}_{\boldsymbol{R}}H(\boldsymbol{R}))\,|n(\boldsymbol{R})\rangle + H(\boldsymbol{R})\,(\boldsymbol{\nabla}_{\boldsymbol{R}}\,|n(\boldsymbol{R})\rangle) = (\boldsymbol{\nabla}_{\boldsymbol{R}}E(\boldsymbol{R}))\,|n(\boldsymbol{R})\rangle + E_n(\boldsymbol{R})\,|\boldsymbol{\nabla}_{\boldsymbol{R}}n(\boldsymbol{R})\rangle \tag{S.12}$$

obtained by differentiating the eigenvalue equation (S.2) with respect to the external parameter $\boldsymbol{R}$, we get

$$\langle m(\boldsymbol{R})\,|\boldsymbol{\nabla}_{\boldsymbol{R}}n(\boldsymbol{R})\rangle = \frac{\langle m(\boldsymbol{R})\,|\boldsymbol{\nabla}_{\boldsymbol{R}}H(\boldsymbol{R})\,|n(\boldsymbol{R})\rangle}{E_n - E_m},\ m\neq n. \tag{S.13}$$

Therefore the integrand of the surface integral (S.8) is given by

$$\boldsymbol{V}_n(\boldsymbol{R}) = \operatorname{Im}\sum_{m\neq n}\frac{\langle n(\boldsymbol{R})\,|\boldsymbol{\nabla}_{\boldsymbol{R}}H(\boldsymbol{R})\,|m(\boldsymbol{R})\rangle\times\langle m(\boldsymbol{R})\,|\boldsymbol{\nabla}_{\boldsymbol{R}}H(\boldsymbol{R})\,|n(\boldsymbol{R})\rangle}{(E_m(\boldsymbol{R})-E_n(\boldsymbol{R}))^2}. \tag{S.14}$$

Here we notice that the vector $\boldsymbol{V}_n(\boldsymbol{R})$ passing through the surface of the integral $S(\mathrm{C})$ is expressed by the differential form of degree 2 of the Hamiltonian $H(\boldsymbol{R})$. Knowledge of the state ket $|n(\boldsymbol{R})\rangle$ itself is usually not needed to calculate $\boldsymbol{V}_n$ as we shall see later. To sum up, the state ket after one complete turn is expressed as

$$|n(\boldsymbol{R}_0),\,t_0 = 0;\,T\rangle = \exp(i\gamma_n(\mathrm{C}))\exp\left\{-\frac{i}{\hbar}\int_0^T E_n\,(\boldsymbol{R}(t'))dt'\right\}|n(\boldsymbol{R}_0)\rangle, \tag{S.15}$$

$$\gamma_n(\mathrm{C}) = -\iint_{S(\mathrm{C})}\boldsymbol{V}_n(\boldsymbol{R})\cdot d\boldsymbol{S}. \tag{S.16}$$

Here, $\gamma_n(\mathrm{C})$ is called the **Berry's phase**, and comes from the adiabatic change of the external parameter. We shall next show that this phase is determined by the *geometric properties* in parameter space.

DEGENERACY POINT AND SOLID ANGLE—GEOMETRICAL PHASE FACTOR

Suppose that the nth energy eigenstate $|n(\boldsymbol{R})\rangle$ is degenerate with the mth eigenstate $|m(\boldsymbol{R})\rangle$ for an external parameter $\boldsymbol{R}^*$. Although the *degeneracy* *point* $\boldsymbol{R}^*$ in the parameter space is by no means on the path of the actual adiabatic change, the existence of $\boldsymbol{R}^*$ is essential for the value of $\gamma_n(C)$. This is because the integrand $\boldsymbol{V}_n(\boldsymbol{R})$ of $\gamma_n(C)$ is singular at $\boldsymbol{R}^*$, as is evident from (S.14).

Let us continue our discussion by taking as an example, the problem of the spin in a magnetic field mentioned earlier (spin magnitude S, and magnetic moment $g\mu\boldsymbol{S}/\hbar$.[†] When the magnetic field $\boldsymbol{B}(t)$ slowly changes its direction while keeping its magnitude constant, we denote the spin component as $S_{z'} = m\hbar$ ($m = -S, \ldots, +S$) along the direction of the magnetic field at each instance (taking this to be the positive direction of the z'-axis). Writing the Hamiltonian of this system as

$$H(\boldsymbol{B}) = -g\mu\boldsymbol{S}\cdot\boldsymbol{B}(t)/\hbar, \tag{S.17}$$

the energy eigenvalue is

$$E_m = -g\mu mB \tag{S.18}$$

and

$$\boldsymbol{\nabla}_{\boldsymbol{B}}H(\boldsymbol{B}) = -g\mu\boldsymbol{S}/\hbar. \tag{S.19}$$

The degeneracy point is at the origin of the parameter space, $\boldsymbol{B} = 0$, where there are $2S+1$ fold degeneracies. When we rotate the direction of the magnetic field $\boldsymbol{B}$ keeping its magnitude constant and bring it back to the initial direction again, the vector $\boldsymbol{B}$ goes around the closed loop on a sphere of radius B. Then the vector $\boldsymbol{V}_m(\boldsymbol{B})$ can be expressed from (S.14) as

$$\boldsymbol{V}_m(\boldsymbol{B}) = \mathrm{Im}\sum_{m'\neq m}\frac{\langle m(\boldsymbol{B})\,|\boldsymbol{S}/\hbar|m'(\boldsymbol{B})\rangle\times\langle m'(\boldsymbol{B})|\boldsymbol{S}/\hbar|m(\boldsymbol{B})\rangle}{B^2(m'-m)^2}. \tag{S.20}$$

In order to make the calculation easier, let us take the z-axis of spin space to be the direction of the magnetic field (z'-axis direction). Since only the terms with $m' = m\pm 1$ contribute to the intermediate states, we use the expression (3.5.41) for the matrix element of the spin operator:

$$\begin{aligned}\langle S, m\pm 1\,|S_x/\hbar|\,S, m\rangle &= \tfrac{1}{2}\sqrt{(S\mp m)(S\pm m+1)},\\ \langle S, m\pm 1\,|S_y/\hbar|\,S, m\rangle &= \frac{\mp i}{2}\sqrt{(S\mp m)(S\pm m+1)}.\end{aligned} \tag{S.21}$$

By substituting (S.21) into (S.20), we obtain the components of the vector $\boldsymbol{V}_m(\boldsymbol{B})$ as follows:

$$(\boldsymbol{V}_m(\boldsymbol{B}))_{x'} = 0,\ (\boldsymbol{V}_m(\boldsymbol{B}))_{y'} = 0,\ (\boldsymbol{V}_m(\boldsymbol{B}))_{z'} = \frac{m}{B^2}. \tag{S.22}$$

[†]Here spin magnitude S is in unit of $\hbar$, while vector $\boldsymbol{S}$ has dimension of $\hbar$.

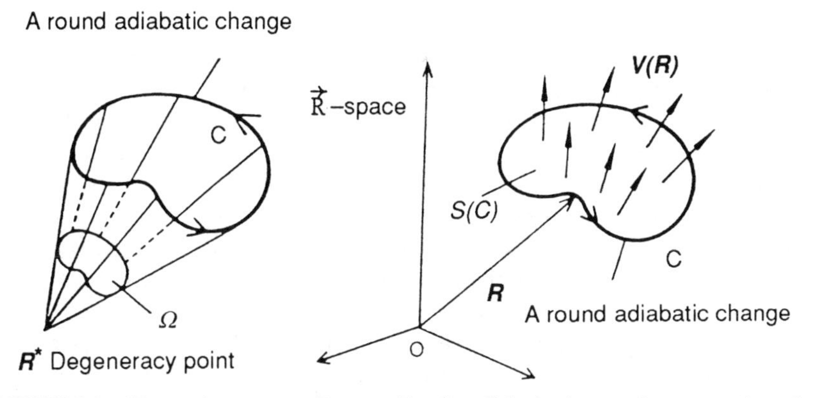

FIGURE S.1. External parameter ***R***-space. For the adiabatic change of one turn along the closed loop C, the *geometrical phase* is given by the surface integral of the vector ***V***(***R***) penetrating through the surface S(C) surrounded by C (right). This is proportional to the solid angle that the closed loop C subtends at the degeneracy point ***R**** (left).

Thus the vector $\boldsymbol{V}(\boldsymbol{B})$ is parallel to the magnetic field $\boldsymbol{B}$ and penetrates normally through the spherical surface of the $\boldsymbol{B}$-space. Hence the phase $\gamma_m(\mathrm{C})$ is given by

$$\gamma_m(\mathrm{C}) = -m \iint_{S(\mathrm{C})} \frac{\hat{\mathbf{B}}}{B^2} \cdot d\boldsymbol{B} = -m \int_{C} d\Omega = -m\Omega(\mathrm{C}) \qquad \text{(S.23)}$$

where $\Omega(\mathrm{C})$ is the *solid angle* which the closed circuit C subtends at the degeneracy point. This simple expression for $\gamma_m(\mathrm{C})$ is dependent neither on the magnitude of the magnetic field nor on the spin which determines the dynamics of the system, but exclusively on the dimensionless quantum number m and the solid angle $\Omega(\mathrm{C})$. The phase factor $\exp(i\gamma_m(\mathrm{C}))$ accompanying the adiabatic change along the closed loop is regarded as a **geometrical phase factor** associated with the degeneracy point that is singular. Figure S.1 illustrates this integral for the geometrical phase in the parameter space.

According to (S.23), the state ket changes its sign if $m\Omega(\mathrm{C}) = \pm\pi$ after one round trip of an adiabatic process. We notice that this is the case not only for fermions with m = half integer but also for bosons with m = integer.

EXPERIMENTAL VERIFICATION

The experiment to verify (S.23) has been performed by letting polarized neutrons pass through a coil which produces a spiral magnetic field inside. The neutrons, having passed through one complete period of the spiral field after time interval T, feel the same field as the initial ones. The solid angle on the r.h.s. of (S.23) can be continuously changed by changing the ratio of the magnitude of the spiral field to the magnitude of another field applied along the direction of the spiral axis. Let us assume that the initial spin of neutrons is in a mixed state of two eigenstates: parallel and anti-parallel to the field. Under an adiabatic change the *ratio* of the numbers of these two states is preserved. The neutrons passing through the inside of the coil proceed while rotating their spins in the plane perpendicular to the field whose direction changes along the course [see (3.2.20)]. After passing through one period of the spiral field, the Berry's phase $\mp\Omega/2$ is added according to (S.23) to the phases of the spin eigenstates parallel or anti-parallel to the field. As mentioned in Section 3.2, twice as many of them contribute to the rotational angle of the spin expectation value. Therefore, besides the dynamical phase, the contribution from geometrical phase is added in the total angle of spin rotation, ϕ, during one period of the field. To wit,

$$\phi = g\mu BT/\hbar + \Omega. \tag{S.24}$$

The value of ϕ is obtained from the measurement of the neutron polarization. The measured value of $\Delta\phi = \phi - g\mu BT/\hbar$ closely agrees with Ω determined by the experimental condition (the ratio of the fields for two directions). Thus, the validity of (S.23) for the Berry's phase was confirmed (see Fig. S.2). This experiment was performed by T. Bitter and D. Dubbers in 1987.

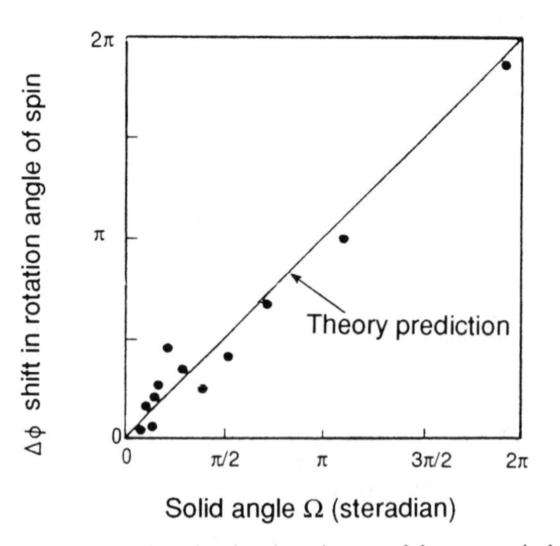

FIGURE S.2. Experimental data showing the existence of the geometrical phase. The shift $\Delta\phi$ in the rotation angle of the spin of neutrons that have passed the spiral magnetic field is almost equal to the solid angle Ω determined by the applied fields. The solid line is the prediction of the theory.

GENERAL CONSIDERATION OF TWO-LEVEL CROSSING

For general cases with external parameter $\boldsymbol{R} = (x, y, z)$, the same situation as the above example of spin in the field can be realized around the degeneracy point $\boldsymbol{R}^*$ if there are some crossing levels. Choosing $\boldsymbol{R}^*$ as the origin of the parameter space, we suppose that the energy eigenstates $|+(R)\rangle$ and $|-(R)\rangle$ are degenerate at $\boldsymbol{R}^* = 0$. Taking each eigenvalue to be such that $E_+(\boldsymbol{R}) \geq E_-(\boldsymbol{R})$, we measure energy from the degeneracy point. We expand the Hamiltonian $H(\boldsymbol{R})$ around the origin up to the first order in $\boldsymbol{R}$, and perform an appropriate linear transformation. Then the general Hamiltonian connecting the two levels is expressed by a 2×2 Hermitian matrix;

$$H(\boldsymbol{R}) = \tfrac{1}{2}\begin{pmatrix} Z & X-iY \\ X+iY & -Z \end{pmatrix} = \tfrac{1}{2}\,\boldsymbol{R}\cdot\boldsymbol{\sigma}. \tag{S.25}$$

The energy eigenvalues are $E_{\pm}(\boldsymbol{R}) = \pm\frac{1}{2}R$ and the two levels *conically* cross at the degeneracy point. $\boldsymbol{\nabla}_{\boldsymbol{R}}H(\boldsymbol{R})$ is equal to $\frac{1}{2}\boldsymbol{\sigma}$. These expressions are the same as those of the case $m = \pm\frac{1}{2}$ for the spin in the field. Therefore, by integrating over surface $\boldsymbol{S}$

$$\boldsymbol{V}_{\pm}(\boldsymbol{R}) = \pm\frac{\hat{\boldsymbol{R}}}{2R^2}, \tag{S.26}$$

the geometrical factor for general two-level crossing is obtained as:

$$\exp(i\gamma_{\pm}(\mathrm{C})) = \exp\left\{-i\iint_{S(\mathrm{C})} \boldsymbol{V}_{\pm}(\boldsymbol{R})\cdot d\boldsymbol{S}\right\} = \exp\left(\mp\frac{i}{2}\Omega\,(\mathrm{C})\right). \tag{S.27}$$

As before, $\Omega(\mathrm{C})$ is the solid angle of the closed loop (adiabatic path) C, at the degeneracy point $\boldsymbol{R}^*$.

We notice here that the expression (S.26) for the vector $\boldsymbol{V}_{\pm}(\boldsymbol{R})$ is equivalent to the magnetic field (2.6.74) produced by a magnetic monopole placed at the origin. The singularity of the degeneracy point in the parameter space is structurally the same as that of a magnetic monopole, and the magnetic field by a magnetic monopole is regarded as determining the geometrical phase. We can exploit further the analogy to the magnetic field; for example, in comparing the two following relations

$$\boldsymbol{B}(\boldsymbol{r}) = \boldsymbol{\nabla} \times \boldsymbol{A}(\boldsymbol{r}) \tag{S.28}$$

$$\boldsymbol{V}(\boldsymbol{R}) = \mathrm{Im}\,\boldsymbol{\nabla}_{\boldsymbol{R}} \times \langle n(\boldsymbol{R})\,|\,\boldsymbol{\nabla}_{\boldsymbol{R}} n(\boldsymbol{R})\rangle, \tag{S.29}$$

we can interpret $\langle n(\boldsymbol{R}) \mid \nabla_{\boldsymbol{R}} n(\boldsymbol{R})\rangle$ as a kind of vector-potential caused by the singularity at the degeneracy point. As we have shown in (S.11), this quantity is gauge dependent, but $\boldsymbol{V}(\boldsymbol{R})$, which corresponds to the magnetic flux density, $\boldsymbol{B}$, is gauge-invariant.

AHARANOV-BOHM EFFECT REVISITED

In Section 2.6 we studied some examples where integrals along paths determine the phase factors. Among them the effect of electric as well as gravitational fields gives rise to dynamical phase factors. On the other hand, the Aharanov-Bohm effect due to the magnetic field can be shown to be just a consequence of a geometrical phase factor.

Let a small box confining an electron (charge $e<0$) make one turn along a closed loop C, which surrounds a magnetic flux line Φ_B (see Fig. S.3). We take as $\boldsymbol{R}$ the vector connecting the origin fixed in the space and a reference point in the box. In this case the vector $\boldsymbol{R}$ is an external parameter in the real space itself. By using the vector-potential $\boldsymbol{A}$ to describe the magnetic field $\boldsymbol{B}$, the nth wave function of the electron in the box (position vector $\boldsymbol{r}$) is written as

$$\langle \boldsymbol{r}|n(\boldsymbol{R})\rangle = \exp\left\{\frac{ie}{\hbar c}\int_{\boldsymbol{R}}^{\boldsymbol{r}} \boldsymbol{A}(\boldsymbol{r}')\cdot d\boldsymbol{r}'\right\} \psi_n(\boldsymbol{r}-\boldsymbol{R}) \tag{S.30}$$

where $\psi_n(\boldsymbol{r}')$ is the wave function of the electron at the $\boldsymbol{r}'$ position coordinates of the box in the absence of magnetic field. Then, in the presence

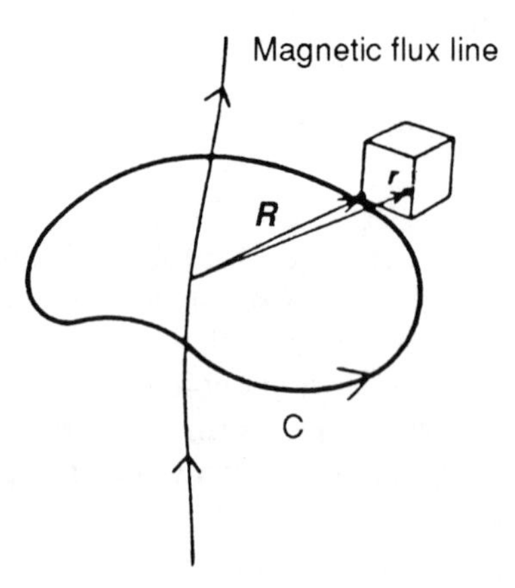

FIGURE S.3. The Aharanov-Bohm Effect. Electron in a box takes one turn around a magnetic flux line.

of magnetic field, we can easily calculate the derivative of the wave function with respect to the external parameter to obtain

$$\langle n(\boldsymbol{R}) \mid \boldsymbol{\nabla_R} n(\boldsymbol{R})\rangle = \int d^3 x \psi_n^* (\boldsymbol{r}-\boldsymbol{R})$$
$$\times \left\{ -\frac{ie}{\hbar c} \boldsymbol{A}(\boldsymbol{R})\, \psi_n(\boldsymbol{r}-\boldsymbol{R}) + \boldsymbol{\nabla_R} \psi_n(\boldsymbol{r}-\boldsymbol{R}) \right\} = -\frac{ie\boldsymbol{A}(\boldsymbol{R})}{\hbar c}. \quad \text{(S.31)}$$

The second term under the integral vanishes for the electron in the box. From (S.6) and (S.31) we see that the geometrical phase is given by

$$\gamma_n(\mathrm{C}) = \frac{e}{\hbar c} \oint_{\mathrm{C}} \boldsymbol{A} \cdot d\boldsymbol{R} = \frac{e}{\hbar c} \iint_{S(\mathrm{C})} \boldsymbol{B}(\boldsymbol{R}) \cdot d\boldsymbol{S} = \frac{e}{\hbar c} \Phi_B. \quad \text{(S.32)}$$

This result is just the expression (2.6.70) of the Aharanov-Bohm effect obtained earlier in Chapter 2.

COMPLEX SYSTEM AND ADIABATIC POTENTIAL

Up to now, the quantity $\boldsymbol{R}$ that changes adiabatically is an external parameter. A more interesting question is to treat quantum mechanically some complex system made up of two constituent systems, where an adiabatically changing quantity is contained in the system as an internal dynamical variable. We shall discuss how the appearance of the Berry's phase in one of the constituent systems reversely influences the other constituent that moves slowly. The former can respond quickly to an adiabatic change caused by the latter. For background, let us first consider the problem of a polyatomic molecule which is composed of electrons and nuclei, a typical example of a complex system. When calculating the energy levels of rotation and vibration of a molecule, we shall apply a method that has been used for many years.

Consider the case of a diatomic molecule; we write the Hamiltonian of the complex system as

$$H_T = H_N(\boldsymbol{R}) + H_{el}(\boldsymbol{r}_1, \ldots, \boldsymbol{r}_N) + V(\boldsymbol{R}, \boldsymbol{r}_1, \ldots, \boldsymbol{r}_N), \quad \text{(S.33)}$$

where $\boldsymbol{R}$ is the relative position coordinate between the two nuclei, $\boldsymbol{r}_1, \boldsymbol{r}_2, \ldots, \boldsymbol{r}_N$ the N-electron coordinates, and the center of mass of the molecule is chosen at the origin. The first, second, and third terms of the Hamiltonian correspond to the kinetic energy of nuclei, the energy of the electron system, and the interaction energies among nuclei themselves and between nuclei and electrons respectively. Since the nuclei with heavy mass move relatively slowly and electrons with light mass move quickly, we can use an **adiabatic approximation** here. We first fix the position of the nuclei and determine the eigenstate of electrons $|n(\boldsymbol{R})\rangle$ corresponding to the fixed configuration of the nuclei

$$\{H_{el}(\boldsymbol{r}_1, \ldots, \boldsymbol{r}_N) + V(\boldsymbol{R}, \boldsymbol{r}_1, \ldots, \boldsymbol{r}_N)\} |n(\boldsymbol{R})\rangle = U_n(\boldsymbol{R}) |n(\boldsymbol{R})\rangle, \quad \text{(S.34)}$$

where $U_n(\boldsymbol{R})$ is the energy of the nth eigenstate. We next approximate the state ket of the complex system in the form of a product of the nucleus ket $|A\rangle$ and the electron ket

$$\Psi = |A\rangle |n(\boldsymbol{R})\rangle. \quad \text{(S.35)}$$

Substituting this into the Schrödinger equation $H_T \Psi = E\Psi$ of the complex system, we obtain

$$(H_N(\boldsymbol{R}) + U_n(\boldsymbol{R})) |A\rangle |n(\boldsymbol{R})\rangle = E_n |A\rangle |n(\boldsymbol{R})\rangle. \quad \text{(S.36)}$$

Since $H_N(\boldsymbol{R})$ is the differential operator with respect to coordinates of the nuclei $\boldsymbol{R}$, it will operate on both the nucleus state ket $|A\rangle$ and the electron state ket $|n(\boldsymbol{R})\rangle$. However, we can ignore the derivative term of electron state ket with respect to $\boldsymbol{R}$, as long as the amplitude of the relative motion of the two nuclei is small compared to the equilibrium distance between them. This approximation is the so-called **Born-Oppenheimer approximation**. Equation (S.36) then leads to a simple eigenvalue equation that involves only the coordinates of the nuclei, as follows:

$$(H_N(\boldsymbol{R}) + U_n(\boldsymbol{R})) |A\rangle = E_n |A\rangle. \quad \text{(S.37)}$$

In the above equation the effective potential $U_n(\boldsymbol{R})$ that determines the motion of the nuclei is called **adiabatic potential**. It is the sum of the Coulomb repulsive force between nuclei and the energy of electrons for fixed R. $U_n(R)$ is obtained for each level of the electron system. The minimum $U_n(R)$ gives approximately the equilibrium distance between nuclei about which the molecule vibrates and rotates. The corresponding energy levels are given approximately by

$$\begin{aligned} E_n &\cong \hbar\omega_n(N + \tfrac{1}{2}) + \frac{\hbar^2 L(L+1)}{2I_n} \\ N &= 0,1,2, \ldots, \; L = 0,1,2, \ldots, \end{aligned} \quad \text{(S.38)}$$

where ω_n stands for the angular frequency determined by the curvature at the minimum of $U_n(R)$ and I_n is the moment of inertia of the molecule. If the two nuclei are identical, L is further limited to be an even or odd number by permutation symmetry.

DYNAMICAL JAHN-TELLER EFFECT

Our object is to investigate the role of the geometrical phase in a complex system. To this end, let us consider the case in which the electronic states are degenerate for a certain space configuration $\boldsymbol{R}^*$ of the nuclei; that is, for a case such that the adiabatic potentials satisfy

$$U_n(\boldsymbol{R}^*) = U_m(\boldsymbol{R}^*) \tag{S.39}$$

and cross one another at $\boldsymbol{R}^*$. Let us remember that an adiabatic change of one turn around the degeneracy point $\boldsymbol{R}^*$ in the $\boldsymbol{R}$-space causes a geometrical phase. Thus for the case of the diatomic molecule where the distance associated with the nuclei is a single one-dimensional parameter in determining the electron state, this is not enough to discuss the Berry's phase. We need to consider tri-atomic molecules (ions) or a group of atoms in a crystal. Here we use the latter one as an example for discussion purpose. Generally in crystals the electronic states in the crystal fields (e.g., a pair of d-orbital states) are often degenerate in energy for certain atomic configurations of high symmetry. Then, in most cases, the crystal becomes slightly distorted to assume a lower symmetric configuration, and thus moves in a more stable state by lowering its electronic energy. This phenomenon is what is called the **Jahn-Teller effect**. To make the argument simpler, consider the electronic state of the central ion surrounded by four atoms in a plane. Suppose that the four atoms are placed in a square configuration, for which electronic states $|a\rangle$ and $|b\rangle$ extending their orbitals in different directions respectively are degenerate. The deformation of the lattice then removes the two-fold degeneracy. Displacements of atoms from the equilibrium position, taken on the square lattice points, are expressed in terms of the normal modes of vibration. Among the several possible normal modes, we look at the two independent modes A and B as shown in Fig. S.4, which couple with the degenerate electronic states. Suppose that A and B have the same eigen frequency, ω, and we denote their normal coordinates as Q_a and Q_b, respectively.† On the other hand, the

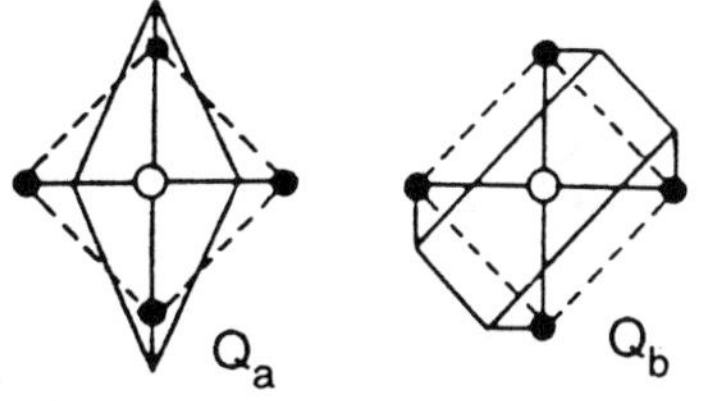

FIGURE S.4. Distortion of lattices causing dynamical Jahn-Teller effect. Two normal modes, Q_a and Q_b, are shown.

†Our explanation for the Jahn-Teller effect is not exact but conceptual.

electron system is described by a linear-combination of the two degenerate levels

$$|\alpha\rangle = C_a|a\rangle + C_b|b\rangle. \tag{S.40}$$

Then the effective Hamiltonian of the normal vibration, which couples with the electron system, can be written in the form of

$$\frac{P_a^2 + P_b^2}{2\mu} + \frac{\mu\omega^2}{2}(Q_a^2 + Q_b^2) - K\begin{pmatrix} -Q_a & Q_b \\ Q_b & Q_a \end{pmatrix} \tag{S.41}$$

where P_a and P_b are the conjugate momenta to Q_a and Q_b, respectively; μ is the effective mass of the normal vibration and K is the strength of the coupling between electrons and lattices. The matrix that represents the interaction operates on the electron state vector $\binom{C_a}{C_b}$. We have expressed the interaction by the first order terms of the normal coordinates, Q_a and Q_b, assuming that the vibration of the atomic displacement is small. For normal coordinates written as

$$Q_a = \rho \cos\theta,\ Q_b = \rho \sin\theta, \tag{S.42}$$

the eigenenergy ε and the eigenket of the electron system are obtained by solving

$$-K\rho\begin{pmatrix} -\cos\theta & \sin\theta \\ \sin\theta & \cos\theta \end{pmatrix}\begin{pmatrix} C_a \\ C_b \end{pmatrix} = \varepsilon\begin{pmatrix} C_a \\ C_b \end{pmatrix} \tag{S.43}$$

as

$$\varepsilon = \pm K\rho \tag{S.44}$$

$$|+(\theta)\rangle = \cos\frac{\theta}{2}\,|a\rangle - \sin\frac{\theta}{2}\,|b\rangle$$

$$|-(\theta)\rangle = \sin\frac{\theta}{2}\,|a\rangle + \cos\frac{\theta}{2}|b\rangle. \tag{S.45}$$

From (S.41) and (S.44) the adiabatic potential for the normal coordinates is given by the two-valued function

$$V_{ad}^{\pm}(\rho) = \tfrac{1}{2}\mu\omega^2\rho^2 \pm K\rho. \tag{S.46}$$

If one illustrates the surface of the adiabatic potential as a function of Q_a and Q_b, a "Mexican hat" shape like Fig. S.5 emerges. The potential consists of two axially-symmetric surfaces and assumes a minimum value at $\rho_0 = K/\mu\omega^2$.

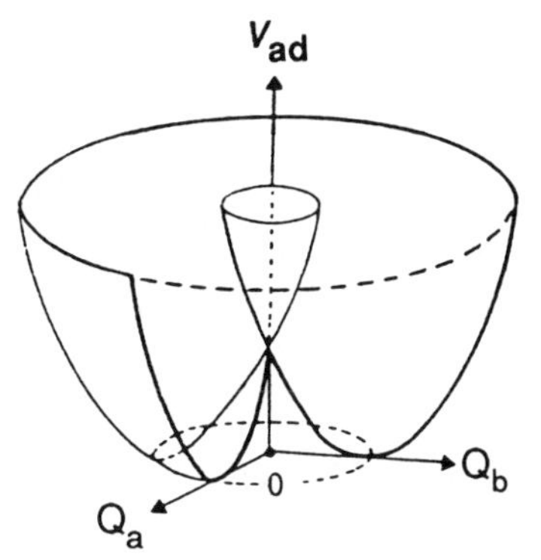

FIGURE S.5. Adiabatic potential surface related to dynamical Jahn-Teller effect (S.46).

Here we notice that the electron state kets (S.45) change sign for rotation by 2π around the symmetry axis:

$$|\pm(\theta = 2\pi)\rangle = (-1)\,|\pm(\theta = 0)\rangle. \qquad \text{(S.47)}$$

Formally, this is the same as the rotation of the spin with $S = \frac{1}{2}$ discussed in Section 3.2, since the sinusoidal dependence of the ket on the rotation in (S.45) is given by just one half of the rotation angle, $\theta/2$. Naturally, in our case, it is the *orbital state* ket of the electron system that has changed sign and thus has nothing to do with spins. The factor (-1) on the r.h.s. of (S.47) is easily shown to be just the Berry's phase factor. Here the origin of the Q_aQ_b-plane is a degeneracy point of the electron system. When the vector (Q_a,Q_b) describing the state of the lattice deformation circulates once around this origin, the corresponding "solid angle" amounts to 2π. Therefore, from the general expression (S.27) the electron state ket is multiplied by the geometrical phase factor $\exp(\mp i2\pi/2) = -1$. It is to be noted that as early as in 1963, this kind of Berry's phase was already taken into account in a very natural way by G. Herzberg and H. C. Longuet-Higgins, when they solved the problem of a coupled system with degeneracy while studying poly-atomic molecules.

As shown here, for such an electron-lattice system in which degenerate electronic states couple to two lattice vibrational modes, a minimum of the adiabatic potential does not form a well but a continuous valley. Thus, the dynamical deformation of the lattice and the corresponding adiabatic state change of the electron system are both incorporated. This is the so-called **dynamical Jahn-Teller effect**. Now, let us consider the dynamical motion of the normal coordinates (Q_a,Q_b) (that is, (ρ,θ)) under the adiabatic potential (S.46) and find the influence of the Berry's phase on the dynamics. The two-dimensional Schrödinger's equation to be solved is

$$\left\{-\frac{\hbar^2}{2\mu}\left(\frac{\partial^2}{\partial\rho^2} + \frac{1}{\rho}\frac{\partial}{\partial\rho} + \frac{1}{\rho^2}\frac{\partial^2}{\partial\theta^2}\right) + V_{ad}^{\pm}(\rho)\right\}\Psi(\rho,\theta) = E\Psi(\rho,\theta), \qquad \text{(S.48)}$$

where the wave function $\Psi(\rho,\theta)$ is expressed by the product of the nucleus piece $\phi_{\pm}(\rho,\theta)$ and the electron state ket

$$\Psi(\rho,\theta) = \phi_{+}(\rho,\theta) \,|+(\theta)\rangle + \phi_{-}(\rho,\theta) \,|-(\theta)\rangle. \tag{S.49}$$

The adiabatic potential written as $V_{ad}^{\pm}$ in (S.48) is to be understood as acting on ϕ_{+} in $\Psi(\rho,\theta)$ for V_{ad}^{+} and on ϕ_{-} for V_{ad}^{-}. Since (S.48) is a differential equation with separable variables we can solve it by choosing the nucleus wave function as

$$\phi_{\pm}(\rho,\theta) = f_{\pm}(\rho)\, e^{ij\theta}. \tag{S.50}$$

Here the most important point is that the state function $\Psi(\rho,\theta)$, which describes the whole coupled system consisting of electrons and lattices, must remain a single valued function. The state ket of the electron system changes sign when θ is rotated by 2π. Therefore, in order to cancel this sign change, $\phi_{\pm}(\rho,\theta)$ must change its sign as well to preserve the single-valued nature of $\Psi(\rho,\theta)$! From this requirement we conclude that $\exp(ij2\pi) = -1$, that is

$$j = (l + \tfrac{1}{2}) = \text{half integer}, \ (l = \text{integer}). \tag{S.51}$$

The presence of the Berry's phase has changed the quantum number that characterizes the motion of atoms, from the usual integer number to *half integer* one.[†]

If two adiabatic levels are well separated, we are allowed to solve the equation (S.48) by ignoring the mixing between $\phi_{+}(\rho,\theta)|+(\theta)\rangle$ and $\phi_{-}(\rho,\theta)|-(\theta)\rangle$. Then the low energy excitations in the potential valley is approximately expressed by

$$E_{nj} \cong (n + \tfrac{1}{2})\hbar\omega + \frac{\hbar^2(j^2 + \frac{1}{4})}{2\mu\rho_o^2}. \tag{S.52}$$

$$(n\text{: integer, } j\text{: half integer})$$

The result is just the sum of the energies of vibration along the radial direction and those of rotation in the plane of the normal mode coordinates. (The factor $\frac{1}{4}$ in the numerator of the second term comes from the electron system.) In the picture of lattice deformation they are the energies related to oscillation of the *magnitude* of the atomic displacement and transformation of the deformation *patterns*, respectively. These kinds of energy levels have been well analyzed experimentally through spectral measurements for ions in crystals and poly-atomic molecular gases and also through several physical measurements on matter. So scientists in these fields have

---

[†]A general discussion is given using path integral method and action in H. Kuratsuji and S. Iida, *Prog. Theor. Phys.* **74** (1985), p. 439 concerning how Berry's phase modifies the quantization condition of dynamical variables.

been familiar with the half-integer quantization which appears in rotational levels for sometime. After Berry's paper, the precise spectral experiment on a tri-atomic molecular gas, Na_3 was repeated again in 1986, and the half-integer quantization was reconfirmed by G. Delacretaz *et al.*

UNIVERSALITY OF GEOMETRICAL PHASE

So far we have discussed a geometrical phase called the Berry's phase by taking quantum-mechanical systems as example. However, let us notice that the Planck's constant is lost in the general expression giving the phase in (S.27). In this respect it is totally different from the dynamical factor (2.6.7) discussed in Chapter 2. Therefore, although the original derivation by Berry is based on quantum mechanics, the geometrical phase factor itself is expected to be universal, regardless of quantum-mechanical systems or classical-mechanical ones. The question is then, where can we find a geometrical phase in a classical system effected by an adiabatic change?

Detection of the Berry's phase has been done for light, which is on the one hand very quantum-mechanical and relativistic, but on the other hand also describable by classical electrodynamics. By letting linearly polarized light pass through the inside of a spirally-twisted optical fiber, one observes the rotation of its polarization plane. According to quantum theory a photon is a particle with spin one, and the spin in an eigenstate is either parallel to the direction of propagation or anti-parallel to it. (This leads to the *helicity* eigenvalue being $+1$ or -1 respectively.) Each eigenstate is a circularly polarized light with a rotational direction opposite to each other. When the linearly-polarized light, which is a combination of these two states, passes through the inside of a once-twisted optical fiber, the propagation vector (the relevant quantity for an adiabatic change here) describes a closed circuit. (At the same time the spin also rotates.) As a result the different Berry's phase is added to each component of the combination state, so that the polarization plane of light rotates. In the 1986 experiment performed by Chiao, Wu, and Tomita, the agreement between the rotational angle of the polarization plane and the solid angle determined by the degree of twist of the optical fiber was confirmed. On the other hand, the effect can also be explained by classical electrodynamics as a result of integration of continuous *parallel transformation* of the electric field vector inside the optical fiber. Although the directions of light at the entrance and the exit are parallel, a difference between the polarization planes is produced for the optical fiber which is twisted in the middle.

Furthermore, even for a classical-mechanical system a geometrical phase factor emerges when we oscillate a conical pendulum on the revolving earth. Seen from the rest coordinate system the time needed for change in the gravity direction at a certain place on the revolving earth is very much larger than the period of the conical pendulum. That is, the direction of the gravity acting on the pendulum *adiabatically* changes as the earth rotates. Let us take the rotational angular velocity of the pendulum in the orbital plane to be ω. In the total rotational angle θ of the conical pendulum after one day (earth period of revolution $= T$) a deviation by $\Delta\theta$ from the simply predicted value ωT, a dynamical angle, is realized. That is,

$$\theta = \omega T + \Delta\theta. \tag{S.53}$$

The shift $\Delta\theta$ is dependent on the latitude of the measurement point but is independent of the gravity and the length of the string which govern the motion of the pendulum. This is after all a geometrical phase and is generally called the **Hannay's angle**.

So the phenomenon is quite similar to the rotation of a quantum spin in a varying magnetic field. Let us compare the two expressions for the total rotational angles in these two cases, (S.24) and (S.53). We may speculate that $\Delta\theta = \Omega$. For the conical pendulum, *this is the case*. The Hannay's angle is equal to the solid angle subtended at the center of the earth by the closed circuit which the gravity vector acting on the pendulum draws following the rotation of the earth. This is a geometrical phase in classical mechanics corresponding to the Berry's phase in quantum mechanics.

In this way geometrical phases accompanying adiabatic changes appear in various systems such as oscillating systems, optical systems, spin systems, molecular systems, and so on, regardless of whether we are dealing with classical or quantum mechanics. In solid state physics, this kind of argument has also been used to explain the quantum Hall effect besides the dynamical Jahn–Teller effect discussed above. Last but not least Berry's phase has also impacted field theory. Without prior knowledge of the geometrical phase, it has nevertheless been possible to accommodate such a phase naturally into theories if one treats precisely Schrödinger's equation, Newton's equation, Maxwell's equations, and the path-integrals. However, we can conclude by saying that our understanding of quantum mechanics has been deepened since the existence of geometrical phases was explicitly revealed.

SUPPLEMENT II

Non-Exponential Decays

Radioactive nuclei furnish us with one of the best known examples of unstable, or decaying, systems. The assumption that the radioactive decay of any nucleus is a random process, and, furthermore, that even in a large sample each nucleus decays independently, leads to the familiar *exponential decay law*,

$$N(t) = N(t_0)e^{-\lambda(t-t_0)}, \qquad \text{(T.1a)}$$

where $N(t)$ is the number of nuclei that have not decayed after a time t, and λ is a constant determined by the properties of the individual nucleus. Since the fluctuations in (T.1a) are small only when the number of nuclei is large, we can rephrase the radioactive decay law as a probabilistic statement: The probability that any particular nucleus will not undergo a decay in a time interval between t_0 and t is given by

$$P(t) = e^{-\lambda(t-t_0)}, \quad t > t_0 \qquad \text{(T.1b)}$$

The quantity $P(t)$ may be called the survival probability. Notice that $P(t_0) = 1$; this is the statement of our knowledge that the nucleus under examination had not decayed at $t = t_0$. Here, we investigate the form of $P(t)$ for quantum mechanical systems. This is necessary because within the framework of quantum mechanics, probabilities are derived from fundamental probability amplitudes, so that it is not *a priori* obvious that the assumptions leading to (T.1) can be consistently incorporated.

In keeping with our intuitive notions, we will say that a quantum system is in an unstable state if the survival probability for that state reduces to zero monotonically with time.† If the state vector of the system at time t is given by $|q_i, t\rangle$ (here, q_i denote all quantum numbers necessary to specify the state), the survival probability $P(t)$ is given in terms of the survival amplitude (c.f. the analogous discussion in Chapter 2 of text),

$$a(t) \equiv \langle q_i, t_0 | q_i, t\rangle = \langle q_i, t_0 | U(t, t_0) | q_i, t_0\rangle, \qquad \text{(T.2)}$$

by

$$P(t) = |a(t)|^2. \qquad \text{(T.3)}$$

†It is important to stress the requirement of monotonicity, since it is possible to find examples where the survival probability exhibits an oscillatory behavior (see, e.g., Rabi oscillations, Sec. 5.5 of text. Notice that the oscillatory behavior persists even when the external potential is time independent).

For an isolated system, the Hamiltonian is time independent, and the unitary operator $U(t, t_0)$ in (T.2) that evolves the states from an initial time t_0 to a later time t is given by $U(t, t_0) \equiv e^{-iH(t-t_0)/\hbar}$. We thus see that if $|q_i, t\rangle$ is a normalized *stationary state* so that it has a definite value of energy, $a(t)$ is a pure phase and $P(t) = 1$.

Although an unstable state of a quantum mechanical system cannot be represented as one of the energy eigenkets, the assumption of completeness of the eigenkets of the Hamiltonian tells us that we can represent the unstable state as a linear superposition over these eigenkets with time-independent complex coefficients; that is,

$$|q_i, t\rangle = S_\varepsilon\, |\varepsilon, t\rangle\langle\varepsilon|q_i, t_0\rangle. \tag{T.4}$$

Here, S denotes a summation (integration) over the discrete (continuous) part of the spectrum of H. We can thus write (T.2) as,

$$a(t) = S_\varepsilon e^{-i\varepsilon(t-t_0)/\hbar}|\langle\varepsilon|q_i, t_0\rangle|^2. \tag{T.5}$$

It is reasonable to suppose that the unstable states are orthogonal to the (normalizable) discrete eigenstates of the Hamiltonian which, as we argued above, cannot decay. We will, therefore, assume that unstable states are associated with the continuum states of the Hamiltonian. This is in keeping with our intuitive expectation from the fact that an unstable particle decays only to continuum states; that is, the unstable state has non-vanishing projections only on continuum states into which it can decay. In this case, the survival amplitude in (T.5) can be written as the Fourier transform,

$$a(t) = \int d\varepsilon e^{-i\varepsilon t/\hbar}\rho(\varepsilon), \tag{T.6}$$

of a real non-negative function

$$\rho(\varepsilon) \equiv |\langle\varepsilon|q_i, t_0\rangle|^2, \tag{T.7}$$

where (for later convenience) we have set $t_0 = 0$.

Notice that $\langle\varepsilon|q_i\rangle$ is just the wave-function for the decaying state in the energy representation, so that $\rho(\varepsilon)$ is just the probability density that an energy measurement of this state yields a value ε. If $\rho(\varepsilon)$ has the familiar Breit-Wigner form,

$$\rho_{BW}(\varepsilon) = (\Gamma/2\pi)/[(\varepsilon - \varepsilon_0)^2 + \Gamma^2/4], \tag{T.8a}$$

we leave it to the reader (c.f. the discussion (5.8.21) to (5.8.24) of text) to show that the corresponding survival probability has the form

$$P_{BW}(t) = e^{-\Gamma|t-t_0|/\hbar}, \tag{T.8b}$$

so that the decay of the state is indeed described by the exponential decay law.

We will now show that with some very reasonable assumptions, Eq. (T.1b) can *never be exact* regardless of the nature of the quantum system. In particular, we will show that for very long times, the survival probability must decrease at a rate that is slower than exponential. To see this, we first remark that any interacting quantum-mechanical system must have a ground state; that is, there is a state $|\varepsilon_0\rangle$, with $H|\varepsilon_0\rangle = \varepsilon_0|\varepsilon_0\rangle$ such that ε_0 is the smallest eigenvalue of H. If this was not the case, it would be possible to extract arbitrarily large amounts of energy from the system by interactions that send it to eigenstates of increasingly lower energy. The existence of the ground state obviously implies that the probability density function $\rho(\varepsilon)$ introduced via (T.7) above vanishes if $\varepsilon < \varepsilon_0$.

To proceed further, we introduce a theorem due to Paley and Wiener (R. Paley and N. Wiener, "Fourier Transforms in the Complex Domain," Am. Math. Soc. Pub. Vol. **XIX**, 1934), which can be stated as follows:

If $a(t)$ and $\rho(\varepsilon)$ are Fourier transforms of each other such that $\rho(\varepsilon) = 0$ for $\varepsilon < \varepsilon_0$ and $\int_{-\infty}^{\infty} dt|a(t)|^2$ is finite, then $\int_{-\infty}^{\infty} dt|\ln|a(t)||/(1 + t^2)$ is also finite.

We will see below (where we follow essentially the argument of L. A. Khalfin, Sov. Phys. JETP Vol. **6**, pg. 1053, 1958) that this remarkable result strongly constrains the long time behavior of the survival probability. First, we note that the fact that $\rho(\varepsilon)$ is real implies $a^*(t) = a(-t)$ so that if the survival probability decays sufficiently fast for large times, $\int_{-\infty}^{\infty} dt|a(t)|^2$ is finite, and the Fourier transform pair $a(t)$ and $\rho(\varepsilon)$ in (T.6) indeed satisfies the conditions for the Paley-Wiener theorem. Thus, for large times, $|\ln|a(t)||$ is required to increase slower than linearly, and hence the exponential behavior for $|a(t)|$ is precluded.

We point out that while our general argument implies that the long time behavior of the survival probability cannot be exactly exponential, it does not shed any light on when the non-exponential behavior sets in or on how much the survival probability deviates from the exponential form. The former requires the specification of a time scale other than the lifetime, and so is probably dependent on the details of the dynamics that causes the decay. We also remark that the Breit-Wigner form $\rho_{BW}(\varepsilon)$ in (T.8a) (which led to the exponential form for the survival probability (T.8b)) is non-vanishing for all values of ε, and so violates the conditions of applicability of the Paley-Wiener theorem. Physically this form of $\rho(\varepsilon)$ violates our assumption of the existence of a ground state, so that the resulting exponential behavior can be, at best, an approximation to the real situation. It will be left as an exercise for the reader to check that if we cut off the Breit-Wigner form of ρ, the survival probability $P(t)$ decays according to a power law for large times.

We stress here that the fact that we have not seen deviations from the exponential decay law at large times does not vitiate our general arguments that do not tell us when the non-exponential behavior sets in. It could be that non-exponential decay occurs only after many lifetimes in which case it would be very difficult to observe. There are, however, equally general arguments that imply deviations from the exponential decay law for very short times (defined precisely below) which we turn to now.

We begin by noting (our discussion essentially follows that of B. Misra and E. C. G. Sudarshan, J. Math. Phys. Vol. **18**, 756, 1977) that the time derivative of the survival amplitude (T.6),

$$da/dt = \int d\varepsilon(-i\varepsilon/\hbar)e^{-i\varepsilon t/\hbar}\rho(\varepsilon), \tag{T.9}$$

is a well defined, continuous function of t if we further assume (as is physically reasonable) that the decaying state has a finite energy expectation value. We can, therefore, evaluate the rate of change of the survival probability,

$$\begin{aligned} dP(t=\tau)/dt &= [da^*(\tau)/dt]a(\tau) + a^*(\tau)[da(\tau)/dt] \\ &= -\dot{a}(-\tau)a(\tau) + a(-\tau)\dot{a}(\tau) \end{aligned} \tag{T.10}$$

where we have used $a^*(t) = a(-t)$ and the dot denotes differentiation with respect to the argument. We thus see that,

$$dP(t=0)/dt = 0, \tag{T.11}$$

which of course means that the survival probability cannot be an exponential at $t = 0$! We remind the reader that in (T.6), $t = 0$ was defined as the latest instant of time when we knew that the system had not yet decayed. This implies that $t = 0$ is also the time when the system was last observed or, in Dirac's terminology, the time of preparation of the state of the system. We remark that while our argument clearly precludes the possibility of exponential decay at $t = 0$, it sets no limitations on when exponential decay may set in.

Taking this analysis a step further, we can consider successive observations of a system. As discussed above, (T.11) must hold after each of these observations. The implications of this are schematically illustrated in Fig. T.1. The survival probability for a system prepared at $t = t_1$ is shown by the curve labeled 1 in Fig. T.1. Notice that, in keeping with (T.11), the slope of the curve is zero at $t = t_1$. If this system is allowed to evolve undisturbed, the survival probability will follow the first curve which may rapidly take an exponential form (of course, for very large values of t, not shown in the figure, we have seen that the decay will be slower than exponential). The size of the flat region at $t = t_1$ is not determined by our general arguments and must be dependent on dynamical details.

If, however, an observation is made on this system at $t = t_2$, our arguments above imply that (T.11) must hold at $t = t_2$, so that the system

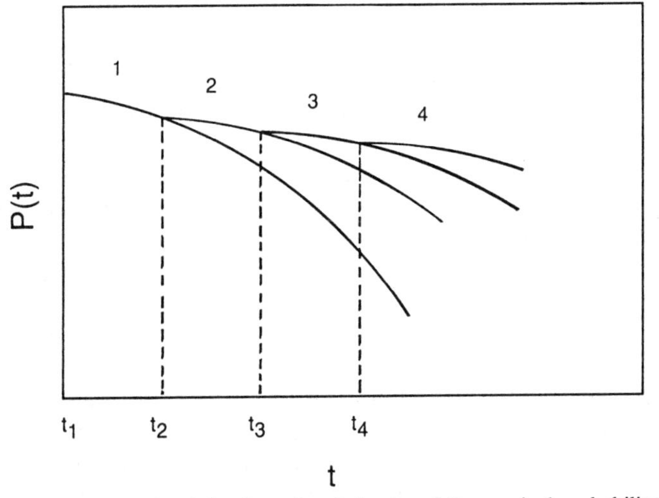

FIGURE T.1. A schematic of the short time behavior of the survival probability $P(t)$ of a state of a quantum mechanical system prepared at $t = t_1$ is illustrated by the curve labeled 1. The curves labeled 2, 3, 4 show how $P(t)$ is modified by observations of the state at times t_2, t_3, t_4. Notice that each successive measurement increases the survival probability from what it would have been had the system not been disturbed by that measurement.

now follows the curve labeled 2. Successive observations at $t = t_3$, $t = t_4$, etc., would make the system evolve along the other curves in the figure. We see that the qualitative effect of these successive observations is to increase the survival probability of the system being observed from what it would have been if the system was allowed to evolve undisturbed after its preparation at $t = t_1$. This inhibition of the decay appears to be a feature of quantum mechanics, once we grant that each observation of the system corresponds to the preparation of the quantum mechanical state.

These arguments imply that it should be possible to slow down the decay of a quantum mechanical system by making successive rapid observations on it. In the limit of continuous observation (provided that this can be operationally defined), an unstable system would not decay. This has been referred to as the **quantum Zeno effect**.

Deviations from the exponential decay law have never been observed for radioactive nuclei or for the well measured K_L mesons of particle physics. W. M. Itano *et al.*, Phys. Rev. Vol. **A41**, 2295, 1990, have however recently demonstrated that the rate for transitions between two discrete levels of a quantum system is indeed inhibited by successive measurements of the state of the system. These authors have further shown that the amount of inhibition increases with the frequency with which the system is monitored. They argue that their results are completely consistent with the expectation from quantum mechanics assuming, as usual, that each measurement is effectively equivalent to preparing afresh the state of the system. This serves to bring home the fact that while this inhibition of transitions may be somewhat surprising, it is a consequence of the structure of quantum mechanics.

At this point, it is fair to warn the reader that we have only touched upon the basic question in our discussion of the quantum Zeno effect. There is considerable (and somewhat controversial) literature on the physical implications of this result. It is not our purpose here to discuss these in detail, nor is it our purpose to enter into various subtleties (both physical and philosophical) of the issues involved. For this, we refer the reader to the original literature.

To sum up, we have presented very general arguments to demonstrate that the decay of any quantum mechanical state must deviate from the familiar exponential decay law, both for very long and for very short times as measured from the instant of preparation of the state of the system. These deviations have not been experimentally observed for systems that exhibit (at least approximately) exponential decay. Since these arguments refer only to asymptotically long and short times, they do not preclude the possibility that a decaying quantum mechanical system may exhibit (essentially) exponential decay over most of its lifetime. Indeed by making suitable approximations, the exponential decay law has been derived (see, e.g., Merzbacher [1970], Chapter 18) for model systems whose decay can be described in perturbation theory. It is instructive to study these approximations and trace why they are invalid for very long and very short times (as they must be in view of our general considerations).

Bibliography

The literature in quantum mechanics is very large and very diverse. We present here a bibliography of those books to which we refer specifically in our text as well as a reference section of more-elementary texts in quantum mechanics (generally at the undergraduate level) that provide the background to our book (addressed primarily to the first-year graduate student). Since many books have been written about elementary quantum mechanics, the list in this reference section is, of course, not meant to be complete. The editor has studied some of the suggested books, read others, glanced at a few, and likely missed others. The editor apologizes in advance if, as is all too probable, there are glaring omissions. Throughout the book little attempt has been made to quote original papers; the few that have been quoted are noted in the footnotes or the text itself. This we believe to be appropriate in a book whose primary purpose is to serve as the textbook to a basic beginning graduate course.

Baym, G. *Lectures on Quantum Mechanics*, New York: W. A. Benjamin, 1969.

Bethe, H. A., and R. W. Jackiw. *Intermediate Quantum Mechanics*, 2nd ed., New York: W. A. Benjamin, 1968.

Biedenharn, L. C., and H. Van Dam. (eds) *Quantum Theory of Angular Momentum*, New York: Academic Press, 1965.

Bjorken, J. D., and S. D. Drell, *Relativistic Quantum Fields*, New York; McGraw-Hill, 1965.

Bohm, D. *Quantum Theory*, Englewood Cliffs, NJ: Prentice-Hall, 1951.

Dirac, P. A. M. *Quantum Mechanics*, 4th ed., London: Oxford Univ. Press, 1958.

Dirac, P. A. M. *Quantum Mechanics*, 4th ed., London: Oxford Univ. Press, 1958.

Edmonds, A. R. *Angular Momentum in Quantum Mechanics*, Princeton, NJ: Princeton University Press, 1960.

Fermi, E. *Nuclear Physics*, Chicago: University of Chicago Press, 1950.

Feynman, R. P., and A. R. Hibbs. *Quantum Mechanics and Path Integrals*, New York: McGraw-Hill, 1964.

Finkelstein, R. J. *Nonrelativistic Mechanics*, Reading, MA: W. A. Benjamin, 1973.

Frauenfelder, H., and E. M. Henley. *Subatomic Physics*, Englewood Cliffs, NJ: Prentice-Hall, 1974.

French, A. P., and E. F. Taylor. *An Introduction to Quantum Physics*, New York: W. W. Norton, 1978.

Goldberger, M. L., and K. M. Watson. *Collision Theory*, New York: Wiley, 1964.

Goldstein, H. *Classical Mechanics*, 2nd. ed., Reading, MA: Addison-Wesley, 1980.

Gottfried, K. *Quantum Mechanics*, vol. I, New York: W. A. Benjamin, 1966.

Jackson, J. D. *Classical Electrodynamics*, 2nd ed., New York: Wiley, 1975.

Landau, L. D.; and E. M. Lifschitz. *Quantum Mechanics*, Reading, MA: Addison-Wesley, 1958.

Merzbacher, E. *Quantum Mechanics*, 2nd ed., New York: Wiley, 1970.

Messiah, A. *Quantum Mechanics*, New York: Interscience, 1961.

Morse, P. M., and H. Feshbach. *Methods of Theoretical Physics*, (2 vols.), New York: McGraw-Hill, 1953.

Newton, R. G. *Scattering Theory of Waves and Particles*, 2nd. ed., New York: McGraw-Hill, 1982.

Preston, M. *Physics of the Nucleus*, Reading, MA: Addison-Wesley, 1962.

Sakurai, J. J. *Advanced Quantum Mechanics*, Reading, MA: Addison-Wesley, 1967.

Sargent III, M., M. O. Scully, and W. E. Lamb, Jr. *Laser Physics*, Reading, MA: Addison-Wesley, 1974.

Saxon, D. S. *Elementary Quantum Mechanics*. San Francisco: Holden-Day, 1968.

Schiff, L. *Quantum Mechanics*, 3rd. ed., New York: McGraw-Hill, 1968.

Tomanaga, S. *Quantum Mechanics I*, Amsterdam: North-Holland, 1962.

Whittaker, E. T., and G. N. Watson. *A Courst of Modern Analysis*, New York: Cambridge University Press, 1965.

REFERENCE SECTION

Dicke, R. H., and J. P. Wittke, *Introduction to Quantum Mechanics*, Reading, MA: Addison-Wesley, 1960.

A very readable but terse book at the junior or senior level. It discusses a few topics, notably quantum statistics, not usually treated in textbooks at this level.

Feynman, R. P., R. B. Leighton, and M. Sands, *The Feynman Lectures on Physics*, vol. 3, *Quantum Mechanics*, Reading, MA: Addison-Wesley, 1965.

This is an introduction to quantum mechanics that approaches the subject from the viewpoint of state vectors, an approach that—together with Dirac's classic—greatly stimulated the spirit and substance of the earlier chapters of our own book. The many fascinating examples (for example, the Stern-Gerlach experiment) discussed with a minimum of formalism are to be noted in particular.

Gasiorowicz, S. *Quantum Physics*, New York: Wiley 1974.

This book is at a level comparable to that of Dicke and Wittke 1960. Gasiorowicz is known to be a very effective teacher, who makes difficult concepts simple and clear even in advanced research areas of particle theory. Hence it is not surprising that his quantum physics book reflects that clarity of purpose that enhances the readability of the book.

Mandl, F. *Quantum Mechanics*, London: Butterworths Scientific Publications, 1957.

This book contains a clear and compact presentation of the foundation and mathematical principles of wave mechanics at a level comparable to Gasiorowicz's book.

Matthews, P. T. *Introduction to Quantum Mechanics*, 3rd ed., London: McGraw-Hill, 1974.

This third edition is lucid, concise, self-contained, and addressed to the undergraduate. It is a mathematically oriented introduction to development in wave and matrix mechanics, based on clearly defined postulates.

Mott, N. F. *Elements of Wave Mechanics*, London: Cambridge University Press, 1952.

A short but delightful treatment of wave mechanics at an elementary theoretical level by one of the pioneers.

Park, D. *Introduction to the Quantum Theory*, 2nd. ed., New York: McGraw-Hill, 1974.

Intended for advanced undergraduates, this attractive book by an outstanding pedagogue has a general introduction to quantum theory, followed by selected applications to various branches of physics, illustrating different aspects of the theory. The applications appear to require little teaching, and hence students can use the book as a reference.

Pauling, L., and E. B. Wilson, *Introduction to Quantum Mechanics*, New York: McGraw-Hill, 1935.

Though one of the older texts, it is still well worth studying. There is a surprisingly complete treatment of the hydrogen atom and harmonic oscillator from the wave mechanics viewpoint.

Rojansky, V. *Introductory Quantum Mechanics*, Englewood Cliffs, NJ: Prentice-Hall, 1938.

Despite its vintage, this book remains a classical text, especially valuable for its detailed coverage of the fundamental principles of wave and matrix mechanics.

Index

Death Benefit

Robin Cook

THORNDIKE PRESS
A part of Gale, Cengage Learning

Detroit • New York • San Francisco • New Haven, Conn • Waterville, Maine • London

Thorndike Press, a part of Gale, Cengage Learning.

Thorndike Press® Large Print Basic.
The text of this Large Print edition is unabridged.
Other aspects of the book may vary from the original edition.
Set in 16 pt. Plantin.

LIBRARY OF CONGRESS CATALOGING-IN-PUBLICATION DATA

Cook, Robin, 1940–
Death benefit / by Robin Cook. — Large print ed.
p. cm. — (Thorndike Press large print basic)
ISBN-13: 978-1-4104-4494-3 (hardcover)
ISBN-10: 1-4104-4494-5 (hardcover)
1. Life insurance—Fiction. 2. Wall Street (New York, N.Y.)—Fiction. 3. Large type books. I. Title.
PS3553.O5545D43 2012
813'.54—dc23 2011046502

Published in 2012 by arrangement with G. P. Putnam's Sons, a member of Penguin Group (USA) Inc.

Printed in the United States of America
1 2 3 4 5 6 7 16 15 14 13 12

To all the foster children

ACKNOWLEDGMENTS

A writer needs a lot of friends: At least this writer does. These are the friends that don't mind fielding a call, often out of the blue, and being asked a question of fact or preference, or being willing to read an outline or character study and lend an opinion. You all know who you are, and thanks. Of course there's always Joe Cox, who knows more about law and business than I'll ever know, and Mark Flowenbaum, a real forensic pathologist. And, of course, there's Jean Reeds Cook, who is a great reader and doesn't allow me any slack. Thank you all.

Prologue

Krasnoyarsk, Russia
March 14, 2011, 4:22 p.m.

Prek Vllasi touched his right forefinger to the scar on his upper lip, where the cleft had been crudely repaired when he was an infant. It was something he did without thinking many times each day, more often when he was under pressure. Right now, standing in a filthy room on the tenth floor of a derelict Soviet-era tower block in the Russian city of Krasnoyarsk, he was getting more nervous by the minute.

Prek checked his watch again and looked over at Genti Hajdini. Genti was leaning against a foldout table, periodically yawning as he worked on a fingernail with his pocketknife. Every time he saw Genti, Prek was slightly taken aback by the straight lines of his lieutenant's beaky nose. From this angle, it looked more like the sharp end of a hatchet. Yes, he was sure the Chechens were coming, Genti had just told him for the tenth time. Good people back home in Albania had vouched for this outfit. Even though he could feel it, Prek checked the Makarov pistol stuck in his belt at the small of his back. The Puma bag containing 500,000 euros was on the floor. Genti had brought the guns and money hidden in a truckload of Turkish fruit he'd driven deep into Russia. No wonder he was tired.

There was nothing else to do but wait.

It was freezing — the temperature had topped out at twenty below and the sun would be gone in an hour and a half. Outside, the sky was the same dirty color as the buildings and the ground. Prek started pacing around the large room, what must once have been a communal area in the apartment block that sat just outside town. Prek was a meticulous man; he'd read up about Krasnoyarsk. Forty miles or so down the Yenisei River was the town of Zheleznogorsk, better known by its old Soviet name, Krasnoyarsk-26. This was a closed city, home to factories handling God-knows-what exotic and dangerous materials to make God-knows-which agents of destruction. Weapons-grade plutonium had been produced in three nuclear reactors there, the last of which had just recently closed. For years the Soviets simply dumped the radioactive waste from the nuclear plants straight into the river until they thought better of it and drilled hundreds of wells to pump the deadly sludge underground. Prek knew there was as much radioactivity as a hundred Chernobyls humming away in the caverns around here, another reason he'd be happy to get out of this place.

Quietly, two men walked in. Wiry and rugged-looking, they wore identical black overcoats. Genti looked up.

"Artur? Nikolai?"

One of them stepped forward until he was ten feet from Prek. "I'm Artur," he said, and gestured back to his companion. "That's Nikolai."

Prek looked back and across to Genti, who nodded. These were the names he had been given: Artur Zakoyev and Nikolai Dudaev.

"This is a lot of money you're asking for," Prek said in Russian.

"This stuff isn't easy to get," said Artur. "If it's easy to get, why do you need us? What do you want this for, anyway? You making a big boom somewhere?" Artur grinned. He was referring to the fact that the substance could be used to make triggers for nuclear weapons.

Prek winced. He'd seen better teeth on a mule.

"What we do with it is our business," said Prek. "How do we even know it's genuine? It needs to be good, not some old shit you had lying around in a warehouse."

"You have to trust us. That's why you pay us. You have the money?"

Prek looked down at the bag and kicked it toward Artur. Prek glanced at Nikolai standing behind his boss and to his right. *That's the second time he's looked at his watch,* Prek said to himself. Artur stepped forward and went down on his haunches, his hands in front of him. Everyone knew the drill: they kept their hands down and in plain sight by their sides. Artur unzipped the bag and pulled out a brick of hundred-euro notes and flicked through it with his thumb. Prek saw Nikolai glance at his watch again.

He's waiting for someone, thought Prek. He looked at Genti, who was watching Artur counting money. *He's waiting for someone and they're late.*

"Now you have to trust me," said Prek. He was in a hurry. "The money's all there so I'll take the merchandise."

Artur stood up and held up his hands.

"Okay, okay." His right arm still aloft as if he were taking an oath, Artur reached into his right coat pocket with his left hand and pulled out a small object. Prek rocked forward and back on his heels — he'd had no time to react but he knew Genti could shoot both Chechens in the head in a second. This wasn't a gun — it was a small aluminum vial about three inches long and an inch around. Prek moved forward, took the vial, and put it in his pants pocket. Nikolai said something Prek didn't understand and without another word, the Chechens turned and were gone, Artur clutching the bag with the money.

"Come on," Prek said in Albanian.

When he reached the doorway, he turned left, the opposite direction to the way they'd come in and where the Chechens were now headed.

"The car's back there," Genti said, but Prek was running now, heading to a stairway on the far side of the building. They could hear loud voices echoing up from the other stairway and the sound of hobnail boots on cement. This was who the Chechens were expecting, and it wasn't the chamber of commerce coming to thank the Albanians for their business. Fortunately, Russian timekeeping hadn't gotten any better since the fall of Communism.

Guns drawn, Prek and Genti raced down

the stairs. Prek saw parked police cars and black vans ahead, doors hanging open. He turned and ran around the back of the building, Genti following close behind. The Chechens were in front of them, running toward a solitary car parked in the corner of a walled-in courtyard. *Fucking amateurs.* Prek saw the opportunity.

The Chechens jumped in the car, Artur throwing it into reverse and backing up and around so the car faced forward. Before Artur could get it into drive, Prek and Genti were on the car, firing into the windshield three times each. Artur was hit and thrown back into his seat, his foot pushing onto the accelerator so the engine raced in neutral. Prek and Genti pulled open the doors and dragged the Chechens out. Artur was dead, his head blown open. Nikolai had been shot twice in the neck, blood bubbling out of his airway as his life ebbed away. Prek slammed the car into drive and took off, looking for another road out of the complex. Prek's heart was threatening to break out of his rib cage and he cursed loudly. He was sitting on glass and he had to lean forward so his head didn't come into contact with whatever of Artur's gray matter was splattered on the headrest behind him.

"What happened?" he yelled.

"They sold us out," said Genti. "There . . ." He pointed ahead to a track off the service road leading away from the apartment complex. He knew they wouldn't last five minutes on a highway in a car with no windshield before being spotted.

"Hey . . ." Genti said as Prek slowed down on the unpaved road. Prek looked over and Genti turned his head to look in the back-seat. Though streaked with blood, the Puma bag sat perfectly safe. Prek banged on the steering wheel and turned to Genti and both men laughed long and hard.

Part I

1

Columbia University Medical Center
New York City
February 28, 2011, 7:23 a.m.

The girl, twelve, awakens with a start. She's lying on a thin mattress on a low, narrow bed and circling around her is a pack of girls. They're older — sixteen, seventeen — and as they shuffle along, they're staring down at her with obviously sinister intentions. Some are suppressing giggles, others are smiling, but these smiles aren't signs of happiness, they're smiles of anticipation. It's still nighttime. There are other cots in the long room and the girl knows the other occupants are awake but they won't move to help her because they know what's about to happen.

Transfixed with terror, the girl is unable to react as the mob falls upon her. As she's being dragged off the bed she sees her chief tormentor, the ringleader's face twisted in a manic grimace. Still, she knows better than to scream for help. Somewhere in the dormitory, there's suddenly a loud banging sound. And again.

Pia Grazdani, twenty-six, woke up in a panic and a cold sweat, unsure for a second of where she was. She breathed out with relief when she realized she was safe and in her dorm room at Columbia University Medical Center. Someone was banging on the door.

Taking another deep breath, Pia leaped out of bed in her flannel pajamas, took three quick strides to the door, flipped the dead bolt, and pulled it open. As she expected, it was George, her fellow fourth-year medical student.

"Pia, do you know what time it is? This isn't a day you want to be late." His tone was not as strident as the syntax suggested. At six-one, George Wilson had seven inches on Pia, but somehow he always felt smaller when he was in her presence. As he explained it to himself, she had what he called a tough, plucky personality and could at times be rather volatile.

Pia held the door open and George took a couple of steps into the small dorm room. Pia let the door close and turned and hurried past George, pulling her pajama top over her head as she did so. George looked at Pia's bare back, at the cut of her shoulder blades framing her flawless olive-brown skin. She stood in front of her low dresser and pulled out some clothes for the day. As she did so she caught George's stare in the mirror.

"Sorry, George, I couldn't sleep and then when I could sleep I was dreaming. You go ahead, I'll catch up with you later today."

With that said, Afrodita Pia Grazdani turned her full attention to getting ready. When she pulled down her pajama pants, George turned his head and looked out the window. He would have preferred to watch her but was afraid to do so. Instead he concentrated on the dramatic view that he and the other medical students had learned to take for granted. He could see the giant George Washington Bridge connecting Manhattan with New Jersey. Its usual morning rush-hour traffic was at a standstill in both directions.

"It's okay, Pia," said George, "I'll wait." Then, searching for something to say, he added, "I guess you still haven't figured out how to use that alarm clock I bought for you. I can't get you up every day — you've got to do something about being on time. You could always use the alarm on your cell phone if you'd prefer."

George stopped talking. He'd turned his attention back to the room and was immediately transfixed by the sight of Pia brushing her jet-black hair. He felt an immediate crushing sadness. The few times Pia and George had slept together, four exactly, Pia had asked George to leave before he fell asleep. And each time she'd stood at that same dresser with her back to him brushing her hair, just as she was doing now. Over the course of time George had become painfully aware they hadn't truly slept together at all those four precious times, they'd just had sex: Wham bam thank you ma'am, in reverse.

George was athletically inclined and good-looking in a preppy, stereotypical fashion with an unruly shock of blond hair and a ready smile. During his Ivy League college days it had drifted back to him that many preppy women found him to be a "hunk." There had never been a shortage of willing girlfriends. But George had set his sights early in life on becoming a doctor and did not want to become involved. As a consequence, George's romantic life had been a string of one-night stands and short flings with little emotional commitment. He'd hurt people, he knew, especially now that the tables had been turned and he was the "hurtee" rather than the "hurter." With Pia it was completely different. She truly didn't seem to care, and it drove him crazy. Numerous times he had told himself to forget her, that she was damaged goods, but he couldn't. Instead he'd become obsessed to an extent. George desperately wanted to have a romantic relationship with this woman, but he had no idea what she wanted and why it hadn't happened. He'd been trying during three and a half years of medical school.

"Come on, what are you waiting for?" Pia barked when she popped out of her tiny bathroom still applying a pale lipstick that was more for lip protection than color. She grabbed her medical student white coat, pulled it on, and draped her medical center ID around her neck. She held open the door behind her as if she were the one waiting.

Emotionally flummoxed, as per usual, George awakened from what could have been considered a petit mal seizure and followed her out through the door. He had to practically run to catch up to her as she hurried down the hall toward the elevators.

Pia continued to walk quickly as they made their way out of the dorm and turned right toward the medical center complex. Columbia University Medical Center lies in Washington Heights on Broadway as it runs north along upper Manhattan's spine. Even at this time of the morning, the place was busy. The more purposeful people in the varying-length white coats making their way along 168th Street were the doctors, students, and staff of the hospitals and research facilities. The patients and their relatives who were arriving were more hesitant, trying to figure out where they needed to be and clearly apprehensive of why they were there and what the day might bring.

George turned up the collar of his jacket against the sharp wind that came off the Hudson River and was funneled into 168th Street by the curve of Haven Avenue. Tomorrow, it would be March, the month in which any day the temperature might be in the sixties or it could snow. At that moment it wasn't particularly cold, but the wind was a reminder that winter still had some kick.

George and Pia were heading for different buildings to start their fourth-year month of elective. Fourth year of medical school was a series of monthlong rotations in various specialties, which included an elective period where the individual students could choose something of their particular interest. That month Pia was to do research, as she'd done during her month of elective in her third year. George was to work in radiology, as he too had done the previous year. These choices were particularly apropos since three weeks earlier, George, Pia, and the rest of the class of 2011 had learned the results of the residency matching program. Both Pia and George had been awarded positions there at Columbia University Medical Center, thanks to their superb academic records and strong faculty recommendations: Pia in internal medicine and George in radiology. By special dispensation, Pia would also be starting a concurrent Ph.D. program in molecular genetics, which would allow her to continue her lab work while fulfilling her requirements as a medical resident.

Awaiting Pia that morning at the William Black Medical Research Building was the noted molecular geneticist Dr. Tobias Rothman, winner of both a Nobel Prize and a Lasker Award. In addition to being known for his accomplishments, Dr. Rothman was even better known around the medical center for being a difficult person to work with, thanks to his renowned total lack of social graces. Rothman did not suffer fools. In fact Rothman did not suffer anyone except his long-term research assistant, Dr. Junichi Yamamoto. Originally Rothman's reputation had made George nervous for Pia when she started her third-year elective in the man's lab, but his concern had been tempered by his personal knowledge that she herself was a handful. From direct experience he knew that she could be counted on to hold her own in most any situation. As it had turned out, to everyone's surprise, even Pia's, she had gotten along famously with the famed and feared researcher. In fact it had been Rothman's suggestion that Pia do a Ph.D. program at Columbia with her practical work taking place in his lab. Until Pia had come along, Rothman had never mentored anyone. For a while it had been a major point of gossip in the entire medical center as people wildly hypothesized what on earth was going on between the exotically attractive medical student and the universally disliked but respected curmudgeon who was the center's major research celebrity.

"Pia! Wait up!" George called out. In her typically self-absorbed fashion, Pia had gotten ahead of George in the crowd. Dodging the phalanxes of medical students in their white coats who were all converging on the Black building, George scuttled ahead to catch Pia just before she entered the building. He pulled her aside. Pia looked up at George with her big brown eyes wide open, as if surprised to see George, whom she was supposed to be walking with.

"Do you want to have lunch? It's the first day, so they might go easy on us. I know for me it's probably going to get crazy after today."

"I don't know, George. Rothman's . . . Rothman's, you know . . ."

"Rothman's an asocial asshole, that's what I know."

"Let's not argue! I know what you and just about everyone else thinks, but the man's been good to me. I don't know what he has in mind for me today or for the month, for that matter. What I do know is that I can't make plans to meet for lunch before I find out what the day is going to bring."

"I can tell you what most people think he has in mind for you."

"Oh, please!" Pia snapped. "Let's not get into that again. I've told you time and again the man's never made one advance or off-color comment in my presence. What he is is a genius who believes he's surrounded by morons, and he might be right, at least comparatively. All he's interested in is his work, and I'm interested in it too. I'm well aware of his asocial reputation, but I'm lucky that he tolerates me. I can't wait to get in there. If I have a moment as the day progresses, I'll call your cell."

For a brief second George saw red. All at once his brain was flooded with unreasonable jealousy of the prick Rothman. Everybody hated the guy and here was the woman he was romantically obsessed with essentially telling him to bug off and that she couldn't wait to rendezvous with the aged sourball instead of planning to get together for what might be the last lunch of the month. George sucked in a lungful of air while he regarded Pia's obviously scornful look. In a flash he wondered anew what the hell he was doing continuing to pursue this woman when she seemed to be merely tolerating his company.

Instinctively George knew he shouldn't put so much importance on whether or not she would make plans to meet for lunch, but he couldn't help it. It was just another episode in a long list of episodes. The last time they had made love, which was the way George wanted to think about the "hookup," as she called it, George had tried to be open about how her dismissing him made him feel. Her response then, as it was now about lunch, was to get irritated. Of course, after he left her room, instead of feeling good about making his feelings known, he'd worried himself sick that he'd scared her off for good. But he hadn't. Instead, a couple of days later, George had received a surprising note from Pia in his mailbox. "Maybe you should call Sheila Brown." It included a cell phone number. George called Sheila Brown and had one of the oddest phone conversations of his life. He was to learn more about Pia's past in that one conversation than she had ever shared with him.

"Hi, my name is George Wilson. Pia Grazdani asked me to call you."

"Hello, George. Pia told me you would be calling. I was Pia's caseworker and her therapist for some time. Pia told me it was okay to talk to you."

"Oh, er, okay . . ." Caseworker? Whatever George had been expecting, it wasn't this.

"I know it's extremely unusual for a therapist to talk to a stranger about one of her patients, but Pia asked me to speak with you."

Therapist? This was going to be interesting.

"Normally I wouldn't be speaking to you as I am since it's against any number of rules of my profession, but Pia persuaded me to do so. If I can help Pia emerge from what she faced in her upbringing, I'm willing to do most anything within reason.

"I worked with Pia for years, since she grew up in foster care, including a stint in what used to be called a reform school. As a result, let us say that she's always found it very hard to form any meaningful relationships. Trust is an issue. She didn't tell me much about you, but I find the fact that she asked me to talk with you very encouraging. I think she wants you to know something about her, but she can't tell you herself. So she asked the person who she believes knows her best to do it. Pia has different ideas about privacy and attachment than most people."

That George knew from painful experience.

Without going into specific details, Sheila encouraged George to "keep trying" with Pia because it was her opinion he would be "good" for her. Sheila concluded by giving him her office number to have in addition to her cell phone number in case he ever wanted to call back. George never did, and despite Sheila's disclaimers, he questioned the professionalism of the conversation. At the same time he appreciated the information. He never brought up the issue directly with Pia by saying that he knew she'd been through foster care but rather tried to get her to open up about her childhood in general. Unfortunately she always responded that it was something she did not want to discuss. It was a no-go zone. That was okay with George; he set it aside and didn't think about it. He was giving her as much time as she needed.

George let out the lungful of air with pursed lips. The slight delay gave him a chance to collect himself and not blurt out something he'd regret later. He even tried to mask the fact that he was upset.

"Well, I hope your day goes as well as can be expected," he said finally. "I know you can handle yourself, Pia. But I still don't know how you can stand to work with him."

"I don't need to get along with him, George. It's not kindergarten. If he tolerates me and I learn from him and he can help my career, that's all I ask. We're grown-ups — we don't have to be friends."

She'd used that line before, and George had to wonder, was she talking about Rothman or about him? George's worry that Pia was going to abandon him resurfaced.

"Okay!" George said simply, while holding up his hands in mock surrender. "Sorry to even mention it."

"Stop apologizing!" Pia said sharply, looking at her watch. "You sound nerdy when you apologize. Now I really am going to be late."

Pia hurried away. George asked himself what time Pia might have gotten up had he not gone to her room to wake her. He couldn't help but notice that she'd not bothered to thank him, much less make a commitment to have lunch. Unfortunately it all was irritatingly pro forma.

Pia showed her ID to the security guard with all the other students who were mostly first- and second-year heading off to their eight o'clock lecture. Instead of following them she took the elevator to the fourteenth floor of the Black Research Building and headed for Rothman's sizable laboratory. He commanded the most space of any researcher in the entire center. As soon as she went through the anonymous-looking metal door and entered the suite, she could sense that the lab's day was already in full swing. The three research technicians, Panjit Singh, Nina Brockhurst, and Mariana Herrera, were loitering around the lab's communal coffeepot having already calibrated all the instrumentation that needed

to be done on a daily basis. Rothman, fastidious about what he ate and drank, kept a Nespresso machine in his office that only he and Dr. Junichi Yamamoto, his senior associate, were allowed to use.

"Morning, Miss Grazdani," said Marsh Langman, Rothman's secretary, from b hind her desk. One overdefined eyebro was raised as she looked at the clock on t wall opposite. "You don't want to make t a habit."

Pia followed the woman's line of vi and glanced at the clock. The second h had just ticked past vertical: it was 7:49 stopped and half turned toward Roth ultraloyal retainer-cum-secretary f inevitable reprimand.

"You know he likes everyone to be Marsha said in an accusatory tone.

"I'm not late," said Pia. Studer supposed to begin lectures and oth ties at eight unless they had bee the night before for specific rota required it.

"Ah, but you aren't early eithe start the month off on the wron I should warn you, you'll be h company in your office. There's maintenance in there trying to

lem in the wiring. The security system is down."

"How long's that going to take?"

Marsha, a middle-aged African-American woman in a lab coat her position didn't call for, made a face that said, *How would I know?*

Pia was exasperated. There was barely room for her in what was generously described as an office.

"Is the chief going to have time for me this morning?" Pia was one of the handful of people who didn't reflexively kowtow to Rothman and wait for him to come to them. As Pia asked the question she turned to ully face Marsha. The research technicians vere quiet. Pia wondered if they'd timed heir coffee run for her predictably not-uite-early arrival and were eavesdropping or any possible gossip.

"You know how pressed for time he always ," Marsha said. "He's being pressured to ish his most current salmonella typhi periment with Dr. Yamamoto. We have to mail the manuscript to *The Lancet* in the xt day or so."

Iarsha always talked as if she were ac-ly involved in the research. It was part her strategy of erecting barriers and ding sand traps for those wanting time

with Rothman. She watched over him like a killer guard dog.

"He's been in since six" — "in" being the biosafety level-3 lab, frequently referred to as BSL-3, where the work on the salmonella strains was being done. "I'll see if I can get word to him that you want to speak to him."

"Thank you," Pia said, her eyes betraying irritation. "Getting word" to Rothman meant flipping a switch and talking to him by intercom. Hating to waste time and having finished the last project he had assigned to her, Pia needed to see Rothman to find out what she was going to be doing that month. And now there was some workman in her office to complicate things.

Pia was fortunate to have an office at all. Few of the other people in the lab had such a privilege. When Rothman's chief technician was fired after getting into an argument with Rothman over some picayune detail of lab procedure, his successor, Arthur Spaulding, took an office nearer the biosafety level-3 area, and Pia took over Spaulding's broom closet. Pia saw that her office door was ajar, and she bristled. There were sensitive files in there, even if only a handful of people on the planet would understand what they meant. As she entered she saw that her bench area that also served as her desk was occupied — an electrical blueprint was laid out on the flat surface and there were tools and wires strewn on top. In the corner of the tiny, windowless room was a stepladder with a human form standing on the top platform, head and shoulders hidden up inside the dropped ceiling. Three panels had been taken down and stacked against the wall.

"Excuse me!" Pia called out. When there was no response, she called louder, "Hey, you up there!"

Pia's sharp words caused the man to flinch and hit his head on a pipe up in the ceiling. The man let out an indistinct curse and slowly emerged from the ceiling. After taking one look at Pia, he climbed down from the stepladder. He was about forty-five with grayish stubble and salt-and-pepper hair, wearing dark blue coveralls. His forehead was deeply lined, and he had the sunken cheeks and pale complexion of a lifelong smoker. His body was thin but muscular. His security tag read "Vance Goslin."

"How long are you going to be?" Pia demanded. She had her arms akimbo.

Goslin was struck immediately by Pia's remarkable and exotic beauty, her glowing, flawless skin, her full lips, and, perhaps most of all, her huge dark eyes. Adding to her allure was her apparent confidence and forthrightness. In Goslin's world, girls who looked like Pia acted significantly differently. He was more than casually attracted to her. He was intrigued.

"Depends when I find the problem," he said. He pointed to two areas on the blueprints lying on the bench. He had a distinct accent that Pia thought she recognized, especially given the name Goslin. "If the problem is here, it's easy to fix. If the problem is there, it's harder, but one way or the other we'll get it done. It might even be done by tonight."

Goslin nodded as he finished talking, eagerly continuing to scan Pia's shapely body as he'd done while he'd been talking. He did it overtly, as if it were a matter of right. Eventually his gaze alighted on Pia's hospital ID. "Grazdani," he voiced, raising his eyebrows questioningly. "Now, that's an unusual name."

Pia didn't respond, making him think she might be hard of hearing.

"Your name is unusual. Is it Italian?" he said, raising his voice. He had assumed a wry smile as if he knew Grazdani was not Italian. It was his way of flirting.

"No, it isn't Italian. And why are you shouting?"

Pia had talked about her Albanian heritage maybe twice in her life and she wasn't about to do it now with this person. There were thousands of Albanians in New York City, and Pia remembered enough of the language to recognize an Albanian accent when she heard it. Once when she was ordering a slice of pizza, two young men behind the counter started a frank appraisal of her physical attributes in their language before Pia asked them in English if they wanted her to talk to the manager about their rudeness.

"Actually I'd guess Albanian," Goslin said, the smile continuing. "I'm of Albanian descent, and I have a lot of Albanian friends here in New York. They work in maintenance like me. We've kinda taken over the business. . . ."

Pia wasn't listening. It was barely an hour since she'd been dreaming about one childhood nightmare and now this man was reminding her of another — her father — which added to her growing irritation. Even though she was offering nothing to this maintenance worker that he could take as encouragement, he was still talking, trying to engage her in conversation.

"So where are you from?" he asked. His eyes narrowed and head tilted, as if he were about to make a guess. Such a situation was not uncommon for Pia. Many people, particularly men, tried to guess her genealogy from her appearance, usually coming up with such suggestions as Greek, Lebanese, or even Iranian, but she wasn't going to play the game with this guy even though he'd been correct about her name. Her father was indeed Albanian, although her mother was Italian.

"I'm American," Pia said. "Hurry up with what you're doing! I'm going to need my office sooner rather than later."

"And what do you do?" Goslin asked, vainly trying to keep the conversation going.

Pia didn't answer. She walked out of the room, pausing only to pick out a couple of files she thought she might need.

To the surprise of the lab technicians, who had moved from the coffee nook to their respective individual benches, Rothman suddenly emerged from the biosafety unit. This was a surprise because everyone expected that he would be closeted in there for the day, as had been the case for the last several weeks. As a stickler for rules, he had gone through the airlock and removed his protective lab garb and donned his street clothes. Without a lab coat on, he resembled a gentleman banker more than a research scientist coming from working with extraordinarily deadly typhoid-causing salmonella. Although asocial to a fault, he was a careful dresser, a disconnect as it suggested he cared what other people thought. But he didn't. The clothes were purely for him and the ensemble was the same day after day: conservative three-button Italian suit, pressed white shirt, dark blue tie with matching pocket square, and black penny loafers. He was not a tall man, but he projected well and appeared taller than he was. Moving quickly, he was an intimidating figure with martially erect posture and an expression that did not encourage conversation. His dark brown hair was cut conservatively to match his suit. His almost invisible rimless titanium glasses were his only concession to current fashion.

As Rothman strode toward his private office, the eyes of the technicians followed him. It was immediately apparent to all of them what had brought Rothman out of the biosafety unit. Catching sight of Pia, he had motioned for her to follow him. As the office door closed, the lab technicians exchanged knowing looks with a tinge of collective jealousy. All of them knew that with the pressure of the upcoming *Lancet* article,

Rothman would never have left the biosafety unit specifically to talk to them. In their minds Pia was a kind of teacher's pet made worse by her not being all that friendly. Similar to Rothman, she was always too busy for small talk and kept to herself. On top of that, they all thought she was a bit too good-looking to be a medical student and thought captiously she would have been better suited to playing one on TV. Pia was an enigma to the laboratory staff, made more interesting by the gossip that had it she was going to be a nun.

If the lab technicians had had the opportunity to view the scene inside Rothman's office, they might not have felt any jealousy whatsoever. It appeared more like Rothman and Pia were engaged in an arcane ritual, rather than having an actual conversation. Neither looked at the other throughout the entire, brief encounter. After Rothman told her he wanted her to edit the current salmonella study for *The Lancet* that day, he picked up one of two copies of the paper from his desk, which he was now studying intently. Pia appeared equally distracted, her arms crossed, looking at her feet. The uninitiated might sense a social ineptness on both sides as the awkward silence extended; a clinical psychologist, given enough time, might talk more in diagnostically precise language.

Finally Rothman, half standing, leaned over his desk and handed a copy of the *Lancet* paper to Pia. "Make sure this is up to snuff. I'll want it by morning. Tomorrow we'll talk again about what you will be doing for the month." He still did not look at her. "Let me say that I know you have always been more interested in my stem cell work than my salmonella work, and I'm fine with that. You've earned it, considering you finally know something practical about genetics rather than the garbage they taught you in class. And one other thing: Two fourth-year students have been foisted on me for a month of elective by the damn dean. So I want you to give some thought to what they can do while they're here. It's not going to be easy. I'm sure they'll be worthless."

"Where are they and how can I meet them?"

"They are supposed to start tomorrow. Dr. Yamamoto will introduce you. The main thing is that I don't want them tying up a lot of Junichi's time as he seems to enjoy that kind of crap. I need him to concentrate on our work."

"I can't do anything with that maintenance man in my office."

"My understanding is that he'll be done sometime today. So, tomorrow then." Rothman was never much interested in the details of running his massive lab. Suddenly he was again absorbed in the *Lancet* manuscript.

Ignoring Rothman's dismissal, Pia said, "There is something I want to tell you. The residency matching results came in. I'm going to be here at Columbia doing a combined program getting a Ph.D. in cellular biology with you, as you so generously offered, and completing a residency in internal medicine. I hope you're pleased."

"Well, I'm not!" Rothman said with emotion, his infamous ire rising. "I'm disappointed. If I told you once I told you a dozen times that for you, doing an internal medicine residency would be a complete waste of time, just like it was for me. I think it is totally apparent that you, like me, are cut out for research, not clinical medicine. You should be here in the lab full-time! I said as much in my letter of recommendation for the Ph.D. program."

A level of tension hung in the air. For a few beats neither spoke, nor did they so much as exchange a glance.

"But I have the sisters to think of," Pia

said. Pia's education had been partly underwritten by the Missionary Sisters of the Sacred Heart, an international religious order situated in Westchester County. Pia had fled to the order for emotional safety after aging out of foster care at age eighteen. Although initially Pia had thought briefly of joining the order as a nun, after finishing her high school equivalency and a portion of college at New York University, she had changed her mind. Consequently the relationship with the sisters, particularly the mother superior, had become more transactional. Although Pia would complete her medical training and still go to Africa to help with the organization's missionary work, she would not become a novitiate.

Although Pia had received full scholarships from New York University and Columbia Medical School, the Sisters' contribution had been considerable. She felt justifiably obligated. "I don't think I can renege on a plan I made ten years ago. Although I've come to agree with you that my personality is more suited to research, I think I have to go through with the original plan to become a doctor and, at least for a time, serve the order's needs."

A rush of mumbled profanity escaped from Rothman's lips. He shook his head in disbelief. “Here I am offering you a part in making medical history with my stem cell research, and I have to be concerned about a bunch of nuns in Westchester.” He paused for a moment to collect his thoughts. “What kind of money are we talking about?”

“I don’t know what you mean.”

“Come on. Don’t be obtuse. What do you think you owe them in dollars?”

“I’m not sure I can think of it in those terms.”

“Let’s not be difficult. Give me a figure however you want to create it.”

Pia thought for a moment. It was not an easy task. She’d never put a figure on how the sisters had nurtured her and given her a sense of protection from the evils of her foster care experience. She shrugged. “I don’t know. Maybe fifty thousand. Something like that.”

“Done,” Rothman said. “You’ll get a loan from my bank to the tune of fifty thousand, and I’ll cosign for it.”

Pia found herself momentarily speechless. Never in her life had someone stood up for her financially, especially to the tune of fifty thousand dollars. She didn’t know how to react. “I don’t know what to say,” she mumbled.

“Then don’t say anything! We’ll revisit this

issue, but for today I want you to jump on this paper for *Lancet.* It needs another set of eyes and the statistics checked. I know you are a whiz with statistics."

Rothman got up from behind his desk. With his attention buried in the piece of paper he'd been intermittently studying, he walked out of his office. Pia was stunned. Rothman had just essentially lent her a large sum of money and asked for her help on a vitally important paper.

"Okay," Pia said to herself, "I've got work to do. Now I just have to get that man out of my work space." Following Rothman out through the door, she headed back to the lab bench where she had set up her temporary work space.

2

Convent of the Sisters of the Sacred Heart
Westchester, New York
February 28, 2011, 7:20 p.m.

Armed with Dr. Rothman's pledge of financial support, Pia made an appointment with the mother superior of the convent of the Sisters of the Sacred Heart for that same evening. It wasn't a meeting she looked forward to. Pia recalled how years before the mother superior had found her as a teenager sitting on the convent wall after getting into a fight with her foster family at the time, who lived a couple of miles away. She had brought Pia inside, and they had talked. The result was that Pia returned the very next weekend, with her family's permission, to help out in an unstructured fashion. The rest was history, culminating in Pia's decision, when she aged out of foster care, to join the convent with the idea of possibly becoming a novitiate.

Pia would be forever grateful for what the mother superior had done for her in the years following her moving in, especially as it was an enormous improvement on what she had experienced in the foster care system. Although it was, in reality, another institution, Pia had finally been at peace. She had found the mother superior to be sympathetic not only in helping her adjust to the convent's community living but also in helping her navigate the tempestuous waters in the real world beyond the tranquillity of the convent. It had been at the mother superior's insistence that Pia turned to academia and became a superior student rather than an adequate one. But obtaining her high school equivalency and attending college had allowed Pia to learn about herself to the extent of realizing that a nun's life was not for her. Instead she decided on a career in medicine, where she sensed she could excel and find equivalent peace. After all, during her entire tumultuous foster care experience, she had always viewed the doctor as the sine qua non of power and control of one's personal destiny. But the decision had had consequences, especially with regard to the mother superior.

About five years earlier Pia had made a similar appointment with the woman. It was then that Pia admitted that she was not going to become a sister but rather a doctor. It had been a difficult meeting as the mother superior had been obviously disappointed and made her feelings known. At the same time she had been encouraging about Pia's new career track and voiced how desperately doctors were needed at their missionary locations in East Africa. Now, as Pia walked into the mother superior's stark office, she knew she faced as difficult a situation as — and maybe even worse than — when she had decided against becoming a nun. The more she thought about her goals, the more she thought Rothman was right about her being uniquely qualified for medical research.

"Pia, my dear, it's a blessing to see you. We have all missed you. All the sisters ask about you day after day."

"And you, Reverend Mother."

Pia kept her eyes glued to her hands as they worked at each other in her lap. Her anxiety had peaked. She hoped it was not reflected in her voice. She had dressed simply in a black dress that broke at the knees and plain pumps. At first glance, the mother superior had looked the same as when she first met her ten years before. The uniform of the order helped with that. But

Pia could tell that age was taking its toll. The mother superior had moved slowly when she walked around her desk to greet Pia. From Pia's perspective her hand had felt bonier and more delicate when she placed it on Pia's shoulder than on Pia's previous visit a month or so previously.

On the short train ride out of Manhattan, Pia had rehearsed what she was going to say. She wanted to be clear so there would be no misunderstanding. She was confident in her decision, more confident then than she'd been in Rothman's office, but she knew the mother superior had a talent for ignoring what someone was saying as she worked the conversation back to a position more in tune with her interests and opinions.

As the pleasantries continued, Pia's mind rapidly played over the extraordinary changes her life had taken since she arrived at the convent in what, at that moment, seemed like a previous life. She was now in her fourth year at Columbia Medical School, as amazing as that sounded even to her. She recalled how difficult it had been to convince Columbia to accept her. She remembered how she'd had to explain why, at age eighteen, she'd decided to join a Catholic African missionary order. Her experience at New York University had been a breeze. From the get-go the college admissions people were convinced, no questions asked, that Pia, as a young woman emancipated from foster care, would make a valuable addition to the rich tapestry of NYU undergraduate life.

Columbia, on the other hand, had expressed early concern about Pia's history and its potential effects on her independence and ability to empathize with patients. They didn't voice their concerns in such a clear fashion, but Pia had gotten the message, especially when she was asked to undergo an interview by one of the medical center's psychiatrists. Recognizing that she wouldn't have been asked to do the interview if they weren't interested in her, Pia had acquiesced. To her surprise, the interview turned out to be more pleasant than she had feared. The psychiatrist had been well versed in the inequities of the New York foster care system and seemed sympathetic when he learned that she had been under its questionable aegis from age six to eighteen. Unfortunately, she had never experienced an adoption or even a final placement.

Although the psychiatrist did not have access to her records by law, Pia was rather open with him and explained her experiences, although she downplayed some of the grittier elements. She fully admitted that in retrospect she knew that she had been abused and that she had had to grow up without a nurturing presence in her life, but she added that rather than hindering her, she believed, her experiences would make her a better doctor. She also downplayed any symptoms she'd experienced such as her mild brush with an eating disorder as a teenager and the recurrent nightmares she still experienced.

As the interview had progressed, Pia's openness apparently won the day as the psychiatrist was equally open with her. He actually told her that he was impressed with how she had been able to cope and that he agreed with her that her experiences might make her a better doctor, especially if she became interested in a specialty like pediatrics. He told her that he was particularly impressed by her near perfect grade point average at NYU, her near perfect MCAT scores, and the fact that she had won acclaim as an actress with the NYU theater group. He said it was all indicative of her commitment to her goal of becoming a doctor and to the adjustment she had achieved to everyday life despite her history. He also told her that he would be strongly recommending her for admittance to the class of 2011.

After the psychiatric interview, Pia had been ecstatically hopeful that she would be accepted. But months later she found out that it had not been enough to convince the admissions committee. There had been a number of people who'd apparently demurred, thinking it was too big a risk despite the psychiatrist's recommendation. It took an unexpected last-ditch intervention by two people to carry the day. First, the mother superior offered to become involved and sent a flurry of carefully worded, beautifully argued, and persuasive e-mails. And the second person was Dr. Rothman, who, at the time, was sitting on the admissions committee for an obligatory three-year term. Pia found out about this surprising twist of events only years later, after working with Rothman during her third-year elective. He'd brought it up suddenly at one of their typically uncomfortable meetings. He admitted to her something that he said no one else knew: that he too had suffered through the New York State foster care system because he had been a difficult, hyperactive child. He said a diagnosis was not made until he was an adult, when he himself recognized he had Asperger's syndrome. Pia had been stunned and was still stunned. Respecting his confidence, she had told no one about the revelation.

"The last time you made a formal appointment to see me," the mother superior continued, "you had sad news for us here at the convent, saying you had decided against joining us by becoming a novitiate. My intuition tells me that you are here today for similar reasons. I hope that is not the case. We love you here at the convent and are very proud of you and your accomplishments."

Pia looked up briefly to engage the mother superior's unblinking stare, but she couldn't maintain it. Almost immediately she looked away, finding herself staring at the crucifix on the wall over the woman's shoulder, thinking of pain, sacrifice, and betrayal. Pia took a fortifying breath. As usual the mother superior was miles ahead of her, seemingly sensing what was coming. "I'm starting another month of research in Dr. Rothman's laboratory."

"He is a gifted man. The Lord has looked kindly on him."

"He is going to make history by personally ushering in regenerative medicine. His work with stem cells will be seminal. I want to be part of it."

"From my perspective, you already are. From what you have shared, he has taken to you. Not that I am surprised. How can I help?"

Pia looked back down at her hands. She felt a tinge of guilt after all that the mother superior had done for her, and here she was immediately offering to do more. "I believe I will want to do medical research full-time, meaning I don't think I want to go to Africa."

There, it's out, Pia thought. She felt immediate relief. For a few moments silence reigned in the room. Pia suddenly realized how cold it was. "I know this is a rather large change since I offered to go to Africa to repay you and the order for all the help you've given me over the years since I aged out of foster care."

"Your going to Africa was to be for you, not us," the mother superior said. "Pia, please don't be rash. I know I'm going to sound very old-fashioned, but is there a man involved? There must be — it is your burden to be so beautiful. I hope to God that Dr. Rothman is being honorable."

Pia suppressed a smile. The mother superior's suggestion was so far from reality as to warrant such a reaction. She and Rothman had trouble making eye contact much less something more intimate. "I can assure you that Dr. Rothman has been quintessentially honorable."

"God has limitless ways to test us," the mother superior continued.

"Reverend Mother, I don't believe God is testing me. This does not involve a man, I assure you. I have made my decision because it pleases me and because God has given me a facility for the work. But I would like to repay the convent. Thanks to Dr. Rothman's generosity, I have access to fifty thousand dollars. I would like to donate this money to the convent."

"I will be willing to accept any donation but not as repayment. For our services you do not owe us anything. After all, your presence was payment enough."

"It would please me to donate the money," Pia said.

"As you will. But I do have another request. I don't want you to forget us. I trust that you will still make it a point to visit us on occasion. If you forget us, that will be a betrayal."

Pia, who had been looking past the mother superior at the crucifix, was stopped short. Suddenly, her suit of armor was dented, and she looked down at her shoes, feeling young and small. *Betray. Betrayal.* When she first encountered the word "betray" in a novel when she was eleven, she looked it up in the school's big dictionary. The definition seemed just right. That's what her family had done, it had *betrayed* her. *Betrayal* was the tragedy that had stalked Pia ever since she was six, on the day when the police burst through the front door of the apartment she shared with her father and uncle and placed her in the clutches of the New York City foster care program.

3

Columbia University Medical Center
New York City
March 1, 2011, 7:30 a.m.

She knows the man's important but she can't remember his name. The girl is standing in front of a long desk, wearing a plain, very loose-fitting, institutional gray shift with her shoulders slumped forward and her hands clasped in front of her, elbows tucked into her sides. Even sitting down, the man is very large, enormous, in fact, and he's leaning forward, talking to her, not looking her in the eye but right at her chest. She can't make out what he's saying. She's been bad, she's misbehaved, she's going to need to be punished, that's all she knows.

She can hear him now. He's even bigger than he was before, telling her to stand up straight, to pull her shoulders back. Why is she wearing that shapeless garment? Pia remembers that she's fifteen or at most sixteen and this is the head of her school, and it's as if she's at the back of the room, watching this girl who is her but not her. He pushes his chair back and stands up. Coming around the desk, he approaches her with a cruel, lustful smile. "Pia . . ." he orders. "Pia . . ."

"Pia . . . Pia . . . !" Pia sat up in bed and breathed a deep sigh of relief. Her T-shirt was sticking to the sweat on her back as she stretched forward, listening to George calling to her from the other side of the door. She let him in and hurried to get dressed, vowing that she'd remember to set the alarm that night. Her internal clock used to wake her up at six without fail but the last couple of weeks she'd had trouble sleeping, suffering from a number of recurrent nightmares. She felt exhausted. She'd not had nearly enough sleep. After visiting with the mother superior, she'd gone back to Rothman's lab. By the time she'd gotten back to her dorm room and crawled into bed it had been 4:23 in the morning.

As she dressed, she found herself mulling over the meeting with the mother superior, and she shared some of it with George on the way to the medical center.

"I'm glad you went," George said as they walked in the crisp morning sunshine. "I mean, you were never going to join the order, and I can't see you in Africa doing whatever missionaries do nowadays. I've never known any nuns, but I can't imagine you're the nun type." From one of their four lovemaking events, an erotic image of a satiated Pia naked in bed flashed through his mind. He noticed Pia shoot him a look, and he cringed. Could she read his mind? It wasn't the first time George had this worry.

"I don't think becoming a nun and taking the vows would have been a problem for me, George. I'm not even saying I'll never do it. I've seen how life is at the convent, and it's very peaceful. It's different from the world out here. The sisters support each other. It's safe."

George immediately felt uncomfortable, like he was patronizing Pia. He couldn't blame her for wanting some security in her life with the little he knew of her childhood. But becoming a nun? It seemed extreme. "I guess what I mean is that it seems like a way of avoiding life. There are other ways of being safe besides going and hiding out in a convent."

"I don't think of becoming a nun as hiding. It's the opposite — they have to give all of themselves to the world they have chosen." *They don't betray each other either,* thought Pia to herself. They had reached the Black research building.

"Actually I think you'd be equivalently hiding if you end up spending your whole career working in there with Rothman," George said, indicating the building with a tilt of his head. His concept of medicine involved helping people directly, one-on-one, having an effect on the lives of people he could see and touch. As far as he was concerned, research was too cold and abstract and populated with asocial martinets like Rothman who were as welcoming and warm as a file full of algorithms.

"So, what about lunch today?" George said, changing the subject and ever hopeful. As he had feared, they hadn't met up for lunch the day before. In the three-plus years that George had known Pia, they'd never had an official lunch date. They had lunched together numerous times, but not as a planned event. During the first two years they had managed to have pretty much the same schedule so it just happened. But now that George was on a radiology elective and Pia was holed up in Rothman's lab, he knew that the chances they'd run into each other by accident were slim. But why he was bothering to ask her he had no idea, since he knew it wasn't going to happen. And why was he always so damned accommodating?

"Sorry, George, I can't make plans," she said. "Yesterday I had to spend the entire day and come back at night to work on one of Rothman's journal articles, and it still isn't done. On top of that I'll be meeting with him sometime to find out what he has in store for me for the entire month. I seriously doubt I'll even be getting lunch."

Pia was unhappy to see that the pesky maintenance man was still in her office. He was up the stepladder again, only facing a different direction this time. The day before, as she had worked on Rothman's paper on one of the benches out in the lab proper, she'd noticed that he'd left at twelve and didn't come back for four hours. At that rate she worried about him being there pestering her and keeping her from her cubbyhole for a week. Her office was small, but it was hers and she could leave her stuff spread out on the countertops, something she couldn't do in the main lab.

Pia made enough noise dropping her bag on her tool-littered desk to ensure Vance knew she was there and not particularly happy. "Hey, you up there," she called out.

Vance pulled his head down into the room and, seeing Pia, climbed down, smiling, rubbing his hands on a rag. "Ah, Miss

Grazdani! How are you today? I missed you yesterday when I left."

"I noticed you took a four-hour lunch. You should have told me you'd be away so long. I could have been working here in my office. Anyway, yesterday you thought you'd be finished. What's up? How long is this going to take?"

"The job is turning out to be more difficult than I had thought. All I can really say is that I'm trying my best. As soon as I figure out what the hell is wrong, I'll knock it right out and be outta here."

Pia merely sighed irritably and lifted her bag.

"Miss Pia, I've got a surprise for you. I made an extra sandwich today for lunch, one for me and one for you. How about joining me for a bite? I make a wicked pastrami sandwich on a ciabatta roll. What do you say?"

He was smiling again. Jesus, men were so predictable. Pia glowered: Was this guy suffering from delusions? She wasn't staying to find out nor did she want to encourage the man.

"Just hurry the hell up with the job. Please!" she snapped. As far as the sandwich offer was concerned, she didn't even want to acknowledge it.

Pia turned and stepped back into the main lab. She put her bag on the bench area where she'd worked the previous day. But instead of jumping right in, she walked back to Marsha's desk to find out where their leader was that morning. To her surprise, she learned that Rothman was in his office and waiting for her. Pleased, Pia hurried in through the open door. Immediately she noticed he was dealing with the same maintenance inconvenience Overhead, a number of ceiling tiles were missing and spaghetti-like wires dangled from the holes. An assortment of tools dotted one of the countertops and a few were scattered on the floor. In the corner was a stepladder leaning against the wall and the security camera was missing from its mounting.

"Good morning, Dr. Rothman," Pia chirped. She never knew what to expect mood-wise but hoped for the best. "Marsha said you were expecting me."

"Miss Grazdani. How do you spell 'catheter'?" Rothman demanded, not even bothering to look up from the sheet of paper he was holding. She could tell it was part of the *Lancet* manuscript she'd worked on.

"C-A-T-H-E-T-E-R. Why?"

"Well, it seems you know how to spell it, so I'm wondering why you felt the need to make up an alternative version for my paper?"

Pia had worked on Rothman's article, making several suggestions for changes in structure and rewriting one whole section she found particularly opaque. Late last night she had been in a hurry to finish, and she hadn't run a spell check.

"One wonders what they taught you at NYU, if anything. There were several spelling errors and two grammatical ones."

From experience, Pia knew how Rothman worked. These jabs at her spelling and grammar almost certainly meant that he had accepted her structural changes. If you lived for compliments and praise, you'd starve to death working for Rothman. Rothman took good work for granted. If you weren't good, you didn't last long, so the only elements worth talking about were the minor faults. Rothman twisted in his seat to face his Mac and started pecking his way around the keyboard. Pia surmised he was adding her changes to the original manuscript. Pia took a seat without being asked. If she waited to be asked, she'd be standing all day.

Pia had enjoyed laboring over the *Lancet* piece. Scientific writing was something she enjoyed and seemed to have a facility for.

Over the previous three years, Pia had collaborated with Rothman on his salmonella studies and had even got credit as one of the authors on several. It had been exciting work. Rothman was continuing his award-winning, landmark research that he had accomplished concerning salmonella virulence, a subject for which he'd won his Nobel and Lasker prizes. Virulence was the microorganism's ability to invade and kill its host cells, something salmonella was particularly good at. Over the years Rothman had found, classified, and defined the five pathogenicity "islands," or areas, in the salmonella genome that encoded for various virulence-related factors such as specific toxins and antibiotic resistance, both of which had contributed to salmonella being by far the largest cause of human food-borne illness in the world. Every year salmonella caused the mortality and morbidity of countless millions of people. Every year typhoid fever alone still killed upwards of half a million people, a situation Rothman had his sights on rectifying and was coming closer to each year.

Initially when Pia first joined Rothman's lab, she had been more interested in his newer area of research, namely stem cells, and had hoped to work with them. But he had had other ideas and wanted her to cut her teeth with his continuing salmonella work. As time passed she'd become as committed as he in the microbiological arena, fascinated by bacteria and viruses in general and salmonella in particular and the microscopic realm they inhabited. Soon she found herself reveling in the involved science while enjoying the thrill of working with one of the greatest minds on the subject. On a daily basis Pia had come to relish refining her knowledge of genetics so that she could one day make her own contribution to basic research. Gradually she had come to realize how exciting research could be and how well it fit with her personality.

Pia watched Rothman type away in front of her. The level of his concentration was truly remarkable. One minute he'd been talking with her, the next he was totally absorbed, as if she were no longer in his presence. Pia did not take any aspect of his behavior personally. After he'd confided about his Asperger's, she'd read about the syndrome and guessed that many aspects of his personality were dictated by it, even ignoring her as he was doing at that moment. Instead of being annoyed, she thought about the content of the article she'd rewritten. It was about studies that Rothman had been doing involving salmonella typhi grown in outer space on the orbiting International Space Station. Rothman had found that growing the bacteria in a zero-gravity environment made it enormously more virulent than control bacteria grown back on earth. It was Rothman's belief that the conditions in space somehow mimicked to a marked extent those present in the human ileum, triggering the bacteria to turn on the genes in the pathogenicity islands to produce effector proteins. Pia was one of the few people who knew that at that moment in the refrigerated storage facility inside the biosafety unit there were three strains of these enormously virulent, space-grown salmonella. She also knew that what Rothman wanted to do was to figure out how zero gravity caused these changes with the hope of learning how to turn them off, not only in space but in the human ileum as well.

Although Pia had learned to be patient in Rothman's presence, she had her limits. After a few minutes had gone by Pia coughed lightly. She'd found by experience that coughing seemed to penetrate Rothman's concentration more than anything else. Almost immediately he peered around the screen of his Mac and pushed a box of tissues in her direction. He had a phobia about people coughing in his presence. He was, after all, a firm believer in the "germ" theory. Pia took one of the obligatory tissues.

"Right. Miss Grazdani, for this month's assignment . . ." He disappeared from her sight again. He resumed his two-finger typing but at least kept talking. She couldn't see his face but she preferred it that way, which he did too as both had trouble maintaining eye contact not only with each other but with everyone else as well. "I want to move you over into our induced stem cell work. You've done a knockout job with salmonella, but it's time you started in the other arena."

A smile of anticipation appeared on Pia's face. Rothman's words were music to her ears.

"We've been making breakthrough discoveries of late involving organogenesis."

Pia's heart picked up speed. It was the first time Rothman had talked to her about his stem cell work. She knew what organogenesis meant as the word was self-explanatory. It was now the cutting edge of stem cell research. It was the last hurdle before creating organs that could be transplanted into patients — organs like hearts, lungs, and kidneys. It thrilled her to think that Rothman was making enormous leaps forward. And the idea that she would become part of the effort gave her chills down her spine.

"At this stage our biggest problem is that tissue culture techniques and fluids have not kept up with the breakthroughs that we're making. Current tissue culture techniques were developed for sheets of cells, not solid organs. I'm sure you can gather what I mean. It relates to oxygenation and removing metabolic waste while maintaining acid base balance within extremely narrow parameters. It's been basically a combination of pushing the limits of biochemistry and engineering. We have come up with some impressive hardware breakthroughs, but the involved fluids have not kept pace. The problem that is now holding us back is the acid base balance. My guess is that the pH is varying too much. We can't figure out why. What I want you to do is to become a tissue culture fluid expert and figure out why we're having this pH problem. Got it?"

"I think so," Pia managed. She had learned that it was never a good ploy to question any of Rothman's directives. Anything and everything could be discussed again, but not on the spur of the moment.

"Good! Get to it! And when I finish making these changes in the manuscript, I'll see that Marsha gets you a copy for your final review. Now get out of here!"

Rothman's typing picked up pace, a few keystrokes followed by several frantic deletions. Pia kept her seat despite Rothman's final comment. She sensed that this was all the information she was going to get about her month's elective at the moment, and it wasn't much. Inwardly, she shuddered a little. She had expected to be working on some aspect of Rothman's salmonella research as she'd done in the past. Tissue culture was a new discipline for her and what she was being tasked to do sounded like an entire Ph.D. project, not a month's assignment. She was going to need a lot of help from Rothman and from the other technicians, especially Nina Brockhurst, whose job it was to take care of the physical plant of Rothman's organ-growing experiments, which would include the baths. In the past Nina had openly resented Pia, claiming Rothman played favorites with her. Pia had taken the situation in stride as she knew there was always intrigue when people were forced to work together, especially when the boss's signals were so hard to read.

But whatever the workload and her colleague's demeanor, Pia knew she was going to find the month fascinating. Even if the fluid bath assignment wasn't, on the face of it, very exciting in and of itself, it was still vital experience she would be gaining, learning the basic techniques for taking care of newly created organs, a key stepping-stone in the journey from studying organogenesis in mice to studying it in people. Most important, the work was in the stem cell arena: the place she believed she really wanted to be.

Pia coughed again, this time into the tissue that she had in her hand. Rothman's face reappeared around the side of his Mac. His expression was one of surprise that Pia was still there.

"I went to see the mother superior at the convent last night," she said. "I told her about my not wanting to go to Africa."

"Good," Rothman said simply. His face disappeared. The typing recommenced.

"She was nice about it, but I could tell she was not happy."

"That's her problem, not yours. You'll be doing a lot more of God's work here in my lab than going to any godforsaken part of Africa."

"She said she did not want to be repaid."

"Good for her. So don't."

"I think I should. Are you still willing to cosign for a fifty-thousand-dollar loan?"

"I am, but I think you're crazy. She doesn't want to be repaid, or so she says. Save your money."

"She used the word 'betrayal,' " Pia said. She knew she was distorting the reason the mother superior had chosen to use the word, but the fact that she had used it at all still bothered Pia.

Rothman gave a short, mocking laugh. "Betrayal! She's just trying to foist some Catholic guilt on you, Pia. For chrissake, pay her the money if you need to and be done with it. I'll have Marsha take care of it with my bank. As a fourth-year medical student, I'm sure your credit is adequate. Remember, it's your life, not the mother superior's. Now, get out of here and get to work."

Pia got up and left Rothman to his typing. Passing Marsha, she thought about hitting the library. Her initial plan was to read everything on tissue engineering that she could get her hands on. She had no doubt it was going to be an overwhelming amount of information.

4

Columbia University Medical Center
New York City
March 1, 2011, 1:15 p.m.

Having loaded up on books and printouts from a Google search in the library, Pia spent the morning reading, sitting deep in concentration at a bench area outside her windowless office. She'd been vaguely aware of the maintenance man doing his thing, but she had ignored him until he came up behind her. Disregarding the fact that she had her iPod buds in her ears, he had the nerve to tap her on the shoulder.

"Pia, darling dear. Want to reconsider my pastrami offer? You won't be disappointed."

"Not on your life," Pia said, with emphasis, hoping he'd get what she intended as a definitive message. He shrugged and smiled and gave a little silly wave as if Pia had been gracious rather than scathing. She was beginning to think that Vance was one of those men who reveled in rejection. Irritably, she repositioned her earplugs and went back to her reading. When she heard her name being called again, Pia was momentarily livid, wondering what she was going to have to do to get him to leave her alone. Tearing out her earpieces, she looked up to see Rothman's assistant, Dr. Yamamoto, standing in front of her flanked by a young man and woman in newly laundered, dazzlingly white lab coats.

"Miss Grazdani," Yamamoto said. He was a slight man with what appeared to be a kind of half-smile frozen on his face. "I would like to introduce our new students with us for the month."

Dr. Yamamoto was often held up within the medical center as a perfect example of how opposites attract. He was well liked, soft-spoken, considerate and communicative, always encouraging people to address him as Junichi: the positive to Rothman's negative. Also in contrast to Rothman, he was casually dressed as always, in a Hawaiian shirt beneath his wrinkled and not-too-clean lab coat. If there was one nod to his playful side, Yamamoto was said to be the originator of elaborate and brilliant practical jokes stemming from his graduate student days involving one particularly pompous medical student's expensive fool's errand to a "convention" in Geneva that never was. On the serious side, Dr. Yamamoto's most important characteristic was his complete and total devotion to Rothman and Rothman's work. What was accepted around the medical center was that Rothman was the brains and Yamamoto was the worker bee. They were yin and yang.

"Perhaps you know Lesley Wong and William McKinley," Yamamoto said.

"Like the president," the young man said. "But call me Will."

Will stepped forward, a big smile on his face, hand outstretched. Columbia University Medical School had about 640 students spread over four years of training. In general the first two years were primarily spent absorbing the science of medicine with progressively more and more time creatively devoted to introducing students to patients. The third year was the principal clinical year with the major attention devoted to internal medicine and surgery. The fourth year was mostly rotations in various clinical subspecialties combined with electives according to each student's personal interests. At Columbia the emphasis was on academic medicine. Lesley and Will were fourth-year students in Pia's class. Both thought they had a new interest in research, which was why they had been assigned to spend a month in Rothman's lab.

Pia took Will's outstretched hand and stood up.

"Pia. Grazdani." She noticed that Will was tall, even a little taller than George, who was above average. Like George, Will had blond, unruly hair.

"You're George's friend, right?" Will said.

"George? Yes, of course."

"Love George, great guy. I often play b-ball with him."

"I'm Lesley Wong," the woman said, shaking Pia's hand in turn.

For a moment there was an awkward silence. Pia briefly eyed the two students, realizing they had to be the students Rothman had briefly mentioned the day before and then never brought up again. He had said something about tasking her to come up with something for them to do, as if she wasn't going to be busy enough. One way or another it was going to be a burden of sorts.

Lesley and Will eyed Pia back. For their part, they weren't terribly excited about meeting her either. For them, finding out they had been assigned to Rothman's lab was the equivalent to being sent someplace in Dante's inferno. Rothman had the reputation of destroying every student's sense of self-confidence by making them feel stupid, which they invariably were in comparison to Rothman's encyclopedic knowledge. And they had heard about Pia as well. She too was known as being over-the-top smart and also strangely detached and had taken an early interest in research in addition to the regular curriculum. For most people being a medical student was demanding enough. Except for being tight with George Wilson, who was one of the more popular students in the class, a mark in her favor, Pia never had had the time or inclination to be particularly friendly with many of her classmates. And all that was on top of the gossip that she and Rothman had something going since she was the only person in the entire medical center that he got along with except for Dr. Yamamoto.

Lesley looked over at Will, but he was staring at Pia. When they'd received their assignment, Lesley had told Will that she'd sat next to Pia in a lab every day for a month the first year but she was sure Pia wouldn't remember. Lesley hadn't made up her mind if Pia was extremely focused on her work or just plain rude, although she thought it was the former. As for Will, he was excited to be finally matched with Pia — he'd wanted to be for three and a half years. He'd been sure to introduce himself to every woman in the student body he deemed attractive, but this was the closest he'd come to her.

"Okay. Introductions done," Yamamoto said. It hadn't been quite as awkward as he feared, and he was relieved. He could get down to business.

"If it's okay, we can go to my office. I'd like you to come as well, Pia. There are a couple of things we need to talk over and then we can all go take a look at the organ baths with Dr. Rothman."

Yamamoto beamed a smile and walked away, the two new students padding closely after him. Pia brought up the rear, reluctant to be leaving her reading but excited at the same time. Even though she had been working there for years, she had never seen the organ baths. Although she had spent more actual time with Dr. Yamamoto than she had with Rothman, she didn't feel she knew him as well. In her mind he was more complicated than Rothman. She thought of him as a kindly man but knew that in his own way, he was just as demanding as the chief. He suffered fools no more gladly, but his reprimands and corrections were delivered more politely and at lower volume. Pia had gathered from experience that the quieter Yamamoto spoke, the more important it was to listen.

"Okay, guys, find a seat."

The condition of Yamamoto's office in comparison with Rothman's was as different as the men's personalities. Yamamoto's looked like a typhoon had passed through it. Books, journals, files, documents, papers lay everywhere, including on each of the two seats in front of the desk. Visitors took it on faith that Yamamoto actually had a desk as every square inch was under paper, including a mountain of academic journals positioned so that a curious passerby couldn't see whether the doctor was sitting behind his desk or not.

"Just move those papers," Yamamoto said, as Pia and Lesley Wong gathered them up from the chairs but searched in vain for a free surface on which to put them down. Yamamoto gestured to the floor, a suggestion the women took. He leaned his posterior against the front of the desk and folded his arms. "You could get a chair from the lab," he suggested to Will.

"No problem," Will said. "I'll stand."

"I thought it appropriate for us to do a small review and go over some introductory material so you people can better appreciate what you're about to see," Yamamoto said. "You are in for a treat today. I've gotten special dispensation from Dr. Rothman to show you our organ bath program, which has been kept a secret of sorts up until now. It couldn't have been kept a complete secret as there are too many people working here, and we're here in a very public medical center. Since the professor and I are close to publication, secrecy is no longer the issue it was since the university has already seen to it that the appropriate patents have been applied for. At the same time, we would prefer that you keep what you see today to yourselves. Deal?"

All three students nodded.

"All right, let's start at the beginning. But I don't want to make this a boring monologue so help me out! Somebody tell me what a stem cell is."

The three students eyed one another. Will spoke up. "In simple terms it's an undifferentiated immature cell that has the potential of becoming a differentiated mature cell."

"Right on," Yamamoto said. "An example is a bone marrow stem that can become an adult blood cell. These cells are often called adult stem cells. What's a pluripotent stem cell?"

Lesley spoke up: "A stem cell that can turn into any of the three hundred or so types of cells that make up the body of a multicellular organism like a human."

"Right on again. You guys are making this easy for me."

Pia felt a wave of impatience wash over her. She was eager to see the organ bath unit. Had it been up to her, she would have preferred to forgo any review session.

"Up until four or five years ago, how were pluripotent stem cells obtained?"

"From blastocysts." Pia spoke up by reflex. She wanted this little talk over with.

"Right," Yamamoto said. "Blastocysts from fertilized eggs, meaning very early-stage embryos. Why was that a problem that led to serious delays in stem cell research?"

"Because it offended conservatively minded people," Pia said. "Particularly here in the United States, limitations were placed on what could and could not be done in stem cell research with government funds."

"Well said," Yamamoto commented. "Here's a harder question. Let's say that the research on embryonic stem cells had been allowed to proceed unimpeded. Can anyone say what the major problem would have been if the research had advanced to a point of using the stem cells to treat patients?"

None of the students moved.

"Let me give you a hint," Yamamoto said. "I'm referring to an immunological problem."

"Rejection!" Lesley called out, her eyes lighting up.

"Exactly. Rejection, meaning that any use of such embryological stem cells would have elicited some degree of rejection reaction. Some techniques would have reduced this problem but not completely eliminated it."

All three students nodded. Everything that Yamamoto was saying they had heard before. "Now, can anyone define 'induced pluripotent stem cells' in contrast to embryological stem cells? These are the cells that Dr. Rothman and I have been working with exclusively."

"They are pluripotent stem cells made from mature cells, usually a fibroblast and not egg cells," Pia explained. "They are 'induced' by particular proteins to revert back from being a mature fibroblast to being a stem cell."

"Exactly," Yamamoto said. "And isn't it a marvel that it works? For a long time one of the tenets of biological science was that cellular differentiation was a one-way street, meaning the process could never revert. But people should have known that this particular tenet was false. After all, it was known that certain animals could regrow body parts, like starfish and salamanders. Also cancer should have been a hint that the process of differentiation could go in the opposite direction, as many cancers are composed of immature cells that arise in organs populated by mature cells."

Pia found herself glancing at her watch and sitting up straighter in her seat. She wanted to speed up the review session but didn't know how. She inwardly groaned when Lesley piped up with a question: "How exactly are the cells changed back from mature to immature?"

"The same way that everything else is accomplished in the cell," Yamamoto said. "By switching on and off genes. Remember, every eukaryotic cell — that is, a cell with a nucleus — contains a copy of an organism's entire genome. Meaning that every nucleated cell has all the information necessary not only to build itself but to build the entire body. How this works is a process called gene expression, meaning the turning on and off of genes in a kind of molecular ballet. I know you learned all this in your genetics courses in college and during your first two years here at Columbia. Anyway, cellular maturation proceeds by a sequential switching on and off of the appropriate genes. It used to be thought that genes functioned by producing specific proteins, sorta one gene for one protein. But now we know it's far more complicated, as there are significantly fewer genes than originally thought. For the cell to go in the opposite direction of maturation, the sequence has to be reversed. Are you all with me so far?"

All three students nodded. Despite feeling impatient, even Pia was now finding Yamamoto's review fascinating. Like all other researchers, Pia was aware that biological science was unfolding its mysteries at an ever-increasing, mind-boggling speed. The nineteenth century had been chemistry, the twentieth physics. The twenty-first was undoubtedly going to be biology.

Yamamoto checked his own watch. As if answering Pia's hopes, he said, "We've got to move this along if we're going to catch Dr. Rothman in the organ bath unit. Let's go back to our discussion of stem cells. Now that we have the induced pluripotent type that are going to avoid immunological rejection problems and be more acceptable to religious conservatives, what is the first step toward making them useful to treat the patient who donated the fibroblast? Anybody?" Yamamoto glanced from face to face.

Will shrugged and offered, "Get them to mature again but into the kind of cell the patient needs."

"Thank you," Yamamoto said. "Indeed, that is exactly what most stem cell researchers have been busy doing for years: finding out how to regulate gene expression such that the stem cells mature into the kinds of cells that make up the body, like heart cells, kidney cells, liver cells, and so forth. Stem cell researchers have now gotten very good at this, including Dr. Rothman and myself. But here is where Dr. Rothman and I have separated ourselves from the pack and are about to usher in twenty-first-century regenerative medicine that's going to extend and improve the quality of life. We have been able to make virtual leaps in the ability to have these mature cells organize themselves into whole organs. In other words, we've managed to stumble on a host of structural genes and other transcriptional processes that are responsible for creating the lattice-like scaffolding that forms the basis of a three-dimensional organ. Once we had the structure, it was relatively easy to get it populated by the appropriate cells.

It's a process called organogenesis. Take, for instance, a liver. Although we and others have been able to make hepatic cells for years, we have never been able to get them to organize themselves into a whole liver with collagen, nerves, and blood vessels, the whole deal. We can do it now. We're doing it with rapidly increasing efficiency. It's phenomenal."

"I assume you've been doing this with animal models?" Pia said.

"Of course! Mostly mice. The whole stem cell field has extensive experience with the murine model."

"And you believe what you've been learning will be applicable to human cells?"

"We do, and not only on a theoretical basis. Concurrently we've been carrying on this research with human cells as well." Yamamoto held up his left arm and pulled his lab coat sleeve down with his right hand. Proudly he pointed to a number of inch-long scars of varying age along his forearm. "I've been the guinea pig for the source of human fibroblasts. Although most of our research is done with mice, we have some human organs functioning equally well, human organs that could be used to treat me if I needed one of them. You'll see in a few minutes. Any questions before we head over to the unit?"

Yamamoto looked at each of the students in turn and then paused. Finally he said, "Okay, let's make our visit. Hope you guys are ready. You are about to visit the future." He pushed himself up into a fully erect posture.

When the women started to replace the papers and journals they had removed from the chairs when they first arrived, Yamamoto motioned for them not to bother. With Yamamoto in the lead, the group exited his office. They walked the length of the sizable lab since the organ bath unit was located at the far end on the opposite side from the biosafety unit. As they passed through the lab, some of the technicians looked up from their work and eyed them questioningly. Visitors to the organ bath unit were not common.

First they entered an anteroom where there were caps, gowns, booties, masks, and gloves. There was a similar room on the other side of the lab that guarded entrance into the biosafety unit, just as this room guarded the organ bath unit. Interestingly enough, the rationale was just the opposite. For the biosafety unit the gear was for protection of the visitors. With the organ bath unit it was called reverse precautions and was for protection of the contained specimens. It wasn't until everyone was suited up and had been checked by Yamamoto that they proceeded.

5

Columbia University Medical Center
New York City
March 1, 2011, 2:00 p.m.

Using a keypad, Dr. Yamamoto punched in a combination, unlocking a simple door. Although she wasn't trying, Pia noticed it was the same code used for the biosafety unit, an alphanumeric sequence that was the time and date of Rothman's birth. Yamamoto stepped aside and ushered the three students into a starkly modern, brightly lit room filled with the gentle, hypnotic sound of running water. Pia felt a slight breeze against her as she advanced. She knew that was evidence of the room having positive pressure, meaning air came out of the room, not in. That was the opposite of the biosafety room, where the laminar flow was into the room.

Stepping beyond the door, Pia shaded her eyes against the sharp bluish light that came from banks of recessed fiber-optic fixtures. She assumed the light had something to do with sterility in the room. Stopping with the others, she took in the scene before her. They had entered a large space painted bright white. Pia was puzzled as to how this lab fit in with the rest of Rothman's suite — it seemed larger than it feasibly could be. At the back of the room, a figure dressed in similar protective garb was hunched over a stainless-steel cart mounted on wheels, making some adjustments on a control panel. There were three rows of such carts and Pia counted thirty of them in all. Set atop each were clear rectangular Plexiglas aquarium-like containers of various sizes. Below were shelves holding various and sundry equipment. Each also had an attached pole supporting a control panel with an LED display. One of the carts was a few feet to Pia's left and she moved over to take a closer look. Lesley and Will followed.

This was one of the organ-growing baths whose contained fluid was what Pia would be investigating for the month. She leaned over and peered into the container at a miniature translucent object suspended in the liquid by a kind of spiderweb that she was later to find out was made of the same material as real spiderwebs. As for the object itself, she could see it was connected by filament-thin lines to a central clamp where the lines were gathered. A thicker cable fed out of the bath and down into the body of the cart, where there was a boxy device with multiple readouts monitoring the conditions in the bath. An attached magnifying device on a movable arm was also connected to the cart. Pia maneuvered it into position so she could see the contained object better. Although it was not too much larger than a pine nut, its appearance was unmistakably that of a kidney, albeit a very tiny kidney. Most of the lines contained red fluid along one larger one that was clear. The red ones were presumably functioning as veins and arteries. The clear one was acting as the ureter to take away the urine the miniature organ was producing. To one side of the container was a jet, like one attached to a pool circulator, only in miniature. It was pulsating at a very rapid rate. Gentle eddies of liquid coursed around the container, causing the organ's surface to pulsate slightly.

"We find that we must keep the fluid in the baths in constant motion despite the organs being perfused internally. But it has to be carefully modulated. Occasionally the rigid bath sets up a wash that can disturb the organ." Yamamoto had stepped over to join the students. He noticed Pia straighten, looking down the length of the room.

"It's quite something, isn't it?" he said, speaking directly to her. "Of course, I see it every day, so I've come to take it all for granted."

"How are the organs started?" Pia asked.

"They are started in tissue culture dishes designed to mimic the mouse uterine environment in terms of temperature and with pulsation waves close to the normal mouse heart rate of around five hundred and fifty beats a minute. As I said earlier, the whole process, first in the tissue culture dishes and then in these organ baths, is a ballet of gene expression with a careful adherence to sequence and timing. It starts with an aliquot of induced pluripotent stem cells held in close proximity by spiderweb-like restraints. Remember, to form a whole organ we have to involve all three germ layers: ectoderm, mesoderm, and endoderm. Once the organ has reached a size that can be manipulated, it is moved into these baths to develop to its full extent."

"Are there other organs in here besides kidneys?" Will asked.

"Heavens yes," Yamamoto said. "We've got all the usual transplantable organs such as livers, pancreases, lungs, and hearts so far. The kidney program is the most advanced, since it was kidneys we started with. To prove we are on track with what we have been doing, we have already transplanted some organs back into the individual mice from which the fibroblasts were taken with complete and utter success. And let me share another leap forward that we are in the process of making. We've found that carrying out organogenesis with multiple organs works even better than growing them singularly, meaning we have preparations in which the developing organs are helping each other, like the heart pumping the perfusing fluid and the kidneys removing waste."

"Do you think sometime in the future you could essentially make a whole new organism?" Pia asked with astonishment and not a little dismay.

"At the rate we're going, I see that as a definite possibility, although I can't imagine what the rationale would be."

Pia reflexively shuddered as she realized that Frankenstein, that nineteenth-century nightmare, could very well resurrect itself to haunt the twenty-first in a frighteningly more plausible fashion. If the Rothman-Yamamoto organogenesis worked well with kidneys, hearts, and pancreases, there was no reason it couldn't work just as well with brains.

"Where are the human organs?" Pia asked.

Yamamoto took a few steps down the kidney line and pointed into a larger Plexiglas container. "This is human, as you might expect considering the size. It's also one of the composite preparations with a human heart to do the kidney's internal profusion."

Pia stared into the bath, transfixed by what she was looking at. The kidney did look human, but the heart did not. She asked Yamamoto why.

"Good question. Since oxygenation of the perfusing fluid is being done by the oxygenator on the lower shelf, we did not need a four-chambered heart as two would do. So we altered the design of the heart."

Once again Pia was amazed. "You have that much control of the organogenesis process to alter the overall three-dimensional architecture?"

"Absolutely. As I mentioned, once we made the original organogenesis breakthroughs, our progress has been truly phenomenal and isn't slowing down."

The figure Pia had seen earlier finished what he was doing, stood upright, and came toward the group. As he neared, and despite the surgical mask, Pia was further surprised. She could tell it was Rothman. Wearing some kind of goggles with thick, tinted lenses, he made for an eerie figure, like the prototypical mad scientist in his lair. Pia knew that what Dr. Yamamoto had said outside was true — this truly was groundbreaking work. In the stem cell race to move from the promising hypothetical to the clinical, Rothman and Yamamoto had advanced much further than any other team in the world.

Rothman moved the goggles to the top of his head as he came to a stop. He looked at Yamamoto. "Have they been given a short introduction?"

"Yes, Doctor."

Rothman nodded. He knew he was going to have to show off his work to any number of interested biotech venture capitalists over the coming years, even though it wasn't something he enjoyed or found easy. Yamamoto had helped him prepare a script that he'd practiced again and again with his wife. The students were to be a kind of dress rehearsal.

"Welcome to Columbia University Organogenesis Laboratory," Rothman said. Yamamoto coughed gently into his hand. Rothman had trouble deviating even slightly from the prepared text.

"It is common knowledge that there are currently more than one hundred thousand people on waiting lists for organ transplants in this country, and these are people with end-stage disease. The list grows at a rate of about five hundred a month. Currently the same number of people, five hundred or so, die every month. On top of this grim statistic there are thousands upon thousands of additional patients who could benefit from an organ transplant even though they are not yet in a life-threatening situation. Obviously in the current environment the supply of viable organs from either a live donor or a recently deceased individual has not come close to keeping up with demand. Even for those patients lucky enough to receive an organ, the match is often far from optimal, meaning they are relegated to a life of immunosuppression with dire health consequences. What we are doing here, in a cost-conscious fashion, is to create organs which will simultaneously solve the supply problem and the immunological issue. This goal has not yet been reached, but we are making significant progress. At this stage we are looking for outside funding to ramp up production at multiple centers across the country.

"What you see in this row of baths are kidneys that have been created from stem cells derived from fibroblasts — connective tissue cells — of specific mice." Yamamoto tried to interrupt to say that Rothman was covering material that had already been mentioned, but he couldn't get Rothman's attention. Rothman was on a roll. "I wear these magnifying lenses so I can work with the lines, but take my word for it, each organ is hooked up to a pump that circulates a blood-like solution into the kidney's main artery and out its main vein. It's connected by a cannula, or thin tube, from its ureter to a port where its urine output can be sampled. That's one of the functions performed by the monitoring unit beneath the bath. All the data is collected in the mainframe so we can see how tiny fluctuations in conditions affect the kidney and its development.

"Each kidney will soon be implanted back into the same mouse that supplied the original fibroblasts. We've already done this twice with no rejection phenomena whatsoever." With his hand, Rothman gestured toward another group of baths. "These vessels contain pancreases, which have quite different needs than kidneys. Initially we had more difficulty getting the organogenesis process to start than we had with the kidneys, but those initial reverses have been solved, and we are now doing equally well. With the pancreases, we have had to be very careful about the integrity of the connections with the pancreatic duct since the pancreatic secretions contain digestive enzymes. Initially some of our preparations digested themselves."

"Have there been any problems with teratomas?" Pia asked. In contrast to the others, she did not feel intimidated by Rothman. She knew that teratomas, a kind of developmental tumor, were something feared by stem cell biologists.

For a moment Rothman faltered. He had not been expecting to be interrupted in his prepared remarks. Except for the sound of moving fluid emanating from all the organ baths, a brief silence reigned.

"No teratomas at all," Yamamoto said, coming to his boss's aid. He was well aware of his boss's quirky personality.

As if he had forgotten the students' and Yamamoto's presence, Rothman's attention diverted to the appearance of a small blinking light accompanied by a pinging noise coming from the control panel of one of the baths. Without a second's thought or explanation, Rothman headed in its direction, slipping his goggles back on as he walked.

"That's an alarm that some aspect of the bath's parameters has started to change," Yamamoto explained.

The students watched him go. Lesley and Will were awed at having been in the famous researcher's presence and having survived without being belittled. Pia was impressed by the alarm: "What would have happened if no one had been in here to hear the alarm?"

"Not a problem," Yamamoto said. "All information is followed in real time by the university's mainframe, and Dr. Rothman and I have apps on our iPhones such that we would have been instantly alerted."

"Earlier Dr. Rothman talked to me about a problem with the tissue culture fluid," Pia said. "Was he referring to the fluid in these baths?"

"I'm sure he was," Yamamoto said. "We've been having a continuing problem maintaining the correct pH balance. Did he ask you to look into the problem? Because if he did, it would be a great help. It's not been a particularly big problem, but neither of us has had a chance to look into it. I know I'd feel a lot better if we could solve the issue."

"I'll give it my best," Pia said. "The problem is I'm starting at ground zero. I've had no experience whatsoever with tissue culture."

"That didn't seem to bother you in relation to salmonella," Yamamoto said.

Pia smiled behind her mask. She took Yamamoto's comment as a compliment. "What about Lesley and Will? Maybe it would be appropriate for them to give me a hand."

"That's a great idea," Yamamoto said. He looked at Lesley and Will. "How does that sound to you?"

Both students shrugged. "Sounds good," they said in unison.

As they left the organ bath unit, Pia turned just before exiting. She looked back at Rothman tending to the bath. The pinging had stopped. Once again the thought of the mad scientist in his lair popped into her mind, and once again she shuddered. She'd visited the future in this room and was excited to become a part of it. At the same time she knew instinctively that there could be a dark side. Biological science was advancing almost too fast, and the problem with science is that it cannot be unlearned.

6

Greenwich, Connecticut
March 1, 2011, 3:30 p.m.

Edmund Mathews went to answer the front door of his waterfront mansion in an extra-exclusive enclave of the already-exclusive Connecticut town of Greenwich. It was unusual that he was alone in the house, but his wife, Alice, had gone to the city with a girlfriend on a shopping expedition, and the au pair, Ellen, wasn't back from school with Darius yet. There was no gardener on the grounds, no workman in the house, nor were there any painters, decorators, deliverymen, mechanics, cooks, or anyone else anywhere on the property. The $10 million house was quiet and unattended, just the way Edmund liked it.

This will be Russell, Edmund thought. Edmund and his partner, Russell Lefevre, had decided to take this Tuesday off because their work was about to get crazy. This was going to be the last free day they would have in months, and it seemed like now he was losing some of his free afternoon. Russell had called a few minutes earlier, sounding upset, and said he wanted to come over immediately to talk about something important. Russell had a habit of insisting on talking about anything sensitive in person. Back at Morgan Stanley, when they worked together in asset-backed securities, their calls had been recorded in case either party misremembered the terms of a trade later on. Edmund doubted very much that anyone was listening in nowadays but the old habit lingered with Russell. He was a worrier and he always had been.

Edmund opened the door and greeted Russell. His partner was a tall, lithe man with a sweep of blond hair tinged with gray. He was wearing tennis whites with a sweater thrown over his shoulders. For a man who was usually quite a dandy, he looked thoroughly bedraggled. When he wasn't in a suit, Edmund preferred to wear old T-shirts and shorts, even in winter. He was thicker in the body than Russell, but not overweight, and he kept his hair short and neat with weekly trips to the barber in town.

Edmund could see that Russell had haphazardly parked his Aston Martin DB9 in the driveway and not over by any of the garages as Edmund preferred. The Aston Martin was a fine piece of automotive engineering but it was too ostentatious a machine for everyday use for Edmund's taste. The garish crimson paint job only exacerbated the feeling. Edmund preferred the in-your-face statement he made in his black Escalade, but for driving enjoyment, he loved nothing better than taking his Morgan runabout on the back roads deep into Connecticut. His true pride and joy he drove only rarely: in his garage was a Ferrari 250 GTO that had cost him millions back in the days when that didn't seem like such an extravagance.

"We've got a problem," Russell said as he entered the atrium.

"So I gather. Let's go into the kitchen," said Edmund, who preferred to keep business discussions out of the house if he could help it. This was going to be one of those days he didn't have any choice.

Russell and Edmund had both worked as derivative traders at Morgan Stanley. Edmund was one of the best traders there, agile and decisive and brilliantly able to find someone to take the other side of a position he was holding. He knew Russell had some limitations as a trader, but he had a quant's mind that could calculate risk quickly, and Edmund could rely on him to tell him if something he was planning was feasible. Russell had seen the potential for making money in CDOs — collateralized debt obligations — exotic financial products that took advantage of the subprime mortgage market to create apparently risk-free investments that could make billions in profits for the company and tens of millions for the traders. With property prices on their seemingly unstoppable upward curve, the investments were safe as houses, as people in the know liked to say.

Eventually it turned out that many of the executives at the brokerages selling CDOs and at the financial institutions here and in Germany and Japan and elsewhere who bought them were completely ignorant of what a CDO actually was. They knew what asset-backed securities were, but the assets here were mortgage bonds packaged together and sliced up and sold in bundles. Many of the individual loans the bonds were backed by were subprime loans that would never be paid off, and only a few loans needed to fail before the whole package defaulted. It was inevitable that this would happen.

When Russell explained to Edmund precisely what the subprime loan crisis was going to mean to CDOs and other financial products and for the system as a whole, Edmund was unnerved and excited at the same time. He immediately, and secretly, used his own money to short his own firm and made bets on the failure of other companies exposed to CDOs. He continued to sell the doomed bonds even when disaster was inevitable. He made staggering amounts of money and after a while he told Russell, a loyal company man who'd never dreamed of acting that way, what he was doing. As Edmund predicted, Russell wanted in, and Edmund gave him some of his action.

As the banking catastrophe unfolded there were many victims: investors who'd lost their money, shareholders who found their stocks worthless, countless workers who lost their jobs. Men like Edmund Mathews and Russell Lefevre were not among them. Amid a clamor that the bankers involved should go to jail, they left the firm with close to $100 million in compensation between them.

Edmund had enjoyed his first weekend of unemployment to some extent, taking Darius to soccer practice without bringing his BlackBerry, having dinner with Alice and another couple in town, reading the Sunday paper. But by 9:05 A.M. on that first Monday, he was bored stiff. In his home office, he had two screens showing Bloomberg and MSNBC and he noodled around, making minor trades for a few tens of thousands of dollars through his online account. At ten, he called Russell and suggested they get back in the game on their own.

"Okay, Russell, what's the problem?" Edmund said, after handing Russell a glass of ice water. The men stood at either end of the island in the middle of the state-of-the-art kitchen. Edmund flipped Russell a place mat before he could put his water-beaded glass on the butcher block.

"I was playing tennis with Teddy Hill —"

"Teddy Hill? He's got to be sixty-five. I hope you went easy on the old boy."

"Ed, this is serious. I play with Teddy because he knows everyone, and he tells me things he hears. As he did today. When he told me, I practically ran off the court, left him standing there."

"Told you what, Russell?"

"We're being shorted big time."

Russell was right. This was serious.

When he called Russell during his first Monday of alleged freedom, Edmund found that Russell was as anxious as he was to get something going. Unbeknownst to Edmund, Russell needed to be earning. In 2008, he found himself personally long on real estate, owning a portfolio of properties in Florida and California suddenly worth a lot less than their outstanding mortgages. When Russell had fixed his problem, he was low on cash and needed to leverage his severance money into something more substantial.

As they had done many times as part of a large corporate group, the two men took a weekend away to a hotel in Boca Raton to brainstorm. Before getting down to business, Russell insisted on going to the local mall to pick up some T-shirts for his four kids. Edmund waited for Russell outside the Gap and watched people passing by.

"Look at the people, Russell," Edmund said when his partner returned. "What do you see?"

"Families, strollers, couples, lots of old people. What's on your mind?"

"Right. Old people. It's Florida, famous for oranges and old people. What do old people have?"

"I dunno, high car insurance premiums?" Russell said.

"That," Edmund said, "but this generation also has lots and lots of life insurance."

And Edmund told Russell his idea. It was called "Life Settlement."

The partners figured they had stumbled upon something big. Russell crunched numbers for weeks while Edmund discreetly got advice from his old contacts: lawyers, traders, bankers, ratings experts, and hedge fund managers. The idea was legal, and it was doable. And Russell said the numbers were watertight.

"The only way this doesn't work is if we have the Second Coming and Jesus stops people dying," Russell said.

"And we know that's not going to happen."

In early 2010, LifeDeals, Inc., was formed with Russell as CEO and Edmund chairman of the board. The start-up money was most of their $100 million take from the subprime debacle, and they used it to buy up life insurance policies from thousands and thousands of Americans desperate for cash. Edmund hired the most aggressive salesmen he knew and told them to hire even more hungry people to go out and pound pavement and buy policies for no more than 15 cents on the dollar. There were millions of Americans who needed money for long-term care, or to finance an operation when they didn't have medical coverage or, as was increasingly the case, even when they did, but the coverage wasn't adequate or the insurance company figured out a way not to pay. LifeDeals had to pay the balance of the premiums, but when the policyholder passed on, as they would as surely as night follows day, the payout was theirs.

Within six months, the LifeDeals board was confident enough to take the company public. Edmund and Russell held options that made them very wealthy once again, but they wanted capitalization to buy more policies. Edmund's favorite statistic was that there was more than $26 trillion in life insurance policies sloshing around out there for the taking. Their plan was to start securitizing the policies, aggregating them, and selling bonds. This time, the assets behind the securities were cast-iron, personally guaranteed by the grim reaper. And thousands of people every day were walking away from policies they had paid into for years because they couldn't afford the premiums. They were waiting to be picked off.

Edmund liked to think his company could someday be worth a trillion dollars.

■ ■ ■ ■

"Who is it?" Edmund asked.

"Teddy doesn't know. He heard it from a friend who heard it from a friend. But he trusts the party. Swears it's true."

"It's just someone being a wiseass," said Edmund.

"No," said Russell. "It's a biggish bet. Whoever it is, they're sure we're going down the toilet."

"Well, we fucking well better find out who it is before we catch a cold."

Russell knew the implications as well as Edmund. They needed a large institutional investor to underwrite their securitized package, and if it was known on the street that LifeDeals was getting shorted, that partner would be very hard to find. Everyone remembered what happened in 2008.

"We need to start going through the 13Fs right away," Russell said, referencing the quarterly statements institutional investment managers had to file with the SEC outlining their holdings.

"And I need to start making some calls."

Russell had left Wall Street with more intact relationships than had Edmund, and he could easily plug into the rumor mill. It was, after all, a very small community. Edmund didn't need to say anything, both men knew what was at stake.

7

Columbia University Medical Center
New York City
March 1, 2011, 7:30 p.m.

Off by themselves, Pia, Lesley, and Will were sitting in the mostly empty hospital cafeteria nursing cups of postprandial tea and coffee. Seeing what Rothman and Yamamoto were working on had left each of them shell-shocked. As medical students, they were well aware from an academic standpoint what was being done in the lab and having seen it with their own eyes made it real and concrete. They'd been to the future, and it was difficult to absorb.

"I can't get over it," Lesley Wong said. "I'm still blown away. Growing organs from a patient's own stem cells. Something like that is going to revolutionize medicine."

"It's certainly going to revolutionize the care of degenerative disease," Pia said. "It's going to provide cures instead of just treating symptoms."

"Down the road we could grow our own organs and freeze 'em for when we need 'em," Will said. "Hey, I wonder how Columbia divides the cash on a medical breakthrough like this. This is going to be huge. Yamamoto said that the university has filed patents, but don't you think that Rothman and Yamamoto have to get some kind of cut? Don't you think, Pia?"

Pia had come with Lesley and Will not because she wanted company per se, but because she still felt unnerved by what she had seen and wanted to talk about it. After Dr. Yamamoto led them out of the inner lab with the organ baths that morning, they'd settled in a corner of the lab with the intention of discussing tissue culture fluid. Instead they couldn't stop talking about the progress Rothman had made in organogenesis. As interested as they were, they realized that reading books from the library and doing Web searches was fruitless. The textbooks on this stuff hadn't been written yet.

"You're asking the wrong person," Pia said, responding to Will's question about Rothman and Yamamoto getting a cut. "Money and I don't have much of a relationship."

"But he's gotta be looking at billions here.

Doesn't he? I'm going to call my dad, he'll know someone who knows."

"Your dad?" Lesley said.

"Yeah, his broker is pretty well connected."

"I don't think you should be talking about any of this to others," Pia offered. "Particularly people outside the medical center. Remember what Yamamoto said. At least during this month while you're working there or until the key publication is released."

"You might be right," Will said. "But it can't be that much of a secret, as Yamamoto admitted. But certainly it's best not to get on Rothman's wrong side, especially with his reputation."

"I'm just pleased to be part of the scene," Lesley said. "I'd be happy just checking bath temperatures for the month."

"After all the terrible stories I've heard about Rothman's treatment of students, I was expecting the worst today," Will said. "But hell, he seemed very nice to us. Maybe he didn't know we were coming today or didn't know who we were?"

"He had to know who we were," Lesley said. "I think he was using us to practice showing off the progress they've been making. But whatever the reason, I don't care.

I'm happy just to have been able to see it."

Inwardly, this was exactly what Pia was thinking. It had been a magical experience for her to visit Rothman's inner sanctum. It had been a long wait, but she didn't care. Nor did she feel any resentment that Lesley and Will had gotten to do it on the very first day of their elective. For Pia, it was as if she had entered a different physical dimension. The room, and what was happening inside it, seemed to belong to a realm quite apart from all that lay outside. What she remembered was a white space, glowing blue, like something from a science fiction movie.

"It was one of the most tantalizing experiences of medical school," Will said. "I loved it."

"Me too," Pia said. "I could have just stood there and watched baths all day."

"Hey, everybody," a voice announced. It was George Wilson, standing at the foot of the table, carrying a cafeteria tray. He'd just emerged from the food table line. "Is this a private party or can a tired radiology extern join you?"

The three students eyed each other. It was Will who spoke up: "If it isn't Mr. Wilson! Hello, George."

"Will, how's it going?" George said. He tried to conceal his displeasure at seeing

McKinley sitting at a table with Pia.

"You know Lesley Wong," Will said, playing the host. "And Pia, of course."

"Lesley, hi, how are you? Pia, how was your day with Rothman?"

George was feeling extremely uncomfortable as he'd not yet been invited to sit down and so stood awkwardly next to the threesome's table. It was late, and he'd managed to get to the cafeteria just before it closed. The last person he expected to see there was Pia. The second-to-last person was Will McKinley, who usually held forth in the medical school dorm cafeteria chatting up all the women medical students.

"We were just talking about it," Pia said, unaware of George's discomfort. Social cues were not one of her strong points. "Lesley and Will are also spending their electives in Rothman's lab. And about the day? It was . . . let's say interesting."

"Crazy Rothman saves the world," Will said.

"What do you mean by that?" Pia said. There was a sharpness in her voice.

"Nothing, nothing," Will said, holding up his hands as if he expected Pia to attack him. "It's known you worship the guy —"

"I *respect* the guy —"

"Listen, it's fine, he's clearly some kind of

crazy genius." The look on Pia's face persuaded Will that he might be better off changing the subject.

"What we were wondering," Will said, "is how much money Rothman stands to make if he finds investors to back him."

"You were wondering," Lesley said.

"Yes, I was wondering. There's got to be big money in this. Rothman's sitting on a gold mine. Don't you think, George?"

"I'm not exactly sure what you're talking about," George said. "But I can tell I'm interrupting a meeting here." He started to leave, but Will grabbed his jacket just above the elbow and restrained him. "Don't go! Take a seat!"

George glanced at Pia, and she motioned with her head for him to sit down, which he did, unsure if it was the right move.

8

Greenwich, Connecticut
March 1, 2011, 9:10 p.m.

Edmund Mathews was back at his front door, and again it was Russell. This time Russell hadn't called ahead; he simply texted Edmund on his BlackBerry to say he was coming over. There was only one reason he'd come back. Edmund figured he must have found out who was shorting LifeDeals.

"Edmund. We need to talk."

"Tell me what you found out."

"I need a drink and so will you. Can you fix me a scotch?"

Edmund knew Russell liked the eighteen-year-old Talisker single malt he kept in the den, so he walked Russell over to the room and closed the door behind them. Edmund had been sitting in there reading some research and had started a fire in the grate. The room smelled slightly of smoke, and when Edmund opened the bottle, the peaty aroma of the whiskey added to the effect that they were in a Highland lodge.

"So what do you know?"

Edmund poured two drinks and handed one to Russell, who stared into the fire, an elbow leaning on the mantel.

"I'm a big boy, Russell, I've heard bad news before. Out with it!"

"Gloria Croft," Russell said, snatching a glance at Edmund, then throwing down all of his drink in one shot.

"Excuse me? I thought for a second you said Gloria Croft."

"I did. That's who it is, Edmund, Gloria freaking Croft. She's doing it plain as day through BigSkies."

"You've gotta be kidding me. You have *got* to be fucking *kidding* me!"

Edmund was yelling nearly at the top of his voice. This was why Russell had been concerned — he knew Edmund was going to flip out. There was a knock at the door, and Alice, Edmund's wife, popped her pretty blond head in.

"Oh, hi, Russell. Edmund, Darius is going to bed. . . ."

Alice took one look at her husband's face twisted in a rictus of unalloyed rage. She could see it would be a hopeless task to get anything out of him in this state.

"I'll say good night for you, then. Bye, Russell," Alice blurted, and withdrew, closing the door behind her.

The interlude cut short Edmund's tirade. He poured himself another drink and then one for Russell, closed his eyes for a second, and took a deep breath. Why did it have to be Gloria Croft? "You'd better tell me everything."

Russell sat down on a low padded stool by the fire. Edmund stayed standing.

"I made some calls. It only took two, actually. I called the guy who told Teddy Hill he heard about the shorting and the guy gave up his source. It was someone I'd met before, at some mortgage convention in Vegas. He runs this bullshit financial newsletter. He got it from the horse's mouth."

"The horse's ass, more like. Did he say why she's doing it?"

"Didn't say much. I think once he'd told me he thought better of it and got off the phone pretty quick. He got nervous. She's a big player. BigSkies has a lot of money."

Edmund was experiencing a less-than-pleasant sense of déjà vu. Through her hedge fund, BigSkies, Gloria Croft had taken huge positions betting on the failure of the CDOs issued by, among others, Edmund and Russell's desk. She'd taken the positions early, in 2006, when no one else was doing it and when it was cheap to do. Hundreds of thousands of dollars could turn into tens of millions. What she was indicating was that she thought the AAA-rated mortgage-backed bonds would fail and jeopardize the futures of Wall Street giants like Bear Stearns and Lehman Brothers. Almost no one agreed at the time: There was no chance their stock could go so low. But it did, and it went lower.

Russell and Edmund were quiet, Russell staring into the bottom of his glass, Edmund at the fire hissing and spitting in the grate. He held back the fire guard and popped another log into the flames.

"She's got some balls," Russell said finally.

"She does."

"This is different, though."

"You're right about that."

Russell and Edmund's thinking was following the same track. Subprime mortgages were a disaster; as assets, they were just terrible. In their "Life Settlement" paradigm, the debtors were the nation's largest insurance companies, some of the richest institutions in the country. The bottom line was solid, encapsulated in one of Edmund's favorite sayings of recent months: What are the insurance policy holders going to do, not die?

"So what is she doing?" Edmund said after another long pause. "It doesn't make sense. We know the numbers, right? We're solid all round. Actuarially, we did the worst case — people living a little longer for God knows what reason — and we've taken those tolerances into account. Unless she's shorting you and me personally. But she's too smart for that. Way too smart. There's gotta be something she sees in the numbers."

"If there was anything in the numbers, *I'd* have seen it," Russell said a little testily.

"I know that, Russell. She's seeing something that isn't there. It really doesn't matter why she's doing it, she's doing it and we might be left standing in the rain with our dicks in our hands. God *damn* it."

"So what do we do?"

"We have to talk to her, find out what she knows," Edmund said. "Try to talk some sense into her. When she understands what the upside is, perhaps we can help her out."

Edmund was talking about offering Gloria an inside track to invest in the company and share in the enormous windfall they were confident was coming.

"Maybe that's what she wants," Russell said. "She's sending up a smoke signal."

"She could call us on the telephone and

ask," Edmund said. He thought for a second. "Let's call her right now."

"Now? It's after nine o'clock at night."

"Call her anyway. She's always working. I'm not sure I'll be able to get to sleep tonight unless we talk to her. Do you have her cell?"

"Yes, but why me?"

"You did business with her. And she's not going to take my call. Pure and simple."

Years back, Gloria had worked for Edmund as a humble analyst two jobs before she went out on her own. Edmund's desk was a boys' club and women had to be very thick-skinned to work there. This much Russell knew. There was more he didn't know, but the bottom line was that it hadn't ended well. Russell happened to have been present the last time Edmund had seen Gloria, when she walked out of a bar full of about-to-be-former traders where she'd gone to commiserate with one of her friends who had been fired. She was walking out because Edmund, who was very drunk, was yelling at her.

"You're the reason these people are out of work," Edmund had shouted. Many people would contend that it was the products Edmund sold that ruined the firms and caused people to be fired rather than the shortsellers who saw an opportunity. But from Edmund's perspective, his own participation was purely short-term expedient. Gloria's role was more causative.

Russell scrolled through his BlackBerry for Gloria Croft's number. He dialed, and Gloria picked up after a couple of rings.

"Gloria, it's Russell Lefevre."

"Russell, how are you?" Gloria's voice was level and deliberate. She didn't act surprised in the slightest to hear Russell's voice despite the hour.

"Good, thank you. I trust you are too. Where are you? I hope you're not still in the office at this hour." Russell could hear background sounds that suggested she was.

"Just watching Asia open. I've been expecting your call. Is Edmund there with you? You can put me on speaker if you like."

"Hang on a second, Gloria."

As Edmund rolled his eyes, Russell fiddled with the buttons on his phone, then propped it up against the bottle of Talisker.

"Okay, Gloria," Russell said.

"Hello, Edmund. How are *you?*"

"Okay, Gloria," Edmund said, trying to sound carefree. He looked imploringly at Russell. This was his call, couldn't he just get on with it?

"Gloria, we'd like to get together with

you," Russell interjected. "We have some things we'd like to discuss."

"What types of things, Russell?" Her voice was carefree, as if she were enjoying herself.

"Stop screwing around, Gloria," Edmund said, all lightness gone from his voice. "LifeDeals, as you know perfectly well."

"Ah, the same charming Edmund Mathews I remember so well. If you want to talk, you're welcome to come and see me at my office."

This was a power move, and Edmund was making a vigorous throat-slashing gesture with his right hand. He didn't want to go to her office and cede her the offensive.

"How about we meet for lunch?" Russell suggested. "You like Terrasini, I remember. And I haven't been there in a while. Would that work for you?" He was suggesting the excellent midtown Italian restaurant, long a financiers' favorite.

"Russell, I'm sorry, I'm booked solid. And I'm going out of town. It's here or it's nothing until next week."

"Hold on, Gloria."

Russell picked up the phone and quickly muted it, just in time.

"Jesus, who does she think she is?" The veins in Edmund's neck were bulging. It was as if Gloria were still working for him, and she were talking back to him.

"Edmund, she's got us over a barrel and she knows it. We need to know what she's looking at. I'll go, if you can't stand it."

"No, I'll go. Like some damned supplicant. She's going to pay for this. Somewhere down the road. And big-time."

For once, Edmund's curiosity had trumped his vanity. Russell got back on the line.

"Gloria, sorry about that. Any chance we can see you tomorrow?"

"How about nine o'clock?"

It was another raised middle finger as far as Edmund was concerned. Nine o'clock meant driving into Manhattan and battling all the commuters. Edmund had an aversion to public transportation so taking the train was out of the question. Edmund was making the throat-slashing motion again.

"Sorry, Gloria, I've got an appointment early on that I can't break. How about ten-thirty?"

"Okay, Russell," Gloria said with amusement. She could imagine how her request had been greeted by Edmund.

"See you tomorrow," Russell said, ending the call. Edmund sighed, grateful for the one small concession.

9

406 Park Avenue
New York City
March 2, 2011, 10:37 a.m.

Edmund and Russell entered the midsize high-rise along Park Avenue in midtown Manhattan. Predictably, Edmund had been in a sour mood on the Town Car ride down from Greenwich. Russell had insisted on sharing the ride — he wanted to see if he could soften his partner's mood before they met Gloria Croft. Edmund had been a bully all his life, and he felt uncomfortable in any situation where he wasn't in charge. Not only was he not controlling events here, it seemed like he was being played, and by a woman, and by a woman who used to work for him. Russell doubted whether his calming words had had any effect.

When they arrived, Gloria used a script she had learned from Edmund, and Edmund knew it. The two men were buzzed into the suite and shown into a glass-walled conference room, where they were left to stew for fifteen minutes. The receptionist was very polite, and they were offered coffee or water. Outside the room, the office seemed calm and peaceful, with only the hum of the air-conditioning system breaking the silence. The image exuded was of quiet authority.

Then Gloria emerged. She'd changed her look since Edmund had last seen her. There was now a slight wave to her mid-length, lustrous brown hair. She wore a well-tailored business suit with a lavender blouse and black heels. There was just the right hint of décolletage. She looked like ten million dollars.

"Gentlemen, I'm very sorry, something going on in Singapore." Russell and Edmund had stood up when Gloria entered, and she walked over to each man and shook his hand. There was a slight, almost imperceptible smile on her face. She was clearly enjoying herself.

"Follow me!" She walked out of the conference room quickly, and Edmund and Russell gathered up their coats and briefcases and hustled after her.

"She's got us trotting along like a couple

of poodles," Edmund muttered under his breath.

Gloria was already sitting at her desk when the men entered her office. Some giant, abstract, and presumably very expensive painting hung on the wall behind her. The desk itself was bare save for several large telephones; carrels behind Gloria were strewn with prospectuses and various files. One entire mahogany wall was inset with the mandatory array of televisions carrying the financial channels. Gloria pressed a button on the underside of her desk, and the office door soundlessly closed. When she spoke, she sounded coy even if Edmund knew that was something she was incapable of.

"I feel like I'm twenty-five again. Back then I was like a suckerfish hanging around the great predators, looking for the little scraps of food they missed when they fed. The ocean was full of blood. It was a lot more fun then than it is now, don't you agree?"

Edmund didn't like the way this had started out. Even he wouldn't have been this bold. Now she was the shark, and they were the suckerfish, and it was their blood she smelled in the water. He bit his tongue until Gloria started talking about the "opportunities" she had lucked into in the subprime field, opportunities she was grateful that the market (meaning Edmund and Russell) had made available to her.

"Well, Gloria," Edmund said, trying to control himself, "you're just not as smart as you think you are. That subprime stuff was never designed to succeed. We knew it was going to fail. We were shorting each other. We were shorting ourselves."

"Maybe you were, but not till the very end. I was buying swaps five years before you" — Edmund snorted — "and you guys were still leveraging your position selling the worthless bonds right up until Lehman went down. Tell me you weren't."

Gloria had at least one of her gloves off now. She felt she held a winning hand against LifeDeals. It might make for a longer game if she held her cards, but she had Edmund and Russell right where she wanted them, and she could enjoy witnessing their reaction if she played her hand now. She'd probably make just as much money in the long run if LifeDeals wasn't a public company. It depended on how far Edmund was going to push his luck. That morning before they arrived she'd looked at herself in the bathroom mirror and said, "Payback time."

Gloria cleared her throat and went on: "The traders who sold those CDOs should have gone to jail. The whole of Wall Street was tarred with that brush. Immoral, greedy, selfish . . . it was *stealing*."

"Nonsense," Edmund said. "You said it yourself: it was an *opportunity*. You destroyed those companies. Your fingerprints are on the bodies. The government mandated that mortgage lenders make subprime deals. Everybody should have a home. No one held a gun to anyone's head. . . . I don't understand why we're raking over all this again. We've moved on. You clearly haven't, but I'd urge you in the strongest possible terms to *get over it*."

Edmund was exerting as much self-control as he could, speaking slowly and evenly. Russell knew the plug on the volcano was shaking, the whole thing was threatening to blow. Robotically Edmund continued.

"We have the utmost confidence that LifeDeals is a winner and is going to prove to be so in the very near future."

"Oh really," Gloria said. "Well, I have half a million in credit default swaps on the line that says it won't. And I'm going to buy more. And you know what, I'll be glad when it fails, because I think you're stealing again, only this time you're stealing vulnerable people's life insurance, and you're paying them pennies. These are old people who are desperate for money because they need an operation and don't want to go bankrupt because our health system has written them off."

Edmund was massaging his temples. They were bankers, they made money, end of story.

"The Supreme Court has ruled that life insurance policies are equity that people can buy and sell," Edmund said.

"You're paying fifteen percent of face value. Ten if you can."

"We are offering a legitimate financial service to older Americans who need cash for whatever reason. We haven't created the need, we're merely filling it. It's of no concern to me whether an individual is paying for a new hip or a cruise to Alaska. Perhaps they just don't want their ungrateful children to get the money. There's nothing immoral or unethical about it. We're helping put money back into the economy. You should thank us."

"Oh, spare me, Edmund. Now that the mortgage market has dried up, some clever analyst zeroed in on life insurance. It's another gold mine and damn the consequences for the people involved."

Russell could see this was going nowhere fast. He moved forward in his seat. "Gloria, with all due respect, Edmund and I didn't drive down here to debate the ethics of life settlements, although I have to say they have been around for years with little protest. We'll have to agree to disagree. We'd like to know why you're so sure we're wrong. I brought along some research as backup."

Russell placed financial statements in front of Gloria along with complicated graphs that showed bell curves for the life expectancy of people whose policies LifeDeals had purchased, separating them by the diseases the holders suffered. He gave Gloria the whole picture, letting her see her more information than was normally presented to prospective hedge fund managers. He then described the plan, how they would securitize the policies into bonds, resulting in vastly increased income that was then used to buy up more policies to make into more bonds. The bonds were weighted, the largest segment based on diabetes, the second largest on cardiovascular disease, and the third on kidney disease. While Russell talked, Gloria glanced at the financial statements and the bell curves. It didn't take her long. When she was finished, she tossed them aside as if she didn't believe any of it.

Finally Russell explained that since the bell curves could accurately predict when the policies would pay out, they could factor in all the other cogent data and determine their cash flow extremely accurately and buy as many policies as the revenue streams allowed. Their actuarial data was enormous, going back fifty years, and even longer if they needed it.

"We've left nothing to chance," Russell said. "It's foolproof, based on real numbers. Sure, a few people are going to have spontaneous remissions, but others will pay out faster than predicted. It's all based on accepted mathematics and the bedrock is the insurance companies. It might be the best investment opportunity ever, backed by the Supreme Court ruling, so there's no chance the insurance industry can lobby Congress to have laws and rules changed. The insurance companies are going to pay every penny the policies have accrued."

Russell suddenly stopped, out of breath. Russell and Edmund looked at Gloria, who stared back for a couple of beats. There was silence.

"Don't you see it?" Russell questioned.

"*I* see it," Gloria said. "You're the ones who don't see it."

"It's real. We've run the numbers up and down and confirmed it with all the actuarial companies. It's real. We're already holding fifty thousand policies —"

Gloria whistled. "How much are the premiums on fifty thousand policies, Russell? You must be paying out about four, five million a month. You're going to run out of capital by the end of next year if you don't start having significant income."

Russell and Edmund knew she was right. That Gloria was smart was not news to Edmund; he wouldn't have hired her otherwise way back when. But in this instance they'd be fine, they were going to be fully capitalized by the end of this year. He wondered if Gloria might be bluffing and was beginning to think she was. So far she'd given them nothing. He was getting tired of this.

"Gloria. All you've told us is that we are mean, heartless bastards taking money from old ladies," Edmund said. "But we knew that already. I think you're fishing. You told some guy you were shorting us knowing it would flush us out, and we'd come down here and explain our business plan. Which we have done. Congratulations. Now we should go and not take up any more of your time. We'll be more than happy to mail you a prospectus in due course."

Edmund's irritated expression had morphed into the insufferably smug look Gloria remembered from whenever he had dressed her down in days of yore. She pulled out the central drawer of her desk and found a red Sharpie on her desk. Looking at Edmund, she took one of Russell's graphs and copied the bell curve, only drawing it shifted to the right of the one printed on the paper. She held up the graph.

"What would it mean to you if this happened?"

Russell squinted at the paper — it was the diabetes chart.

"That's not going to happen."

"Humor me. Hypothetically."

"You're projecting chronically sick diabetes patients living about ten years longer than they're going to live. As I said, it's not going to happen."

"Let's say forty percent of your policies are diabetes patients. If we have a curve like mine instead of the curve like yours, I reckon that's twenty thousand policies you're stuck with for ten additional years. That's, um, two hundred and forty million in premiums you weren't expecting to pay. Kind of cuts into your model, doesn't it? Perhaps they're half your policies. I think the curve needs to move a little more.

Fifteen years and it's four hundred and fifty million. Your biggest source of revenue becomes a sinkhole of toxic assets."

"That's hypothetical, and it flies in the face of fifty years of actuarial data. Fifty years!"

Edmund was yelling but Gloria was looking at Russell. He looked worried.

"Yes, you have fifty years' worth of *old* data. But you're not looking at the future. Technology can make a monkey out of a table in a minute. If you have any more great ideas, please share them with me, I will be happy to take a position on them too."

"What the fuck are you talking about?" Edmund demanded.

"Do you know what an iPS cell is?"

"I've heard of them, yes," said Edmund. "Something to do with stem cells. But I don't see —"

"Induced pluripotent stem cells," said Gloria. "If you looked to the future and not to the past, you might know that iPS cells are going to have a huge impact on regenerative medicine."

"You mean stem cell therapy?" said Edmund. "That bubble burst ten years ago, all those biotechnology start-ups? Penny stocks today."

"Edmund, you're still talking about the past." Gloria had come this far. "You're ignoring the future."

"Okay, Gloria, what do you see in your crystal ball?"

"Have you heard of the Nobel laureate Tobias Rothman? Or Junichi Yamamoto? What they're doing at their research lab up at Columbia Medical Center?"

"No," Russell said, feeling that the ceiling was pressing in on him.

"Through a contact who follows biotech patents I've learned that Rothman has mouse organs, entire organs, that he has grown from iPS cells that he's transplanted back into the same mice that donated the cells. Any day now he's going to do it with human iPS cells if he hasn't already. He'll be able to grow pancreases. For humans. To make insulin. Pancreases that are custom made for a patient so no rejection. You know what that's going to do?" Gloria pointed to the graph she had drawn over and dragged her finger from Russell's bell curve to her red version.

"That."

Gloria sat back.

Russell had done the real math in his head. Thanks to some particularly aggressive salesmen in Texas and Florida, they were extremely long on diabetics' policies. Gloria had in fact undershot — they were almost two-thirds of their policies. Meaning they might be on the hook for almost $600 million in additional premiums. Who knew if the science was going to work and when, and not every patient was going to be helped, but still, if she was right, their paradigm would be in tatters. Was there any way they could dump those policies? Could they securitize them anyway? Would anyone invest in the company with this much doubt about the nature of the risk? These questions were occurring to Russell; Edmund just wanted to get the hell out of there.

"Think of LifeDeals as a swimming pool," said Gloria. "There's water pouring out already and there's going to be a lot less of it pouring in than you planned in the near term. You guys are going to be left high and dry with no life preserver."

Gloria was enjoying herself.

"You want some advice? I doubt it but I'm going to give it to you anyway. Hurry up and securitize and sell those tranches ASAP, before others start to see that the ground under LifeDeals is going to be more like quicksand than bedrock. Once that happens, your bonds are going to go begging. Some of the money that comes in from the bonds you might be able to squirrel away if you're clever, and I know you are, but you certainly aren't going to get back your seed money. Unless you want to break the law. Which brings us neatly full circle. Perhaps you'll end up going to jail this time."

"Russell, we gotta go," Edmund said, as Russell gathered up his papers. Edmund and Gloria held each other's glare. Gloria had played her hand, and she could see it had hit home.

"So sorry you guys have to run, but I have to go to lunch anyway," she said.

Gloria handed Russell some more of his papers. She'd already decided to strengthen her position against LifeDeals significantly later that day. Edmund was right in part — she had wanted them to tell her about their business plan, and she assumed Edmund would be arrogant enough to tell her too much. Now she'd seen their model, and it was even worse than she could have hoped. Or better. Maybe she'd cost herself some money, but she already had more than she could reasonably spend in three lifetimes.

That look on Edmund's face was priceless.

Edmund and Russell were silent as they waited for the elevator. Russell stole a look at Edmund's face, and it bore an expression he had never seen. It looked like grief. They got in the elevator.

"Hold these a second," Edmund said to Russell, handing him his case and his coat. Edmund stepped forward and slammed the elevator door hard with the fist he'd made of his left hand. He cried out and grabbed his hand. The pain, when it came, was a relief.

10

Columbia University Medical Center
New York City
March 2, 2011, 1:00 p.m.

Pia had quickly learned to feel at home in what Dr. Yamamoto liked to call the "bathroom," the organ bath facility where the mouse kidneys, hearts, lungs, and pancreases were being nurtured. She had spent the morning in there, harvesting reams of data on the pH levels in the baths and using a handheld tablet to look at the histories of a few organs that had failed. They were later found to have very subtle variations in acidity or alkalinity from the rest of the samples. Pia's task was to monitor the baths, and she tried to figure out how she might rig up some kind of electronic alarm to her cell phone like Rothman and Yamamoto had that would alert her when a bath developed a slight variation in pH.

Dr. Rothman had come and gone a couple

of times. Pia knew from speaking with Dr. Yamamoto that the team was running complex and time-consuming studies concurrently, both here with the baths and in the biosafety level-3 lab on the other side of Rothman's complex. Rothman's work on salmonella had made his reputation, and he wasn't about to abandon it, even if it meant working at superhuman levels of energy and concentration. He treasured his access to the highly virulent strains that NASA provided him with, and with the space shuttle program winding down, he didn't know when he might get more.

Lesley and Will had left the room to find Dr. Yamamoto. It had been decided that in addition to helping Pia, they would initiate their own study of the effects of slight variations in the temperature of the baths. Unfortunately their study had reached a quick impasse, and they preferred to consult Dr. Rothman's associate rather than the man himself.

Dr. Rothman entered the room, moving to the last row of baths.

"We seem to have a problem with number nineteen," he said, apparently into thin air. Pia joined Rothman, who was fiddling with the monitoring unit under the bath.

"The blood flow is compromised. There's

a blockage, so we may have to section the organ to see if the problem is developmental or some kind of embolus. There are few journeys longer from *in vitro* to *in vivo*."

"How long before you can start human trials?" Pia said.

Rothman flinched a little and looked around at Pia, apparently in surprise. Had he been talking to himself?

"We're a little closer with the kidney than with the pancreas. The kidney is basically a filter. Quite simple. But the pancreas is very complicated. It's fascinating to me that one gland would have so much to do, and such important tasks."

"Hormones and enzymes," Pia said.

"The islets of Langerhans. I always loved that name. They were discovered by a twenty-one-year-old German named Paul Langerhans in 1869. I remember when I was a teenager and first heard the term I thought they were named after some actual islands someplace."

Pia had rarely heard Dr. Rothman sound so jovial. He seemed to revel in his lair. Pia thought it fitted his temperament to enjoy the name of the hormone-producing cells of the pancreas that pumped insulin and glucagon into the bloodstream to regulate sugar levels. Or at least, they were intended to.

"Of course it was necessary to locate the pancreas adjacent to the duodenum so it could inject its enzymes into the digestive system. The ampulla of Vater, another of my favorites."

Rothman was referring to the junction of the bile duct and the pancreatic duct where food passing through the intestine was mixed with the agents necessary for its digestion and to control the level of acidity.

"But it's so deeply buried in there. It's very elusive. That's why pancreatic cancer is so hard to detect and so lethal. The organ has such a large blood supply cancers tend to spread very quickly."

Rothman's mind was wandering. He seemed so uncharacteristically relaxed.

"Its organogenesis is very elusive too. All the hormone- and enzyme-producing cells have to be genetically coded to create the gland, and we're just coming to grips with the process."

Rothman had moved to a different bath.

"The mouse pancreas is remarkably similar to ours. We're making strides here, but I want to speed things up."

Some scientists were working on implanting glucose sensors and insulin pumps into patients. Others were examining gene therapy solutions, with patients ingesting a medication containing a virus to cause the production of insulin in the presence of glucose. Rothman was tackling the issue the only way he knew: by swinging for the fences. Pia loved that confidence and ambition. She felt some of it had rubbed off on her in the time she had spent with Rothman over the last three years. She also knew how other people saw him. They saw that confidence as arrogance of the worst kind, but it could be arrogance only if the conceit was deliberate. It wasn't just that Rothman didn't care what other people thought, he didn't notice either.

"I wanted to thank you, Doctor," Pia said.

"For what?"

"For offering to lend me money to pay the Sisters."

"The Sisters helped you in the past, but the past is the past. You don't need them anymore. You need to move beyond all the problems foster care caused you, just like I did."

"I'm trying to," she said, referring to rising above the legacy of her childhood experiences. But about not needing the Sisters anymore, she wasn't so sure.

"My sons are not as healthy as I would

like. I feel very guilty," Rothman said out of the blue, shocking Pia. He rarely said anything personal, especially something so very personal. The only other time was when he'd admitted having Asperger's.

"I'm so sorry," Pia said. "I had no idea."

"Nobody does," Rothman said, uncharacteristically wistfully. "I never talk about it. But it's a big part of my race with stem cells and stem cell science."

Pia didn't know what to say. What was suddenly clear was why Rothman had made such a deviation in his scientific pursuits after such success with his salmonella work.

Rothman continued to watch the tiny pancreas suspended in the bottom of the bath. Pia could only imagine what flight of hope his mind was taking him on now. She could see him almost physically shake it off. He took one more look at the figures on the monitor and wordlessly left Pia's side. It was amazing and distressing how he could turn on and off.

11

New York City
March 2, 2011, 1:30 p.m.

After leaving Gloria Croft's office, Edmund Mathews, still steaming mad and nursing a sore left hand, turned east onto Lexington Avenue and found a Duane Reade pharmacy. He bought a bottle of Motrin and took four. His hand throbbed sharply, but he was certain he hadn't broken any bones when he slammed the elevator door. Had he used his stronger right hand, he surely would have. Russell Lefevre often found Edmund's behavior during crises to be unpredictable and alarming, but he knew Edmund had a laser-like focus that he could bring to bear at times like this. Edmund had the ability to break down complex problems and attack them piece by piece until he had it licked.

Edmund had the car brought around and the two men sat in it double-parked on East

Fifty-eighth Street. Russell swore he could hear Edmund thinking.

"We have to get a head start on securitizing what we have," Russell said.

"Yes. For certain. And look at the diabetics whose policies we own," said Edmund. "See if it mightn't be cheaper to cancel some rather than carry them. And we should lay off some of our sales force until further notice."

Edmund had Russell call a contact at Goldman Sachs, a man named McDonald in the asset-backed securities division. McDonald had been interested in LifeDeals but was wary. Russell was still confident they would get one of the major players on board, and he hadn't spoken to McDonald in a while. It happened that McDonald had a few minutes to spare an old client, and Edmund and Russell headed down to West Street in Battery Park City and Goldman's global headquarters.

"That guy's small-time, no vision," Edmund said, after the unsatisfying meeting. Russell had answered all the questions the traders asked about securitizing their portfolio of life settlements sooner rather than later. But the traders couldn't see what the hurry was. From their perspective the more policies they had to bundle, the better their product was going to be. And they hadn't done the grueling legal legwork to create the complex CDOs to be in a position to take them to market. What Russell and Edmund had been looking for was reassurance after their meeting with Gloria Croft. As they left Goldman they admitted to each other that the reaction they'd gotten wasn't exactly negative, it just wasn't wildly positive either.

In the car Russell and Edmund made another decision. As Gloria Croft had so devastatingly demonstrated, the main problem they were facing was with the mortality-rate bell curves that LifeDeals' viability was predicated on and the damage that medical advances might do in terms of moving the curve to the right.

"We need to go see Henry Green," Edmund said. Henry Green was CEO of Statistical Solutions LLC, the company that had produced all the actuarial data, including the bell curves. Edmund got out his BlackBerry. This was a call he wanted to make.

"Henry Green, please. . . . Okay, well, tell him Edmund Mathews is in the city and on his way over and needs to see him right away. . . . Well, I'm sure he'll understand when you tell him. Our key data has flaws.

There's new information. We need to fix it."
Edmund hung up.

"He'll see us," Edmund said.

LifeDeals had put Statistical Solutions on an expensive retainer. Russell wanted to be armed with the best available statistical analysis when they went on the road selling their product. One supposed lesson from the subprime debacle was that investors wanted to know exactly what they were investing in. It might appear to be self-evident, but it wasn't. Russell wanted to be able to show an investor the latest data, even individual policies, if they cared to see them.

For his part, Henry Green was less than pleased to be hearing from Edmund Mathews at all, let alone when he was demanding a face-to-face meeting without notice. While Russell Lefevre wanted the full breadth and depth of research that Statistical Solutions offered, he gave the firm time to accomplish what was asked for. In contrast, Edmund Mathews would call up and want immediate answers to complicated questions. Green had had to drive his people hard, squeezing every last cent's worth of work out of the company to comply. Edmund expected Henry to drop everything whenever he called.

Edmund and Russell arrived at the Statistical Solutions offices in Chelsea in short order and within a couple of minutes were sitting with Green in his office.

"Edmund, I gather you mentioned something on the phone about 'new information,' " Green said hesitantly.

"That's right," said Russell, who wanted very much to avoid having Edmund yell at Henry Green, which had happened in the past. "There's new material and we need your expert opinion to see if we need to be concerned about it."

Edmund sighed at the understatement.

"What my colleague is trying to say, Henry, is that you may have been wrong in some of your forecasts by amounts that could put us out of business. So I'd be very grateful, Henry, if you could get those geniuses you told us about who could have gotten jobs at Google to come in here and prove to us that they are in fact smart enough to tie their own shoelaces after all."

Edmund's voice was rising in volume but the dam held, just. Henry Green pressed a number on his phone pad and picked up the receiver.

"Yes, Laura, would you have Tom and Isabel meet us in the conference room right away?" Green hung up the phone.

"Gentlemen, shall we?"

Everyone Russell and Edmund could see around the office was young. Henry Green at least affected the look of a businessman with his dress slacks and dark shirt, but his messy hair was at least two inches too long in back. The numbers geeks, dressed in black, looked like they had just come in from an all-night party. Statistical Solutions was gaining a reputation for all kinds of data collection and algorithm solving, and a lot of their employees did indeed go on and work for Silicon Valley giants that paid for their dry cleaning and accommodated their dogs at work. To keep them, Henry Green had to be equally tolerant and generous. As long as they gave him six months of hard work, Henry Green didn't mind. Statistical Solutions was strongly in the black.

Anticipating that Russell wanted to speak, Edmund dove right in, addressing Isabel and Tom directly.

"What do you know about stem cell research in relation to the treatment of diabetes?"

"I know what stem cells are," Isabel Lee said.

"Did you build it into your projections?"

"Build what in?"

"The fact that a professor up at Columbia is making strides toward creating human pancreases outside the body to be used as transplants. If he succeeds, he'll be prolonging the life of diabetes patients."

"Which is a good thing, of course," Isabel said. Neither Isabel nor her colleague Tom Graham enjoyed working on mortality statistics for LifeDeals, even less when they found out what LifeDeals was doing with them. They'd mentioned their misgivings to Green, but he'd waved them off, saying they were not being paid to make ethical value judgments. Yeah, the concept of making money from people's deaths is creepy, but the pay is good.

"Yes, it's a wonderful day for medicine and fat people, less good for my investors," Edmund said.

"Listen, we gave you full license to draw up parameters for us, using actuarial data and cross-referencing it with our cash-flow projections, and nowhere did we see any information about this," Russell said, getting a nod of agreement from Edmund.

"Russell, we built in increases in life expectancy and added tolerances for unexpected developments, but they were capped at five percent," Henry said. "As we discussed, as you agreed. If there is about to be a major breakthrough like custom-made transplantable pancreases from stem cell research or fallout from the human genome project, we can't be held responsible. You can't anticipate once-in-a-century events."

"Then all this fucking statistical research is useless," Edmund snapped, throwing up his hands in frustration. "It's all nothing but mental masturbation."

"Na-ha," Isabel said, not at all intimidated. "It's good data for what we had. If there's a paradigm shift, then numbers change and graphs have to be adjusted to reflect it. Simple as that." She shrugged and sat back in her chair.

Tom Graham was looking at a fingernail and didn't respond.

"That's it! That's what we get! Whoops, sorry, wrong number. We didn't think of that! We were paying you to think of everything. It's not as if stem cells popped out of the blue. What sort of company are you running anyway?"

"Okay, let's not get worked up here," Russell interjected. "Henry, Edmund apologizes —"

"Don't apologize to me, apologize to them," Henry said, indicating Isabel and Tom. Isabel glowered at Edmund, who eventually raised a hand to speak. It was about as contrite as he was going to be.

"Henry, listen, we need some fresh models

run based on some new assumptions that I can e-mail to you in an hour or so, soon as we get back. We need to see how our cash flow is affected in these new scenarios. You'll have to make assumptions as there is no real data available. We'd be very grateful if you could do it for us. And we need it right away. As you know, there's provision in our contract —"

"Yes, Russell, I know," said Henry. "In actual fact, I happened to take a look at our contract when you were on your way over. We'll do the work for you, have it for you tomorrow, as soon as we can. As *you* know, Russell, there's a mutual twenty-four-hour cancellation provision in the contract, under certain circumstances. I believe circumstances such as these cover the provision more than adequately. So consider this your notice."

More deflated than they had been on arrival, Edmund and Russell didn't have it in them to protest. They got up to leave.

12

Castle Towers Retirement Home
Phoenix, Arizona
March 2, 2011, 11:50 a.m.

At her home outside Phoenix, Sally Mason sat on a bench next to the entrance making the most of the last of the morning air before the sun made sitting outside unbearable. Even though Sally had been born in the state and lived here all her life, the heat had always got to her. Sally was proud of the fact that she was a lifelong Arizonan. There'd been only about 450,000 people living in Arizona when she was born in 1933, and that was about the current population of Mesa, a place that was barely a dot on the map when she was growing up.

Today, Sally was scheduled to receive another visit from Howard Essen, the salesman she'd met a couple of times and spoken to on the phone frequently in the last few weeks. Sally appreciated the fact that Essen hadn't been too much of a hard-seller, not as relentless certainly as the man who sold her husband the life insurance policy in the first place. She'd actually enjoyed talking to Howard about his family, a wife and three children, to whom he was clearly devoted. He'd also shown interest in her story, about Arizona when she was a girl when they still tied up horses outside stores in downtown Phoenix, about her Preston, now twenty years gone, and about their only daughter, Jean, and her son. This was one of the good days, a day she didn't have to travel forty-five minutes for hours of tedious and uncomfortable dialysis.

Sally had decided that today she was going to tell Howard Essen she was accepting his proposal.

Howard agreed to Sally's request to come at noon, as she wanted to keep the afternoon free. She checked her watch — it was about ten of — and Sally closed her eyes and thought about Preston as she did most days. She'd been so young when they met, barely eighteen, and he looked so dashing in his Air Force uniform when he came into her pop's convenience store. The fifth day in a row he came in he ran out of things he needed to buy and he had no ready excuse: He was here to see Sally. Life with Preston hadn't always been easy, but he was always a caring man. Toward the end, he set up the life insurance policy for Sally and funded an annuity to make the payments with a little left over. Preston wanted to make sure their daughter, Jean, was taken care of, and he hoped that would be one less worry for Sally.

The money going to Jean under the policy always seemed like such a huge sum to Sally. That was until Jean's husband died suddenly, leaving her with a mountain of bills and debts she had no idea existed. The money that Sally had squirreled away after selling the house Preston bought in 1965, the best year he'd had in his plumbing business, had been diminished to help Jean pay her bills. Now Jean was having to give up most of her inheritance to help her mother. Sally protested a little, but Jean insisted, and Sally knew she was right. Preston Mason would never have hesitated — he'd have done whatever it took to help his wife live the best life she could.

Sally's kidney condition was at stage five, the end stage, and she needed a new organ. But there were thousands of people on the waiting list and the state had just decided to stop paying for lung transplants altogether, as well as some heart and bone marrow procedures. How long would it be before kidney transplants were added to the list? Sally didn't want to wait to find out. She didn't want to spend her last years chained to a machine; she wanted her freedom back but it came at a price. She needed at least $250,000 for the operation. Above and beyond the money she'd need to keep her place at Castle Towers, she had some savings and there was a little money Jean had promised she could have. But she was still tens of thousands of dollars short, which was the reason that when the idea of selling her life insurance policy was presented to her, she was receptive to it.

The call from Howard Essen came at a particularly apposite moment. It wasn't a coincidence, although Sally would have been upset if she'd known how it came about. Howard found potential clients through an informal network of contacts he'd established at more than two dozen retirement homes and nursing facilities. He paid a mix of orderlies and supers and front-desk staff to tip him off when a resident told them about certain medical or personal issues, like starting dialysis or visiting a heart specialist or not being able to afford to help with their grandchild's college tuition. Howard found it distasteful, but he felt he had little choice. These were tough times, and he had to find a way to keep his own family's heads above water. In this case, Sally had told a friendly orderly her predicament, and he mentioned it unthinkingly to the superintendent, who in turn called Howard.

For ten years, Howard had made a decent living selling starter mortgages to young Arizonans. When things were going well he'd been caught up in the all-pervasive hysteria of home ownership. Everyone was selling mortgages with no supporting documentation, so why shouldn't he? There was no one who said it was a bad thing. After more than six months out of work he'd found this job with LifeDeals. In fact, they'd come after him, looking for once-successful mortgage salesmen and offering a job paid almost wholly on commission. The cheaper Howard bought a policy for, the better his remuneration. It helped him sleep at night when he didn't squeeze that extra percentage point out of the policyholder. Not that he thought it would have been so easy to get Sally Mason to capitulate.

The first time he paid her a visit, Howard had introduced himself and Sally said, "Essen, like the city in Germany?"

"Yes, ma'am. Exactly." This was a sharp

one, he could tell at once.

Howard presented his spiel, showing Sally the graphs and tables indicating how much money Sally would save if she didn't have to make the premiums and how much money she might make if she invested it wisely.

"So if I stop paying into the policy and use the annuity money, I can have this much cash when I'm, let's see, a hundred and two?" Sally pointed at a very large figure at the outer edge of one of the projections.

"That's right. Who's to say you're not going to live twenty years with your new kidney? And we base our projections on a historically average rate of return on a sensible mixture of investments. I can give you the name of a great investment specialist who could help you with that."

"I'm sure you can, Howard. And what rate of return might you expect?"

"As I said, using historical averages, about eight percent, give or take."

"Oh, Howard, I wish you'd called me thirty years ago. If you had, I wouldn't be in this position."

A couple of minutes before the hour, Sally saw Howard pull up in his Ford truck and park. She waved to him, and he walked over.

"Hello, Mrs. Mason," he said.

"Good morning, Howard. Let's go do some business," and Howard smiled at her.

Sally's room was small so Howard and Sally sat in the dining room of the home where she felt more comfortable. Howard had brought all the paperwork, and he laid it out in front of Sally for her to sign. Sally picked up her pen and put it down again.

"You know, Howard, when Preston bought this policy, he said it was going to set up our daughter for life. But instead of that I'm using it to give me another ten years because I can't trust the state I've lived in all my life to help me anymore. I'm almost out of money, my daughter's almost out of money. There's just my grandson, George, up there in New York at medical school who's always said he wants to make some money so he can help his mom out. He doesn't know anything about this because he's already further in debt than my kidney's going to cost me. No one's got any money, they've just got debt. How'd it get this way, Howard?"

Howard Essen looked down at his feet. They'd talked a little about Howard's previous career and the mortgage craziness and about how Sally's pop sometimes gave customers a little bit of credit at the store before the end of the week when they'd spent the previous paycheck. And how he almost always regretted doing it.

"I swear I don't know, Mrs. Mason."

"Oh, I think we have some idea, Howard."

Howard watched as Sally Mason signed the paper that turned over her half-million-dollar life insurance policy in exchange for just over $75,000, 15 percent of its value. Sally had gone quiet and didn't say much to Howard after the transaction was completed. Howard would come by in a few days with a cashier's check and Sally's copy of the agreement. When he finished up and said his goodbye and left, Howard felt like going back home and taking another shower. Sally decided she'd wait a couple of hours and call her grandson, George, and leave a message. She wanted to make sure things were still going well for him in New York.

13

Greenwich, Connecticut
March 3, 2011, 6:45 a.m.

Edmund Mathews was sitting at his kitchen island with a cup of coffee when the house phone rang. Edmund picked it up in the middle of the first ring. It was Russell.

"Sorry if I woke you."

"God no, I've been up for hours. You hear anything?"

"Henry Green e-mailed me a couple of minutes ago. His team has put together some numbers, and they want to show us at nine this morning. What's the earliest I can pick you up?"

"You can pick me up now. What did he say about the numbers? Did you call him?"

"No, his message said not to call, just come by."

"So we have no idea what they came up with. Great. Well, just swing by whenever you can. I'm ready when you are." Edmund hung up the phone.

Nothing he'd thought about since the meeting at Statistical Solutions the previous afternoon had offered Edmund much solace. He wasn't a numbers-cruncher like Russell, but he understood how heavily they were exposed by the insurance policies they'd purchased from diabetics. These people had looked like such a solid foundation for their business: a widespread and chronic condition with severe complications and a lot of lower-income policyholders. He'd seen more than a few e-mails from salesmen saying they'd reached someone whose policy was in arrears just as they were about to lose it. These were the perfect candidates, people happy to settle for ten cents on the dollar for something that to them was worth nothing.

Edmund wasn't a man who spent a lot of time in regret or recrimination. If something was broken, you fixed it. The key was, you had to get ahead of the problem before it got serious. Edmund's favorite historical character, predictably enough, was General George S. Patton. Edmund appreciated such a man of action. If Patton had been allowed to reach Berlin first in 1945, and then been allowed to keep going to Moscow, how different would things have turned out in the world? Such great men of history were always thwarted by the weak and the small-minded.

Edmund hated nothing more than feeling powerless, which was where the events of the previous day had left him. Gloria Croft had fired the first shot, and then Henry Green provided the coup de grâce. Edmund felt completely blindsided. He hadn't seen it coming, and neither had Russell. Russell was supposed to be the details guy who knew people who knew what was going on, the one who had his ear to the ground. Edmund had said as much on the long car ride home from Statistical Solutions the previous evening. Gloria would have been gratified at how long they spent stuck in traffic.

By the time he got home, Edmund was done sniping at Russell, and a dark cloud had made its way across his face and stationed itself there. Alice spent another evening staying out of her husband's way and again, Edmund had few words for their son. With his scotch bottle as a companion, Edmund had taken up sorcery. He was trying to bore his way into Henry Green's simulation software, willing it to come up with some way of limiting the damage that was being threatened to LifeDeals' future. Short of relying on magic, he was certain there was something he could do. He had to get himself and his company to Berlin.

14

Columbia University Medical Center
New York City
March 3, 2011, 7:15 a.m.

The call early the previous evening from his grandmother Sally had jolted George back to a place nearer his usual moorings. One day before that, he had found Pia in the hospital cafeteria with Will McKinley. Lesley Wong had been there too, but George had fixated on the fact that Will McKinley had worked his way to Pia's side for a month's elective. Although George considered himself reasonably facile with women, in that he got along with most, Will was a more practiced seducer with fewer scruples. George wouldn't have imagined Pia would be interested in a guy like Will but what did he know? Jealousy was a cruel emotion and while the four of them had sat together he'd suffered through believing Will was reveling at the scale of George's discomfort. Will had never made any secret of just how attractive he found Pia. More than once, he'd asked George what he thought Pia saw in him. George secretly enjoyed Will's rude question because it implied that he as well as others saw him and Pia as some kind of couple.

George knew his grandmother well enough to know she would disapprove of Pia. Or, rather, she would find his continued interest in her to be unhealthy for him. With all that he had on the line, George wondered for the thousandth time what he was doing continuing his pursuit of Pia's affections. It wasn't so much the time that it took up, though it took up enough, it was the amount of emotional energy he expended on her, dissecting her words and actions, thinking of strategies to win her affections, worrying about her well-being. He needed to conserve that energy for his studies. More than anything, he wanted to be the best doctor he could be.

George knew how much his family was pulling for him to succeed. They had suffered so many reverses, and it seemed like they were sliding down a one-way slope like so many middle-class families. If he didn't make it, he knew his mother and grandmother might put a brave face on it but be devastated inside.

George also knew his mother, Jean, had money issues. She'd moved into a much smaller house in the same Baltimore neighborhood a few years before and still didn't seem to have any extra cash. Jean had had the misfortune or lack of foresight to work in decaying industries after George's father died. She had been a bookkeeper at the Bethlehem Steel works at Sparrows Point for a while, then she found and lost a job at the General Motors plant. She always said she was fine when he asked about her finances, and refused to let him look at her bank statements. Although George was on a full scholarship, she never failed to send him a twenty-dollar bill whenever she could.

"You're a student, George," she told him. "You concentrate on your education!"

When Sally had called it was five o'clock in the East, and she didn't expect George to pick up his cell phone. Her intention was to leave him an encouraging message without taking up any of his precious time. She had an exaggerated sense of how busy George was every minute.

"Busy day?"

"Not too bad. They're not killing us just yet. Actually it was a good time for you to call since I'm on a break. I'm taking a radiology elective, and they don't work themselves to death like some other specialties. How is your day going?"

"Oh, you know. It's pretty quiet around here these days. Have you spoken to your mother recently?"

"Not lately. What's up?"

"Something interesting did happen yesterday. I sold your grandfather's life insurance policy to a very nice gentleman. I'll be paid in a few days. How are you doing moneywise? Are you okay? I could send you some."

"I'm doing fine," George said, even though he was constantly short of cash. He couldn't wait for July 1, when he'd start his residency. Instead of money going out, he'd be getting a salary. It wasn't going to be great, but anything was better than it was at the moment. Even with his scholarship, he'd assumed a sizable debt.

"If you need any money, let me know."

"I will," George said, although he had no intention of asking his grandmother for money. "I've never heard of someone selling a life insurance policy. Is that common?"

"Mr. Howard Essen, the man who bought it, said it's very common."

"Oh," George said simply. He told himself he'd try to remember to investigate online such a scenario when he got back to his room. At that point he'd switched the conversation with his grandmother back to her health, which he knew was not good as she was being kept alive by kidney dialysis.

Later when George had looked up "Life Settlements" and read about the issue, he wasn't happy. It seemed to him that it was one more way that the elderly could be victimized, this time by the financial world. He couldn't help but worry that his ailing grandmother had been taken advantage of, and such a thought had helped rearrange his priorities.

Grabbing his jacket from the closet, George headed down to the elevators. Once there he briefly thought about Pia, and wondered if he should go up to her room to make sure she was awake. But he hit the down button. Hell, if she was going to be spending the day with Will, she could get herself up. Instead he decided to grab a coffee, rewarding himself with a more leisurely entry into the day than usual.

15

Statistical Solutions LLC
Chelsea, New York City
March 3, 2011, 9:17 a.m.

As was his habit, Edmund made sure he and Russell were fashionably late for their meeting at Statistical Solutions headquarters. They were greeted coolly by Henry Green and hustled directly into the same conference room as the previous day. The mood was somber, if not funereal. The room was occupied by a half-dozen people, including the slacker Tom dressed in a plaid shirt, wrinkled, low-riding Bermuda shorts, and flip-flops. Isabel was not to be seen. Two of the other people in the room — a young man and a woman — were dressed in a similar casual vein as Tom; two additional men who were a couple of years older wore dress shirts without jackets, pleated pants, and striped ties. Their haircuts were neat, conservative. The final man was dressed in a full dark suit and had an ostrich briefcase at his side.

Henry Green spoke first. In front of him were several copies of what looked like a bound report.

"Thank you, gentlemen, for coming in today. As I mentioned yesterday, Statistical Solutions LLC has decided to exercise its option to terminate its consultancy agreement with LifeDeals, Incorporated, as of close of business today, March third, 2011. In doing so, we are acting without prejudice, and we adhere to the articles of our initial agreement —"

"Yes, yes, blah, blah, blah," Edmund said, interrupting Henry rudely. "We get it, Henry, you're reading us the legal fine print so we think twice about suing you for incompetence. 'If our information is useless, don't blame us.' Now, let's play a game. Hands up, who in this room's a lawyer? You?" Edmund asked, pointing quickly at Tom. Tom stared back at Edmund and didn't flinch. "Don't think so," Edmund commented with a snicker. "How about you two?" Edmund gestured toward the two men dressed in shirts and ties. "Wrong again. Looks to me like a couple of accountants."

"Mr. Mathews, I am trying to do this as

painlessly and professionally as I can. Yes, I have asked our legal representation to join us, as you have so astutely observed —"

"He's calling me Mr. Mathews now," Edmund said, swinging around to address Russell. "He's definitely been talking to his lawyers."

"Okay, Edmund, that's enough," Russell mumbled. He was tired of making excuses for Edmund when he blew up in public like this. It was like traveling around town with a rebellious and obnoxious teenager.

In fact, Russell and Edmund had discussed whether they should bring their own lawyer to the meeting. Russell contended that if each side lawyered up, it was more than likely that the meeting would be aborted before it began. One lawyer would speak, the other would object, and LifeDeals and Statistical Solutions would be advised to leave the matter to their legal representation. What Russell had not anticipated was that Edmund saw red the moment he spotted Statistical Solutions' legal counsel, who stood out from the others like a sore thumb. On edge from everything that was going on, Edmund took the man's presence as a personal affront.

"If I may make a suggestion," Russell said. "We came here this morning for the report that you promised. Let's deal with the legal issues of our relationship after the fact."

"Okay, Russell, thank you," Henry said, glancing at Edmund, who appeared reasonably composed. *Well, you won't look so composed in five minutes,* Henry thought. "We'll proceed with our presentation. There's a copy of our termination letter in the packets we'll give you at the end of the meeting, as well as a completely nonprejudicial note from our legal department reiterating the reasonable scope of our services, which is what I was trying to do a few minutes ago. But I appreciate you're eager to hear the results. I want to assure you that we had our best people working on this. Isabel Lee, whom you met and who is unable to join us today, put in a solid shift on this. So did Tom Graham, who graduated two years ago from MIT. . . ."

Edmund rolled his fists one over the other, like an umpire indicating for play to proceed. He wanted the facts, not the support behind the facts, and the longer it took getting the information implied that it wasn't going to be to their liking.

". . . and Paul, who had more than five years' experience at the Department of Defense." Edmund drummed his fingers on the desk.

"Okay. The work we did last night — all night, in fact — was to estimate how a shift to the right of the bell curves we previously created on the timing of the redemption of the life insurance policies LifeDeals holds would affect cash flow. We'll have the formal report in a day or two but we can give you a preliminary one today. I have to say, we were surprised."

Henry paused and took a sip from a glass of water.

"We were surprised how quickly even the slightest shift of the bell curves would affect the company's financial situation. It would create a period of time in which the payment of premiums would need to be continued to maintain policy values with limited income. This effect is predicted because of the steepness of the bell curves' slopes. As we know, once the policies start to be cashed out by LifeDeals, income can be expected to rise very rapidly, which is why we had strongly recommended that you maximize your purchase of life settlement policies in relation to capitalization. Everyone clear on this?"

Russell nodded vigorously. There was nothing new here.

"Okay, good. Next we looked at the lifespan statistics of individuals lucky enough under current procurement and distribution protocols to get a new organ, be it a lung, a heart, a liver, a kidney, or a pancreas, depending on which degenerative disease is involved. We found that getting an organ alters these people's life expectancies to a marked degree. Understand that we already factored in standard organ replacement rates in the preliminary data that we all approved, and it was at that time a small variable. But the new circumstances, the potential new circumstances, I should say, caused us to drill down into those statistics further and find different research that previously wasn't relevant."

Henry paused again to take a drink. Edmund was fit to burst.

"There are newer statistics which show how well new organs work for patients over a long period. The older figures implied that organ recipients still had the propensity, whatever it was, to affect the new organ and often adversely. But to a greater extent than we initially would have expected, new organs, or at least a large percentage of them, do very well over many years provided it's a good match. Of course newer antirejection drugs have helped as well. In many or most cases, ten to fifteen years of life expectancy can be added to patients' lives.

In lay terms, it appears that you can put a new radiator in a beat-up car engine, and it doesn't matter so much how you drive. The radiator's going to hold up.

"We applied these newly developed statistics to the bell curves of LifeDeals' policyholders and there's no way around it — the implications are pretty catastrophic for cash flow if a percentage of policyholders get new organs. The higher the percentage, of course, the more catastrophic the effect."

"What percentage?" Edmund barked.

"Well, the problem seems to be that the cash flow issue starts almost immediately with just a small shift to the right. Just a few percentage points."

"What do you mean by 'a few'? Five? Ten?"

"Er, five's not good, ten would be, as I said, catastrophic."

"So we'd need five to ten percent of diabetics to get a new pancreas," said Edmund. "What are the chances of that happening?"

There was silence.

"These are not rich people. They won't be able to afford it. It's fucking pie in the sky."

"Not necessarily," Tom Graham said, speaking up for the first time, his voice surprisingly deep. "What you said about people not being able to afford it. Look at the statistics. There might well be thirty-five million people with diabetes in this country. It costs something like a hundred and fifty billion dollars a year to treat them. You think insurance companies won't want to jump on this opportunity? How about state programs that have to carry the cost of these people for decades? Not to mention Medicare, Medicaid. Even far-right politicians are going to find a way around their qualms about stem cells because who won't want to help get twenty, thirty million Americans healthy? It's a magic bullet. If it works, it cures people. And once the organ is accepted, it costs *nothing.* Pancreas doesn't work? We'll grow you a new one and you're welcome. People have been searching for decades for a way to reduce healthcare costs. Regenerative medicine is going to be the answer."

This was worse than Edmund had dared think. He was a salesman; he knew that even if not every diabetic got a new pancreas, the idea of buying diabetics' life insurance policies in a climate in which people could receive a new organ was suddenly looking very outdated, like investing in steam engines after the Model T Ford was produced.

"This is all laid out in the final report?"

Russell asked.

"It will be," Henry said. "It's summarized in the report we'll be giving you today."

"Of course, this is all confidential."

"Of course."

"But it's all dependent upon the date induced stem cell organs will be available," Edmund said. "It's not happening next week. At least I assume it isn't. What is it — two years? Five years? Have you looked into that?"

"Perhaps Ginny here might speak a few words in that regard," Henry said. Sitting next to Tom Graham, a tall woman with long black hair nodded to Henry. She shared Tom's fashion sense and was wearing a T-shirt with a bright image of a robot on the front.

"I read the journals I could find online and tried to figure out some kind of timeline, but the articles in this arena aren't very speculative. It's a new technology so there are no statistics to predict such a big leap in something like regenerative medicine," Ginny said.

Ginny proceeded to talk about the rapid developments already accomplished for stem cell maturation into specific cell lines by researchers around the world. "The next step would be to turn these cells into organs or organ-like apparatuses by a process called organogenesis. This work is going on in Russia, in China, in Germany, but it's having the most success at Columbia University with Doctors Rothman and Yamamoto. Rumor has it that these two researchers have already formed whole organs, which have been transplanted back into the mice that donated the cells from which the organs were formed. Supposedly in the next month or so there's going to be an article in *Nature* about it with all the supporting data. Apparently they have had such success that they've already requested clearance from the FDA to do it on a human. They're waiting for FDA clearance for the next step," said Ginny.

"And when might that happen?" asked Edmund.

"I did talk to one scientist friend last night," Ginny said. "He told me no one knows, but the best guess is in the next couple of months."

"As far as the business angle on this, the numbers strongly suggest that partial remedy would be provided in circumstances like this if the party holding the policies were to raise capital immediately as a hedge against this newly raised eventuality." Henry was taking over in the hope of wrapping up the meeting. Edmund's fuse was getting shorter by the minute, he could tell. Henry was now practically reading verbatim from a script.

"We already ran revenue models for you based on the idea of securitizing tranches of life insurance policies, and while it's difficult to build into the models the prospect of degraded assets, in the final report there will be a recommendation that the securitization proceed immediately and that a significant portion of the funds obtained be set aside to satisfy premiums on those policies that will have to be carried longer than expected. As far as additional life settlements, it would make sense to buy only those life settlements involving individuals with clearly terminal illnesses like metastatic cancer . . . ALS . . . things like that."

The list was a lot longer but Henry's distaste had finally got the better of him.

Edmund thought that all this meeting had done was to confirm what Gloria Croft had told them less than twenty-four hours earlier. It figured: Gloria was one of the best analysts he ever had. And she knew about this before he did. Edmund was confounded by the fact that his great moneymaking scheme could be jeopardized by two geeks he'd never heard of.

"Let me ask you a question," he growled

at Ginny. "You find this research and you find a scientist to call in the evening who says the FDA is going to green-light this project that could revolutionize medicine or whatever. So why am I not reading about this in *The New York Times?*"

"Because researchers and their universities have become much more sophisticated about patent issues. There used to be a rush to print because they yearned for the notoriety, but they're much smarter now. There are fortunes to be made in biotech, and this area of organogenesis might be the biggest yet. It will probably overshadow every other technological milestone in the history of medicine. Believe me, when the Rothman work hits *Nature,* it's going to be all over *The New York Times, The Wall Street Journal,* and every other media outlet."

Edmund and Russell rode the elevator in silence. It was the same car that Edmund had assaulted the day before. He'd taken a prescription painkiller from his cache, so he was feeling only a dull throb in his left hand. He looked closely and thought he could make out a slight indentation in the metal elevator door. The trouble was he felt like doing it again.

Only when they reached the street, away

from any prying ears, did they speak.

"What do you think?" Russell said.

"I think we paid those idiots too much money. And we will sue them."

They met their car in front of the building and got in. Edmund thought for a moment. His mind was racing.

"Okay, here's what we're gonna do. Today. No more diabetics, obviously. Tell the sales force, even if they're hooked on the line, even if they're in the boat and you have the billy club in your hand, put 'em back in the water. Any contracts in the works — cancel. Checks being cut — cancel.

"Have someone go back and look at contracts of people who have diabetes and something else. We like the something else now. Get the lawyer to draft a letter in his best incomprehensible legalese to say we're no longer interested in diabetes and send it to these people. Scrub those people out of the stats. They never had diabetes. And we need new policies. Target smokers. I know they're the worst because none of them thinks they're going to get sick and when they do, they die too fast. Find out if we can target smokers or ex-smokers with policies who've missed a couple payments. They're all lying anyway. And offer them twenty-five percent —"

"But the model —" said Russell.

"Fuck the model!" roared Edmund. "Don't you understand? As of today, there is no model. We don't have a *business* if this shit goes down, never mind a *model.* Jesus. These are Band-Aids I'm talking about and we've got a head wound."

"Potentially," Russell said.

"Right, yes, there's a chance this research may not get anywhere. But still, we're up to our necks in this either way. Gloria Croft is already shorting us, and she won't be shy talking about it. We have to do something. It's not like we can sell our stock and walk away."

"Perhaps we should go and see Jerry Trotter," Russell said, after an awkward pause.

"Yeah, I was thinking the same thing," said Edmund.

Dr. Jerred L. Trotter, an old friend of both Russell and Edmund, ran a very successful hedge fund. Trotter was a man who enjoyed outthinking people, which meant he wasn't above a certain amount of sharp practice if he was confident he could get away with it. There were so many areas where the regulatory authorities were lax, or there just weren't any regulations. It was by using Trotter's good offices, and through a number of disguises of Trotter's creation, that Edmund had purchased credit default swaps on his own company while continuing to sell subprime bonds to his corporate clients. It was just the kind of caper that Trotter relished. Especially when his cut was so generous.

Edmund had sounded Trotter out very early in the planning stage of LifeDeals. He sent Russell over on what looked like a fact-finding mission. Did Jerry think a model like this might work? Did Jerry think these were good assets to spin off bonds from? Did Jerry think enough investors would want to buy such a product? Russell never mentioned that they were seeking investors.

Three days later, Jerry had called, practically biting Edmund's arm off over the phone. He'd run the numbers, and he wanted in. Edmund, after pretending to think about it, pretended to reluctantly allow Jerry to invest $25 million of his own money and more of his fund's. Trotter would definitely want to hear about these new market conditions. He wouldn't be pleased, to put it mildly, but he was always a man of action and he'd think of something.

The driver had been awaiting instructions as the two men sat in the back of the car.

Now Edmund gave the driver an address, and they set off.

16

Terrasini Restaurant
Midtown Manhattan
March 3, 2011, 12:45 p.m.

"There was no way we could meet you guys today, no way. And then Russell said the magic word: 'Terrasini.' " Everyone laughed, even Edmund, who was making an effort to calm down.

Dr. Jerred L. Trotter, former plastic surgeon, current hedge fund maven and LifeDeals angel investor, was holding court alongside his longtime number two, Maxwell Higgins. There was always a table at Terrasini for Trotter, and the four men — Higgins, Trotter, Mathews, and Lefevre — were nestled at a table in a corner of the restaurant's front room.

"I thought it couldn't hurt," Russell said, attempting to sound breezy. Edmund and Russell were glad Trotter could see them at short notice, even if it had meant killing a couple of hours until lunchtime.

"You know it's my favorite place to eat. Unless I'm picking up the tab, right, Edmund?"

"There's no danger of that, Jerry," Edmund said. "Everyone knows it's against your religion to pay for lunch."

Trotter guffawed. "Damn straight," he said loudly, and waved expansively toward no one in particular and a waiter quickly came over.

"That glass of Barolo, make it a bottle. I trust that's okay, gentlemen? Second thought, bring the glasses and then the bottle in a few minutes. I need a real drink after the morning I had."

Edmund looked on, suppressed irritation sticking in his craw. That little sideshow just set him back at least a couple hundred bucks. But scenes like this were part of the cost of doing business with Jerry Trotter.

"No serious problems, I hope, Jerry," Edmund said.

"Everything's a fucking problem," Jerry said in his usual penetrating voice. A man in a Cucinelli sweater who'd brought his family to lunch turned around and shot Jerry a look.

"Whoops, apologies, forgot where I was," Jerry said to the man.

"I thought we should meet in the office," Edmund said. "You know, easier to speak freely." Jerry Trotter, like a lot of financiers, swore like a stevedore. Edmund didn't care about that but he wanted to discuss LifeDeals' predicament without having to hold back in front of the other diners.

"Got to eat, Edmund," Trotter said, picking up the menu.

"Of course, Jerry."

Edmund thought that Jerry Trotter probably knew the menu better than the restaurant's own waitstaff. Everyone insisted on playing these childish games, thought Edmund, aware that, once more, he was the one being toyed with. Edmund studied Trotter's face. He knew the man was at least sixty, but he looked closer to forty-five with his sweep of gunmetal hair, a few trace laugh lines, and bright blue eyes still a shade of azure that caused more than a few people to take a breath when they saw them. If he'd had any cosmetic work done, it was very good.

"Max, what was that special again?"

The waiter was handing out the glasses of Barolo.

"The pasta?" Higgins questioned with his upper-crust London accent. "Orecchiette with sweet sausage, broccolini, touch of ricotta. Sounds marvelous."

"Ah, gives me a hard-on just to think about it. Of course. Times four, what do you say, guys? Cold day, calls for some soup, I think. The other special, squash soup, little bit of cream, am I right, Max? All around, sir, if you don't mind."

The waiter agreed it was a great selection. Edmund made a point of looking at his menu a couple of beats longer, then folded it and handed it over. He was too upset to protest. Now Trotter was swirling the red wine around in his glass. It was a lovely ruby color and on another occasion, Edmund would have been rhapsodizing about it.

In the often flamboyant world of hedge fund management, Jerry Trotter was something of a celebrity. He'd already enjoyed one very successful career as a plastic surgeon, catering mostly to the moneyed ladies of New York's Upper East Side, for whom he performed face, eye, and buttock lifts. In truth, Trotter was a better showman than he was a surgeon. He knew, like everyone in the medical profession, that during medical training grades counted more than the need to display particularly good physical skills in obtaining training positions in such surgical specialties as the eye or brain, or in plastic surgery. Trotter had made sure he always got good grades to make up for his hand-eye coordination, which was poor enough for him to find surgery a taxing grind. But that was in the past, and he didn't need good hand-eye coordination to manage money.

Trotter had always enjoyed looking after his own money and taking the odd risk here and there. After years of working six days a week at his practice, he had a lot of it to manage. Trotter had known Max Higgins for years and well enough to know he was keen to strike out on his own from his trading position at Goldman. Trotter made a proposition to Higgins: we'll set up a fund, you run it and teach me what you know, and I'll supply the money. It worked, and Trotter quickly found that his patients trusted him and were grateful and many were happy to let him invest for them. Trotter was a quick study and soon his fund, the immodestly and unimaginatively named Trotter Holdings, was heading toward the middle rank of name funds.

"So, Edmund, Russell, what was so urgent that we had to meet today?"

Russell looked over at Edmund before he started. The delay in Trotter being able to meet with them had at least allowed Edmund and Russell to decide how they were going to present their news to Trotter. They'd sat in a diner on Lexington Avenue for an hour and strategized.

"We have a potential public relations issue we want to give you a heads-up about. We thought you might be able to help us get out ahead of the possible publicity — nip it in the bud, so to speak."

"Edmund, has your checkered past finally caught up with you?" Jerry asked, not sounding as though he was joking at all.

"We've heard about some medical research that's taking place," Russell continued, to spare Edmund from having to respond. "It's at an experimental stage, with no guarantee it's going to work or anything, but in some quarters it's being taken more seriously than in others."

"How does that affect LifeDeals?" asked Jerry, who glanced between Edmund and Russell even though it was Russell who was speaking. All the joviality had left his voice. The two waiters brought over the soup and left it quietly, detecting the tension at the table.

"It may not mean anything," Russell added. "As I said, we want to get ahead of any potential bad publicity."

Jerry Trotter picked up his spoon and tasted a mouthful of soup. It was delicious, of course, but when he sensed he was about to hear something unpleasant, his appetite faltered. "Russell, you're going to have to tell me a little more clearly what's going on."

"Okay, Jerry, sure. There are a couple of researchers at Columbia who think they can grow artificial organs using human stem cells that they make from a patient's own cells. Obviously I don't know the details, but the process is called organogenesis. It's supposed to start what will be called regenerative medicine. If they can grow new pancreases, for example, they can help people with diabetes live longer. But at this point, it's a huge 'if.' "

"I read something about that idea in some research we did on our due diligence for LifeDeals, but it sounded like science fiction," Jerry said.

"Apparently not anymore. The future is now, as they say. Or it might be," Russell said.

"How many people know about this?" Max Higgins asked.

"Not many," Russell said. "Outside Columbia and the stem cell field, very few is my guess."

"How did you hear about it?"

There was a pause. This was where it was

going to get a lot trickier.

"Gloria Croft told us," Edmund said.

Max Higgins quickly figured out the implications and, way ahead of Trotter, asked Russell the main question.

"Is she taking any action?"

"Yes."

"And knowing Gloria, she's shorting LifeDeals, am I right?"

"Yes."

"Hold on, hold on, wait a second," Jerry said. "LifeDeals is being shorted because of some work being done in a lab at Columbia?"

"I'm afraid so, Jerry."

Higgins went on, "Am I to assume that some numbers have been run looking at cash projections based on a potential breakthrough in diabetes treatment? And that the projections don't look so good?"

"Is he right?" Trotter asked. Higgins had got right to the heart of the matter in seconds.

"Broadly, but look —"

"So he is?" Trotter questioned.

"As I said, it's early in the —"

"Would you mind telling me how this . . . this fucking *disaster* is what you would describe as a public relations issue?" Jerry was hissing, using his soup spoon to point first at Edmund, then at Russell. The table was silent until Max Higgins spoke up again.

"The science may fail eventually, but with Gloria Croft taking a position, investors are going to be alarmed when they find out. Then it becomes as much about Gloria as about the science. She's a barometer. So from that perspective, they're right, Jerry, it's a PR problem first."

"And if the science succeeds?" Jerry said.

"Then we have a bigger problem," said Edmund.

Jerry put down his soup spoon and took a large drag on his $30-a-glass Barolo.

"You guys didn't see this coming?"

"Obviously not," Edmund said. "It's a once-in-a-century breakthrough if it happens. You can't do projections for being hit by an asteroid."

No one was eating now. The waiter, coming by for a second time, asked if anyone was still working on their soup and no one was. Yes, the soup was fine, but everyone was preoccupied. The uneaten soup disappeared.

At Jerry's insistence, Russell walked him through what they knew of the research. He made a point of saying that there was no guarantee it would succeed; the odds had to be that it wouldn't because in the vast majority of research projects there was always some major unanticipated obstacle that would arise to thwart the hoped-for result.

"So what are the chances it will succeed?" Trotter questioned.

Edmund said there was no way to know. He then had Russell talk about the effects such an eventuality would have on LifeDeals' cash flow. As Edmund and Russell had agreed before meeting with Trotter, Russell erred on the conservative side.

"What's our position vis-à-vis the public offering?" Jerry directed his question at Max, his partner. He didn't care that he might offend his lunch hosts.

"The lockup period expires May thirty-first," Max said.

When an investor takes part in an initial public offering, they can't sell their shares for a certain amount of time, in this case 180 days. Trotter Holdings was halfway through the mandated blackout period, meaning that Jerry Trotter and his fund were stuck with his shares for another three months.

"Shit. What are the chances of Gloria Croft keeping her mouth shut for three months?"

"Listen, Jerry, life insurance is still a twenty-six-trillion-dollar business," Edmund said. "There's plenty of money to be made. These diabetes policies are basically a drop in the bucket. It's not a question of dumping LifeDeals shares, it's a problem that needs a solution. That's why we came to you, you're the man who deals with problems, everyone knows that." Edmund was being purposefully flattering and Trotter wasn't unhappy about it. It's true, he was a guy who could solve a problem, but this was a big problem and a new problem.

"Gloria Croft is so full of it you wouldn't believe," Edmund continued. "She was carrying on, telling us that our life settlement bonds were a bad product. Blah, blah, blah. She even told us that we should have been prosecuted over subprime."

"She is one sanctimonious bitch, I know that for sure," Jerry agreed.

"She was enjoying telling us about the research, twisting the knife."

The pasta was set at each man's place. There was slightly less tension around the table now — the problem had been identified, there was a common enemy, and the four of them were on the same side. Soup could be resisted, but the pasta was a different proposition, and each of the four men ate.

"Of course, there is a medical angle here, Jerry, that we think you can help us with. And Statistical Solutions is drawing up projections of the effect on revenue if we have to pull the diabetics' policies. We're going to need more capital. We can see if we can fill in these capital shortfalls with different initiatives. We've already directed our sales teams to go back and beat the bushes for metastatic cancer patients with big policies. It costs more money to buy those policies, but it's completely risk-free."

"More money, how so?" Higgins said.

"We've authorized the salesmen to offer more than the standard fifteen percent. Metastatic cancer patients are not going to cause us any problem about dying on time. It's just that we have to be more aggressive to find them."

"Okay, Edmund, I'm hearing you. Needless to say, Max and I will have to put our heads together and talk about this research issue. We want to see that Statistical Solutions data as soon as you have it, of course. And we have a bunch of meetings this afternoon, so I'm sorry but we have to eat and run."

Jerry downed another forkful of his lunch

and chased it with the remnants of his third glass of wine. The leave-takings were quick and less effusive than the greetings had been when the four men met. Trotter and Higgins left Edmund and Russell behind, picked up their coats and got into their waiting town car on Fifty-fourth Street.

"So?" Trotter asked when he was settled in his seat. They were heading south on Park Avenue.

"Gloria Croft," Higgins said. "She's a pit bull."

"I sense there's history there with Edmund — perhaps it's more about that than about the product."

"Maybe it's both. She's found a flaw in the model, and it just happens to be Edmund Mathews, so it's a bonus."

"Be that as it may, we have to do something," Jerry said. "And we can't leave it to those two. They're floundering, you can tell. There's too much time till we can sell the stock."

"I agree. But Edmund has a point: it's still a good business even if there is a bump in the road with the current model. It will just take some juggling and deep breaths on our part. It's not a time to panic. Besides, since we can't sell the LifeDeals stock, we can't panic even if we wanted to. I also agree that in an ideal world, he should have seen this coming. But our due diligence didn't pick it up either. It's a function of the times. Technology's changing too fast just like markets are changing too fast. It's getting harder and harder to factor this kind of thing in."

"Well, I think we need to do some of our own research. A little more street-level. We can't rely on Edmund et al. for that, clearly. Let's get an investigator up to Columbia, sniff around a little. And someone can do some digging on Gloria Croft. A woman like that doesn't get where she is without pissing a lot of people off. And maybe cutting some corners. We need to get some angles to work with, get some leverage."

"Okay, I'll get right on it. This is a new one on me — shorting stocks because of a medical breakthrough."

"It's new all-round. We may have to get creative," Trotter said, watching the Park Avenue traffic move slowly in a steady afternoon drizzle.

Jerry Trotter knew how to make an entrance into a room, and he also knew how to make an exit. Edmund didn't know what to make of what had just happened. Sure, Jerry was royally pissed when he first heard about the problem, but he seemed calmer when he left. The problem was that the departure had been so sudden.

"What do you think?" Edmund asked Russell. They had ordered cappuccinos.

"For me, Higgins is the brains of that operation, he sees the bigger picture," said Russell. "It didn't take him long to zero in on the heart of the problem. They definitely got the message."

"Do you think they'll come up with some suggestions?" Edmund asked.

"I do," Russell said. "I think it was right to tell them early in the game. It feels good to have Jerry Trotter and his team working on this with us. I just hope they get back to us soon."

"My guess is that he will," Edmund said. "Jerry isn't the kind of guy who's going to watch while his sixty million dollars turns into pocket change. But I do have one reservation."

"What's that?"

"I never can be a hundred percent confident Jerry's working *with* us."

17

Columbia University Medical Center
New York City
March 3, 2011, 6:23 p.m.

Pia glanced up from her reading to see that it was almost six-thirty, and she wondered how long she was going to have to wait. She was sitting in the deserted lab outside her office, which continued to be off-limits. She was waiting for Rothman, who was still in the biosafety unit trying to finish up the data for the *Lancet* article. Yamamoto had told her to wait because Rothman wanted to speak with her. What about, she had no idea. Whatever it was, it made her feel uneasy.

The day had been interesting in some regards although it had not been one of her best. She had spent the morning doing the lab equivalent of KP duty, washing and rewashing glassware. For the second day in a row Pia had been actually late, not just late by the secretary Marsha Langman's standards. For the second day in a row George had failed to appear at her door, and she wondered why. It irked her to a degree because she'd come to count on it, even though she was the first to agree that he wasn't obliged to come and wake her. She'd slept badly again, as she'd been doing for a week or more. As per usual it was nightmares involving childhood memories she'd battled to suppress and which surprised her in their lucidity. Even after years, they still possessed ample power to shock and disturb her.

The head technician, Arthur Spaulding, had seemed amused by Pia's punishment. He had made a point of detouring when in the lab to pass by Pia, saying nothing. He didn't have to, his smirk was enough. Spaulding had been at the research center for years, and he liked things done just so. He'd started working for Rothman only in the last eighteen months, and Spaulding got on with him about as well as his predecessor had before he was fired. Spaulding seemed to take exception to any irregular request from Rothman, and Rothman being Rothman, most of his requests were irregular.

"Do you need something, Mr. Spaulding?"

Pia said finally on about his fifth pass. "You seem to go out of your way to come over here."

Spaulding said nothing and left, as Yamamoto had suddenly appeared and come over.

"Miss Grazdani, we would like to see you in the organ bath lab."

Dr. Yamamoto hadn't directly said anything about Pia being late, and he didn't need to; Pia had gotten the message. She suited up quickly and went in the lab to join Lesley and Will, who had been in there all morning. They were standing in a side room with Rothman, who was holding a thick paperback book in one hand and pressing impatiently at the display of a newly installed machine that looked like a large inkjet printer.

"Dr. Yamamoto, it appears we have bought a lemon," Rothman said.

Will gestured down to the power strip in the wall, and Yamamoto bent down and flicked on the surge protector. The machine wheezed and babbled into life.

"Anyone know what this piece of machinery is?" Rothman said, not missing a beat and bending forward to stare into the innards of the device. As she moved closer, Pia could see how intricate the engineering on the device was, with tracking bars and a large array like the inkjet of a printer. Pia opened her mouth, but Lesley beat her to it. "An organ printer?" She had seen the cover of the book Rothman was holding.

"Yes, for three-D bioprinting. We have an older machine, but this one is new. Someone from the manufacturer is coming tomorrow to show us how to use it."

"Perhaps we should wait till then to turn it on, Doctor," Yamamoto said.

"No harm in warming it up. Mr. McKinley, what do you know about three-D bioprinting?"

"I think it works like a regular printer, spraying living cells onto a sheet of . . . a sheet of something. It goes back and forth building up the layers into a three-dimensional structure. The cells can often organize themselves to function collectively. So far it has applications making skin and cartilage."

"Indeed. You can print a spinal disk. Which I may need to try by tomorrow," said Rothman, straightening up and rubbing the small of his back.

"We're all going to get up to speed on this machine together. As this engineering matures it might be quicker than growing organs. At this stage I'm thinking of using it to fix defects in the organs that we're growing. But who knows? The value of this technology is that an organ would be fashioned as the mirror image of the patient's by using data from MRI studies. As is often the case with organogenesis, the hardest part is not in replicating the function of the organ, or gland, but in connecting it to the rest of the body — veins, arteries, ducts, and so on have to be oriented appropriately to make the surgery feasible."

Rothman had found a switch and he toggled it on and off. He then reopened the instruction book and immediately became engrossed.

Yamamoto ushered the students away. Will looked back at Rothman.

"I hope that thing's still working when the guy comes tomorrow."

The students spent the rest of the day monitoring the organ baths. They'd found that some did experience small changes in temperature or pH spontaneously, and the three talked with Yamamoto about designing an alarm system even for those minor fluctuations. It seemed to them that even such small changes might impact the results. It was real science, demanding and exciting for each student. Will and Lesley worked closely, keeping a respectful distance from Pia, who was in her own world.

After Dr. Yamamoto dismissed the students for the day, he asked Pia to stay behind for a moment. "Dr. Rothman would like to speak with you later, after we're done in the level-three lab. I hope you don't mind."

"Of course not," Pia had said. What else was she going to say?

18

One Central Park West
New York City
March 3, 2011, 8:30 p.m.

Jerry Trotter and Max Higgins left Edmund Mathews and Russell Lefevre with a souvenir of their lunch together at Terrasini. It was a healthy check, and as Edmund paid up, he hoped that Jerry and Max were going to do something to justify it. He was reasonably confident, as he knew them to be guys who did not sit on a problem. They were known to do whatever it took.

And Jerry and Max did not disappoint. Within an hour of leaving the restaurant, Trotter and Higgins had two of their best investigative guys working on the two separate cases, one on the Columbia regenerative organ situation, the other on the hoped-for dirty little secrets of Gloria Croft in hopes of having something to control her with. Jerry knew that everyone, especially

Wall Street people, had their secrets. The PIs were also given Edmund's and Russell's full names so they could Google them for background information.

Jerry couldn't use his very best guy, who was in fact a woman named Jillian Jones, because she was already involved in looking at a company that Higgins thought might be tanking its results to pave the way for a takeover. But as PIs, Tim Brubaker and Harry Hooper were separated from Jones only by a sliver. They'd do a thorough job, and they were very quick.

This was the first time Jerry ever had three PIs on his payroll at once. Strictly speaking, Jones, Brubaker, and Hooper weren't on the Trotter Holdings payroll at all; each was employed off the books in a strictly cash-only arrangement. There was no paperwork — no pay stubs or invoices or receipts, the latter because Trotter trusted the three not to pad expenses. The PIs undertook enough on-the-books domestic surveillance cases to report to the IRS to show sufficient income to justify their lifestyles. They were usually able to piggyback a cash job for someone like Jerry Trotter on one that was recorded by the business accountant.

Jerry Trotter loved this semi-legal, clandestine work because it was so far removed from anything he'd done as a plastic surgeon or a money manager. He loved everything about it and even saying "Brubaker" and "Hooper" gave him a little thrill; to his mind, they were perfect PI names. For him it was like being in his own movie. Brubaker and Hooper were ex-cops and had seen everything; Jillian Jones had seen everything too but no one knew what work she had done prior to this and no one dared ask. In contrast to most PIs, she always acted anxious and quick to take offense. She also had a black belt in karate and was always armed.

As usual, Higgins took care of all the practical arrangements. He bought three use-and-lose cell phones and used one to place a prearranged marketing cold call to Hooper and Brubaker's offices. The call purported to be from a firm interested in talking to the business manager about the company's metered postage. This identified the source of the call. The first digit of the call-back number times one hundred was the rate being offered; the last four specified a time for a meet at the usual location. Brubaker and Hooper had a habit of checking their messages regularly and both picked up Higgins's call within a half-hour. The promise of a $300-an-hour rate drew them to the four o'clock meeting place in a back booth at Flanagan's bar on Second Avenue. It took Higgins five minutes to give them their marching orders, a cell phone for each to call in on, and a $1,200 down payment on their work.

Another aspect of the game that Trotter relished was asking his guys how they'd found whatever information they dug up. At first neither Brubaker nor Hooper wanted to talk about their MOs, but they'd come to indulge Trotter's whims. He did sign the checks, after all.

When the phone rang at eight-thirty, Trotter was fixing his second Glenlivet in his apartment near the summit of the Trump International on the corner of Central Park West and Columbus Circle. Trotter lived so high up he didn't bother with drapes in the living room — he didn't want anything to interfere with his view of Central Park. Tonight, banks of low clouds and rain were all he could see. He was pleased to see it was Brubaker. The number on his LED was the cell phone Higgins had given him.

"It's B," Brubaker said. It was the code name Trotter insisted on his using.

"So soon?" Trotter could barely conceal the childish excitement in his voice.

"Yes, I got to speak directly with the

laboratory's secretary by posing as a journalist. Couldn't shut her up. Thought she was doing her boss a favor talking him up. Thinks he's shy and needs the pub. She even said as much."

"So she was informative."

"Very. Don't understand half the stuff she was saying, but she was knowledgeable. I'm transcribing the tape myself word for word. I don't want any transcription service looking at this stuff."

"Of course, very prudent. So give me the headlines."

"Okay, the two names you mentioned to me are the guys for sure. . . ."

"Rothman and Yamamoto," Trotter said, talking over Brubaker.

"Shit, what's the point of the code words and all the cloak-and-dagger you insist on if you don't follow it yourself? Yes, those are the guys. The first one is the big cheese."

"Sorry," Trotter said, inwardly cursing himself.

"Okay, so she tells me all this stuff that they're doing, and I'm supposed to be a science reporter and able to follow it. So I asked her at the end for the Cliff Notes version that I can use for the readers."

"What paper does she think this is going in?"

"No paper. I told her I was doing the research to see if there was a story and if there was, I'd sell it and call her again."

"What if she calls you?"

"She doesn't have my number. I told her this was very hush-hush on my end, and I asked her not to tell anyone we'd spoken because this is such a big story that other reporters are going to be on it soon, and I want to get a jump on it. I'm actually thinking of writing it up for real — I wasn't lying to her, this is going to be way big."

Jerry's joy at playing Dick Tracy evaporated.

"What do you mean, 'big'?"

"Well, according to her, these guys are close commercially to growing organs outside the body, organs that will be perfect matches for the person who needs them. The trials have worked with animal subjects, and they want to move on to using human stem cells."

"When?"

"She did get a little cagey there. Not because she wouldn't tell me — I think she didn't know and didn't want to let on that she didn't know. But it's months that they will be moving to human cells, maybe even weeks and certainly not years."

"Weeks or months? The difference is important."

"Well, I guess I need to make a few more calls. But it's happening. And soon. He's working on something else too. Something about growing salmonella strains that cause typhoid fever on the space shuttle. Can you imagine? To think where our tax dollars go. It makes me sick."

"Tell me about it," Jerry said. "Okay, thanks, B. Keep me posted."

"Got it, boss."

After hearing from Brubaker, Trotter was impatient to know what kind of progress, if any, Hooper was making. Although it was technically against protocol, Trotter called Hooper's new cell.

"Yes," Hooper said after one ring. He was between calls on the Gloria Croft assignment and thought it might be one of the contacts he had made calling him back.

"Hi, it's the boss."

"Hi, boss."

"What's happening? Any dirt?"

"I'm only three hours in. Not even."

"What's the setup?"

"I'm a headhunter looking for someone for a major bank CEO job. The board wants a woman for appearances' sake. I'm asking around about people on my supposed list."

"Our friend doesn't need a job, she makes

seven figures–plus a year," said Trotter with disappointment. He purposely avoided using Croft's name.

"I know that. They know that. But people like showing off how *much* they know. I think someone might tell me just how much she doesn't need a job in a bank. Or need the scrutiny of running a public company, more like it."

"So you want someone to brag about what they know."

"Sure, everyone does it. Most everyone. And the finance world is like a small, competitive club which feeds on gossip."

Okay, that was more like it. Jerry was struck again by how much Hooper and Brubaker sounded alike. They sounded like Brooklyn cops, which is what they both had once been.

"So you shake anything out of the tree yet?"

"Just spoke to a guy who knew her at Morgan, back in the day. I said someone mentioned him as a possible reference, and he laughed. Real asshole, thinks I'm a moron having gone to Brooklyn College nights. I don't like these Ivy League types. But he has something, I'm sure. Trying to fuck with me a little. Hope he doesn't push it 'cause he's messing with the wrong guy. I can get his nice car towed tonight, and it ain't goin' to the pound."

"Yes, I'm sure you can. That's what keeps me honest in our relationship."

Hooper laughed, then added, "One other thing. He mentioned that I might ask one of the bankers Higgins mentioned when we talked this afternoon. He said if I wanted dirt on our friend to ask him, because he thought he had been literally and figuratively fucking her back in their Morgan days."

Trotter frowned. "Which one?"

"The thick guy with the short hair," Hooper said.

"Now, that *is* interesting," Trotter said. "Don't call and ask him directly. Make it part of your investigation. It could be interesting."

"Got it," Hooper said.

As Trotter hung up the phone, he smiled. "Edmund, you rogue."

19

Columbia University Medical Center
New York City
March 3, 2011, 9:02 p.m.

Pia waited more than two and a half hours for Rothman and Yamamoto to finish their work in the BSL-3 lab. She spent the time productively reading papers on tissue engineering and organ printing on the Internet, which is what she would have been doing anyway if she'd gone back to the dorm. As the time had passed and her empty stomach growled, she became progressively concerned it had something to do with her being late two days in a row. Eventually, Rothman and Yamamoto appeared. Yamamoto immediately left. Rothman wordlessly waved for her to join him in his office, where he got straight to the point.

"I want to talk to you about the future. Your future. I need to know that you are committed to this work."

"I am, truly," Pia said. She was panicked. "I know I was late this morning —"

"You were late two mornings in a row, at least according to Miss Langman."

"I'm sorry . . ." Pia stammered. Her fears were coming to pass.

"It doesn't help to be sorry," Rothman shot back. "I'm concerned about what it implies."

"I will make sure it never happens again," Pia offered.

Rothman waved her off. "Let me speak while I'm inclined to do so. As you know, I'm not accustomed to talking too much about this kind of nonsense. I don't have the time. Last year I confided in you some information about myself because I had been progressively confident that you were turning out to be the person I thought you could be. Remember, as I told you, I played a role in getting you admitted when others on that damn admissions committee where I was forced to serve were reluctant because of your foster care experience. Since I had the same experience, I thought you might have promise of being a researcher."

"I've come to the same conclusion," Pia blurted.

"Don't interrupt!" Rothman snapped. "Last year when I told you those secrets about me that are only known by my wife, bless her soul for putting up with me, concerning my foster care history and my Asperger's, I wasn't completely open. I said my sons were not as healthy as I would like. To be more specific, not only are they too on the Asperger's spectrum, but even worse, they have type 1 diabetes. Having passed on the Asperger's was reason enough for guilt and depression. The diabetes has put it over the top. The main reason I've turned to stem cell work is to see that my boys are cured in my lifetime. It's a quest that pulled me out of a serious bout of depression. Depression has been my bête noire."

"I'm so sorry to hear about your boys," Pia offered.

"I'm not telling you this to elicit any sympathy. I'm telling you this so that you understand me better. I have never ever agreed to mentor anyone, and it is not just because my Asperger's puts me at a social disadvantage. I feel I don't have the time for other people's nonsense, and this includes Ph.D. students as well as medical students. You were a first. I thought your foster care experience would make you thrive in the lonely pursuit of science and that you should have been given a chance."

"I think you're right," Pia said. "I know I

struggle with social issues as well."

"Pia, commitment to research has to be total. Two days ago you came in here and told me that, yes, you were going to take me up on my offer to do your Ph.D. here in my lab. At the same time you told me you were going to do a simultaneous residency in internal medicine. And then you expected me to be pleased. Pia, that old myth of the doctor doing both clinical medicine and research at the same time is totally passé. It wasn't even true when it was current. Research is more than a full-time job."

For one of the first times in their three and a half years of knowing each other, Pia and Rothman held each other's eyes. It was a kind of Mexican standoff. Both were conflicted. Pia had struggled hard and surmounted considerable obstacles in her drive to become a doctor. And now she was so close. In months she would be getting her M.D. degree. The problem was she wouldn't yet be a doctor, one that can get a license from the state. A resident was on his or her way to becoming a real doctor. If she didn't take a residency, she'd always be a medical student with an M.D.

Both individuals looked away.

"I realize that I'm hard to read," Rothman said, breaking a short silence, "or at least my wife tells me so. She advised me to have a conversation with you."

"She knows about me?"

"She knows about everything in my life. It is the only way we could have survived as a couple. I'm not easy to live with."

Rothman had been rehearsing this speech for days so he was relatively comfortable once he had started talking.

"What I seek in a colleague is commitment. A med school graduate knows nothing. No offense. But if they're able to pass through medical school it means they have the basic intellectual wherewithal to do research. After the initial flash of inspiration that tells you what to look for, most of the success in research lies in doggedness. In covering every angle, tracking down every lead. I'm mixing metaphors but you know what I mean. Dr. Yamamoto was in fact a rather mediocre student, but he exhibited more application than ten other men who had better grades. I can already tell what kind of doctors your colleagues will be. Ms. Wong is desperate to help sick people, and she'll be very good at it. Mr. McKinley will probably end up doing something flashy but unchallenging, like surgery, and worse yet, plastic surgery."

Pia didn't know what to say.

"I'm pleased you made the right decision about the nuns. It shows me that you're orienting yourself in the direction of research, but I don't want you to make the same mistakes I made. For me, doing a residency in internal medicine was a big waste of time."

Rothman paused.

"I think we're similar in a number of ways."

Pia's eyes opened wide, and she flushed and looked down at her feet. She didn't share Rothman's newfound comfort in discussing such personal matters.

"I need more help here. Dr. Yamamoto can't do everything, and I can't rely on med students rotating in and out of the lab. No offense. The two of us are spread thin, working on salmonella and organogenesis at the same time. But the university is committed to helping us, and I hope we'll be able to add laboratory staff and take over more lab space to ramp up our organogenesis work. We need at least another researcher. Which is why I'm talking to you about commitment. Although I usually have no time for excuses, tell me why you were late the last couple of days."

"There's no excuse, really," Pia admitted. "But I've been having trouble sleeping."

Silently she cursed George for her getting accustomed to him waking her up even though she knew it was totally unfair.

"Why have you been having trouble sleeping? Anxiety?"

"Nightmares."

"About what, if I may ask?"

"Childhood memories. Ancient history."

"Pia, I think I need to know more about you. You've told me little about yourself even though I've tried to be open with you about me."

"What do you want to know?"

"Everything!"

Pia took a deep breath. Occasionally in life there is a key crossroads moment that can be identified as such even as it's taking place. Pia understood she couldn't hide or keep quiet. The time to talk was now. Except with her caseworker, Sheila Brown, she'd never talked about her childhood. She took another breath. She felt as if the room were closing in on her. The only light in the room was a small library light on the desk, which put Rothman's face in shadowy relief.

"I do remember my mother — not much, just little things. She was also a Pia. My first name is actually Afrodita, but we were both called Pia. Sometimes out of the blue, a scent on someone or a gesture will make me think of her. But she died when I was little. I don't know how, and I don't even know how I know she died, but I do. I lived somewhere in the city with my father, Burim, and his older brother, Drilon, who stayed with us. They were Albanians, real Albanians, just off the boat, very rough around the edges. My father was away a lot, and I had to spend time with my uncle, who was a real creep. Have you got any water or anything?" Pia's throat was dry.

Rothman got out a bottle from a small fridge under the Nespresso machine. He pushed it and a cup across the desk's leather surface.

"He hit me a couple of times, my uncle. He used to touch me, never when my father was around, he was very careful. He made me touch him inappropriately, to say the least. He took pictures of me, you know, and developed them himself, and I think he used to sell them to other creeps like him. It developed into a side business for him. One night I had enough, and I went for him."

"What did you do?"

"I tried to stab him in the penis with a pair of scissors. Unfortunately I missed the target. It bled a lot. I remember later during first-year anatomy when the professor was talking about the femoral artery and joked about how you should avoid puncture wounds there. I suppose that's where I must have stabbed my uncle.

"When he got out of the hospital, he beat me up pretty well, and I guess someone called nine-one-one about a little girl with a bruised face because the police came and grabbed me. My uncle and my dad both got arrested. So I became the property of New York City Child Protective Services."

Pia took another drink of water.

"And your father?"

"He vanished. Years later I made Sheila Brown, my final and best caseworker, tell me what happened. She managed to find out that the two of them made bail and disappeared into the city. I've no idea where he is, if he's even alive. I guess I convinced myself that my father wasn't to blame for what his brother did to me. When I was in foster care, I used to fantasize that he would suddenly appear and rescue me from those places I was in, but he never did. I gave up hoping after a few of years."

"How old were you when you were taken into care?"

"Six. I was a problem for the foster care people from the get-go. My father hadn't officially given up his parental rights, so I couldn't be adopted. He had to register with Child Services before — I wasn't in school, I guess — and genius that he was, he said I was a Muslim, which, as an Albanian, he may have been. Sheila said he must have thought I'd get better treatment if I was a minority, but it just meant none of the religious agencies wanted me. Back then the system was dominated by Catholic, Protestant, and Jewish organizations. So instead of having a chance at some reasonably appropriate placement, I was put in a kind of temporary group home, and a janitor there tried to abuse me like my uncle did, but I fought him off. I complained about it, and I guess that marked me down as a troublemaker.

"I ended up in juvenile court a couple of times. I was lucky not to get locked up, I suppose, but I was designated a 'person in need of supervision' — I love that phrase. So the first dump I got sent to was called the Wilhelmina Shelter for Troubled Children. I got in trouble a lot. They'd write me up for not looking the staff in the eye, which they interpreted as being insufficiently remorseful, that kind of thing."

Pia looked at Rothman. His face was impassive. He nodded almost imperceptibly.

"When I was twelve or so, I got shipped

to the Hudson Valley Academy for Girls, in a town called Eden Falls. It sounds nice, right? Eden Falls. The supposed school was a nineteenth-century institution. They used to send teenage prostitutes there, you know, to reform them, but when I was there it was all the hard cases who'd been thrown out of everywhere else — the girls no one wanted to adopt or deal with appropriately. The girls lived in these cottages. The paint was peeling, the plumbing didn't work. Roasting in summer, freezing in winter. That was bearable. It was the other girls that were the problem.

"The place was run by these girl gangs that were like crime families. Well organized — 'daddies' at the top, then 'housewives,' 'cousins,' 'uncles,' and so on. They'd take your money, make you do their chores, beat you up for no reason. They came out at night. There was a lot of abuse — you know, physical stuff, sexual stuff sometimes. The staff knew, but they didn't care — the girls kept much better order than the staff ever could. I tried to stay out of the way, but they got everyone in the end. They found me hiding one night and attacked me in a bathroom. . . ." She paused.

"Anyway, I found an old book on boxing in the library and a couple of the younger girls knew some karate so we did a makeshift self-defense class. I was determined. Every time they came looking for something from me, I fought back. So I never joined their 'shadow society,' as one social worker called it. I got into a lot of fights, spent quite a lot of days and nights in the little punishment room, which was the moniker for solitary confinement. I spent a week there once. What I did was get the girl who was the leader of the group that attacked me in the bathroom off by herself and gave her one hell of a beating."

Pia worried that she was saying too much, but Rothman simply nodded again when she looked at him.

"I worked really hard at the lessons that were available at Hudson Valley Academy. It was my escape. There were some teachers who cared. I was determined not to end up pregnant on welfare or in jail like most of the girls did when they left. For the most part the staff didn't care what happened or what anybody did as long as we didn't actually kill each other. Excuse my English, but it was like a goddamn garbage dump. Leave the trash there till it's eighteen and had developed a taste for drugs and then release them into the world with no supervision. Good luck.

"The superintendent there, he knew it was the system that created as many problems as it fixed. Papitano was his name. He tried to get better therapists and teachers. He even tried to have the place shut down, but he was threatened so he stopped trying.

"I knew about that because the superintendent lived in a big mansion on the school grounds and a few of us used to clean his house and make his meals. He helped me out by putting me on that detail and getting me the good teachers. But he was a real sad sack and abusive — I guess his wife had left him, and he never saw his kids. I was sixteen, and I was around his house a lot and he got it into his head that I was interested in him even though I had resisted all his advances. One night he got really drunk, and I was in the library reading — it was the only library in the whole school — and he came down and said he loved me. He was really pathetic, but he got me in a corner, and I don't like being cornered. I think he said he got the black eye falling in the shower.

"I think Papitano was more embarrassed than mad at me because nothing happened to me. Except he wanted to get me out of there, which was a good thing. In retrospect I don't think it was for me, but rather so he wasn't tempted again or whatever his problem was. But his intercession worked by finding me a good and competent caseworker."

"Sheila Brown," Rothman said.

"Sheila Brown. She was very persistent, and she went to court, and Child Protective Services agreed to move me to a group home so I could possibly get my high school diploma before I was 'emancipated' out of the system. Emancipation, which is a very well chosen word. So I left Eden Falls, thank God. I was happy to leave, but that Papitano guy, the superintendent . . ."

Pia's voice trailed off and she paused to collect herself. When she resumed she talked very quietly, leaning forward, practically addressing Rothman's desk.

"You know, I really had grown to trust him. I thought we had a connection. But before I got shipped out, he got drunk again, and he was a big man. He caught me alone in the library again. I'd let my guard down, and he betrayed me."

Pia stopped talking. They were sitting so still that the motion-detector-operated switch in Rothman's office doused the lights. The sudden pitch darkness caused both Pia and Rothman to jump. The lights came right back on.

"Jesus, I thought they fixed that," Rothman said. "Used to happen to me all the time."

"You must think I have an antisocial personality disorder," Pia said, regretting saying so much about her violent past. It had been like a dam bursting. "I've never really talked about it straight out like this. Not to anyone except maybe Sheila. But with her it wasn't all at once. It was over time."

"I don't think you have a personality disorder in the slightest," Rothman said. "You did what you had to do. I admire you. My foster care experience was nowhere near as bad. With Jewish parents, I got a decent placement right off. It wasn't a picnic, and I had to do without much nurturing, but I was older to boot. I was eleven at the time. Also I got to spend vacations with an aunt who wasn't the warmest of souls but at least was family. Even though my label then was only ADHD, my parents couldn't deal with me, so they had given up. In their defense, I was a handful. They had four older children, and I just figured that my mom and dad had run out of love by the time it got to me.

"Listen, Pia, I'm not trying to make you feel grateful to me, or feel any different about me because of what I've told you. I'm just saying that I understand some of what you went through — more after what you've been willing to share tonight. It's no wonder you have nightmares, and to tell you the truth, your being late doesn't bother me as much as it bothers Marsha and Junichi, especially with what I know. Ironically, it bothers them because they believe it bothers me. The major point I want to leave you with is that research is the calling where I found myself despite my past and despite my Asperger's. I think it could be for you as well, but you're going to have to make a decision. It can't be halfway. It's got to be either research or clinical medicine. It can't be both."

"Reactive attachment disorder," Pia said. "One of the social workers at Hudson Valley Academy told me I had reactive attachment disorder. It's supposed to mean I can't establish a relationship with anyone."

"Well, I guess we make quite the pair then," Rothman said, and smiled. Pia had never seen Rothman smile before and his whole face lit up for a second. "You think about your future. You don't have to say anything now. But I have to know soon so I can prepare. Once our article for *Nature* comes out, things are going to accelerate."

Rothman stood up. "Now I have to go back in the biosafety unit for another hour or so." Typical of his Asperger's syndrome behavior, he didn't comment further, just left.

Pia remained sitting after Rothman had departed. Except for the sounds emanating from some of the automatic equipment out in the lab, there was silence. Even the desk lamp went out again until she waved her hand in the air. She'd been taken aback by the evening's events. She felt exposed, emotionally naked, and found herself worried that Rothman might decide on further thought that she was much too big a risk. She sat in her chair for at least ten minutes before she got up and left. As she descended in the elevator, she started to feel better, relieved to a degree that she'd been as open as she had. Having been in foster care himself, Rothman understood her. All at once Pia felt confident all was going to work out. It was her opinion that you can trust only a man whose actions match his words, or better still, who acts without asking for anything in return. The only person she knew who fit that bill was Dr. Rothman. She knew that even George, as generous as he was, had his own agenda.

Exiting into the cold night, Pia didn't

know exactly what she was going to do, but she had to admit that Rothman had made a lot of sense. And as unbelievable as it was, she felt he'd morphed into the father figure she'd never had.

20

One Central Park West
New York City
March 4, 2011, 8:05 a.m.

When his home office phone rang just after eight in the morning, a very anxious Jerry Trotter snapped it up. He'd been hoping it would ring, and he was hoping it would be Harry Hooper.

"I just had breakfast with that Morgan guy I was telling you about last night," Hooper said, launching right in after Trotter picked up.

"You met with the guy? Sat down in front of him and ate breakfast?" Trotter was surprised. Brubaker and Hooper were usually more indirect, avoiding face-to-face meetings.

"He wouldn't say anything more over the phone. He wanted a meeting, insisted on it. At six-thirty in the morning. He thinks I'm a headhunter for real, and he wants a new job, like yesterday. I couldn't see the harm in it. It's not like I'm going to see the guy again."

"But what do you know about being a headhunter?"

"What's to know? I just asked the guy to tell me about himself, about his strengths, where he sees himself in five years, all that BS. I said I didn't know of anyone who needed someone exactly like him, but I'd keep an ear to the ground, keep him in mind."

"Didn't he ask you for your business card?"

"Said I was all out," Hooper said. "Said I'd been meeting with a lot of bankers the last couple of weeks and underestimated demand. Almost convinced myself I was that busy. Anyway, we finally got around to talking about your woman friend. She and the thick guy who she was working for at the time definitely slept together. More than once. Not just some drunken hookup at a convention but an actual affair — hotel rooms in the afternoon, that kind of thing."

"And he knows this how?"

"He was going out with this woman who was good friends with your girl. Real good friends, like girlfriends who told each other everything. So your friend tells this woman she's seeing this guy who's married. Then she says it's her boss. She swore the friend to secrecy, made her swear she'd never tell anyone, all that. But the woman told my guy. Information's valuable, as you know, and it all depends on the circumstances. This woman thought it would help in her relationship with my guy, bring them closer, having a shared secret. It didn't work. They broke up after a while."

"So why'd he tell you?"

"Like I said, information can be valuable. I was asking about your girl, he knew something. Maybe he wanted the supposed job I was checking out your friend for. I don't really know. I guess I may have led him to believe I'm better connected on Wall Street than I am."

"He's going to be pissed when you disappear all of a sudden."

"What's he gonna do, tell his boss? Anyway, I plan to call him next week, start letting him down slow. It looks like I'm going to be downsized myself. Sure is a cruel world."

"Okay," Jerry said. "Give me a second to think."

Jerry held the phone in both hands. This was good — Gloria Croft and Edmund Mathews had slept together ten, twelve years ago. And clearly it hadn't ended well because Gloria was apparently enjoying trying to ruin Edmund. But for what Jerry had in mind, there had to be more. This was good, but it wasn't enough.

"Okay, I like this, but I need more. Keep digging. Try and figure out why it ended between them, and why it ended so badly."

"All right, got it."

Jerry sat back in his chair. He was a man with a lot of secrets, which is why he assumed everyone else had them. Some of Jerry's secrets concerned the fact that he was unfaithful to Charlotte, his wife of twenty-two years. He had had affairs with some of his patients, one of which continued after Trotter ended his medical practice and went into finance. It was still going on, with trysts at an apartment Trotter maintained in the Village for that express purpose. Trotter didn't feel any guilt about Charlotte. He thought of it as a kind of deal even though Charlotte had never been approached about it. He played around, and she lived the high life. Shopping was her sport.

From Jerry's perspective risk was a big part of life. Everybody handled risk differently. He thought he handled risk well, which was what made him a good hedge fund guy. Others handled risk poorly. The real question that dogged Jerry's mind at that moment was how much would have to be on the line for someone to do something truly desperate. He was just beginning to think there might be a way to solve the problem that Edmund had tossed into his lap.

Jerry Trotter had another secret, one that weighed on his mind more heavily than any other. It had nothing to do with women. Not only had Jerry taken a very sizable personal stake in LifeDeals, in addition to the position his fund had acquired publicly, but he had made a third and completely clandestine investment that was larger than the other two stakes combined. Jerry had studied what Edmund and Russell had set up with LifeDeals, read the business plans, and pored over the sales reports. He had commissioned his own secret research and paid lawyers hefty fees to set up financial instruments ready to be sold at a few days' notice. And then, masked by a series of offshore shell companies, he had set up the bare bones of a parallel company that would mimic LifeDeals, right down to the type of policies it went after. As Edmund never tired of saying, life insurance was a $26 trillion business in the USA alone. There was plenty of money to go around.

Edmund and Russell's bad news about regenerative medicine had hit Jerry Trotter like a hammer blow, much more than Edmund could have guessed. His due diligence had completely missed it, as had Edmund's. To his partner and his firm, LifeDeals' predicament was unfortunate but it hardly threatened the hedge fund's success, even in the short term. But Jerry stood to lose much more. His personal stake was very large but also survivable. But if the shadow company that he was rolling out went down, he was probably ruined. The various subsidiaries were already buying policies. Individually, each was tiny compared with LifeDeals'. Together, Jerry had once been proud to think, they were larger.

Over the course of approximately eighteen-plus hours, from the moment he'd left the Terrasini restaurant, Jerry Trotter had become an extremely desperate man. He hadn't slept all night, instead using his old calculator and various files and portfolios to try to figure out ways in which he could emerge from this intact. He knew he was clutching at straws with Harry Hooper, but he was hoping against hope that Edmund Mathews had something more than just money at stake, something that would mean Jerry didn't have to try to fix this mess all on his own. Jerry had few qualms, but he much preferred to delegate the truly dirty stuff, the stuff that could get you thrown in jail or worse.

21

One Central Park West
New York City
March 4, 2011, 11:55 a.m.

By noon Jerry was near to being a basket case. After finishing the call with Harry Hooper, he went back to doing what he'd done at the end of the night: surfing the Internet just to have something to do. Jerry was buzzing on the amphetamines he'd taken to keep him awake and he knew he had thirty-six to forty hours until he crashed. Every couple of hours he drank a Red Bull, and he chugged Diet Cokes constantly. His wife, Charlotte, had no idea what was going on but was familiar enough with the routine to keep well out of her husband's way. For Jerry the Internet was a wonderful resource and babysitter, so to speak. You could find out anything you wanted to know on it, as well as plenty of things you didn't know you wanted to know.

It couldn't help much with finding the Fountain of Youth or proving the existence of God, but otherwise, it was golden.

The Internet was particularly useful when it came to providing practical solutions to all manner of problems. Jerry had recently discovered how to tune his universal remote so that it operated the controls on his TV, and he was grateful for that. This was a different kind of problem. As he sat alone in his darkened study with the shades drawn, he stared at the screen on his Mac, following threads in obscure discussion groups, piling up memberships in esoteric organizations, clicking on links that took him to some tortured recesses of our collective consciousness as represented on the World Wide Web.

Some of Jerry's on-screen reading reminded him of being at medical school. What he wouldn't have given to have had this resource back then! The dry phrasing of the medical material hadn't changed in thirty years. Jerry thought he'd perhaps spent a couple of hours reading about salmonella when he was a student. He'd always been slightly germophobic, especially when it came to the more powerful microbes, and reading about this one made him uncomfortable. But Dr. Rothman's first specialty, the one that brought him his first Nobel, was fascinating.

It was such a versatile and dangerous bacteria.

The longer he sat at his desk, the more convinced Jerry became that only one course of action was open to him. He was initially horrified by the thought, but it looked like there were no other options, and he hated to be backed into a corner. Whenever he got squeamish, Jerry pondered the prospect of being broke and disgraced. If it all came crashing down, he'd be a laughingstock. Some ambitious hack would write a book about him, and he'd come across like a buffoon, an idiot. He would avoid that fate at any cost.

Once Jerry had the idea percolating in his mind, really all he needed was the resolve and the money. Spending hours researching certain specialized activities on the Internet had convinced him of something else: Money really could buy you anything. He had the money. He just had to convince himself he could follow it through.

Now, toward midday, the throwaway cell phone rang again. Trotter was hoping for Hooper, but he got Brubaker.

"What do you have?" Trotter said.

"Confirmation that those two guys are

definitely the leaders in this organ-making field. Way out in front. Independently confirmed beyond that source I mentioned. And no one can be precise on the timeline because it depends on the results of tests that no one can predict. They might do a test and it doesn't work, which sets them back a week, a month. Or it does work and it's on to the next one."

"But eventually it's going to work?"

"That's what I'm hearing."

"Too much to hope that it blows up in their face."

"If you're looking for them to fail, doesn't look like it's gonna happen. From every source, they're very confident."

"How do you know?"

"So I'm told. Plus they've formed a private company to control the patents that have been applied for. And it's not one patent. It is a whole series of patents to be sure they'll control the whole field."

"Thanks, I figured. That means they're close."

"Not necessarily — just means they're confident they're going to get there."

"How'd you find out about the company?"

"You really wanna know?"

"Indulge me."

"Okay, boss. I have a friend in the New

York State Division of Corporations. Can find out when people register corporations or LLCs. Comes in handy when guys set up limited liability companies to hide money from their wives."

"I'll bear that in mind."

"Rothman Medical they call it. So it wasn't hard to find. Registered two weeks ago. They probably registered it overseas too, in better tax locations. As I said, they're being thorough."

"And who are the partners?"

"The members of the company? Just the same two guys."

Jerry ended the call. Rothman and Yamamoto. It seemed like the two of them were piloting the whole ship on their own. Jerry checked his watch. It was nearly twelve-thirty, almost four and a half hours since he'd spoken with Hooper. Suddenly Jerry felt crushingly tired. It was vital to him that Hooper find something he could use as leverage on Edmund Mathews. His brain was close to fried; he had to have someone help him with this. He knew Hooper would call him the second he had anything, but like the previous evening, he couldn't resist calling.

"It's me," he said redundantly when Hooper picked up.

"Is there a problem?"

"Just checking in," Jerry said, trying to control his voice.

With his antennae constantly up, Hooper sensed there was a problem, and the problem was Jerry. Jerry had said only five words, but it sounded to Hooper like Trotter was tweaking on crystal meth. Having been a policeman, he'd had to deal with all manner of drugs. "You don't sound so good."

"I'm tired is all."

"Well, I got some lines in the pond," said Hooper. "Just waiting for a bite. Just try and relax."

Sure, thought Jerry, as he broke the connection. *That's easy for you to say.*

22

Columbia University Medical Center
New York City
March 4, 2011, 12:35 p.m.

The night before, Pia had taken the time to set the alarm that George had given her as well as her own cell phone and had awakened refreshed and ready to go at 6:30. She'd slept like a rock. First time in more than a week. After showering, Pia had taken a trip to the cafeteria and knocked on George's door bearing a toasted bagel with cream cheese and a cup of coffee.

"Wow, this is a first," George had said when he answered the door. "And you brought breakfast. Come in!"

"I'm just returning the favor. Or favors. But where were you yesterday and the day before? I was late both days, and yesterday I ended up scrubbing beakers for two hours as punishment."

"Oh, I . . ."

"Doesn't matter. I have some news."

"Good news?"

"I think so."

George had continued getting ready while Pia sat on his bed.

"Rothman wants me to work in his lab full-time when I graduate."

George had come out of the bathroom holding his toothbrush. His mouth was agape and foaming.

"Can he do that?"

"I think around here, he can do pretty much whatever he wants. All he has to do is threaten to go to Harvard or Stanford."

"So what did you say?"

"I didn't say anything. For one thing, I was too stunned, and for another, he told me to think about it. But it's a no-brainer. I'm going to say yes. I'll talk to the dean about postponing my residency. I imagine I can still qualify as a Ph.D. student. But the important thing is to work in his lab. You can't believe what's happening in there. He's going to become even more famous. I wouldn't be surprised if he wins another Nobel."

George had returned to the bathroom and stood in front of his mirror. He looked himself in the eye and bit his tongue.

"That's great, Pia. Congratulations." He

had tried to sound convincing, but didn't think he managed it.

"I thought you were going to give me one of your rants about Rothman."

"Hey, if this is what you want to do, I think you should do it." He bit his tongue again.

"My thoughts exactly. Come on, George! Hurry up, we're going to be late."

The early morning had been dark and a cold drizzle was in the air. March was not one of the best months in New York City. George and Pia had hurried to their assignments, talking about how their first days had gone.

"So what's it like working with Will McKinley?"

"He's a bit of an ass and into himself. Rothman thinks he'll go into plastic surgery. Anyway, he's holding his own since he's smart enough. I do like Lesley."

"I'm sure that Rothman's comment was not meant as a compliment. I've heard he's never said anything positive about anybody."

Pia had merely raised her eyebrows without commenting.

"McKinley reckons he's God's gift to women. I'm sure you figured that out."

Pia had shrugged, as if to say, "And?"

"Is he leaving you alone?"

"I can handle Will McKinley, believe me. He is quite cute though."

George had caught up the stride after momentarily lagging behind Pia. He looked across at her. She was smiling, having a laugh at George's expense. He couldn't help but laugh along with her. Silently he chided himself for being such a wimp.

The morning passed uneventfully for the students. They spent their time on their respective projects in the organ bath unit. Pia also spent several more hours reading about pH buffers designed to be used for tissue culture. The maintenance man had still not finished in her office or in Rothman's office either and wires still hung out of both ceilings. The electrical blueprints that had been in Pia's cubbyhole were now in Rothman's office. Pia had gone in to see if the worker was there as he wasn't in her space. She wanted to give him an earful about the job not being done. But he wasn't in there either, and after starting her reading, she forgot about him altogether.

As if taking a cue from jealous George, Will showed up and tried to engage Pia in small talk. Pia wasn't sure if he was hitting on her or not but couldn't have cared if he was. She answered his first few questions but then told him directly she wanted to concentrate on her reading. He took the hint and vanished.

At twelve thirty-five Rothman and Yamamoto appeared from the depths of the BSL-3. Pia couldn't help but notice that they were acting out of character. They were actually talking excitedly to each other. Pia didn't stare directly but watched out of the corner of her eye. The lab was quiet, which was why she had heard them emerge. It seemed that everybody was at lunch.

All at once Yamamoto came toward her as Rothman disappeared into his office but without shutting his door. Even that was out of the ordinary. Pia sensed that something was going on.

"Where are the other students?" Yamamoto asked when he reached Pia's side. His voice had what Pia would have described as an anticipatory ring.

Pia looked up. "I believe in the organ bath unit," she said.

"Good," Yamamoto said. "I want you in there too. Rothman and I want to show you students something."

Five minutes later all five people were in the organ bath room attired as per usual in caps, gowns, masks, and booties.

"Okay," Rothman said, clasping his gloved hands together in excitement. After Pia's surprising talk with him the previous evening and now his excited behavior, she felt she was seeing a side of Rothman that she never imagined existed. "Dr. Yamamoto and I want to show you something but in the strictest confidence. You will be here a month, Pia longer, but we would be very grateful if you could keep what you're about to see to yourselves in that period and thereafter. Agreed?"

The three nodded their assent.

"Good. We don't want anyone to get excited prematurely. The stakes here are very high."

As he talked, Rothman edged toward the back of the room. Set in the wall was a door with another security pad like the one on the main door outside. Rothman shielded the code pad with his body, punched in a code, which Pia assumed was the same as for the other security doors, and pulled open the door. Dr. Yamamoto held it as Rothman stepped over the threshold, followed by the students. Yamamoto stepped in and closed the door behind him.

They were standing in a room that was maybe ten feet square, an identical but smaller version of the one they had just left.

The five people made the room feel crowded. The same bluish light filled the room, which had its own HVAC system that hummed a little louder than its larger neighbor. The recessed ceiling light illuminated two carts like the ones they had been working with next door. They stood side by side, but only one was operating.

Rothman gestured toward the bath atop the one cart. It was similar to the ones out in the main room. In it was a kidney much larger than the mouse organs. Soon they learned that it was a human kidney, and like the human kidneys in the outer room, it had been made from Yamamoto's fibroblasts. It was a pale color and appropriately kidney-shaped. The difference was that this organ had ports through the Plexiglas wall that were connected with Y connectors to the organ's artery, vein, and ureter.

"What you're looking at is what is going to be the world's first human organ exoplant made from induced pluripotent cells. This morning we received official sanction from the FDA to go ahead and attach this organ to Dr. Yamamoto's cannulated inguinal artery and vein. We will be allowing the organ to function as it would if it were transplanted into Dr. Yamamoto's abdomen."

"You've volunteered for this?" Will asked Yamamoto.

"Yes, of course," Yamamoto said with enthusiasm. "It is a great honor for me."

"When will you do it?" Pia asked. It seemed as if she was being overwhelmed on a daily basis in Rothman's laboratory.

"As soon as we can schedule it with the surgery department. It will be done in one of the main operating rooms for safety's sake. We'll allow the organ to function for several hours while we monitor it carefully. It's going to be a big day. A milestone really."

But what Pia could see before her seemed like a destination in itself. The dreams inherent in what was growing next door were being made into a reality in this tiny secret chamber. Pia felt a sense of awe, that she was present at the creation of something immense and extraordinary. No one in the room was saying a word. Pia stared at the artificial kidney sitting in its nutrient solution, the blue light reflecting in the bath and flickering over her face. She had been working in Rothman's cathedral but now she had seen the shrine. She knew that Rothman would have preferred the organ to be a pancreas, but she knew he knew that would not be far behind.

She could hardly wait to see that happen.

23

One Central Park West
New York City
March 4, 2011, 1:20 p.m.

Jerry Trotter was slumped in his study, his head twisted awkwardly as it rested on his desk next to his slender Mac keyboard. As he snored fitfully, Jerry was having a particularly lurid dream. He is sitting as far back in a chair as he can while a man is yelling right in his face. In the dream, Jerry is late for a vital appointment, but he doesn't know what it's for, and he can't find out until the man stops yelling at him and gets out of the way. Jerry twitched himself half awake but didn't move. He'd drooled on the desk and his head was pounding. A phone was ringing somewhere nearby.

After an hour that morning, Jerry had turned off the ringers on all the phones in the house and on his regular cell phone. It appeared that there were plenty of people who wanted to talk to him. He hadn't shown up at work so they figured he was at home or at least somewhere where he could pick up his cell phone. But the only calls Jerry cared about would come in on the device that Max Higgins had given him. So that must be the phone that was ringing.

Jerry sat bolt upright and pulled something in his neck. A quick spasm of pain ran up into his head as he scrambled for the phone. It stopped ringing.

"Shit, shit."

Although he was one-part asleep and couldn't orient himself properly, Jerry found the phone and pushed at the unfamiliar buttons. A 917 number came up and he pressed the green button. The phone redialed. It wasn't Higgins, and he couldn't remember Hooper's number or Brubaker's.

"Let it be Hooper," he said quietly, with purpose. "Let it be Hooper."

Someone picked up the call.

"Where've you been? I called twice."

It was Hooper.

"You got something?"

"Bingo."

"What is it? You gotta tell me. . . ."

"We need to meet. The Starbucks on the corner of Sixtieth. Opposite the Mandarin Oriental."

"That's right across the street."

"See you in ten minutes."

Jerry Trotter looked at his Rolex again and then around the Starbucks. Harry Hooper had said ten minutes and that was almost a half-hour ago. Jerry had come straight down to the street from his skyscraper aerie and hurried across Columbus Circle, and he'd reached the place in four minutes flat. Max Higgins would be at the apartment at any moment too. Jerry had quickly called him and asked him to come up from the office with the car. Things seemed to be moving.

As usual, the Starbucks was jammed. There was a line of people snaking around the store, all waiting to order, and customers to their left waiting for their beverages to be delivered. Most of the two-tops were occupied by individuals with laptops and a lot of notebooks. Who were these people? Jerry wondered. Didn't they have homes? Or offices? A homeless guy had wedged his shopping bags and himself into a corner. He had a cup of water and as long as he stayed awake, he could sit there for the rest of the year.

What sort of venue was this for a meeting? Jerry wondered. There was little chance of finding somewhere to sit and less chance of talking discreetly. Jerry took out the phone and was ready to call Hooper again when he felt a hand squeeze his elbow, and not lightly. He twisted around. Hooper.

"Let's take a walk," he said.

Hooper guided Jerry out of the store and across the street so they skirted the front of the Time Warner building. Hordes of people were coming in and out of the doors.

Hooper turned right on Fifty-eighth Street and walked toward Columbus, steering Jerry through traffic to the south side of the street. They entered a pair of light green glass doors and took an escalator up to the lobby of the boutique hotel on the corner. Hooper led Jerry to a quiet section of the expansive lobby and sat at a table where a drinks menu was resting.

"A little quieter in here," Hooper said.

"What was all that about? We could have just met here."

"You sounded very tense on the phone," Hooper said. "I'd say nervous was more like it. And nervous people make me nervous. Just basic precautions."

Jerry looked at Hooper. How old was the guy, fifty-five? He was smaller than Jerry remembered, no more than five-eight, with dark hair that might be dyed but which was all his. He had a pinched, smoker's face and friendly eyes. Trotter trusted him not at all.

"Shall we have a drink?"

"Sure," said Jerry, who was running on fumes. Hooper waved a hand and a waiter came over from a bar.

"Scotch, rocks, a little water," Hooper said.

"Vodka martini with a twist," Jerry said. "Thank you."

"You look a bit rough there, boss."

"Didn't sleep well," Jerry said. "Nothing some good news won't cure. I'm presuming it's good news because you couldn't tell me over the phone."

"I wanted to have a little word with you, face-to-face."

"Oh, yes."

"I'm kinda wondering why you're so interested in this guy."

"Well, what's it to you, Harry? I asked you to find out some information and you seem to have found something. Obviously I want to have some leverage over this person, but it's nothing you need worry about."

"I'm curious how valuable the information might be."

Jerry paused while the waiter served their drinks. Was this little asshole trying to shake him down? The waiter left, and Jerry picked up his glass slowly.

"Cheers, Harry." Jerry knocked back half of his drink and put it down. "I'd say the information is worth the three hundred an hour I'm paying you. That was our agreement, I believe. A very generous one."

"Agreements are made to be renegotiated," Hooper said.

"What do you have in mind?"

"Another ten grand."

"Ten grand? Are you kidding?"

"Not in the slightest."

Jerry laughed, he couldn't help himself. Ten thousand was chicken feed. Anticipating that Hooper would try something like this, although not quite so unsubtly, he had brought fifty thousand dollars with him and was willing to spend it.

"Let me think about it," Jerry said, pretending to look pensive. "You must think I'm an idiot," he added, taking another gulp of vodka. "You and Brubaker both. Do you call each other and say, 'What an idiot that Jerry Trotter is, thinking he's some kind of spy'?"

Hooper looked at Jerry coldly. He didn't say no.

"I'm an idiot, but I'm not a total idiot."

Jerry reached into the pocket on the front of his leather jacket and took out a small digital recorder of the type Hooper was familiar with.

"What's that?" Now Hooper was smiling.

"I taped our calls, Harry. Not on this machine but another one just like it. What did you say — 'basic precautions'? I prefer to think of it as insurance. Ha, me and insurance." Jerry finished his drink and held up his glass for the hovering waiter. Hooper hadn't touched his drink.

"There's nothing on there. I never say anything on the phone."

"Oh, really? Then you've got nothing to worry about."

Hooper's eyes darted around the room briefly, and he took a sip of his drink.

Jerry had got him thinking, he could see that.

"We're in this together, my friend. I have no intention of doing anything with the recordings. As you say, there's probably nothing there. But we've definitely entered a new phase in our relationship. You were very honest with me — you want more money. Okay."

Jerry reached into his jacket again and took out a thick manila envelope. He threw it down on the table next to Hooper's drink. Hooper picked up the envelope, held it below the level of the tabletop, and opened it with a finger. He flicked through the bills and looked up at Jerry. Jerry thought that if Hooper had ever seen that much money before, it was evidence he'd seized in an investigation, and it was going under lock and key.

"I don't get it," Hooper said. "That's a lot more than ten."

"Yes, it is. That's fifty."

"Fifty grand! Holy shit."

"Ah, Mr. Hooper, your grim exterior is slipping." Jerry finished off his drink. He was feeling a lot more like his old self.

"What do I gotta do?"

"You tell me two things, and I give you another one of those envelopes in a couple of weeks. That's all. First, I'm going to tell you what I think. I think you're a greedy little man. I know you pad your bills for me — that's fine, everyone does it. But this is real money. And I have more real money that I intend to keep giving you as long as we can help each other out. Because we really are in this together. I also think you don't know exactly what I have on tape. Hmm?"

Hooper had regained his composure and was looking Jerry right in the eye.

"I notice you already took the money. I also think you're thinking, Screw it, I want the money. It's easy money too, Harry, because I know you already know the first thing — that's why we're here. And I really think you'll find out the second one very quickly, a man of your experience."

"You're playing a dangerous game. You're an amateur."

"I know." Jerry closed his eyes and smiled. "But I'm a quick learner. So tell me what you found out about Edmund Mathews and Ms. Croft."

In a few sentences Harry Hooper told Jerry Trotter what he'd been told and about the source of the information. There was no doubt in Hooper's mind that it was true.

"Thank you, Harry. That might just be enough for me."

"So what's the other thing you want to know?"

Jerry leaned in toward Hooper.

"I want you to tell me how I get my hands on some polonium-210."

24

Greenwich, Connecticut
March 4, 2011, 3:23 p.m.

It was thirty miles, give or take, from Columbus Circle in Manhattan to Edmund Mathews's house in Greenwich and by some miracle, Jerry's driver, a former New York State highway patrolman, made the trip in just over fifty minutes. After leaving Harry Hooper in the hotel bar, Jerry had found Max Higgins waiting in the limo in front of his building. He got in and called Edmund Mathews immediately, pretty much ordering him and Russell to leave their Greenwich office and meet at Edmund's house within the hour. Jerry had told Max nothing. Max thought Jerry looked terrible — red-rimmed eyes, unshaven cheeks, hair in disarray, and wearing a strange and rumpled shirt-and-khakis combination under an old leather jacket such as a biker might wear. And he could smell the alcohol on his breath. Max would have to wait for an explanation because as soon as he'd spoken with Edmund, Jerry stretched out in the limo's generous backseat and fell into a noisy and fitful sleep.

In the hours since their lunch with Jerry and Max the day before, Edmund and Russell had done nothing significant in terms of solving their problems. Russell had busied himself overseeing the implementation of some of Edmund's ideas about buying different types of life insurance policies and legal staffers had started combing through existing diabetics' policies looking for what Russell had called "anomalies." Anyone who'd used a middle initial on one document and not on another, they were to see if that was grounds for termination of the policy. Any agreements in progress were halted pending investigation. But these were stopgaps. If there was to be a macro solution, Edmund and Russell hoped it would come from Jerry.

When Edmund received Jerry's summons, he was optimistic that salvation was at hand. Jerry had sounded hoarse, and he was even more abrupt than usual. But no matter. Russell had been positively giddy as the two men waited for Jerry, Edmund more reserved. From experience Edmund knew that if Jerry had thought of something, it wouldn't be pain-free. There'd be a price to pay somewhere down the line.

Jerry's limo pulled up to Edmund's front door. As Edmund watched from a second-floor window, the driver hopped out and held the door open for Jerry, who slowly stepped into the chilly winter air. Even from this range, Jerry didn't look so hot. As Edmund made his way downstairs, his wife, Alice, ever the good hostess, opened the front door.

"Alice!" Jerry said jovially. "I was hoping I'd see you. You look as lovely as ever." And she did, her blond bob tucked back behind her ears, her light green eyes set off by a mint-green sweater, her gym-toned legs setting off a sharp, knee-length skirt.

"Hello, Jerry, how are you?" Alice grabbed Jerry's elbow and leaned in to kiss his cheek. Jerry had tried to tame his hair and had quickly polished off half a roll of breath mints, but he hadn't completely overcome his dishevelment. Nor had he done anything about a subtle ripeness that hovered around him like an invisible cloud. Alice recoiled a little.

"I was just saying to Max," Jerry continued as Edmund joined them, "what a wonderful couple Alice and Edmund are. And little Darius makes three. Beautiful wife, a healthy heir, this stunning house. Edmund, you are a lucky SOB. The man who has everything. Wasn't I just saying that, Max?"

"Absolutely, Jerry, and who could disagree?" Max had no idea what Jerry was talking about, but he played along. Two minutes earlier Jerry had been all but dead to the world.

Jerry had his arm around Alice's shoulder as the group filed into the house. Edmund wondered what on earth was going on. Jerry had never shown the slightest interest in Alice, nor Edmund in Charlotte Trotter. They didn't have that kind of relationship — it was all business.

"Russell here? Ah yes, there you are," Jerry said, spying Russell emerging from the library.

"Can I get you gentlemen anything?" Alice asked, extricating herself from Jerry's grip. Jerry moved to lean against a wall. From Edmund's perspective it appeared as if Jerry was having trouble standing up.

"I'd love a coffee, thanks, Alice. You have one of those fancy machines, right? As strong as it comes, and in a large mug if you wouldn't mind. Didn't sleep so well last night."

Alice moved toward the kitchen and the four men stood in Edmund's expansive entryway.

"We don't have all the paperwork from Statistical Solutions corroborating the concerns we have about the bell curves," Edmund said, eager to get the ball rolling.

"I don't care about that," Jerry said. "It's as bad as you thought. Actually, it's probably worse than you feared. We have to preserve the capital we've invested, and the only way to do so is to act quickly and decisively. Like now."

"Well, shall we go into the library and sit down and talk about it?" Edmund asked. "Or the living room?"

"No, Edmund," Jerry said, suddenly sounding more focused. "You and I are going for a little walk outside."

"A walk? It's freezing out there! It's going to snow later."

"Don't worry, Edmund, you're not going to freeze to death. Go grab a coat."

As Russell and Max moved into Edmund's library, Edmund and Jerry stepped outside, Edmund fortified by a woolen overcoat, Jerry by the coffee Alice had made. It was five shots of espresso staining the inside of a Syracuse University mug.

"They've set up a company to control the patents for the organogenesis techniques," Jerry said. "Rothman and Yamamoto. These are the guys, no doubt about it. They're the problem."

"I'm glad you're taking the issue to heart," Edmund said. They walked along an ornamental path in the front of the house, past rosebushes that had been severely pruned back for the winter. Patches of snow lay on the lawn in the shadow of the hedges. This was as barren as Edmund's garden ever looked.

"We have to act at once. Those bell curves move at all to the right, it's a disaster."

"I'm pleased you see the same problem we do."

Jerry stopped walking just short of the lawn.

"Unfortunately, we don't see a simple financial solution, like selling ourselves short through an intermediary or securitizing our policy holdings immediately. With Gloria Croft shorting big-time, we probably couldn't find any institutional buyer."

"I agree," Edmund said. "But the life settlement concept is still sound. It's maybe the best business opportunity I've ever come across. It would be a shame to have to give up at this early stage."

"I agree," Jerry said. *And more than you know,* he thought, more than even Max knew. "Which is why I've come up with another plan."

There was a silence, then Jerry went on.

"It's a little unorthodox, but it's the best plan that serves all of our interests. Believe me, I've thought about nothing else over the last twenty-four hours. But it's not for us to do — it's for you to do. It *was* your idea, this whole thing. Your mess to get out of. Just you and I will speak of it, nothing will be in writing."

Edmund nodded. He didn't expect anything different. Not from Jerry.

"There's only one solution, and it's the way it has to be because this guy Rothman has got himself out there so far ahead of the pack."

Another silence ensued.

"I think Rothman's momentum has to be stopped. If it is, I think we'll have a good five years before the rest of the research community catches up to where Rothman is today."

Neither man said anything. Jerry's words hung heavily between them as if they were written in the air. Finally Edmund broke the excruciating silence.

"How do we stop Rothman's momentum, Jerry?"

"Easy," said Jerry. "You kill him."

Edmund turned and walked away from Jerry, back toward the house. He took a path on the side of the building and Jerry set his empty coffee cup down and followed him to the rear garden, where Edmund sat on a bench with a view of Long Island Sound. Jerry sat down next to him.

"Murder, Jerry? Like having him shot?" Edmund was appalled. At the same time he didn't think he had the luxury of dismissing any idea out of hand no matter how preposterous it sounded.

"No, not at all. The two of them should die in a way that doesn't invite suspicion of homicide. It must look like an accident. There shouldn't even be an investigation, although I suppose that would be inevitable. But there can be nothing that makes this look deliberate. Because it wouldn't be beyond the realm of possibility that any semi-competent murder investigation would lead right to LifeDeals. You sat there yourself at Statistical Solutions and talked about what this could do to the company's bottom line."

"Do you have any specific suggestions,

Jerry?" Although the proposal was outlandish and terrifying, Edmund wanted to find out what Jerry was proposing. It wasn't as if Edmund had any plan B waiting in the wings.

"As a matter of fact, I do."

Edmund continued to stare out at the water.

"I'll tell you," said Jerry, "most of the medical people will know that Rothman's first interest in research, before he became involved in regenerative medicine, was salmonella, which is the number-one cause of food-borne illness in general and typhoid fever in particular. He's investigating the virulence of the bacteria — what causes it to be a tremendously deadly bacteria on the one hand, and a bothersome but nondeadly cause of gastrointestinal distress on the other. Why does one type give you the runs but another kills you? We did a little research. He's found that growing salmonella in outer space produces a very lethal strain. He should be fed some of this particular strain.

"A lot of people don't care for the man — they're jealous of his Nobel Prize, and they think he's got an attitude. If he dies from the bacteria he's studying, a lot of people are going to say, 'Oh, that's terrible,' and then smile at the irony of it later."

Jerry made it sound so easy.

"I suppose that would be clever," Edmund said. He felt he had to say something.

"That's not the half of it. The typhoid fever he'd immediately develop might or might not kill him. There has to be something else that will kill him quickly and definitively, but it's got to be something you can't easily detect. There's a substance called polonium-210 — very radioactive and deadly if you ingest it but not harmful otherwise. We'd use it because it produces many of the same symptoms as typhoid and would be masked by it. It's what killed Alexander Litvinenko in London a few years ago."

"I remember that. That was just a theory, surely, the polonium."

"I think it was more than that," Jerry said.

"Why do we need it?"

"To make sure the guy dies. It's very potent. The challenge is that Rothman and his sidekick work in one of the premier medical centers in the world. The salmonella, no matter how virulent it might be, cannot be counted on by itself. One or both of them could be saved. That's a chance that cannot be taken. We need to be sure. One-hundred-percent sure, ergo the polonium, and a massive dose of it to boot."

"So where the hell do you get this stuff? Who's going to buy it? Russell?"

"You hire the right people. Professionals."

"You've been watching too many movies," Edmund said. "So tell me, Jerry, who is going to procure this deadly radioactive poison for us?"

"Albanians."

"Albanians?" Edmund's voice betrayed his skepticism.

"There's an Albanian Mafia that's grown big in New York in the last twenty years. Very violent, very ruthless. But also very reliable, if you do business with them. Their word is their bond and all that. The FBI put a crimp in their operations in the nineties, but they've grown back and they're looking to make names for themselves again. You're going to ask, how do I know? I got this from a man who spent years of his life trying to put these guys in jail. He gave me a name."

Jerry held out a piece of paper, folded in half, for Edmund. Edmund thrust his hands in his coat pockets and looked at Jerry.

"You're out of your fucking mind."

Jerry let Edmund stew for a couple of minutes. Edmund had moved down to the edge of his property and was standing, looking out at the gray water of the Sound. Jerry could imagine Edmund's state of mind — part of him horrified to even consider such a thing, another part telling him he had no option but to consider doing it. Which side was winning? Jerry decided to play his trump card. He didn't want to have to do this either, but again, there was no choice. He walked down to join Edmund and stood about four feet to his right, looking ahead.

"I know about you and Gloria Croft."

"What about me and Gloria Croft? You mean personally?" Edmund waited a beat, then turned to look at Jerry, who was stony-faced and staring straight ahead.

"What's that got to do with anything?"

"So you know what I'm talking about?"

"Yes, yes. Gloria and I had a . . . a thing when we were working together."

"When you were her boss."

"Yes, Jerry, *Jesus,* what does that have to do with *anything?*"

"You got married young, I believe."

"I was married at the time. I admit it, I was a bad boy. I got carried away, and I wasn't the only one who's ever done that. You tell me you never did. But I learned my lesson. I steer very clear of bitches like her."

"So no harm, no foul is what you're saying, right, Edmund?"

"Jerry, I swear I have no idea what the relevance of this is. You just asked me to *kill* two people, for Christ's sake." Edmund turned his head as he said it, checking that no one had joined them. "Are you trying to put pressure on me with *this?*"

"There's something I don't think you're aware of. I was hoping I wouldn't have to bring this up, but it seems like you leave me no option." Jerry looked at Edmund. He had crossed one bridge with Edmund a few minutes ago. Now he was going to burn it down.

"When you were sleeping with Gloria Croft, she got pregnant —"

"Oh, bull*shit,* Jerry —"

"She got pregnant, Edmund, and she had a termination, and it didn't go well. She used a good clinic, I can give you the name, but the procedure had some serious complications. I can give you details, if you need them. She survived, but it left her sterile, so she can't have children. And I would imagine it also left her with a lot of resentment for the man involved."

"Why should I believe this?" Edmund's face was dark with fury, his hands bunched into fists still thrust deep into his coat pockets, his left hand throbbing from punching the elevator door. He leaned toward Jerry, almost goading him.

"You're trying to blackmail me? I can't believe you."

"The information came up quite by chance," Jerry said. He was surprised how remarkably calm he felt. "We were looking for dirt on Gloria when we heard about this. I know someone who has contacts in the records department at certain hospitals, and he found the relevant file. The timing's right, we checked, and there's even a note in there saying that she only had one sexual partner. They were ruling out some STDs, so they asked. My guess is that partner was you."

"Bullshit."

"You use a condom every time, Edmund? She miss a bit of work around the time you stopped seeing each other? You might not remember, but I doubt you liked your analysts taking a lot of time off. And perhaps she left the company soon after that, am I right?"

Edmund sighed. He felt deflated, almost literally, as if the air had been sucked out of his lungs. He stared back out at the water again.

"So what will you do with this information? And I'm not saying it's true."

"I said just now what a lovely wife you have, what a beautiful home. It's true, of course. I'm just pointing out to you what's at stake here, Edmund. You might not see things as clearly as I do. We've all worked so hard to get what we have, and there are so many people who are jealous of us, who say we don't deserve all this, but we both know the truth. We earned everything we have. Without us, this country would be starved of innovation. Nothing new would be created. Okay, so someone's going to grow organs outside of the body, but not now, not when they'll destroy this wonderful product of yours. It's a fantastic idea that you had. And you have to protect it."

Jerry paused.

"Now you say what I just told you is bullshit. It's not all bullshit, is it? It can't be. And what's Alice going to say if she gets a note that says her husband slept with his analyst, and she got pregnant? I doubt she's going to be placated that easily, just by telling her it's all bullshit."

Edmund said nothing.

"I'm telling you, these Albanian guys can make all of this go away. I assure you they have done more difficult things than this. It turns out that it's true what they say: money really can buy you anything. Just look around you, Edmund, you just have too much to lose."

"What about Gloria Croft?"

"Don't worry about her," Jerry said. "She'll get hers when LifeDeals share price skyrockets."

Jerry held out the piece of paper again. This time, Edmund wearily put out his hand and took the scrap, unfolded it and read it. Jerry touched him once on the shoulder with his left hand and turned and walked back toward the house. Edmund stood where he was, staring at the name written on the paper, a name that meant nothing to him, and everything.

Part II

25

Columbia University Medical Center
New York City
March 23, 2011, 12:02 p.m.

Tobias Rothman was happiest when he could work uninterrupted in the safe confines of the lab, with Dr. Yamamoto at his side. Yamamoto was like Rothman's right arm. He could hold out a hand, and Yamamoto would know what he wanted without him having to ask for it. The two men communicated by looks and pointed fingers and sometimes, Rothman swore, by intuition. If Rothman could intuit anything as they worked together under the hood in the biosafety level-3 lab today, it was that his colleague didn't feel so good because a couple of times he'd uncharacteristically missed Rothman's cues. In truth, Rothman hadn't been feeling particularly well himself for the last hour or so. He had some mild gastric distress but worse was a kind of lightheadedness, as if he were walking on eggshells. It had started about an hour after their coffee break at nine. They'd been in the unit since six.

Rothman looked over at Yamamoto. He was facing the wall, resting his hands on the lab bench, breathing hard. Yamamoto turned to look at Rothman, and Rothman could see that he was shivering. With a hood and mask, all Rothman could see of Yamamoto's face were his eyes, which reflected fear. Suddenly Rothman felt it too, and began to shiver himself. It was as if he'd just jumped into a bath of ice water, yet he was sweating, and he felt nauseous. It was impossible that what flashed through his mind could be happening — they'd taken all their usual precautions, and their safety record was perfect.

The next moment Yamamoto's eyes rolled up in his head, and he collapsed limply onto the floor. Rothman tried to steady himself before going to Yamamoto's aid, but he felt suddenly much worse. The room swam before him. He knew he was going to black out and just before he did, his hand reached for the red button on the wall.

Pia was sitting in her office comparing notes with Will and Lesley. It was crowded but quiet. They'd taken refuge in there despite its diminutive size as there was yet another workman in the lab proper, again working up in the ceiling with all the electrical wires. He'd briefly been in her office as well as Rothman's but thankfully had left both. Luckily it wasn't the same guy, Vance, who'd been such a pain weeks earlier.

The three students had formed an effective unit in their three weeks together and were making good progress with the temperature and pH issues with the organ baths. They'd spent almost all of their waking hours including weekends in the lab, but none of them begrudged a minute of it.

Then, in an instant, it was as if a riot had broken out at the door to the lab.

"What the hell?" Will said, as the three students piled out of Pia's office.

From her vantage point, all Pia could see were people barging in through the door. The place was being invaded — she must have seen twenty people dressed in gowns, hats, masks, and booties rush toward the biosafety unit. Bringing up the rear was a pair of gurneys sprouting IV poles with plastic bags of IV fluid slapping against the metal poles, pushed by more gowned-up figures. The gurneys disappeared into the biosafety unit, whose door had been propped open with a doorstop. Pia felt a terrible sensation growing in the pit of her stomach.

One man stopped by Marsha's desk and stood next to the terrified secretary, who had a hand clasped over her mouth; another blocked off the entrance door, which was again closed, denying access to the corridor and the rest of the medical center. The laboratory staff crowded into the center of the room and there was a ripple of loud conversation and shouted questions.

"Is this a drill?" Lesley said. "What's going on?"

The figure by Marsha's desk pulled down his mask. He was a fifty-something African-American man with skin as black as ebony; his voice was calm yet commanding.

"Okay, folks, this is not a drill. We have a situation, and I need you all to stay right here, right where you are. Is everyone in the lab accounted for?"

People looked around, checking for coworkers among the fifteen or so technicians and support staff standing around. Pia could see the maintenance man in his coveralls standing at the back of the room, gaping like everyone else.

"Everybody here? Okay. My name's David Winston. I'm from hospital security.

These other people are a mixed group from the hospital ER and the Department of Infectious Disease. I'll give you more information when we have it. I am asking you to please remain in this area. Thank you for your cooperation."

The staff stood in small groups and talked among themselves. Pia, unable to stay still, walked around in a small, tight circle. Whatever was happening, she knew it wasn't good. A wave of anxiety washed over her.

The lab door opened abruptly and a tall, distinguished-looking man walked in quickly and made his way through the cordon toward the biosafety unit, conspicuously avoiding eye contact with anyone. He was dressed in protective clothing like the others except his mask hung down on his chest. Under the gown was a suit, not scrubs like the others. Pia knew this was the chief of Infectious Disease, Dr. Helmut Springer, as she had attended several lectures he'd given during second-year pathology.

The background buzz of conversation grew louder. Most recognized Dr. Springer. Everyone in the lab was well aware that they worked with highly virulent and contagious microorganisms. Was it possible there'd been some contamination of the lab? Where were Dr. Rothman and Dr. Yamamoto?

Springer's appearance only heightened the tension. The man by the door was on a cell, apparently quarterbacking whatever was happening. "We're on our way, ETA five minutes," he was heard to bark into his phone.

Quickly tying his mask in place, Springer pulled the biosafety unit door fully open. As if on cue, the gurneys reappeared, the one in front carrying Dr. Rothman, Dr. Yamamoto in the rear. Both men had IVs and were wearing oxygen masks. Rothman passed right in front of Pia, who pushed forward to take a look. She could see he was deathly pale and shivering violently. His eyes were fixed straight ahead, staring at the ceiling. He looked like death.

As fast as they had come, the cavalcade of medics was gone. Only Dr. Springer and Winston remained. Springer addressed the shell-shocked staff, a few of whom were clutching each other for comfort, others holding their hands over their mouths in disbelief at what they had just witnessed.

"As you can see, Doctors Rothman and Yamamoto have been taken ill. At first guess, we have to consider it to be severe typhoid fever. Both men are presenting the classic symptoms — fever, sudden prostration, abdominal distress, delirium, right lower quadrant borborygmi." Springer counted off the symptoms on the fingers of his left hand as if he were on formal grand rounds. Once a professor always a professor, thought Pia. "Obviously, they were working in the biosafety unit. But can anyone tell me what they were working on exactly?"

Lab technician Panjit Singh stepped forward. "They were working on salmonella strains grown in the space station lab. I know that for a fact because I set everything up for them this morning. They've been working on it for weeks."

"Okay, thanks, that's very useful. Do you know if there are any antibiotic-sensitivity studies available for these special strains?"

"Yes, lots of them. I can get them for you."

"That's good, I'm going to need them, thank you. Mr. Winston here will talk to you about procedure a bit later, but here's a thumbnail: no one is to go into the level-three lab until it's cleared. The Rothman lab itself will be off-limits until further notice. I've already put in a call to the CDC to get their help on the epidemiology side so we can find out how this contamination occurred. Right now, everyone needs to follow me to the Infectious Disease Clinic, where you'll be screened for typhoid fever.

Everyone will also need to take a prophylactic course of antibiotics. This is very important. For the next week you'll have to monitor your own temperature twice a day. Anything unusual, come in right away. A degree either side of normal, I want to see you. Any questions?"

"Who raised the alarm?" Singh asked.

"There's a panic button in the biosafety lab," Springer said. "One of the doctors must have hit it. We'll check the tape."

"Does everyone need to come to the clinic?" Pia asked. "Even people who haven't been in the biosafety unit today?"

"Absolutely. And Mr. Winston will also be gathering names of everyone who's been here delivering supplies or takeout or whatever. We want to see everyone who has set foot inside this lab. That's it. Thank you for your cooperation."

The din of conversation erupted again.

"Oh my God!" Lesley said. "Did you see how they looked? It must have come on fast."

"Dr. Yamamoto told me he didn't feel so hot this morning," Will said. "But yeah, I saw how they looked. I guess we better go do what the man said."

Pia looked around. The maintenance man was hanging back, and though Pia didn't want to talk to him, she knew he needed to follow the protocol.

"There's a medical issue," she said to the man. His temporary name tag read "O'Meary." "You have to come to the clinic with everyone else."

O'Meary looked nervous and didn't say anything. Winston called out to Pia.

"Time to go," he said. "We're locking down." There was clearly no room for argument. Pia waited for O'Meary to leave and exited in front of Winston. As the last person out, Winston pulled the door shut and talked to two figures in full hazmat suits standing outside.

"No one gets in," Winston said. "Put up the caution tape." The men in the hazmat suits nodded and set to work.

As they made their way to the elevator, Pia could see that the whole floor was being cleared, with other personnel being led down the stairs. There were more people in hazmat suits that looked like robots. On the elevator ride down, Pia could feel her heart beating too fast, and she had to concentrate on breathing deeply. She felt some dizziness from her shallow breathing. As she walked along the sidewalk she was gripped by what felt like panic — everything around her felt very close and incredibly far away at the same time. She had stopped walking and was holding on to someone. Voices were loud in her ear.

"Come with me," a woman is saying. It's a hot sunny day but Pia's freezing cold. The woman has a nice smile and she's holding Pia's hand. This is a new place, Pia knows that. She hasn't been here long. This is the first smile she's seen, though it's odd now — grown-ups don't keep smiles on their faces the whole time. Pia and the woman have come inside and they're walking up to a large door. It feels like they're walking uphill. "This is the headmaster's office," the woman says. She opens the door and pushes her in. Pia can hear the lock being turned. "Hello, Pia," says the man. He's smiling too but it's a twisted smile, not a smile of welcome. . . .

Pia looked up. She was sitting on the ground on 168th Street with traffic passing by. Winston was supporting her with his hand, looking down at her.

"You okay?"

"Yes, I think so."

"You fainted. Or almost fainted. You're not perspiring, so I don't think you have a fever. I think you're okay. Ready to get up?"

Pia waited a second and allowed Winston to pull her to her feet. Then she remembered where she was and what had just happened.

With disturbing clarity she saw Rothman lying on the gurney, his face looking like death, and the image terrified her. Over the course of three and a half years she'd come to rely more and more on the man's strange friendship, particularly after their heartfelt conversation a few weeks ago. Up until then, their relationship had been akin to two people comfortably wandering around in a darkened room, occasionally sensing each other's presence but not much else. But after the conversation and the personal revelations, she felt they'd moved to another level. Rothman had become the ersatz father she'd always pined for. Most important, she'd allowed herself to begin to trust Rothman despite having learned not to trust anyone, not to allow anyone into a position where they could betray her, like so many had done.

Now, as she stumbled along the street, Pia was overwhelmed by the thought that just when she'd allowed Rothman into her world, he was going to abandon her. Why was he doing this? And why now? It was irrational to think so, but did he do this to spite her? Did he purposely set her up? After all, he'd admitted to being depressed. She was almost paralyzed with anxiety.

At the Infectious Disease Clinic, Pia was

shaking when she was handed the Z-Pak prophylactic antibiotic. She sat down in the waiting area and her head started to clear. She was aware that several people had tried to talk to her, but she didn't hear them.

"Miss Grazdani!" a nurse called out sharply, standing directly in front of Pia. She was on the brink of calling the ER if the young woman continued in her fugue-like state, thinking Pia might have to be admitted.

Not quite awakening, Pia sat up straighter and focused on the nurse's face.

"I'm here," Pia voiced. "I'm sorry. What did you say?"

"I said that you can't go back to the lab. It's going to be closed until the CDC epidemiologists get here from Atlanta and declare it clean. What you should do, as we have advised the others, is go home, start your antibiotics, and watch your temperature. Is there someone we can call who can meet you there? Miss Grazdani? Are you okay, Miss Grazdani?"

"I'm just fine," Pia assured her.

26

Columbia University Medical Center
New York City
March 23, 2011, 2:37 p.m.

It wasn't a warm day, but Pia had wanted to sit outside. She had found a bench set in a small rectangle of public cement, what in New York City is called a park, and sat down, hands in her coat pockets, her chin down, hood pulled over her eyes. Her mind played over the scene she had just witnessed several times. There was a surreal quality to it, like it was one of her nightmares. Unfortunately it was real.

After she had gotten herself reasonably calmed down, Pia got up from the bench and started to walk toward the dorm. She got halfway, changed her mind, and turned and headed back to the hospital. There she took the first two of her antibiotic tablets at a water fountain before heading to the internal medicine floor.

At the main nurses' station, she asked for Dr. Rothman and was directed to the infectious disease wing a floor above, where Rothman and Yamamoto had been admitted. She wanted to check on Rothman's status, hoping that he'd revived with treatment, and if so, she wanted to ask him if he knew how he and Yamamoto had become contaminated. Pia knew the epidemiologists would certainly be asking the same questions, but she had a personal reason for finding out — the crazy thought that he'd done it on purpose, an idea she knew to be irrational but which demanded, in her mind, to be investigated.

Pia had another concern. Experience had taught her to absolutely distrust authority in any institution and to assume that nothing would happen the way it should. She knew Rothman was disliked by almost one hundred percent of his colleagues in the medical center. He was rude, seemingly arrogant, and antisocial. While medical protocol and simple human decency demanded that each patient receive the undivided attention of medical staff and the best care available, she couldn't help but think Rothman's reputation might degrade the standards.

Pia used her medical student credentials

to get on the floor and found that the two researchers were in adjoining, negative-pressure rooms where air flowed in but not out. They were in strict isolation but there was no one guarding the rooms. Pia started to put on the isolation gear in the anteroom — the gown, hat, mask, gloves, and booties — but just as she was about to put on her mask, Dr. Springer emerged from Rothman's room. He undid his mask and stared at Pia.

"What on earth are you doing here? You're Rothman's student, aren't you? You're supposed to be home."

"I took my antibiotics and my temperature's fine. I know I'm clean — I wasn't in the biosafety unit today, or even in contact with Dr. Rothman or Dr. Yamamoto. It's very important I speak with Dr. Rothman."

"Good God! Of course you can't speak with him. The only people allowed in are medical staff assigned to his case. No family, no friends, and certainly no medical students."

"There's no one in there looking at him now. Are you sure of the diagnosis? Is this the best place to treat his condition?"

"What do you mean, 'is this the best place?' " Springer shook his head in disbelief.

"I know what people around here think of Dr. Rothman —"

"Young lady, I don't know what you're implying but everyone at Columbia Medical Center gets the same superb care as everyone else, friend or foe, rich or poor. It makes no difference whatsoever. And I happen to like Dr. Rothman."

"Okay, okay. Sorry, I'm just upset." Pia didn't want to get thrown off the floor. "I've been working with both men for more than three years on the salmonella strains that are probably involved, and I thought I might be able to help."

"Okay," said Springer. He relaxed a degree. He sensed Pia's intentions were good even if totally unrealistic.

"I have to tell you that both men are delirious. Even if I let you in there, you wouldn't get anything out of Dr. Rothman. Follow me." Springer took off the protective gear and tossed it into the covered hamper. Pia did the same.

Springer took Pia back to the nurses' station and, sitting down, itemized the laundry list of tests that had been ordered, including a complete blood count, electrolytes, blood cultures, urine cultures, stool cultures, stat DNA microbiology tests, and the appropriate X-rays. At that point the tests that had come back confirmed that the infectious agent was one of the salmonella strains Rothman and Yamamoto were working on, which Rothman had named the alpha strain, the most virulent of the three grown in space. He also mentioned that the white cell count showed a mild leukopenia, meaning the white cell count was mildly depressed, something often seen with typhoid fever. He declared the electrolytes, meaning primarily sodium, chloride, calcium, and potassium, were normal. Springer concluded by telling Pia that Rothman's and Yamamoto's temperature, heart rate, blood pressure, degree of oxygenation of blood, urine output, and central venous pressure were being monitored, and that at the moment the only thing abnormal was the temperature.

"They're both in bad shape, especially considering how quickly the illness came on," Springer added.

"What antibiotic are they on?" As Pia knew from her studies with Rothman, there was debate about which was the best to use in serious salmonella cases.

"That's a good question," Dr. Springer said. "Actually Dr. Rothman briefed me recently on findings he had made in his studies of antibiotic sensitivity on these zero-gravity strains. All three strains he was working on are very sensitive to chloramphenicol. That's an antibiotic that at one time was considered the best choice for typhoid, but it went out of favor in the seventies because newer strains of salmonella were becoming resistant. Dr. Rothman said that because these strains were grown in space, they were more virulent but somehow they'd also lost their chloramphenicol resistance. He was interested because drug resistance is a big problem with salmonella."

"Have you thought of trying ceftriaxone?" Pia asked, referring to a newer antibiotic.

Springer hesitated, giving Pia a once-over look. He'd been trying to be nice to her, as she was obviously concerned about her mentor. When he resumed speaking his voice and syntax had changed. There was an edge. "I actually wasn't requesting a consult by speaking with you. It's purely as a courtesy that I'm filling you in on Dr. Rothman's condition and course of treatment. But to answer your question, if it was a question, there is some sensitivity to ceftriaxone but significantly less than there is to chloramphenicol."

"Chloramphenicol can cause aplastic anemia," Pia said, missing Springer's signal that she was pushing it.

"Yes, we've taken into account the side effects, of course. Excuse me." Springer suddenly got to his feet. Abruptly he left Pia, talked briefly with one of the residents helping with Rothman's and Yamamoto's care, then left the floor.

Pia waited for a few minutes before wandering over to the same resident, who was reading a chart.

"What do you think of Dr. Springer? Do you think he's qualified?"

"What do I think? He's the best in the country. I wouldn't be here otherwise."

Puzzled by the question, the resident walked away, leaving Pia standing by the nurses' station, alone.

27

Columbia University Medical Center
New York City
March 23, 2011, 7:38 p.m.

News of the Rothman/Yamamoto event spread rapidly through the Columbia medical community. George Wilson, like everyone else, had heard about it, and he could only imagine the effect it was having on Pia. Concerned, he had looked for her. It took some searching but George finally managed to track her down. She didn't answer her cell phone and neither Will nor Lesley had seen her, so he had had to physically find her. George struck gold in the library stacks, a place he knew she found comforting. After some cajoling, Pia agreed to go with him back to the dorm cafeteria.

Pia was as distraught as she could ever remember being. She was especially upset because her emotions were so conflicted. Usually in her tumultuous life, distress had a definitive cause, but now she didn't know whether to be upset about Rothman's dire condition or angry at his carelessness in getting infected with the bacteria he'd been working on. And there was an undercurrent: Pia was terrified about her own future, which she thought she'd been so careful about but now seemed to be in the balance. She was also furious with herself for allowing Rothman to penetrate her well-constructed protective shell. And now she had the added distraction of George, who was trying to be solicitous but making things worse with all his questions.

"I can't sit here anymore," Pia said suddenly, interrupting George, but she didn't care.

"You haven't eaten anything," George said, looking down at her tray. "You've got to eat."

"I can't eat," Pia complained. "Feeling I'm in control is important to me. I don't feel in control. My life is coming apart. I've got to see Rothman. I have to."

Security was essential to Pia, as was control. At the moment she felt neither.

"Is he allowed to have visitors?"

"I don't even know if he's conscious. But I'm not a visitor, I'm concerned about the course of treatment he's on."

"I'll come with you," George said.

Pia didn't know whether she wanted him to come or not.

"Don't you have things to do?"

"Nothing important. I want to help you."

"Whatever!"

Pia jumped up from the table, leaving her tray of food untouched. George stuck her turkey sandwich, still in its wrapper, in his jacket pocket and hustled after Pia. As she marched to the hospital, George trailed along in her wake. He tried to talk to her but gave up when she wouldn't answer. She was on a mission.

The floor housing Rothman and Yamamoto was bustling with staff and orderlies. There were few patients in evidence. Most were too sick to be up and about. Pia found the resident on duty, Dr. Sathi De Silva. As the sole infectious disease resident, she had her hands full, not just with her two celebrity patients but a ward full of others, and several more people in the emergency room awaiting her attention. Pia and George were in their medical school white coats so Dr. De Silva accepted them as students most likely on their internal medicine rotation. Dr. De Silva took her teaching responsibilities very seriously, so when Pia started asking questions about Dr. Rothman, she stopped what she was doing. "To answer your question, both Dr. Rothman and Dr. Yamamoto are dangerously ill. They're both delirious and uncommunicative."

"I understand they're on chloramphenicol. What's your feeling about such a choice?"

Dr. De Silva shrugged. "I think it's a good choice. Yes. It's a unique situation because there are newer antibiotics, but in this case we have sensitivity studies that show the involved strains of salmonella to be uniquely sensitive. Dr. Springer believes it's our best hope. We're monitoring for side effects, but we haven't seen anything. If there are any problems, we can always switch to one of the newer, third-generation cephalosporins."

"Strange case," Pia commented.

"One of the strangest," Dr. De Silva agreed. "And not a little ironic."

"Do we know how they got infected?"

"If we do, I haven't heard anything. I know the CDC epidemiologists went through the lab and particularly the level-three containment area where the salmonella strain was kept. I think their initial concerns were about a malfunction of the hood, but apparently it was working fine. There was some bacteria in the hood itself, but you'd expect that. They took cultures, I know, and we'll get results in twenty-four hours. I'm hearing all this secondhand. My job is to look after them."

"Of course," said Pia. "Has the CDC finished with the lab?"

"Dr. Springer said an hour ago that most of them were already headed back to Atlanta."

Dr. De Silva's cell phone beeped, and she glanced at a text message. "Oops, gotta go! Nice talking with you."

"Can we see Dr. Rothman?"

"I don't see any harm, but you're not going to be seeing much," Dr. De Silva said, already walking away. "As I said, he's delirious. If you do go in, just make sure you put on all the gear and don't bring anything out!"

Eagerly Pia set off toward Rothman's room. George stumbled after her.

"What are you doing?" George complained. "You can't go in there. He's sick, he can't tell you anything. Why take a chance?"

Pia didn't answer. She suited up as per the Universal Precautions established by the CDC, which were posted on the outer door. George continued to try to talk Pia out of the visit, but she ignored him. He found a set of protective gear for himself and followed Pia into the room. As they passed through the door they could feel air entering with them.

Pia walked directly over to the bed. Several IVs were running, each laced with antibiotic.

"Dr. Rothman? . . . Dr. Rothman?"

Rothman stirred and half opened his eyes.

"Dr. Rothman, can you hear me?"

"What are you *doing?*" George's nerve was failing him on many counts. Neither of them was on an internal medicine rotation, so they had no business or excuse to be there, and why was Pia trying to talk to Dr. Rothman? The man was delirious. Apart from the trouble they could get into, George was nervous about the salmonella that was making Dr. Rothman sick. The man looked gravely ill, with an ashen coloring and loose strands of hair matted to his pale forehead.

"He doesn't look good at all," Pia commented.

"Tell me about it," George said nervously.

"My gosh, look! He's losing some hair." Pia pointed to tufts of hair on Rothman's pillow, but George wasn't interested. Rothman had become agitated now, twisting against his restraints while mouthing some words. Pia grabbed his chart and was flipping through the pages.

"His temperature's up — not a lot, but up nonetheless."

"Pia . . . let's *go!*" George stage-whispered.

"You go, George, I'm not leaving. Not yet." From working with Rothman, Pia had learned a great deal about typhoid fever and its cause, salmonella typhi. She knew the danger signs of the illness and the fact that the disease attacked the small bowel, concentrating in lymphoid tissue in the small intestine called Peyer's patches. Rothman's gown was pulled over to one side, and Pia exposed Rothman's abdomen a little more. She slowly pushed in on his upper abdomen, and Rothman squirmed and moved his head from side to side.

"He's definitely showing signs of discomfort, maybe pain in his abdomen," Pia said. "This is not a good sign."

George was beside himself. He could see a few people passing by in the outer hall through the wire-embedded windows in the two doors of the isolation room. He walked over and closed the blinds, hoping to buy Pia some time. When Pia suddenly let up on the pressure she was exerting, Rothman reacted slightly, to Pia's surprise, as if that caused more discomfort.

"Did you see that? He recoiled. Would you say he recoiled?"

Pia repeated the maneuver and got the same result.

"He definitely recoiled."

"Whatever it is you're doing, it's going to get both of us kicked out of school if we don't leave right now. We're pushing the limits on a couple of celebrity patients."

"It's rebound tenderness," she said. "It's a sign of peritonitis, inflammation of the lining of the abdominal cavity. It means the bacteria have penetrated the lining of the small intestine."

Pia reached over and punched the intercom button. The nurse at the station picked up.

"Is Dr. De Silva available? If she is, get her in here stat. The patient has developed rebound tenderness."

George was hopping from one foot to the other. *Now she's really done it,* he thought.

At once, Dr. De Silva came in the room, palpated Dr. Rothman's abdomen and confirmed Pia's finding.

"And look, he's losing some hair," Pia said.

"That could be the chloramphenicol. But regardless, the rebound tenderness suggests the chloramphenicol is not controlling the infection. We'll have to change the antibiotic. I'll call Springer and get his suggestion. Thanks for your help."

Dr. De Silva ducked out of the room.

"He's getting worse," Pia said, looking at Rothman forlornly.

"Rebound tenderness isn't a good sign, I know that," George said. "But you've done all you can do. Let's go. You heard her, she's calling Springer."

By the time George and Pia took off their gear and got back to the nurses' station, Dr. De Silva was on the phone with Springer. Pia stood where she could hear Dr. De Silva's half of the conversation. It sounded like Springer was doing most of the talking.

"Okay, ceftriaxone . . ." she said. ". . . And the hair loss . . . Right, of course we'll stop the chloramphenicol. . . . Okay. I'll see you soon, and I'll call Dr. Miller."

Dr. De Silva turned and saw Pia. She hung up the phone and redialed immediately. She covered the receiver with her left hand and talked to Pia as the phone rang.

"Dr. Springer's on his way in. He wants to check the rebound tenderness for himself — Oh, hello. I need Dr. Miller. . . . Dr. Miller, this is Dr. De Silva in Infectious Diseases. I'm treating Dr. Rothman and Dr. Yamamoto. Dr. Springer would like a consult. We're seeing rebound tenderness in Dr. Rothman and may have to remove the infected bowel. . . . No, just Dr. Rothman at the moment . . . His temperature is up slightly. Other levels — blood pressure, pulse, oxygenation — are the same. Okay, thanks."

Dr. De Silva hung up the phone and exhaled. She was a small woman of Sri Lankan descent who prided herself on running a tight ship. She was embarrassed that a medical student had picked up an important sign that she'd missed. "I just checked him a few minutes before you two showed up. Temperature was holding steady," she said, half to Pia, half to herself. She turned to Pia.

"It can come on very quickly. Dr. Miller, the chief surgical resident, is coming in. And Dr. Springer's on his way over. So, who's your preceptor? I should at least give you credit for what you found. And how did you know what to look for? I'm impressed."

"Actually I'm not on internal medicine at the moment."

"Are you on an infectious disease elective? If you are, I haven't heard your name."

"I'm not on an infectious disease elective either."

George was desperately trying to get Pia to shut up. Out of Dr. De Silva's line of sight he was frantically making a time-out gesture like a football official.

"Well, what brought you here?" Dr. De Silva asked.

"I just happen to know a lot about salmonella."

"Really? From whom?"

"Dr. Rothman," Pia said, as George grabbed her arm and literally pulled her away, angling her toward the elevators.

George felt a sense of relief as they left the hospital. With as busy as Dr. De Silva was, he hoped she wouldn't say too much about the two mysterious med students, one of whom had been very helpful. Actually he doubted she would. He knew that there hadn't been any negligence on Dr. De Silva's part, but he knew that in the competitive atmosphere of the academic center, she was probably chagrined that she'd been, in a fashion, upstaged by a medical student. Pia had detected the change in Dr. Rothman's condition before she did. But George's relief was short-lived.

"I want to go back to the lab," Pia said, stopping suddenly. They had just reached the corner where 168th Street turns into Haven Avenue. "I want to see if there are any clues as to why or how he got infected. He's so careful, I don't understand it. He's so detailed and compulsive about his work, his organization, his technique, it's all flawless. It doesn't make sense."

Still lurking in the back of Pia's mind was the thought that Rothman had infected himself intentionally. But why would he involve Dr. Yamamoto? It couldn't be the case, or could it? What she wanted to do was completely eliminate the idea as even a remote possibility. If Rothman died, it was going to be a betrayal of sorts, but she didn't want it to be his betrayal. Betrayal by fate she thought she could ultimately handle. Personal betrayal by Rothman would be something entirely different.

George groaned inwardly. Visiting Rothman had been bad enough. Visiting a lab that was off-limits by order of the CDC was something else entirely. "The lab is closed," George said in a fashion that wouldn't brook discussion. "Order of the CDC. Let's head up to your room. I saved the sandwich you didn't eat." To prove his point, George pulled the food from his pocket.

"I'm going," Pia said.

"What on earth do you think you're going to turn up that the CDC hasn't?"

"I don't know, but I can't do *nothing.* You can come with me or not, I'm going anyway. Of course two sets of eyes may be better than one."

George realized Pia was asking for his help, however indirectly, which was a first. Still, it wasn't an easy decision. He was okay with bending the rules, but not breaking them like this. He couldn't afford to get kicked out of medical school. It had been his goal as long as he could remember, and he had his family to consider. But George had no time to ponder his decision. Pia had already turned around and was heading toward the research building.

"You're not worried about getting typhoid fever?" he asked, catching up to her.

"I was in there this morning. And there's protective gear we can put on just like we used in Rothman's room."

Pia entered the building. George followed. It was like making a decision without making a decision. They showed their IDs to the security man and headed for the elevators.

As George had expected, the door to the lab was crisscrossed with yellow caution tape. "See, just as I expected. We can't go in."

Pia didn't respond. She merely peeled back the necessary tape and tried the door, which was locked. It didn't deter her. Many times over the last three and a half years,

Pia had been asked to take a reading in the lab at night, or monitor an automated experiment. She took the key she'd been given for those eventualities, opened the door, and stepped over the threshold.

"Pia, this is crazy," George said. Reluctantly he came in after her. It was dark, and very quiet.

"Relax. The security cameras are out, they've been working on them again for days. Who's going to come in now? I just want to check the refrigerated storage facility in the biosafety lab and take a peek at the logbook. And before you say it, I know the CDC has probably investigated all that. They might have even taken the logbook. Be that as it may, I need to make sure they didn't miss anything."

Pia turned on the minimal light necessary. It was a small lamp by the communal coffee machine. She then quickly checked her own office, and Rothman's. George trailed after her like a shadow. As far as she could tell, nothing had been disturbed in either office. Pia pointed out Rothman's desk to George. The in-tray, the few files, the pictures — everything was just so.

"See how orderly he is?" said Pia.

All George could think about was getting out of there. An air circulator kicked on and

George jumped half out of his skin. He followed Pia to the biosafety level-3 room and they donned the protective gear once more. Pia used the coded punch pad to reach the lab itself. Since there were no windows, Pia turned on the overhead lights. The ventilation system was still running and there was an eerie calm in the place. Pia checked the logbook, which the CDC had not taken. There were the usual entries; the next to last one was Panjit Singh, when he went in that morning to set up. Then there was Rothman and Yamamoto's entry. There was nothing abnormal. She then went to the refrigerated storage unit. Using a separate keypad, she was about to open it when she heard a noise that caught her attention.

"Did you hear that?" she whispered intently to George.

"Hear what?" George said nervously.

Pia held up a hand and went to the door and cracked it open. There were quiet sounds but unmistakable — voices in the lab outside. Voices getting louder.

"In here . . . c'mon," she said urgently.

"Shit," George said under his breath. He'd heard the voices. "Shit on a brick," he mumbled to himself.

Silently but urgently Pia waved for him to follow her. George saw where they were heading and beat it out through an emergency exit in the far corner of the lab. The door complained when he pushed it open as it hadn't been opened since it was installed back when the lab was last renovated. It had also been made airtight.

Pia followed close behind George. She might have stayed and faced the music had she been there by herself, but she was well aware of George's utter fear of authority. Where that had come from, she had no idea.

The unit's emergency door led to the lab storeroom where Pia and George pulled off their protective gear and stumbled out into the main part of the microbiology department that housed Rothman's lab. Staff on the evening shift at the microbiology clinical lab were curious to see two young people running by, then stunned to see them followed a minute later by three figures in full hazmat gear.

Microbiology led into the anatomy department and George and Pia crashed through the connecting doors and into the familiar surroundings. As first-year students they had spent a good deal of time in the department. George was leading, but he didn't know exactly where he was going. All he knew was that he wanted to avoid getting caught. He ducked into the darkened anatomy room, dimly illuminated by night-lights. For the benefit of the current first-year students who were taking anatomy at the time, the room was well stocked with cadavers, most covered with oilcloth shrouds. Several torsos sat upright on the head teaching table. They'd been cut across the upper chest and then halved in a sagittal section so that half the gullet and half the brain were visible. George was eye-level with the torsos, the exposed whites of their eyes seeming to glow in the half-light.

George and Pia ducked behind the long teaching table, but there was nowhere to hide. A moment after their arrival, the banks of ceiling lights flickered and came on. Three security guards in hazmat suits stormed into the room. Pia stood up and George, very reluctantly, followed suit.

The security men were angry, demanding Pia's and George's identification cards. They then made several calls on their radios before turning back to the students. George was cowering, Pia taking it all in stride. "You're coming with us," said the nearest figure to George, grasping his arm and marching him out of the room. Pia was escorted out behind him.

The group wended their way past the few onlookers in the clinical microbiology lab and down to the street via a service elevator. George's mind was racing but he couldn't think of any way Pia could talk her way out of this. As they walked across the campus, the group attracted a lot of stares and comments from passersby. Some of them wondered if they were watching some med-student prank.

George and Pia were taken through a featureless corridor in the hospital bowels to the security department. They walked past a bank of TV screens being monitored by two bored-looking men, down another corridor and into a small office with a handwritten sign on the door: DUTY OFFICER. Standing up, watching a couple of monitors mounted on the wall, was David Winston, the man who'd taken charge in the lab earlier that day. He recognized Pia, having helped her when she fainted in the street.

"Ah, you again. I see you're feeling better than when I last saw you."

"Mr. Winston," Pia said. "My friend and I were just retrieving some of my belongings from my office."

Winston referred to a list on a clipboard resting on his desk.

"Miss Grazdani, and . . ." He looked at George.

"George Wilson."

"George Wilson. Not on my list. You a fourth-year student as well?"

George nodded.

"Well, you'll be taking antibiotics too," Winston said. "Folks, there's a protocol we use in these situations. You broke into a secure, potentially contaminated area. I actually saw you do it myself, sitting right here. The cameras might not be operating inside the lab, but outside they work just fine. So I see two people go into the lab, and I have to send three of my guys in full body gear to go in and find you. And it turns out to be you two. So the protocol is, I make a call to the dean of students, who loves to hear from me, as you might imagine. It's just a heads-up because my next call is to my friends at the Thirty-third Precinct, and I'll have a full and frank conversation about criminal trespass."

George was aghast. If the police got involved, he was screwed.

"I don't know why you guys went in there, and I'm not going to ask. The CDC might have cleared it, but the caution tape was still over the door. Especially you, Miss Grazdani, as you were specifically told the lab would be off-limits. Frankly, I'm dumbfounded. But I've never understood medical students since I took over this job heading the center's security."

Pia started to speak, but Winston held out his hand to silence her and called the dean of students. He explained the situation. He then listened for a good two minutes and hung up the phone.

"She's coming down. I don't know who I'd rather deal with if I were you, the dean or the Thirty-third."

Winston showed George and Pia into a small side room and closed the door. George was too agitated to speak; Pia started pacing around the room. She couldn't sit still. After what seemed like an age but was in fact thirty minutes, the door opened and a tall, dark-haired woman in sweats and a ski jacket came in and shut the door behind her. Her name was Helen Bourse. She had been dean of students for almost a decade and was well liked but hardly a pushover.

"What the hell did you think you were doing? You two have made me cash in more favors than I actually own, stopping Mr. Winston from having you arrested. I want you to convince me I did the right thing."

"I'm so sorry, Mrs. Bourse," George said. He took one look at Pia's defiant face and decided he should speak for the both of them.

"*We're* very sorry."

"So what in God's name were you doing there? In a lab that was sealed and potentially contaminated."

"The only part that might have been contaminated was the biosafety unit," Pia said, interrupting George, who'd started to respond. "We took the necessary precautions. I wanted to look for myself. I just can't understand how Dr. Rothman managed to get infected, knowing him as I do."

"So you weren't picking up your stuff as you told Mr. Winston. And what, you're suddenly epidemiologists? We had a team of actual epidemiologists check out the lab today both from here and from the CDC. They combed the place, including the biosafety unit."

"What did they find?"

"Nothing, but that's not the point."

"I've been working in there on and off for over three years. I wanted to check it out. If something was different, I might have been able to see it, probably better than strangers from Atlanta."

Some of Bourse's vinegar lost its acid. She realized that Pia had a point. Still, it didn't justify what these two otherwise gifted students had done, something totally foolish and out of character. After a pause she asked, "Well, what did you find?"

"Nothing, but we were interrupted. Do you have a report from the epidemiologists?"

"Certainly not from the CDC. Not yet. But I spoke to the head of our own team. Apparently nothing was found amiss."

Dr. Bourse knew that Dr. Rothman was closer to this student than to anyone in the whole medical community. She knew quite a bit about Pia, more than she guessed Pia surmised. Bourse had had access to all the deliberations of the admissions committee, which she had pored over in great detail. Up until the call from Winston, she'd had high hopes for her, hopes she wanted to maintain. For Pia, Bourse's intent was to try to keep the damage from the evening's escapade and poor judgment to a minimum. Such was the burden of being dean of students. Earlier that evening Bourse had had to deal with an even more difficult issue: A third-year student had been caught stealing prescription drugs from the medical floors. Bourse turned her attention to the second delinquent. At least he met her eye, which she couldn't get from Pia. "So what's your excuse?" she asked George, with a certain resignation in her voice.

"No excuses. I was helping my friend," he

said as evenly as he could.

Dr. Bourse studied George. He too was a top student, more liked in general than Pia, who could be considered standoffish. Bourse was well aware of George's apparent infatuation with Pia, so she took his excuse at face value. Once again she marveled at how such an apparently accomplished young man like George could be reduced to such a lovelorn teenager that he'd risk his future like this. If Bourse had allowed Winston to have him arrested, it could have affected his becoming a doctor.

"All right," Bourse said. She took a deep breath and regarded the ceiling for a moment to clear her head. "Here's what we're going to do. You're going to go back to your rooms and stay there. You won't fraternize with anyone or talk about this episode with anyone. You'll monitor your temperatures and take your antibiotics as directed. George, I'll make sure you get some. And I'll see both of you in my office at seven tomorrow morning. We'll discuss your elective then, Ms. Grazdani. Mr. Wilson, tomorrow you will return to Radiology. Both of you will also say a prayer for me and thank the Lord that I'm in a benevolent mood. I'll now go and square this away with Mr. Winston. If I can."

With the dean out of the room, George let out a deep sigh and sat back in his chair. "Oh my God, I thought we were dead. If the police aren't involved, it's just an internal thing. It won't be on our records. It'll be like this never happened." George looked at Pia, who didn't respond. Her face was a blank, her mind clearly still back at the lab.

"You can't let this drop?" George questioned.

"Of course I can't let it drop," Pia shot back. "Something had to have happened. Something out of the ordinary."

"What about one of the technicians messing up, either by accident or design? I mean, Rothman was like a bull in the proverbial china shop. I imagine there are a lot of people who aren't crying tonight about what happened to him."

Pia shook her head. "There were people in the lab who found him unpleasant. But the same people admired him greatly. I can't imagine anyone in the lab being involved in any underhanded way."

"So what are you thinking?"

"I don't know what to think," Pia said. Her mind was swirling. Her first concern was whether or not Rothman would pull through. At the same time she was reconsidering the two possibilities for what had happened: Rothman contaminated himself by accident, or he did it deliberately. But then, another idea started to take shape in her mind. She realized there was a third possibility she hadn't yet considered.

28

Columbia University Medical Center
New York City
March 24, 2011, 5:05 a.m.

Although it was only a little after five in the morning, Pia finally quit trying to go back to sleep and got out of bed. The previous night she'd returned to her dorm room from the security office exhausted mentally and physically. Before she and George went to their respective rooms, he gave her the turkey sandwich that he'd saved. It was flattened to a degree but still recognizably a sandwich. Once back in her room, she'd eaten a corner of it, then threw the rest in the trash and went to bed, hoping to get some rest. She'd not slept well, but at least she couldn't remember her dreams.

Pia showered quickly and dressed. She understood how essential it was to her future that Rothman pull through. Despite the hour, she knew she had to go over to the hospital to check and make sure he was okay. Her hope was that the new antibiotic had worked wonders and was controlling his infection. In that case, she further hoped that his delirium had cleared up and she could have a word with him. She wanted to ask him if he had any idea what had happened in the biosafety lab the previous morning.

Emerging from the dorm onto Haven Avenue, Pia felt conspicuously lonely. It was morning although not yet light. She felt like the only person in the world as she made her way over to the hospital. Once inside, it was different as the hospital never slept. Quickly she made her way up to the infectious disease wing.

When she got to the ward she was perplexed. She felt she must have gotten turned around because the room she thought was Rothman's was being disinfected in preparation for its next inhabitant. But no, this was his room. So Rothman had been moved, perhaps because he was showing signs of improvement following the new course of treatment. Pia didn't allow herself to think that anything else could be the case. She checked Yamamoto's room. His room was also being cleaned. He'd been moved as well.

Pia turned around and went back to the nurses' station to find out where Dr. Rothman and Dr. Yamamoto had been sent. The station was humming, even at that time of the morning, in preparation for the shift change at seven.

"Excuse me," Pia said to one of the night nurses standing at the counter filling out one of the millions of forms they were required to do. "I'm looking for Dr. Rothman and Dr. Yamamoto." Suddenly Pia felt nauseous and an overwhelming sense of panic rose within her. *They weren't moved because they got better.*

"Please, tell me where they are," Pia pleaded, hoping against hope.

"And who are you? Are you related to Dr. Rothman?"

"I'm Dr. Rothman's medical student. Please, where is he?"

The nurse took Pia by the arm and walked her away from the busy nurses' station into the waiting room, which was empty at that hour. She didn't turn the light on and the two women stood in the semidarkness. Pia was worried her legs were going to give out and she would collapse on the floor like a rag doll.

"Listen, we only just told the families," the nurse said. "I'm sorry, they both passed.

Dr. Rothman first, then Dr. Yamamoto about an hour ago."

"What do you mean they 'passed'?" Pia asked, but intellectually she knew what the nurse meant. But just maybe . . .

"They died, sweetheart, I'm sorry. Dr. Rothman died being prepped for surgery. That's all I know. That's all I can tell you. Look, I have to go." The nurse put her hand on Pia's arm and left the room.

Pia went down on her haunches, her mouth open in a silent scream. She clasped her arms around her knees and pulled herself tight into a ball, as if she were trying to hide somewhere within her own body. Pia felt like she'd been punched in the gut. She was disoriented and angry — angry at the hospital, angry at the world, angry at Rothman himself. If she was going to judge a man by his actions, what had he done? She'd been abandoned. Betrayed. Pia stumbled out of the room and off the ward, descended in the elevator, walked outside in a daze. The sky was now light in the east, but the sun had yet to make it up over the horizon.

29

Columbia University Medical Center
New York City
March 24, 2011, 6:30 a.m.

Pia staggered down Fort Washington Avenue like a drunk, clutching her stomach. Thinking she was about to throw up, she stopped by a garbage can, but she was only able to bend over the rim and dry-heave a couple of times. Pia stood up and breathed deeply in and out. She had to get a grip on herself, but what was she going to do? There was an immediate issue to contend with, and she was almost grateful for the distraction: She remembered that she and George had to see the dean of students at seven, so she went to the dorm to find George.

George had set his alarm for 6:15. He was accustomed to sleeping soundly, even if the rigors of medical school meant it might be for only three or four hours at a time. But last night he'd been unable to sleep save for brief snippets, constantly turning over the events of the evening in his mind. At one point he'd given up and sat for almost an hour in the semi-comfortable armchair by the window and nursed a glass of Jack Daniel's from a bottle that had lasted through his whole medical school career. Through the window he could see a portion of the George Washington Bridge and a bit of the New Jersey Hudson River waterfront. He was totally confused and humiliated about his motivations.

Last night, Pia had come desperately close to wrecking all of George's plans. George stopped himself; it really wasn't Pia's doing that he had tagged along with her. He had to take some responsibility. The problem was that he cared about her too much to let her blindly charge into these predicaments by herself, and whether she acknowledged it or not, she needed his help. Still, he knew there was a point where he'd have to stop and put his own interests first. He just didn't know where that point would be.

He'd been asleep for only a little more than an hour when the alarm went off, and George unknowingly swept his hand over the snooze button. Nine minutes later he did it again, and he would have done it a third time if Pia's knocking hadn't woken him up. He was pleased to see her until he saw the look on her face and realized there was something wrong.

"What happened . . . ? Wait, what time is it? We've got the dean . . ."

Pia walked into the room like a reanimated zombie and fell full-length on George's bed. She mumbled something into George's pillow.

"What's wrong?" George saw what time it was and started to get dressed. After a few seconds, he went over to the bed and sat on the edge and moved a few strands of hair from Pia's cheek. She was a basket case.

"Tell me," he said quietly.

"He's dead. They're both dead."

"Who? Rothman? Yamamoto?"

"Yes."

"Oh, Pia, I'm sorry. I'm so sorry." George laid his hand on Pia's shoulder. "Pia, I don't know what to say. God, this is really a tragedy. From what you've told me, they were on the verge of something huge. What a setback for regenerative medicine, probably years, maybe even a decade. There's no one to fill their shoes."

Pia was silent. George removed his hand. She turned to look up at him. Her face no longer looked blank, just angry.

"Right now, I don't give a flying fuck about the future of regenerative medicine. Jesus!"

Pia sprang up from the bed and stormed out of the room. George followed her, tucking in his shirt. At first he couldn't see her but heard footfalls receding down the stairs.

"Pia, wait up!"

George ran down the street after Pia, who was speed-walking toward the dean's office. He caught up with her and jogged alongside.

"Pia!"

Pia waved her hand, dismissing George.

"Look, I'm sorry."

She stopped dead in her tracks, clenched her fists with her arms by her sides, and let out a low scream in exasperation. She turned and looked George right in the eye.

"George, don't tell me you're sorry. Please just *shut up!*"

With that she walked off and left George standing in the street like a just-dumped lover. George's shoulders fell. Condolences clearly weren't his strong suit although he hardly thought he deserved such a brush-off. He thought back to his long night of self-examination. If this woman needed him, especially now, she sure had a funny way of showing it.

30

Columbia University Medical Center
New York City
March 24, 2011, 7:00 a.m.

Pia and George arrived separately at the suite of offices of the dean of students. They were each buzzed in by the dean herself as the secretary, due in at eight, had yet to arrive. They sat at either end of the leather couch outside the office proper, avoiding eye contact, not saying anything. For Pia, such silences came naturally, while for George, who'd happily talk to anyone, it was a strain not to communicate. At the same time he had no wish to be told to shut up again by Pia, which was what he was certain would happen since all he wanted to do was apologize, again, for inadvertently upsetting her. It was part of his nature to feel responsible.

A few minutes after the hour, Helen Bourse emerged from her office.

“Thank you for being on time this morning,” Bourse said, beckoning Pia and George to follow her. She gestured to a couple of straight-backed chairs for them to sit in. She had seen a note in Pia’s filc, which she’d read overnight, written by a particularly zealous clinical preceptor during her second-year introduction to surgery. It said that Pia had difficulty arriving on time for early-morning appointments, even after being admonished that such behavior was not tolerated in surgery. Although it was usually only about five to ten minutes, it was consistent, and the preceptor indicated that she thought it was a serious lapse.

Dr. Bourse sat down and regarded her two students. “To begin, I’m afraid I have some very bad news.” Her voice had the gravity appropriate to the situation. “There’s no way to say this other than just to say it. Doctors Rothman and Yamamoto both died this morning. They were being prepped for surgery for rapidly developing peritonitis, but they didn’t make it.”

“I know,” Pia said.

“How do you know?” Dr. Bourse was confused. She had just learned it herself.

“I went to the infectious disease unit this morning. I woke up early. I thought maybe the new antibiotic would have had an effect, but I was told they had died."

Dr. Bourse stared at Pia, whose voice sounded as if all the fight had gone out of her. She could see that the young woman's eyes were rimmed with emotion and fatigue. Dr. Bourse sighed. Here was yet another example of Pia's marked willfulness, as she herself had told both Pia and George in no uncertain terms to go back to their rooms, check their temperatures regularly, and stay there until this meeting. Yet Pia had willfully ignored the order.

Dr. Bourse sighed again, still looking directly at Pia, whose eyes were, as usual, diverted. "Okay, I'll try to overlook the fact that I told you to stay in your dorm room. I gather you went to the hospital because of your closeness with Dr. Rothman?"

Pia nodded. She had an urge to admit that for her, Rothman had morphed into the father she never had, but she held her tongue. It wasn't in her to be forthright with her secrets.

"At least you made no attempt to go back to the lab, or did you?"

George looked quickly at Pia, worried. The thought that Pia might try to go back to the lab without him had not occurred to George.

"No, I didn't," Pia said quietly, and

George exhaled.

"Did you both check your temperatures as I asked?"

Both students nodded, although George had had to abandon his thermometer when Pia flew out of his room that morning.

"And I assume everything was fine. Okay. Rothman's and Yamamoto's deaths are a blow to everyone here at the medical center, particularly the school. I knew Dr. Yamamoto a little, and he was a fine colleague. Dr. Rothman I knew better, of course, and I understand you got along well with him, Miss Grazdani. He certainly took a keen interest in your progress and afforded you more privileges than he did other students." And any fellows, thought Dr. Bourse. "I viewed that interest as a compliment to your abilities as a researcher and to the potential that Dr. Rothman recognized in you."

Pia was staring at the ground.

"Of course it's terribly ironic that Dr. Rothman, who spent so much time uncovering the pathogenicity of salmonella, should die from the same organism he'd come to understand so well. . . ." Bourse let the thought drift.

"So, Miss Grazdani, I have arranged for you to do research, starting today, with Dr. Roselyn Gorin, who is one of the most talented people on campus. She has a Lasker Prize, as you may know, and she's doing absolutely groundbreaking work on the differentiation of stem cells into specific adult cells. Roselyn is a friend of mine. She's a wonderfully warm and understanding person. I just talked with her ten minutes ago, and she's very happy to take you on. *Happy*'s not the right word in these circumstances, but she's very willing to help."

Dr. Bourse smiled hopefully.

"Today? I can't possibly start today," Pia said.

George winced as it was apparent to him that Dr. Bourse's first reaction was of intense irritation.

Dr. Bourse paused in order to gain control of her emotions, as she was as irritated as George had sensed. Roselyn was a friend, but in fact she hadn't been overjoyed about taking on a new student, especially Rothman's student, who had built up a reputation of her own, fairly or not. Dr. Bourse wanted to tell Pia to get her act together, but she held her tongue.

"I'm very grateful for what you've done," Pia said quickly, trying to sound sincere. "Really," she added, as if she sensed she was pushing it. "But I just got this news an hour ago, and I really can't think straight. I need a couple of days. To get my head together."

Dr. Bourse sighed again. Pia was not the easiest individual to get along with. At the same time what she was saying was undoubtedly true. Everyone at the center was going to be shaken by the deaths. Dr. Yamamoto was a very popular presence and even if few could tolerate Rothman on a one-on-one basis, his death was still a blow, especially under these circumstances. He was, after all, the center's scientific celebrity.

"Okay, Miss Grazdani. Today is Thursday. Monday morning, first thing, I shall expect you to resume your responsibilities as a fourth-year medical student. I am also reminding you to stay away from Dr. Rothman's laboratory. This is compassionate leave, not an opportunity for you to play epidemiologist again. We have real epidemiologists who are qualified to do the work. Do you understand?"

Pia nodded.

"Please say, 'I understand,' " Dr. Bourse said. She wanted to be absolutely clear.

"I understand," Pia said, almost inaudibly.

"Mr. Wilson. You'll return to Radiology today —"

"Absolutely, Dean," George said, cutting her off.

"And you will also cease enabling Miss Grazdani. Perhaps you might want to ask yourself why you, with a hitherto spotless record, are drawn into the kind of behavior we saw last night. *Gnothi seauton* — do you know what that means? It means 'Know thyself,' and it's something that we as physicians need to remember always.

"I doubt that it was your idea, Mr. Wilson, to break into Dr. Rothman's lab, and I hope that in the future you will let your actions be guided more by your intellect than by your id — by your cerebrum more than your hypothalamus."

George eagerly nodded agreement.

"Everyone clear?" Dr. Bourse said. Pia and George nodded in unison.

"Well, thank you. You can go."

Dr. Bourse watched George hold the door open for Pia, who left without acknowledging George. She acted as if George holding the door for her was a matter of right.

Dr. Bourse sat at her desk for a few minutes of contemplation. Since a large part of her job was to get to know her population of medical students at the Columbia University College of Physicians and Surgeons, she thought about the strange relationship between Pia and George. Of course, fraternization between students was not necessarily encouraged, nor was it discouraged, provided it did not interfere with performance. In this romance, it was pretty obvious what he saw in her as she was the source of considerable gossip around the center as a particularly beautiful, intelligent, but enigmatic young woman. What wasn't so clear was whether there was any reverse attraction.

Relations between staff and students, on the other hand, were officially frowned upon, but it was difficult to enforce a ban when the parties were all consenting adults and the students mostly in their late twenties. There had been persistent rumors concerning Pia Grazdani and Dr. Rothman. Again, Pia's exotic beauty and obvious intelligence were lost on few people, but what she might possibly have seen in him was beyond most people's comprehension. But nothing was ever substantiated and while there was every reason to believe that Dr. Rothman had given his student significant responsibilities and privileges, there was never any evidence that he had done so inappropriately. And now, Dr. Bourse thought, the conundrum of their relationship would just have to remain one of life's little mysteries.

31

Columbia University Medical Center
New York City
March 24, 2011, 12:10 p.m.

After leaving the dean's office, Pia spent the rest of the morning stewing in her room. Her feelings were in absolute turmoil, emotions and thoughts swirling in and out of her consciousness so that one minute she felt utterly depressed, the next she was sharp and motivated. She was still angry at Rothman for getting sick, and also at Springer, the chief of Infectious Diseases, for letting Rothman die, which she believed he did by selecting an outdated antibiotic despite the supposed sensitivity testing that Rothman himself had done. And why did it have to be her, a lowly fourth-year medical student, who made the diagnosis of the incipient peritonitis, the onset of which had heralded the man's death?

But most acutely, Pia knew she was devastated by the simple fact that Rothman was dead. She was depressed at the thought of what it meant to her future. Pia had a lot of practice at dispassionately breaking down a situation and identifying how it affected her. It was the emotional aspects of her state of mind that she struggled with the most.

Pia had been completely convinced by Rothman's argument that she wasn't suited to clinical medicine, since she actually didn't like most people, especially when they were sick and complaining about it. She had little sympathy for illness and none at all for any kind of complaining. At one point during her residency, after thirty-six hours without sleep, she had had to take blood from a patient, a tough-looking young policeman who was deathly afraid of needles. As the man wriggled and squirmed and Pia couldn't for the life of her find a vein, she told him to "quit whining like a baby." Fortunately, no one overheard her, and the man didn't complain, although the staff did wonder why he made such an effort to avoid Pia for the rest of his stay.

With Rothman gone, she didn't know if she had it in her to follow through with getting a Ph.D., which was a requirement if she was serious about pursuing a future in research. Most of all, the nagging issue of what had happened to Rothman wouldn't leave her alone, and she mulled it over continuously. It could have been an accident, she knew, but it seemed very unlikely knowing Rothman as well as she did. He was too careful, too compulsive. And the two of them getting sick simultaneously? It didn't make sense. But the alternatives seemed equally unlikely, especially the idea that he did it deliberately. The only other possibility, namely the idea that someone else did it deliberately, like Panjit, who might have had the opportunity, seemed even more unlikely.

The longer Pia sat in her room, the more agitated she became and the more motivated to take some action. But what? She paced around the room as best she could, given the close confines of the living area. She lay on her bed, but that was intolerable. She thought about calling Will or Lesley, but she didn't know what she would say to them. She walked down the corridor to the soda machine, but she didn't want anything to drink. Her mind was buzzing, racing, overheating.

Suddenly Pia knew what she could do to get herself centered, to regain some focus. George had been hanging around, offering to help. She thought there was something he could do for her, as he'd done a few times in the past. George wasn't so different from any other man she'd ever encountered. But each time it had happened, when Pia thought George would have been put off by her needs, there he was, back again the next day.

Pia imagined that George would be on his lunch break, one of the benefits of a rotation in something predictable like radiology or pathology. There was a predictable schedule. She wanted to call him, but she couldn't find her cell phone. And when she did find it, in the pocket of her coat, she saw that the battery was dead. She plugged the phone into the charger and called George, catching him as she had hoped on his way over to the cafeteria.

"I was going to swing by later to make sure you were okay," George said. They hadn't parted on the best of terms after meeting with the dean, and George's perennial insecurity about Pia had surfaced again.

"You offered to help. Is that still the case or are you still mad at me for getting you in trouble?"

"I'm not mad at you, I'm just worried about you."

Pia rolled her eyes.

"So you'll help me?" This was awkward.

Pia wanted George to say yes, I'll be right over. Instead, he said, "Not if it means going back up to the lab."

"No, George. What I'd like you to do is come over here for a few minutes."

"Right now?"

"Right now, George! I assume you're on lunch break?"

"Okay," George said. "I'll be right there."

Pia prepared. Almost to the minute of his expected arrival there was a knock on Pia's door. She pulled it fully open.

George's eyes sprang open to their fullest extent. He was clearly taken aback. Nervously he glanced up and down the dorm hallway to make sure no one could see what he could. Pia was standing in the open doorway buck naked.

"This isn't quite what I expected," he managed, as Pia pulled him into the room. She was shockingly deliberate, as she had been on previous occasions, and again, as on those previous occasions, he didn't resist. Under the circumstances she was a force greater than him, and he was powerless. Pia grabbed at the belt in George's pants, and he obliged. She then pulled his sweater and T-shirt over his head. Pia pushed him onto the bed and handed him a condom just as she had on the other occasions. He was ready — achingly so — and Pia got up on him immediately. She closed her eyes and looked up, rocking herself rhythmically and hard against him. He knew it was simply sex, that she was looking for the endorphin rush, and she found it fast, shuddering slightly as she did so.

As soon as she was finished, Pia put her hands on George's chest and slipped off him. She looked right at him, but it was like she didn't see him. "Thanks. I needed that," she said. She walked over to her bathroom, turned on the shower, and, after a couple of seconds, jumped in.

George put his hands behind his head and looked down at himself for a few beats. He then slipped off the condom, walked into the bathroom, and flushed it away. From a birth control perspective, it had been a waste. Pia had finished showering and was toweling off. George couldn't help but admire her athletic body, exquisitely shaped breasts, and deep, flawless, honey-colored skin.

"Would it kill you to kiss me?" George was bemused; he didn't know what to think. He was being used, he knew, and didn't understand why.

"I don't like kissing. Doesn't do anything for me."

George could tell Pia's mind was already elsewhere. There was no point in him saying, "Well, what about me?" He could hear her reply: "What about you?" George didn't know what else to say. Each time they had sex, George hoped it meant they'd made a breakthrough, that their relationship had climbed out of its curiously stalled state into a level of true intimacy. But it had never been the case. Nor was it now. She was a train running on a totally separate track. In many respects his role was irrelevant, as if it could have been anyone lying there.

"Thanks," Pia repeated airily as she passed him coming out of the bathroom. There was no modesty, whether pretend or real. In her upbringing there had never been an opportunity for even pretense.

"What for? I didn't do anything."

"No, you did! Really. You've given me a reboot like what needs to be done with a modem once in a while. You've made it possible for me to see what I have to do, rather than sit here paralyzed."

"Is that what it was? I want . . . I want us . . ." George felt like that hopeless teenager again. Pia was dressing quickly. George was standing naked and felt very self-conscious. He slipped on his boxers. "So tell me. What are you going to do?"

"Get in more trouble, I expect."

"What does that mean?"

"You should just leave, George. My problem is I don't think Rothman was treated correctly, whether people believe it or not. There was something wrong about how he got sick and there was something wrong about how he was treated. Chloramphenicol? It's almost never prescribed these days. Third-generation cephalosporins are where it's at now, so why give him something old that potentially causes truly catastrophic side effects?"

"You told me yourself. They used it because of Rothman's own sensitivity studies."

"That's what they said. He shouldn't have died, period, yet he was dead within what, fifteen, sixteen hours? He got sicker in the hospital — there was no delay in treatment, he was taken straight to the ward right after he showed the first symptoms. I think the treatment made him worse."

"Okay, I understand you're frustrated," said George, "but the dean told you directly not to interfere. Not to play epidemiologist. Do you want to get kicked out of here in your fourth year?"

"I've got time off, I'm not sitting here, I'll lose my mind. I'm going to talk to Springer about the treatment and why it didn't work. No one said I couldn't talk to him."

"Springer! Everyone knows he hates med students. By reputation he was second only to Rothman. You pull him as a preceptor for your rotation in internal medicine and half the students try to switch within a week. And the other half are lining up on the roof to jump off the building. Not to mention the fact that you've already pissed him off."

"Don't worry, George, I'll be my normal diplomatic self."

"That's exactly what I'm worried about."

32

Columbia University Medical Center
New York City
March 24, 2011, 2:05 p.m.

As Pia sat and sat in the narrow waiting room of Dr. Helmut Springer, her determination to see him didn't waver. Her tryst with George had succeeded in establishing in her mind what she needed to do. She had a burning need to know two things. The reason why Dr. Rothman became sick was one issue; another was why the vaunted and lauded Columbia medical staff had, in her mind, apparently screwed up his treatment. She knew she was only a fourth-year medical student, but from her perspective she couldn't come up with a compelling reason why Dr. Rothman and Dr. Yamamoto should have died at all, let alone died less than a day after the men were admitted to the infectious disease ward at the hospital. It wasn't as if they were in some backwoods operation — this was one of the absolute premier medical institutions in the world.

Though Springer probably wouldn't be happy to see her, she was hopeful that if she talked with him he could aid her quest to find out what had happened. He was, after all, a world-renowned infectious disease specialist. She knew his reputation of not treating medical students with anything close to respect and she knew their meeting the day before had not ended well; still she was optimistic. If he didn't know that she was the one who first recognized Rothman's incipient peritonitis, she was going to tell him herself, thinking it should count for something.

After forty-five minutes of waiting, Springer's receptionist finally announced to Pia that the doctor could see her now. Pia quickly entered his office. Springer was at his desk facing into the small room. There were no other chairs; it was Springer's way of keeping meetings short.

"Dr. Springer, I'm sorry to bother you again, and I know I annoyed you the last time we met. I apologize for all that. But I'm a medical student, and if I can't learn from my experiences, then I'm a pretty poor excuse for one. And I apologize for questioning —"

"Yes, yes," Springer said, cutting Pia off midstream. Her apologies sounded rehearsed and there was nothing resembling contrition in her eyes. Worst of all, his schedule was completely full with residents, at that moment, awaiting his arrival in the emergency room. He cleared his throat. "From our last chat I suspect you believe you know better than some of the foremost authorities in the land what has taken place here. Well, I want to disabuse you of that notion. Also I'd like to say that I wouldn't have even taken the time to see you this morning were it not for the fact that you discovered the early signs of peritonitis in Dr. Rothman. Dr. De Silva told me about a medical student who she assumed was on rotation catching rebound tenderness in Dr. Rothman's abdomen, which had not presented itself previously. We'll overlook the fact that this medical student was not, in fact, on rotation and had essentially broken into the ward and was wholly unauthorized to approach the patients. Of course, I later learned that this medical student was yourself."

It took Pia a couple of seconds to realize that Springer was paying her a slight compliment, even if it was cloaked in a sardonic reprimand. Pia took it as an opening. "I fully admit it was, and perhaps I shouldn't have been there," she said. "But it was an important discovery with important consequences. The man was clearly getting worse, which makes one wonder why the original antibiotic was chosen."

"Please," Springer said, his face empurpling. "This is where we left off last time. I just indicated that we were grateful for your help, and now you're back with this nonsense. I can't win. Again, there is nothing to indicate that chloramphenicol wasn't doing the best job under the circumstances. And, as we have said about fifty times, we were informed by the sensitivity studies carried out by Dr. Rothman himself that it was the correct choice of antibiotic. We are working under the assumption that Dr. Rothman's work in the study was as thorough and accurate as was customary."

Pia couldn't believe what she was hearing. Was Springer attempting to shift some of the responsibility onto Rothman? In this case it seemed especially crass to even suggest blaming the victim. "So how come, considering those sensitivity studies, neither Dr. Rothman nor Dr. Yamamoto showed any sign of response to the chosen antibiotic?"

Springer closed his cycs for a moment.

"The answer to your question is simple. The virulence of the involved strain of salmonella overwhelmed both the antibiotic and the patients' defenses. Remember, antibiotics, contrary to myth, do not cure. It is the patient's immune system that cures. Obviously with Rothman and Yamamoto, their immune systems were completely overwhelmed. Simple as that."

Pia started to respond, but Springer cut her off. "Listen, we've been over this issue. And let me add that a department head at this hospital does not have this kind of conversation with a medical student. A department head does not have this kind of conversation at all — there are protocols to be observed, there are panels that are convened if there are questions about the diagnosis or treatment. It's not clear in this case that there are any questions. Jesus, why am I justifying myself to you? This is not how we conduct business around here."

Pia wasn't picking up on Springer's rising sense of outrage. She had him in the room and she wanted answers. "Why weren't Rothman and Yamamoto being monitored more closely?"

"They were being monitored extremely closely. Each had his own nurse."

"Closely? How did it happen that a medical student had to pick up on the signs of developing peritonitis?"

"That was a fluke. It would have been picked up very quickly. Trust me. Now, is there anything else I can help you with, any other hospital policy you might want to critique for me?"

Springer's sarcasm was lost on Pia.

"This case confuses me," Pia continued. "In fact, it's one of the worst cases of salmonella or typhoid that I've ever come across."

"In your vastly broad experience," Springer said.

"In my experience, yes."

"Well, what are you alluding to? I'm sure you're alluding to something. So enlighten me, please."

"One of the first things they told us when we got here was about diagnosis. 'When you hear galloping hoofbeats, you should think of horses, not zebras.' "

"Yes, of course, it's the oldest saw in medicine. What about it?"

"Should we be looking for zebras here, Dr. Springer?"

"*We* are not looking for anything here, Ms. Grazdani. But I am dying to know what it is that you are looking for. So enlighten me again."

"Okay. Is it possible that this case represents some exotic form of an antibody/antigen reaction the body can have, like a Shwartzman reaction? In which case would it not have made sense to use Decadron or some similar anti-inflammatory agent, something potent, to try to head it off at the pass?"

"If that is your great revelation, well, I'm sorry to say it's not much of one. Because we used Decadron in the evening when it became clear that the two researchers were approaching extremis. Perhaps you should review the patients' charts before making accusations like that."

"Of course. If I had been given access to the charts I wouldn't have made the mistake. But I'm not making accusations, I just want to get to the truth, Dr. Springer."

"We all do, Ms. Grazdani."

Springer was suddenly overcome with fatigue. Talking with Pia Grazdani was frustrating, and he had more people he was going to have to deal with that afternoon who were going to be even more of a burden. There would be the inevitable press and the patients' families. It was not going to be a good day, since ultimately, it was the patients he cared about.

"Do you think perhaps there could have

been yet another bacteria involved besides salmonella, a bacteria or a virus that was being covered or camouflaged by the salmonella? And maybe it was this bacteria that was totally resistant to chloramphenicol and was the real killer?"

There was silence while Springer tried to control his anger. This was simply too much. His eyes drilled into Pia's while she maintained her composure, waiting for an answer while looking down at her feet. Finally Springer exploded with bottled-up emotion.

"I cannot, for the life of me, imagine a more ridiculous scenario. We made the diagnosis by fulfilling Koch's hypothesis. The illness was caused by salmonella, whose presence we ascertained in multiple ways, but most convincingly from blood culture. We classified the strain in multiple ways as well, particularly by DNA analysis. The offending organism was, without an ounce of doubt, the alpha strain of salmonella typhi that Rothman himself had had grown in space with the cooperation of NASA. There was no other pathogen, for Christ's sake. The blood cultures only grew out the salmonella. Nothing else! Nothing at all!"

Undaunted, Pia changed the subject on a dime. "What about the hair loss? Does serious salmonella infection cause hair loss?"

Springer was having difficulty controlling himself, yet the woman seemed completely calm. "The stress of almost any severe illness, particularly one presenting with high fever, can cause hair loss. Anyway, what hair loss are you talking about?"

"I saw hair loss with Rothman before I discovered the rebound tenderness. The resident suggested it could be attributed to the chloramphenicol."

"That's not something I am aware of," he said. And then, suddenly and angrily, "Oh, for God's sake. You wait here!"

Springer bounded out of his desk chair, pushed past Pia, and disappeared. Pia stood in the room and waited. Within a few minutes, Springer reappeared and sat down, giving Pia a nasty look. Thinking she had probably maximized what she was going to get out of the conversation, Pia eyed the door.

"I told you to wait," Springer said. "Stay there!" Confused, Pia did as she was told. There was silence except for Springer's labored breathing. *The man's boiling,* she thought. *I'm not getting anywhere.* Pia eyed the door again.

"Dr. Springer, I sincerely thank you for your time."

"Stay where you are!" Springer said brusquely.

Pia rolled her eyes, confused. *First he can't wait to get rid of me, now he wants me to stay. . . .* Then, bursting through the door came Dr. Helen Bourse, dean of students.

"Ah, Dean Bourse, it's simply not possible for me to do my job if I am to be hounded by a medical student who thinks she should be running my department. She goes onto the floor and sees patients with no authorization, which I am sure could open us up to all manner of liability issues. She repeatedly questions my medical ability and second-guesses decisions that were made, and now she's come up with an outlandish suggestion that we might have completely missed another organism which was responsible for Rothman's and Yamamoto's untimely deaths. First it was the choice of antibiotic, now it's a second pathogen. This is outrageous and it has to stop."

Pia looked at Springer, and she was unable to conceal the contempt she was feeling. He had run off like a coward and got the dean to come tell her off. She glanced at Bourse, who was standing arms akimbo, a hard expression on her face. She was angry and dumbfounded.

"I would like Dr. Springer to understand that I'm not trying to do his job," Pia said in her defense. "I'm just trying to answer some questions that I would have thought were important ones. My sense here is that something is wrong."

Neither Springer nor Bourse could believe the gall of the young woman. The question in both of their minds: Who the hell does she think she is? Springer found his voice first.

"Do you see what I'm talking about? This woman is off the wall. I'm going to talk with Groekest about the advisability of rescinding the position she was offered here as a resident/Ph.D. candidate. This is absurd."

At the mention of the chief of the internal medicine department, Helen Bourse signaled with a snap of her head that Pia should leave Springer's office. Pia was happy to oblige. Bourse then nodded to Springer to indicate that she had the situation under control. "I'll get back to you. Sorry about this." Bourse then followed Pia out of the office and into the hallway. Pia might be temporarily unbalanced, but Springer was a bully, and he'd made his point. Before Pia had a chance to say anything, Bourse lit into her.

"What the hell do you think you're doing?

When we spoke this morning, and I gave you time to get your head together, I don't believe I said you should go see Dr. Springer and belabor the head of Infectious Diseases about his patients or his diagnosis. Where is your social sense? Good grief, woman! It's common knowledge that Springer is not a fan of medical students in general, but this episode has pushed him over the edge. I have never heard him as exasperated as he was when he called."

Pia started to speak, but Bourse wasn't done with her.

"You're fast developing a reputation as a troublemaker, Ms. Grazdani, and that will not look good on your résumé if it gets recorded. You are here, essentially, as a guest of the institution, and guests do not behave like this. If they do, they're usually asked to leave. I gave you a couple of days to get over the tragic death of your mentor and that wasn't supposed to be time for you to come in here and stir things up again."

"But don't you think these questions need to be answered?"

"No, I don't, not if he doesn't," Bourse said, gesturing at the door.

Pia started to speak again, but the dean had had enough. "Have you shown any signs of a fever?"

"No."

"Then get yourself back to your room. If I hear you're causing any more trouble over this unfortunate affair, I will think seriously about rescinding your welcome here as a medical student. Which would be something of a tragedy for you, considering you only have a couple of months left before you graduate. And it would be a tragedy for us because we'd be admitting we made a mistake in taking you in the first place. I don't think Dr. Springer will go to Dr. Groekest on his own, but he might. So be careful, young lady. You are now officially on very thin ice. I must not have made myself clear last time we spoke. Am I making myself clear now?"

"Yes," said Pia. "Perfectly."

33

Columbia University Medical Center
New York City
March 24, 2011, 4:45 p.m.

For Pia, being in trouble was as natural as breathing in and out. She'd spent most of her life under some kind of probationary supervision undertaken by people who didn't know her, care to know her, or understand her situation. Long ago, Pia had wondered how it was that she was the one who ended up in front of some panel or other. She never instigated the trouble, she was always reacting to someone older and more powerful trying to take advantage. Somehow that fact got lost in the paperwork. More often than not, it was only she who suffered the inquisition and the punishment that followed. For her, injustice belonged with pain on the same one-way street.

By the time she reached age twelve, Pia

had simply stopped questioning the way of the world as it concerned her. This was just how it was and how it was going to be. Over the years she'd come to know how the individuals of influence in her life operated. Dr. Springer was a familiar type. He was fiercely protective of his own reputation and would adopt any position that protected him, even at the expense of reason and fact. Quick to take offense, Springer had no backbone. When Pia pushed back and kept pushing, Springer literally ran away. He went and found someone who did have some fortitude — Dr. Bourse — and he hid behind her. Bourse was a different proposition. She wasn't afraid, Pia could see that, and she wasn't willing to take the easy way out and simply dispense with the problem — Pia — which she could have done.

Pia had spent the afternoon anxiously mulling over Springer's behavior. She'd learned nothing. There was also the fact that no one seemed concerned about the medical issues she was raising, which couldn't help but fan Pia's semi-paranoid belief that the medical center, and the Infectious Diseases Department in particular, hadn't looked after Rothman and Yamamoto properly. And how could anyone prove that some medical center personnel had nothing to do with their getting sick in the first place? Pia was starting to consider the idea that some kind of cover-up might be under way, orchestrated by Dr. Springer.

And Rothman was still emotionally in her head. If she hadn't let him take such an influential role in her life, she wouldn't have found herself in her current predicament. If you let people into your life, she thought, you were bound to hurt sooner or later.

A knock on her door jolted Pia out of her agitated state. It was George. Who else?

"What happened with Springer? I was so worried I couldn't concentrate all day."

"It was a disaster."

"I'm sorry. I'm also sorry I didn't offer to go with you, really I am. You shouldn't have to do this all by yourself."

"George, stop saying you're sorry, please! Besides, I never expected you to come with me. In fact, I never gave it a thought. And after what happened, I'm glad you didn't come. Springer was a lot angrier at me than he was the first time. He went and fetched the dean to tell me to stop interfering. And he threatened to go to Groekest if I didn't."

"So, are you going to?"

"Going to what?"

"Stop interfering."

"How can I? They're the ones interfering, covering things up, not me. They're sitting on something, I'm sure of it."

"If you don't mind me saying so, that sounds rather paranoid."

"So be it. And remember, even paranoid people have real enemies."

"So you got confronted by the dean again?"

"Afraid so."

"What did she do?"

"Bawled me out big-time. Gave me a lecture about being a troublemaker. Threatened to have me kicked out of school."

"Shit!" George commented.

Pia checked her watch. "Actually I was just about to go back to Rothman's lab. I'm just waiting until it's late enough. I don't want to run into anyone, especially not the dean."

"Pia, the last time I looked, the dean isn't working security. They have a whole staff for that, and they caught us in about five minutes the last time we went to the lab. Bourse made it pretty clear you're not supposed to go back to the lab. Now she's spoken to you a second time. Maybe they're right. You *are* crazy."

"I think I have an aptitude for science, George. There are facts here, evidence that doesn't add up. No scientist is going to just walk away from that."

"Then tell me this: What are you going to do when you get kicked out of this place? That would make you an ex-scientist. Or not even. More like an ex-almost-qualified scientist. I don't think there's a great deal of demand for them in the job market today. You're going to graduate in a couple of months, if you're lucky. Yes, Rothman's death is a bad experience, a terrible one, but you might be compounding it and throwing away a career before it's even started."

"Career? Right now I don't see that I have a career. And I couldn't live with myself if I gave up now. Do you know if Rothman's lab is still officially closed?"

"How would I know? Well, I do know it's closed to you."

"The epidemiologists must be done by now," Pia said, ignoring George's point. "If they're not still checking the place out, there's no reason I can't go. I do have stuff I left in there. The dean was upset that we went in when it was still officially closed. If it's still officially closed, I won't go in, I promise, but if it's not off-limits, I will. At the very least I need to check the contents of that storage freezer in the level-three biosafety unit, which we didn't get a chance to do last night, remember? I'm one of the few people who knows the code that Spaulding uses in the logbook for the storage freezer. I want to be sure that all the samples that should be in the freezer are in the right place."

"Who's Spaulding?"

"The head lab technician. Rothman and Spaulding used to argue about the state of the storage freezer. Rothman thought the freezer was a mess, Spaulding disagreed. Rothman was thinking of sacking him. But that wasn't unusual — everyone thought they were about to get fired. Spaulding was the only one who pushed back."

"This is all very interesting and maybe you might find something is amiss in the storage facility. But even if you did, then what? Remember, it's not Rothman's lab anymore. All that is history. Unfortunately. And *you're* going to be history if you keep doing what you're doing. And are you really suggesting that the senior lab tech might have had something to do with Rothman's death? That's crazy."

"Actually, I don't know what I'm thinking. I do have some wild ideas, like these two scientists plotted together to carry off a dual suicide."

George looked at Pia with consternation.

"I'm kidding. I'm kidding. But there are so many things bouncing around inside my head right now, so many theories, and I can't rule anything out. Maybe it's something someone didn't do rather than what they did do — what do they call that? A sin of omission, not commission. The only thing I *know* is that something about this whole situation is not right."

"Of course the whole situation isn't right, Pia — two people died. That can never be right. But it doesn't mean there can't be a simple and logical explanation for why it happened."

Pia thought for a moment. She considered opening up to George and talking about herself and her state of mind, but that was something she had always been loath to do. That was what she had done with Rothman and look where it had gotten her. She glanced at George's face. He'd been looking at her the whole time; she had been mostly looking at the floor. He looked less eager than usual and more serious. Pia took a deep breath. She decided she'd at least try.

"I don't want to think Rothman had anything to do with his own illness. But I'd like to be sure. If he did, it will mean that he let me down. In a real way, he betrayed me. Rothman was very important to me, and it's hard for me to admit that anyone has such an influence on my life. Now he's gone, I feel like I'm starting at square one. And I don't want it to be his fault."

George nodded, but he was having a very hard time understanding Pia's reasoning. Even if Rothman accidentally infected himself, why would that mean she should think less of him, that he "betrayed" her?

"It was Rothman's idea to start the research track of my training, and it was going to be under his direction. Who's going to do that for me now? I was going to be working in his lab for my Ph.D. Where am I going to go now? Once again I've been abandoned."

George was a little taken aback at what sounded like self-centeredness in the face of Rothman's and Yamamoto's deaths. "I'm sure the university will find you another lab," he said. "They found you another rotation. Will and Lesley are already doing theirs."

"Maybe they'll find me one, maybe they won't."

George hesitated for a moment. He knew there was a risk Pia would take what he was about to say the wrong way. But he decided to say it anyway. "Pia, I'm sorry, but I don't understand how Rothman could have 'betrayed' you, as you put it. He got sick and died. It's tough for me to understand you sometimes. I don't think you should insert yourself into this where you don't have to. If you're now thinking Rothman's death wasn't an accident and that there's a cover-up going on, I don't see any other way for this to end but badly."

"Unless it's true."

"You're talking about murder. Who would want to murder one of the best research teams in the country?"

Turning it over in his mind, George could think of only one reason why someone would be so willing to risk their career without blinking an eye. He knew for sure this line of reasoning was going to get him into trouble. "Look, it's none of my business, and I've never said anything to you that would make you think I was jealous of any other guy, er, getting close to you, but your relationship with Rothman, well . . ."

George was cut off by Pia's loud laugh.

"Oh, God! Is that why you think I'm caught up in all this? You think I was sleeping with Rothman?"

"No. Well. Yes. Maybe. I don't know. It might explain why you're so worked up about it. It's what some people are saying on campus."

"So I have to be sleeping with a guy to care about how he died? Thanks a lot, George. I did let him get close to me but not like that. Typical male. I'll say it if it helps: No, there was nothing physical between me and Dr. Rothman. Zip. Believe me, I can tell when a man is interested in me like that, and he wasn't. He was actually happily married and devoted to his family, despite how asocial he seemed."

Pia was steamed and George didn't know what to say. The thought he'd expressed had taken on a life of its own in his mind. But as soon as he had voiced it, he knew it was very unlikely. Now he was just embarrassed he'd even mentioned it.

"All right, that does it, I'm going to the lab," Pia said. "I really do have stuff there that I need to get. I worked there for more than three and a half years. And don't worry, if it's off-limits, I'll come right back like a good girl."

"And if it isn't?"

"Then I'll check out the storage freezer and get my stuff."

"I'll walk over to the med center and wait for you in the library."

"You don't have to do that."

"After Springer, it's the least I can do. Really."

"I guess I can't stop you."

George knew that was as close to an invite as he was going to get.

34

Columbia University Medical Center
New York City
March 24, 2011, 5:07 p.m.

Pia and George walked over to the Black research building and passed through security with their medical student IDs. They were bucking the tide, as it was past five o'clock and most of the staff was streaming out, having finished work for the day. Pia and George parted at the elevators, with Pia telling George she would come find him in the library when she'd finished her visit.

In the elevator, Pia was glad George wasn't with her. She'd be able to do what she wanted more quickly without him. She was pleased but not surprised that the caution tape was gone from the lab entrance. More good news: The door was unlocked, meaning the lab was officially back to normal. But the positive feelings were short-lived when she saw that a few of the familiar denizens of the lab had taken the same opportunity to show up and get on with whatever business was pressing. Marsha Langman was tidying up her desk, its previous pristine neatness a victim of CDC investigators who had gone through most of the lab's records. Unfortunately the head technician, Arthur Spaulding, was also there, for similar reasons, trying to get everything back to normal.

Seeing Spaulding was a disappointment. His presence precluded her plan to visit the biosafety lab. If he saw her, particularly if he saw her in the refrigerator storage facility, he would undoubtedly make a scene. Pia cursed under her breath that she hadn't gotten into the lab before any of the others. Neither Marsha nor Spaulding, nor any of the other technicians who were there, greeted her, or obviously snubbed her — it was like she wasn't there at all. It surprised her, because they were all going through the same trauma involving their bosses' deaths. It was like they were a group of automatons.

Pia headed toward the open door leading into her small office, thinking she might have to get her stuff, leave, and then return later that night to check out the biosafety microbiological storage freezer. In the process she practically collided with the maintenance man from the day before, O'Meary.

Obviously he knew Pia's name. "Miss Grazdani! Nice to see you again. We just heard ten minutes ago that we can get back in here tomorrow morning to finish up. I'm just checking the site, making sure all the tools are here." He then leaned toward Pia and whispered, "I'm not a hundred percent happy about being in here after what happened yesterday. But the job's gotta get done. You think it's safe in here now? Our boss says so."

"I think it's safe," Pia said. "I don't think it was ever unsafe."

"Good to hear." O'Meary straightened up to his full height. He jerked his thumb toward Pia's ceiling. "I think we've isolated the problems with the short. It's up there, so we should be outta your hair by lunch tomorrow."

Pia didn't respond. She doubted the problem would ever get solved. Besides, she wouldn't be there at noon tomorrow or any other day.

"I hope we won't be bothering you too much tomorrow," O'Meary said, trying to be considerate. He tried to go around her. Pia stopped him.

"I know you've only been here a couple of

days, but did you see anything unusual yesterday morning? Before all the excitement started. Anything that struck you as odd?"

"That Springer guy asked me already, and the Disease Control people. They took a long time about it."

"I'm sure they were very thorough, but I'm wondering, because you were in here all morning and in different parts of the lab with your wires, if you saw anyone early on who you didn't see in the lab later. Anyone who looked like he didn't belong."

O'Meary narrowed his eyes, but playfully. "What are you, a cop now?"

"No, I'm not a cop."

"I wasn't working in that bio unit where they got sick, so I wouldn't know about anyone in there. Are you sure this place is safe? The disease people were asking about contamination, like 'Before the contamination . . .' and so on. Is it really safe in here?"

"I'm sure it's safe. I came back and I'm not about to take a risk with that bacteria."

"So why the questions? You're making me nervous."

"I'm just making some of my own inquiries. Supposedly nothing abnormal was found here in the lab or in the biosafety unit. Did you see Dr. Rothman or Dr. Yamamoto at all?"

"I didn't even know which one was which. There was a lot of people in and out of this lab, dropping stuff off."

"Do you know Arthur Spaulding, the head technician?"

"Yeah, he was introduced to us when we first came on the job."

"Did you see him when you were in Rothman's office?"

"Sure, a few times. Quick ins and outs."

"Anyone else more than once?"

"The secretary, Martha."

"Marsha."

"Whatever, yeah. You know, you sound an awful lot like a cop."

"I'm not a cop, just a student who has a few questions. Sorry to keep you. But if you think of anything unusual, just find me."

"Are you gonna be here?"

"Actually, no. Let me give you my cell phone number. If you remember anything, please call. I rarely use it, but I'll pick up a message." *A message that I can ignore if it's not germane,* thought Pia. She usually made it a point not to give out her cell number. O'Meary wrote it down.

"Okay, I got it."

Over O'Meary's shoulder Pia saw Spaulding say good night to Marsha and leave. She silently applauded. She was now free to go and check out the storage freezer.

To be on the safe side, Pia did a circuit of the lab to see who else was around. A couple of the other support personnel were tidying up in the main lab area, and Marsha was busy around the front desk, but there was no one there who might go into the biosafety unit. Stepping into Spaulding's office, she took the microbiological storage freezer log from where she knew Spaulding kept it in his desk. In the anteroom of the unit she quickly suited up, and once inside the unit itself, she used her own key to enter the large walk-in storage freezer. The door closed automatically behind her. She was surprised the interior light was on, which was odd, because Spaulding was scrupulous about turning it off when he left. As Pia started to contemplate what that might mean, the door was yanked open. Pia's heart leaped in her chest. She found herself face-to-face with an equally surprised Arthur Spaulding.

"What are you doing here?" Pia asked quickly, feigning indignation.

"I came to turn off the light. More to the point, what are *you* doing here? This is off-limits to anyone but Nina Brockhurst, Panjit Singh, Mariana Herrera, and me. You know that. And how did you get in, damn it?"

"I have a key," Pia said, producing it and dangling it in front of her masked face. "Dr. Rothman gave it to me, and he gave me authorization to come in here."

Spaulding snatched the key roughly out of Pia's hand, causing her to start again.

"Maybe you didn't hear, but Dr. Rothman isn't around anymore to give you his authorization."

"I bet you're happy about that," Pia blurted. As soon as the comment left her mouth she regretted it.

"For now I'm in charge of this lab, and from this point on, you no longer have authorization. And I'll take that too." Spaulding snatched the logbook that Pia was holding.

Pia stood there for a couple of beats, deciding what to do. She'd recovered from the shock of Spaulding's unexpected appearance and now was just angry. She'd never liked the man. She brushed past Spaulding heading toward the freezer door.

"You're not the princess anymore, Pia darling. Your access to this entire lab is revoked, as I'm sure the dean would be happy to confirm if I asked."

Pia said nothing. In the anteroom she took off her protective garb and passive aggressively left them where they fell. Seething, she walked back into her tiny office and packed up the few belongings and files she'd accumulated in three-plus years. She found an empty box and put them in it. Without looking back, she closed the office door and walked toward the entrance to the lab. Marsha Langman didn't look up as she passed by. What a bunch of jerks, Pia thought.

In a huff, Pia headed back to the dorm, but then she remembered George in the library. Reversing her direction, she quickly found him and got his attention. She then left. He quickly closed the journal he was reading and followed her out into the hall. He had to jog to catch up with her. It was no secret she was angry.

"Can I ask what happened? You weren't gone long. Was the lab still off-limits?"

"Maybe it would have been better if it had been," Pia said. "I hope you're hungry because I'm starved."

"I'm hungry enough. Let's go to the cafeteria."

"Fine," Pia said.

They exited the hospital. It was cold and dark. They walked quickly.

"You still haven't told me what happened," George reminded Pia.

"What happened was that prick Spaulding surprised me at the storage freezer and ended up kicking me out of the lab for good. The dummy left the light on and came back to turn it off. My bad luck he's so anal. Figures, he's a huge asshole."

"You can't be surprised he got mad. And why do you care if he kicked you out of the lab? You're going to another lab come Monday. You can strike another staff member off your Christmas card list."

"I don't send Christmas cards."

"It's a figure of speech, you know what I mean. So now the lab's really off-limits, right?"

Pia nodded. "I guess so."

"So there's not much more you can do, unless you want to break into Springer's office."

Pia looked over at George with confusion.

"I'm joking. Seriously, there's nothing more you can do. You've talked to the doctor in charge of Rothman and Yamamoto's care and now you can't go back into the lab. You have to let it go and let the authorities do their work. You can be sure there's an investigation going on. So you should drop it. Yes?"

Pia wasn't hearing George's plea at all.

"Pia, are you listening to me?"

Far from dropping it, Pia was now wondering whether Spaulding was hiding something. Yet what could she do? And what was she going to do with the rest of her life? Without her mentor and his program, was research still a possibility? Becoming a physician had seemed to offer safety in her life, something she coveted. But Rothman had made her realize that her discomfort with patients, with other people in general, might not bode well for such a career. She was at a crossroads in her life and had no answers. Thinking about it made her anxious.

Pia let out a sigh and George asked her what was up. Pia ignored the question. She suddenly knew that her preoccupation with finding out what had happened to Rothman was allowing her to avoid thinking about her career and the decisions she was facing. It was her first line of defense. The future could wait. Pia stopped walking and pulled George to a stop just outside the dorm.

"I'm not giving this up. I've got to find out why this tragedy happened. There are too many questions. Every time I stop and think, there are more questions; more people acting strangely. The infectious disease people insisted on using a fifty-year-old antibiotic and the patients died within hours despite being diagnosed and treated. And no one, I mean no one, liked Rothman. All his colleagues were jealous of him having a Lasker and a Nobel and possibly being in line for another. Okay, Spaulding was mad that I was in the storage facility, which he has always been weirdly proprietary about, and he knew I'd been in his desk since I was holding the logbook. But he was acting bizarre, like the lab was now his. The moron is only a fucking technician. He's not a researcher. And what about the fact that I, a mere medical student, was the one to pick up the peritonitis? Maybe if they had gotten him to surgery sooner, he would still be alive."

"What do you want me to say?" George asked. He tried to look Pia in the eye but she avoided it.

"Does it really sound like I'm paranoid? Don't answer. Whatever, I'm not finished looking into all this as there doesn't seem to be anyone else doing it."

"How many times do I have to remind you that this is one of the premier medical centers in the world? Do you really believe you have anything to add? You're going to flame out, Pia. Is that what you want? Do you have a career death wish?"

Pia thought for a moment. "Maybe."

"Even if you persist in this self-destructive investigation or whatever you call it, I don't see what your options are. Springer, Bourse, Spaulding — you're on a last warning from all of them."

"I'm not afraid of Spaulding. He took my key, but I still know where Rothman kept his. He was acting like he had something to hide."

"As I said before, I think this is crazy. Look, if you insist on pursuing it, why don't you check out the autopsies? There's probably a simple pathological explanation for Rothman and Yamamoto's clinical course. Or even a complex explanation, I don't know. But that's where you're going to find the answers, not pissing off everybody in the hospital. The autopsies were probably done today so they could dispose of the bodies because they're still hot."

"Hot?" Pia said.

"Yeah, hot, as in contaminated with salmonella," George said sharply. Sometimes Pia could be a little slow on the uptake. She still looked bemused, as if her mind wasn't making a connection. "You understand what I'm saying?"

"Like you'd say someone full of virus is 'hot.' "

"Exactly. No Pathology Department wants to keep hot bodies around. I tell you what you should do: Find one of the pathology residents tonight and ask what they know, or what they can find out. You haven't worn out your welcome in that department yet. Have you?"

"Actually, that's a good idea. I didn't think of that."

"If you want to look into that, I'll come with you, so you stay within the lines." George smiled as he said it. He wasn't reprimanding her, he was joking because he knew he wouldn't be able to keep her out of trouble if she was determined to get into it. She'd more than proved it.

"But I'm not going back to the lab," he added. "If you want to do that, you're on your own. Spaulding is bound to have alerted security. And what does it mean if there's a sample missing in the freezer? What does that prove? Other than the fact that Spaulding's not as good at his job as he'd like to think. Which is another thing you knew already."

"I don't think Spaulding would alert anyone. He doesn't have that kind of authority, despite his big talk. But at the moment

I'm not thinking about going back to the lab. I think I'll follow your suggestion and check with Pathology. As I said, I hadn't thought of that. It's a good suggestion."

"Then I'll tag along. I know, I'd feel guilty if I don't go and you get carried away again and get yourself kicked out of school."

"I don't care one way or the other," said Pia. She was intrigued. She wondered why she hadn't thought of it.

35

Columbia University Medical Center
New York City
March 24, 2011, 8:50 p.m.

The wind blew fiercely down Haven Avenue and into the canyon of 168th Street. Pushed from behind, Pia and George made rapid but chilly progress from the dorm toward the medical center. It was raining, and George struggled with a cheap umbrella until it finally turned inside out for the third time. He stuffed it headfirst into a trash can. Pia strode on, head down, the hood of her sweatshirt pulled tightly over her head.

Once inside the hospital, they descended into the labyrinthine bowels of the nearly century-old building to the morgue. It was malodorous and poorly lit and filled with outdated equipment. The morgue served as a way station for the dead: In simple cases, bodies would be picked up by a funeral home. A van from the OCME, the Office of the Chief Medical Examiner, would come if there were complicating factors.

Both Pia and George found it difficult to reconcile this run-down, dirty place with the hospital and medical center as they knew it. Some campus buildings were slightly shabby on the outside but pristine and modern within. The morgue was unkempt inside and out, and at this time of the evening seemed to have been deserted by anyone from the realm of the living. Old-fashioned wooden meat-locker-style doors with metal signs proclaiming that only authorized personnel were to be admitted were the order of the day. The only noise in the godforsaken place was a low-grade electrical hum and the drip of water onto cement floor.

Following their noses, they walked into the tiered amphitheater of the abandoned autopsy room, which looked like the set of a horror movie depicting Victorian times. Some of the seats were broken. The pit area, with its two ancient autopsy tables, was being used as storage for a random collection of pipes, old sinks, and discarded toilets. With the constant fight for space in the medical center, George wondered out loud why the area hadn't been retrofitted.

Finally Pia and George walked into the

morgue itself. Arranged along the walls was a series of walk-in coolers. The fifteen or so corpses in the room were on separate gurneys, some covered by sheets, others fully exposed. Still in place on a few of the bodies were the various tubes and wires used to treat and monitor them while they were still alive. A couple of the bodies were clothed, others naked. Most still wore the hospital johnnies they expired in. Mixed in were a couple of long black body bags.

Pia and George were wondering why they couldn't find anyone working there when the night diener, or mortuary attendant, startled them.

"What do you want?" the man asked. It was obvious he was unhappy at being disturbed. He was perhaps fifty or sixty, small of stature, wearing a stained lab coat. He had a bulbous head too large for his body with a badly maintained comb-over wedged on top. He wore small oval glasses and squinted through them at his uninvited guests. The casting department for the Victorian horror movie had done a good job.

"And how did you get in?" he added before Pia or George could answer his first question.

"We came in that way," Pia said, indicating their entrance route.

"That's the back entrance. Visitors are supposed to come in the front. No one ever comes in the back way."

"We're here to ask about a couple of autopsies," George said. "Autopsies that might have been performed down here today. The patients would have been two staff members who died early this morning, Dr. Rothman and Dr. Yamamoto?"

The attendant laughed cynically, as if this was the funniest thing he had heard in ages.

"There hasn't been an autopsy done down here in fifty years. I never heard of either of those patients. They're not here, if that's what you want to know. And if there was an autopsy, it would be done in the anatomy department of the medical school, where they still do them. On account of the teaching. You need to get in touch with the pathology resident on call. And you can go out the front way." The man pointed in the direction of the usual exit. He then stood there implacably.

George looked at Pia, who didn't seem in the mood to argue.

"Okay," George said. "Thanks."

As Pia and George waited for the elevator to rumble its way down to them, George snuck a look back toward the chamber of horrors.

"Can you believe that guy?"

"I've been in some creepy places, but that's the creepiest."

"Do you think he ever leaves?" George said.

"You get the impression that he lives down here."

"I'll be happy if I don't see that face again."

"I should think so," said Pia. "If we come back down here, it means we're dead."

Back in the land of the living, George called the on-call pathology resident. Dr. Simonov agreed to meet with them and asked that they come up to the clinical path lab. When Pia and George found him, Simonov was taking a break in a small windowless office with a giant mug of strong coffee on the desk in front of him.

"So what can I do for you guys? It's not often I get called by medical students. What's up?"

Simonov was Russian but had lived in the West long enough to almost completely lose his accent. Only the occasional dropped article gave him away. He'd gone to both college and medical school in the States.

"We're wondering if there was an autopsy

performed today on either Dr. Rothman or Dr. Yamamoto or both," George said. He had suggested to Pia that he do the talking this time. She didn't care. "They died early this morning when —"

"Yes, I know who they were," Simonov said. "Everybody in the medical center knows about them. Why are you asking?"

"There are questions about how quickly they died," Pia said before George could speak. "It was a relentless downhill course despite maximum treatment so that we —"

"No autopsies were done on them here," Simonov said, cutting Pia off. "Not a lot of autopsies are done anymore in general. It's a pity, but it is reality. There's no money. But Rothman and Yamamoto, they would not have been done here under any circumstances. Having died like they did of an infectious disease while working meant that they are definitely medical examiner cases, pure and simple. All we did here was put the bodies in body bags, seal them up, and decontaminate the exterior. They were then picked up by the OCME." He spelled out the acronym, explaining it stood for Office of the Chief Medical Examiner.

"I know what the OCME is. So do you know the results?"

"Results!" Simonov laughed at Pia's question. “Maybe in three weeks or more,” he said. “They get a lot of bodies down there, and they generally take their time.”

“Down where?” Pia said. “Where exactly is the OCME?”

“You gonna go down there now? I wouldn’t advise it. But, okay, whatever. It’s on East Side, First Avenue and Thirtieth Street, near NYU Medical Center.”

“Thanks. If we call them up, do you think they’ll answer our questions?”

“Now?”

“Tomorrow.”

“How would I know? Maybe they never had medical students asking questions. But then again it is affiliated with NYU Medical Center, which is a teaching hospital. For all I know they may have a medical student elective.”

“Who should we call? Should we ask for someone in particular?”

“I knew one of the MEs, but he’s no longer there. But they have a PR department. I’d call them. Maybe call the medical legal investigator on the case.”

“Do you think they’d tell us the results if we called?” George asked.

“You mean call the ME’s office?” Simonov smiled and let out a quick, knowing laugh. “You think in this great big city bureaucracy you can just call and they jump and tell you results? Not in a million years. This case is important, they were important guys. It's going to be a media event. There's probably going to be lawsuits about safety, that kind of thing. Since it's an infectious case, the autopsies have probably already been done, but they're not going to release any results for three, four weeks after the toxicology screens are completed. But there's not going to be general access to the information, and they definitely won't give the results to a couple of green medical students."

"You're probably right," Pia said. She knew more than most people about city institutions.

"If I were you, find something else to do. But it's your life. If you insist on looking into the case, I'd go down there. I wouldn't try calling on the phone. If you're there in person and meet with someone who more or less takes pity on you or likes you, you might learn something." Simonov winked at Pia. She got his inference but ignored it.

"So if you're really committed," Simonov continued, "go to the OCME. Just don't count on getting any answers. As for calling, you might as well call three-one-one." Simonov was referring to the citizens' help line — people called to report a cat stuck up a tree or a loud movie set on the street. Simonov checked his watch and picked up his coffee.

"If you decide to call three-one-one, tell them there's still a big pothole on my street. Been there since Thanksgiving."

Back out in the rainy night, Pia and George slogged along 168th Street, keeping as far away from the curb as they could. Every time a yellow cab zipped by, it splashed water up onto the sidewalk.

"Well, that was almost useless," Pia managed against the wind.

"I'm not sure I'd write it off as useless. He reminded us about the politics involved. He also emphasized that there's undoubtedly going to be a thorough investigation as a prelude to any legal action. I think that's information you should take to heart. It's time to drop all this, Pia."

"Dreamer," Pia said. "I'm in this until I get some answers."

"You are impossible," George commented, as a sudden gust of wind came down from Haven Avenue, halting their forward progress for a moment. They had reached Fort Washington Avenue. Looking to the side, Pia realized they had come abreast of the Black research building.

"What time is it?" Pia asked.

George managed to glance at his watch. "It's after ten. Time for us to be in bed." For George the idea of bed had immediate appeal. It brought up the fact that they had had sex that day, or at least Pia had had sex. Ever the optimist, he wondered if just maybe, after his accompanying her back over to the hospital to check on the autopsy, she might consider a continuation. George closed his eyes and screwed up the courage to speak.

"Do you want to come to my room? Stay the night? Or we could go to your room, whichever you prefer."

"What for?" Pia asked, blankly.

"Well, for one thing, we ended things a little quickly this afternoon. Maybe if we had more time . . ."

"That's a thought," Pia said in a preoccupied fashion. "Have you noticed where we're standing?"

George looked up. In truth he hadn't been particularly aware of the immediate surroundings.

"We're just outside the Black building," Pia added. "It's after ten, as you said. I want to go up to the lab for another quick visit to check out that damn micro storage locker.

I'm not going to be satisfied until I do it, and this is the best time. I've been in there fifty times at night like this."

"No, Pia!" George said firmly. "It's too big a risk."

"I don't think there's any risk whatsoever. You head back to the dorm. It'll only take me twenty minutes at most."

George looked ahead at the dorm looming in the misty night. It beckoned as a haven of warmth and security. He looked back at Pia. She was smiling up at him, confident as usual. Most important, she hadn't said no to his suggestion that they sleep together. "You really think it'll be safe, that no one will suddenly pop in?"

"Absolutely. Twenty minutes it'll take me. I'll call you as soon as I get back to the dorm."

"And you remember that whatever you find out, it won't prove anything?"

"I'm aware of that."

George's mind went into overdrive. Maybe it was a good idea. Maybe if she got the damn micro storage locker out of her mind, she'd give up on her self-destructive investigation.

"All right," George said with sudden resolve. "I'm coming with you. Maybe I can help speed things up." He grabbed her hand

and started to pull her toward the Black building entrance.

Pia resisted. "Are you sure?"

"I'm sure," George said. In his mind's eye what he could see mostly was them climbing into bed, holding each other tightly.

Pia shrugged. "It might be quicker with two people. All right, let's do it."

Without another word, Pia and George ducked into the Black building. The security man knew Pia well and didn't bat an eye. Pia used her key to open the main door, a key Spaulding had not asked for. The logbook was back where she expected it would be, in Spaulding's desk. Inside the biosafety unit she used Rothman's spare key from his office to open the storage locker. They worked quickly and efficiently.

George wouldn't have wanted a physician to check his blood pressure at any point during the visit, but Pia seemed icy cool and focused.

Pia had George read out to her how many of each sample were recorded in the logbook while she counted the actual samples. As Pia suspected, there were three samples missing from the storage freezer, at least according to the book. There were supposed to be thirty samples of the zero-gravity salmonella typhi, divided evenly between what was called alpha S. typhi and beta S. typhi. One of the missing samples was from the beta salmonella strain and the other two were the alpha strain, which was the strain that had infected Rothman and Yamamoto. Out in the main part of the unit near the hoods, Pia found a small collection of six labeled petri dishes in the incubator. Each was labeled with either an alpha or a beta.

After Pia and George had left the biosafety unit and removed their protective clothing, Pia found two of the same type of stoppered containers used in the storage facility without labels sitting next to Spaulding's sink.

After replacing the logbook and the spare key, Pia said to George, "Okay, we're done."

George's heart rate calmed down once they had exited the lab without incident.

"What does all this mean, Pia?" George asked as they rode down in the elevator.

"I don't know," Pia admitted. "It might not mean anything, but information is information. What I'd like to do is go over the discrepancies with Spaulding if I can figure out how."

"Good luck with that," George said.

Pia and George fought their way through the weather back to the dorm. Though he was exhausted, George felt strangely exhilarated. He and Pia had actually worked together. George knew he'd been useful and was acutely sensitive to Pia's gestures, like the way she had put her hand on the small of his back to encourage him to precede her through the outer dorm door. She was obviously pleased with what they had accomplished. They stopped in the lobby and pushed the elevator button. Both cars were on upper floors.

Pia stared at the elevator's slow-moving floor indicator. George cleared his throat to speak, but Pia didn't want to hear what he had to say. She just wanted to get to her bed and try to sleep.

"Pia, you must know how I feel about you. I've tried to tell you a hundred times. More than that. Pia, would you look at me?"

Pia reluctantly turned to George. He had that earnest look.

"You know I worry about you because of my feelings for you. I love you, you must know that. I think about you constantly."

On hearing the words, something in Pia's brain fizzed into life. A laboratory animal learns to stop engaging in a certain behavior, like touching a red button, if it gets a painful shock every time, even if previously it got a reward, like a piece of food. In Pia's mind, there was a connection between protestations of affection and pain. She had learned that the people who said those words would cause her pain, and should be avoided, like an electric shock.

Pia pressed the elevator button again, as it appeared that the car was stuck on the eighth floor. She said nothing.

"Our relationship can't be totally one-sided."

"What do you mean, 'relationship' . . . ? Look, George, this isn't the time or the place for this."

"When is the time, Pia? I've wanted to tell you I love you for years."

The elevator finally arrived, the doors opened, and a cluster of students noisily piled out. A party had started in someone's room and now was moving to a bar up on Broadway.

George pulled Pia aside as the door closed. She rolled her eyes.

"George, come on. Not now."

"I'm sorry, but I have to say it. I know you don't want to hear this, but I don't understand you."

"That makes two of us."

"But we need each other, don't you think? I know I need you."

"I don't know what that means — to 'need' someone. Someone needing me, I don't want that responsibility."

The second elevator arrived with a straggling student who hurried to catch up with his friends. Pia got in the car and held the door for George.

"Get in, George, Jesus."

Pia punched the eleventh floor for her room and seven for his. Message delivered. George reluctantly got in. Pia's mind was already full of competing problems — Rothman, the Sisters, Africa, the rest of her life — and now here was another one. She wondered what it was like to think about someone constantly as George said he did of her. It was an alien concept. She glanced at George, who was looking at the floor. She had no idea what he was thinking or feeling. The elevator stopped on seven and Pia reached out and pressed the hold button. George hesitated for a moment, than stepped out.

"Good night, George," Pia said.

George just nodded as the doors closed. To Pia he looked pathetic.

36

Columbia University Medical Center
New York City
March 24, 2011, 11:15 p.m.

George knew something about loss. His own father, Morgan Wilson, died when George was three, and no matter how hard he tried, George couldn't really remember anything specific about him other than a vague sense of contentment. He did have a few vague memories, but they'd been pieced together from photographs shown to him by his mother, Jean. There was one silent home movie of a time Jean and Morgan took George to see his grandparents, Sally and Preston, in Arizona. George had watched the film over and over and his father always looked impossibly young and endearing. In the short film Morgan is holding George on his lap and alternately kisses him on the cheek and hugs him. Morgan's absence had caused George a degree of melancholy similar to the melancholy he was feeling at that moment.

George got up from his bed where he'd flopped after Pia's rebuff. He needed to get out of his room if only for a short interval. There was always the vending machine room on the first floor. He needed to see people, normal people, and there were usually students getting sodas or bags of chips.

As George headed toward the elevators, he tried to concentrate on how much he was loved by his family. He'd always had that to fall back on whenever he felt lonely. He knew that Pia did not have an equivalent situation, which made her behavior even more confusing. Why did she so consistently reject the love that he wanted to share with her and finally had the courage to voice? It just didn't make sense.

George slapped the down button. Almost as if the car had been sitting there waiting for him, the elevator doors opened. Inside was Will McKinley, perhaps the only person in the world who could have made George feel even lonelier.

"George!" Will said. "What a coincidence. You heading down for a snack? Hop in!" Will took George's arm and pulled him in. The ground-floor button had already been pressed. George lacked the strength to resist.

"What's the matter, George? You look terrible."

"I'm just tired. It's been that kind of day."

"How's Pia? Have you seen her? She must have taken what happened to Rothman real hard."

"She did."

"Lesley and I tried to call her but she's not picking up."

"She's not great at staying in touch with people," George said.

The elevator reached the ground floor and Will guided George off.

"Listen, George, if there's anything I can do to help Pia, just let me know. Really. We want her to get through this in one piece, she's a great girl."

George simply nodded. Will walked away toward the vending room. When he realized George wasn't accompanying him, he turned and waved to George to follow.

"Come on! My treat."

George sighed, wearily turned, and pressed the button once more to call the elevator. He wanted company but not Will's company.

37

Columbia University Medical Center
New York City
March 24, 2011, 11:30 p.m.

The moment the elevator doors had closed on George, Pia had already relegated him to his proper place in her mind, well down the list of her immediate concerns. She didn't like being abrupt with him, but she didn't want to get into a long-drawn-out conversation either. She was exhausted from not sleeping well the previous night. Unfortunately, when she reached her floor, she had a bit of bad luck. She had run into Lesley and had had to have a conversation about Rothman and Yamamoto. Curious if Lesley had any interesting thoughts, Pia had tolerated the chat, but after ten minutes or so, when it was evident Lesley was not about to add anything significant, Pia broke it off.

Pia put the key in the lock and opened

her door, walked in, and hooked the door with her right heel to slam it shut. In complete darkness, she felt the wall with her left hand to find the light switch and flicked it on. With her right hand she tossed her keys in the general direction of the desk. All Pia wanted to do was take a quick shower before going to bed. She'd been on the go all day and tomorrow wasn't going to be any quieter with a planned visit to the OCME.

Pia walked over to the window and closed the blinds. She took off her lab coat and tossed it over the arm of her reading chair. Next she removed her sweater and laid it on top of the jacket. She opened her closet door and kicked her shoes directly inside, first the left, then the right. Next she shucked off her black skirt and then her bra and let them fall to the floor. Pia couldn't wait to get into a hot shower. She put her hand on the bathroom door and thought it was odd that it was closed — she never closed the bathroom door, even when she was using the toilet.

Before Pia could process another thought, the door was yanked open away from her and the handle wrenched out of her hand. A tall figure materialized in the doorway and put the heel of his hand on Pia's breastbone and pushed her hard to the ground. Pia's head snapped back and smacked against the floor. A reflex cry rose in her throat, but it was choked off by the man, who was straddling her now with his knees on her arms, his left hand over her mouth. Pia tried to clear her head, but her ears were ringing. The man kneeling on her was wearing a balaclava, and she could see a second figure, obscured by the first. He was also hooded. Both wore hospital security uniforms.

The first man struggled to keep Pia still. He reached behind him with his right hand and the second man handed him a roll of duct tape. The first man looked back again and waved the tape.

"Cut me a piece," he said in heavily accented English, and his colleague obliged. He freed Pia's mouth, took the strip of tape in both hands, and clamped it down over Pia's mouth before she could let out more than a stifled screech.

"Stop moving. We're not going to hurt you," the first man said. Pia wriggled once more but stopped. She was struggling to get enough oxygen through her nose and her head was throbbing where she had hit it on the floor. Her arms were getting numb where the man was kneeling on them. She looked into his eyes and nodded.

"Okay. I'm going to let you up. Don't do anything stupid."

The man got to his feet, digging his knees into the fleshy part of Pia's arms as he did so. He stood back and she got up. She felt small. She was wearing only a pair of panties, and even though they were wearing balaclavas, she knew the eyes of the two men were moving up and down her body. She was going to be raped, she was certain. Pia raised her arms to rub her sore triceps and cover her breasts.

Pia thought, *It's only ten feet to the door.*

Pia thought, *They're not expecting me to do anything.*

Pia thought, *I don't want to be raped. Not again.*

Pia looked from one man to the other and then down at the floor. She wanted them to relax, even slightly. Then she got up on her tiptoes, tapped her right foot on the floor behind her, and in one move, using her arms first for balance and then for forward thrust, she drove her right foot heel-first into the front man's groin. He doubled over and staggered back and into his colleague, and Pia stood square at once, moved forward and reached over and hit the second man in the face twice, boxing jabs that fit the narrow space she was in. Both men were hurt but not enough. Pia got in a couple more kicks that would leave bruises, but the two men quickly regained momentum and charged her. The first man, his groin aflame, feinted a couple of times, then smashed Pia in the jaw with a right hook, knocking her unconscious.

When Pia woke up her head hurt like hell, and she couldn't move her limbs. She understood why: She was duct-taped to her chair, her arms bound behind her and her ankles fixed together. Pia's eyes were barely open, but she could make out one of the men coming toward her, his arm moving back and then quickly forward. She flinched, then took a face-full of cold water that the man had thrown over her using the Tupperware container she sometimes fixed oatmeal in.

"That's what you do when I say do nothing stupid, huh?" The man's concealed face was very close to hers. His blue eyes bored into Pia's. She tried to speak, or at least snort.

"You are a good fighter, but we are more experienced, and there are two of us. We have respect for you because we are family men. But we know some young men who are less, what is the word? Civilized. They are, in fact, animals. If they were here now and not us, then God help you."

The man was speaking in a whisper. The fight and the overturned furniture that resulted had prompted the upstairs neighbor to bang on her floor to ask for quiet. The men didn't want to try the neighbor's patience.

"I say this only once. We are here to give you a message. Stop what you are doing. Stop asking questions. Your doctor was careless and got himself and the other doctor infected, and he put the whole medical center in a compromised position. It will be dealt with quickly and quietly and everyone will move on."

Pia was rocking back and forth in her chair, her eyes wide with fury. It was her rebelliousness surfacing.

"Stop rocking!"

Pia didn't stop. The man slapped her in the face, not hard but well placed enough to make her jaw throb even more severely than it had been. Pia sat still.

"You will be watched. Not by us, by our friends. If you keep meddling, if you call the police, our other friends, the animals, will come and take you away and you will ask them to kill you after a couple of days.

You will beg them. You understand?"

Pia stared at the man. He moved even closer than before and the rough material of the balaclava touched her skin. She could feel his hot breath through the damp wool. He spoke in barely a whisper.

"You understand?"

Pia waited a beat, then nodded.

"You will tell no one we were here. If you talk with anyone here, like that boy you are with, they will be killed too. If you go to the police or the medical authorities, you will be killed. It's easy. Just stop, go about your life, and all this goes away."

The man stood up. His colleague stepped forward and jabbed the needle of a syringe hard into Pia's thigh. She gasped at the pain, then fell unconscious almost at once. The men tore off the duct tape binding her, leaving her skin red and swollen where it had been in contact with the tape. When they pulled the tape from her mouth it tugged at the wound on her jaw and further opened a tear in her lip. Blood trickled down her chin. The first man wiped it off with a tissue he got from a box on Pia's dresser. He pocketed it after using it. He picked her up and laid her on the bed with her head over the side. He knew that the drug he'd given her had the tendency to cause vomiting.

The men removed their balaclavas and prepared to leave. If Pia had been conscious she would have seen at once that the face of one of the men, the leader, was marked with a cleft lip. The other man had a peculiarly and memorably pointed nose. The first man cracked the door open and, seeing an empty hall, quickly exited the room, followed by the second man. They put on their official hats and, adjusting their uniforms, made their way quickly to the stairs.

38

Columbia University Medical Center
New York City
March 25, 2011, 8:07 a.m.

Pia woke up in stages. First, when it was still night, she skimmed the surface of consciousness but quickly fell back into the darkness. Then, later, it had become light outside, and she was aware of her own breathing and a sharp pain in the back of her head and a throbbing along her jaw. Finally, she awoke and was hysterical — there were men in her room, chasing after her, she had to get away. She tried to get up, but her body wasn't obeying her commands. She slumped back on the bed and closed her eyes.

Then she remembered. Men had been hiding in her bathroom and had attacked her. The last thing she remembered was getting jabbed with a needle. She felt her leg at that spot, and it was sore. She looked down at the puncture wound. So she had been drugged, and hit. No wonder she felt so bad. She reached down and felt between her legs: nothing; she experienced a modicum of relief.

Dazed with the fog of her drug hangover, Pia was unsure of what to do. Her mind clicked over to George. Pia remembered the conversation they'd had in front of the elevator, the confessions George made to her and the look on his face when Pia said she wasn't thinking about those kinds of things right now. Last night she wanted George to leave her alone; now she wished he were here with her.

As the drug gradually started to wear off, the pain in Pia's jaw intensified. She stood up. She was dizzy. She managed to get herself into the bathroom. She looked at her face in the mirror, and it was a mess. A livid red welt with a small laceration covered much of the left-hand side along the jawline. Pia's lip was swollen and bloody, and there were red marks where the duct tape used to gag her had ripped at her skin. She remembered the fight, how she'd kicked one of the men in the groin and been smashed in the face in return. Pia leaned in and looked at her eyes. She saw that they were puffy and ringed with dark circles. She hadn't had a normal good night's sleep in an age. Being unconscious for hours didn't count. Pia looked at herself again and hoped to get an answer to the question: What was she going to do now?

She washed her face with cold water and took a long, hot shower. She put on her most comfortable sweatshirt and pajama pants. She located a bottle of Advil in her travel bag and took four tablets, washing them down with two glasses of water. Then she called George on his cell phone. When he didn't answer, she didn't leave a message, fearing she wouldn't be able to say anything coherent. She sent a text message instead: "Something's happened. Please come over. Urgent. P."

Pia lay down on the bed and waited.

George's phone vibrated in his pocket. He hung back a little from the group doing radiology rounds and read the message. He had a coffee break coming up, and he figured he could wait until then to reply. Pia had probably found a way of indicting someone else in her conspiracy theory, and he had been enjoying being a regular medical student for the last couple of hours. George put the phone back in his pocket and caught up with his group.

■ ■ ■ ■

George called Pia after finishing a cup of coffee in the X-ray technician lounge. It was 9:45. At first, George thought he had a bad connection because he couldn't understand what Pia was saying. He moved out of the room with its background chatter, into a corridor, and stood by a window.

"Pia, can you hear me? You're very faint. What is it?"

What Pia was trying to say was "Please can you come to my room?" But it didn't sound like that at first.

"Say that again, Pia, I can't understand you."

Pia repeated herself.

"You want me to come?"

"Yes."

George was confused by the sound of her voice and wondered if Pia's state of mind had anything to do with the way their conversation had ended the previous night. It crossed George's mind that Pia was drunk, but it sounded more like she had a mouth full of cotton.

"Okay, I'll be right there."

George asked one of the other students to tell the resident he'd been called away on hospital business and headed to Pia's room. He found he was less eager to see her than usual. The previous night he'd made a decision, recognizing that he was probably barking up the wrong tree with Pia. George wasn't confident he could follow through with his decision, but he was going to try. It was for his own peace of mind.

Fifteen minutes later George knocked on Pia's door. When she opened it and he saw her face, all of his plans, doubts, and recriminations were washed away. He was instantly morphed back into the slavish dog he'd been for three years.

"Oh my God, what happened to you?"

Pia shook her head and pointed to her jaw. George fetched Pia's desk chair and sat her down.

"Take your time, tell me what happened."

"There were two men in my room. Last night," Pia said. She spoke slowly and deliberately.

"Last night? This was last night? Why didn't you call?"

"They drugged me. I just woke up."

"Jesus. Who were these men? What did they do? Did they . . . ?" George hesitated, not sure he wanted to hear.

"No, they didn't rape me, if that's what you're asking. They warned me to stay away from the Rothman case."

"Jesus, Pia. Do you want to lie down?"

"No, I'm okay."

"I'm going to call security. And then the police."

"No! Don't!" Pia said. She shook her head vigorously, an act that hurt a great deal. She was still dazed from the sedative, but the clouds were clearing.

"No security and no police. I have to take what they said seriously. They were waiting for me in my room. They said they'll be watching me. I mean, they already had been watching me. You see what this means, George? It means I was right. There's a conspiracy behind Rothman's and Yamamoto's deaths."

"Wait a minute, Pia, slow down," George said. "These two men, who were in your room, who obviously beat you up, they said specifically, 'Stay away from the Rothman case'?"

"Not like that, but they said it."

George was horrified, but his first instinct was one of skepticism.

"Did you recognize the men at all?"

"They had ski masks on. But they were dressed as hospital security. Shit, George, maybe they *were* hospital security, you know? That would mean the hospital is covering it up. Spaulding, the dean, Springer, all those guys . . ." Pia stood up as if she wanted to flee.

"Oh, come on, Pia. This is New York City. In America. Maybe in a movie or some Third World dictatorship they kill off their own doctors and beat up medical students, but not here. I can't believe you could think that. Get a grip."

"Well, someone did this!" Pia said, pointing at her face, shaking, partly with rage, partly with fear. "I know what institutions can do, George, what people can do to someone they're supposed to be looking after. If you grew up in the system I grew up in, maybe you'd be a little more cynical. I know one thing: Everyone has their own agenda. If you're in the way, things like this happen to you." Pia sobbed once and her shoulders heaved.

"Okay, Pia." George stood and reached out to Pia, and she went into his arms. He held her tightly.

"I think we should call the police. You need an ambulance as well —"

"No!" Pia pushed George away. "I need to think about what this means. If we call the police, they'll call the administration and security here, and for all I know, they were the ones who attacked me. I need to think." Pia grabbed both sides of her head and shook herself. "The drug, I can't think straight."

"Maybe we should go to my room," George said.

"They know all about you, George. It won't be any safer there. They won't do anything now, I'm just sitting in my room."

George looked around. "You think they're watching you that closely?"

"Well, think about it. Every time we moved, we got caught. It happened twice in the lab."

"One time it didn't."

"But we didn't find anything important then, remember? And we were allowed to go through the morgue just fine because there wasn't anything there to find."

"I'm having a hard time believing all these people are in on a conspiracy. Bourse, Springer . . . Dr. De Silva, who was treating Rothman. Why, Pia? What are they conspiring about? And there's no proof the deaths were anything other than accidents."

"Let me remind you again. You have no idea how much people hated Rothman. I saw it every day I was in his lab. No one liked him — he was rude, disrespectful, mean. And they were all jealous of him, how he got special treatment from the hospital, how he got a Nobel and might well get another. He had a lot of enemies, all sorts of reasons, including people in his own lab."

"Okay, but you don't kill someone because you don't like them. It's too much, it's so theatrical!"

"Well, how do you explain this?" Pia gestured to herself. "I was attacked," she yelled. "I was ordered to stay away. Now I'm sure Rothman was killed. His death wasn't accidental, it was deliberate. The only thing I'm not sure of is why they didn't kill me too last night rather than just warn me. They must be more afraid of how people would react to my disappearing than afraid that I wouldn't respond to the warning. As they said, if I keep quiet, all this goes away. If I disappear, they talk to you and find out what I was thinking."

George felt a sudden chill. If Pia was right, he might be next in line for a visit. But how could she be right? It was so far-fetched. George needed some time to think too.

"Can I get you some ice for your face? I'll just be down the hall."

"Sure, thanks."

George went to the ice machine at the end of Pia's corridor, but it was out of order. He could go down to the cafeteria where he knew ice was always available, but that meant leaving Pia on her own in her room for a few minutes. George walked back to her door and opened it, startling Pia.

"Shit, George, can't you knock?" she said.

"Sorry. The ice machine's out. I'm going downstairs to get some. I'll be right back."

39

Columbia University Medical Center
New York City
March 25, 2011, 11:20 a.m.

George returned with the ice. Pia was sitting at her desk, writing furiously on yellow legal paper, trying to make sense of what had happened in the past forty-eight hours. George took some ice and wrapped it in a towel for Pia to hold to her face. The rest he put in her sink. He then sat on the bed and watched her write on page after page.

Pia tested her prodigious memory, trying to isolate fact from speculation. She worked backward from the undeniable fact that she'd been attacked and threatened in her own dorm room. That was obviously a criminal act — but what else of all the events of the past two days had constituted unlawful activity? While she worked on the what, she also worked on the who. She tried to piece together a cast of characters using the information she was certain about. There had been two men in her room — but who else was involved and how broad was the conspiracy she now knew existed?

After an hour, Pia stopped.

"This is getting me nowhere. It could be anyone. And there's so much that's happened, and I bet we don't know the half of it."

"What about trying to establish a timeline? Isn't that what they do in those cop shows on TV? They use a whiteboard: 'six forty-two P.M., suspect seen in O'Leary's bar . . .' "

"But we don't know who the suspects are unless we just include everyone. And we can't really investigate anything. Say we think Springer was involved somehow. The only times we know what he was doing are when I was with him. I can't pick up the phone and demand he answer any questions about his whereabouts at any other time."

"This is why we should call the police," George said. "They can investigate to their heart's content and ask him anything they please."

"Given that there's a conspiracy, one of the things we don't know is why."

"Only, as you keep saying, Rothman was hated by half the human race. Of course that raises the question, why kill Yamamoto as well? He wasn't unpopular, was he?"

"Not at all, people loved him. He was devoted to Rothman. They were like two peas in a pod, working together. If they weren't working together in either the bio-safety unit or the organ bath unit, Yamamoto was in Rothman's office. They even ate together if they took time out for lunch, which wasn't always the case. Yamamoto was the only one Rothman allowed to use his private coffee machine or drink the bottled mineral water from his private office fridge. They were like Siamese twins."

"So there's much more we don't know than we do know, as far as what other people were thinking and what they were doing," George said. "So what *do* we know, other than the fact that you were attacked last night and told to stop involving yourself in this?"

Pia turned back to her desk and picked up her pen and underscored a couple of lines on the page.

George looked at his watch. He was concerned about getting back over to the hospital but decided he was more worried about Pia. The resident to whom he'd been assigned for the day was rather laid-back, to say the least, and probably didn't even realize George wasn't around. Besides, George wanted to stay and humor Pia for a while. He was worried that she might have a concussion from the attack, and he wanted to make sure her mental status didn't change. In addition, he reasoned, she couldn't get into any more trouble while they were there in her room.

Suddenly Pia turned back around. "You know what we know the most about?"

George shrugged.

"We know the most about Rothman and Yamamoto's illness even without the autopsy results and even without seeing their charts. I was in the lab when it presented, I saw them in the hospital, I talked to the doctor who was treating them, I examined Rothman myself, I diagnosed new symptoms, I spoke with the department head involved."

"Good, yes," George said. They had gone over all this before but under the circumstances, George was happy to do it again. Pia tore the sheets she had written on off her legal pad, crumpled them into a ball, and threw the ball in the general direction of the trash basket. She missed. Pia began writing again, more slowly this time.

"Okay," she said while she worked. "We do have a timeline of the infection. The onset was extremely rapid. Rothman or Yamamoto pressed the panic button and there was a medical team in the lab almost at once. I saw them arrive. Rothman and Yamamoto knew what to look out for, so from the first symptom to the medical team arriving may have been only ten minutes, at the outside. Springer showed up, and he went into the lab. Then he stayed and talked to the staff while Rothman and Yamamoto were taken directly up to the infectious disease floor and put in isolation, and treatment was started. I'd say they were there in five or six minutes. And Springer told us it was classic typhoid fever — high temperature, delirium, and so on, so it was diagnosed immediately. No delays. They got antibiotics within an hour of the initial symptoms."

Pia had the pad on her knee.

"So Rothman and Yamamoto got all the symptoms straight off. It apparently wasn't the usual sequence where a patient gets one symptom initially and then another a few hours later. It happened like a bolt of lightning. As far as I know, that's not the way typhoid fever develops. Then the patients got the more ominous rebound tenderness by that evening. It's all so speeded up."

"You said this was a particularly virulent strain," George said.

"True. One of the zero-gravity strains. The alpha strain. But still."

"And you also said that Rothman's own sensitivity studies suggested that the strain should have been knocked out by the antibiotic he was given."

"That's right, the chloramphenicol and later the ceftriaxone."

"So what are you saying? Are you suggesting that it can't have been that strain of salmonella?"

"No, I'm not. The strain had to be involved since Koch's postulates were satisfied."

"Meaning that they managed to grow the culture from samples taken from the patient."

"Or by using more modern DNA techniques, yes."

"Pia," George complained, "you're confusing me. What's the bottom line here? What are you trying to say?"

"I suggested to Springer that there might be a second bacteria involved, a bacteria or a virus that was actually more virulent than the salmonella typhi and that was resistant to the antibiotics. It could explain the shockingly fast clinical course Rothman and

Yamamoto experienced."

"What was Springer's reaction to your suggestion?"

"He went bonkers on me," Pia said with disgust. "That was the end of the interview because he went out and called in reinforcements, meaning the dean."

Pia put down her pad and pen back on the desk.

"So you think there might have been two bacteria involved," George said.

"Right at this moment that's the only thing I can think of. The clinical course was just too fast, especially in the face of two antibiotics given within hours of initial symptoms and known to handle salmonella. I know it's contrary to recognized diagnostic rules, the major one being that one should look for a single causative agent even with seemingly multiple symptoms. But it's the only way I can explain what we've seen with Rothman and Yamamoto."

She turned back to her desk and read from her notes.

"We have all the symptoms right here: fever, delirium, prostration, temperature elevation, sweating, low white cell count known to be associated with salmonella, all leading up to abdominal rebound tenderness. From intestinal perforation and finally death."

George got up from the bed and headed into the bathroom. Pia was overwhelming him. He was amazed she remembered Koch's postulates from second-year microbiology. He certainly didn't. He put some of the ice that was melting in her sink into a fresh towel, rolled it up and brought it to her desk. He exchanged it for the first one he'd made. Pia was staring at the paper, her back to him.

"Here's some more ice," he said.

Pia swiveled around in her chair, and George winced when he saw her jaw at such close quarters.

"How's it feel?"

"It's not too bad. A bit better with the ice."

Pia took the fresh towel and held it to her face. An image flashed into her mind: Rothman lying on his deathbed, sweating into his pillow, delirious . . . Suddenly she stared directly up into George's eyes with a fierce intensity that made George look away.

"The hair loss!" Pia said slowly, with emphasis. "What about the hair loss?"

40

Columbia University Medical Center
New York City
March 25, 2011, 12:15 p.m.

Pia got up from her chair, put the ice-filled towel down on the desk, and started pacing her room, circling George. First he'd been unnerved by the intensity of Pia's stare when she had her epiphany, whatever it was; now she was stalking around the room like a cat closing in on a mouse.

"What hair loss?" George questioned.

"Rothman's! Remember I saw there was some hair on his pillow before we found the rebound tenderness."

"Yes, I remember. You pointed it out to the resident, and, as I remember, she suggested it might have been due to the chloramphenicol."

"Exactly," Pia said.

She stood still. "Can I use your computer?" Pia's laptop was old and slow; the year before George had invested in a new Dell with a much faster processor.

"Sure. Let's go!" George picked up the wet towel from Pia's desk and gestured toward her with it. Pia shook her head. George took the towel into the bathroom while Pia changed from her pajama pants into sweats. She also quickly downed more Advil.

Gathering up her notes and heading for the door, Pia paused and looked back. She'd experienced a flash of anxiety. Although her room was where she'd been attacked, she still felt safer there than outside. Her attackers were lurking out there somewhere. Perhaps they really were watching her, as they'd threatened. George sensed her trepidation and put an encouraging hand on Pia's shoulder and squeezed lightly. They exchanged a reassuring glance. Pia breathed deeply and walked out of the room, shutting off the light as she went.

"Let's take the stairs," Pia said, and she and George descended the four flights, walked along the corridor, but then stopped in front of George's door. Both had the same thought: If they knew about George, they probably knew his room number.

"What do you think?" he asked. It wasn't crazy to imagine that the two men might have it in their mind to pay George a visit too.

"Now I think we're being paranoid," Pia said.

"But as you correctly said, even paranoid people have real enemies. Wait there!" George unlocked the door and pulled it fully open. He and Pia were prepared to flee if anything looked amiss. Nothing did. George entered his room, making sure nothing had been disturbed before throwing open the bathroom door. "All clear," he said with a sigh of relief.

"Let's get to work," Pia said.

George booted up his laptop and checked the wireless signal before giving up his desk chair to Pia. He went and sat on the bed. His room was a mirror image of hers.

Pia quickly went online, typed "hair loss" and "chloramphenicol" into the search engine and scrolled through the results for a few minutes. "There's nothing that lists hair loss among chloramphenicol's side effects. Actually, there are some alternative healers who sell chloramphenicol to reverse hair loss. Wow, De Silva was so wrong when she said chloramphenicol might be the cause of the hair loss."

Pia continued surfing. "Springer attributed it to fever and stress," she said as she read. "It seems that stress can cause hair loss. But I don't think it could be involved in this case. I mean Rothman and Yamamoto were certainly being stressed with their fevers and all, but for stress to cause hair loss I think it has to be over a period of months, not hours."

Pia continued her search. George couldn't see the screen from his vantage point, but he could see the light flickering on Pia's face as page after page flashed by. Suddenly there was a steady light, and Pia leaned forward in her chair. "Yeah, here we are. Hair loss and stress. Yup, I was right." Pia read out loud: " 'Unless the stressed patient is pulling his own hair out, severe stress merely changes the hair follicle from an active state to a resting state. The hair doesn't fall out immediately but rather over a period of months.' "

Pia looked over at George. "Clearly Springer's suggestion wasn't much better than De Silva's."

"So what are you thinking?"

"Since I've never heard of salmonella causing hair loss, we have to think of something else to explain it, bringing us back to the second-agent idea, like another bacteria or a virus. But if there was another microbe involved, it would have to be one whose clinical symptoms mimic typhoid fever because all the other symptoms were consistent with typhoid fever. Are you following me?"

"I think so."

"I'm saying that we have to come up with an agent that mimics typhoid fever symptomatically but which also causes hair loss and can kill in hours in the presence of chloramphenicol and possibly ceftriaxone. Of course, without access to the charts, I can't be sure they ever got the ceftriaxone, but for the sake of argument, let's assume they did."

"You know what I wish?" Pia said after a few minutes of silence. "I wish that we examined Yamamoto as well as Rothman. Just to be sure he was suffering from the same signs and symptoms."

"Maybe we can ask Dr. De Silva's opinion."

"I don't think she's going to want to hear from me. Let's keep going."

Pia looked up at the corkboard on the wall behind George's desk. A business card for a taxi service was tacked next to a picture of George's mother and grandmother. There was a postcard from Hungary alongside it. Suddenly Pia snapped around again.

"What are the usual causes of hair loss

besides what I've mentioned?"

"This sounds like internal medicine rounds, which I'd like to forget. That was one venue where I did not shine in the slightest."

"Come on," Pia said. "What causes hair loss?"

"Er, hormonal changes, alopecia areata, stress like you said." Pia motioned for George to come up with more. He thought harder.

"Dermatological diseases of the scalp, particularly cicatricial diseases. Wow, that's a good one. That's the kind of response that would have gotten me kudos on rounds. Trouble is I always choked up."

"What else?" Pia commanded. She waved her hand, indicating she wanted more.

"Okay, certain drugs."

Pia nodded and looked expectantly at George, as if she knew the answer and was waiting for him to get it. It was like a game of charades.

George became impatient, ready to give up before he remembered something else.

"What about chemotherapy and radiation?" George sounded uncertain. Sure, they caused hair loss, but what could be the relevance?

"Right on!" Pia exclaimed. "Radiation!

You saw people undergoing radiation when you were on oncology during internal medicine."

George nodded.

"Chemotherapy and/or radiation destroys the hair follicles and the hair falls out immediately."

"What are you getting at?" George noticed that Pia's face had brightened considerably.

"I said I was wondering if Rothman could have been infected with another bug besides the salmonella, another bug that was not sensitive to chloramphenicol or the third-generation cephalosporin he was given."

"Right, the ceftriaxone."

"I suddenly don't think there was another microbe," Pia said. "Damn it, George, you said it yourself, remember? You said they had to autopsy the bodies the same day the men died because they were 'hot.' I thought at the time it was a strange word to use, but I think you were more right than you knew. I don't think they were hot because they were full of bacteria. I think they might have been hot because of radiation."

41

Office of the Chief Medical Examiner
520 First Avenue, New York City
March 25, 2011, 12:32 p.m.

Laurie Montgomery had been sitting in her office at the Office of the Chief Medical Examiner catching up with an old friend, Detective Captain Lou Soldano, when the phone rang. She saw it was her boss, apologized, and took the call. She was soon rolling her eyes, and Lou smiled.

Laurie Montgomery had been back at the OCME for eleven months, since the harrowing events of the infamous Satoshi Machita case, involving both the New York mob and the Japanese Yakuza, which led to the kidnapping of her infant son, John Junior. The story had been plastered all over the media for several days as the details of the case unfolded. After JJ's rescue, Laurie had come back to work but only after she had found a live-in nanny, Paula, who immediately proved to be a godsend. With Paula looking after JJ, Laurie felt secure. Right now, her husband and fellow ME, Jack Stapleton, was working in the same building, and JJ was safe with Paula at the couple's home on 106th Street. It didn't hurt that she and her husband had friends like Detective Captain Lou Soldano either. Right after the kidnapping, he'd insisted on a twenty-four-hour security detail for the Stapleton home.

From Laurie's side of the conversation and from knowing Laurie's boss, Dr. Harold Bingham, Lou sensed he'd be in for a wait. He took his copy of the *New York Post* from his briefcase and flipped through it until he saw the story: IVY SPACE GERM DOCS DIE. He quickly reread the first few paragraphs. He had wanted to show the article to Laurie, which was one of the reasons he'd stopped by.

"Sorry, Lou," Laurie said, hanging up the phone. "That was Bingham."

"I assumed as much. No problem. Did you see this article?" He held up the paper.

"Yes, but not that one specifically. There was the same story in the *Times*."

"Crazy and scary at the same time. It says that two researchers at Columbia contaminated themselves in a lab with some virus grown in the space station or something. The bodies were supposedly brought here to OCME. Is this all true?"

"Most of it. But the contamination agent wasn't a virus. It was a bacteria called salmonella typhi that causes typhoid fever. Jack did both autopsies yesterday. Very sad. I understand they were stem cell researchers who were making huge strides growing human organs."

"That's what I understand," Lou said. "Anything unique about the autopsies? There were some wild theories about the deaths in the article. Apparently one of the guys was a big-deal researcher who was not particularly liked by his colleagues."

"Jack didn't mention anything other than that he was impressed with the pathology. He'd never seen an entire gut in both patients in such bad shape. Typhoid fever isn't usually so generalized. Anyway that was the case I was just talking to Bingham about. He expects there will be some political fallout. If there's a press conference scheduled, he wanted to give me a heads-up that he might want me to host it. He knows Jack hates doing it and isn't the most diplomatic."

Lou laughed because Jack was one of the most undiplomatic men he knew. "You guys make a good pair because you complement each other." Changing the subject, he said, "What about a bite of lunch today? Do you have time for a quick one?"

"I'm sorry, Lou, they're dropping like flies out there."

Lou laughed again. He was glad the public couldn't hear the black humor that was often engaged in within the OCME walls.

"I hear you."

Lou levered his stocky frame out of Laurie's chair, put his trench coat back on, and made his farewells.

42

Columbia University Medical Center
New York City
March 25, 2011, 1:18 p.m.

The first time George heard Pia's theory, he thought she had lost it for sure. She said she believed that Rothman and Yamamoto were murdered using a radioactive agent, polonium-210, that was masked by the salmonella bacteria they had also been given. George had sarcastically asked which James Bond movie she'd gotten that from, but Pia was, as ever, deadly serious.

"George, it's actually happened before so it's a copycat crime. Someone was actually murdered in this fashion. Truly. The man's name was Alexander Litvinenko, a Russian. He was killed in London in 2006. Don't you remember? It was all over the news."

"I don't remember," George admitted.

Pia waved George over to the desk so he could see several of the newspaper stories she had found online, then she filled him in on the basics of the case.

"Litvinenko was in the KGB, then the FSB, which is what replaced it. He fled Russia and was given political asylum in England. He wrote a couple of books that were critical of the Russian president, Putin. You'd think he would have been super-careful, knowing what he did. He meets these guys, ex-KGB like him, in a London hotel bar for tea. Within hours he's sick and hospitalized. After several days he's diagnosed with radiation poisoning, which is later figured out to be polonium-210. He gets progressively worse, since there was not a lot the doctors could do, and dies about three weeks later."

"Three weeks. That's a lot longer than Rothman and Yamamoto."

"Yeah, I know. But polonium's effects are dose-related. We don't know how much polonium was used and when Rothman and Yamamoto were poisoned."

We don't know if *they were poisoned,* thought George, but he kept it to himself. Pia was on a roll.

"So the Brits investigate and find out about the bar and the tea, and there was radiation all over the place, especially in the teapot. Ultimately it was proven that he died from deliberate poisoning. They did an autopsy and the pathologists had to wear hazmat suits. Litvinenko's GI tract was very hot, to use your word. The guy had to be buried in a lead-lined coffin."

"Okay, I can see why one spy might use something like this to kill another spy, but why use it on doctors? If you want to kill them, why go to all that trouble? Why not just shoot them?"

"That's the clever part. Whoever did this did not want anyone to know it was an assassination. They wanted it to appear like an accident. The symptoms of radiation poisoning are camouflaged by the salmonella: fever, prostration, delirium, GI symptoms, low white count. Everything's the same except the hair loss. They were counting on the fact that no one would think to look for this kind of agent because of the typhoid fever diagnosis. Polonium is unique in that it decays by only emitting alpha particles, which would only be detected if someone thought to look specifically for it, but nobody would because the diagnosis of typhoid was so obvious."

Pia was picking up steam again. It all seemed to fit.

"Nor would the alpha particles make anyone else sick, which obviously has been the case. Alpha particles can only travel about a centimeter, and they can be blocked by as little as a sheet of paper. It's only if the polonium-210 is breathed in or ingested that it's dangerous, and then it's *really* dangerous, especially in a large dose, which can be rapidly fatal. Even as little as a millionth of a gram can kill you."

Pia sat back in her seat with a look of triumph on her face. "What do you think, George?"

George was overwhelmed both by the amount of information Pia had thrown at him and by her enthusiasm. Things fit, but he couldn't help but wonder if Pia was getting ahead of herself. "You have to assume the hair loss had no other cause," George said. He thought a little more. "But I guess this would explain why the antibiotics didn't work. Or maybe they were working, but the radiation overpowered them."

"Exactly," Pia said. "It is diabolically fiendish. Whoever is involved is smart — probably a doctor or a scientist who knows a lot of medicine."

George thought about it some more. He started pacing the room.

"I guess it's possible," George said. He could see no insurmountable problems with the theory. "So let's tell the authorities, let them figure it out."

"No, we can't. We don't know who did it."

"I guess you have to assume that the guys who attacked you had something to do with it."

"No doubt, but this has to be a major conspiracy. You know what this stuff is used for? They use it to make firing mechanisms for nuclear weapons. No one admits to making it, although the major source is supposed to be Russia. I just read about it two minutes ago. So the FSB can just call someone and get it. But how do you get it in New York City? There has to be a lot of people involved. Serious people with access to this material. And I believe those guys when they warned me not to go to the police. I'm not going to the authorities till I have enough proof to go to the media at the same time."

"The media?"

"I told you, I don't trust 'the authorities.' " Pia made quote marks in the air with her fingers. "If I give this story to the newspapers, whoever is involved won't be able to bury it."

"So what proof do you need?"

Pia turned back to the computer and made another search.

"Okay! Polonium-210 has a half-life of a hundred and thirty-eight days, meaning that it takes a hundred and thirty-eight days for it to lose half of its radioactivity. So if that's what they used, there has to be some trace around someplace, either in the lab or in the rooms Rothman and Yamamoto occupied in the hospital. Even if someone was very careful giving it to them, there's bound to be some residue just like there was in London in 2006."

George joined Pia at the computer, looking over her shoulder. "How is polonium detected?"

"Here it is," Pia said, pointing at the screen. "Alpha particles can be detected with a Geiger counter. Pretty simple."

"Where are we going to get a Geiger counter?" George said. "Oh, let's use mine. It should be in the bottom drawer on your right."

"Very funny," Pia said. "Geiger counters are not all that uncommon, especially around a medical center like this. They must have them in Nuclear Medicine. We'll go over there and see if we can't borrow one."

"I couldn't help notice that you said 'we,' " George said. "So is this a formal invitation?"

"Of course it is," Pia replied.

"Well, thank you," George said. Actually there was no way he would have allowed Pia to go over to the hospital without him. He reached out and felt along Pia's jaw. "Maybe we should get you X-rayed while we're there. I know a technician who would do it as a favor."

Pia pushed his hand away. "I don't want to take the risk."

"All right, here's the plan," George said. "I'll help you look for a Geiger counter, but first I have to talk to my resident and come up with an excuse for not being around."

"Fine," Pia said. "I wouldn't mind a few minutes of rest. Whatever drug I was given has me feeling sleepy again. I could use a bit of rest while you take care of business. Do you mind if I just lie down here on your bed? Call me when you're ready." Pia moved from the desk chair over to the bed and lay down. She closed her eyes and let out a sigh. She was tired and wired at the same time.

George went over to her and, taking out his pocket light, made her open her eyes. He quickly checked pupillary light reaction. It was normal.

"Jesus, that's bright," she complained, turning her head to the side. "What the hell are you doing?"

"I'm just worried about a concussion," George said.

"That's a happy thought," Pia said.

"It's something to think about, especially if you're sleepy."

"Good point, but I think I'm fine, just tired."

"I'll go back to Radiology and make my excuses. I'll be right back. Wedge the chair under the door handle. Do you have your phone?"

Pia nodded.

"Make sure it's charged. My charger's on the desk. And call me if you need me."

George would have been far happier calling the police or renting a car and driving Pia as far out of the city as he could. But he'd come this far with her. He just had to cover himself, and he'd be clear to stay with her for as long as it took to resolve this thing. They'd get a Geiger counter and find the proof if it existed. Or they wouldn't find the proof, and Pia would have to drop this theory as she had all the others. Perhaps then she would quit sleuthing.

43

Columbia University Medical Center
New York City
March 25, 2011, 2:01 p.m.

Pia's brief sleep was interrupted by George knocking gently on his own door. She woke up with a start and recognized that she was in a room that was not her own. She sat up feeling like she had a hangover. All at once she realized where she was, and the whole previous night's experience came flooding back. Pia got up from the bed, swayed for a moment, and went to the door.

"Who is it?"

"It's George." Pia took the chair away from under the door handle. She then hurried into George's compulsively clean bathroom and checked out her face, wincing when she touched her bruised left jawbone and wincing again when she noticed for the first time that she also had the makings of a good black eye. Reflexively, she closed her uninjured eye to make sure she could see normally. She also examined her cut lip, and then wiped away a bit of blood encrusted in her nostrils. Then she began to fill the sink to wash her face.

"What time is it?" she called through the open door. Her mind was clearing, and the dizziness she'd experienced when first getting off George's bed had totally disappeared.

"It's about two, a little after," said George. "You want something to eat?"

"No! There's no time. We have to get moving. The longer we wait, the less chance we have of finding anything. As we learned, polonium has a relatively short half-life and you can just wash it off normally, like dirt."

"So you're sticking to the polonium idea?" George had half hoped Pia would have cooled to her rather outlandish scenario by the time he got back.

"Absolutely. It fits so well. You agreed it fit, didn't you?"

"It seems to," George said. "Provided we can't think of another reason for the hair loss. But the practicalities seem so daunting. And you're sure there was hair loss, right?"

"Oh God, yes, George, I'm sure. You saw it yourself."

George looked at Pia as she emerged from the bathroom. She had an intent expression on her face. It seemed that Pia was pleased with the power of her deductive reasoning, or else with the elegance of the plot to murder her mentor.

"Do you have any idea how hard this must have been to pull off?" she said. "It makes the Kennedy assassination look easy."

"I think Oswald acted alone."

"Okay, bad example. This has to be a sizable conspiracy, with a number of people involved. Once we confirm the polonium, I can't let the authorities spin the story, which they will. I need to make sure my version of the story, which is the truth, gets out."

"But if there's proof, the police will protect you."

"Bullshit. It's the police I'm most afraid of. Listen, the more I think about it, the more I think it has to have been other researchers or doctors. The science behind it is impressive. I mean, it had to be someone with a medical background who thought all this through. Otherwise, as you said, why not just shoot them?" Pia stopped herself.

"I'm jumping the gun. We've got to look for radiation left in the lab. If it's going to be anyplace, that's where it will be, I'm sure.

We need that Geiger counter. But let's make a quick detour back to my room. I need to dig out some concealer. The fact that I look like I got run over is going to raise some eyebrows."

"Let's make it quick," George said. "I've only managed to wrangle a couple more hours. I have to be back in the Radiology Department for an important lecture at four o'clock."

Pia and George were able to borrow a Geiger counter from a resident in the department of nuclear medicine with ease. It was an out-of-commission machine awaiting recycling that was actually better at detecting alpha particles than the newer models. With the detector in hand, they hustled over to Rothman's lab to check for any leftover radiation.

Once at the lab's outer door they hesitated. "The only person I'd rather not run into is Spaulding," Pia said. "He's the only one who might cause us trouble. I never got the impression that any of the other technicians liked me much, but I can't imagine they would physically stop us."

"Want me to duck in and ask if he's around?" George questioned.

"Good idea," Pia responded.

It took George less than a minute. When he reappeared he said that the secretary told him Arthur Spaulding was taking a late lunch.

"Lucky us," Pia said. "Let's do it."

The pair entered the lab with Pia in the lead. Marsha Langman looked up. Pia said she was just coming by to get some personal items. Marsha shrugged and went back to her work, whatever that was.

Pia made a beeline for the biosafety unit. They quickly donned protective clothing. They were in a hurry and didn't want to be interrupted. Pia wanted to start in the unit because it was there that Rothman and Yamamoto had spent the entire morning on the fateful day, as well as the day before.

The Geiger counter was a small yellow box about the size of a large flashlight, with a handle on top. Pia held the main instrument in her left hand and ran the sensor, much like a microphone, over the bench surfaces. The machine made a slight crackling sound from background radiation every few seconds. To Pia's chagrin, they found nothing, even under the hood itself.

As they removed their protective clothing they didn't talk. Emerging back into the lab, they detoured to Pia's small office for Marsha's benefit. Pia had said she was there for some personal items so it was a command performance. As per usual, O'Meary was still there, half up in Pia's ceiling space. He poked his head down when he heard the students enter.

"Miss Grazdani, you back? My God. What happened to your face?"

Pia said nothing.

"Good news. I found the short after all this time. It was between here and the doctor's office. We'll be out of here today. Sorry about the inconvenience."

Pia ignored him.

"Is that one of those Geiger counters?"

"We did some radioisotope labeling in here," Pia said. "We're just checking the place is clean. Which it is."

"How does that thing work, then?"

"Look it up online. Like I did."

It made George uncomfortable that Pia was being so short with this guy. As his family was blue-collar, George felt a kinship with people like maintenance workers.

Disappointed at not finding any contamination in the biosafety unit, Pia was beginning to feel a big letdown. Yet there was one other place she wanted to check: Rothman's office. Besides the biosafety unit and the organ bath unit, that was the only other place where both Rothman and Yamamoto spent any time. The problem was Marsha and her guard-dog mentality. Even with Rothman gone, she suspected Marsha would be Marsha.

As Pia and George returned to the lab proper, Pia was running through her mind possible ways to handle Marsha. Luckily the problem solved itself. Marsha was no longer sitting at her desk. Pia guessed and hoped she'd taken a late lunch like Spaulding.

With Marsha no longer standing guard, Pia and George hustled into Rothman's inner office. Inside, there had obviously been some packing going on because scattered around were open cardboard boxes half full of books and papers. Pia ran the probe around the desk, the shelving behind the desk, the couch, and the coffee table where Rothman's guests, usually journalists, would sit and hope that they would be the one to breach Rothman's famous guard. They were inevitably disappointed. Next they tried Rothman's private bathroom, which his celebrity status had afforded him; no other lab had a bathroom like this. But the Geiger counter stayed mostly silent except for the background crackle, just like in the biosafety unit.

Pia almost forgot about it but there was

one more room — not so much a room as a storage area, where Rothman used to keep supplies scientific and secretarial. The place was stacked with toilet paper and paper towels, cases of beakers and test tubes, reams of paper, and old files. And here also was Rothman's beloved Nespresso machine. Well, maybe, thought Pia. Just maybe.

There were a few wayward clicks from the Geiger counter next to the coffee and cappuccino maker that made Pia's pulse pick up speed. Next to the coffeemaker was a dish towel, folded in half and carefully spread out in a small space between the coffeemaker and the coffee fixings themselves. The towel was supporting four white porcelain coffee cups: two espresso size and two regular size. They were sitting upside down. There were a few more clicks as Pia ran the probe over the bottom of the two regular coffee cups. Then she held the counter in her left hand and turned the cups over. She put the probe into one cup, then another. There was definite activity. It wasn't off the chart but there was more activity in the cups than in the rest of the lab.

"They got Litvinenko with his tea," Pia said excitedly. "Maybe they used the coffee here. It would explain how they both got hit at the same time and no one else was affected."

"Doesn't sound like much radiation. You think it's significant?"

"It's not much, but it is registering alpha particles. The cups were probably washed but there's still something left. Anyway it's definitely significant. Let's get out of here."

Pia took one of the coffee cups and held it gingerly by the handle. She took a padded envelope, slipped the cup inside, and put it in the reusable shopping bag in which she'd brought the Geiger counter.

George and Pia reversed their steps and emerged into the main part of the lab. Unfortunately they were in for a surprise. Marsha had apparently not gone for a late lunch, and Spaulding had returned from his. Both of them, with indignant expressions, blocked their way. Spaulding in particular had his hands on his hips and glared at Pia. "How dare you!" he said haughtily. "I told you not to come back here. And what are you doing with that?" He pointed at the Geiger counter.

Pia motioned for George to follow her. Her intention was not to engage them in conversation. She started to skirt Spaulding, but he grabbed her arm. Protectively, George moved to get in between them.

"It's all right, George," Pia said in a calm voice. "Arthur, let go of my arm or I'll file a complaint with the medical center authorities for sexual harassment."

Spaulding let go of Pia's arm. "Whose Geiger counter is that? Does it belong to this lab?" He was sputtering.

"Don't worry, Arthur, we signed it out from the appropriate department."

"But what are you using it for in my lab? I demand you tell me."

"It's a bullshit detector, Arthur. Oh, look." Pia held the sensor up to Arthur's face and it crackled with its background noise. "It appears to be working just fine after all."

Pia pushed past Spaulding and shot Marsha an indifferent look before leaving the office.

44

Columbia University Medical Center
New York City
March 25, 2011, 2:48 p.m.

George was relieved to get away from Spaulding and the lab unscathed and hoped their trespass wouldn't be communicated to Bourse. He hustled to catch up with Pia and found her waiting for an elevator.

"Third strike you're out with that guy, and you're at oh-and-two. Now you really should stay away from him."

"That's okay," said Pia, "I have no reason to go back to the lab. We picked up some clicks from the coffee cups, but it was hardly what I was hoping for. I don't know if that's going to be conclusive enough. We need more evidence."

"I was worried you were going to challenge Spaulding about the freezer log."

"I thought of it. He's such an ass. He has no authority to bar me from anywhere. I have no idea if what we found in the storage facility has any relevance whatsoever. If someone used a sample of salmonella from the freezer to infect Rothman and Yamamoto, there's no way we're really going to know."

"So do you really think Spaulding had something to do with Rothman's death?"

"If he did, he was a very small part of something larger. He's not smart enough to think that up all by himself."

Pia thought about what that might imply, about who might possibly be involved if there was a broad conspiracy. If Spaulding had been recruited, anyone could be a threat to her, a thought that made her shudder. She felt extraordinarily vulnerable, as much as or more than anytime during her childhood. As difficult as it was, she just had to keep her nerve and find some more evidence.

"George, can I stay in your room tonight? I don't want to be alone."

"Of course you can," George said. He was pleased she was asking. He only wished the circumstances were different.

When they reached the street, George expected Pia to head back to the dorm. Instead she accompanied him toward the hospital. It was raining and the wind was blowing sideways. People were moving along with their heads down, collars turned up against the cold.

"So you're not satisfied with the readings from the coffee cups?" George said. "You don't think it's enough to go to the media with?"

"I really don't think so. The few clicks we heard might not be that uncommon. I really don't know. The cups obviously were rinsed out, I think, but not scrubbed clean. I want to check the infectious disease ward. The killer may have been able to clean up after himself at the lab, but they wouldn't have been able to get onto the ward. Unless they have people on a cleaning crew."

Which was entirely possible, thought Pia.

"So that's where we're going?" George asked. He checked his watch. He still had time before the lecture he had to attend.

"Yes."

They reached the ward and quickly saw the futility of the mission. There were new patients in the rooms that had been occupied by Rothman and Yamamoto. An infectious disease ward had to be kept spotless because of the nature of the illnesses and infections treated there. The hospital was always aggressive with its general precautions, here even more so.

After a couple of minutes of scouting around, Pia looked at her watch.

"We're wasting our time here. Let's go back to the morgue."

They took the elevator down to the basement this time, taking the approved route. In daytime, there was slightly more activity than they had found on their visit the night before, when the place was presided over by the mortuary diener from central casting. The place itself had gained no luster. Without windows it could be day or night, still bleak and decrepit. There were several quite normal-looking men in their fifties whose job it was to handle the comings and goings of the corpses. They were helpful, seemingly appreciating a visit from live people. Pia and George asked them if they remembered handling Rothman's and Yamamoto's bodies. They did, because there had been quite a bit of fuss, including a warning about the possibility of typhoid and instructions about general precautions.

"We disinfected the body bags on the outside after the bodies were put inside," the first man explained. "Of course we used full precautions the whole time."

Pia and George could see no ID tags on the men and thought it prudent not to offer their own names.

"So which gurneys did you use? And were they treated afterward?"

"Sure they were treated," the man said. "And they're still where they were treated."

"Do you mind if I take a look?"

The attendant took the two students to another old autopsy room. This one had a special ventilation system because it was used for "dirty," or decomposing, cases. Pia proceeded to run the radiation detector around the gurneys, but she found nothing.

"What are you looking for, exactly?" the diener asked.

"One of the patients had had radioactive isotope therapy," Pia said, thinking quickly. "We want to make sure there wasn't any leakage. Which there hasn't been, so thank you."

As Pia and George made their way back to the elevator, George complimented Pia on her quick thinking in reference to the radioactive isotope idea.

"I had to come up with something. Maybe I'll use the same story when I go to the ME's office." Then she added, "Nix that! I couldn't get away with it with a medical examiner. From the autopsy they would already know neither of the men had cancer."

Pia pondered again. "I know. I'll say that

Rothman and Yamamoto were using a radioactive alpha emitter in their research, and we want to be sure they didn't contaminate themselves with it as well as contracting the salmonella. I'll tell them that it's a safety issue."

"Sounds good," said George.

By now they were inside the hospital lobby.

"I have to have something to eat before I go downtown. The ME's office closes at five, I think, but I'm going to faint if I don't eat something. I don't even know if I ate anything yesterday. I know I haven't today. How are you with time?"

"I've got a few more minutes before my lecture."

They went to the hospital cafeteria. It was busy, even at off-hours like this, as illness and its treatment didn't operate on a convenient nine-to-five schedule. Staff, patients, and visitors alike had to grab something to eat when they could. Pia got something substantial while George grabbed a coffee and a cookie.

"So," George said, "what about revisiting the idea of going to the police and telling them about the clicks we got with the Geiger counter? I just can't believe it's a large conspiracy. Even if it was murder with polonium, I'm sure the explanation is going to be something much more banal. Something we haven't thought of. I don't know, perhaps he had gambling debts?"

"I couldn't disagree more. If it was polonium, it would have to be a high-level operation of some kind. Polonium is impossible to get. It would mean there would have to have been intense planning for weeks — months, even. There has to be powerful people involved, there has to be. I essentially *lived* inside a conspiracy for years. I told you, George, if you saw what I've seen, you'd know people are capable of anything."

George was aware that a couple of other diners, people he didn't recognize, were looking over at them from other tables. Pia's makeup couldn't conceal all her bruises, especially in the harsh fluorescent light of the cafeteria. George leaned in toward her and lowered his voice.

"So what do you intend to do exactly before you give up on this thing?"

"There's only one more place to check for significant radiation — the bodies at the OCME. If there is none or I can't get in to check, then I'll give up my investigation. Is that good enough?"

"How are you going to get in the door at

the ME office?" George asked, remembering their conversation the previous evening with the pathology resident.

"I don't know that yet," she said, with a touch of irritation. "If I just tell them what I believe, they'll think I'm crazy. 'Remember those research guys from Columbia that were brought in here? Well, I think someone put polonium in their coffee. . . .' I may go with the alpha particle emitter idea I mentioned. I don't know. I'll see how the land lies. I'm just going to have to wing it."

"Maybe you should just go and hang out in your room. Hold off on going to the medical examiner's for today. You've been running around like crazy and you got beat up last night, for Christ's sake. I can't go with you this afternoon, but perhaps we can go together in the morning?"

"Tomorrow's Saturday. I have no idea if the ME's office is even open. I don't know how long they keep the bodies there, either. I would imagine the office encourages families to arrange to take their relatives to the funeral homes as quickly as possible. Plus, if we go tomorrow, they're likely to be short-staffed. They'll probably say, 'Come back Monday.' And I just can't sit around. I have to know."

"You're probably right about Saturdays,"

George said. “But —”

“No, George, I’m going today. You go to class. There’s no way I can get in trouble at the OCME.”

“You’re very resourceful,” George said with a wry smile.

Pia ignored the comment. “I’ll leave now. Would you mind hanging on to this coffee cup? I’ll keep the Geiger counter and hope I need it.”

George took the bag. “Not at all. We’ll meet up as soon as you get back.” He watched a focused Pia leap up and lift her cafeteria tray. “And be careful! Try to stay out of trouble!”

Pia merely glowered at him before leaving.

George watched her wend her way among the tables, deposit her tray at the window, and then leave. The idea occurred to him to call the police now that Pia was safely out of sight. But he knew if he did that, whatever the outcome, Pia would never talk to him again. He was certain that she would consider it an act of betrayal.

Tossing back the remains of his coffee, George got to his feet. At least he was going to be on time for the lecture.

Part III

45

Belmont Section of the Bronx
New York City
March 25, 2011, 3:28 p.m.

Aleksander Buda ended the call on his cell phone and held the device in his hand, then used his spatula-like index finger to tap the end-call button. There was now a problem in an operation that had gone so smoothly up to now, and he hated problems. Problems gave him terrible heartburn. He found the container of different-colored antacid tablets he always carried, took a handful, and chewed them quickly, one after another. Buda was fifty-something — fifty-what he didn't know for sure, because his family had left Albania with a few pots and pans and a little money but no documentation indicating when he was born. Over time, first in Italy and then in the United States, he had acquired the paper trappings of the immigrant, including a date of birth in 1958, but he had no idea if it was true.

Buda wasn't a big man, perhaps five-nine, but he had close-cropped hair with a scar that ran into his hairline on the right side of his broad face and enough prison tattoos on his arms, should he care to show them, to make anyone think twice about approaching him. Buda dressed unobtrusively, today in a tan long-sleeved shirt and jeans and sneakers. One might imagine he actually was a handyman for a group of East Side co-ops, work he showed up for every now and again, rather than what he really was: the head of a crew in the Albanian mob.

Buda's crew, or clan, was less hierarchically organized than a Cosa Nostra family, and leadership was often fluid and based strictly on results. Through a combination of caution and brutality, Buda's power had remained unchallenged for years. The members of the crew shared their compatriot's reputation for ruthlessness and violence, gained over more than twenty-five years of aggressive criminal activity. The Albanians had come late on the New York scene and they had been keen to make up for lost time. They took low-level positions in Italian organizations only to rise up and challenge the longer-established Mafia stalwarts.

In Europe, Albanian groups established a

strong presence in hard and soft drug trafficking, dominating the heroin trade in many countries, running the raw materials from Afghanistan through Turkey to Albania. Processed heroin and other drugs could then be distributed anywhere in the world, through hubs like the terminals at Port Newark, New Jersey. Heroin was just one business the crews were involved in. They also had interests in such prosaic activities as extortion, loan-sharking, and illegal gambling. Aleksander Buda had lieutenants working such operations so he could keep a low profile and take on more lucrative projects, such as the one he was working now, the one with a problem.

Buda was very aware that Albanian crews had developed a profile. One based in Queens had been taken down by the FBI a few years before; another in Staten Island was broken up in 2010. There were now more than two hundred thousand Albanians in the New York metro area, maybe three hundred thousand. The vast majority, all save a couple of hundred, were hardworking and law-abiding. Buda and his men drifted in and out of this Albanian diaspora, hiding among them in plain sight. The mob groups were clannish and secretive, hypersensitive to any kind of insult, and quick to use violence for the sake of vengeance. Under the Albanian code of *besa,* a man's word was his bond and a handshake was a cast-iron seal. Buda had an agreement to complete this task, and he realized he was going to have to expose some of his men to take care of this particular problem. And doing work in public was another thing that made him nervous.

After Jerry Trotter made his proposal to Edmund three weeks previously, it had taken Edmund two days to call the number Trotter gave him. Ten, fifteen times he had told himself that he wouldn't call, that he'd throw the piece of paper in the fireplace and forget all about it. Other times, he convinced himself that this was a test of his resolve set by Jerry, that if he called the number, Jerry himself would answer the phone. But at times, usually in the dead of night, when he sat by himself in his study drinking whiskey, Edmund ran through what such a call would be like to make. Say this guy actually was a killer for hire; how do you introduce yourself to someone like that? What do you say? He figured that if you called on business like this, you didn't use your own phone.

Edmund finally called the number from a

pay phone in a Laundromat on Second Avenue in the Sixties in Manhattan, a busy spot without any obvious security cameras trained on it. Edmund steeled himself, inserted his money, and dialed the number. Someone picked up but didn't speak, and Edmund ran through his rehearsed lines.

"Hello. I got your number from a friend. I have a proposition for you. This isn't a joke."

Edmund didn't say any more; the phone line quickly went dead.

An hour later, Edmund called again, from the same pay phone.

"Can we meet somewhere? I think you'll want to hear what I have to say —"

Click.

The next day, on the fourth attempt, at ten in the morning, a thickly accented voice said, "Call in an hour. Pay phone bank at Grand Central. Main-level concourse."

Edmund did as he was told.

"Take six train to Morrison Avenue, exit platform north side and wait."

Edmund was at a point of no return. All he'd done was make a few phone calls, but now he was going to meet someone he knew was a killer. He looked at the commuters walking through Grand Central, ordinary people like him. If he went ahead with this, he would no longer be ordinary. In the recent endless days and sleepless nights, Edmund had weighed the possible costs of doing what Jerry demanded and doing nothing. If he failed to act, he'd be ruined financially and personally. But Jerry's terrible scheme gave him a chance.

Another thought had occurred to Edmund and was proving impossible to ignore. These doctors were destroying his business. It was their fault he was in this position, and he was damned if he was going to let them get away with it.

Edmund rode the subway north to a section of the Bronx he'd never visited before. He got off the train on a windswept elevated platform. There was hardly anyone else around at eleven in the morning, just two men who had alighted at the station — one who had sat in Edmund's car, another from the car behind his. Edmund left the station, walked down to the street, and stood by the exit. He checked his phone, and crossed and recrossed the street, looking for some sign of life.

Suddenly, a dark blue panel van drew up, and the back doors opened. A voice from within told Edmund to get in, and he did. The van drove away, and immediately Edmund's arms were seized, tape was secured over his mouth, and a cloth bag was roughly forced over his head. His body was patted down, hands thrust into his armpits and groin. Then his clothes were removed, all of them, and he was left naked, bound, and gagged on the floor of the van, first as it rattled along the street and then, for what seemed like an age, parked someplace.

"Okay, Mr. Edmund Mathews, rich banker man from Greenwich, how did you get that phone number?" The voice came from the front of the van somewhere.

Edmund tried to talk but his mouth was taped shut. He mumbled and the voice said, "How rude of me. Let the man speak."

The tape was ripped away crudely, and Edmund reeled from the shock.

"A friend of mine gave it to me. He wouldn't say where he got it."

"We'll see. So what you want?"

Edmund laid out what he wanted. It didn't take long, but he had to explain a couple of times the need for using polonium to effect the killings.

"Okay, this is what we do. You come to Middletown Road subway station, eleven tomorrow. Bring a deposit for me. As a gesture of goodwill. Say, fifty thousand dollars in Ben Franklin notes. Nonrefundable. Give the man back his clothes."

Edmund's arms and legs were freed, and

he dressed quickly. The van moved again and stopped after a few minutes, and the doors opened. Edmund got out in a bleak parking lot behind an abandoned building. He figured out where he was, less than a half-mile from where he had been picked up, and he took the subway back to Manhattan.

More than at any point throughout the whole ordeal, Edmund's flight reflex was strongest that night. If he called the FBI, surely he could give them Jerry and this guy, whoever he was, and at least he would be free from this crazy plot. But he wouldn't be free of LifeDeals and Gloria Croft and his own imminent destruction. The Statistical Solutions data had finally come in, and merely underscored what Russell and Edmund already knew. Their model was shot to pieces the moment regenerative medicine became a reality. His need to stand and fight kicked in.

Edmund traveled again to the Bronx, and was again driven away in a van, this time of a different color. Again, he was bound and stripped, but his clothes were returned more quickly this time and his mouth wasn't taped, a small mercy for which Edmund was grateful. He could feel that the envelope with the $50,000 was no longer in his jacket pocket.

"Thank you for the money," the same voice said. "A more cautious man would throw you out of the van now and be happy with good takings for one day's work. But I read about you, Mr. Mathews, and I am intrigued. Then I read about the people you say you want to die and I think, *What are they doing? I don't understand, I am a stupid peasant.* Then I think, *This guy must be for real.* I don't know why but I do. I also think this is a very expensive idea. Someone has to go to Russia, buy this radioactive material from some very bad men and not get caught. They have to give this material to the marks, plus the bacteria, and not get caught. We can do it, but not for one million dollars."

"How much, then?"

"Two. And a half."

"Jesus."

"Mr. money guy, I see where you live, how much money you make. You are not doing a trade, here, on Wall Street. I don't negotiate — that is the price. And tomorrow, the price is more."

"Okay."

"Sorry, speak up, please."

"Okay," said Edmund.

The two men met once more, three days

later. Edmund told Russell he needed a huge amount of cash but not what it was for. Russell asked once, and Edmund bit his head off so Russell just did what he was told. It took Russell two and a half days to assemble one and a half million dollars from various business and personal accounts. Edmund packed it into a large baseball equipment bag and drove to the address he had been given on the phone. It was the same parking lot where he had been let out the first day. Once more, Edmund got in the van and went through the same degrading procedure.

"I guess you trust me," the voice said. "I now have one-point-five-five million dollars from you and I haven't done shit. But I am a businessman, and I will fulfill my end of the deal."

The man gave instructions for paying the rest of the money once the job was finished. The job would be done sometime in the next month. Edmund said nothing.

"But one more thing first. Something I need to know, otherwise I am afraid I won't be able to go ahead."

Edmund said nothing.

"Who gave you my phone number? Was it your partner, Mr. Russell Lefevre?"

"No."

"So, who was it?"
Edmund was silent.
"I really want to know."
So Edmund told him.
"Okay, thank you. Now release Mr. Mathews."
From the front seat of the van, a man turned back toward Edmund. He was wearing dark glasses and a baseball cap but Edmund could see a livid scar on the man's forehead running up toward his scalp. The man was holding out his right hand.
"Shake my hand, then we have a deal."
The men shook hands, and Edmund heard nothing more, until March 25.

Aleksander Buda thought some more about the information he'd just received, and with his phone still in his hand, he called Edmund Mathews.
"Yesterday we followed to the letter the course of action you recommended, and it didn't work. I didn't think it would. Someone I have on-site tells me he saw her going around today with a Geiger counter, her and that guy she hangs out with. You know what that means? It means someone's seen through your brilliant plan."
"Shit."
"Yes, shit. As in what we are standing in

up to our knees. Unless we do something right now, it's going over our heads."

The conversation was much too specific for Buda's liking. But he felt he had to get the okay from the banker and make sure Edmund realized the price had gone up. The job itself shouldn't be difficult, the girl was hardly being discreet, but he had a troublesome task to take care of. When he heard the girl's name was Grazdani, he paused. It sounded Albanian to him, and he wanted to be very sure that killing an Albanian girl wasn't going to step on anyone's toes. He didn't want to be the cause of a blood feud like there'd been in the 1990s. He'd have to snatch her, hold her, and put feelers out to see if there were any Grazdanis in the crews in the neighboring areas. But what were the odds?

"Okay," Edmund said finally, feeling the same numbness he had felt when he agreed to the deal in the first place.

"There are two of them, actually," Buda was saying. "It will be another ten percent."

"Ten percent of the total or the balance?"

"Ah, ever the money guy," said Buda. "The total."

Buda ended the call and summoned Prek Vllasi and Genti Hajdini to his office in a trailer parked inside a low-ceilinged warehouse. Buda dressed his senior lieutenants down severely in Albanian.

"You were useless last night. She wasn't put off at all. Didn't you do anything to her?"

"You see," Genti said to Prek, "we should have done her when we had the chance. Like I said last night, knocking her around wasn't going to be enough." He turned to Buda. "She's a tough bitch."

"I can see that," Buda said. Genti had been nursing a black eye all day. "Now this whole thing is about to go down the drain because of her. She knows what happened, God knows how. You gotta get back over there right now and take out her boyfriend and grab her off the street."

"Boyfriend?" Prek said. "What boyfriend? You mean the kid she was hanging around with last night? Unless he's with her when we grab her, I don't know if we'd recognize him."

"He's the guy with his tongue in her ear!" Buda snapped. He was steaming, but he quickly realized Prek was right to be cautious. Killing the wrong man would be counterproductive.

"I'll get a picture of him from the medical school database and send it to your phones.

His name is . . . George Wilson," he said, referring to a note.

"And remember, pick up the Grazdani woman," Buda said. "And don't touch her, you animal, unless she's not related to anyone important, in which case she's all yours. Understand, Genti? Word is she was in Rothman's lab just a few minutes ago, poking around. Take her to the summer house and call me when you get there. And take Neri Krasnigi with you. It seems like the two of you can't handle her."

Krasnigi was relatively new to the crew, younger, inexperienced, and more vicious than either Genti or Prek. The two men were affronted by the order but didn't show it.

As the men exited the trailer, Buda shouted after them, "Use the white van for the pickup and then dump it. Take the blue one to the house."

Prek gave a thumbs-up sign and walked away.

Prek and Genti found Neri Krasnigi sitting in a battered old armchair at the back of the warehouse reading a German *Playboy.* Prek told him to follow, and the three men got into the white van. The plates were obscured with what appeared to be dried mud but which was actually cleverly painted plaster.

As they pulled out into Lorillard Place, heading quickly toward East Fordham Road, Prek filled Neri in on the afternoon's operation. What they had in mind to pull off was a pair of Albanian specialties: a blindingly fast hit and snatch, in broad daylight if necessary. In the Albanian mindset, it didn't matter. Neri was excited; this would be his first official hit. They checked that their automatic pistols were loaded. Duct tape, blankets, balaclavas, two Columbia University Medical Center police uniforms, and a can of Ultane, a volatile, rapid-induction anesthetic, were piled into the back of the van.

The white van pulled into a garage and Genti got out and climbed into a blue van. Starting it up, he followed Prek in the white van. They parked the blue van near the George Washington Bridge and set out again in the white van toward Columbia University Medical Center.

46

Columbia University Medical Center
New York City
March 25, 2011, 3:54 p.m.

Unbeknownst to George Wilson, at the same time he was thinking about contacting the police, Pia was too. Not seriously, but it had crossed her mind. It was certainly true that if an investigation needed to be done, they were far more capable and could turn over rocks she couldn't even approach. But then there was the issue of what she would tell them. Would she say that she'd picked up a few lonely alpha particles in a coffee cup and thought it was evidence of a grand conspiracy? Of course not. There was no doubt in her mind that she would not be taken seriously, which ultimately would increase her vulnerability rather than decrease it. The police would think she was crazy and probably call the dean, believing they were acting in her best interest. Of course, going to the police had another downside. They might be tempted to look her up in their computer, and even though bad stuff about her teenage years was not supposed to be there, it might be. No, she would not go to the police. Instead, as planned, she'd go to the OCME in a final attempt to figure out the affair. If that didn't yield overwhelming evidence of wrongdoing, she'd give up, just as she'd told George.

When Pia had reached the street coming from the hospital cafeteria, her intention had been to head down to Broadway to take the subway downtown. But feeling the temperature and noticing that the rain had increased, she decided to take a quick detour back to the dorm for a better coat and an umbrella. She knew the subway would get her only so close to the OCME and that she'd end up walking. How far, she had no idea.

In the dorm she hesitated outside her door, just as she and George had done outside his earlier. Being attacked in her room the previous night made her paranoid. She didn't know how the men had gotten into her room.

Repeating what she and George had done, after silently unlocking her door, she opened it suddenly, prepared to flee if needed. She also carefully checked her bathroom to make certain it was empty. It was.

With a warmer, rain-resistant coat and an umbrella, Pia set off on her way to the subway. She had put the Geiger counter in another shopping bag to make it easier to carry. She checked the time. It was almost four, so if she was going to make it to the OCME before it closed, she had to get moving.

Walking quickly in the darkening day, Pia passed the Black building. She went another fifty feet along West 168th Street when she saw two hospital security police up the street, walking toward her. She stopped. She couldn't see them well in the fading light with mist rising off the pavement, but she could see their uniforms well enough. They were the same uniforms her attackers had worn the previous night. To make matters worse, they seemed approximately the same height and build.

Fighting the urge to flee, Pia stopped dead in her tracks. Ahead to her right was a porte cochere for the hospital. Pia considered racing to it and into the hospital where she could vanish into the crowds, but she'd hesitated too long. She'd have to pass the security guards before she got to the entrance.

She looked behind her and saw there were surprisingly few people on the street. She thought about dashing back to the Black building but thought that if the men wanted to catch her, they probably could before she got inside. Turning back around, she eyed the approaching guards. They seemed to be staring intently at her. She froze, suddenly remembering a similar scene imbued with the same fear and dread.

She was thirteen at the time and had been at the Hudson Valley Academy for Girls for maybe a year. The stress of constantly being on guard, the fear of being attacked at any time, wore on Pia. A couple of times she'd cracked and tried to run away from the school, and she did so again. This time she got lost and had to spend a harrowing moonless night in the woods surrounding the school grounds. The night wore on and on and hummed with threat. Pia tried and failed to make her way back to the academy in the hope of breaking back in before anyone realized she was gone.

Pia had spent the hours before dawn slumped against the trunk of a tree, dozing fitfully. She arose at first light and walked east toward the sun until she found herself on an unfamiliar street winding downhill. It was then that she saw two policemen in the middle distance. They were implacably and threateningly walking toward her, staring at her unblinkingly, silent like automatons. Pia froze, as if by standing as still as possible they might not see her. When they were within ten feet of her, they parted, one walking to her left, one to her right. Perhaps they hadn't seen her! Perhaps they weren't looking for her at all! But when the men came alongside her, they suddenly lunged at her, roughly grabbing an arm each. Once again she was a prisoner of the state, totally vulnerable.

Now the same feelings flooded Pia's brain. The security guards were coming at her with the same silent intensity, drilling her with their beady eyes. Pia stood stock-still and closed her eyes. Just like the Eden Falls policemen, the men parted right in front of her and went on either side . . . and moved on. The man to her left brushed past her upper body, then turned and said something — was he apologizing or saying "Get out of the way," or making a lewd remark? Pia didn't know. She breathed out with relief, unaware that she'd been holding her breath.

Pia was embarrassed by the depth of her paranoia. She shivered and a chill ran down her spine. The unnerving episode had taken only seconds. She wiped perspiration from her forehead and then hurried on toward Broadway. By the time she passed the hospital entrance, her breathing and pulse were almost back to normal. On Broadway she relaxed further as there were significantly more people. She felt safer there — she was safer there. In the near distance loomed the subway entrance swallowing pedestrians like an insatiable monster. A gust of wind blew the umbrella she was holding almost out of her hand, and she fought with it for a moment to get it under control. Once she did, she closed it and hurried forward to the stairs.

47

168th Street and Broadway
New York City
March 25, 2011, 3:56 p.m.

"Holy shit, there she is!" Genti shouted, pointing, as Prek made the right-hand turn from Broadway onto 168th Street running between the Columbia University Medical Center on the left and the Armory on the right. Genti's glance had alighted on the woman carrying a shopping bag when the wind blew her umbrella, threatening to lift her into the air. He saw her face clearly, and it was the Grazdani girl, no doubt about it. She was walking quickly toward the subway entrance. "Stop!" Genti shouted as Prek slowed.

"Who's with her?"

"No one, I didn't see no one," shouted Genti. "Let me out. Wait for me here."

Genti, who was riding shotgun, leaped from the van the moment Prek was able to stop. He ran after Pia, who he guessed was maybe twenty yards ahead of him, making her way toward the subway entrance. As he ran, Genti checked that the gun was secure in his jacket pocket. He wasn't certain what he'd do if he caught up with her — should he shoot her in the street? Grab her and bring her back to the van?

He just knew the one thing he couldn't do was lose her.

Genti watched Pia disappear from his view as she descended quickly into the station while he weaved in and out of the cars, gypsy cabs, buses, and vans on the busy corner of Broadway and 168th Street. He reached the subway entrance and rushed down the stairs but couldn't see her. Was she catching the A train or the 1? Probably the express train, the A, he thought. He was desperately looking ahead for her, half pushing people out of the way.

"Excuse me, excuse me." He didn't want to get too aggressive: New Yorkers were liable to get aggressive right back at you. Genti rarely rode the subway and didn't have a Metro card that he could swipe to get through the turnstiles, and he certainly didn't have time to stop and buy one from a machine. Hoping there were no cops looking out for fare dodgers, he followed a schoolkid and pushed through the turnstile behind him.

Genti had to choose, the A or the 1. Changing his mind, he opted for the 1, and as he neared the giant, aging elevator that took riders to the platform deep underground, he caught a glimpse of Pia at the head of a group of passengers who'd just got on. The doors began to close. He saw she was standing to the side, ready to be first off. He raced forward.

"Hold the elevator!" he shouted. "Hold it."

Genti reached the doors as they were almost closed and frantically tried to stop them. For a moment his hand was trapped and he was forced to pull it out quickly. He looked around. Stairs. Genti wasn't careful anymore; he pushed past an old woman and leaped down the refuse-strewn stairs three or four at a time. He didn't know that the platform was the equivalent of eight stories below and pressed on, dodging the few passengers walking up or down, yelling at everyone to get the fuck out of the way. He was breathless but reached the bottom only to find that the elevator had already discharged its load of descending passengers, and new ones were already boarding.

Genti took in lungfuls of air, his hands on his knees. He was the first to admit not being in the best shape. Then he heard nearby the high-pitched squeal of a subway train's brakes. Uptown or downtown? He guessed she was going downtown, as the majority of people were doing. He pushed forward and heard the mechanical sound of train doors opening. He entered a barrel-ceilinged passageway leading from the elevators to the station itself. Suddenly there was a crowd of people coming toward him, filling the tunnel from one side to the other. They'd just disembarked from the train, and he had to fight his way through them. When he reached the platform, he looked up and down its length, spotting Pia farther down the platform.

Genti saw her clear as day. She was right there, maybe thirty feet in front of him. She stepped into a car.

Then Genti made a mistake. While he waited for the "Please stand clear of the closing doors" announcement that always preceded the train's departure, Genti marched down the platform intending to board the train through the same door she did. Then, as he drew level, the doors closed without warning. Genti banged on the door and looked back toward the conductor about twenty feet away. "Hey, man, the door!"

The conductor ignored him, and the train's brakes released with a hiss of air. Genti looked into the car. Pia's and Genti's eyes locked for a brief second before the train began to pull away from the station. All Pia saw was another guy trying to push his way onto the train.

Genti turned to stare down the conductor, who drew his head into his cab with a slight smile on his face as the train picked up speed. Genti watched it disappear into the tunnel, keeping his eyes on its taillights until they disappeared into the gloom.

He'd failed.

Genti walked back to the elevator. In some ways he was embarrassed that he'd missed the girl, but he rationalized that it probably was for the best. Maybe he would have had trouble getting her out of the station without interference. Besides, he reasoned, if they were supposed to get the boy too, it would be easier to get them together and deal with them at the same time. If they'd taken the girl, the boyfriend might have gone to ground and been hard to find.

By the time Genti had ridden up in the elevator he was feeling much better about having missed Pia. And he had been reminded of what a beauty she was. He looked forward to snatching her off the street and taking her to Buda's isolated summer house where no one could hear anything going on inside.

Genti stepped onto the street and looked for the white van but couldn't spot it. He called Prek, who told him he was down beyond the Neurological Institute where he'd been able to snag a great parking place between the medical school and the dorm just beyond where 168th Street turns into Haven Avenue.

Genti walked west and soon saw the van. He got in and related how he'd just missed her in the elevator and then missed her a second time, as she was getting on the subway car. He said he was so close it was frustrating.

"It's not the worst thing," Prek said, echoing Genti's earlier thoughts. "This is the perfect spot right here. Hopefully, if we're lucky, when she shows up it will be with the boyfriend, and we'll be waiting. I'm thinking we need to get them together."

"How will we know which one he is?" Genti said. "I bet she has a lot of boyfriends."

As if Aleksander Buda were reading their minds, Prek's cell phone buzzed. It was an e-mail from Buda with an attachment. When Prek opened the JPEG he found himself staring at George's medical school admission photo listing his height and weight.

"He's six-one, one-ninety, with blond hair," said Prek. "Taller than most. We shouldn't have too much trouble picking him out."

"Perhaps he'll come by on his own," Genti said.

"No, my sense is they'll be together since they seem to be spending a lot of time together recently. I know I would if I was him. I imagine they'll meet up when she comes back from wherever she's going."

48

Corner of First Avenue and Thirtieth Street
New York City
March 25, 2011, 4:40 p.m.

As the last of the daylight threatened to fade away completely thanks to the low clouds and rain, Pia paused outside the Office of the Chief Medical Examiner, gazing at the front of the half-century-old building. It wasn't inviting in the least, with its weird façade of blue glazed tiles that reminded her of the Ishtar Gate of ancient Babylon. She'd seen pictures of the latter in an almost equally ancient *Encyclopædia Britannica* at the academy. She glanced at the tile and then at the outdated 1960s-style aluminum-framed windows. World-class ugly.

Pia had been worried about making it in time, but she'd been lucky with the trains. Now that she was there, she wasn't as confident as she had been about going in cold like this. She had no contacts there, no one she knew she could trust, no one she knew on any level, and this wasn't a feeling she liked. She was very aware that this was the *New York City* OCME and she'd had a lot of bad experiences with various city agencies as a child. The state may have provided her with food and shelter, but it had also fed and sheltered her enemies and abused her. There was not a lot of reason not to worry that this city agency wouldn't be equally as nasty.

Pia had another thought. What would happen if she went in there and actually succeeded in her mission? What if she asked to see Rothman's body and somehow got permission and then found that he had in fact been irradiated, proving that he and Yamamoto were murdered? Her mind raced. There'd be a huge police action, and she'd be at the center. It would be a media circus. She and George would be questioned, she'd have to voice her myriad suspicions and assist with inquiries and make statements and possibly appear in court. But then she touched her tender jaw and remembered the beating and the warning she'd gotten the night before. She had no choice. The answers to her questions lay in this building, or they were nowhere at all.

Taking a deep, fortifying breath, Pia tucked her umbrella under her arm and went in the front door.

The reception area showed every bit of its fifty years. It was dark and somewhat dingy with a worn dark brown leather couch and a few other worn, mismatched chairs. The linoleum floor was cracked and scarred. On a low coffee table were some dog-eared magazines with the address labels torn off the covers. There was a crowd of people sprinkled on the furniture and some leaning against the wall. Soon it became apparent they were all together, members of a grieving African-American family with at least three generations represented. Two teenaged girls were hugging in a corner, crying softly.

"Excuse me," Pia said, approaching the receptionist. A security guard sat at another table off to the side next to a set of glass doors. The receptionist, carefully dressed and pleasantly plump, was wearing a name tag that read "Marlene." She looked up. She had a friendly smile.

"Yes, honey."

"Hi. I'm a medical student and I'd very much like to talk to one of the medical examiners."

"Are you here to find out about teaching opportunities at the OCME, like the elective for medical students?"

"Maybe," Pia said, wanting to keep her options open.

"Maybe?" Marlene questioned, with a smile. "What is it exactly you want to talk to the medical examiner about?"

Pia hesitated.

"Actually, I'd like to discuss a specific case. Two cases, to be more precise."

"Are they relatives of yours?"

"No."

"Perhaps it's best if you talk to the department of public relations if it's information about a specific case you want, seeing as you're not a relative."

Pia saw she was losing ground. The last thing she wanted was to be shunted off to the public relations department, where she surely wouldn't get access to Rothman's corpse. Fearing she might be turned away, Pia studied the receptionist's face, trying to think of an approach. Marlene seemed to be a congenial person and for a brief moment Pia toyed with the idea of telling at least a partial truth. Quickly she decided against it. Any way she explained the reason she was there that approached the truth would sound too bizarre.

"Actually, I also want to talk about the teaching opportunities here. The two cases I mentioned are teaching cases that I've heard about. I've come all the way down from Columbia." Pia smiled to try to cover up any potential inconsistencies. "I've become interested in forensics. Very interested."

Marlene was confused — what did this girl really want? She was also impressed that she had come all the way from Columbia Medical Center way up in Washington Heights. That took some effort, especially late on a Friday afternoon. Marlene didn't have the heart to send her away without talking to someone. Besides, she was a pretty thing, and Marlene knew exactly who would be more than happy to speak with her.

"Okay then. I'll call Dr. McGovern."

"Is he a medical examiner?"

"He's a medical examiner and he also happens to be the coordinator for teaching at OCME."

"Thank you." Pia was delighted.

Marlene put in a call to Chet McGovern and motioned for Pia to take a seat. Pia stepped away from the receptionist's desk. She didn't sit down as there were no chairs available. It was close to five o'clock, so she had to make a quick impression on this McGovern guy. A moment later the glass doors opened and a heavyset woman in a lab coat appeared, holding a clipboard. She introduced herself to the grieving family as Rebecca Marshall, an ID coordinator, and asked them to follow her. Dutifully the entire clan disappeared through a door marked ID ROOM.

Pia took one of the newly vacated seats and tried to be patient. While she waited she tried to decide what approach to take with the medical examiner. Should she be aggressive or coy? Eventually she decided she'd have to wait and see what kind of man Dr. McGovern turned out to be. She was hoping for someone on the youngish side, someone she could flirt with to a degree. Over the years she'd learned she had an effect on most men, and she thought that this was one of those situations where it could work to her advantage. Usually it was the opposite.

A few minutes later her prayers were answered when a youngish man came through the inner doors in a long white lab coat with the confident air of a doctor. When he took one look at Pia, who was the only person in the waiting room, his face lit up. Pia recognized the reaction. She'd seen it too many times not to. He appeared to be in his forties, early fifties at most. He was blond and good-looking in a masculine, all-

American way, somewhat like George, and Pia could tell he was in good shape.

He came right over to Pia like a bee to honey and introduced himself. Pia did the same, avoiding his stare. She recognized the type immediately: an irrepressible Lothario who undoubtedly saw every attractive single woman younger than him as a challenge. She was encouraged.

After the introductions, which included him repeating proudly that he was indeed the current coordinator for teaching at the OCME, he said, "Let's go up to my office and see if we can't help you out. Thanks, Marlene."

McGovern winked at Marlene behind Pia's back, and Marlene rolled her eyes.

As McGovern escorted Pia to his third-floor office he peppered her with questions about where she was at medical school, what year, and what she thought she might like to specialize in. He mentioned how interesting medical forensics was and offered up his credentials.

Pia played along, answering McGovern's questions, acting as if she were interested in his life story and achievements. They entered McGovern's small office and sat on either side of McGovern's untidy desk.

"Sorry about the mess. So what can I do

for you, Miss, er . . ." McGovern's eyes shone with the struggle he was having remembering her name.

"Grazdani. Thank you for seeing me with so little warning."

"My pleasure, I'm sure."

"I want to know about autopsies that were performed here on Dr. Tobias Rothman and Dr. Junichi Yamamoto from Columbia University Medical Center. They died the morning of March twenty-fourth — yesterday — of typhoid fever." Pia was all business, taking McGovern aback. "I'm assuming autopsies have already been done," she said.

"Er, well, er, I wasn't involved in either one, and I haven't heard much scuttlebutt around the office other than one of them was the famous Nobel-winning researcher. All I heard was that he died of an extremely aggressive infection. But let me check what we have."

McGovern keenly wanted to be helpful. He looked at Pia, and she half smiled back at him. Using the computer on his desk, McGovern looked up the names to get the OCME accession numbers and then brought up the individual cases.

"Here we are. Yes, it says the autopsies were performed the afternoon of the twenty-fourth, so that was, er, yesterday. The day they died." McGovern scrolled through one file, then the other.

"They certainly were both serious infectious cases with severe erosion of the gut, both small intestine and large. Wow! Anyway they're considered OSHA cases, which was the main reason they were autopsied."

"OSHA cases?" Pia questioned. She'd heard the acronym but couldn't remember what it was for.

McGovern looked up. "The Occupational Safety and Health Administration. It's a government agency that gets involved when there are deaths in the workplace involving public safety issues. The autopsy results will be reported to OSHA as the OCME is required to do by law."

McGovern looked back at his screen.

"Okay. Both cases were done by Dr. Jack Stapleton. He's our super-doc who does more cases than anyone else. He's never satisfied, always pushing for more, works hard like he doesn't have a life.

"Let's see. The cause of death for both cases is listed as infectious disease — typhoid fever — and the manner of death is accidental. Let me ask you, do you know why the manner of death is considered accidental?"

Pia said that she didn't, not adding that she might be challenging that official verdict.

"If the two researchers had come down with typhoid after eating at a restaurant, like the hospital cafeteria, then their deaths would have been labeled natural, since typhoid is a food-borne pathogen. But since they contracted the disease in a laboratory, or in a workplace setting, then it's accidental because it certainly couldn't be considered a natural process."

McGovern was trying his best to sound authoritative.

"And if for some reason the researchers infected themselves on purpose, then the manner of death would be suicide. And last but not least, if someone purposely infected them, then it would be homicide."

McGovern laughed and held his hands out wide as if to say, "See what a good teacher I am."

Pia didn't laugh with him or even smile. For her he was acting stereotypically transparent. *He's talking to me like I'm a college coed,* she thought.

After a slightly awkward beat because of Pia's lack of response, McGovern said, "Do you have any specific questions about the autopsies? If you do, I can call Jack and ask him directly. I know he's still here."

Chet McGovern would have liked nothing better than to have Pia indebted to him for his help. An hour earlier he'd learned his Friday-night plans had fallen through, and he hated spending the best night of the week on his own. He was about to ask her if she was free and if she might like to have a bite of dinner when he noticed she was lifting her bag up onto the desk. She then reached into it and pulled out a yellow instrument, a lead, and mike-like device attached. It took McGovern a minute but he recognized it as a Geiger counter.

"Well," Pia said, "to be honest, what I'd really like to do is check if Rothman and Yamamoto might be emitting a small amount of radioactivity. I mean, if that would be all right."

"I suppose," McGovern said, not wanting to say "no" but confused by the strange request. There was obviously something she wasn't telling him, but he decided to play along. "Why do you think they might be emitting radioactivity?"

Here was the thousand-dollar question. She still hadn't decided how she was going to respond, even though she had been reasonably certain it would come up. She could go for broke and voice her suspicions or be more prudent and try to be obtuse about it. On the spot she decided on the latter.

"I'm involved in a project for a thesis involving radioisotopes used for research," she said. Pia decided this wasn't the time to raise suspicion about why she was really there at the OCME. She didn't want to show her hand just yet. She didn't want the OCME calling up the medical center and talking about her visit because it would reveal to whoever was involved in the conspiracy that she hadn't stopped her meddling.

"I worked in Dr. Rothman's lab for more than three years, and I know that certain isotopes were used in that period for various experiments. I just want to be sure there hasn't been any contamination to the personnel. I checked Rothman's lab and there was a very small amount of what we want to believe was rogue radiation in the office by his coffeemaker. I hope you can help. It's for everyone's peace of mind."

Pia stopped. She knew what she had just said didn't make total sense, but it sounded good. She smiled as pleasantly as she could. She hoped her smile didn't look as fake as it felt. She could tell that McGovern was suspicious and hesitant, but that he hadn't ruled out granting her request.

"Is that what you told Marlene downstairs?" he asked.

"I told her I was interested in a couple of particular cases."

"Oh, okay. She said you wanted to know about the OCME electives. Never mind. Listen, we have radiation detectors down in the mortuary area just in case, and nothing has sounded recently, especially not yesterday. I know that for a fact."

"Well, that's not surprising because the isotopes we've been using in the lab were all alpha emitters for targeted alpha therapy such as bismuth-213 and lead-212, which wouldn't be picked up by general radiation detectors made for beta and gamma radiation."

Pia smiled again and McGovern nodded knowingly, even though he had no idea what she was talking about. The last time he read much about radioisotopes was over a decade earlier when he was studying for his boards. McGovern looked pensive. Pia thought he was thinking about alpha particles. In fact McGovern was running a mental checklist. At first he'd questioned himself, but no, he was certain. He'd never seen a better-looking medical student, which was saying something as they were, in his opinion, getting better-looking every year, at least at NYU, which was where most of the medical students he met in his position as OCME teaching coordinator were from. He should spend more time at Columbia, he thought.

"So you just want to make sure Rothman's and Yamamoto's bodies are not emitting alpha radiation?" McGovern asked, just to be certain he understood.

"That's right. That's why I brought this Geiger counter. It's specially programmed for detecting alpha particles."

McGovern went back to his monitor.

"Let's see. There might just be a problem. The bodies of infectious cases like these don't stay around here very long, for obvious reasons. . . . Yup!" he said suddenly, tapping the screen with a forefinger. "Just as I thought. There's a problem. As I said, in serious infectious cases like typhoid fever and a few other communicable diseases, the bodies aren't held here in the OCME. After the autopsies are completed and the cause and manner of death corroborated, the bodies are released to the families and the respective funeral homes and cleared for cremation. In other words," McGovern said, "the researchers' bodies are no longer here. You're about twenty hours too late."

Pia mouthed a repressed "shit," which

McGovern caught and appreciated. He associated colorful language with feistiness, and he loved feistiness in a woman. It was his hope that now that he'd ascertained the bodies were no longer at the OCME, perhaps they could move on to more interesting topics, like Friday-night plans. Meanwhile, Pia stared into the middle distance, thinking. She could hardly reproach herself; twenty-four hours ago, when the bodies left, she'd never even heard of polonium-210.

Watching Pia's expression, Chet suddenly was afraid that after hearing the news she might get up and leave. She was clearly disappointed. In his mind, her leaving at that point would be a major tragedy because so far he'd not gotten either cell phone number or an e-mail address from her.

"The guy who did both autopsies is just down the hall," Chet reminded Pia. "And he's a friend. So if you have a specific question about what he found, I'm happy to go ask him."

Pia was disappointed. It had never occurred to her that Rothman's and Yamamoto's bodies would have already been sent to funeral homes. She thought briefly about trying to find out the names of the funeral homes, but she didn't know how she would do that without raising a lot of suspicion.

As for talking to the ME down the hall, what would possibly be the point?

49

Columbia University Medical Center
New York City
March 25, 2011, 5:25 p.m.

Earlier, while Prek and Genti had sat in the van, Neri Krasnigi, the recruit foisted on them by Buda for God knows what reason, had been continually walking along 168th Street and the small portion of Haven Avenue from the medical school entrance back to the van. He'd been ordered to have his cell phone in the radio mode to act like a walkie-talkie to stay in contact. Neri was dressed in one of the security guard uniforms, which now looked pretty wet from the rain. Prek knew he was taking a chance that Neri might bump into one of the real security guards, but he had been willing to risk it. Prek wanted as much notice as possible when either Pia or George appeared, heading in their direction on their way to the dorm. Nevertheless Neri had been ordered back inside the vehicle.

Prek was as content as he could be given the situation. He was certainly keyed up as he always was before a hit, especially with a couple of Red Bulls under his belt. He had another beside him just in case he needed it. The radio in the van was playing heavy metal with the volume turned low. As he sat and waited, Prek methodically rubbed the scar on his upper lip. It was a habit that he wasn't even aware of. It was now almost five-thirty.

Aleksander Buda had called to check in at five o'clock, and Prek had to explain that they had spotted the girl but lost her in the subway. When Buda exploded with a string of choice expletives in Albanian questioning the virtue of Prek's mother and parentage, Prek had held the phone away from his ear and even blushed slightly. Neri, who could hear Buda clearly even though he was sitting on a milk crate in the back of the van, let out a chuckle before he could stop himself, earning a scowl from Prek. As soon as Buda's volcano subsided, Prek held the phone to his ear again.

"Was she carrying anything, like an overnight bag?"

"No. Just a shopping bag and an umbrella. I'm sure she's coming back."

"She better be. . . . What about the guy?"

"No sign of him yet. He may be in class or whatever it is medical students do. They're just getting out now, streaming by the van. Of course he could be in his dorm room having passed by before we got here. But from watching her like we did, I'm sure they'll meet up. And we'll be here."

"Don't fuck this up," Buda said, and ended the call.

Prek looked around the floor of the van by his feet and picked up one of the empty Red Bull cans he'd dropped there and hurled it into the back of the van in the direction of Neri.

"You ass, you think this is funny? Get the dry hospital security uniform on. You're going back out for a walk."

The command-performance radiology lecture George Wilson had made a point of attending had finally finished. Unfortunately it hadn't been great. The speaker had a soporific voice, and George and the rest of the attendees had had a difficult time staying awake. Late lectures were a problem in that regard for most people, especially when the lights dimmed for the de rigueur slides. Halfway through the talk, George's mind had wandered to what Pia was finding downtown and whether or not she was safe and keeping out of trouble. George knew that if she caused trouble and the OCME called Bourse to complain, it would probably be the end of Pia's medical school days, at least at Columbia. As the lecturer had droned on, George found himself wishing he'd gone with her.

George got his stuff together and exited the lecture hall. He certainly hadn't learned anything. Reaching the street, he donned his coat and turned up the collar. It wasn't raining so much as drizzling. He had a knot in his stomach from worrying about Pia. He was worried that he'd allowed her to go on her own and wondered when he'd hear from her.

Along with a large clot of first- and second-year students, George walked through the early-evening air toward the dorm and past a young security guard who seemed to be patrolling the front of the building. George looked at him quickly, as he had no umbrella and his black fake-leather coat with its fake-fur collar appeared soaked. He looked about seventeen, and George paid him no mind. He walked into the dorm building and waited with the throng of students for an elevator. For the fiftieth time, George checked his phone.

There was no text message from Pia, no call or e-mail.

When he reached his room, George flopped down on the bed, exhausted and hungry. He suddenly felt alone and afraid. He knew he wasn't nearly as tough as Pia. Armed with what little he knew of her upbringing, he was aware of how much she'd been through in her life. It was so much more than he had ever experienced. Sure, his dad died when he was young, and there hadn't been a lot of money around when he was growing up, but his mother had always made sure he was loved and looked after. She paid attention to his education — made sure he studied and guided him through high school and college and on to medical school. She was always there, making sure he worked hard enough to justify the scholarships he needed to attend Arizona State University and then Columbia Medical School. All in all George had had support and security all his life, exactly the opposite of Pia. Vaguely he wondered where he would be today had he shared Pia's experiences. Probably in something like a hamburger joint slinging hash.

Suddenly George missed hearing a friendly voice. He called his mom but got the ancient answering machine she still insisted on using. He didn't leave a message. Then he looked at his watch and called his grandmother Sally Mason in Phoenix. He thought the middle of the afternoon would be a good time to catch her, but it was not meant to be. This time he left a message.

After George had passed by and entered the dorm, Neri went to the driver's-side window of the van. Prek lowered the window and looked at the rookie and felt sorry for him. He looked bedraggled with his dark hair plastered against his forehead. "Okay," Prek said. "Get back in the van but stay in the uniform."

"Thanks," Neri said, and meant it. He quickly entered through the van's sliding door.

Prek watched him in the rearview mirror as he pulled off the wet jacket. Genti was tapping a pencil against the dash in time to the music.

"Did the George fellow look at you?" Prek asked, still watching the underling in the mirror. It was easier than turning around.

"He did. He looked me right in the eye. Why do you ask?"

"Just curious."

"It doesn't matter, does it?"

"Can't imagine," Prek said. "I was hoping they'd be together, but what can you do. When you finish getting settled, bring that milk carton up to the front and sit between us. I want you to look out the windshield for the Grazdani girl along with us. Six eyes are better than four." The trickle of medical students had swelled to a horde. Like livestock returning from the pasture on their way back to the barn.

The van was parked on the west side of Haven Avenue facing southeast. Prek et al. were facing the medical students coming from the medical center complex, passing the van on the passenger's side.

"We have one bird in the nest. Now, where the hell is the other one? Where the devil did she go?"

50

Office of the Chief Medical Examiner
520 First Avenue, New York City
March 25, 2011, 5:30 p.m.

Chet McGovern waited with anticipation for the beautiful med student sitting in front of him to tell him if there was anything she wanted him to ask Jack Stapleton, who had performed the autopsies on the dead researchers she was interested in. McGovern tried to read her face. She'd looked crestfallen a few minutes earlier when he told her the bodies were gone, but now she seemed to have brightened. After a few moments' contemplation, something seemed to have occurred to her.

"Well, maybe there is something you could ask," Pia said.

"What? What would you like me to ask him?" McGovern said. He tried to mask his eagerness, worried it might scare her off.

Pia remembered Rothman's rebound

tenderness, which she had been the first to find. It heralded the peritonitis, which bore witness to what was going on in Rothman's gut. Typhoid's target organ was the small intestine. From her recent research she knew that the gut was also sensitive to radiation, particularly the cells that lined the gut. But it was the whole gut, not just the small intestine. If polonium was involved, then the whole gut would have been damaged.

"I'd like to know if the autopsy findings were typical for typhoid fever."

"Let me go and find him," McGovern said with alacrity. "Not a problem. Don't move!"

He jumped up before Pia could change her mind and dashed out of the room, heading down the hall to Jack Stapleton's office. He knocked on the door and barged in without waiting for an answer. To his dismay, the office was empty.

"Damn!" He then ran down to the office of Jack's wife, Laurie, whose door was ajar as usual. To his delight both Laurie and Jack were there.

Laurie and Jack liked to hang out in one of their offices after the rush of the day and go over their cases and maybe make plans for the evening. The rush-hour traffic, particularly bad on Fridays, would subside a little if they waited to leave until six, and with the live-in nanny at home tending to JJ's needs, there was no rush. This was their quiet time, and they relished it because it was in short supply considering how busy they both were.

"Jack. Thank God. Oh, hi, Laurie, how are you? Jack, listen." Chet was talking excitedly and loudly at first but got mock-conspiratorial. He looked behind him and pushed Laurie's office door almost closed so no one else could hear.

"Jack, I have got the best-looking fourth-year medical student in my office that I've ever seen. I mean ever. I need you to keep her interested until I can get her info. I had nothing to do tonight, but then she showed up. It's like a sign. You gotta help me, man."

As usual, Jack was amused by Chet, his former office mate and longtime friend. Jack had heard innumerable episodes of McGovern's social antics. Laurie, on the other hand, had wearied of Chet's incessant womanizing. She couldn't resist baiting him a little.

"Chet, how old are you?" she said.

"I know," he said, pretending to be sheepish.

"No, seriously, how old *are* you?"

Jack thought he should step in at this point, between his wife and his friend.

"How can I help you, Chet?"

McGovern stuck his head around the door and looked down the corridor to make sure Pia hadn't left.

"Listen! This Columbia med student just came in asking about those two typhoid cases you worked on yesterday. Actually, she came in supposedly interested in an elective, but I guess that was a cover story. For some reason she wants to check the corpses for evidence of alpha radiation because they'd been using some alpha emitter radioisotopes in the lab where your two patients worked. She even brought in her own Geiger counter. When I told her the bodies had already gone she was disappointed. Thanks for being so over-the-top efficient with the death certificates and signing out the cases, Jack!"

"You're welcome, buddy."

Jack and Laurie smiled knowingly at each other. This was typical McGovern behavior. Each week there was a new hot prospect. It used to be Laurie felt badly for the man because she thought he was lonely. But that had changed. She was now convinced Chet did not want to find a mate. It was the chase he wanted, and he never tired.

"When I told her the bodies were gone, she wanted to ask you whether your findings were typical for typhoid."

"Tell her the findings were indeed typical for typhoid, but a very serious case of typhoid from a remarkably virulent strain."

"How about coming and telling her yourself? She'll be more impressed."

Jack looked at Laurie, who shrugged as if to say, "It's okay by me." Jack heaved himself to his feet, told Laurie he'd be right back, and followed Chet back to his office.

Chet made the introductions, and Jack could understand Chet's enthusiasm. Grazdani was fetching. He noticed the Geiger counter. He quizzed Pia about her interest in his cases. She gave him the same story she'd told Chet, and Jack purposely didn't challenge her although he was tempted. Instead he said, "My understanding is that you're interested to know if the autopsy findings were typical for typhoid fever. Yes, they were: a very virulent form of typhoid fever. The gut, the target organ of the disease, was in bad shape, which is why they died so quickly. There were multiple perforations into the peritoneal cavity."

Pia sat up straighter in her chair. "Have you seen anything like that before?" she asked.

"Well, no, not to that extent. But you have to remember that typhoid fever, and especially such a bad case, is rarely seen these days. It's no longer the scourge it used to be before we had antibiotics."

Laurie suddenly appeared. She'd decided not to be left out. Chet introduced her to Pia. Pia shook her hand and then turned her attention back to Jack and said, "The strain they were working with and which caused their infections was particularly virulent because it was grown in space, under a NASA program."

"Really?" Jack said. He made a mental note to ask why no one had mentioned that fact.

"Was the involvement just in the small intestine or was it the whole intestine?" Pia asked.

"It was the whole intestine," Jack said. "From the duodenum all the way down and including the rectum. In that sense it was unique. Usually it's just in the small bowel. It was unique enough that I saved some rather large specimens in formalin. I thought they could be used in the future for teaching purposes. We take our teaching responsibilities very seriously around here, right, Dr. McGovern?"

The dig got Chet McGovern to mumble something, and Jack laughed. Pia looked confused, but in actuality she was giddy.

She hadn't even heard Jack's sarcastic comment. All she had heard was that he'd saved sections of the gut! The bodies were gone, but pieces of the involved intestine were still available.

"I mean, I can't show you any slides because the specimens haven't been processed yet because the autopsy was only yesterday. But if you want to view the gross specimens, I'd be happy to show them to you. As for slides, if you provide your contact information, I'll either tell you when they're ready and you can come back, or, if you'd prefer, I could send some slides up to you at Columbia Medical School."

"Oh, absolutely, I want to see the gross specimens," Pia said. "And I want to see the slides too, when they're ready."

Jack looked at McGovern with a smile. "Dr. McGovern, make sure you get Miss Grazdani's contact information."

"I'll be happy to do so," McGovern said, beaming.

"Well, let's go up," Jack said, and all four trooped out of Chet's office and headed for the stairs. Pia carried both her umbrella and the shopping bag holding the Geiger counter.

On the fourth floor they all filed into the histology lab. The supervisor, Maureen

O'Conner, was still on duty. Jack could swear that since redheads had become very cool recently, Maureen's red curls had gotten redder.

"So what do we have here on a Friday night?" Maureen quipped. "Is this a party or is it work?" She looked from Jack to Laurie to Chet to Pia. Chet made the introductions and Maureen shook hands with Pia.

"I want to look at some samples, if it's okay, Maureen," Jack said. "I know it's late."

"Ah, it's never too late for you, Jack," Maureen said, and Laurie rolled her eyes. Maureen had taken an early liking to Jack and babied him with special attention. Jack's slides were always available a little quicker than everyone else's.

Under Jack's instruction, Maureen took out a number of formalin-filled sample bottles from the sample storage area and put them on an available and reasonably empty bench.

After donning gloves, Jack took out the pale intestine samples and put them on the countertop. He showed Pia the perforations and the marked erosion of the internal, mucosal epithelium that lined the organ. When she saw that Jack was ready to put them back into the sample bottles, Pia asked a question as casually as she could.

"Would you mind if I checked the sample with my Geiger counter?"

Jack shrugged. "It's okay with me."

Pia pulled the Geiger counter out of the shopping bag. After opening up the mica port specifically designed for alpha particles, Pia turned on the machine and positioned the Geiger counter as close to the intestinal sample as possible without touching it. Immediately the counter started giving off the clicks that announced the presence of radiation. As Pia moved the instrument even closer the clicks intensified until they were a continuous noise. Then the needle on the counter's gauge went off the scale.

"Whoa," said Jack. "What's that about?"

Pia said nothing and moved the counter away from the sample and then back. It was unmistakable, the sample was emitting radiation, a lot of radiation. She did it again just to be certain, then turned off the Geiger counter and slipped it back into the shopping bag.

The three shocked MEs looked at one another and then at the young med student. Something wasn't adding up at all. The sample of intestine had come from a man who had been signed out as having died of salmonella poisoning, yet the sample was emitting extremely high levels of alpha-particle radiation. This student had said that they had used radioisotopes in the laboratory as part of the experimental regimen, but could that have caused this radiation?

"What's going on here?" Laurie asked, addressing Pia. Her voice was even, unchallenging. "This is all rather surprising. Do you have any explanation?"

Pia's heart was racing, and she felt as though she might actually be in shock. She was not prepared to face the reality that Rothman's and Yamamoto's deaths might be a copycat of Alexander Litvinenko's in London. Pia was terrified of not being able to get to the truth. Now, when it appeared that she'd found it, all she could feel was a rush of anxiety and paranoia. All she wanted to do at that moment was to get the hell out of the OCME, go back to the dorm, and give herself an opportunity to think about the implications of the discovery and what her next step should be.

"Miss, we need you to tell us what you think is going on," Laurie said, her voice hardening to a degree. "This is an unexpected and very significant finding."

Pia said nothing. She could feel the eyes of the MEs boring into her. She'd never had any reason to trust anyone in a position of authority. These three weren't the police or hospital security, but they did work for the city. Who are the bad guys and who are the good guys? She didn't know. The bigger question was, are there ever any good guys? She had to get away.

Jack was as flabbergasted as anyone else. "You mentioned isotopes, radioisotopes being used in Dr. Rothman's lab?"

"Um, I'll have to find out for sure," Pia said. "I can get back to you in the morning. Do you come in on Saturdays?" She picked up her umbrella and hooked the shopping bag over her shoulder. She eyed the door to the hall.

Chet McGovern was trying hard to think of what Pia had told him about alpha emitters. "Earlier you mentioned something about lead and bismuth, something like lead-213 and bismuth-212, was that it?"

"It was the other way around: lead-212 and bismuth-213, actually. But yes, I did mention those isotopes, and now I have to go back and check to make sure they were the ones being used. I really need to leave." Pia checked her watch. "Oh my goodness, it's almost six o'clock. I promised to be back by six and it's a forty-five-minute subway ride up to Washington Heights."

The MEs could sense Pia's acute anxiety.

No one was convinced by her display of surprise at the time.

"I think you need to stay here until we get to the bottom of this," Laurie said. "You might have been exposed yourself. Alpha emitters are dangerous if either ingested or breathed in. There might be other people who need to be checked out."

"Thank you so much for your help," Pia said nervously, looking at Laurie and Jack but not quite looking them in the eye. She was itching to get away. "I can be in touch about the isotopes tomorrow."

Pia didn't want to be trapped there when the MEs called the authorities, which she knew they would do shortly. She had to finish this on her terms.

"Young lady, what's going on?" Jack said. "You show up with a Geiger counter in a shopping bag and bruises on your face. Are you a medical student at all? Who sent you here?"

"No one sent me," said Pia. "I can see how this must look, but I am a medical student. You have to trust me — no one else was contaminated, I'm sure of it. But I can't stay here, I have to get back, I'm sorry."

Pia started backing toward the door, and Jack stepped toward her.

"You can't hold me here if I want to

leave," Pia said. "And I want to leave. Right now!"

Laurie touched Jack on the shoulder, and he paused. Pia turned and walked away quickly. Chet followed and looked back at Jack, puzzlement written on his face. He had no idea what to do. He didn't even have her cell number. Pia and Chet disappeared. Maureen was confused too, wondering if she should call security.

"She's right, Jack, we can't keep her here. She said she's at Columbia, so she won't be hard to find."

"If she wasn't lying about that too."

Part of what Jack and Laurie liked about their work as MEs was the unexpected. This was something very new.

"What do you make of it?" Laurie asked.

"I dunno," said Jack. "There's a lot she isn't saying. She suspected there'd be radiation in the bodies. Of course she did, she brought her own Geiger counter! But when she found what she was looking for, she was completely spooked. More like terrified."

"Definitely," Laurie said. "We need to get someone to track her down."

"I agree."

Jack thought for a second.

"Let's check the other guy quickly."

Maureen was glad to have something to

do. She fetched Yamamoto's specimens. They looked for all intents and purposes the same as Rothman's, mirror images in fact. But whether they were radioactive they had no idea. Pia had taken her Geiger counter with her.

"Should we call DeVries to find out how we can determine what the radioisotope we're dealing with is?" Jack asked, referring to the OCME head toxicologist.

Suddenly Laurie remembered a disaster kit that the OCME had put together after the agency had recovered from 9/11, the events of which had caught them, and most of the city agencies, completely unprepared. The concern was that if 9/11 had been a nuclear terrorist event, OCME would have been completely unable to cope. So as not to be caught unawares, the disaster kit had been put together. "I think there's an instrument in the disaster kit that detects radiation," Laurie said. "And it should be able to identify the radioisotopes involved. You remember? Bingham insisted on getting it."

Jack didn't recall, but he trusted Laurie's memory. When she left to see if she could find the device, Jack called John DeVries, the toxicologist, and asked him how they could identify the radioactive material.

"I honestly have no idea, Jack. My whole

career I've never had to, thank goodness. The only radioactive cases to come through the OCME in my experience were patients being treated by nuclear medicine so we already knew the identity of the radioisotope. I guess you'll use atomic absorption somehow but I'll have to get back to you. It's Friday night, you know, Jack."

"I know it is, John. Many thanks."

That was a dead end for now. Then Laurie returned. She'd found the disaster kit and in it a Berkeley Nucleonics Corp. handheld Model 935 Surveillance and Measurement System, capable of identifying individual isotopes. Together Jack and Laurie read the directions and then used the machine to measure Rothman's intestine's emissions. After about five minutes, the result was available. Although mostly alpha particles were being emitted, there was also a low level of gamma radiation. It was the gamma radiation that yielded the result. It was polonium-210!

"The death certificates are wrong, both of them," Jack said. "Damn it, I missed this completely. This was no accident."

"Obviously. Do you know much about polonium?"

"I happen to know a little bit about it. First of all, there are no medical uses for it.

In fact, you know what it's mainly used for? It's mixed with beryllium such that the alpha particles from the polonium cause the beryllium to release neutrons to act as a trigger for nuclear weapons."

"Good God!" Laurie exclaimed. "How do you know *that?*"

"I don't know how I know it, but I know it," said Jack. He remembered something else. "It was used to kill that Russian guy in London, you remember that?"

"Oh, yes, the defected former KGB officer?"

"Right."

Laurie and Jack had taken a professional interest in the case a few years back as did most forensic pathologists.

"We have to report this to Homeland Security," Laurie said.

"Yes," Jack said. "It doesn't mean Rothman and Yamamoto were making nuclear weapons, but it does mean that they didn't die from typhoid fever alone. They had typhoid fever from the salmonella, but they obviously had radiation sickness on top of it. My guess at this point is that the typhoid was a mask for the polonium, which, in retrospect, was probably the lethal agent. I should have questioned the fact that the entire gut was involved."

"Don't be hard on yourself," Laurie said. "I can assure you that no one could have made this diagnosis."

"I suppose you're right," Jack agreed, although he wondered for a moment if he wasn't just making an excuse for himself. "I have to say, it's a rather devilishly ingenious way of murdering someone. And whoever did it nearly got away with it. It fooled me. It would have got past everyone if not for that girl. What happened to Chet — maybe he talked her into staying?"

Jack got out his cell phone and called Chet.

"Chet, that girl, is she still here?"

Jack listened for a second.

"All right. You better get back up here." Jack disconnected. He looked at Laurie. "She's gone. According to Chet, no matter what he said, she literally ran out of the building. And he didn't get her contact information."

"She's got to be found. She could be in danger," said Laurie.

"You've got that right. If whoever is involved in this knows what she knows . . ." Jack didn't finish his sentence. Laurie instinctively knew what he meant. Instead Jack said, "I'll call the chief. This is going to be a bombshell and a media circus."

"And I'll call Lou. And then Paula. It looks like we're spending more of our Friday night here."

Jack nodded. He looked over at Maureen. "Sorry about all this," he said. "A bit of an emergency. Would you mind getting the rest of the specimens? They're going to have to be put in a shielded container of some kind."

"Will do," Maureen said. She'd picked up on Jack and Laurie's anxiety.

Jack and Laurie then ran out of the histology lab, down the stairs, and back to Laurie's office. As he punched in the numbers for Dr. Harold Bingham, the OCME chief, Jack could already see the problems that lay ahead: It was a high-profile case involving prominent medical researchers and the cause of death and the manner of death had been missed. At least they'd found it now, but that was unlikely to appease Bingham. It was Bingham who would have to report the findings to the various government agencies and deal with them, a job that Jack was thankful he did not have to do.

While Jack was calling Bingham, Laurie called Lou Soldano on his cell.

"Lou, it's Laurie. Can you talk?" She dispensed with any pleasantries.

"Hey, Laurie, nice to hear your voice,"

Lou said, his tone becoming wary. "What's happening?"

"We have a situation here at the office. It seems we have a case of polonium poisoning. Remember that case in London four or five years ago?"

"Of course I do!" Lou said gravely.

Laurie took Lou through what she knew — about the mysterious medical student arriving with her Geiger counter, about how she was upset not to find the researchers' bodies but able to test the tissue Jack had saved, and about her extreme reaction at the findings.

"If you're right and it's a copycat of that Russian guy and you mention the words 'KGB' and 'radiation,' this is going to be a shit storm, pardon my French. All the alphabet agencies, the media . . . And if the Russians are involved, it's serious trouble. You gotta keep a lid on it."

"Jack's on the phone to Bingham right now. I'll call the PR people and get them in lockdown mode."

"Appreciate it, Laurie. Now we gotta find this girl. Any reason to think she's not going back up to Columbia?"

"Hold on a second, Lou, Chet just walked in." Laurie turned to McGovern.

"Chet, the woman, she say where she was going?"

"No, she half jogged down Thirtieth Street, heading west. I assumed she was heading back up to Columbia. Why are you asking? What's going on, Laurie?"

Laurie ignored Chet.

"I'd guess she's heading back to the subway, Lou. Ten, fifteen minutes ago now."

"Lou? Are you talking to Lou Soldano?" Laurie waved him off. Jack was standing in the corner, on the phone with a finger in his other ear and saying a lot of "yessirs" and "nosirs."

"I'll put out an APB if you can give me a description. You said she seemed scared?"

"Very. She couldn't wait to get out of here," Laurie said.

"Sounds like she knows more than she should. So how would you describe her?"

"Maybe five-six, slim, about what, one-ten? Black hair, just down to the shoulder. Lovely skin."

" 'Lovely skin' ain't a description, Laurie."

Jack had finished talking with Bingham, and he piped up. "She's incredibly attractive. Maybe French/Moroccan/Slavic. Chet McGovern here was panting like a dog."

McGovern took the phone from Laurie.

"I'd say she's more likely Italian. Dark

skin, like olive color. Soft features, dark brown eyes. Gorgeous, like a supermodel. She told me her name is Grazdani, that's all. You think she's in danger?"

Laurie snatched back the phone.

"Lou, it's Laurie again. Remember I said she's a fourth-year medical student at Columbia."

"Good point," Lou said. "If we're lucky, we might be able to get a picture from Columbia, if she actually goes to Columbia. Now, you guys have to zip your lips. Keep me posted if anything happens. I gotta go, Laurie. I'll be putting a task force together including the NYPD organized-crime unit. This is serious, Laurie. This has organized crime written all over it. And it has to somehow involve the Russians. Jesus, Laurie, that polonium stuff is associated with nuclear weapons."

51

Broadway, in Front of Columbia University Medical Center
New York City
March 25, 2011, 6:45 p.m.

Pia emerged from the subway at the same entrance she'd used on her way to the OCME, pausing near the MTA area map in the shadow of the Columbia Medical Center's art deco buildings. She'd raced out of the medical examiner's office with her head spinning. It was dark, the streets were wet and slick, and there seemed to be hundreds of people on the sidewalks. Pia couldn't face going all the way across town in the rain despite her umbrella so she took a different route, walking to Park Avenue South and the Twenty-eighth Street station on the 6 line, then traveling to Grand Central and taking the S train, the shuttle across town to Times Square. There she had taken the A express train up to Washington Heights.

Throughout the entire unpleasant train ride Pia had acted like a zombie, seemingly impervious to her environment. A few people, mostly men, tried to talk to her, but she didn't respond in the slightest. She was in a daze, going over and over the events since Rothman and Yamamoto had fallen ill. It was as if she were experiencing a living nightmare. Having her suspicions corroborated at the OCME afforded her no satisfaction in the slightest. All it had done was cement her fears and sense of dread. She didn't know specifically if the lethal agent that had been given to Rothman and Yamamoto was polonium, but her intuition told her it was. What to do now was the question for which she had no answer. Maybe she should just run and hide someplace until all the pieces fell wherever they were going to fall. The reality was that she had certainly opened the floodgates at the OCME. Whether she liked it or not or intended it or not, the police were now going to be involved, along with every other law enforcement agency. In her vernacular, the shit was about to hit the fan.

Pia's intention when she got out of the subway was to hurry back to the dorm. She felt her only resource was George. Even though she was under no illusion that

George would know what to do, she hoped she could use him as a sounding board. The fact was, she had no one else. She'd thought briefly about involving the two other stalwarts in her life — Sheila Brown and the mother superior — to get their advice, but the story was much too long and complex and more important, Pia was reluctant to put either of them at risk. In the current situation, knowledge was dangerous.

Although Pia was desperate to get to the dorm, she was also terrified. The moment she'd emerged from the relative safety of the subway, she felt inordinately vulnerable. The men who had attacked her said they would be watching her, and she believed them. It meant that they were there, lurking somewhere in the darkness surrounding the medical center. Although where she was at that moment, near the corner of Broadway and 168th Street, was lit and crowded with commuters, looking west down 168th Street was neither.

Holding her umbrella in the crook of her neck, Pia got out her cell phone, which she'd turned off before going into the OCME. She switched it on. Immediately she saw she had more than ten missed calls and three voice mails. She called George, but he didn't answer. Instead she left a voice message of her own: "George, it's me. It's about six forty-five. I'm at the hospital entrance to the subway at 168th Street. Can you come get me so we can walk back to the dorm together? Okay, I'll be waiting here."

52

Haven Avenue, Columbia University Medical Center
New York City
March 25, 2011, 6:59 p.m.

George had not meant to fall asleep, but he had. It was a legacy from the soporific lecture. There was also the fact that he hadn't been sleeping as well as usual with everything that was going on. Not only was he asleep, he was in the deepest stages such that he didn't hear his cell phone emit its cricket chirping. The phone was on his desk not ten feet away. He didn't hear it again when it chirped fifteen minutes later. But that call brought him up from where he'd been such that when the phone chirped a third time, he got up and answered: "Hello."

"George, it's Grandma. I tried you before, but I missed you. How are you?"

George was suddenly wide awake. He didn't know how long he'd been asleep, and he fumbled for his watch, checking the time. It was almost seven, and he panicked. Where the hell was Pia?

"Grandma, I'm good, but I'll have to call you back, okay?"

"Oh, okay, George. Make sure you do now. We haven't talked in a while. Is everything good?"

"Everything's good. I'll call soon! Gotta go!"

George saw he'd missed two calls and listened to the message Pia left the first time. He checked his watch. Shit, she'd been waiting fourteen minutes. As he pulled on a pair of shoes, he tried calling her but got her voice mail. He then raced out into the hall, heading for the elevators.

Prek and Genti sat in the front of the van anxiously scanning passersby through the windshield. Neri was perched uncomfortably on the milk crate, just a little behind and between them. What had been easy earlier, checking out the students as they passed, was now much more difficult. There was a streetlamp at the corner of Fort Washington and Haven, but it was far enough away not to be of much help. It was still raining and much darker. They had been there long enough to be stiff and sore, and in foul moods.

"Where the fuck is she?" Prek questioned morosely. He didn't expect an answer and didn't get one. "This is turning into a bitch."

Neri, as the most inexperienced, was suffering the most. He'd been so keyed up with excitement, and now that he had had to wait he felt let down, depressed. Although his role was going to be the easiest in that he was going to do the hit, he'd never actually killed anyone before. He had his right hand in his jacket pocket holding his military-issue Beretta M9 semiautomatic pistol with the thumb safety on. He'd fired the gun hundreds of times in practice and considered himself a good shot. But shooting a man in the head at point-blank range was a very different proposition from hitting static targets at twenty-five, fifty, or one hundred feet. Yet he knew he had to do it to rise within the crew. Like Prek and Genti, he had his wool balaclava in his lap, ready to pull it over his head and swing into action.

An NYPD cruiser drove by, and all three reflexively crouched down. Prek watched it disappear in his side mirror. Then another NYPD cruiser went past, and Prek tensed up further. He watched that one disappear as well.

"You see that?" he said.

"Of course," Genti said. "It's Friday night. I wouldn't give it much thought."

"I don't like to see cops in the area when we're doing a job. Where the hell is this bitch?"

"It's getting harder and harder to see these kids' faces until they're right on top of us," Genti said.

A group of three students in lab coats walked past the van, followed by a couple of people walking alone. One of them caught Genti's attention, and he leaned forward and picked up his balaclava. A minute later he slumped back in his seat. It was yet another false alarm.

Pia had been walking back and forth along the side of the subway stairs, waiting for a call from George, trying to figure out where he could be. They had definitely planned on getting together when she got back from the OCME. More than once, she had resolved to quit waiting for George and walk back to the dorm by herself, at least until she looked down 168th Street and saw that it was darker and more deserted than it had been fifteen minutes earlier. Pia had been about to call George for the third time when she'd been frightened by a hand on her shoulder. Spinning around, she had had to restrain herself from lashing out at her attacker. But it wasn't an attacker. It was Will McKinley, who'd come up from the subway and caught sight of her pacing back and forth. After initial small talk and mutual sympathies about Rothman's and Yamamoto's passing, Pia had latched onto him for company back to the dorm. As an enticement, as if she needed it, she had offered to share her umbrella.

After reminding each other about the deaths, they'd each regressed into their own worlds. They walked in silence until beyond the 168th Street hospital entrance. Pia wondered what Will might say if she told him what she now knew. She thought he probably wouldn't believe her.

"I was surprised to see you," Will said. "Did you come out of the subway like I did?"

"I did," Pia admitted. She tried to think of what to tell him if he asked where she'd been, so she changed the subject. "Have you seen George at all today?"

"Wilson? No, but I haven't been around since lunch. Lesley and I still haven't found a home for our month's research elective. I took the opportunity to do a little shopping." He held up a shopping bag.

Pia spotted a police cruiser heading in

their direction along 168th Street. She tilted down the edge of her umbrella to keep her face from being seen, which Will noticed immediately as it bumped into his forehead. She had reacted reflexively. She wouldn't have been surprised if the police were looking for her. Although being picked up by the police was surely not something she wanted to happen, at least not yet, she was the first to admit it would be far better than being confronted by her attackers again.

"What, are you a fugitive from justice?" Will quipped, unknowingly interpreting Pia's gesture correctly.

"Hardly," Pia said with a fake laugh. Another police car was coming, so she kept the edge of the umbrella angled down.

They reached Fort Washington Avenue and waited for the traffic light to change. There were only another couple of hundred yards to the dorm. Pia relaxed a degree. They had not seen any men in Columbia Medical Center security uniforms. She was looking forward to getting to the relative safety of George's room.

Genti was the first to spot Pia and Will coming around the corner heading right for them, backlit by the corner streetlight.

"There, in front, fifty yards."

"Both of them!" Prek voiced with delight. "Fantastic! D-day. Wait for my word. You okay, Neri?"

"Sure!" Neri said with more bravado than he felt. He slipped off the safety on his pistol and pulled on his balaclava as Prek and Genti did the same.

Neri looked out the van's rear window to see if anybody was coming in the opposite direction.

"Wait!" he said. "Who's this coming from the dorm? Is that him?"

"Who?" said Prek. He swung around to look at George. Then he flipped open his cell phone and studied the picture Buda had sent him. The only illumination in the van was from the dim streetlights and it was a small photo. He swung back around to look at the man walking with Pia. They could have been twins in the misty half-light. "It has to be the guy with Pia. What is this, the Swedish Olympic team? Everybody is blond."

Prek waited a couple of beats as the couple got closer. "It's him. Hell, he has his arm around her. How close is the guy coming from the back?"

Neri checked again. "About two hundred yards."

Prek picked up a rag from the dash and

soaked it with a hefty slug of the Ultane anesthetic, which they were going to use to take down Pia. He shot a look at Genti, and Genti nodded back.

"Okay, go!"

Just as Pia and Will came alongside the van, the three masked men leaped out, Prek and Genti from either side of the front, Neri from the back. Neri stepped around from the back of the van as Will McKinley stood stock-still in front of him, his mouth open wide. Neri pointed the gun at Will's head and a fraction of a second later Will reacted by turning toward Pia, who'd let out a scream. Neri fired, sending a nine-millimeter bullet into the side of Will's head. Simultaneously Genti grabbed Pia in a crushing bear hug while Prek slapped the Ultane-soaked rag over her face. Almost immediately the fight went out of Pia, and she lost consciousness.

Neri ran around to the front of the van and got into the driver's seat while Genti dragged Pia to the back of the van and pulled her inside. As Prek ran to help Genti, he caught sight of Neri's spent casing. He picked it up from the sidewalk just before jumping into the van behind Genti and slamming the doors shut behind him. Neri already had the engine started, and the moment he heard Prek's "Go!" he accelerated from the curb, made a rapid U-turn, and headed north on Haven Avenue. The hit-and-snatch had taken about seven seconds.

There were three witnesses who saw everything and eight more who heard the gunshot and saw the van drive away. One of the witnesses had his own reasons for not wanting to talk to the police that evening so he kept right on walking as if nothing had happened. The second was a male med student who had been walking back to the dorm, twenty yards behind Pia and Will. He had watched in horror as the raid unfolded. At first he thought he might be watching a movie being filmed, but it was dark and there were no cameras. And the blood coming from the shooting victim's head was very real. He called 911 and tried desperately to remember what, if anything, he'd learned in the past two years about gunshot victims.

The third witness was George. He'd seen Will and Pia before the event and had stopped, waiting for them to come to him. He was relieved to see Pia, but his relief was short-lived. In the next second he saw the men leap from the van, shoot Will, and snatch Pia. It happened so quickly he didn't have a chance to move. He blinked, as if blinking would reset the scene back to when Pia and Will were walking toward him. But it didn't. It was only then that he ran forward to where the other student was kneeling over Will McKinley's body.

Inside the van, Prek used a prepared syringe to give the barely conscious Pia an injection of Valium, enough to knock her out completely.

"Don't drive too fast," he yelled ahead to Neri. "Keep it steady." Prek and Genti then struggled to roll Pia up in a threadbare carpet. It wasn't easy in the rocking and swaying van.

Neri's hands were shaking, and it was all he could do to stop himself from throwing up. The mark had looked right at him. Neri blinked rapidly and concentrated so he wouldn't drive right off the road.

"And, Neri," Prek said.

"What?"

"Good job."

On a quiet side street just north of the George Washington Bridge, Neri pulled over behind the waiting dark blue van, and the men quickly transferred their cargo. With that done, Prek returned to the driver's seat with Genti riding shotgun. Neri was told to keep a watch on Pia.

Abandoning the white van, Prek drove ahead to where he could get on the Henry Hudson Parkway to loop around and get onto the George Washington Bridge, heading over to New Jersey. As they expected, the bridge was chockablock with rush-hour traffic. But the crew didn't mind in the slightest. The hit-and-snatch had been flawless, and they were giddy with their success. It was, as Prek said, a tribute to the Albanian mob tradition.

"I even got this," Prek said proudly, as he pulled Neri's bullet casing from his pocket and held it up. "Are we good, or what?" He then handed his cell to Genti and told him to text Buda that the operation went smooth as silk.

53

Belmont Section of the Bronx
New York City
March 25, 2011, 8:05 p.m.

Aleksander Buda had been happy to get the text from Prek. He'd been mildly concerned about the operation even though it was a relatively simple job. But he knew from sore experience that "shit could happen." Be that as it may, the pesky girl was unconscious in the back of the van, and they were on their way to the agreed-upon location, Buda's summer home, and the boyfriend was dead. The white van they had used had been abandoned. Now the only suspense that remained was the fate of the girl.

Buda was confident Pia Grazdani wasn't connected with any of the prominent Albanian mafia crews in the immediate area; he would have heard the name, which he knew was undoubtedly Albanian. The problem was that if she was related to anyone in a crew, anywhere up and down the East Coast and as far west as Detroit, custom dictated that she be accorded a degree of protection. Even so, Buda had debated with himself whether or not he would have been justified in simply disposing of the girl at the same time as her friend. It would have been neat and efficient. She'd certainly become a serious pain in the ass, especially having somehow, on her own, figured out the polonium issue. But Albanian mob bloodbaths had been fought over even less of a provocation. Buda had decided he had to be sure.

A cautious man, Buda had made it a point to investigate Pia Grazdani in a discreet manner. He was known to the FBI, of course, and he knew how the FBI loved patterns and didn't believe in coincidences. If the head of one Albanian crew, like himself, suddenly called all the other local heads in quick succession, Buda knew there was a good chance the feds would find out about it and come snooping around.

So Buda had sent live emissaries to crews in Queens and Staten Island and asked one associate to call a crew in Pennsylvania just in case. Manhattan and Brooklyn had also been dealt with, and since he controlled the Bronx, that was covered. He'd received negative rcports all the way around, even from Detroit. There was no connected Grazdani. The future was not looking promising for the girl.

But there was one unit left unchecked: Berti Ristani's crew based in Weehawken, New Jersey. Ristani was a particularly nasty customer, willing to do just about anything to make a name for himself. Buda realized he hadn't seen the guy in a year. He thought it wouldn't be a bad idea to make the visit himself for political reasons in addition to providing an alibi for tonight, just in case. Buda grabbed his car keys and set out for Weehawken. He knew he didn't need to call ahead. Ristani could always be found in the same place.

54

Columbia University Medical Center
New York City
March 25, 2011, 8:31 p.m.

Detective Captain Lou Soldano was frustrated. He was standing inside the taped-off area of the street that marked the McKinley crime scene. A full crime-scene unit had combed the area for clues, but none were to be found. They hadn't even found the shell casing from the gun that was used to shoot the student. All they had was a shopping bag with a device one of the techs identified as a Geiger counter that the woman had been carrying when she was abducted.

Officers were taking statements from witnesses, hoping to get information on the perpetrators and on the van they had used. The reports he'd heard were wildly contradictory, from the men's heights to their clothing. One witness swore that only two men had been involved, while all the others claimed three. The one thing they agreed on was that all the men had been wearing ski masks. The van was described consistently as dirty white, but they had nothing on the make or license plate.

George Wilson had given the most detailed account, and even told Lou something important and most likely related. He said that Pia had been assaulted in her dorm room the previous night and threatened with violence, which now had come to pass. When asked why she didn't report the incident, George said she was afraid of the police because of her childhood experiences. He said that he had encouraged her to go to the police on multiple occasions. When asked why he didn't report it himself, George said because he respected her wishes and privacy, and she had asked him not to.

Lou's frustration was not only over the paucity of evidence at the scene. He was also frustrated that the local precinct of the NYPD hadn't blanketed the area with personnel after Lou had specifically sent out a message to do so. He had given them the description of Pia from Jack and Laurie with the added information that she was carrying an umbrella and a canvas shopping bag. He had hoped that the police or the

Columbia hospital security would have been able to pick her up. He had wanted to detain Pia not just to learn exactly what she knew but also for her protection, and now the bad guys had beat him to the punch. If the local boys had done what he had asked, the snatch-and-killing might have been avoided.

The only part of the operation that seemed to be going right was the radiation angle. Lou knew that the ME's office had notified the relevant authorities about the possibility of alpha radiation at four sites in New York City: Columbia University Medical Center, the OCME itself, and the two funeral homes where the bodies of Dr. Rothman and Dr. Yamamoto were taken. But the ME wasn't able to mobilize law enforcement. Soldano had had to do that, and his task force was still more on paper than in the field.

One of the other things that had frustrated Lou was how long it had taken to get a photo of the Grazdani woman. Lou himself had called Columbia hospital security to confirm that Pia was indeed a medical student. Lou had also asked for Pia's details, as well as a recent photo, which the hospital security had trouble getting because it was locked up in the office of the dean of students, and the dean was unavailable. So it wasn't until after the abduction that the photo was sent out to law enforcement. That was akin to locking the barn door after the horses were stolen.

"God damn it," Lou said out loud for the umpteenth time. Nothing about the case seemed to be going well. The sole positive development was that a white van had been located that was believed to be the vehicle involved in the abduction. At that moment another team of forensic experts was going over it. Lou had no idea if it would lead to any clues, but he was hopeful. Meanwhile an APB had been put out in New York, Connecticut, and New Jersey. Lou was hoping they'd get lucky. But he doubted that would be the case. Lou's intuition was telling him that organized crime was involved big-time. He knew for sure that this wasn't a kidnapping for money — meaning he feared for Pia's life.

All of a sudden several news vans showed up and parked just beyond the crime-scene tape. As their antennae rose, their doors burst open and a bevy of cameramen and journalists alighted.

Lou groaned. He knew this was going to be a media circus, and he wondered how long it was going to take before the mayor got involved.

First Will McKinley was unlucky — twice, in fact — and then he got lucky twice. Will was unlucky to have been involved in the Rothman case to begin with — to be found on the street with Pia and mistaken for George, resulting in his being dealt with as an annoying tagalong who knew too much. Will was also unlucky that Neri Krasnigi's gun had fired at all. When Neri had cleaned and loaded his gun earlier that day, he hadn't been as careful as he thought. A few sizable pieces of grit had been stuck to the first bullet he loaded and lodged inside the chamber. Under different circumstances, or if the pieces of grit had been just slightly bigger, perhaps the gun would have blown up in Neri's face rather than misfiring slightly and dispatching the bullet at perhaps fifty percent of usual velocity. That was lucky.

Will was lucky again, if someone shot in the head can be said to be lucky. Will had turned his head so that he was hit in the temple, not the forehead, meaning the bullet made a complete transit through his frontal lobe, a kind of injury that had seen miraculous recoveries in the past. It might also be considered fortuitous that he'd been shot a hundred yards from a major trauma center, where expert help was immediately available. A superb team of doctors had treated Will within minutes of his being shot and continued to monitor him closely. He was now in a medically induced coma, hooked up to an array of monitors and life-giving machinery. Everyone was hoping Will's luck would not run out.

55

Green Pond, New Jersey
March 25, 2011, 8:45 p.m.

Prek drove carefully into Green Pond, a private summer lake community in Morris County in northern New Jersey. What could have been an hour's drive had taken nearly two, partly because of traffic, partly because Prek kept the van well under the speed limit, even on the open stretches of road. Some of the items in the van would have taken a lot of explaining if the vehicle was involved in a traffic stop.

The Green Pond that lent its name to the town was actually a lake. This night its surface was dark as the moon had yet to rise. Prek navigated the twisting, hilly eastern shore road where the few homes, sitting at the end of long driveways, were mostly hidden by dense, leafless hardwood forest. There were just a handful of homes on the cliffs on the western side, which could be reached only by boat. The village itself lay to the north. After a mile or so the road wound closer to the lake and to the dwellings along the shore. Some of the homes belonged to year-round residents and their windows glowed with warm incandescent light and the occasional flash and flicker of the ubiquitous big-screen TVs. At the southern end of the lake were a number of summer cottages and they were dark, with their docks stacked in neat piles and their boats under tarps.

Prek pulled into the driveway of Aleksander's waterfront cottage, which was situated on a peninsula at the very southern end of the lake facing out onto a cove a couple of hundred yards in diameter. Aleksander had been drawn to the house by its private location on a spit of land. He had been delighted to find out that for seven or eight months of the year, the houses and cottages on either side of his on the eastern cove were dark and deserted with their heating systems off and pipes drained. There were only two houses on the western side of the peninsula, and they were also empty except during the summer months. Buda loved the place because of its serenity, particularly in the winter when the lake froze up solid.

Prek parked in back of the house facing the road, found his key and opened up, turning on the lights and the oil-fired heat.

"Honey, I'm home," Prek called out, laughing.

The closer they had got to this sanctuary, the brighter the mood in the van had become. Prek was positively gleeful that everything had gone so well. The boss was sure to be very pleased.

The three men unloaded the rolled-up carpet that contained Pia and brought it quickly inside. The front door opened directly into the living area, where there were two leather couches, one black, one brown, facing each other in the center of the room in front of a fieldstone fireplace. Prek moved a low coffee table scattered with automobile magazines to the side so that Genti and Neri could set down the carpet roll containing Pia. They then unrolled her until she lay sprawled, facedown, on the room's massive fake oriental.

Prek called Buda. He wanted to tell his boss that they'd arrived and all was well. He also hoped to get the okay to finish up with Pia. It was a perfect night, quiet and dark, to dump a body in the large swamp that extended for almost a mile from the southern end of the lake into a protected virgin forest surrounding a government arsenal called Picatinny. It was the kind of wilderness whose remoteness would come as a surprise to most residents of New York City. It had certainly proved useful to the Buda crew on more than one occasion.

To Prek's irritation, Buda didn't pick up. Prek didn't leave a message; Buda would see the missed call and know Prek was trying to reach him.

Prek's annoyance was exacerbated when it dawned on him that despite all their planning, they'd forgotten to bring any food. There was a store up the road about five miles north, but Prek didn't think it was a good idea for any of them to show their faces in a public spot, not when they were going to be disposing of a body. Prek went into the kitchen, such as it was, and looked in the fridge. There was one carton of milk with a past-due date. The cupboards were even more depressing. There was an open box of cold cereal, but one corner was chewed off and mouse droppings could be seen.

Discouraged, Prek reentered the living room. There was a sudden silence. Prek could tell Genti and Neri had been talking about something and had stopped abruptly.

"What is it?" Prek asked.

The two men were looking at Pia. There had been a disagreement.

"How long before she wakes up?" Neri said.

"She's got ten milligrams of Valium in her, so she's going to be sleeping it off awhile," Prek said. "She'll start to wake up, but she'll be very groggy. We can always give her another shot if need be. Buda didn't answer his phone."

"She's beautiful," said Neri.

"Our young friend here was just telling me what he'd like to do to her," said Genti. "I suppose we could all take turns, maybe even be interesting to watch. What do you think, Prek? We should have done it when we were in her room last night."

"I like my girlfriends to participate," Prek said. "And anyway, we're not doing anything till I hear from Buda that she's safe to get rid of. Remember, she's got an Albanian name. We got to be sure we wouldn't be stepping on someone's honor, if you know what I'm saying."

"Oh, come on, Prek," Genti said, "what are the odds of that? There's two hundred fifty thousand Albanians in the area, and this is one girl. I've never seen anyone that good-looking related to any of us. She sure doesn't look like your sister."

Genti and Neri laughed. Prek didn't. He had a premonition there was going to be trouble after everything had been going so smoothly.

Neri and Genti were sitting on separate couches. Neri looked like a dog in heat, practically panting while adjusting himself. He was looking alternately at the supine Pia and at Genti, who he believed was in his corner as far as having sex with the woman was concerned. Genti Hajdini had come up through the crew's ranks at the same time as Prek, but Buda entrusted Prek with more responsibility and bigger jobs. As a consequence, Prek made more money than Genti, and when Buda wasn't around, Prek called the shots. He knew this was something Genti resented, but it rarely was an open problem between them. Prek knew that Genti was still miffed about not being allowed to have his way with Pia the previous night.

"The boy did good tonight," Genti said, pointing at Neri and making a shooting sound. Genti and Neri laughed again, then stopped and Genti looked at Prek. "Maybe we should reward him. Maybe we should reward all of us." There was a heavy silence in the air.

"Who made you the boss, anyway?" Neri

said quietly.

Prek looked from Neri to Genti and back again. Neri was still wearing the black jacket he'd had on when he shot the boy in the street and Prek assumed he still had the gun in an inside pocket. Genti might be armed too, for all he knew. His own gun was in the glove compartment of the van. Did he really think Neri and Genti would gang up on him and take him out? In this crew, as Prek and Genti both knew very well, stranger things had happened. Turning his attention back to Neri, Prek held his frankly impertinent gaze.

"Buda said I'm the boss when he's not around."

Pia let out a groan.

"Listen, you assholes! Buda told me we wait until he's sure there is no family connected with this woman. You go and mess with this woman and Buda finds out she's someone's daughter or niece and you two haven't been able to keep it in your pants? The uncle or father, whoever he is, is not going to be happy. He's going to be unhappy with Buda and that means Buda is going to be very unhappy with you."

"She's unconscious," Neri said. "Out of it. She won't even know, at least not for sure. It's such a waste. Like it's a crime."

"She'll know, you asshole."

"You not interested in girls anymore, Prek?"

Now it was time to stare down Genti. Prek knew that Genti's comment was meant to rile him, but he decided to ignore it. "She's a pretty girl, sure, but there are a lot of pretty girls."

"I don't see any others in the room," Neri said. He was looking at Genti, hoping for support.

"You don't want to be the reason for a blood feud. Trust me."

"Unfortunately, he's right," Genti said. He got up from the couch and stepped over to Prek. He draped an arm around Prek's shoulder and jostled him.

"We're just screwing with you. If we get the all-clear, Neri gets his, okay? I might take a turn, why not?"

Genti stepped over to Pia's body and with his forefinger lifted her skirt.

"Not bad, not bad at all. Whaddaya say?"

"I say we don't touch her until we get the green light to whack her. When that happens, you two can do what you like. For the time being, help me put her on the bed to get her out of sight. You two are like teenagers."

Prek went over to Pia and grabbed both

ankles. "Come on! Give me a hand!"

With Genti and Neri taking an arm each, they carried Pia into the bedroom. There they tossed her onto the bed.

"Now leave her be," Prek said, motioning his two colleagues to precede him back into the living room. As Prek followed them he wondered what the hell was keeping Buda.

56

Weehawken, New Jersey
March 25, 2011, 8:48 p.m.

"So, Berti, what about this Grazdani girl?" Buda asked. "Anybody in your organization might be related to her? I've been told she's in her mid-to-late twenties and gorgeous. A real looker."

Berti Ristani was sitting behind a desk in his office in a small industrial building in Weehawken. Berti's office looked for all the world like the workplace of a reputable building contractor. Supply catalogs were piled on the desk, the room was ringed with file cabinets, and a storage chest for architectural plans stood in the back of the room. Ristani was a contractor, Buda knew, but not all his contracts involved construction.

Berti leaned back in his office chair, his huge body causing the frame to complain bitterly. Berti's florid face, tracked with broken blood vessels, creased a little as he pondered Buda's question.

"Ah, yes, the business you came for. But I never see you, Aleksander, do we need to talk business? How about a drink?"

"It's a loose end, Berti. It's something I need to take care of soon and I'm trying to do the right thing. I can't leave the situation as it is for too long."

Berti Ristani had no business agenda he wanted to bring up with Buda, and he was mildly offended that Buda kept bringing up this Grazdani issue. He'd been enjoying talking to Buda about old times when they both first arrived from Albania. Back in those days it was no easy thing getting to America. Both had been lucky. On top of a common past, Aleksander Buda was one of the crew leaders Berti respected, and he'd been pleasantly surprised when Buda showed up unannounced.

"Okay, let's find out. I know no one by that name specifically, but I do have two of my best guys with a similar name. But it's not Albanian. It's Italian. Anyway, I might complain, but I appreciate your concern like this. There's been far too many blood feuds. Thanks for coming to talk to me."

"It's nothing, Berti. It would be foolish to act any other way."

Ristani shifted his weight forward and the

chair complained again. He placed his fleshy arms on the table and punched a button on an intercom.

"Drilon, can you come in the office for a second?"

Ristani looked at Buda.

"Drilon, one of my most loyal guys. He and his brother, who's out on a job."

"Anything special?"

"Not really. He runs a bunch of books in South Jersey, down to Philly. Friday nights, he likes to collect. He's smarter than a whip, in contrast to Drilon, who, as the saying goes, is not the sharpest knife in the drawer. Oh, Drilon, come in here."

Drilon was used to being called into his boss's office from his perch near the building entrance numerous times each evening he was on duty. Usually Ristani wanted something to eat, and this was what Drilon expected on this occasion. He walked into the office and saw the back of a figure seated in front of Ristani's desk.

"Drilon, Mr. Buda has a question."

Buda? Had Drilon heard correctly? The man twisted in his seat and Drilon saw the scar on his forehead. It was Aleksander Buda, a serious dude. What did he want?

"Simple question," Buda said without emotion. "Do you know anyone named Pia

Grazdani?"

"Say that again," Drilon said. He thought he was hallucinating.

"Pia Grazdani."

He'd heard correctly. The name triggered a movie that played in fast-forward in Drilon's head. Twenty or so years ago, at least, Drilon had been drinking, drinking a lot. He goes home where he lives with his brother, Burim, and his brother's wife, Pia, and there's Pia looking as hot as you like almost without a stitch on, and Burim's out of the apartment running some errands like he was always doing trying to rise up in the Rudaj organization, one of the most notorious early Albanian mafias. But the bitch rebuffs his advances even though she was asking for it and digs her nails into his chest, like deep, and Drilon sees red. Goes berserk. What happens next, he certainly didn't plan. He grabs his gun in a rage and shoots her in the forehead. Blam. Story's over. But her kid's there — little Pia. He considers shooting her too except he can suddenly hear people next door, so instead he hits her in the head with his gun, trashes the apartment, and takes the $500 stash that the brothers had hidden in the stove. He goes back to the bar where he was drinking, drinks more, stays until it closes, sleeps an hour on a park bench, and goes home to raise the alarm that his sister-in-law was murdered by intruders.

The fallout turned out to be a breeze. Burim accepted the intruder story as he was happy enough to be done with his wife and had been thinking seriously of dumping her, and the Rudaj organization and crew took care of everything. There was no investigation, no nothing. As far as anyone was concerned, Pia had just disappeared, leaving her daughter behind.

Did Buda mean either of those Pias?

"So?" Buda said. He'd noticed Drilon's pause and blank expression. "Do you know a Pia Grazdani?"

Drilon felt the hair on the back of his neck stand up. His face flushed, he could feel it. Three questions immediately crossed his mind: One, was he talking about Afrodita, his little niece, or Pia his sister-in-law? Drilon hadn't given either of them a thought in more than twenty years, but one was dead and the other, who knows? Two, had Buda asked Burim the same question? And three, what the fuck was the correct answer?

"Er, I don't think so," Drilon said. "Why do you ask?"

Drilon had spoken to Buda, but it was Berti Ristani who spoke up.

"He's asking because he wants to know. I'm asking you too, Drilon. I haven't called your brother, because he's busy at the moment. I assume you'd know if either you or your brother is related to this girl. The surname is pretty similar."

"Let me think, we got a big family," Drilon said. So Burim didn't know — that was to his advantage. And perhaps it wasn't her at all. But Drilon feared it was the girl — whose mother he murdered, a secret he'd managed to keep all this time. But if Aleksander Buda was asking about her, she probably was halfway in the ground already. Drilon could see no reason to rock the boat. There surely couldn't be anything connecting him to the girl. "I don't think I've ever heard the name."

"You certain about that, Drilon? You took long enough thinking about it."

"You know me, boss, I'm not the brightest guy. Like I said, we got a real big family, but mostly back in the old country."

"So I've heard," Ristani said conversationally.

"That's right, boss."

"What's your family name?" Buda said. The expression on his face had changed not at all the whole time.

"Graziani," Drilon said. It had been

Burim's idea to drop Grazdani after the Rudaj crew had been broken up and many of its members sent to jail. Graziani was the name Burim came up with when he asked Ristani for work years back. It was the surname of one of his favorite Italian soccer players, and he always liked the fact that it was close to his name too, off by only a single letter.

"It's close but different. Italian instead of Albanian," Berti said. "Close but no cigar. Thank you, Drilon."

Drilon left the room. He was sweating and the color had faded from his face. He wanted to hide as far away from Buda as possible until the man left.

"Anyone else you need to ask?" Buda said.

"No, I know all the other guys' families and have never heard of a Grazdani."

The men hugged briefly, Buda having difficulty getting his arms around Berti.

"Let's not be strangers," Berti said, waving.

Buda got in his car, but before he drove away, he called Fatos Toptani, his most trusted man back in the Bronx. In the Buda organization, Fatos was number two.

"It's me. I need to get hold of someone right now. Name is Burim Graziani, one of Ristani's crew. He's working down in South

Jersey. . . . No, no, nothing heavy, I just need to ask him a question. Yeah . . . Something ain't right here."

From his office, Ristani waited a couple of minutes till he thought Buda had left the premises, then made a phone call of his own.

57

Green Pond, New Jersey
March 25, 2011, 8:52 p.m.

Prek was pleased to leave the tense atmosphere of the house even to undertake the menial task of cleaning out the van, something Neri should be doing. Prek was going crazy just sitting around, waiting for Buda to call to give either a thumbs-up or a thumbs-down. The other two numskulls could not stop talking about sex and who was going to be first even though the woman was no longer in their sight.

In a closet in the house he found a mop and bucket and an inventory of cleaning supplies, including a battery-operated hand vacuum mounted on the wall. Prek filled the bucket with water and poured in a generous quantity of Pine-Sol, loaded up the gear, and walked out to the van. In the morning, they would take it to a car wash and take care of the outside of the vehicle;

for now Prek wanted to make sure the interior was free of any trace of the girl. He vacuumed the cab and washed down all the surfaces with Windex. When he was finished he moved to the back to continue his work when his cell phone rang.

At last, he thought.

Neri Krasnigi got up from the couch in the living room, went to the window, and watched Prek get in the back of the van with his wet mop. *There's Prek doing women's work,* he thought to himself. It seemed appropriate. He hadn't been able to get Pia out of his mind, and he also felt slighted by Prek's attitude toward him. Through the whole job, Prek had ordered him around, throwing stuff at him in the van, giving him the toughest part of the mission but showing him no respect at all. Even when Prek told him he did a good job it was in the tone of voice you'd use with a puppy who was taking a piss outside.

Neri had forgotten the initial terror he'd felt after he shot the student. Now he was feeling bold, accomplished, entitled — feelings that had heightened when he chugged down the last of the Red Bull. Perhaps he had a chance to show up Prek and have some fun in the process, even if it would have to be a private triumph.

Neri went to check on Genti, who'd fallen asleep on one of the sofas. With Prek out of the house and busy and Genti asleep, Neri thought he could risk a quickie. After all, what could Prek do? Neri slipped a hand into his jacket to briefly fondle his gun. It gave him courage. He was his own boss.

Being careful not to make a sound that might wake Genti, Neri slipped into the master bedroom and went over to the bed, took a good look at Pia. She was again supine, clearly breathing, but she looked as dead to the world as she had when they'd first brought her into the house. Tiptoeing, he went back to see what Prek was doing and when he walked to the kitchen window, he could see the van's doors were open with someone inside. With Prek attending to his housekeeping duties, Neri thought he was safe for at least ten or fifteen minutes. With mounting excitement, Neri went to the front door and threw the dead bolt. Then he quickly tiptoed back into the master bedroom, literally shaking with excitement. He closed the door.

"So what does that mean for us?" Prek asked Buda. He wasn't sure he understood what his boss was trying to tell him.

"It means we don't do anything with the girl till I figure out what's going on with this moron Drilon."

"You're sure he was lying?" Prek couldn't understand why someone might lie when asked a direct, simple question about whether someone was related to them or not.

"I'm pretty sure. Everything about his behavior told me he was lying. When I asked him a direct question, he hesitated and then started stammering that he didn't know if he was related or not. It was obvious to me he knows the name. And it's practically his name. If you're gonna change your name, change your name."

"And his boss didn't say anything?" Prek asked, meaning Berti Ristani.

"Nothing. Either he didn't notice or he didn't want to say anything with me sitting there. I bet it's the latter because he's not stupid. It's this Drilon guy who's stupid."

"Why would he lie about something like that? He must know what the implications are."

"I would assume as much," Buda said. This, of course, was the question that was nagging at him. If Drilon Graziani was lying, it meant whatever he was trying to conceal from his boss was more important to him than this girl's life, even if she was a relative. Paradoxically, that fact made the girl suddenly more valuable to Buda, even if he didn't know why. This was the reason it was so important for him to talk to Burim Graziani, if that was actually his name. Buda figured that Ristani had also realized Drilon was lying, which had an entirely separate set of consequences.

Buda himself didn't appreciate being lied to, especially by a subordinate, and he wouldn't want to be in Drilon's shoes if Buda was right about his supposition. It also put Buda in a tricky position. His visit was probably now the cause of a problem within the Ristani crew. He hoped Berti didn't blame him for that.

"I don't want to talk about it anymore," Buda said, meaning that he wasn't comfortable having even this guarded a conversation over a cell phone line. "I'm coming up there to the summer house. Just make sure our guest is treated as a guest until this is cleared up," Buda said.

"Will do," Prek said, ending the call. He'd left Genti in charge in the house, and he trusted him, for the most part. But he thought he'd better check.

Immediately after talking with Prek, Buda

got another call on his headset.

"Aleksander, it's Berti. Sorry to bother you."

"No bother, Berti," Buda said.

"I talked with Burim," Berti said. "I asked him about Pia Grazdani, wondering if he'd ever heard the name. And you know what? He said he did. Can you believe that?"

"No," said Buda, but he could.

"Then Burim called back and said one of your guys tried to call him." Berti said nothing more and left the statement hanging in the air. Buda thought he'd better play it straight.

"I did have one of my guys call Burim," Buda admitted. "You know as well as I do, Berti, Drilon acted strangely to my question. My sense was that he was lying. I figure that's your business, him lying to you, but he lied to me too. If I could ask the brother, maybe I wouldn't have to bother you directly. But I have to find out so I can deal with the woman I'm holding without starting a blood feud."

"I appreciate that, Aleksander. Of course, none of us want another blood feud: Albanian brother against Albanian brother. Of course I noticed Drilon was lying, and I called him back after you left and asked him again. I said, 'No fucking around,' and he said yeah, well, maybe he did know a Pia Grazdani. He tried to say he'd forgotten because he hasn't heard the name or seen the girl for twenty-some-odd years."

Buda was relieved that Berti was seeing it his way.

"So what do we do, Berti?"

"You hold on, I can conference you through to Burim."

"I need to make another quick call first," Buda said.

"Okay. Do what you need to do, then call me right back."

Buda was navigating a complicated course but calmly made his call to Prek. When Prek picked up, Buda spoke and didn't give Prek a chance to respond. He told Prek that he had to talk to a man named Burim Graziani before he could say yea or nay about Pia Grazdani. He said he was about to talk with him, so he'd be getting back to Prek straightaway with a final answer. "Hold the course with our guest for another half-hour or so," Buda said. "I thought I'd also let you know I'm only about a half-hour away. I'm in Wayne, on Route 23. I'll be back to you shortly."

As soon as Prek hung up from Buda the second time, after being told to hold the course, he jumped out of the van. For a moment he stood and listened. He had expected to hear muffled conversation from his two sex-starved underlings, but he heard nothing, which was disturbing. A half-hour earlier the men had been unable to stop talking. With gathering urgency, Prek headed for the front door, hearing in his mind Buda telling him the woman was to be treated as a guest.

With his intuition setting off alarm bells, Prek reproached himself: He should not have left the two alone no matter how much he had wanted to get out of the house. He went to open the front door and found it locked.

"What the . . ." he said. He ran around the corner of the house directly to the window of the master bedroom. Neri hadn't even bothered to close the drapes. Prek banged twice on the window, then ran back to the van, grabbed his gun from the glove compartment, ran back to the window, and smashed it with the butt of the gun. He was furious. Reaching in awkwardly, he fired off a single round.

58

Turnoff on Route 23
Wayne, New Jersey
March 25, 2011, 9:19 p.m.

"I understand you're trying to reach me," Burim Graziani said.

"Berti, are you still on the line?" Buda questioned.

"I'm getting off. You two men talk." There was a click when Berti hung up.

"Yes, I need to talk to you," Buda said to Burim. "We haven't met each other, right?"

"No, I don't believe so. But I know who you are, of course."

In their line of work, everyone knew Aleksander Buda. This was going to be a complicated conversation, Buda could tell. He wanted to make sure it wasn't also too compromising. Cell phones could be hacked, even new cell phones like the one Buda was currently using.

"For that reason, we need to be careful."

"I understand."

Neither of them was willing to start. Burim had been shocked to get Ristani's call. He had been in his car, driving back to Weehawken from South Jersey where he'd concluded his business early. Ristani's question had shaken him so much he nearly rear-ended the truck in front of him. "Pia Grazdani?" he'd repeated out loud, and he thought of his wife, not his daughter. He remembered her fiery personality, the fights, how Pia stayed out all night to party, leaving him alone with the baby. His sudden fury meant he wasn't listening properly to what Berti was asking him.

"She's about twenty-five," Berti had said. "Apparently quite beautiful. Burim, shit, can you hear me?" The connection had not been good, going in and out. It was at that point that Burim realized Berti wasn't talking about his late wife, but rather about his daughter, Afrodita Pia Grazdani.

Buda cleared his throat. "Berti told me you recognize the name Pia Grazdani. Is there any relation?"

"I remember her by a different name," Burim said. "Afrodita, which is what I called her. Her middle name was Pia, like her mother. She was my daughter."

Afrodita. The kid had been a pain in the

ass almost as much as the mother since she'd inherited her mother's personality. Drilon had been the only one who got along with her. A miserable little thing, very demanding at a time when Burim had been too busy trying to make the grade in the Rudaj organization. He'd had no time for a kid. After she'd been taken away by city services, Burim told himself that he'd go and get her back when he was legal in the country, but when he got his green card, he decided he was happier as a man without the burden of responsibility. Then he had to drop out of sight as Burim Graziani and he never got around to establishing his new identity beyond getting a driver's license in case there was ever a traffic stop. He imagined he'd now have to explain all this to Berti Ristani, something that was a bigger issue to him than the fate of his daughter.

"So you think this girl might be your daughter?" Buda asked, not wanting to believe this was happening.

"It's possible, for sure. It's hardly a common name, and the age is about right, mid-to-late twenties." Try as he might, Burim couldn't remember Afrodita's birthday — neither the day nor the year.

"What's the story with the change of the family name?"

Burim related that issue. Since Buda, like all Albanian mafia, knew the details of the Rudaj debacle, he understood. When the FBI came bursting in, lots of people had to go underground.

"So you lost touch with your daughter a long time ago?"

"Yes, you know how it is in this business."

Having been at the time a gofer in a neighborhood crew that was heavily involved in the drug business didn't make him ideal parent material. Buda and Burim both understood. Burim didn't think it was necessary to fill in the details. That the cops had come and taken the kid away and put her in foster care, that he hadn't bothered to stay in contact, was all understood. Burim went quiet again.

"You think she'd remember you?"

"She was six, I believe, when she went away, and I guess a kid can remember back that far."

Burim couldn't help wondering why a man like Buda cared about this woman who might possibly be his daughter. "So how did this Pia Grazdani show up? How did she get involved with you?"

"She's associated with a job I was asked to do," Buda said vaguely. "She's a medical student at Columbia University, doing work with some researcher who had an accident and died."

Burim was shocked once again. Could his daughter be a medical student? And at such a famous university? It seemed incredible. If pressed, he would have thought the girl would end up on a similar path as her mother, would have been with a guy like him or maybe even out on the street. A medical student? He was surprised to feel something like pride.

"And is she pretty, like Berti said?"

"I haven't seen her, but I'm told she is quite beautiful. And, er, scrappy."

"You mean she likes a fight?"

"You could say that."

"That sounds right," Burim said ruefully. "Her mother was a tigress. So what is this about?"

"Where are you?" Buda asked. "With this development, we need to talk in person."

It turned out that Burim was only about fifteen miles from where Buda was parked, near the Lincoln Tunnel exit on the New Jersey Turnpike.

"Do you know the Swiss House Inn?" Burim asked, and Buda did. The restaurant was just off Route 80, convenient for Burim and Buda and not far, as it happened, from Green Pond either.

"I want my brother to come," Burim said.

"Okay," Buda said, curious. The two brothers seemed to be like night and day. Why he'd want his moron brother there, Buda couldn't imagine, but he didn't care. It was, after all, a family affair.

"I will have an associate with me as well," Buda said, thinking of Fatos Toptani. If he could get Fatos Toptani to get there in time, he thought.

"About thirty minutes," Buda said, and rang off. He wasn't happy that the call had taken so long, but his hand had been forced somewhat. What were the odds that Berti's guy Burim was this Pia's father? From that perspective, he was very glad he'd thought to look into the issue. Killing the daughter of a connected man, even a long-lost daughter, even a daughter the father was ambivalent about, would have been a serious matter, especially for a man associated with the Ristani crew. More than any other crew Buda knew, they were addicted to violence. For them it was like a sport.

Buda quickly phoned Berti back and gave him a synopsis of the conversation. "As strange as it may seem, this Pia Grazdani may be Burim's long-lost daughter." Berti was as surprised as anyone. "We're getting together in person," Buda added.

"Good," Berti replied. "I appreciate the care you are taking with this. I wouldn't want anything to come between our organizations."

"Nor would I," Buda responded, and meant it.

Buda then made one more call before heading off to the restaurant rendezvous with Burim. He called Prek. It was now more important than ever that Pia be treated with kid gloves. Her fate was going to have to be in Burim's hands.

59

Green Pond, New Jersey
March 25, 2011, 9:24 p.m.

Prek's phone rang again. Again it was Buda, just as he expected. He took the news saying little until Buda had finished. He then reassured Buda that everything was fine at the cottage, and before he rang off, he asked Buda to bring them some takeout if it was convenient. Buda agreed, saying he'd bring it from the Swiss House.

"What is it?" Genti said after Prek had disconnected.

"The Buda is on his way with Fatos to a restaurant not far from here to meet with two of Berti Ristani's guys who are brothers. It appears likely that the girl is related to them — the daughter of one and niece of the other. If it turns out to be true, my guess is they're going to talk for a while and figure out how they will vouch for her silence just like we vouched for her safety. Then they're going to drive here and find out that our promise was worth nothing. Then they're going to shoot Neri in the head and you in the legs. If you're lucky."

Prek indicated first Neri and then Genti. Neri had crammed his body into the corner of the couch. His hands were thrust down between his knees and his body was slumped forward although his head was up and he was looking at Prek. His eye, where Prek had smashed him in the face with the pistol, was a livid red. It would turn into a big black shiner, if he lived that long. Genti was sitting at the other end of the couch. He wanted to be sitting up with Prek, who was perched on the back of the couch opposite with his feet on the seat cushion, but he understood the symbolism. He was in the doghouse almost as much as Neri was.

"You fucking idiot," Prek said to Genti.

"Hey, I trusted him too," Genti said.

"I trusted *you!* And you egged him on with all your sex talk."

"Well, you didn't say, 'I'm going out to the van now, Genti. You make sure that Neri keeps his pants on.' You said, 'The boss says he doesn't know yet, leave the girl alone,' and you said it to both of us. I figured he heard it as good as I did."

"So you went off and took a nap."

"You were right outside, Prek. If you were so worried about her, why didn't you stay in here? It's as much your fault as it is mine."

"My fault?"

"Okay, Prek, not your fault, but I didn't touch her. What about this little fucker?" Genti waved his hand in the direction of Neri.

"I never did nothing," Neri said very quietly.

"What did you say?" Prek said.

"He said he never did anything," Genti said. "He said you stopped him before he got anywhere. He said that already."

"Didn't look that way to me."

"Me, I tend to believe the guy. I would if someone near shot my head off like you did, then hit me in the face with a gun. This guy, I tell him the truth. Listen, Prek, the girl ain't gonna remember anything either way."

"That's not the point." Prek was yelling now. "He's saying he didn't do anything and I'm saying that is not what I saw."

"Which part, Prek?"

"Which part what?"

"What did you see?"

What Prek had seen through the window was Neri with his pants off lying on top of Pia — there was no other conclusion, he was raping her. When Prek smashed the window and fired the gun, it was with the intention of scaring Neri, but the shot had been too close, passing barely three inches above the young man's head and slamming through a cheap wardrobe into the brick wall behind. But it had the desired effect. The shot awoke Genti, who let the apoplectic Prek in the front door. Prek immediately went to Neri and hit him in the face as he stood there, his pants around his ankles.

Neri was thoroughly humiliated and more than a little scared. He was telling the truth, he didn't rape the girl, but it wasn't for lack of trying. He'd found he had the same issue with a sleeping beauty as he did with any hooker — he couldn't get an erection. He'd wanted to prove something to himself, but he'd failed. And right now, that wasn't the worst of it. The girl had squirmed pretty good underneath him. He was worried that she wasn't nearly as out of it as she looked.

Pia gradually felt more and more awake. Her head was pounding, but she remembered a few things. She remembered being at the subway station near the hospital and she remembered George was coming to get her. And Will was there. Something had happened to Will. She remembered that the street was wet, because it was raining. She'd hurt her knee a little when she'd fallen. Why had she fallen?

The room she was in was a mess of shapes she was having trouble deciphering. She was lying on a bed, she could feel that. She could hear voices from a room nearby. Who did the voices belong to? She shook her head. That's right, she was in a car. No, a van. And someone stuck a needle in her thigh. Ouch. She listened. The voices belonged to the men who had stuck her with a needle. They were the voices of the men who had attacked her in her dorm room. They were the men who had done something to Will. *And one of them was just doing something to me. I have to get out of here,* she thought to herself.

Pia could move her arms and legs. She was surprised to discover she wasn't tied up. She looked around the room and saw a closed door on one side and a broken window on the other. There had also been a really loud noise. *And a man on top of her.* Something in Pia's brain kicked in. She had to get out of that room, even if it was just to the other side of that door or out the window — either would be better than in there. The men's voices were loud again, and they were coming from the other side of the door. The window, then.

Pia swung her legs over the side of the bed and tried to stand but she flopped down onto her knees and then her hands. Avoiding shards of glass, she crawled to the window and pulled herself upright. The old-style window had a handle that Pia held on to. As she turned the handle, the empty sash flew open, and she fell halfway out of the frame. It took some effort to reach down until her hands were on the earth in front of the window and she walked herself forward, using them until she could lift one leg then the other out and over the sill. She collapsed on the ground. Pia assumed she'd made so much noise, whoever was shouting in the other room would have heard. But she could still hear the voices, fainter but still there, still raised in disagreement.

"So what are you going to tell Buda?" Genti said. He was now scared.

"What do you think I should tell him?"

"Tell him nothing happened. Look him in the eye and tell him nothing happened."

"I don't want to lie to Buda. Why should I lie? It's you guys who are at fault."

Neri looked over at the door to the bedroom where Pia was sleeping. *She better not remember anything,* he thought, *or I'm dead.*

■ ■ ■ ■

Now sitting on her haunches, Pia resisted the urge to close her eyes, lie down, and go back to sleep, although every bone in her body was telling her to do it. Not for the first time in her life, an adrenaline rush drove her on. She looked over at the van, but she realized it was probably directly in the line of sight of where the voices were coming from. And she was in no state to drive; she'd crash into the first tree she came across.

Pia had no idea where she was, so she tried to size up the situation. She was cold, she knew that. She had been in a house, and she couldn't see any other houses around, or any lights. She stared into the dark in front of the house. Was that water? Yes. A river? A lake? Was it the ocean? She had no idea. She saw the light of a midsize moon, mostly obscured by clouds, but she couldn't tell if it was rising or setting. Pia kept low but started to move to her right, away from the van. She could see better now, and on the other side of the body of water she saw a single house showing one light. There were other homes, but those that she could see both across the water and the immediate neighbors were just dark geometric shapes.

Leading away from the house in a curve was a pea stone driveway, and Pia walked unsteadily along its edge, trying to keep off the stones. She felt like she was getting her legs beneath her now. Reaching a stretch of pavement, she didn't know which way to go, left or right. She noticed she was somewhere out in the country, with forest all around. Pia made an arbitrary decision and turned right. On the pavement, she tried to up her speed to a jog, but she staggered along like a drunk. Pia guessed she'd been given some drug. Again she remembered the stabbing pain in her thigh.

The road was flat and straight and Pia passed driveways on her right but none on the left. The trees kept the road dark; there were no lights on in any of the houses she passed. Pia listened intently for the sound of the van back at the house starting up. Suddenly, the road stopped and split into a starburst of driveways leading off into the darkness. Shafts of moonlight had broken through the clouds and in a gap in the trees, Pia could make out water to her left. Water on her left, water on her right. She got the uncomfortable feeling she was heading down to the end of a peninsula.

Pia turned and retraced her steps, but then, to her dismay, the quiet of the woods was shattered by the raucous sound of an automobile engine starting. It was coming from outside the house she'd escaped from. The light from the headlights bounced as the van made its way quickly down the driveway. If the van turned right, she was a sitting duck. Pia swung around and ran down a driveway to her left, trying to make as little noise as possible on the gravel. Reaching the house, Pia made a detour on flagstones set into grass around the house, quickly coming upon a small sandy beach. Now she could see she was at the edge of a circular two- or three-hundred-yard-long cove with its relatively narrow neck to her left leading out to a large lake. At this point, the opposite shore was just a couple of hundred feet away and along the shore was the house with the light on.

Pia weighed her options. If she yelled, Pia knew the men in the van were more likely to hear her than anyone else. She could hide, but she would have to move eventually, and when it got light, for all she knew, she'd be plainly visible to the men who had taken her. Noticing a pile of rocks breaching the water's surface in the middle of the narrow expanse between where she was standing and the opposite shore, Pia wondered if the water might be shallow all the way across. Although she knew the water would undoubtedly be freezing, she thought that crossing the cove represented her best chance.

Pia took off her shoes and her shirt and bunched them against her chest and stepped into the water. As she had expected, the water was numbingly cold, and Pia breathed in sharply. She looked behind her, but there was no sign of the van lights. There was sand on the bottom of the cove, then a slick mud and the occasional rock. As the water came up to her waist, there were mostly rocks and Pia slipped, exposing more of her skin to water. Regaining her balance, she continued forward. Suddenly, the headlights flashed across the water in front of her, then again, twenty feet to her right. Pia slowed as she reached the rock pile. Her legs and feet were totally numb and felt more like stilts than legs. Skirting the rocks, she had only fifty feet or so to go.

Without warning, the bottom dropped away and Pia's feet slid down an underwater slope covered with slimy silt. In the next instant, she was trying to tread water using one hand, holding her shoes and clothes over her head with the other. She tried to swim, holding her breath in the icy water. Almost immediately she felt her muscles begin to lose some of their function. She gasped for breath: now she was numb all over except for her face. Pia gave up holding her clothes over her head; she dropped her shoes and tried to swim a few strokes. She was able to move forward a little faster even though she felt like she was barely moving.

Eventually, her right foot touched a sandy bottom. She stood up with the water up to her neck and pushed toward the shore. She was shivering so much it was hard to hold her soaked clothes. Within a few feet, the water was back to waist-deep. The lighted house was about a hundred feet to her left. She tried to call out, but the loudest sound she could make was a whisper. She staggered; her legs weren't hers. Finally she was out of the water, on a kind of point with the shoreline falling off on both the lake side and the cove side. But the route to the lighted house along the edge of the lake was blocked by large boulders, underbrush, and numerous trees. She'd have to reach the house via the road.

Pia found a path of sorts through the uneven ground, and she saw there was a long dark driveway from the road down to the house. She made her way toward the driveway so she could reach the road. There were sharp stones under her still totally numb feet, and she was carrying her dripping wet clothes. What would these people think? She reached the pavement and turned left. Walking was difficult, but the house was getting closer.

Then, behind her, she heard an approaching vehicle. How far away, she couldn't tell. With mounting panic, she looked behind her into the darkness, and she could see the glow of approaching headlights. She had no time to hide, and she knew she wasn't able to run. She tried to call out, but the feeble noise was drowned out as she was bathed in bright light. Maybe it was someone else, part of her brain was telling her. Shielding her dark-adapted, squinting eyes with her free hand, Pia stared back. The vehicle braked and came to a halt within inches of her near-naked, shivering body.

Please, please, please.

Pia's heart sank. It was a van, a blue van.

60

Swiss House Inn
New Jersey
March 25, 2011, 10:09 p.m.

Aleksander Buda waited in the parking lot of the Swiss House Inn for Fatos to arrive. It was a standing joke among the crew that the thinnest guy they knew was Fatos, although no one ever said it to his face more than once. Fatos was slender and wiry as a greyhound, with quick hands that made him very proficient with a knife. He was rarely seen without his baseball hat, worn jauntily backward like a hip-hop devotee. When Buda wanted backup like he wanted that evening, he always called Fatos.

As dependable as ever, five minutes after Buda arrived, Fatos pulled his black Cadillac sedan into an empty space next to Buda's. Both cars were off by themselves at the back of the lot. Neither man got out. Buda barely acknowledged Fatos with a nod. They didn't need to talk a lot.

Buda's eyes swept around the half-full parking lot. A guy he thought might be Burim was sitting in a new Chevy Camaro in a slot facing out no more than twenty yards to his right. The driver sitting behind the wheel fastidiously ignored Buda. Then another car drove up, an Escalade, and Buda recognized Drilon riding high at the wheel. Drilon flashed his lights at the Camaro.

"Gang's all here," Buda said to himself.

Drilon parked, and first Burim, then Buda, then the other two men got out of their cars, met in the middle, and exchanged greetings.

"Let's get something to eat," Buda said. "I'm starved."

The restaurant and bar was situated in a rather ordinary-looking two-story wood-frame house painted a deep green with white trim. Except for the lighted sign announcing SWISS HOUSE INN in front of the building next to the road, the structure didn't look much like a restaurant, more like another house along the road only in better shape. The parking lot was fairly full, so it was obviously a popular place on a Friday night. Burim walked in the door first and the lady at the door made a fuss of him, asking after his health and saying his table was ready. Buda guessed he'd called ahead and was a frequent patron. Other diners who had been waiting to be seated took one look at the group and to a man and woman decided not to make an issue with the men who jumped the line. They were all in oversized leather jackets, the accepted attire of the Albanian mafia.

In the back of the busy restaurant was a single booth situated partway between the kitchen and the main room. The only traffic was from staff coming out of the kitchen, several of whom made a point of saying hello to Burim.

"So they know you in here," Buda said. He was mildly disappointed. If he'd known Burim was a regular, he wouldn't have agreed to the location.

"I'm a big tipper," Burim said, winking at Drilon across the table.

Buda studied the two men. Drilon was sweating and looked distinctly uncomfortable. Burim was relaxed and exuded calm confidence. Burim got the looks and the brains, Buda thought.

"What can I get you guys?"

It was the hostess doing double duty. They ordered four beers and the special, schnitzel, all around.

“So here’s the deal, gentlemen,” Buda said. He dove right in, dispensing with small talk. “I’m doing a job for someone and halfway through, there’s a problem. The problem is this girl who is sticking her nose in, investigating a situation and making things very difficult for me. We tried to dissuade her, but it didn’t work. The sensible course is to eliminate the problem, so I clear it with my client, who is willing to pay a hundred grand for the extra work, meaning to take care of the meddling bitch.”

Buda paused to take a sip of water and he glanced at Fatos. Fatos knew the contract was for $250,000, but there was no reason these guys had to know.

“If, as might be the case, this girl is your daughter, then you have to take responsibility for her, and you and me, we’ll share the money. But that means it’s your responsibility to make absolutely sure she desists from investigating this case, from poking around, from talking to people, from thinking about it, from dreaming about it. And if she isn’t your daughter, then we will fulfill the contract and you must promise not to mention anything about what we have done. In that case you get a quarter of the money for your inconvenience.”

“Where is she right now?” Burim asked.

At that point he was anticipating meeting her.

"Not far from here. She's perfectly safe."

"And what was she sticking her nose into, as you put it?"

"I'd rather not say. Ah, our food already."

The steaming plates of schnitzel and noodles arrived. Remembering Prek's request, Buda ordered four more to go.

"The key point is," Buda said, poised with a forkful of veal, "that she stops doing what she's doing. I have to tell you, she's making things very dangerous for me. It would be good if she went on a long vacation."

"I'm sure we can arrange that," Burim said. He had no reason to believe he could make that happen, but he liked the idea that there was to be a payday in this for him. "Right, Drilon?"

"Sure." Drilon was facing his own set of demons, but he had little choice but to play along. He pushed a noodle around the edge of his plate with his fork.

"We're happy to help you, Mr. Buda, but we are incurring some more expenses here," Burim said.

"Don't worry," Buda said, "we'll take care of dinner."

"No, really, your money is no good here. There will be other expenses, for the girl."

Burim had already decided the girl had to be his missing daughter. The coincidences were too extraordinary. And if the girl's personality resembled that of his late wife, he was going to need some cash to even hope to control her.

"Of course," Buda said. He had been expecting this. "I think ten thousand is a fair amount for the girl."

"For her inconvenience," Burim said. Buda nodded at Fatos, who leaned back and shuffled some bills under the table and presented Burim with a thick wad folded in half.

"Either way," Buda said, "daughter or no daughter. If she is not your daughter, you can keep the cash for your expenses."

"You have been very thoughtful," Burim said.

"And you have been very cooperative," Buda said.

Buda and Burim shook hands across the table, and in turn all four men shook hands, making the agreement as binding as any legal document in Albanian tradition and logic. The men finished their meal quickly, and with the takeout in hand, they left. Burim, having left three twenties on the table as a tip, was detailed to follow Buda, who would lead the way in a motorcade of sorts.

As Burim walked out of the restaurant, he patted his pocket with the fresh bulge of cash and smiled, thinking it was going to be an interesting and profitable evening.

61

Green Pond, New Jersey
March 25, 2011, 11:15 p.m.

There was nothing the occupants of the summer house could do but wait. Prek sat next to Neri on one couch with Genti and Pia sitting opposite. He had thought about tying her up, as he should have done before, but he didn't want to make a worse impression than he had to. Besides, the girl wasn't going anywhere. Prek had found an old T-shirt in a closet and gave it to Pia to wear along with one of his own New York Jets sweatshirts he kept at the house. The jersey came halfway down her thighs. She also wore a pair of soccer socks pulled up to her knees. She had draped a towel over her shoulders, but she was still shivering.

Pia sat and glared at Neri. That was the guy who had touched her, she was sure of it. She stole glances at the other guys too — the one who seemed to be in charge with the thick scar on his upper lip, and the guy with the dominating nose. She wasn't entirely certain but believed they were the men who had attacked her the night before. She recognized their voices.

Prek cradled a gun in his hand. He wondered if he would have to use it that night, and if so, who would be the target. He could make a case for any of them: Neri for brazenly disobeying an order, and Genti for failing to stop him. The only person Prek wasn't mad at was Pia. He admired her for trying to escape and for getting as far as she had. If he had to shoot her, it would not be emotional. It would just be business.

At least the fiasco was going to have a conclusion soon, thought Prek, as he heard a group of cars pull into the driveway one after the other. A moment later they heard car doors opening and then slamming shut in quick succession.

"Go wait in the bedroom," Prek said to Pia.

As soon as Buda, Fatos, Burim, and Drilon walked into the house, they could all tell that something was wrong. The atmosphere among the three men inside was clearly strained. Neri was sitting on the couch, staring at the floor and didn't stand up. Genti wouldn't make eye contact, and Prek acted fit to be tied. Buda had to find out what had happened and fast.

"Gentlemen," he said, turning to Burim and Drilon. "I left the food we picked up in the back of my car. Would you mind? I'd just like to have a quick word with my guys."

Burim and Drilon left the room and closed the door. Buda lit into Prek.

"What the fuck is going on here? Stand up, Neri! Genti, look at me when I'm talking! Where's the girl?"

"She's in the bedroom," Prek said. "She got out the window, jumped in the lake, and swam off."

"What? Are you serious?"

"Yes, but we found her right away."

"Did anyone see her?"

"No, I'm certain. There's no one here but us."

"You quite sure about that?"

"Yes."

Buda's three men were standing like guilty schoolboys in front of the principal.

"What happened to you?" Buda asked Neri, whose eye was closing rapidly. Neri didn't speak but looked across at Prek.

"Did she do that?"

"No," Prek said. "I did."

"What for?" Buda leaned forward, his

hands on his hips. Fatos was standing by the door with his arms crossed. The message was clear: No one out or in.

"Prek," Buda said, "you better tell me what happened right now, or we're going to have a major problem."

"He attacked the girl," Prek said. Neri's face fell. He'd hoped Prek would make up some story on his behalf.

"Was that before or after she escaped?"

"Before."

"And where were you?" he asked Genti.

Silence.

"All right, I'll deal with this later. It all depends on whether or not this girl is the daughter. Let's hope for you she isn't. Fatos, let 'em in."

"Burim, Drilon," said Buda in as friendly a tone as he could manage. "What happened here is that the girl tried to escape, but she didn't succeed. My men are very embarrassed, as they should be."

Burim looked at Neri but no explanation was forthcoming regarding his injury.

"My wife was certainly a tigress," he said. "Perhaps this woman is too. Mr. Buda, I am ready to meet her."

Buda showed Burim into the bedroom and left. Pia was sitting on the bed, facing the window, shivering.

"Afrodita. Pia," Burim said. "Is it really you? I am Burim. Burim Grazdani. I think I'm your father. Pia, look at me, please."

Pia sat for a second and then turned, glowering at the man, her face clouded with unadulterated fury and loathing. Burim's expression went from disbelief to pure amazement.

"Oh, God," he said. "You're exactly your mother's image." Burim knew the look she had on her face, from the first Pia, a beautiful woman full of hatred. Burim had feelings he'd never experienced and couldn't come close to articulating.

"I'm told you're a student at Columbia Medical School. That's amazing. You must be very intelligent."

Pia had turned around again, and Burim continued talking to her back.

"You look like your mother, you know that? Probably you don't. The same hair, the same eyes, it's amazing."

Pia said nothing. Could it possibly be him?

"I feel this is a miracle, our meeting. Pia, please say something."

Silence.

"Your uncle Drilon is here."

Now Pia reacted. She hacked up some spittle and spat loudly on the floor by the bed. Burim was disconsolate.

"Pia, I'm sorry I never came for you. I was young and stupid. I meant to come, so many times, but I knew that if I came forward they would find out I was illegal in this country and send me home, and then I would never have a chance of seeing you. I was working with these guys, the Rudaj group, you know, and the organization fell apart, and Drilon and I had to go underground. Then when we started working for Ristani, we had to change our names and leave our pasts behind. I wish we didn't have to do it, but we did. Pia, please."

As soon as Pia saw Burim's face, she knew who he was. This was the man she had waited years for, the man who put her through torments while she fervently hoped that he'd come back to save her. He never did. Now he was showing up, and for what? And he had brought that monster with him? What were they going to do, kill her? At this point, Pia barely cared.

"Listen, I know I abandoned you, but suddenly, now that I see you, it's important to me that you are my daughter and that you're safe."

"Safe? Do you have any idea what it was like for me in foster care?" Pia snarled.

Burim was startled by the sound of her voice.

"Do you?"

"But you're going to be a doctor, look how it all ended up!"

"*This* is how it ended up, you moron. Guns, gangsters, *murderers.* That's what I remember from being a kid. And my mom was there, and then she wasn't. What happened to her?"

"I don't know."

"You're a liar!" Pia turned and screamed the words. Buda opened the door — he must have been standing right outside.

"Everything okay?"

"Leave us, please," Burim said. Silent tears were tracking down Pia's face. She turned back around and faced the wall. Pia couldn't make sense of what was happening. How was her father involved with the people who murdered Rothman, Yamamoto, and Will McKinley? They had been waiting for him to show up, which meant that he might be able to stop them killing her too. When she spoke again, Pia's voice was quieter.

"That's all I know about you. That you're a liar."

"I'm here now."

"Are you here to finish the job they

started?"

"I understand why you say these things, but you have to believe me, I am here to save you."

"You and your white horse."

"What?"

"Whatever."

"What I am telling you is no lie. Those guys in the other room, they have been paid money to stop you because you were looking into some deaths. And they want you to stop looking."

Pia said nothing.

"They know you have an Albanian name and they asked around if anyone knows you, and I said, 'Yes, maybe.' It's the case that Albanian cannot kill Albanian: not in our business unless the killer wants to die too. If you weren't Albanian, if you weren't my daughter, you would be dead already. Do you understand?"

"That's very nice of them."

"Actually, it is, yes."

"They murdered my teacher and another doctor by giving them typhoid fever and a massive dose of polonium. Tonight they murdered my friend by shooting him in the head because he was helping me. I should be grateful to them because they're sparing me?"

"I can't do anything about the other people. What I can do is save you."

"And how will you do that?"

"I guarantee them that you will give up your investigation. And that you won't mention their involvement to the authorities. Take a vacation. Something. We can work it out."

"You? You haven't seen me since I was six. They'll take your word?"

"If I give it, yes. I have shaken their hand, and my family honor is at stake."

"Or they'll kill me."

"Or they'll kill you."

"And you'll take my word that I'll give up?"

"If you give me your word, yes."

Pia snorted. It seemed that the only person who could save her was her father, the least likely person on the planet, the person she trusted the least and hated the most, the man who was the cause of all her travails. In a situation that defied comprehension, Pia tried to think dispassionately. The drug wasn't completely out of her system, she could tell; she was more fatigued than she could ever remember being, and frightened and upset and angry. Yet she had to think.

In order to live, Pia would have to promise

to stop investigating, but could she do that? There was very little left to investigate. At the OCME, she had proved that polonium was involved in Rothman's and Yamamoto's deaths and she was certain the MEs would be looking into what she had found. The police would surely be all over Columbia, searching for Will's killers and her kidnappers. There was nothing more she could contribute to the investigation, other than providing evidence, and he hadn't mentioned anything about that. Her work was finished.

"As if you care about family honor," she said.

"I do. But if you don't believe it, take my word that I care about my honor."

"And that's all I have to do, stop investigating?"

"But you really have to stop — maybe go away for a while. You have to believe me, they will kill you otherwise. Whatever you think of me, you have to look at the alternative. You have to stay as quiet as you can. If you lead the police to Buda, there's no chance that you would ever testify against him."

Pia realized she had no choice. But perhaps there was something her father could do for her and right some of the wrongs she had suffered. Pia turned to face him.

"Okay. But you should know, not all these men have been exactly honorable with me."

"I'm glad you agree, Pia. But what do you mean?"

"When I got here I was drugged. But I remember one of them at least forced himself on me, the young one for sure. Maybe all of them."

Burim reacted the way Pia had hoped. He looked at her for a beat, with his face empurpling, then leaped up and threw open the door.

"Mr. Buda! I need to talk to you."

Buda could see Burim was spoiling for a fight, flashing angry looks at Neri. The girl must have told him what had happened in the house. Everyone in the room stood and the tension was immediate. Buda took Burim by the arm into the kitchen. The packages of takeout sat unopened on the stovetop. Burim spoke quietly but with suppressed fury.

"She is indeed my daughter. And she says she was raped. By the youngest one for certain, maybe more of them. Did you know about this?"

"Listen, I was told that one man did lose control of himself briefly but there was no sex —"

“But —”

“I understand this is shocking to you, all of this, but there was so little chance she was your daughter —”

“That is no excuse. Perhaps it is better if she was killed rather than be shamed like this. I gave my handshake, but perhaps I have to take it back.”

Buda looked Burim in the eye. Was he serious or was he just shaking him down for more money? Ten minutes ago the guy didn’t even know he had a daughter, and now he was concerned about her honor? Some of these guys really were peasants.

“I will punish the men, you can be assured of that.”

Burim shook his head and pulled back his jacket, exposing his shoulder holster.

“It can only be put right if I get to do the punishing. Do you want me to call Berti?”

“No, of course not. The reason I called you was to avoid this kind of situation. A killing will only lead to more killing — that is always the way. Punishment, yes. Killing, no. I will apologize to her myself.”

“I doubt she is going to accept any apology. That’s how I knew it was her, she has the same temper as her mother.”

“Listen, I will apologize. I will pay money to her and to you, money I will take from the three men in there. But I will not have a blood feud over this. It shouldn't have happened. I regret the situation. Ultimately, I am to blame. But I need you, Burim, to live up to your handshake and for her to give up her investigation."

Burim paused to think. Buda wouldn't allow a man from another crew to punish his own men. A blood feud was in no one's best interests, and he didn't want to be the cause of a dispute between Aleksander Buda and Berti Ristani.

"Okay. Let me talk to her."

Burim went back to the bedroom. Pia knew she had to accept Burim's help, however distasteful it was to her. Now she wanted more than anything to get out of there, to go and find George. Burim closed the door and relayed what Buda had said. Would she be willing to forgo the revenge she was entitled to? Pia knew justice was being twice denied — she was being prevented from implicating Rothman's killers and also from seeing some street retribution brought down on the person who attacked her.

"If that's the way it has to be, I want to talk to those men outside," Pia said.

"Okay," Burim said. "But I want to shake on our agreement: an Albanian shake."

Burim thrust out his hand. Pia eyed it. She didn't care. She shook hands, and her skin crawled when she touched his.

They walked into the living room where everyone was still standing, although in slightly more relaxed poses.

"I am going to accept the offer," she said to Buda. "I will do what you ask and drop the investigation. But I have a couple of things to say." Pia walked over to Neri and stood right in front of him. Neri started to shake, looking first at Prek, then at Buda, then at Burim.

"You are a piece of trash."

"I swear I didn't do anything. I can't, it's impossible —"

Pia jabbed Neri hard in the sternum with her forefinger.

"You're not so tough when the girl is awake, are you, huh? You know what my father is going to do with you? He's going to cut off your tiny little prick and shove it up your ass."

"No, no, I didn't —"

"I'm sorry?" Pia jabbed Neri again. He was crying now, great shuddering torrents of tears pouring out of his eyes. He held his hands together, pleading with Pia.

"You see how much stronger than you I am? You're a pathetic little boy." Pia poked him once more, and Neri collapsed backward onto the couch where he sat whimpering.

"And you," Pia addressed Drilon. "You will never speak to me or ever come near me again."

Drilon looked at Burim and raised his hands as if to say, "I don't understand." Quickly Pia went on.

"Now I have a question for you." Pia looked at Buda, who raised his eyebrows.

"Me?"

"Some men paid you money to frighten me?"

"Yes."

"Some men paid you money to kill me?"

"Yes."

"Are these the same men who asked you to kill Dr. Rothman and Dr. Yamamoto?"

Buda paused.

"Yes."

"Why did they do it? When I realized the deaths weren't accidental, I couldn't figure out why anyone would want to go to such lengths to kill two medical researchers. The work they were doing — they were about to change the world."

Buda looked at Burim. Was it possible to control this woman at all?

"Some people had investments that were

threatened by the research."

"Investments? You mean they did this for *money?*"

"I guess," Buda said. Why does anyone do anything? he thought.

Pia was incredulous. She thought back to her heart-to-heart talk with Rothman and how it had seemed to be the start of something meaningful in her life: the father she never had. She recalled Yamamoto's kindnesses, small and large. And Will, his life snuffed out too. Then she remembered standing in the blue-lit room looking at the pulsating baths of artificial organs and the enormous excitement she had felt. And the even greater joy she experienced at the awe-inspiring sight of the artificial pancreas. Now, it was very likely those two rooms were being closed up and put in mothballs. The research would continue, but not at Columbia and not with her. Pia felt empty and bereft.

It was likely that Rothman and Yamamoto's killers were in this room. Pia couldn't touch them, she knew that; her life depended on their getting away with murder. But she wasn't completely powerless.

"In that case, there's something I want you to do. And then I promise I will stay off the trail and hold my tongue."

Pia told the men her idea. Buda liked it — this job had far too many loose ends. Burim agreed that it would satisfy his daughter's honor. The men shook hands again, and then each in turn shook hands with Pia.

Buda was happy with the resolution, although he was left with more work to do and he'd have to decide what to do with his men, especially Neri, who seemed to have fallen to pieces completely. Prek and Genti were eating the lukewarm takeout, but Neri was still cowering on the couch.

Buda found an old pair of his wife's sneakers, which were too large for Pia but would work for now. He took them into the bedroom where she was resting.

"What will you do now?"

"You think I'm going to tell you?" she said.

"Listen, I'm sorry it happened this way."

"It's a bit late for that. Please, leave me alone."

When Pia came back into the room, it was filled with cigarette smoke. The men were standing around talking and a couple of them were laughing. Pia went over to Buda.

"Where's my cell phone?"

Buda looked at Prek, who shrugged.

"May as well let her have it. Just don't turn it on till we're done here."

"I won't."

Prek took Pia's cell phone, student ID, and wallet she used for her credit card and cash out of his jacket and gave them back to her.

"I'll be outside," Pia said. "It stinks in here." Without another word, she went outside, slamming the door behind her hard enough to shake the house.

Burim shook his head. "She is her mother, exactly."

"We should go out there — she might call someone," Prek said.

"She won't," Buda said. "She's Albanian, she promised."

"She's half Albanian," Burim said. "And half Italian. I better go."

The men laughed.

Standing on the other side of the van, Pia had turned on the phone and it flooded with messages and e-mails and texts. She saw there was a text from Lesley Wong.

"God bless you," it read. "Praying for Will's recovery."

"Pia?"

It was Burim. She shut off the phone and emerged from behind the van.

"We're leaving," Burim said.

Pia had but one thought. Recovery? Could Will possibly be alive?

62

Green Pond, New Jersey
March 26, 2011, 12:03 a.m.

Buda gave his men their marching orders. He would drive back to the Bronx with Prek and Genti, while Neri would remain at the house and clean it and the van thoroughly to erase all traces of Pia's stay. Buda was quite specific about what products Neri would use and how long he should spend on each part of the task. Buda emphasized what a good job he wanted Neri to do and that it would take him the entire weekend to complete. That would give Buda time to figure out what to do with Neri. Before he left the house, he put the van's keys in his pocket. Fatos had to drive Drilon back to the parking lot at the restaurant to pick up his car because Pia flatly refused to get into a car with Drilon. She wasn't about to explain why.

Pia sat in the front seat of Burim's vehicle

and stared straight ahead as the men said their goodbyes in the driveway. Burim and Pia set off, heading for Weehawken. Burim turned up the heat for Pia's benefit.

"What's your problem with Drilon?"

"I'm not going to talk about it," Pia said.

"I hope you will later. So, tonight we'll go to my house."

Is he kidding? thought Pia. She was desperate to get away from this man.

"No, I want to go back to the hospital."

"I can't let you do that," Burim said.

"Sure you can," Pia said. "I promised not to meddle any more and I'm not going to. You'll have to trust me. It's the same tonight as it will be in a week, in a month. I have to check on something."

"The place will be swarming with cops."

"I'll have to talk to them eventually. Or do you think I'll move in with you and live in New Jersey and play happy family? Because that's not happening. You can't just walk back into my life, don't you understand that? We have an arrangement, that's all. You have to trust me, I have to trust you. We shook hands, remember?"

"You can't tell the cops anything, obviously, you know that. Anything about Buda or his men or about seeing me and Drilon."

"Don't worry, it won't be difficult to

forget you."

Burim ignored the barb.

"So we have to come up with a story for what happened to you," he said.

"The police will know as much as I do, about the polonium. But I don't know who did the killing, I just know the why."

"The less I know the better too."

"They'll find my system was full of drugs, I imagine," Pia said. "So I'll say I was drugged, then I was held in a house outside the city, but I escaped."

"So how did you get back to New York?"

"Okay, I woke up in New York, and I don't know where I've been."

"Where did you get the clothes?"

"I don't remember where I got the clothes and that's the truth."

"So it's this: You were out of it, drugged. Some guys drove you around, but you never saw their faces. Then they stopped in a house somewhere, and you were given different clothes. Then they drove again and let you off in Manhattan. I can't drive to the hospital myself, I can't risk being seen. You better get in the back, stay out of sight of the cameras on the bridge. I'll drop you at the top of Manhattan, on Broadway somewhere. You can take a cab from there."

"All right." Pia climbed into the backseat

and curled up. She was exhausted and still shivering.

"Pia, we have to stay in touch. What's your cell phone number?"

Pia figured he could find out if he wanted to so she told him, and Burim said he would remember it. He didn't bother telling Pia his number.

Burim continued to talk, telling her little anecdotes about times he remembered from when Pia was a child. Burim convinced himself that his memory was correct, that these things had happened the way he remembered them. He concentrated on the road, and he knew Pia probably wasn't listening. He would try to reach out to her, but he wasn't confident she'd respond. After a while he stopped talking, and they rode in silence.

After forty minutes, Burim reached Broadway at the very tip of Manhattan. In the middle of a quiet block, he slowed down and Pia hopped out of the car without saying a word and didn't look back. Burim stopped the car and watched as Pia walked to an intersection and held out her hand to hail a taxi. A gypsy cab pulled over, and Pia leaned toward the window and told the driver something. Before she got in the car, Burim thought she looked small and vulnerable in her crazy mismatched outfit. But he had a feeling she'd be okay.

63

Columbia University Medical Center
New York City
March 26, 2011, 1:00 a.m.

Pia asked the car to drop her off as close as he could get to her dorm building on Haven Avenue. There was still a police presence with artificial lighting at this location of the abduction and shooting. The fare was $12, and she gave the driver the twenty Burim had given her and didn't stop for change. On the ride down, Pia had concentrated her thoughts on Will, ignoring her father, who was jabbering away in the front seat. She tried not to think about her ordeal; at least she was safe now. Pia had no thought about whether or not she was going to try to establish a connection with her father, but she did know she'd have nothing at all to do with Drilon. The few things she remembered about him were all painful.

Pia focused. She wasn't worried about

talking to the police — after all, it wouldn't be the first time. She'd build a wall around what happened at the house and not recount any of it; in all other aspects, she could be truthful. And there were some truths she was still determined that everyone know. There would be no possibility of a cover-up.

Pia walked up to the front desk in the dorm. There were two uniformed cops by the elevator, but Pia hoped that her strange garb and the fact that she'd bunched her hair up under a baseball cap would throw off a casual onlooker. Despite the late hour, students were coming in from studying at the Health Science Center or from a night out. A few others were on their way out, having been called to the hospital for emergencies.

Pia knew the person staffing the front desk and asked him about Will McKinley.

"Pia, is that you?" the young man said. "Cops are looking for you. They said you got *kidnapped* or something crazy."

"No, I'm fine. Will — tell me about Will." Pia gestured with her index finger over her lips to silence the man from calling attention to her presence.

"Oh, man, I heard he got shot in the head, but he survived. He was taken over to the

Neurological Institute, and he had surgery. One of the other fourth-years said he's in Intensive Care."

Without another word, Pia turned and walked away from the desk and made her way over to Neurosurgical Intensive Care. She saw plenty of cops and security guards, but they were on the lookout for a woman with long black hair, not someone wearing a New York Jets sweatshirt to mid-thigh, soccer socks, and a baseball cap. She looked like a cheerleader.

At the doors of Intensive Care, there were more police. Pia was stopped by the nurses, who eyed her less than appropriate clothes and the bruise on her jawline. Pia explained she was a medical student and flashed her student identification with her finger over her name. She hoped that everyone there had been on duty the whole night and hadn't seen or heard the news. The head nurse said she wouldn't let Pia into the intensive care unit, but she paged the resident.

When the resident arrived he looked quizzically at Pia. Still, he was considerate after hearing that she was a medical student interested in the case. He assumed she was a girlfriend of the young man.

"Mr. McKinley is being maintained in an

induced coma post-surgery," the resident, Dr. Hill, said. "He received a gunshot to the head, but the bullet made a complete transit through the frontal lobe. It's an injury that people have recovered from in the past. But I would emphasize that anyone suffering this kind of injury may not be exactly the same person he was before being shot and having brain surgery."

"He's a friend of mine," Pia said. "I was there when he was shot."

"So it's very important that you understand he will be different even if there is a seemingly complete recovery."

"Different how?"

"It would take too long to go into now. Look up the case of Phineas Gage, from 1848, which involved much more severe trauma to the frontal lobe. It was the first recorded case about how penetrating head trauma can affect personality."

"Can I see him?"

"I don't see why not. His family is on the way. You have to wear a gown and so on."

"Of course."

Pia went off to don her protective garb.

Only then did Dr. Hill remember something about being on the lookout for a young woman.

■ ■ ■ ■

In Will McKinley's room, Pia found George standing by Will's bed.

"Pia, my God!" George said, and grasped her in an embrace. "Are you okay? What happened to you?"

"I'm fine. I'll tell you later. Will . . . how is he doing?"

"No one knows. I have to go back and talk to some more cops, but I wanted to see him. I saw the whole thing. I saw him get shot and you taken. I can't believe he's alive. And you too. Thank God. What happened?"

George stared at Pia as if she were an apparition, but she turned to look at Will. His breathing was being handled by a machine, there were tangles of wires and tubes enveloping him, and he was surrounded by banks of devices with illuminated readouts. Will's face looked calm and peaceful and his color was normal. Except for all the medical equipment and the beeping and clicking, he might have simply been sleeping. A nurse hovered nearby. Pia looked around the room and caught her reflection in the unit's large window. She looked terrible, like something the cat dragged in. She turned her attention back to George.

"George, I'm so sorry I got you into this. Please forgive me," she said. "If I had listened to you, then this would have turned out differently, I know that."

"Pia, I feel as terrible as you do about this. I was sleeping while you were waiting for me at the station. I slept through your calls. I should have come to you. It should be me lying there."

"That doesn't make me feel any better. Will had no idea what was going on and I didn't say anything. I don't know what's going to happen to me so there's a couple of things I want to say while I have the chance.

"I want to say thank you for going out of your way to help me. I don't really understand why you'd do that for someone without her asking, and without her appreciating what you're doing. But there are a lot of things I don't understand.

"I guess the main thing I don't understand is myself. I think you do know yourself, which is why you're able to say you love someone, like you did to me. And I'm sorry for not listening then either. I'm jealous that you can do that, and I wonder why I can't. I think there's something broken in me or something that was never there, and it's taken until now for me to see that. For a lot of reasons, I find it very difficult to trust people. As if I need to tell you that. . . . But I don't know how to love someone either, or how to accept their love. It's a big responsibility, being loved, and you should think hard before rejecting someone's love.

"But you've made me want to learn more about myself, to see if I can't fix that broken part. I think we studied that course together, in first-year psych, the part about people with personality issues who never accept that they're the ones who are different. So if they're marching, if they lead with their right foot while everyone else uses their left, they say with unshakable belief that it's everyone else who's out of step, not them. I think I'm like that."

Pia looked around. She hadn't realized the nurse had left, nor had she seen or heard the man enter the room. He was stocky, in a cap and gown, just like she and George were wearing, over his streetclothes. He was standing at the back of the room as she stood with George by Will's bed. The man waved his hand as if to say, "Don't mind me! Go on!"

"I never understood people's feelings, George. I sneered at people who said they were in love because I never knew what that meant. I don't know if I can change, and I don't know if someone can be taught how to love. But I do know I want to try to change."

Pia reached out and touched George's cheek with one fingertip.

"Please try to forgive me."

George closed his eyes.

"Pia, there's nothing to forgive. I'm just so happy you're safe."

Pia stepped back and studied Will's peaceful face, then turned toward the visitor. She sensed he was there to talk to her.

"Miss, I'm Detective Captain Lou Soldano. You're Pia Grazdani, I assume?"

"Yes."

"You have to come with me now."

"I understand. Do you mind if I use the bathroom first?"

"Of course not," Lou said.

After Pia told George she'd see him later, she and Lou walked out of the intensive care unit.

"I'm glad to see you," Lou said. "Are you okay?"

"I'm fine," Pia said, before disappearing into the women's room near the elevators. After locking the door, she took out her smartphone. Quickly she tapped out an e-mail, forwarding a sizable message she'd already written. After making certain it had gone, she used the toilet. She then looked at herself in the mirror over the sink and said: "Now the shit hits the fan." Taking a deep breath, she composed herself to go out and meet Detective Lou Soldano who represented her old nemesis, the City of New York.

64

East Tenth Street
New York City
March 26, 2011, 2:13 a.m.

The man was aware of the buzzing of a phone right next to his ear. He went immediately from deep sleep to partial consciousness but it took him a few beats to realize where he was. He picked up the phone, saw his device, didn't recognize the number but accepted the call just to stop the noise.

"McGovern. This better be good, whoever you are."

"Is this Chet McGovern?" a female voice said.

"I believe so, ask me tomorrow. What time is it anyway?"

"About two-fifteen, sorry about that."

"Do I know you?"

"My name is Jemima Meads. I'm calling from the *New York Post.*"

"The *Post?*"

The mention of the paper made McGovern sit up. He looked across at the redhead lying fast asleep on the other side of the bed. Her bed, he remembered, somewhere in the Village. What was *her* name?

"Dr. McGovern, we're looking at a story that has two researchers at Columbia being killed by the radioactive agent polonium-210, just like the KGB colonel in London. Do you have a comment?"

"It's two-fifteen in the morning," McGovern said groggily.

"And I do apologize, but we want to be first and make sure we have the story right."

"But I thought we weren't releasing the cause of death," said McGovern.

"So you can confirm it?"

"That's not what I said."

"It kind of is."

"Look, speak to my colleague, Jack, he did the autopsies. But I recommend it be tomorrow during normal business hours."

"Jack Stapleton, the ME?"

"Yes, him."

"Okay, thanks. And sorry for disturbing you."

The woman ended the call, and Chet lay back in bed. What was that about?

Epilogue

Greenwich, Connecticut
March 26, 2011, 6:05 a.m.

Even though it was Saturday, Russell Lefevre had set his alarm for 5:45. He clamped down on the buzzer, before it woke his wife. Lefevre padded into the bathroom and then downstairs to make coffee and to check on events on the Internet. As the coffee was brewing, Lefevre scanned the online headlines of *The New York Times, The Wall Street Journal,* and *The Washington Post.* Russell had always been fastidious about keeping up with the news, but in the past few weeks he'd become obsessed, especially since Edmund had become less and less communicative.

Even though Russell had asked him numerous times, Edmund had never told him what he and Jerry Trotter had talked about at Edmund's house a few weeks before, even though Edmund had looked thoroughly shaken afterward. A week or so later, Jerry Trotter disappeared. When Russell called Max Higgins, Max said Jerry had gone on a fact-finding trip to Asia, and he had no idea when he'd be back. Edmund had nothing to say about that. Then Russell read about Gloria Croft being attacked while out running one morning in Central Park, and Edmund told Russell he had no idea what had happened then, either.

Two days earlier, all the newspapers carried the story about Rothman and Yamamoto, first about their being sick. Then they reported that the pair had died in a tragic accident in the lab. Russell didn't know what to feel or what to think. First Jerry disappeared, then Gloria was attacked, then Rothman and Yamamoto died. On its own, each of the latter two events was a piece of good fortune, but together, they were surely more than a coincidence. Did Edmund have anything to do with it? Could these events have been what he and Jerry talked about? It seemed impossible to comprehend that Edmund was involved, but Russell couldn't bring himself to confront his partner.

Russell made coffee and looked for the *New York Post*. When he saw the newly updated headline on the home page he nearly choked:

COLUMBIA MEDS IN KGB COPYCAT SLAY?

Under Jemima Meads's byline was an exclusive about Rothman and Yamamoto. Hedged with "allegedly" and "reportedly," the story said that acting on an anonymous tip, the reporter had contacted members of the New York Office of the Chief Medical Examiner who were working on the theory that the exotic radioactive agent polonium-210 was involved in the deaths of the two prominent Columbia University researchers. The find was made by the husband-and-wife team of Drs. Jack Stapleton and Laurie Montgomery, who, having been reached by the reporter at their Upper West Side town house, refused to confirm or deny the story, referring the reporter to the OCME's public relations department.

The discovery was immediately reported to the FBI, the CIA, Homeland Security, and the NYPD Joint Organized Crime Task Force because of its significant implications and similarities to the 2006 murder in London of a defected Russian FBS agent by the Russian FBS, the current incarnation of the Soviet KGB.

Polonium-210, the article said, is a remarkably poisonous compound millions of times more deadly than cyanide if swallowed or respired. It's also extraordinarily difficult to come by because of its association with triggering nuclear weapons and is thought to be available only in Russia, Pakistan, and North Korea.

At this time it wasn't known if the deaths were connected to a shooting reported outside the Columbia Medical Center that evening.

Russell dashed to the phone and fumbled to call Edmund. He knew he was waking him as the phone rang for the sixth time.

"Russell, what the hell?" His voice was thick with sleep.

"Edmund, go online, look at the *Post.* It says the researchers were murdered, with some nuclear poison. Oh my God, Edmund."

"All right, Russell, calm down. You better get over here." Edmund hung up. Russell wanted to throw up, but he composed himself, went back upstairs, and got dressed.

He started driving toward Edmund's house, his mind racing, trying to make connections, thinking about the coincidences and how they now looked like something so much more. Like murder. As he drove, Russell failed to see that a beat-up old Toyota Corolla had pulled out and was following him through the twisty Greenwich back roads.

Edmund had opened his gates and Russell drove directly into the walled courtyard in front of the waterfront mansion. He leaped from the car and bounded up the front steps and impatiently leaned on the doorbell, whose muffled chords he could just make out coming through the massive door. Where was Edmund? He rang the bell again. The only other sound he could hear was the gentle cacophony of songbirds.

At last, Russell heard a bolt being drawn back on the heavy door, then another sound, of a car coming quickly up the drive. He turned and watched bemused as a tan sedan skidded to a halt inches from his own vehicle and two figures jumped out and ran toward him. They were wearing hoods and holding guns. The door opened and Russell twisted his head back and said one word. "Edmund."

"They sold us out," Edmund said.

Then the men opened fire, both with pistols muffled with silencers. Russell fell forward the way he was facing, across the threshold of Edmund's house. Edmund had no time to process what he was seeing, that there were two men firing weapons at him, that he'd gambled on this venture and this was how he lost. He fell backward, propelled by three bullets in the chest. He fell straight back, only the soles of his most comfortable pair of slippers visible.

The first man walked up the stairs, looked at Edmund, leveled his gun, and shot him once more in the forehead. The second man kicked over Russell's body and did the same to him. The men looked at each other and nodded. They found the spent cartridge cases and picked them up, then walked back to the car, got in, and removed their balaclavas before driving off.

At the wheel, Prek Vllasi navigated the drive out of the gate and turned onto the road. Prek turned to Genti Hajdini and banged once on the steering wheel. Both men smiled.

ABOUT THE AUTHOR

Dr. **Robin Cook** is the author of thirty previous books and is credited with popularizing the medical thriller with his wildly successful first novel, *Coma*. He divides his time between Boston and Florida. His most recent bestsellers are *Cure, Intervention,* and *Foreign Body*.

The employees of Thorndike Press hope you have enjoyed this Large Print book. All our Thorndike, Wheeler, and Kennebec Large Print titles are designed for easy reading, and all our books are made to last. Other Thorndike Press Large Print books are available at your library, through selected bookstores, or directly from us.

For information about titles, please call:
(800) 223-1244

or visit our Web site at:
http://gale.cengage.com/thorndike

To share your comments, please write:

Publisher
Thorndike Press
10 Water St., Suite 310
Waterville, ME 04901